THE ROUTLEDGE HANDBOOK OF METAMETAPHYSICS

Philosophical questions regarding the nature and methodology of philosophical inquiry have garnered much attention in recent years. Perhaps nowhere are these discussions more developed than in relation to the field of metaphysics.

The Routledge Handbook of Metametaphysics is an outstanding reference source to this growing subject. It comprises thirty-eight chapters written by leading international contributors, and is arranged around five themes:

- The history of metametaphysics
- Neo-Quineanism (and its objectors)
- Alternative conceptions of metaphysics
- The epistemology of metaphysics
- Science and metaphysics.

Essential reading for students and researchers in metaphysics, philosophical methodology, and ontology, *The Routledge Handbook of Metametaphysics* will also be of interest to those in closely related subjects such as philosophy of language, logic, and philosophy of science.

Ricki Bliss is Assistant Professor of Philosophy at Lehigh University, USA. Her research focuses primarily on issues in foundational metaphysics with a particular interest in how history, in particular the history of Christian thought, might bear on those issues.

J.T.M. Miller is a philosopher based at the University of Nottingham, UK. From July 2020, he will be Assistant Professor of Philosophy at Durham University, UK. His research focuses on issues at the intersection of metaphysics, and philosophy of language and linguistics, including on the nature of metaphysical disputes, how to assess metaphysical theories, and the metaphysics of language.

ROUTLEDGE HANDBOOKS IN PHILOSOPHY

Routledge Handbooks in Philosophy are state-of-the-art surveys of emerging, newly refreshed, and important fields in philosophy, providing accessible yet thorough assessments of key problems, themes, thinkers, and recent developments in research.

All chapters for each volume are specially commissioned, and written by leading scholars in the field. Carefully edited and organized, *Routledge Handbooks in Philosophy* provide indispensable reference tools for students and researchers seeking a comprehensive overview of new and exciting topics in philosophy. They are also valuable teaching resources as accompaniments to textbooks, anthologies, and research-orientated publications.

Also available:

The Routledge Handbook of Collective Responsibility
Edited by Saba Bazargan-Forward and Deborah Tollefsen

The Routledge Handbook of Phenomenology of Emotion
Edited by Thomas Szanto and Hilge Landweer

The Routledge Handbook of Hellenistic Philosophy
Edited by Kelly Arenson

The Routledge Handbook of Trust and Philosophy
Edited by Judith Simon

The Routledge Handbook of Philosophy of Humility
Edited by Mark Alfano, Michael P. Lynch and Alessandra Tanesini

The Routledge Handbook of Metametaphysics
Edited by Ricki Bliss and J.T.M. Miller

For more information about this series, please visit: www.routledge.com/Routledge-Handbooks-in-Philosophy/book-series/RHP

THE ROUTLEDGE HANDBOOK OF METAMETAPHYSICS

Edited by Ricki Bliss and J.T.M. Miller

LONDON AND NEW YORK

First published 2021
by Routledge
2 Park Square, Milton Park, Abingdon, Oxon OX14 4RN

and by Routledge
52 Vanderbilt Avenue, New York, NY 10017

Routledge is an imprint of the Taylor & Francis Group, an informa business

© 2021 selection and editorial matter, Ricki Bliss and J.T.M. Miller; individual chapters, the contributors

The right of Ricki Bliss and J.T.M. Miller to be identified as the authors of the editorial material, and of the authors for their individual chapters, has been asserted in accordance with sections 77 and 78 of the Copyright, Designs and Patents Act 1988.

All rights reserved. No part of this book may be reprinted or reproduced or utilised in any form or by any electronic, mechanical, or other means, now known or hereafter invented, including photocopying and recording, or in any information storage or retrieval system, without permission in writing from the publishers.

Trademark notice: Product or corporate names may be trademarks or registered trademarks, and are used only for identification and explanation without intent to infringe.

British Library Cataloguing-in-Publication Data
A catalogue record for this book is available from the British Library

Library of Congress Cataloging-in-Publication Data
Names: Bliss, Ricki, editor. | Miller, J. T. M. (James T. M.), editor.
Title: The Routledge handbook of metametaphysics / edited by Ricki Bliss and JTM Miller.
Description: Abingdon, Oxon ; New York, NY : Routledge, 2020. | Series: Routledge handbooks in philosophy | Includes bibliographical references and index.
Identifiers: LCCN 2020006616 (print) | LCCN 2020006617 (ebook) | ISBN 9781138082250 (hardback) | ISBN 9781315112596 (ebook)
Subjects: LCSH: Metaphysics.
Classification: LCC BD111 .R684 2020 (print) | LCC BD111 (ebook) | DDC 110–dc23
LC record available at https://lccn.loc.gov/2020006616
LC ebook record available at https://lccn.loc.gov/2020006617

ISBN: 978-1-138-08225-0 (hbk)
ISBN: 978-1-315-11259-6 (ebk)

Typeset in Bembo
by Wearset Ltd, Boldon, Tyne and Wear

CONTENTS

Notes on contributors ix

Introduction: what is metametaphysics? 1
Ricki Bliss and J.T.M. Miller

PART I
The history of metametaphysics 11

1 Metametaphysics in Plato and Aristotle 13
 Vasilis Politis

2 Kantian meta-ontology 23
 Ralf M. Bader

3 Rudolf Carnap: pragmatist and expressivist about ontology 32
 Robert Kraut

4 Quine's metametaphysics 49
 Karl Egerton

5 Metaphysical realism and anti-realism 61
 Jussi Haukioja

6 From modal to post-modal metaphysics 71
 Nathan Wildman

PART II
Neo-Quineanism (and its objectors) — 83

7 Ontological commitment and quantifiers — 85
   Ted Parent

8 Quantifier variance — 100
   Rohan Sud and David Manley

9 Verbal disputes and metaphysics — 118
   Brendan Balcerak Jackson

10 Absolute generality — 130
    Agustín Rayo

11 The metametaphysics of neo-Fregeanism — 143
    Matti Eklund

12 Easy ontology — 159
    Amie L. Thomasson

13 Defending the importance of ordinary existence questions and debates — 171
    Jody Azzouni

14 Ontological pluralism — 184
    Jason Turner

PART III
Alternative conceptions of metaphysics — 197

15 Grounding — 199
    Alexander Skiles and Kelly Trogdon

16 Fundamentality — 211
    Ricki Bliss

17 Metaphysical explanation — 222
    Naomi Thompson

18 Truthmaking and metametaphysics — 233
    Ross P. Cameron

19 Essence — 245
    Jessica Leech

20	Fictionalist strategies in metaphysics *Lukas Skiba and Richard Woodward*	259
21	Global expressivism *Stephen Barker*	270
22	Hylomorphic unity *Anna Marmodoro*	284
23	Feminist metametaphysics *Elizabeth Barnes*	300
24	Social ontology *Rebecca Mason and Katherine Ritchie*	312
25	Natural language ontology *Friederike Moltmann*	325
26	Phenomenology as metaphysics *Dan Zahavi*	339

PART IV
The epistemology of metaphysics — **351**

27	A priori or a posteriori? *Tuomas E. Tahko*	353
28	The epistemology of modality *Sonia Roca-Royes*	364
29	Ideology and ontology *Sam Cowling*	376
30	Primitives *Jiri Benovsky*	387
31	Conceptual analysis in metaphysics *Frank Jackson*	395
32	Contingentism in metaphysics *Kristie Miller*	405
33	Is metaphysics special? *Thomas Hofweber*	421

PART V
Science and metaphysics — 433

34 Science-guided metaphysics — 435
    Kerry McKenzie

35 Methods in science and metaphysics — 447
    Milena Ivanova and Matt Farr

36 Thing and non-thing ontologies — 459
    Michael Esfeld

37 Moderately naturalistic metaphysics — 468
    Matteo Morganti

38 Metaphysics as the 'science of the possible' — 480
    J.T.M. Miller

Index — *492*

CONTRIBUTORS

Jody Azzouni has published *Talking About Nothing: Numbers, Hallucinations, and Fictions* (Oxford University Press, 2010), *Objects Without Borders* (Oxford University Press, 2017), and many articles in metaphysics, philosophy of logic, mathematics, science, and language. He is Professor of Philosophy at Tufts University.

Ralf M. Bader is Chair for Ethics and Political Philosophy at the Université de Fribourg in Switzerland. Previously, he was Associate Professor at the University of Oxford, as well as Bersoff Assistant Professor and Faculty Fellow at New York University.

Brendan Balcerak Jackson is an assistant professor of philosophy at the University of Miami. His research focuses on numerous issues at the intersections of the philosophy of language, metaphysics, and epistemology.

Stephen Barker did a PhD at Melbourne University, and post-doctoral fellowships in UNAM Mexico and Monash University, Melbourne. He then took up a research position at the University of Tasmania, Australia. He has been at the Philosophy Department, Nottingham University, UK, since 2002.

Elizabeth Barnes is Professor of Philosophy at the University of Virginia, USA. She works primarily in social philosophy, feminist philosophy, and metaphysics, and is especially interested in places where these topics overlap.

Jiri Benovsky (University of Fribourg, Switzerland) specializes in metaphysics, metametaphysics, aesthetics, and philosophy of mind. He is the author of several books, including recently: *Meta-metaphysics* (Springer, 2016). *Eliminativism, Objects, and Persons: The Virtues of Non-existence* (Routledge, 2018), and *Mind and Matter: Panpsychism, Dual-Aspect Monism, and the Combination Problem* (2018, Springer).

Ricki Bliss is Assistant Professor of Philosophy at Lehigh University, USA. Her research focuses primarily on issues in foundational metaphysics with a particular interest in how history, in particular the history of Christian thought, might bear on those issues. She is also interested in aspects of Buddhist metaphysics, as well as issues surrounding what she doesn't know how to better describe than as the ethics of reading.

Ross P. Cameron is Professor of Philosophy at the University of Virginia, USA. His recent book *The Moving Spotlight* (Oxford University Press, 2015) defends the objective passage of time. His next book *Chains of Being* is about ontological dependence, infinite regress, and metaphysical explanation.

Sam Cowling is the John and Christine Warner Associate Professor of Philosophy at Denison University in Granville, Ohio. He is the author of *Abstract Entities* (Routledge, 2017). He works primarily in metaphysics and the philosophy of science.

Karl Egerton works at the intersection of epistemology, metaphysics, and the history of analytic philosophy. Much of his research critically revisits Quine's methodology for metaphysics and its relations to contemporary metametaphysics. He is currently a lecturer at the University of Hertfordshire, UK.

Matti Eklund is Chair Professor of Theoretical Philosophy at Uppsala University, Sweden. He has published widely, in for example metaphysics, philosophy of language, philosophy of logic, and metaethics.

Michael Esfeld is Full Professor of Philosophy at the University of Lausanne (Switzerland). His main area of research is the metaphysics of science and the philosophy of mind. His latest book publication is *A Minimalist Ontology of the Natural World* (Routledge, 2017, with Dirk-André Deckert).

Matt Farr is a lecturer and researcher in the Department of History and Philosophy of Science at the University of Cambridge. His work addresses philosophical problems about time and causation across the sciences, particularly what it means for time to have a direction.

Jussi Haukioja is Professor of Philosophy at the Norwegian University of Science and Technology. He specializes in philosophy of language, focusing especially on the theory of reference and its consequences for issues in metaphysics, epistemology, and philosophy of mind.

Thomas Hofweber is a professor of philosophy at the University of North Carolina at Chapel Hill. He works in metaphysics and the philosophy of language, and is the author of *Ontology and the Ambitions of Metaphysics* (Oxford University Press, 2016) as well as numerous articles.

Milena Ivanova teaches at the Department of History and Philosophy of Science and holds a Bye-Fellowship at Fitzwilliam College at the University of Cambridge. Her research concerns how scientific theories uncover truths about the world, the nature of scientific principles, and the role of values in science.

Frank Jackson is Emeritus Professor at the Australian National University. He is the author of books and papers in the philosophy of mind and language and in ethics.

Robert Kraut is Professor of Philosophy at The Ohio State University. His primary interests are metaphysics, aesthetic theory, and the philosophy of language. His *Artworld Metaphysics*

(Oxford University Press, 2010) explores neo-pragmatist theories of artworld interpretation, description, and evaluation. He has held visiting positions at Pittsburgh, Rutgers, and the Stanford Humanities Center.

Jessica Leech is Senior Lecturer in Philosophy at King's College London, UK. She was jointly awarded her doctorate from the Universities of Sheffield and Geneva in 2011. Her research interests are primarily on issues to do with possibility and necessity, both contemporary and historical.

David Manley is Associate Professor of Philosophy at the University of Michigan, Ann Arbor. His research has mainly been concerned with semantics, ontology, probability, and evidence; he is currently thinking about conditions for rationality and well-being for people, groups, animals, and other cognitive systems.

Anna Marmodoro holds the Chair of Metaphysics at the University of Durham, UK and is concomitantly a research fellow at Corpus Christi College at the University of Oxford. Her research interests are in metaphysics; ancient, late antiquity, and medieval philosophy; philosophy of mind; and philosophy of religion. She has published monographs, edited books, and journal articles in all these areas. Anna is also the co-founder and co-editor of the peer-reviewed journal *Ancient Philosophy Today: DIALOGOI*, published by Edinburgh University Press.

Rebecca Mason is Assistant Professor of Philosophy at the University of San Francisco. She specializes in feminist philosophy and metaphysics, especially social metaphysics. She also has interests in philosophy of language, epistemology, and social and political philosophy.

Kerry McKenzie is Associate Professor of Philosophy at UC San Diego, USA. She specializes in the metaphysics of science, especially in the areas of fundamentality, structuralism, and methodology, and in the issue of what it means to be a naturalist in philosophy.

J.T.M. Miller is a philosopher based at the University of Nottingham, UK. From July 2020, he will be Assistant Professor of Philosophy at Durham University, UK. His research focuses on issues at the intersection of metaphysics, and philosophy of language and linguistics, including on the nature of metaphysical disputes, how to assess metaphysical theories, and the metaphysics of language.

Kristie Miller is Associate Professor of Philosophy and Joint Director of the Centre for Time at the University of Sydney. She works primarily in metaphysics, on the nature of composition, persistence, and time, and on the associated modal status of views in these areas. Her most recent work focuses on the error theory of time, and timeless theories of quantum gravity, and on the nature of metaphysical explanation.

Friederike Moltmann is Research Director at the French Centre Nationale de la Recherche Scientifique (CNRS) and in recent years has been a visiting researcher at New York University. Her research focuses on the interface between natural language semantics and philosophy, in particular metaphysics. She has published over seventy articles and is author of *Parts and Wholes in Semantics* (Oxford University Press, 1997) and *Abstract Objects and the Semantics of Natural Language* (Oxford University Press, 2013). Moltmann received a PhD in 1992 from the Massachusetts Institute of Technology and taught both linguistics and philosophy at various universities in the USA, the UK, France, and Italy.

Matteo Morganti is an associate professor at Rome Tre University, Italy, Department of Philosophy, Communication and Visual Arts. His research focuses primarily on the philosophy of science and metaphysics. He is the author of *Combining Science and Metaphysics* (Palgrave Macmillan, 2013).

Ted Parent Nazarbayev University (in Nur-Sultan, Kazakhstan) writes on epistemology and metaphysics, broadly construed, with publications in *Philosophical Studies* and *The Journal of Philosophy*, among others. His first monograph is *Self-Reflection for the Opaque Mind* (Routledge, 2017), and his second (in preparation) is *A Critique of Metaphysical Thinking*.

Vasilis Politis teaches at Trinity College Dublin. He has written extensively on Plato and Aristotle, including *The Structure of Enquiry in Plato's Early Dialogues* (Cambridge University Press, 2015), papers in *Phronesis* and other journals, and, co-edited with George Karamanolis, *The Aporetic Tradition in Ancient Philosophy* (Cambridge University Press, 2018).

Agustín Rayo is a professor of philosophy at MIT and a professorial fellow at the University of Oslo. His research is at the intersection of the philosophy of logic and the philosophy of language.

Katherine Ritchie is an assistant professor at the CUNY Graduate Center and the City College of New York. Her research focuses on questions about how we think and talk about social groups, the nature of social entities, and the normative and political upshots of both. Ritchie has published articles in journals including *Australasian Journal of Philosophy*, *Philosophy and Phenomenological Research*, and *Philosophical Studies*.

Sonia Roca-Royes (PhD University of Barcelona) is currently a senior lecturer at the University of Stirling. She specializes in metaphysics and epistemology of modality, with broader interests in epistemology, ontology, philosophy of mathematics and logic, and philosophy of language.

Lukas Skiba is a research associate (*wissenschaftlicher Mitarbeiter*) at the Institute of Philosophy of the University of Hamburg, Germany. His research interests cover (meta)metaphysics, the philosophy of logic and language, and the history of analytic philosophy. He obtained his PhD from the University of Cambridge.

Alexander Skiles is an assistant teaching professor of philosophy at Rutgers University, New Brunswick, USA. His research mainly focuses on three topics: non-causal explanation in metaphysics and the sciences; essence, identity, and individuation; and the nature of existence.

Rohan Sud earned his PhD at the University of Michigan and is currently Assistant Professor of Philosophy at Ryerson University in Toronto. His research interests include metaphysics, philosophy of language, and metaethics, with a particular focus on issues related to vagueness.

Tuomas E. Tahko is Reader in Metaphysics of Science at the University of Bristol, UK. He is the author of *An Introduction to Metametaphysics* (Cambridge University Press, 2015) and the editor of *Contemporary Aristotelian Metaphysics* (Cambridge University Press, 2012), and has published numerous articles on metaphysics, philosophy of science, and philosophical logic.

Contributors

Amie L. Thomasson is the Daniel P. Stone Professor of Intellectual and Moral Philosophy at Dartmouth College in Hanover, NH, USA. She is the author of *Fiction and Metaphysics* (Cambridge University Press, 1999), *Ordinary Objects* (Oxford University Press, 2007), *Norms and Necessity* (Oxford University Press, 2013), *Ontology Made Easy* (Oxford University Press, 2014), as well as over seventy book chapters and articles.

Naomi Thompson is a lecturer in philosophy at the University of Southampton, UK, and is currently a member of the Metaphysical Explanation project at the University of Gothenburg, Sweden. She works mostly on questions about structure, fundamentality, and explanation.

Kelly Trogdon is an associate professor at Virginia Tech in Blacksburg, VA, USA. Recent publications include "Grounding-Mechanical Explanation" (*Philosophical Studies* 175/6, 2018: 1289–1309) and "Prioritizing Platonism" with Sam Cowling (*Philosophical Studies* 176/8, 2019: 2029–2042).

Jason Turner is Professor of Philosophy at the University of Arizona. His work lies at the intersection of logic and metaphysics, focusing on the metaphysics of logical concepts such as existence and identity, despite being secretly interested in everything.

Nathan Wildman is an assistant professor at Tilburg University, the Netherlands, and a member of the Tilburg Centre for Logic, Ethics, and Philosophy of Science (TiLPS). His research concerns various topics in metaphysics, philosophy of language, logic, and aesthetics.

Richard Woodward's research focuses on a cluster of connected issues in metaphysics and aesthetics, including the nature of possibility, the methodological foundations of ontology, and the connection between fiction and the imagination. He presently works at the University of Hamburg.

Dan Zahavi is Professor of Philosophy at University of Copenhagen and University of Oxford, and Director of the Center for Subjectivity Research in Copenhagen. Zahavi's primary research area is phenomenology and philosophy of mind, and their intersection with empirical disciplines such as psychiatry and developmental psychology.

INTRODUCTION
What is metametaphysics?
Ricki Bliss and J.T.M. Miller

Metaphysics sometimes seems to have a particularly special or important place within the wider philosophical enterprise. For some, it is the centre of philosophical work, devoted to trying to answer the fundamental questions that could be asked about the world around us. For others, it is nonsense, mere sophistry masquerading as serious intellectual inquiry. Naturally, between these two extremes are various middle-ground positions – perhaps metaphysics does ask substantive questions, but it has been taken over in importance by empirical science, or perhaps metaphysical answers reveal more about us and the conceptual schemes we employ than about the ultimate nature of reality.

All philosophers (or at least all metaphysicians) think about some variant of these questions, either implicitly or explicitly, at some point or another. Throughout the history of philosophy, different answers to the questions, and hence differing views about the importance of metaphysics, have also been an implicit or explicit part of the work of major philosophical figures. Few would argue that we must provide answers to all meta-level questions in order to do first-order metaphysics, but to think that the answers to metametaphysical questions will have no impact on first-order theories is to misunderstand the scope and focus of metametaphysics.

The above-mentioned questions are all within the scope of metametaphysics, at least as the domain will be understood in this volume. Given this wide focus of metametaphysics, whether you are a knee-jerk metaphysical realist, Kantian, Carnpian, Aristotelian, or subscribe to (almost) any other general view about the value and purpose of philosophical inquiry, (we hope) that there will be something in this volume of interest to you. For better or worse, metaphysics is part of the philosophical landscape, and what we think about metaphysical questions and theories will affect how we think about a range of other topics across philosophy.

For those wanting a more formal definition, on a first pass, we can characterise metametaphysics as the domain of inquiry that is concerned with methodological issues that arise within metaphysics. If correct, then we can say that metaontology is the analogous domain concerned with methodological issues that arise within ontology. However, whether this is the right characterisation of either metametaphysics, or metaontology, is a matter of much debate within the recent literature. It is, itself, a metametaphysical (or metaontological) question, with many differing views about the proper scope of these 'meta-' domains and their relationship with first-order topics.

Given this, the aim of this introduction, or indeed this handbook, cannot be to settle what are the proper and correct topics to be classified as part of metametaphysics and/or metaontology. Rather, here we simply propose that the topics of metametaphysics include, but are strictly not limited to, the topics covered in the chapters of this volume. In light of this, when choosing what to include we have tried to cast the net wide; to include core topics that all (or at least most) would accept as being part of what counts as metametaphysics, and some topics that are less clearly core metametaphysical topics but have the potential to inspire new avenues of research.

The structure of this volume

This volume contains thirty-eight chapters, organised into five thematic sections, and the remainder of this introduction is devoted to providing a brief guided summary of the topics contained within it and its structure. For those entirely new to metametaphysics, this will be useful as a way to begin to understand the main issues; for those with more experience in the domain, this will hopefully serve as a way to see at a glance the focus of the chapters covering both familiar and novel topics alike. This will also, we hope, serve to help readers who are planning to dip into only certain sections of this volume see where to begin.

Part I: The history of metametaphysics

This section is devoted to the history of metametaphysics. Each chapter considers one (or two) central figures within the domain. Though all of them wrote before the term 'metametaphysics' had gained widespread usage, they each made significant contributions to metametaphysical topics and continue to have a lasting effect on the contemporary literature.

Going back to the beginning, as it were, Vasilis Politis argues in Chapter 1 that, in at least one important juncture in their work, both Plato (in *Sophist* 242c ff.) and Aristotle (in *Metaphysica* IV.1–2) argue that in order to properly consider the question, 'What is there?', we need to consider the question, 'What does "to be" mean?' Unlike typical modern metaphysicians, however, Politis argues that neither Plato nor Aristotle think that the question 'What does "to be" mean?' is prior to or foundational of the question, 'What is there?'.

In Chapter 2 of the volume, Bader develops a Kantian approach to meta-ontology. He contrasts first-level and second-level construals of existence with Kant's modal interpretation of existence and then identifies the problems of modal representation as the central issue of Kantian metaontology. He shows how this problem can be overcome by non-conceptual resources. In Chapter 3, Robert Kraut makes the case that Rudolf Carnap is a Pragmatist and an Expressivist about ontology. According to Kraut, Carnap's conception of ontological practice is not the 'deflationist' view – often attributed to him – that ontological disputes are vacuous and/or inconsequential. Contrary to common belief, Carnap does not seek to dismiss arguments about the existence of various sorts of entities as 'merely verbal', or to endorse the eliminativist view that ontological arguments should be expunged from our repertoire. Carnap's goal, rather, is to show that continued participation in ontological theorising is consistent with empiricist scruples. Ontological discourse for Carnap is a device that serves to express commitments to adopting certain linguistic/conceptual resources, hence his expressivism. And the acceptance or rejection of linguistic frameworks is to be decided by their 'efficiency as instruments', hence his pragmatism.

In Chapter 4, Karl Egerton discusses Quine. He explains why a significant part of the enduring value in Quine's work should be accorded to his contribution to metametaphysics. This involves revisiting the core Quinean ideas that inform the methodology of metaphysics, but also includes showing how some less regarded, more contentious aspects of Quine's thought can be seen as indispensable to that thought, problematising the widespread belief that one can extract core aspects of Quine's metametaphysics in isolation without eroding their warrant.

In Chapter 5, Haukioja presents an overview of the debate between metaphysical realism and anti-realism, as it has been carried out during the last decades. Metaphysical realism is characterised as a combination of metaphysical, semantic, and epistemic theses: first, the world consists (mostly) of mind-independent things and properties; second, truth involves a correspondence between words and the world; third, even an ideal theory could be radically false. Arguments against metaphysical realism typically aim to show that this combination of views is internally incoherent or unstable; that the metaphysical realist 'cannot say what he or she wants to say' (as recently put by Tim Button). Haukioja reviews the main challenges to metaphysical realism: Hilary Putnam's model-theoretic argument and his 'brains in a vat' argument, as well as Philip Pettit's theory of global response-dependence of concepts. He also discusses the content of metaphysical anti-realism. Critics of this view typically want to defend common-sense realism, and the anti-realist should therefore reject either the semantic or the epistemic component of metaphysical realism (or both). However, it has turned out to be difficult to give clear positive content to such a view.

In Chapter 6, Nathan Wildman discusses the more recent history of the now waning philosophical obsession with the notion of modality. From the early 1960s until the turn of the twenty-first century, metaphysics was dominated by broadly modal issues. But times have changed. Metaphysics has outgrown its modal myopia, as a number of hyperintensional notions have emerged and now play many of the roles that metaphysical modality formerly did. In his contribution to the volume, Wildman charts the rise and fall of modal metaphysics. In so doing, he also demonstrates that the main argument modal metaphysicians used to motivate the general shift of focus from the metaphysics of what is to the metaphysics of what must be equally motivate a further shift, towards a metaphysics centring on the various post-modal, hyperintensional notions. The upshot of Wildman's discussion is that even modal enthusiasts should agree that modality is not (and should not be) the central metaphysical notion anymore.

Part II: Neo-Quineanism (and its objectors)

The chapters in this section focus on the 'neo-Quinean' conception of metaphysics. Quine has (probably) had one of the largest ongoing influences on (Anglo-American analytic) metaphysics in the twentieth century, and his conception of the central question of metaphysics as being the question of 'what exists?' continues to be defended and critiqued in contemporary work. In this context, metaphysics becomes the domain interested in providing a list of all the things that exist, or in Quinean terms, all the things that are the value of a bound variable within our best theory. This section covers both defences and criticisms of this conception of metaphysics, and broader considerations about the nature of the existential quantifier on which such a conception is founded.

In Chapter 7, Ted Parent offers a slightly opinionated review of the three main factions in metaontology: Quineans, Carnapians, and Meinongians. Discussion of the first faction includes consideration of Quine, van Inwagen, and Sider; discussion of the second includes Carnap, Thomasson, and Hofweber; and of the third Meinong, Priest, Berto, Zalta, and Crane. Particular attention is devoted to how quantification is interpreted by different philosophers. Parent also offers some pragmatist remarks about the legitimacy of ontology.

Rohan Sud and David Manley take on quantifier variance in Chapter 8. Quantifier variance is the view that there are several distinct notions that are relevantly similar to our notion of unrestricted existence, and none of these notions are metaphysically distinguished. They offer an overview of that thesis. They explore some challenges to the claim that there are several distinct existence-like notions and to the claim that these notions are metaphysically on a par. They also consider what implications the view has for first-order ontological debates, particularly common-sense ontology.

In Chapter 9, Brendan Balcerak-Jackson discusses verbal disputes in metaphysics. Certain disagreements in metaphysics and ontology are sometimes thought to be defective. One oft-cited type of example has to do with the special composition question, the question of when some things compose a thing. According to the Universalist: There is an object composed of my nose and your ear. According to the Anti-universalist: No, there is no such object! Many philosophers are tempted to diagnose disputes like this one as merely verbal: the parties don't really genuinely disagree, but are merely talking past each other. Importantly, the diagnosis of mere verbalness is thought to have deflationary consequences. If a dispute in ontology or metaphysics is merely verbal, this is taken to show that there isn't really anything substantive at stake in the dispute; it is really just a disagreement about how to use language. Or it is taken to show that the dispute should not be regarded as being about a genuinely metaphysical matter, but, instead, about what is most interesting or what we should care about more. Balcerak-Jackson aims to elucidate these issues by addressing such questions as: What is it for an apparent dispute to be merely verbal? If a dispute in a certain area of metaphysics turns out to be merely verbal, what (if anything) does this imply about the possibility of substantive metaphysical theorising in that area?

Chapters 10 and 11 come from Agustin Rayo and Matti Eklund respectively. Rayo discusses Absolutism: the thesis that there is sense to be made of absolutely general quantification. Rayo aims to clarify what exactly the metaphysical debate is between absolutists and some of their detractors. Matti Eklund provides an overview of the so-called metametaphysics associated with neo-Fregeanism in the philosophy of mathematics. Neo-Fregeanism – championed most prominently by Bob Hale and Crispin Wright – as the name indicates, is inspired by ideas found in the work of Gottlob Frege. One characteristic of neo-Fregeanism is that it seeks to combine a version of platonism (there exist mind-independent mathematical objects) with a version of logicism (the ontologically committing mathematical truths are akin to truths of logic). Eklund discusses how platonism and logicism can reasonably be combined. Another characteristic of neo-Fregeanism is its focus on abstraction principles, and the chapter discusses the role of such principles as well.

The 'easy approach to ontology' – best associated with Aimee Thomasson – maintains that many existence questions that are contested in metaphysics are actually easy to answer via trivial inferences that take us from an uncontroversial premise, via a conceptual truth, to a conclusion about what exists. Such arguments have often been dismissed or treated as paradoxical – in part because they conflict with the neo-Quinean way of addressing existence questions by seeking the best 'total theory' and determining what it must quantify over. Easy arguments are important not merely because they seem to answer questions about whether properties, numbers, propositions, and things of many other sorts exist, but also because they call into question a way of approaching existence questions that was so dominant as to be virtually unquestioned for decades. Chapter 12, authored by Aime Thomasson, reviews the history of easy arguments and explains their importance both to first-order ontological debates and to meta-ontological questions – where they seem to bring the prospect of demystifying metaphysics, and turning attention away from existence questions. It also reviews major objections to easy arguments, along with paths that have been used in replying to these objections.

Ontology is a subject matter directed at questions such as: Are there numbers? Are there objects that are empirically undetectable? Are there properties? Do tropes exist? Are events real? What about fictional beings such as Sherlock Holmes? But questions of this nature appear to motivate debates amongst scientists as much as they do philosophers: Are there molecules? Does Bigfoot exist? What about extraterrestrials? In Chapter 13, Jody Azzouni evaluates different metaphysical positions on whether (and in what ways) they leave intact our ordinary practices of debating about metaphysical claims and disagreeing over such claims. Some positions – Carnapian and neo-Carnapian positions, such as the one held by Amie Thomasson – explicitly rule out the kinds of metaphysical claims that occur in philosophy. Others philosophers, such as W.V. Quine and proponents of quantifier variantism, such as Eli Hirsch, characterise our (collective) logic resources in such a way that debates can only occur subject to the successful translation of the claims of opponents into the 'languages' of one another. The translations possible, however, undercut important aspects of these debates, such as the fact that the disagreement is often over just the existence of an entity or kind of entities, and not over descriptions of them. Only one position, Azzouni's own quantifier neutralism, leaves metaphysical debate – as we ordinarily do it – intact.

The final chapter of Part II of the volume deals with ontological pluralism. According to the ontological pluralist, there are different ways, modes, or kinds of being. In Chapter 14, Jason Turner discusses recent treatments of the doctrine of ontological pluralism, which tie it to the thought that there are different metaphysically special quantifiers. After discussing some attempts to characterise this thought more precisely and some pitfalls they must avoid, he responds to two arguments against the viability of ontological pluralism. The first of these is Trenton Merricks' argument that disagreements between pluralists about how many modes of being there are must be either trivial or inexpressible. The second is Bruno Whittle's arguments that ontological pluralism and ontological monism are notational variants of each other.

Part III: Alternative conceptions of metaphysics

Part III includes various alternative ways of conceiving of the purpose and methodology of metaphysics that have blossomed in recent years, many in direct response to the apparent failure of the neo-Quinean conception of metaphysics. This includes chapters that defend and those that criticise the idea that metaphysics is a substantive domain of research.

This section opens with a topic of much contemporary interest, grounding. Metametaphysics concerns foundational metaphysics. Questions of foundational metaphysics include: What is the subject matter of metaphysics? What are its aims? What is the methodology of metaphysics? Are metaphysical questions coherent? If so, are they substantive or trivial in nature? Some have claimed that the notion of grounding is useful in addressing such questions. In Chapter 15, Alexander Skiles and Kelly Trogdon introduce some core debates about whether – and, if so, how – grounding should play a role in metaphysics. They consider how grounding might be relevant to whether metaphysical questions are substantive, how to choose between metaphysical theories, and how to understand so-called 'location problems'. Following on from this in Chapter 16, Ricki Bliss discusses what, in the contemporary debate, has become a related notion – that of fundamentality. She considers several different ways in which fundamentality appears to be contemporarily understood; along with the reasons often cited for believing the view in the first place.

In Chapter 17, Naomi Thompson takes on a topic also enjoying an enormous amount of contemporary attention, metaphysical explanation. Thompson distinguishes the different things we might mean by the term 'metaphysical explanation', discusses metaphysical explanation in contrast to the more familiar notion of scientific explanation, and motivates the view that the two are to be distinguished. There is little consensus on what is the best way to model metaphysical explanations, but she presents the most prominent alternatives. She ends with a discussion of realism and antirealism about metaphysical explanation.

In Chapter 18, Ross Cameron explores how questions in metametaphysics interact with questions concerning truthmakers. He defends an account of ontological commitment in terms of truthmaking. Whereas Quine said that the ontological commitments of a theory are those things that must be in the domain of quantification if the theory, regimented in a first-order language, is to be true, the truthmaker account says that the ontological commitments of a theory are those things that must exist to make the theory true. He also argues that the truthmaker account is the best way of defending ontology as a deep and distinctively metaphysical project in the face of challenges from those who argue for a deflationary metaphysics.

Jessica Leech, in Chapter 19, addresses the question, 'What is essence?' In rough terms, the essence of a thing is what it is to be that thing. So, perhaps what it is to be Socrates is, in part, to be human. Or what it is to be the number 2 is to be the successor of 1. Or what it is to be the Mona Lisa is, in part, to have been painted by da Vinci. Such a phrase has its origins in Aristotle, and the notion of essence can be found throughout the history of philosophy. This chapter focuses primarily on attempts to give an account of what essence is, and its relationship to other (meta)metaphysical notions such as necessity, real definition, grounding, and identity more contemporarily. Leech reviews attempts to define essence in terms of necessity, and how these might be refined in the light of some well-known purported counterexamples to such attempts. She considers alternative accounts of essence, for example, given in terms of real definition, grounding, or identity. She considers the breadth of the notion of essence, introducing the distinction between constitutive and consequential essence, and considering what has an essence (objects? properties? facts?). She also takes a critical look at essentialist theories of modality; where rather than giving an account of essence in terms of modality, an account of modality is given in terms of essence.

Chapter 20 looks to a discussion of fictionalism. Fictionalists try to reconcile our linguistic practices with our ontological scruples. They want to speak as if there are, say, numbers, possible worlds, or properties, without thereby accepting the existence of the entities in question. In this chapter, Lukas Skiba and Richard Woodward offer a minimal and inclusive characterisation of fictionalism that tries to do justice to the large variety of projects associated with this label. The authors provide an overview of what they take to be the core choice points facing the fictionalist, a survey of some of the main issues facing the viability of fictionalist strategies, as well as pointers to how these issues might be addressed.

Stephen Barker considers the prospects of globalising expressivism – a position in the philosophy of language that questions the central role of representation in a theory of meaning or linguistic function. An expressivist about a domain D of discourse proposes that utterances of sentences in D should not be seen, at the level of analysis as representing how things are, but as expression non-representational states. So, in the domain of value-utterances, the standard idea is that speakers are expressing affective-states, such as approval or disapproval focused on objects or conditions. Global expressivism is the thesis that for *all* domains of discourse, we treat utterances, at the level of analysis, as expressing non-representational states. In Chapter 21, Barker sets out several conceptions of how one might formulate this programme, advancing his preferred account. Having set out key aspects of his favoured approach, Barker then addresses the question of metaphysics. What metaphysical conception of reality goes with global expressivism so understood?

One of the greatest metaphysical insights that Aristotle contributed to the history of philosophy is that objects may be partitioned in two ways: into parts and into abstracta. The latter kind of division has not received due attention among contemporary extensional mereologists (who advocate division into parts only), and even in neo-Aristotelian quarters. In Chapter 22, in addition to clarifying which type of part is relevant for understanding Aristotle's hylomorphism, Anna Marmodoro urges that we need to critically re-examine certain assumptions we make in our study of Aristotle's theory of substance. Among the questions she raises in this chapter are these: Is there a primary matter-to-form 'relation' in a substance? Is the 'relation' between matter and form in a substance analogous to that of potentiality to actuality? Does Aristotle's theory of substance deliver a sound account of substantial unity? She argues that it doesn't, and supplies an account, which derives from principles within Aristotle's metaphysics, but differs from the account given to us by Aristotle. She argues that Aristotle came very close to having a full account of the oneness of a substance, but fell short of it. He unified matter and form in a substance definitionally, but did not explain and justify the oneness of the definition.

The next chapter in this section is from Elizabeth Barnes. When looking at contemporary analytic feminism, there are three main themes that can be well-described as metametaphysics: the metaphysical importance of the social, the political significance of metaphysical claims, and the link between metaphysics and social progress. The bulk of Chapter 23 focuses on the third of these themes, as Barnes considers this to be the place where feminist philosophy becomes the most metaphilosophically distinctive.

As Rebecca Mason and Kate Ritchie point out, in Chapter 24, traditionally, social entities (i.e. social properties, facts, kinds, groups, institutions, and structures) have not fallen within the purview of mainstream metaphysics. In their contribution to the volume, they consider whether the exclusion of social entities from mainstream metaphysics is philosophically warranted or if it, instead, rests on historical accident or bias. Mason and Ritchie examine three ways one might attempt to justify excluding social metaphysics from the domain of metaphysical inquiry and argue that each fails, thus concluding that social entities are not justifiably excluded from metaphysical inquiry. Finally, they ask how focusing on social entities could change the character of metaphysical inquiry. They suggest that starting from examples of social entities might lead metaphysicians to rethink the assumption that describing reality in terms of intrinsic, independent, and individualistic features is preferable to describing it in terms of relational, dependent, and non-individualistic features.

In Chapter 25, Friederike Moltmann gives an outline of natural language ontology as a subdiscipline of both linguistics and philosophy. She distinguishes natural language ontology from both folk metaphysics and foundational metaphysics, as well as other sorts of cognitive ontologies. She offers a characterisation of the subject matter of natural language ontology drawing a distinction between the core and periphery of language, and argues that part of the constructional ontology reflected in natural language is, in significant respects, on a par with syntax (on the generative view).

The relationship between phenomenology and metaphysics is controversial. One reason for this is that the meaning of both of the central terms is equivocal. In Chapter 26, Dan Zahavi focuses on phenomenology in the Husserlian sense. He addresses such questions as, 'How did Husserl view the relation between phenomenology and metaphysics?' 'Is phenomenology metaphysically neutral, does it disregard all questions pertaining to being, or is it on the contrary committed to a specific kind of metaphysics, say, a form of subjective idealism or robust realism?' The answer to these questions is controversial, and there is no obvious consensus in the literature. He presents and discusses some of the prevailing interpretations before finally turning to Husserl's own texts for some answers.

Part IV: The epistemology of metaphysics

Part IV takes up the question of how metaphysical knowledge is possible, if it is at all. Providing an epistemology of metaphysics is crucial to any full defence of the domain of inquiry. This section will consider the pros and cons of recently defended accounts of metaphysical epistemology, the correct methodology of metaphysical inquiry (for example, whether metaphysical inquiry is an a priori or a posteriori enterprise), and the role that various notions such as 'primitive' and methods, such as conceptual analysis, play in metaphysical theorising.

In Chapter 27, Tuomas Tahko asks what the significance is of the distinction between the a priori and the a posteriori in metametaphysics. The primary relevance of the distinction concerns the source of metaphysical knowledge and our epistemic access to that source. For instance, how can we gain knowledge about things like numbers and sets, which are abstract objects? What about possibility and necessity and modal truths more generally? In many cases, it seems that we have to resort to a priori methods. But this raises an immediate concern, because many contemporary metaphysicians are in favour of naturalistic metaphysics, whereby a priori knowledge is often regarded as problematic, if not altogether impossible. Tahko discusses all of these issues, focusing especially on the distinction between a priori and a posteriori methods and the links to modal epistemology, naturalistic metaphysics, and the relationship between metaphysics and science more generally.

In Chapter 28, Sonia Roca-Royes discusses, in the form of an inconsistent set of three claims, the main challenge in the epistemology of modality. Each of the three main responses to it – Rationalism, Empiricism, and Scepticism – can be seen as consisting in denying one of those claims. This chapter focuses on two meta-epistemological distinctions that are becoming more and more central in the literature. The first one classifies the epistemologies of modality on the basis of the amount and nature of the suggested routes to modal knowledge: uniform accounts suggest only one route to modal knowledge whereas non-uniform accounts suggest more than one. The second one classifies epistemologies on the basis of their suggested epistemic priority relations: accounts that deny epistemic priority relations are known in the literature as symmetric accounts. Within the asymmetric ones, necessity-first accounts take (at least some) knowledge of necessity to be prior to any knowledge of possibility, while it's the other way around for possibility-first accounts. While the primary aim of the chapter is to survey recent literature and identify the directions that current research is taking, it also motivates the need for a non-uniform and symmetric epistemology of modality.

In Chapter 29, Sam Cowling discusses ideology and ontology. Metaphysical theories can differ without differing in their ontological commitments. This is due, in part, to the fact that theories accrue ideological, as well as ontological, commitments; where ideological commitments concern what kinds of truths obtain or what concepts are expressible in theories. Familiar kinds of ideological commitments include the introduction of primitive modal and temporal operators or various primitive predicates satisfied by the putative ontologies of theories. Cowling's entry surveys various kinds of ideological commitments, strategies for characterising the ideology-ontological distinction, and the principles of theory choice that might guide inquiry into the world's ideological structure. He focuses on the respective contributions of David Lewis, W.V. Quine, and Theodore Sider to the notion of theoretical ideology.

All metaphysical (and other) theories have something in common: they all contain primitives. In Chapter 30, Jiri Benovsky explores the role primitives play in theories, emphasising the fact that they do most of the theoretical work. He also discusses the nature of primitives, as well as taking in a discussion of the notion of 'explanatory power'.

Introduction: what is metametaphysics?

In Chapter 31, Frank Jackson discusses how conceptual analysis – understood as the clarification and explication of 'what it takes to be a so and so' questions, of how something has to be to be a so and so – bears on issues in metaphysics, where metaphysics is understood as concerned with what there is and what it is like. He argues that conceptual analysis tells us when accounts of how things are in terms of one set of properties (or concepts) are or are not in potential conflict with accounts in terms of a putatively different set of properties; it can alert us to properties that might otherwise have escaped our attention, or to properties we might have thought incoherent; it can alert us to patterns in data that are important for prediction and explanation, and, relatedly, suggest when it would be wise to make one or another modification to the concepts we employ in prediction and explanation; and that conceptual analysis, in its 'rewriting sentences guise', can show us how to avoid unwanted ontological commitments.

Focusing principally on metaphysical contingentism rather than entity contingentism, in Chapter 32, Kristie Miller outlines three positions: global metaphysical contingentism, global metaphysical necessitarianism, and metaphysical moderatism. The first of these is the view that all metaphysical principles are contingent, the second that all metaphysical principles are necessary, and the last that at least some metaphysical principles are contingent. Arguments for each view are considered, and it is suggested that jointly, these arguments tend to militate in favour of metaphysical moderatism.

In the final chapter of this section, Chapter 33, Thomas Hofweber asks the question 'Is metaphysics special, either in a positive or a negative way?' His contribution to this volume aims to clarify the question and critically discusses several proposals about why one might think that it is special. Among such proposals is the view that metaphysics is especially glorious since it is the queen of the sciences or the true revealer of reality, as well as the view that metaphysics is especially problematic, since it is either attempting to answer meaningless questions or is engaged in unjustified speculation.

Part V: Science and metaphysics

The last section of the volume, Part V, connects in various ways to the topics in Part 4, but with a particular focus on the relationship of metaphysics to the empirical sciences, most centrally physics. Chapters in this section will consider various accounts about how to distinguish metaphysics and science (if we can at all), whether metaphysics should be (merely?) the 'handmaiden of science', whether metaphysics is prior to science or science relies on metaphysical work, and whether metaphysical claims can disagree or even overrule empirical theories.

In Chapter 34, Kerry McKenzie considers the thorny issues of how science is to guide metaphysics in her piece, 'Science Guided Metaphysics'. It seems ludicrous to deny that serious metaphysical enquiry ought to be 'informed by science' – the basic tenet of naturalistic metaphysics. But while seemingly so unobjectionable, McKenzie discusses three problems that must be addressed by any metaphysician defending a naturalistic approach. First, that the very notion is ill-defined without a clear demarcation of what counts as 'science' – a task that was largely abandoned as hopeless in the 1980s. Second, that there is no consensus on what the methodology of naturalistic metaphysics is – in particular on the role of supra-empirical virtues in it – and as such no clear place to start assessing the claim that its methodology is superior to that of its rival. Third, that the 'problem of theory change' that vexed philosophers of science has yet to be addressed in the context of metaphysics, and there are extra reasons to be pessimistic about the idea that metaphysical theories can be said to 'make progress' through shifts in their underlying scientific paradigms. In sum, while on the face of it a naturalistic approach to metaphysics seems an attractive, even compulsory, position, much work remains to be done both in characterising it and in securing its superior epistemic credentials.

While science is taken to differ from non-scientific activities in virtue of its methodology, metaphysics is usually defined in terms of its subject matter. However, many traditional questions of metaphysics are addressed in a variety of ways by science, making it difficult to demarcate metaphysics from science solely in terms of their subject matter. Are the methodologies of science and metaphysics sufficiently distinct to act as criteria of demarcation between the two? In Chapter 35, Milena Ivanova and Matt Farr focus on several important overlaps in the methodologies used within science and metaphysics in order to argue that focusing solely on methodology is insufficient to offer a sharp demarcation between metaphysics and science, and consider the consequences of this for the wider relationship between science and metaphysics.

In Chapter 36, Michael Esfeld takes on the topic of thing and non-thing ontologies. He first considers the main versions of ontologies of discrete objects, most notably atomism and Aristotelian substance ontology, along with ontologies of discrete events. He then considers non-thing ontologies, most notably the ontology of one continuous stuff. In assessing these ontologies, the chapter lays stress on how they fare with respect to physics. In Chapter 37, Matteo Morganti provides an overview of the meaning, motivation, and available forms of naturalised metaphysics. He gives arguments in favour of what will be called 'moderately naturalistic metaphysics', an approach that, while acknowledging the priority of science in providing knowledge of the natural world, also preserves the autonomy of metaphysical inquiry. In particular, he argues that, due to the fact that any explanatory hypothesis is underdetermined by the empirical data, radical naturalism – the view that metaphysics should be made entirely dependent on science or even discontinued – is untenable; and that science and metaphysics should be developed in parallel, the former providing substance and support to metaphysical hypotheses, and the latter making the interpretation of scientific theories possible. In the closing chapter of the volume, Chapter 38, J.T.M. Miller considers the view that a central concern of metaphysics is what is possible. That is, that unlike science, metaphysics studies not only what is actual, but also the ways that reality could be. This view, if right, provides metaphysics with a distinct subject matter from that of science, and, depending on what modal epistemology we adopt, a distinct methodology too.

Of course, much more can be said about each of the issues discussed in this volume. Indeed, we could have doubled the number of its contributions and still not exhausted the spread of valuable and rich topics that fall within the scope of metametaphysics. Alas, thousand-page volumes are a relic of the past, and editors with the skill to corral eighty-plus authors are in short supply in the contemporary academy. Suffice it say, however, that we hope that readers interested in topics in metametaphysics might find something of value in these pages.

Editors' acknowledgements

Editing a book, let alone one as long as this, requires help from too many people to mention them all individually. We would, though, as editors, particularly like to mention a few people. First, thanks to our authors for producing excellent chapters for the volume. Second, thanks to Adam and Tony at Routledge for their help throughout this process, and Penny, Allie, and Pip for helping in the production and copy-editing. Third, we would like to thank Matthew Duncombe and Alessandro Torza for help reviewing. Fourth, thanks to colleagues at Trinity College Dublin and Nottingham University for help and advice about how to approach editing such a large volume. Lastly, but certainly not least, thanks to Anna Bortolan for continued and constant support.

PART I

The history of metametaphysics

1
METAMETAPHYSICS IN PLATO AND ARISTOTLE

Vasilis Politis

Introduction

To consider whether there is in Plato and Aristotle a philosophical enquiry and mode of argument comparable to what is today called metametaphysics, it will, I think, be useful and important to distinguish two very different ways in which one may conceive of such a philosophical enquiry and mode of argument. Both ways are present, it seems to me, in the current metametaphysics literature, but, if I am not mistaken, one of them is dominant.

On a radical conception of metametaphysics, a philosophical enquiry and mode of argument may be characterized as metametaphysical, if the philosopher, in preparation for engaging in metaphysical enquiry in general, and any metaphysical enquiry in particular, deems it necessary to conduct a different enquiry of a semantic-conceptual character.

On a modest conception of metametaphysics, a philosophical enquiry and mode of argument may be characterized as metametaphysical, if the philosopher, in preparation for engaging in a particular metaphysical enquiry, deems it necessary to conduct a different enquiry of a semantic-conceptual character.

It seems to me the difference between the radical and the modest conception of metametaphysics is very important. On the radical conception, it is possible to separate semantic-conceptual enquiry from metaphysical enquiry; and semantic-conceptual enquiry is, in the order of philosophical enquiry, prior to metaphysical enquiry. Not so on the modest conception. The modest metaphysician may, in preparation for engaging in a particular metaphysical enquiry, deem it necessary to conduct a different enquiry of a semantic-conceptual character; but he may think that that metaphysical enquiry is part of or tied up with other metaphysical enquiries, and there is no implication that he will deem that in preparation for engaging in them, too, it is necessary to conduct a different enquiry of a semantic-conceptual character.

In this chapter I argue that we can find in Plato and Aristotle significant metametaphysical enquiries of the modest variety, and the modest variety only. I shall demonstrate this with reference to a particular juncture in Plato's enquiry into being in the *Sophist* and a comparable juncture in Aristotle's enquiry into being in *Metaphysics* IV. Demonstrating that these are metametaphysical enquiries, but of the modest variety only, does not directly establish that there are not radical metametaphysical enquiries in Plato or Aristotle. But it does establish this, if we may suppose that these two particular metametaphysical enquiries in Plato and Aristotle are paradigmatic and exemplary examples of such enquiries. For reasons of space, I cannot defend this supposition in this chapter.

While the commitment to a modest form of metametaphysics seems sensible, I confess that I am puzzled at why anyone would want to commit to the radical form of metametaphysics. At the same time, I recognize that, as metametaphysics is characterized today, it is commonly the radical variety that is being invoked. For, as metametaphysics is commonly characterized, it is 'the foundation of metaphysics'.[1] And it is clear that, on the modest variety, metametaphysics is not foundational.

Plato takes some time out to reflect on the meaning of the expression "to be"

At a particular juncture in the dialogue *Sophist*, Plato turns to the question of what there is. He introduces this juncture as follows: 'It seems to me that Parmenides has communicated with us in a slapdash fashion, as has anyone that ever rushed to judgement in the matter of distinguishing how many the things that are, are, and of what sorts they are' (242c4–6; trans. Rowe). The question at issue, evidently, is "What is there?"; or, in the precise terms in which Plato formulates it, "How many things are there, and of what sort?" But it appears that Plato intends a complaint about the way in which this question has been taken up by his predecessors. What is the complaint?

Having provided a short and colourful summary of a variety of answers that past thinkers have defended in response to the question of what there is (242c8–243a1), Plato says the following:

> Now as to whether any of this, as said by any of them, is true or not, it would be harsh and inappropriate to rebuke men of fame and antiquity for failings on such a scale; but there is that one aspect that no one could begrudge our pointing out ... they have been too inclined to look down their noses at us ordinary mortals and treat us with contempt; each pursues his own project without caring at all whether we're following what they say or being left behind ... Whenever anyone utters something to the effect that many, or one, or two are, or have come into being, or are coming into being, ... by the gods, Theaetetus, do you ever understand a single thing they're saying?
>
> (243a2–b7; trans. Rowe)

His complaint, he says here, does not concern the truth or falsity of past theories about what there is; it concerns the intelligibility of those theories. But what is supposed to be unintelligible about them? And why is it supposed to be unintelligible?

The Eleatic Visitor directly spells out what he finds unintelligible in all available theories, to date, about what there is: it is their use of the expression "to be". He says things like: 'we need first to investigate what is (*to on*), and look for exactly what those who use the expression think they are indicating (*dēloun*) by it' (243d3–5) '"So come on, you people who claim that all things are hot and cold or some other such pair of things: what exactly are you uttering (*phtheggesthai*) about them both, when you say both and each of them *are*?"' (243d8–e2). 'So since we're quite puzzled about it all, it's for you to clarify for us what exactly you intend to designate (*sēmainein*) when you utter (*phtheggesthai*) the word "is"' (244a4–6; translations Rowe, with minor changes).

It is less clear why the Eleatic Visitor finds their use of the expression "to be" unintelligible and puzzling.[2] It seems to me that two very different answers suggest themselves. According to one answer, Plato is articulating a puzzle about the use of the expression "to be" in theories about what there is, because he is in general puzzled about the use of this expression. According to a different answer, Plato is articulating a puzzle about the use of the expression "to be" in theories about what there is, because of something peculiar about how this expression is used in such theories.

I do not believe the first answer finds support in Plato's text; the second answer, rather, is the right reading. The choice between the two answers is critical for our present purposes. Suppose we accept the first answer. This lends itself to the suggestion that what Plato intends at this point is to step out of metaphysical questions altogether, such as the question "What is there?", and intends to take up the semantic question, "What do we mean by the expression 'to be'?", for its own sake purely. This would be a general turn from a metaphysical to a semantic question, and a semantic question that is not intended to be related, either directly or through the justification for asking it, to metaphysical questions. While this is not by itself a turn to radical metametaphysics, it prepares for such a turn.

Suppose, on the other hand, that we accept the second answer. In this case, we may expect that Plato intends to take up, and to stay with, the metaphysical question, "What is there?"; but he thinks that, to advance the enquiry into this particular question, it is necessary to take some time out and investigate how the expression "to be" is used. In the first instance, this means that it is necessary to investigate how the expression "to be" is used in available theories about what there is. As we shall see later, Plato thinks this may lead to a wider investigation, into the use of the expression "to be" more generally. This investigation will, evidently, be semantic-conceptual; but it will, no less, be part of a metaphysical enquiry. Apparently, it will not be possible to separate the semantic-conceptual enquiry from the metaphysical enquiry; because, first, that which is puzzling is the use of the expression "to be" specifically in theories about what there is; and, second, the desire to resolve this puzzle is, in large part at any rate, rooted in the desire to take up and to search for an answer to the metaphysical question, "What is there?"

I think it is beyond question that the second answer is the correct one: what Plato intends to be the source of the puzzlement and impression of unintelligibility is the expression "to be" as used in theories about what there is. Here are my reasons for thinking this.

First, in order to state and articulate this puzzlement and impression of unintelligibility regarding the expression "to be", Plato refers to a variety of theories about what there is; *and he does not refer to a use of this expression outside the context of such theories*. He does so, first, by distinguishing several such views (242c8–243a1). He then divides them into two main varieties: pluralism (243c10–244b2) and monism (244b2–245e5). Finally, he distinguishes two further ways of answering the question "What is there?": the view of 'the giants', or what we would call physicalism or materialism; and the view of 'the gods', which includes both Platonists and Parmenideans (245e6–250b11). He begins by stating and articulating the puzzlement and impression of unintelligibility in regard to all such views (up to 243c9). He then repeats this, with greater care and precision, in regard to a simple variety of pluralism, namely, the dualist view that all things ultimately consist of the hot and the cold; and in regard to monism. Finally, he repeats it in regard to physicalism and Platonism-Parmenideanism. He concludes that we are as puzzled about the use of the expression "to be" in regard to the last views as we were in regard to the first, and for similar reasons (see 249d9–250b11). In all of this, appeal to the use of the expression "to be" outside the context of theories about what there is, is entirely absent.

Second, Plato is emphatic in making the hyperbolically sounding statement that the use of the expression "to be" in such theories is unintelligible to us. This statement would be outrageous hyperbole, or plain absurdity, if it were intended to concern any and every use of the expression "to be", such as, for instance, in the statements "There is something in the larder" or

"There is nothing in the larder". But it is not obviously absurd to maintain that, as things currently stand, this word is unintelligible as used in such theories.

Third, having first made the claim to the unintelligibility of the expression "to be" in regard to a variety of such theories (242c8–243b7), Plato goes on to defend this claim, with rigorous argument, when he distinguishes two main varieties of such theories: (a simple variety of) pluralism; and monism. He argues, in what is clearly intended as a rigorous and demonstrative way, that neither of these theories have the resources to explain what they mean by the expression "to be", in contradistinction to what they mean by, e.g., "to be hot" or "to be cold" (for the pluralists) or "to be one" (for the monists). It is hard to see how anyone could intend to argue, in such rigorous and demonstrative fashion, that the use of the expression "to be" is unintelligible quite generally.

These arguments are quite complex and difficult to analyse, but let me indicate briefly what they look like. What resources, Plato asks first (243c10–244b2), does a simple dualist have, who thinks that all there is ultimately consists of the hot and the cold, to explain what he means by the expression "to be"? He has, it appears, precisely three options: he can explain "x is" as meaning "x is hot"; or as meaning "x is cold"; or as meaning "x is both hot and cold". It is not so difficult to see that none of these options will work: the first option implies that, contrary to what the dualist supposes, cold things cannot be or exist, since "x is cold" means something different from "x is hot"; the second option founders for the same reason; the third option would, at best, allow for lukewarm things only!

What resources, Plato asks next (244b2–245e5), does a monist have, who thinks that all there is is a single unitary whole of parts, to explain what he means by the expression "to be"? Again, the conclusion of his argument is that the monist does not have the resources to explain this. But the argument is more difficult to follow. The gist of it is that such a monist must distinguish between what the expression "to be" means and what the expression "to be a single unitary whole of parts" means; and that he does not have the resources to do this, since he is committed to these two expressions referring to one and the same thing and he wants to explain what the expression "to be" means simply in terms of what there is.

What is most important to observe in regard to both these arguments, it seems to me, is that it would be misplaced to think that, in and through them, Plato is committed to a purely reference-based account of what the expression "to be" means; in particular, that he thinks what the expression "to be" means must be explained in terms of what there is. If anything, Plato is arguing the opposite: that what the expression "to be" means *cannot* be explained in terms of the resources provided by an account, *any* account, of what there is.

I think we may conclude that, while it is proper and potentially illuminating to think of Plato as having significant metametaphysical moments, but of the modest variety, it is wrong to suppose that such moments mark a turn to radical metametaphysics.

One might push back in two ways.[3] First, one might think the semantic enquiry in the dialogue really just is a metaphysical enquiry: asking what "to be" means really is just a way of asking what it is to be, and Plato just talks loosely. This is a common way of reading the *Sophist*. In this case, the metametaphysics of the *Sophist*, even the modest kind, just is metaphysics. Second, one might think that the enquiry breaks down into *aporia* just before the semantic discussion comes in. That might make the semantics foundational to the new enquiry, and hence the *Sophist* is doing radical metametaphysics.

In response to the first pushback strategy, I think it is correct that Plato does not separate issues about the expression "to be" from issues about what there is (and perhaps also, about what it is for something to be). But we may not infer that he does not distinguish between these two issues, or that he is only concerned with the latter. We have seen that, at 243a2ff., a shift is marked from the question "What is there?" on its own, to this question in conjunction with the question "What does the expression 'to be' mean?"

In response to the second pushback strategy, I think it is correct that Plato presents the enquiry as breaking down into *aporia* just before the semantic discussion comes in (again, see 243a2ff.). But, as we saw, he does not think it will be possible to separate the semantic-conceptual enquiry from the metaphysical enquiry.

It might be said that it is still possible to separate the semantic-conceptual enquiry from the metaphysical enquiry. For Plato may think that the *aporia* about the use of the expression "to be", as motivated by considerations about what there is, implies the complete breakdown of metaphysics and the need to provide a new foundation for metaphysics on the basis of semantics. But this, it seems to me, is pure invention and anachronism – as if this particular *aporia* in the *Sophist* did to Plato what Kant's reading of Hume did to him. Plato thinks this is a very serious aporia, *in* metaphysics; but there is not reason to suppose that he thinks it is an aporia *about* metaphysics and its very possibility.

To see this, it is sufficient to observe that, after having at some length (from 242c to 250b) articulated how thoroughly puzzled we are about what those who theorize about what there is mean by the expression "to be", Plato goes on to develop and defend a positive solution to this puzzle or these puzzles; or at least what he intends as the beginning of a solution. Most important, this solution involves an account both of what the expression "to be" means and of what being is. It is impossible in this extended part of the dialogue (250–264) to distinguish two separate accounts, one semantic, one metaphysical; just as it would be wrong to deny that Plato distinguishes semantic from metaphysical elements in this account. Plato's account of what being is basically consists in distinguishing a minimum of five basic kinds ('most important kinds', *megista genē*), and working towards determining both what each of them is in its own right (*kath'auto*) and how it is related to each of the others (*pros alla*). The five basic kinds are: *change (kinēsis), changelessness (stasis), being (to on), sameness (to tauton)* and *difference (to heteron)*.

It is important to recognize that it is true *both* that Plato's account of being in the *Sophist* (250ff.) is developed and defended in response to puzzles that relate specifically to how the expression "to be" is used in theories about what there is *and* that this account of being is intended to account for the use of the expression "to be" not only in such theories but generally. How can Plato have it both ways? How can he exploit problems rooted in a narrowly metaphysical use of the expression "to be" to develop and defend a general account of being and the use of the expression "to be", which is not restricted to the narrowly metaphysical use of the word? And how can he do this, while staying with the question "What is there?" and not intending to step out of this or related metaphysical questions?

One necessary condition for having it both ways, and doing so while staying with the question "What is there?" and related metaphysical questions, is that Plato should not think that there is a firm boundary between a philosophical use of a word, such as the expression "to be", and its everyday use. A further necessary condition is that Plato should think that the philosopher's task is not only to describe the actual use of a word, but also, if called for, to reform and revise it.

A philosopher may be motivated to revise the actual use of a word, if she can show that its actual use gives rise to serious and genuine problems, or *aporiai*. To show this, she need not show that its actual use is immediately, and as the word is used in all or most everyday discourse, subject to serious and genuine problems. Rather, she may show that its actual use gives rise to problems, if it is suitably extended in a certain direction. For example, she may argue that the actual use of the expression "to be" leads to problems as soon as its use is extended from relating to *some* things to relating to *all* things. She may argue that this extension reflects a problem even in the original, non-extended use of the expression "to be", because this original use allows for the extension but without anticipating it or having the resources to provide for it.

I believe it is along these lines that Plato proceeds in the *Sophist*. It is striking how, when he formulates the various theories about what there is, he formulates them as being not only about 'that which is' (*to on*) and 'the things that are' (*ta onta*), but also about 'the totality of things' (*to pan*) and 'all things' (*ta panta*).[4] The combinations, in Plato's text, of the terms *to on* and *ta onta* with the terms *to pan* and *ta panta* are so frequent that it is hard not to suppose that Plato thinks that what is distinctive of the use of the expression "to be" in theories about what there is, is that this word is used of the totality of things (*to pan*) and of all things (*ta panta*).

I may add that, irrespective of what metametaphysicians may think, it is not true to say that only metaphysicians use the expression "to be" for all things and the totality of things. I recall the Hollywood blockbuster *The Theory of Everything*, which would hardly have drawn much of a crowd, had it been titled: *A Theory in Physics about all Physical Things*. In Plato's day, as he was well aware, people generally were keen on thinking about all things; a kind of thinking they articulated in cosmic myths. Perhaps we today are not so different.

Aristotle takes some time out to consider whether the multivocity of the expression "to be" is compatible with an enquiry into being

Suppose, as many critics do,[5] that the enquiry in Aristotle's *Metaphysics* gets properly under way only with the introduction, in the first two chapters of book IV, of the question "What is being?" In that case, it will be plausible to suppose that Aristotle's metaphysical enquiry in the *Metaphysics* is premised on a semantic premise, namely, that the expression "to be" is said in many ways. For, after the very short first chapter of book IV, in which he basically states that the enquiry is addressed to the question "What is being?", Aristotle begins the second chapter with the claim that "to be" is said in many ways and that this presents an apparent problem for the possibility of an enquiry into being. He then dedicates the first part of this chapter (IV.2, up to 1003b19) to proposing a solution to this problem. Basically, the solution consists in distinguishing between primary being (*to prōton on, ousia*) and other kinds of being, and positing a relation of ontological dependence between the two. And this distinction will emerge as central in the remainder of the *Metaphysics*.

Now, if it is true that Aristotle premises the enquiry into being, which is recognizably what we would call a metaphysical enquiry, on a semantic premise, which says that the expression "to be" is multivocal (i.e. said in many ways), then Aristotle may be likened to a current metaphysician who is at the same time a radical metametaphysician who thinks that semantics is the foundation of metaphysics.

What, if anything, is wrong with this account of Aristotle's project in the *Metaphysics*?

One response would be to say that, when Aristotle begins the second chapter of book IV with the claim that 'being is said in many ways' (*to on legetai pollachōs*), he does not intend this as a semantic claim, a claim about the expression "to be". One might defend this response by appealing to the *Categories*, to which the claim that being is said in many ways appears to go back; and by arguing that, in the *Categories*, this claim is not intended as being about the use of the expression "to be".

But I do not want to avail of this response. For it seems to me that, as stated and articulated in the opening of *Metaphysics* IV.2, the claim is indeed that the expression "to be" is said in many ways; just as, as Aristotle directly goes on to spell out, the expression "to be healthy" is said in many ways. I recognize that this reading is controversial. But I think there is reason to by-pass the controversy, for our present purposes. For I want to argue that, *even if we suppose that* the statement *to on legetai pollachōs* is understood to mean 'the expression "to be" is said in many ways', and hence as a semantic claim, *still* it is wrong to think that Aristotle intends to base the enquiry into being on a semantic premise.

It is one thing to say that (i) a particular semantic supposition (e.g. that the expression "to be" is said in many ways) presents a problem for the possibility of an enquiry into being; quite another to say that (ii) the enquiry into being is premised on and based on a particular semantic supposition. There is no immediate inference from (i) to (ii). The question is whether Aristotle uses (i) as support for (ii).

It is plausible to think that Aristotle uses (i) as support for (ii), *only if* it is plausible to think that he is committed to the following view:

> *The semantic-based reading of Aristotle's enquiry into being*
> The solution (as spelled out in the first part of *Met.* IV.2, up to 1003b19) to the problem whether the multivocity of the expression "to be" is compatible with an enquiry into being, serves to establish the very possibility of the enquiry into being, its structure and aim and end.

On *the semantic-based reading of Aristotle's enquiry into being*, this metaphysical enquiry will be heavily informed by the semantic supposition that the expression "to be" is said in many ways. For, it will be true to say that the very possibility of the metaphysical enquiry, and its very structure and aim and end, is based on a suitable engagement with the semantic supposition. If this much is the case, it may not be going too far to conclude that there is an important sense in which Aristotle's enquiry into being is premised on and based on a semantic premise.

Let me indicate what I think is at issue in considering whether *The semantic-based reading* is right or not. On any reading of *Metaphysics* IV.2, Aristotle's solution to this apparent problem (i.e. the problem whether the multivocity of the expression "to be" is compatible with an enquiry into being) consists in distinguishing between primary being (*to prōton on, ousia*) and other kinds of being; with some kind of relation of dependence between the two. And on any reading of the *Metaphysics* as a whole, a general distinction between primary being and other kinds of being serves, in the remainder of the *Metaphysics*, to structure and provide a distinctive aim and end to the enquiry into being. As Aristotle memorably states at the end of the first chapter of book VII:

> So it is, indeed, that the question "What is being?" (*tí to on*), which has forever, both in the past and in the present, been a question of enquiry and a source of puzzlement, is precisely the question "What is primary being?" (*tís hē ousia*) ... And this is why we too must, principally and primarily and practically exclusively, attend to the investigation of that which is in this way (*peri tou houtōs ontos*, i.e. that which is in the primary way).
> (1028b2–7)

What is at issue is, I believe, the following question:

> Q: Does Aristotle think that the general distinction between primary being (*to prōton on, ousia*) and other kinds of being, which, on any reading, serves, in the remainder of the *Metaphysics*, to structure and to give a distinctive aim and end to the enquiry into being, *is precisely* the distinction between primary being and other kinds of beings that is established (in *Met.* IV.2.1003a33–b19) through engaging with the semantic premise that "to be" is said in many ways?

If the answer to this question is YES, then *The semantic-based reading* will be right; if NOT, NOT.

Let us consider, therefore, what is required for answering YES to this question; hence for defending *The semantic-based reading*; hence for thinking that there is a sense in which Aristotle's metaphysics is based in semantics.

It seems to me that two suppositions are required for answering YES to question Q. First, the supposition that, when Aristotle introduces the question "What is being?" in *Metaphysics* IV.1, he does so as the starting-point of a metaphysical project. Second, the supposition that the distinction between primary being and other kinds of being that he makes in the first part of *Metaphysics* IV.2, and which he makes in response to the claim that the expression "to be" is said in many ways, *is precisely* the general distinction between primary being and other kinds of being of which he will make crucial use in the remainder of the *Metaphysics*.

It should be clear that both suppositions are required for answering YES to question Q. For suppose, contra the first supposition, that the question "What is being?", as introduced in IV.1, is a question that he considers it necessary to raise in order to advance a metaphysical project that, rather than taking its starting point here, is already under way. In that case, there will be no plausibility to the view that when, in IV.2, he articulates, and proposes a response to, a semantic-based problem regarding the question "What is being?", *that* problem is intended to be *the* basic problem of, and hence potentially foundational to, his metaphysical project. It will be plausible, rather, to suppose that that problem is but one of the many problems that (as book III, the book of *aporiai*, is testimony to) will have to be addressed in order to advance a metaphysical enquiry that is already under way.

And suppose, contra the second supposition, that the general distinction between primary being and other kinds of being, of which Aristotle will make crucial use in the remainder of the *Metaphysics*, is intended to be different from, and significantly wider than, the semantic-based distinction between primary being and other kinds of being that he introduces in the first part of IV.2. In that case, there will be no plausibility to the view that semantic-based distinction between primary being and other kinds of being is *the* basis of, and hence potentially foundational to, Aristotle's metaphysical project in the remainder of the *Metaphysics*.

It seems to me that both suppositions are mistaken. I conclude, therefore, that *The semantic-based reading of Aristotle's enquiry into being* is mistaken. I have (with Jun Su 2017, and with Philipp Steinkrüger 2017) argued against both suppositions. The following is a compressed version of those arguments.

Contra many critics and a prominent tradition of interpretation of the *Metaphysics*,[6] it is mistaken to suppose that, when Aristotle introduces the question "What is being?" in *Metaphysics* IV.1, he does so as the starting-point of a metaphysical project. On the contrary, it can be shown that the question "What is being?", and the concepts *that which being is (to on hēi on)* and *that which determines that which being is (to prōton on, ousia)*, so central in the first two chapters of book IV, are introduced in book III; and introduced for the purpose of engaging with certain *aporiai*, especially the Fifth Aporia (997a34–998a19) and the Twelfth Aporia (1001b26–1002b11). Furthermore, it can be shown that book III serves to develop the project of book I.[7] This is clear especially from the Fifth Aporia, when read together with certain passages at the end of book I (I.8, 988b22–26; 989b21–29; I.9, 992b18–22). Most important, in I.9, 992b18–22 Aristotle provides an argument for the claim that there are *ousiai*; and this claim is presupposed in IV.1–2. This argument relies on the idea of searching for the elements of things; elements being a kind of primary causes. This shows that Aristotle's argument for the claim that there are *ousiai* relies on his very project as he has been characterizing it from the beginning of book I (esp. I.1–2); which is the project of searching for the primary causes of all things.

In this way, the concept of *ousia*, so central in IV.1–2, goes back first to certain *aporiai* in book III; from there it goes back to the end of book I (esp. I.9, 992b18–22); and from there it goes back to the beginning of the *Metaphysics* (I.1–2) and the distinctively metaphysical characterization there of the whole project. This shows that the famous two chapters at the opening of book IV are a further step in a metaphysical enquiry which is already well under way. And this shows that, even if the enquiry in those chapters is significantly premised on a semantic premise (that the expression "to be" is said in many ways), we cannot conclude that Aristotle wants to premise metaphysics in general on a semantic premise.

Likewise contra many critics and a prominent tradition of interpretation of the *Metaphysics*,[8] it is mistaken to suppose that, in IV.2, when the distinction between primary being and other kinds of beings is introduced, it is introduced only in response to the problem whether the multivocity of the expression "to be" is compatible with an enquiry into being. It is true that this is *one* problem in response to which the distinction between primary being and other kinds of being is introduced (this problem is addressed up to 1003b19). But it is not the only problem.

A second problem (addressed at length after 1003b19) is this: How can a plurality of different kinds (they include identity, similarity, their opposites, and such opposites in general) be the subject-matter of a single science and, in particular, the science of being *qua* being? Whereas the first problem is based on the semantic supposition that the expression "to be" is said in many ways, the second problem is based, not on any semantic supposition, but on what is clearly a metaphysical supposition. This is the supposition that *being* is not the only property that is true of *each and every thing*; rather, there are other properties that are true of *each and every thing*, such as: *being one, being either changing or changeless, being identical with itself and different from other things*, and the like. Aristotle's second problem, therefore, is this: Why suppose that the enquiry into being and all things should be focused on the question "What is being?", any more than on the questions: "What is unity?", "What is change/changelessness?", "What is identity/difference?", and the like?

While it is true that Aristotle responds to both these problems by introducing *a* distinction, in both cases, between primary being and other kinds of being, he does not think that both problems are solved by *one and the same* distinction between primary being and other kinds of being. Aristotle is content to introduce a general and flexible way, or ways, or distinguishing between primary being and other kinds of being, and he leaves it for later in the *Metaphysics* to consider how this distinction will need to be worked out.

It is, I think, very important to recognize that this is how Aristotle proceeds. Had he based the distinction between primary being and other kinds of being solely on the semantic-based problem, he would have excluded from his metaphysical project any constructive discussion with anyone who does not share the view that the expression "to be" is said in many ways. He would thus have excluded Platonists and natural philosophers, for they have never heard of this view and may not accept it once they hear of it. But it is clear (both from IV.2 and from, e.g., the end of VII.1 (quoted above) and VII.2 especially) that Aristotle intends to enlist such very different thinkers as Plato and natural philosophers into the enquiry into being and primary being. Aristotle, in the *Metaphysics* at any rate, is a dialectical thinker, not a card-carrying Aristotelian.

I conclude that it is proper and potentially illuminating to characterize Aristotle as a modest metametaphysician. For, as this juncture of the *Metaphysics* shows, he is a philosopher who, in the course of a particular metaphysical enquiry and argument, and for the purpose of advancing that particular enquiry and argument, deems it necessary to take some time out from the particular enquiry and argument in order to address a semantic-based problem with which that enquiry is faced. But to suppose that Aristotle thinks that a semantic enquiry, or a semantic-based problem facing a metaphysical enquiry, occupies a basic and foundational role in metaphysical enquiry is, I think, no better than a mistake.

Notes

1 Manley, in the Introduction to the 2009 collection of papers on metametaphysics (suitably sub-titled *New Essays on the Foundations of Ontology*), begins on p. 1 with the statement: 'Metametaphysics is concerned with the foundations of metaphysics.'
2 For the role of *aporia* ('puzzlement', 'a particular problem') in Plato's *Sophist*, see Brown 2017.
3 I am grateful to the anonymous referee for suggesting that I consider these pushback strategies.
4 For the (clearly deliberate) move from *to on/ta onta* to *to pan/ta panta*, see: 242e5, 243d9, 243e3, 244b3, 243b6, 245b9, 245c8, 249d1, 249d4, 250a2.
5 For references, see Politis and Su 2017.
6 For references, see Politis and Su 2017.
7 I am supposing book II is a later insertion.
8 For references, see Politis and Steinkrüger 2017.

References

Brown, L. (2017) '*Aporia* in Plato's *Theaetetus* and *Sophist*', in G. Karamanolis and V. Politis (eds.), *The Aporetic Tradition in Ancient Philosophy*. Cambridge. 91–111.
Manley, D. (2009) 'Introduction: a guided tour of metametaphysics', in D. J. Chalmers, D. Manley and R. Wasserman (eds.), *Metametaphysics. New Essays on the Foundations of Ontology*. Oxford. 1–37.
Politis, V. and Steinkrüger, P. (2017) 'Aristotle's second problem about a science of being *qua* being: a reconsideration of *Metaphysics* iv 2', *Ancient Philosophy* 37, 1–31.
Politis, V. and Su, J. (2017) 'The concept of *ousia* in *Metaphysics* Alpha, Beta, and Gamma', in R. Polansky and W. Wians (eds.), *Reading Aristotle. Argument and Exposition*. Brill. 257–276.

2
KANTIAN META-ONTOLOGY[1]

Ralf M. Bader

1 Meta-ontology

Ontology (as conceived of nowadays) is concerned with the question of what there is.[2] The task of ontology is to provide an inventory of the world. It is supposed to address first-order existence questions and identify what (kinds of) things exist. Meta-ontology, by contrast, is concerned with second-order questions regarding existence. On the one hand, it examines the nature of existence and attempts to explain what it is for something to exist. On the other, it examines the methodology and epistemology for addressing first-order existence questions. It attempts to explain how one can settle existence questions and adjudicate ontological disputes. These two projects are related, in that understanding what the nature of existence consists in is likely to help one get a better understanding as to how one can find out what (kinds of) things do in fact exist.

A prominent contemporary approach to existence proceeds via quantificational resources. Existence is understood in terms of the existential quantifier. What it is to exist is to fall within the range of an existential quantifier. Existence is understood in terms of either 1. being identical to, or 2. being instantiated by some member of the domain of quantification:[3]

1. a exists iff $\exists x(x = a)$
2. F's exist iff $\exists x(Fx)$

When operating with an ontologically loaded existential quantifier, one can define both a first-level and a second-level existence property.

FIRST-LEVEL:

existence is the property that some thing has iff it is such that there exists something that is identical to it, i.e. Ea iff $\lambda x[\exists y(y = x)]a$

SECOND-LEVEL:

existence is the property that some concept or property has iff it is such that there exists something that instantiates it, i.e. EF iff $\lambda X[\exists x(Xx)]F$

When existence is understood in this way, it is natural to read off the ontological commitments of a theory by looking at what this theory quantifies over. To settle existence questions, one has to determine which theory is true and then identify the ontological commitments of that theory.

2 The nature of existence

The Kantian approach rejects this approach to meta-ontology. Most importantly, existence is neither to be understood in terms of the existential quantifier, nor in terms of a first-level or second-level property.

2.1 The particular quantifier

Kant does not countenance an existential quantifier. There are three quantifiers in the table of the logical functions of judgement, namely

1   universal: 'all'
2   particular: 'some'
3   singular: 'the'

None of these quantifiers involve existential commitments. Quantification is existentially non-committal. Instead of using an existential quantifier, Kant operates with the particular quantifier: 'some'. This quantifier is not ontologically loaded and does not have existential import.[4]

We can see this clearly when noting that, for Kant, universal quantification implies particular quantification, i.e. 'all' implies 'some'. One can infer 'some' from 'all', e.g. 'some unicorns have horns' follows from 'all unicorns have horns', without any commitment to the existence of unicorns. Neither of these judgements is existentially committing. Universal judgements can be non-vacuously true when they do not have any instances. Correspondingly, particular judgements can also be non-vacuously true when they do not have any instances. Such judgements are made true, not by their instances, but by the concepts involved in the judgement.

Analytic judgements, in particular, are non-vacuously true even when they do not have any instances. This is possible because there are two ways in which judgements can be true.[5] On the one hand, logical truth is purely a conceptual matter. It is a question of concept containment. Analytic judgements are logically true, independently of whether they are universal or particular judgements. Material truth, on the other hand, is a matter of the world being the way that it is represented to be. The connection that is represented by the judgement has to obtain not (only) at the level of the concepts involved in the judgement, but (also) at the level of the things corresponding to the concepts. Material truth does require existence. For a judgement to be made true by its instances, it must have instances.[6]

2.2 Existence as a modality

Existence is not a matter of the quantity of a judgement. When making existence judgements, one does not achieve existential purport by means of the quantifier. Nor is existence a matter of the predicate that is employed in a judgement. This is because existence is not a property. It is neither a first-level nor a second-level property. As Kant notes in his critique of the ontological argument, 'existence' is not a real predicate. Instead, it is a merely logical predicate. It functions grammatically as a predicate. However, there is no property that corresponds to this predicate. As a result, it cannot be used in determining objects.

Rather than being part of the matter of what is being judged (which would be the case if existence were to be a real predicate) or being part of the form of the judgement (which would be the case if existence were a quantifier), existence is a modality.[7] It is one of the three categories of modality: 1. possibility, 2. existence and 3. necessity (alongside their correlates: impossibility, non-existence and contingency). None of these modal categories are real predicates and none of them contribute to the content of judgements.[8]

Whereas the subject and the predicate constitute the matter of a judgement, its quantity, quality and relation constitute its form. Together they constitute the content of the judgement. They determine what it is that is being judged. Modality, by contrast, concerns the manner in which it is judged. The different modalities constitute different modes in which one and the same content can be judged. Modality is not part of the content but applies to such a content. As such, it functions as an operator, i.e. existence can be construed as: EXISTS(Fa).

This operator applies to judgeable contents. It is predicational complexes, with both a subject- and a predicate-component, to which EXISTS applies. This is because it is property instantiations or facts that exist and that are represented by predicational complexes. This means that, on Kant's conception, it is strictly speaking a category mistake to apply existence to objects or to properties, as is done by the quantificational approach as well as by approaches that construe existence as a first-level or second-level property of objects and properties, respectively. More generally, it is a category mistake to speak of modal properties. A property is something that an object instantiates. Modality, however, does not apply to objects. Instead, it applies to the connection between the object and the property, namely to the property instantiation. In Kant's terms, modality concerns the "value of the copula" (A74/B100). Accordingly, one needs a copula, i.e. a predicational complex, in order for modality to apply.

The category mistake involved in ascribing existence to objects or properties is analogous to the category mistake that is involved in ascribing truth to concepts. Existence is the metaphysical analogue of truth, i.e. the logical modality corresponding to the category of existence is truth. In the same way that truth applies to judgements, so existence applies to property instantiations. It is not the subject that is true, nor the predicate, but instead the predication. Likewise, it is not the object that exists, nor the property, but instead the property instantiation.[9]

The modalities are thus to be understood in terms of operators that do not contribute to the content of judgements but apply to such contents. An important upshot of this is that there is no room for iterated modalities. Whilst much of contemporary modal logic is concerned with iterated modalities and the accessibility relations on the domain of possible worlds that give rise to them, these questions do not make any sense from a Kantian perspective. To get an iterated modality, one would have to make the modality part of the content, so that one could then apply a further modality to that content. The first/inner modality would have to be part of the content to which the second/outer modality could then be applied. This, however, is ruled out if modal predicates are not real predicates and cannot enter into the content of judgements.[10]

Modality is nowadays often construed as being primarily concerned with possibility and necessity (and their correlates impossibility and contingency). On the Kantian approach, by contrast, existence is as much a modality as are possibility and necessity. Once existence is recognised as a modality, one can conceive of ontology as not only the study of what there is, but more broadly as the science of being that encompasses all modes of being. It deals not just with what exists, but also with what is possible and what is necessary, i.e. what there can be, what there is and what there must be.

This approach also renders reductive theories of modality incoherent. One cannot reduce modal notions to non-modal notions by reducing modality to what exists, for instance to concrete possible worlds à la Lewis 1986, given that existence itself is a modality. At most, one can reduce some modal notions to other modal notions, not however to non-modal notions. Similarly, theories that attempt to banish modality from the fundamental level of reality, as done by Sider 2011, are a non-starter when operating with the Kantian construal of existence. Given that existence is a modality and given that the fundamental level of reality exists, the fundamental level cannot be non-modal. Likewise for fictionalist, conventionalist and eliminativist approaches to modality. One cannot adopt these approaches to modality, unless one is a fictionalist, conventionalist or an eliminativist about existence. There is, however, no such thing as a non-modal or pre-modal characterisation of the world, since characterising the world requires one, at least in part, to give an account of what exists. Nor can one reduce modality to essences. Though proposals to make sense of possibility and necessity in terms of essences have some plausibility, one cannot reduce or explain existence in terms of essences. After all, the upshot of Kant's critique of the ontological argument is that nothing is such that it exists by its very essence.[11] Given that existence cannot be reduced, modality cannot be reduced.[12]

3 The problem of modal representation

Kant advocates a distinctive conception of the nature of existence. Existence is to be understood neither as a real predicate (= property) nor in terms of quantificational resources. Instead, it is a real (= metaphysical) modality. This account of existence gives rise to a distinctive way of thinking about meta-ontology. The crucial issue for Kantian meta-ontology is to explain how we can represent existence (as well as the other modalities). How can we employ the modal categories to make possibility, existence and necessity judgments?

The problem of modal representation constitutes the core of Kantian meta-ontology. It arises because none of the modal predicates are real predicates. They do not contribute to the content of what is judged/represented. The content is determined by the component representations together with the way in which they are determined with respect to quantity, quality and relation. Modality, by contrast, is not part of the content. It applies to such a content but is not itself part of any content. As a result, one needs to explain how modality enters into representations and how it is that we can represent different modalities. In particular, one needs to explain how we can make existence, possibility and necessity judgements and how these various modal judgements differ from each other.

This focus on the question as to how existence (as well as the other modalities) can be represented forms part of Kant's representational turn and its associated faculty-based meta-ontology. This involves focusing on the ways in which one can represent the world and on the faculties that can be employed in generating judgements, distinguishing the various cognitive faculties and the different roles that they can play. This faculty-based approach makes use of two distinctions that are at the heart of Kant's theory of cognition. On the one hand, there is the distinction between the three cognitive faculties: understanding, judgement and reason, to which corresponds the contrast between possibility, existence and necessity judgements. On the other, there is the distinction between the two ways in which a given faculty can be employed, namely the contrast between the logical and real employment of our faculties, which gives rise to the corresponding contrast between the logical and the real modalities.

3.1 Relative positing

As long as we remain at the level of what Kant calls the logical employment of our faculties, we will not be able to give our judgements existential purport. The logical employment is existentially neutral. This is because it consists in putting together concepts to form simple judgements

(= categorical judgements), as well as putting together simple judgements to form complex judgements (= hypothetical and disjunctive judgements). No matter how we put them together, we will be unable to achieve existential purport. It does not matter whether we are forming a universal, particular or singular judgement, an affirmative, negative or infinite judgement, a categorical, hypothetical or disjunctive judgement, or a problematic, assertoric or apodictic judgement.

Existence is not a function of the logical form of the judgement. How a given judgement is determined with regard to the logical functions of judgement has no bearing on its ontological commitments. Logic is free of existential commitments.[13] It has nothing to say about existence (or non-existence).[14] It operates entirely at the level of concepts and judgements, not at the level of the world. Existence does not have a place in logic and is not to be found in the table of the logical functions of judgements. Instead, it belongs to metaphysics and, correspondingly, features in the table of categories.[15]

Nor does it matter what concepts we are combining in our judgements. Existential purport cannot be achieved by appealing to the concept <existence>. Given that existence is not a property and that 'existence' is not a real but a merely logical predicate, existence judgements do not involve an attribution of the property of existence. No such property exists. One is neither attributing a first-level property to objects, nor a second-level property to concepts. This means that the contribution that the category of existence makes to existence judgements is not at the level of the predicate.

More generally, the problem is that concepts are general representations that can be empty. Such general representations might not have any objects corresponding to them. There might be nothing in the world that falls under them. As long as one is working with concepts, one is representing things only mediately. The representation relation between concepts and objects is not immediate. Concepts (immediately) represent properties and thereby (mediately) represent those objects that instantiate the properties in question. One is representing various features that objects may (or may not) have. One is not representing objects directly, but only representing the properties that they might instantiate. This means that existential purport cannot be achieved when merely employing concepts.

All that we are doing when combining concepts is to engage in relative positing. One concept is posited relative to another. Representing existence and giving our judgements existential purport, however, requires absolute positing. Doing so requires a switch from the merely logical employment to the real employment of our faculties.

3.2 Absolute positing

The key question is how we can move beyond the conceptual, how we can move from the level of thought to the level of the world, from logic to metaphysics. When concerned with what there is, we need to get outside the realm of the conceptual. We need to move from the logical to the real employment of our faculties and, correspondingly, from the logical functions of judgement to the categories.

We need to posit something, not in thought, but in the world. Rather than merely connecting concepts, we have to immediately represent the world as being a certain way. In order to do so, we need to immediately represent objects. This allows us to represent property instantiations in the world. Rather than representing the connection between the predicate concept and the subject concept, we are then representing the connection between a property and an object. This means that we need to make use of objectual representations. Only then can we achieve existential purport. Only then can we represent existence.

In the case of a logical judgement one combines a subject concept S and a predicate concept P. The former mediately represents those objects that fall under it. The latter represents a property that is predicated (or denied, depending on whether it is an affirmative or negative judgement) of these objects (either of all of them or only some of them, depending on whether it is a universal or particular judgement).

This model, however, does not apply when it comes to existence judgements, such as 'God exists' (or more clearly in subject-predicate form 'God is existent'). On the one hand, the predicate 'existent' does not represent a property that can be ascribed to the subject of the judgement. On the other, one would not be able to achieve existential purport but only make a claim to the effect that whatever falls under the concept <God> (if there should happen to be any such objects) also falls under the concept <existent>. That way one would merely be making a claim about the way in which the (possibly empty) extensions of these concepts relate to each other. An existence judgement, however, is meant to have existential purport. It is not meant to represent a relationship between possibly empty extensions, but instead represent the world as being such that the relevant extensions are not empty.

The question then is: what is the correct logical form of existence judgements, if not in terms of a subject-predicate judgement whereby <existent> is predicated of the subject concept? The answer is that existence does not come in at the level of the predicate but instead at the level of the subject.

The concept <God> contains various predicates, namely the divine predicates. Something that has the properties corresponding to these predicates is something that falls under the concept <God>, i.e. what it is for something to be God is for it to instantiate the divine properties. This means that we can represent God as existing by representing an object as instantiating the properties corresponding to the divine predicates. Whereas <God> appeared to be the subject of the judgement, this concept in fact provides the predicates that are being predicated in the judgement. The existence judgment 'God exists' is then to be understood as predicating the divine predicates of something in the world.

The representation that functions as the subject of the judgement, accordingly, has to immediately represent an object in the world. This means that it has to be a non-conceptual/objectual representation by means of which we can immediately represent particulars. The existence judgement 'God exists' represents an object as instantiating the divine properties. It attributes these properties to a thing that is immediately represented by means of a non-conceptual representation, i.e. its correct logical form is EXISTS(Da), where D are the divine predicates and a is the object that is non-conceptually represented.

Accordingly, we are no longer dealing only with concepts but are also operating with non-conceptual representations that function as the subjects of existence judgements. This is precisely what the real employment of our faculties consists in, namely operating on non-conceptual representations (paradigmatically on intuitions). That way we are not merely combining concepts, namely general representations, but are synthesising non-conceptual representations that have existential purport by representing particulars. It is thus not by means of the predicate but instead by means of the subject that we represent existence. Existential purport is achieved, not by means of concepts, but by means of non-conceptual resources, most notably by means of intuitions which immediately represent particulars.[16]

4 Cognising the real modalities

Doing ontology requires one to move beyond the merely logical to the real employment of our faculties. One has to not only operate with concepts but bring in non-conceptual resources.

By employing non-conceptual resources one can solve the problem of modal representation. Understanding how the real modalities can be represented helps us to make progress with the epistemological and methodological part of meta-ontology, namely the question how we can cognise existence, as well as the real modalities more generally. It thereby allows us to get clearer on the methodology that is to be used in settling ontological questions.

When we are concerned with the real modalities rather than the logical modalities, we are dealing not only with analytic conditions that are based on logical principles but with synthetic conditions that are based on metaphysical principles. As a result, a crucial gap opens up between the logical and metaphysical modalities. Most notably, being non-contradictory is no longer enough for being possible. Whether something is logically possible can be determined by inspecting the concepts involved. Analysis of concepts suffices for the logical modalities. All that one has to do is to see whether they satisfy the relevant analytic conditions. However, one cannot do so when it comes to real possibility. Real possibility, existence and real necessity cannot be established on the basis of conceptual resources alone.

The key question then is how we can determine whether our concepts have objective reality, i.e. whether the objects that they represent are really possible. Unless objective reality can be established, it may turn out that our concepts are empty figments of our imagination that lack any basis in reality. This is precisely the issue that Kant raises when noting that "thoughts without content are empty" (A51/B75). Nothing is cognised by means of such thoughts. Even though one is thinking something, i.e. one is combining concepts in thought, one is not cognising anything. If the concepts that one is employing lack objectively real content, then the resulting judgements do not amount to genuine cognitions but are instead idle speculations that lack any basis.

This problem arises in particular when dealing with synthetic judgements. To ensure that such judgements have objectively real content, one needs to bring in non-conceptual resources. This is relatively straightforward when dealing with a posteriori judgements. In that case, experience can establish objective reality. One can cognise the real possibility as well as actuality of that which is encountered in experience. Experience involves the combination of intuitions and concepts and constitutes the paradigm example of the real employment of our faculties. One is moving beyond the purely conceptual level and is instead synthesising intuitive representations.

Difficulties arise, however, when dealing with synthetic a priori judgements. Such judgements are not established on the basis of conceptual connections alone, nor are they based on experience. This makes it difficult to understand how one can establish their objective reality. This concern is particularly pressing when it comes to metaphysical theorising. Such theorising is neither based on experience, nor on the concepts themselves. In short, the problem is that of making sense of synthetic a priori judgements. How can cognition be extended beyond that which is given in experience?[17] This is where Kant's transcendental account comes in. Cognition is not restricted to what is given in experience (= empirical cognition), but also includes what makes experience possible (= transcendental cognition). Even though synthetic a priori judgements are not based in experience, they are justified on the basis that they make experience possible. This means that we can cognise the objective reality of metaphysical concepts, such as substance and causation, on the basis that they are required for experience to be possible. Metaphysics thus acquires a transcendental justification.

5 Conclusion

The Kantian approach operates with a distinctive conception of the nature of existence. Existence is neither a first-level nor second-level property but is instead a modality that applies to predicational complexes, rather than to objects or properties. Ontology on this approach is not only concerned with what there is, but also with what there can be and what there must be. The central problem of Kantian meta-ontology is the problem of modal representation. This problem arises because none of the modal predicates are real predicates. They do not represent properties and do not contribute to the content of our judgements. This renders it difficult to explain how the real modalities can be represented. Doing so requires one to go beyond the conceptual. Real modality cannot be represented as long as one remains at the level of the logical employment of our faculties. One has to bring in non-conceptual resources, most notably intuitions. Intuition thus plays a central role in making modal representation possible. It enables us to immediately represent the world as being a certain way (= absolute positing), rather than merely representing various connections amongst concepts, i.e. combining them in thought (= relative positing). Once the real modalities are distinguished from the logical modalities, it becomes imperative to give an account as to how we can ensure that our representations are not empty, that they are not mere figments of the imagination but instead represent real constituents of the world. We can do so either empirically on the basis of what is given in experience, or transcendentally in terms of what is required for experience to be possible.[18]

Notes

1 This chapter outlines the central commitments of a Kantian meta-ontology. For a systematic exegetical treatment of these issues cf. "Kant's theory of modality" (Bader: forthcoming).
2 Cf. Quine 1948 for the classic statement of this view.
3 An important problem for this approach is that it presupposes a domain of *existing* things. The domain of quantification needs to be interpreted as consisting of all and only those things that exist. This, however, presupposes a prior understanding of existence that cannot be spelled out by means of a quantificational account.
4 Cf. Priest 2008 for some interesting historical observations about how the particular quantifier came to acquire existential import.
5 The contrast between logical and material truth corresponds to that between truth understood in terms of the logical extensions of concepts and truth understood in terms of the non-logical extensions of concepts. (For a helpful discussion of the role of logical and non-logical extensions in Kant's logic cf. Anderson 2015: chapter 2.4.)
6 The relevant distinction is thus not between analytic judgements and synthetic judgements, but between logical truth and material truth. Synthetic judgements are special because they cannot be true solely in virtue of facts about the concepts involved. Hence, for them to be true, they have to be materially true.
7 Kant uses existence interchangeably with actuality. Since there is no distinction between existence and actuality, non-actual things cannot exist, i.e. there cannot be any mere possibilia.
8 This holds not only for the real but also the logical modalities. The logical modalities also do not add to the content of judgements. For instance, existence in metaphysics is the analogue of truth in logic – in the same way that truth does not add anything to the content of what is thought, so existence does not add anything to the content of what is represented.
9 Loosely speaking, one can say that objects or properties exist, on the basis that they feature as subjects or predicates in a suitable predicational complex. This is analogous to the way in which one can say that P is true of something or that something is true of S. Yet, S and P, strictly speaking, are not themselves true. Instead, it is the predicational complex 'S is P' that is true. Likewise, it is not F and a that exist or are actual but the property instantiation Fa that exists. This loose way of speaking leads to confusions since it suggests that there can be existence claims that do not involve any predicate components, such as 'something exists' or 'everything exists' (cf. McGinn 2000: 26–28). Such claims, however, are ill-formed and do not follow from ordinary existence judgements.
10 Importantly, the negative correlates of the modal categories are not to be understood in terms of applying a negation operator to their positive counterparts but instead in terms of the logical division of their higher genera, i.e. negation comes in, not at the level of the operator, but at the level of the differentiae, e.g. NON-EXISTS does not result from applying negation to EXISTS but from dividing POSSIBLY.

11 Even if one rejects the Kantian critique and considers some things to be such that their existence can be explained in terms of their essence, it is implausible to hold that this applies to everything that exists, since it would imply a modal collapse whereby possibility, existence and necessity become co-extensive.
12 Whilst one might try to reduce actuality by adopting the indexical account proposed by Lewis, this option become unavailable once existence and actuality are identified.
13 This applies not only to general logic but also to what Kant calls transcendental logic.
14 Importantly, not even logical necessity implies existence. This implies that metaphysical modality is not a restriction of logical modality. One cannot represent these modalities in terms of possible worlds, whereby the set of metaphysically possible worlds forms a subset of the set of logically possible worlds, such that if something holds in all logically possible worlds it also holds in all metaphysically possible worlds. Logical necessity does not imply existence, yet metaphysical necessity does imply existence. This is because logical and metaphysical modalities have different domains and apply to different things.
15 This means that the existential quantifier approach not only mistakenly attributes existential commitment to non-universal quantification but also confounds logic with metaphysics. Not only does it place existence under the wrong heading, given that existence belongs to modality rather than quantity, it also, more troublingly, places it in the wrong table. Whilst quantifiers belong to the table of the logical functions of judgement, existence belongs to the table of categories. Existence is a real modality, not a logical modality. Accordingly, it belongs in the table of categories, not the table of the logical functions of judgement.
16 Intuitions are to be understood in a broad sense that does not imply object-dependence and allows for non-veridicality. For a helpful discussion of this notion cf. Stephenson 2015.
17 This is not restricted to absolutely a priori judgements, such as those of metaphysics, which are entirely independent of experience, but also includes comparatively a priori judgements which proceed from something that is given in experience to something that is not given in this way, which raises the question how we can extend cognition beyond what is given in experience. This is what Kant addresses in the Postulates of Empirical Thought (A218–235/B265–287).
18 Thanks to audiences at Oxford, Keele and St Andrews. I am grateful to Mario Schärli, Erica Shumener and especially to Andrew Stephenson for helpful comments on an earlier draft of this chapter.

References

Anderson, R. L. (2015) *The Poverty of Conceptual Truth: Kant's Analytic/Synthetic Distinction and the Limits of Metaphysics*, Oxford: Oxford University Press.
Kant, I. (1900) *Kants gesammelte Schriften*, Reimer/de Gruyter.
Lewis, D. (1986) *On the Plurality of Worlds*, Oxford: Blackwell Publishers.
McGinn, C. (2000) *Logical Properties – Identity, Existence, Predication, Necessity, Truth*, Oxford: Oxford University Press.
Priest, G. (2008) "The closing of the mind: how the particular quantifier became existentially loaded behind our backs", *The Review of Symbolic Logic* 1: 42–55.
Quine, W. V. (1948) "On what there is", *The Review of Metaphysics* 2: 21–38.
Sider, T. (2011) *Writing the Book of the World*, Oxford: Oxford University Press.
Stephenson, A. (2015) "Kant on the object-dependence of intuition and hallucination", *The Philosophical Quarterly* 65: 486–508.

3
RUDOLF CARNAP
Pragmatist and expressivist about ontology

Robert Kraut

Question: If a horse's tail were called a leg, how many legs would a horse have?
Answer: Four: because calling a tail a leg does not make it one.

The moral of this familiar conundrum – use/mention conflations and missing quotation marks aside – is that matters of language are distinct from matters of fact. Calling a tail a leg does not make it a leg. Likewise – by parity of reasoning – calling the world a place that contains natural numbers does not make it a place that contains natural numbers. And – to push the point further – the pragmatic utility of calling the world a place that contains natural numbers does not make the world a place that contains natural numbers. Analogously: arguments about whether Finns are Nordic, Slavic, or Scandinavian are not arguments about the suitability of descriptive predicates; nor are they instances of negotiating the appropriate use of a bit of language. They are arguments about culture, ethnicity, and/or geography.

So goes the folk wisdom on such matters. If Pluto is a planet, then trendy shifts in taxonomic practices cannot render Pluto a non-planet: semantic facts do not determine astrophysical realities. At most they determine the way we describe such realities.

Carnap's views on ontology – as customarily construed – appear to contradict such wisdom. In claiming that ontological questions are "formal mode" questions (concerning linguistic frameworks) misleadingly formulated in the "material mode of speech," Carnap appears to conflate questions about the way the world is with practical questions about the advisability of deploying one or another "linguistic form" in describing the world. His rhetoric smacks of such conflation ("To accept the thing world means nothing more than to accept a certain form of language …" (Carnap 1950, 208).) But any such conflation appears inconsistent with a language/world dichotomy and with Carnap's profound commitment to a mind-independent reality that "pushes back" on our empirical inquiries. The contrast between how things are and how it is convenient to talk must be preserved.

Carnap's views – properly understood – do not threaten the contrast. His theory of ontology leaves intact the notion of a mind-independent reality that constrains practical decisions about linguistic/conceptual resources. Nor is his conception of ontological practice the "deflationist" view – often attributed to him – that ontological disputes are vacuous and/or inconsequential. He does not seek to dismiss arguments about the existence of various sorts of entities as "merely verbal," or to endorse the eliminativist view that ontological arguments should be expunged from our repertoire. Ontological questions are neither patronized nor slated for disappearance.

Carnap's goal is rather to show that continued participation in ontological theorizing is consistent with empiricist scruples. This he does by suggesting that ontological discourse is not a descriptive, fact-stating mechanism, but rather a device that serves to express commitments to adopting certain linguistic/conceptual resources ("linguistic frameworks"). Thus his expressivism. Moreover, the acceptance or rejection of linguistic frameworks "will be decided by their efficiency as instruments ... the ratio of the results achieved to the amount and complexity of the efforts required" (Carnap 1950, 221). Thus his pragmatism.

The goal in what follows is to clarify Carnap's account of ontological discourse, articulate its consequences, and correct some common misunderstandings. Several caveats: (1) Choice of examples and case-studies is critical: fragments of metaphysics tied to semantic theory, e.g., strike Carnap as worthy of legitimization, whereas metaphysical speculations that appear idle and/or unrelated to empirical scientific inquiry are deemed worthy of dismissal. Not all metaphysics is on a par. (2) It is likely that Carnap's views on these matters shifted and evolved. It is no goal of this discussion to put forward a unified account of Carnap's views over time. The current discussion seeks to develop the position articulated primarily in "Empiricism, Semantics and Ontology," though supporting text is often drawn from works of different periods. (3) Occasionally it is unclear whether Carnap advocates deflation, elimination, expressivism, revision, error theory, analytical reduction, explication, and/or some other strategy in accounting for ontological practice. Interpretive caution is required: certain methodological contrasts (e.g., between "eliminativism" and "expressivism") might not have been evident to Carnap, given the intellectual environment.

One terminological caveat is especially urgent. It is not uncommon to hear Carnap's approach characterized as "deflationary." Amie Thomasson, e.g., speaks of "Carnap's original form of ontological deflationism" (Thomasson 2016, 122). Rasmus Jaksland notes

> the recent surge of interest in deflationary approaches to ontology in particular and metaphysics in general, often claimed to follow a tradition after Rudolf Carnap. Deflationists regard metaphysical debates to be misguided.
>
> *(Jaksland 2017, 1196)*

Such characterizations are prevalent but misleading. 'Deflationary' – as a term of philosophical art – denotes a family of theories (not any specific theory) about the semantics of '... is true.' These theories share the basic assumption that 'S is true' serves not to attribute a property to a sentence but to do something else. There are many varieties of such theories: redundancy, performative, prosentential, semantic ascent, minimalist, robust, disquotational, etc. Details vary considerably. Some are friendlier toward metaphysics than others; some illuminate Carnap's strategies, some not, and some are irrelevant.

Perhaps these terminological caveats (or reprimands) are beside the point. Those who describe Carnap's approach to ontology as "deflationary" do not, perhaps, intend to gesture toward deflationary theories of truth, but rather intend to portray Carnap's theory as serving to "deflate" ontology.[1] To *deflate* is to devalue, deprecate, minimize, depict as vacuous or misguided, weaken confidence in, depict as shallow, portray as eliminable without substantial loss, or otherwise downplay. If this is what 'deflationary' signals when applied to Carnap's theory of ontology, then serious interpretive errors have occurred: Carnap's goal is not to let the air out of ontology and minimize it, but rather to portray it as legitimate in the face of empiricist misgivings. He wishes to earn ontology the right to go on.

In light of such considerations, it is best to leave the word 'deflationist' behind when articulating Carnap's views.

The data

Carnap invokes several traditional metaphysical disputes as a motivating backdrop for his theory of ontological practice. One such dispute concerns the reality of mind-independent objects; another concerns the existence of abstract entities. Here we focus upon the latter as a case-study.

A venerable tradition in metaphysics – Platonism or *ante rem* realism – explains facts about resemblance, classification, laws of nature, and truth in terms of universals (or properties) and relations among them. Universals are *abstract entities*: eternal, immutable, non-localizable in space-time, multiply instantiable, and not dependent for their existence upon mental or linguistic activity.

The question is whether such entities exist. It seems an intelligible question. If the question is puzzling it is not because *existence* is puzzling, but because *universals* are puzzling. Surely the concept of existence is familiar and straightforward: either universals exist or they do not (however difficult it might be to determine). Questions about the existence of neutrinos, prime numbers greater than ten, or golden mountains are meaningful, intelligible, sufficiently precise, and offer no special conceptual problems; questions about the existence of universals are no exception.

That's one view. It is not Carnap's view. Carnap doubts that all existence questions are equally clear and intelligible; he wonders, e.g., what it would BE for universals to exist (or to not exist). He questions the very concept of *existence* ("the concept of reality") as deployed in this metaphysical context. Presumably a world that contains universals differs, in some way or other, from one that does not: but it is unclear how to characterize the difference without begging questions. Carnap does not formulate his skeptical misgivings with precision and clarity. He says, e.g.,

> I cannot think of any possible evidence that would be regarded as relevant by both philosophers [viz. the realist and the nominalist concerning the existence of numbers], and therefore, if actually found, would decide the controversy or at least make one of the opposite theses more probable than the other.
>
> *(Carnap 1950, 219)*

Here Carnap's skepticism is obviously grounded in matters of verification and confirmation: if no possible evidence is relevant to resolving a controversy, then the controversy is illusory.

Here, however, Carnap displays a slightly different (though related) perspective:

> [concerning disputes about the reality of the external world] Realists give an affirmative answer, subjective idealists a negative one, and the controversy goes on for centuries without ever being solved. And it cannot be solved because it is framed in a wrong way. To be real in the scientific sense means to be an element of the system; hence this concept cannot be meaningfully applied to the system itself.
>
> *(ibid., 207)*

Here the claim is that concepts are being deployed illegitimately–i.e., applied in ways that outreach their meaningfulness.

Speaking of philosophical disputes about the existence of numbers, Carnap says

> Unfortunately, these philosophers have so far not given a formulation of their question in terms of the common scientific language. Therefore our judgment must be that they have not succeeded in giving to the external question and to the possible answers any cognitive content. Unless and until they supply a clear cognitive interpretation, we are justified in our suspicion that their question is a pseudo-question, that is, one disguised in the form of a theoretical question while in fact it is non-theoretical; in the present case it is the practical problem whether or not to incorporate into the language the linguistic forms which constitute the framework of numbers.
>
> *(ibid., 209)*

Admittedly, much of this looks like heavy-handed radical empiricism. Nevertheless, the upshot is that Carnap does not understand the philosophical disputes in question. He seeks conceptual clarification. Our interpretive task is complicated by the fact that Carnap's remarks on this front already deploy his formidable theoretical apparatus – linguistic frameworks, internal vs. external, quasi-syntactic, etc. – which presumably were introduced by Carnap *to solve a problem*; here we need to get a clear view of the (alleged) problem. For now, let us simply regard Carnap as dumbfounded and frustrated with ongoing, seemingly insoluble ontological disputes; he wants to know what is going on.

More specifically: Carnap challenges the metaphysical disputants to specify the *content* of their disagreement about the existence of universals. He thinks the disagreement is vacuous (pending further directives). He does not know what they are disagreeing about, or what would constitute a compelling argument for one side or the other.

It is not easy to sympathize with Carnap's skepticism: the content of existence disputes – whatever the subject matter – is surely clear and obvious. Carnap's misgivings appear to flow from an unduly restrictive empiricist epistemology – perhaps an obsessive verificationism.

Not clear. Puzzlement about certain occurrences of 'reality' and 'existence' talk is not unique to Carnap, and need not be grounded in positivist epistemology. Here, e.g., is Murray Gell-Mann reflecting in conversation upon Hugh Everett's "many worlds" interpretation of quantum mechanics:

> As interpreted by some people, Everett's work has two peculiar features. One is this talk about many worlds being "equally real," which has confused a lot of people … What does it *mean*, 'equally real'? It doesn't really have any useful meaning. But what people mean is that there are many alternative histories of the universe – many alternative course-grained, decoherent histories of the universe – and the theory treats them all on an equal footing, except for their probabilities. Now if *that's* what you mean by 'equally real,' OK – but that's *all* it means.[2]

Yet another instance of concern about basic ontological notions occurs in Einstein-Podolsky-Rosen's classic paper:

> A comprehensive definition of reality is, however, unnecessary for our purposes. We shall be satisfied with the following criterion, which we regard as reasonable. If, without in any way disturbing a system, we can predict with certainty (i.e. with probability equal to unity) the value of a physical quantity, then there exists an element of physical reality corresponding to this physical quantity. It seems to us that this

> criterion, while far from exhausting all possible ways of recognizing a physical reality, at least provides us with one such way, whenever the conditions set down in it occur. Regarded not as a necessary, but merely as a sufficient, condition of reality, this criterion is in agreement with classical as well as quantum-mechanical ideas of reality.
>
> (Einstein et al. 1935, 777–8)

It is hard to tell from EPR's text whether their concern is epistemic – viz., how to recognize a physical reality – or metaphysical – viz., *what it is* for something to be a physical reality. Certain forms of empiricism and/or operationalism might lurk in the background, blurring the contrast between epistemology and metaphysics. Nevertheless, the above remarks suggest that Carnap is not alone in wondering *what it is* for an entity or magnitude to exist and/or be real. The content of the existence/nonexistence contrast (or: real/non-real contrast) is not always clear.

Return to the ontology of universals. Carnap's puzzlement is perhaps prompted by the fact that universals carry no fund of sensory experiences on their sleeves (such is the sting of radical empiricism). But there are, contra Carnap, strategies for addressing such skepticism: i.e., strategies for showing that the touted difference between the existence and nonexistence of universals is, in fact, a genuine difference that makes a difference. For the record, we briefly rehearse several such strategies, lest it be assumed that we endorse Carnap's misgivings about the content of ontological claims:

1. Perhaps the content of the disputed ontological contrast can be found on the level of *truths*. Universals are among the truthmakers for a variety of sentences. The argument invites us to consider such sentences as

    a Some human properties are inherited, whereas others are acquired.
    b There are properties of microparticles that have not yet been discovered.
    c Similar objects share properties.
    d Redness is a sign of ripeness.
    e Patience is a virtue.

Such sentences, it might be argued, depend for their truth upon the existence of universals. Therefore the nonexistence of universals entails the falsity of such sentences. That gives clear sense to the disputed contrast between the existence and nonexistence of universals.

This is not a compelling argument. Even allowing that the logical form of (a)–(c) involves higher-order quantification, higher-order variables need not be construed as ranging over universals: such quantification might be understood in substitutional rather than objectual terms. As for the abstract noun phrases in (d) and (e), nominalistic paraphrases might dispel the sense that reference to universals is involved.[3]

So this effort to address Carnap's bewilderment about the alleged contrast between the existence and nonexistence of universals is not conclusive. It seeks content for the contrast on the level of sentential truth. It is not likely to be found there. Other possibilities, however, are more promising:

2. Various objects resemble one another in significant ways; they have similar causal powers. The explanation of such resemblance might turn on an appeal to universals. It is because the objects in question instantiate certain universals, rather than others, that they have the powers and resemblances they do.

If this is right, then the problematized existence/nonexistence contrast is manifest in the contrast between a world that contains resemblances and one that does not. Surely this would address Carnap's skepticism.

3   Laws of nature, unlike mere generalizations, support counterfactuals; and laws of nature, unlike generalizations, figure in the explanation of their instances. One possible explanation of the contrast is that laws are singular statements of fact describing relationships between universals or magnitudes. It is because predicates embedded in lawlike statements denote universals, and because certain connections hold among those universals, that certain lawlike statements are true.

If this is right, then the partisan can insist against the skeptic that the contrast between the existence and nonexistence of universals is manifest in the contrast between a world in which laws of nature hold and one in which they do not. Surely this would address Carnap's skepticism.

4   Certain predicates correctly apply to certain individuals. Linguistic usage is normatively constrained. The explanation of such normativity might lie in the fact that the universals expressed by those predicates are exemplified by those individuals.

If this is right, then the partisan can insist that the contrast between the existence and nonexistence of universals is manifest in the contrast between normatively constrained behavior and behavior that lacks such constraints. Surely this would address Carnap's skepticism.

Each such strategy purports to locate an area in which Carnap's skeptic can be reassured that disputes about the existence of universals are substantive.[4] But Carnap's skeptic – driven by an overly restrictive epistemology, a commitment to empiricist operationalism, and/or unreasonable standards of intelligibility – remains skeptical.

The point of this exercise is to emphasize that the forces driving Carnap toward his theory of ontology are *not* irresistible. If he seeks a definition of 'exists' and cognate expressions, it is not clear that any such definition is forthcoming: some concepts are so basic as not to be susceptible to analysis in terms of other concepts. 'Existence' and 'reality' might be among such concepts. On the other hand, if he seeks empirical, sensory-perceptual confirmation for the existence of universals, it is not clear that such a demand is reasonable.

For the present we put aside the persuasiveness of Carnap's motives. Analogue: Hume's theory of causation and causal concepts is rooted in theories of experience, impressions, epistemic justification, and concept acquisition that might or might not be plausible; nevertheless Hume's theory is of considerable philosophical significance. Carnap's theory of ontology is analogous. However persuasive (or not) his motives might be, Carnap does not wish the entire practice of metaphysics to grind to a screeching halt: ontology is an essential aspect of semantic theory, an enterprise he deems significant. His task at hand is *not* the rejection of ontology; his task is rather to earn ontologists the right to do what they do. Thus Carnap asks – with admirable pragmatist resolve – "What are people *doing* when they engage in ontological disputes?" His theory is an elaborate response to this question.

The theory

It is undeniable that Carnap regards the bulk of metaphysics as consisting of pseudo-statements devoid of meaning, and thus worthy of elimination. But not all metaphysics is thus condemned: ontology – the study of What There Is – is an essential part of semantic theory and thus worthy of retention. Because there are several distinguishable strains in Carnap's thinking about metaphysics, it is unwise to seek a unifying insight. One strain is dismissive and eliminativist; another is revisionist; another is reductive. Yet another – less publicized and often overlooked – is expressivist.

In *Philosophy of Logical Syntax*, Carnap draws an explicit distinction:

> We have here to distinguish two functions of language, which we may call the expressive function and the representative function.
>
> *(Carnap 1935, 27)*

Carnap thus countenances a bifurcation between descriptive, fact-stating discourse and discourse which serves some other purpose. The contrast is especially familiar (though controversial) in metaethics, wherein emotivism and other strains of noncognitivism draw heavily upon it. Thus equipped, Carnap stresses the expressive function of metaphysics: "[ontological] propositions in the metaphysical books obviously have an effect on the reader, and sometimes a very strong effect, and therefore they certainly express something" (*loc. cit.*). Note that here Carnap neither patronizes nor downplays ontological propositions; he rather shifts attention away from one kind of meaning to another. Granted, such a shift might be deemed a variety of "downplay": but this assessment would illegitimately assume that expressive discourse fares lower than representational discourse on the scale of discursive significance. It does not.

As a matter of historical precedent, it is relevant that noncognitivist explanations of moral discourse were part of Carnap's intellectual climate: in 1935 he notes that "a value statement is nothing else than a command in misleading grammatical form" (Carnap 1935, 24). And the metaphysical discourse that constitutes the bulk of Heidegger's *Was ist Metaphysik?* is construed by Carnap as expressive. Carnap says

> The (pseudo)statements of metaphysics do not serve for the description of states of affairs. … They serve for the expression of the general attitude of a person toward life.
>
> *(Carnap 1959, 78)*

Here we find Carnap's expressivism made explicit. But even as an expressive mechanism, most metaphysics is deemed defective. It is deluded and self-deceived:

> … through the form of its works it pretends to be something that it is not.
> The metaphysician believes that he travels in territory in which truth and falsehood are at stake. In reality, however, he has not asserted anything, but only expressed something, like an artist.
>
> *(ibid., 79)*

Thus metaphysics is more like art – a mechanism aimed toward expression – and less like science – a mechanism aimed toward articulation of truths about the world. But metaphysics is *bad* art: inferior to poetry, music, and other expressive art forms:

> lyrical poets do the same without succumbing to self-delusion …
> The harmonious feeling or attitude, which the metaphysician tries to express in a monistic system, is more clearly expressed in the music of Mozart. … Metaphysicians are musicians without musical ability.
>
> *(ibid., 80)*

Here Carnap treats the bulk of metaphysics as a defective tool, an inadequate expressive substitute for art: it achieves – in misleading "theoretical" form – what art achieves more honestly and effectively.

Not all metaphysics is regarded with dismissiveness: statements about the existence of systems of entities are essential to semantic theory – a vital enterprise. Carnap's "metalinguistic" pragmatism aims to legitimize ontology – at least, those portions of ontology required by semantic inquiries. The meaning of ontological claims is to be understood in expressivist terms: not as expressions of commitment to a way of life, but to the value and/or pragmatic advisability of deploying specific linguistic/conceptual frameworks.

The human condition tolerates myriad commitments: to forms of life, linguistic resources, religious perspectives, artistic genres, styles of human companionship, philosophical methods, or whatever. Pragmatic consequences of any such commitment vary: some are adaptively beneficial, some conducive to communal solidarity, some enriching, and some stultifying. But once the bifurcation between descriptive and expressive discourse is acknowledged, Carnap's basic conception of the practice of ontology is easier to fathom.

Expressivism is a prominent theme in Carnap's thinking, and ought not to be neglected. The central thesis of this discussion is that Carnap recommends a "non-descriptivist" or "antifactualist" account of ontological discourse, analogous to noncognitivist accounts of moral discourse advocated by emotivists. Insofar as this interpretation captures Carnap's theory, it is clear that he seeks neither elimination nor analysis of ontological claims: he rather wishes to explain them – while preserving their integrity – within the larger context of human commitment, as a mechanism that functions to achieve certain nondescriptive ends. Note that this strategy does *not* constitute a "deflation" of ontology. Any appearance to the contrary derives from the prejudice that departures from descriptivism constitute deflation (in the sense adumbrated earlier). But such prejudice is misguided.

Return to the traditional dispute about universals. One strain of realist insists that such entities exist; one strain of nominalist denies their existence. The disputants engage in sufficient discussion to ensure agreement about the properties universals would have, were they to exist, thus establishing that the dispute is non-illusory and concerns the same subject matter. How, according to Carnap's account, should we understand what is happening here?

The central ingredient in Carnap's explanation is the notion of a *linguistic framework* or *linguistic form*. There is room for disagreement as to how these notions are best understood. Frameworks are basically formal systems: to "introduce a framework" is to "introduce new ways of speaking, subject to new rules ..." But "ways of speaking" are *meaningful*: they have semantic content. Often it is unclear whether Carnap intends frameworks to be semantically interpreted structures or merely syntactic systems (i.e., sets of vocabulary items combined with recursive formation rules and "customary deductive rules"). As a case in point, his specification of the *system of natural numbers* is helpful:

> The framework for this system is constructed by introducing into the language new expressions with suitable rules: (1) numerals like 'five' and sentence forms like 'There are five books on the table'; (2) the general term 'number' for the new entities, and sentence forms like 'five is a number'; (3) expressions for properties of numbers (e.g. 'odd', 'prime'), relations (e.g. 'greater than') and functions (e.g. 'plus'), and sentence forms like 'two plus three is five'; (4) numerical variables ('*m*', '*n*', etc.) and quantifiers for universal sentences ('for every *n* ...) and existential sentences ('there is an *n* such that ...') with the customary deductive rules.
>
> (Carnap 1950, 208)

For the present we leave open the question of how robust a notion of *framework* is required to capture Carnap's strategy.

Thus equipped, the dispute about the existence of universals involves – in a sense yet to be articulated – a specific linguistic framework: viz., *the system of thing properties*:

> The thing language contains words like 'red,' 'hard,' 'stone,' 'house,' etc., which we used for describing what things are like. Now we may introduce new variables, say '*f*,' '*g*,' etc., for which those words are substitutable and furthermore the general term 'property.' New rules are laid down which admit sentences like 'Red is a property,' 'Red is a color,' 'These two pieces of paper have at least one color in common' (i.e., 'There is an *f* such that *f* is a color and ...').
>
> (ibid., 211–12)

Call the system of thing properties 'Lp': this is the linguistic framework the adoption of which lies at the foundation of the controversy about universals. As Carnap sees it, the realism/nominalism issue manifests a question about the advisability of adopting Lp; that question, in turn,

> is not a theoretical question but rather the practical question whether or not to accept those linguistic forms. This acceptance is not in need of a theoretical justification (expect with respect to expediency and fruitfulness), because it does not imply a belief or assertion.
>
> (ibid., 218)

The realist has made "the practical decision to accept certain frameworks" (viz., the framework of thing properties); the nominalist embraces an opposing decision. The conflict is real – as real as that between a strategist who decides to go to battle and one who reaches the opposing decision that appeasement is the optimal strategy. The conflict involves a clash of commitments rather than one of beliefs: viz., commitment to the pragmatic advisability of adopting a given discursive framework. But the conflict is *real*, drawing upon beliefs, commitments, agendas, goals, and interests.

Not only is the conflict real, but its resolution has ramifications. Every framework comes with benefits and liabilities; disputes about the pragmatic advisability of their adoption can be complex and difficult. In the present case, adoption of Lp involves risks as well as rewards. Lp offers tremendous advantage in expressive power; but specifications of logical form and inferential connections require deployment of a second-order logic, which presents formal and philosophical complexities. Here is how Stewart Shapiro describes the situation:

> Some philosophers argue that second-order logic is not logic. Properties and relations are too obscure for rigorous foundational study, while sets and functions are in the purview of mathematics, not logic; logic should not have an ontology of its own. Other writers disqualify second-order logic because its consequence relation is not effective – there is no recursively enumerable, sound and complete deductive system for second-order logic. The deeper issues underlying the dispute concern the goals and purposes of logical theory. If a logic is to be a calculus, an effective canon of inference, then second-order logic is beyond the pale. If, on the other hand, one aims to codify a standard to which correct reasoning must adhere, and to characterize the descriptive and communicative abilities of informal mathematical practice, then perhaps there is room for second-order logic.
>
> (Shapiro 1998)

For present purposes, the point is that any pragmatic cost-benefit analysis of "the system of thing properties" must acknowledge not only that quantification over second-order variables brings immense increase in expressive power, but that additional, perhaps unwanted complexities arise. If the monadic second-order variables range over sets, questions arise as to whether they range over the full power set of the specified domain: if so – i.e., if the semantics deploys full models containing all subsets of the first-order domain – then completeness fails, compactness fails, and Lowenheim-Skolem fails. There is no sound and complete proof system for second-order logic. Moreover decisions must be made about the form of a comprehension scheme: i.e., whether every open higher-order sentence specifies a property or whether type restrictions apply.

And the complexities do not end here. If sufficiently ambitious we might wish to explore Second Order Number Theory – that is, the theory of the integers together with all sets of integers; questions then arise which are not solvable from ZFC. One solution is to add axioms; but choice of axioms is notoriously complex and delicate (see Woodin 2001).

Perhaps such complexities are not prohibitive (depending upon one's goals); but clearly the decision to adopt a linguistic framework which requires a higher-order logic has strings attached; adoption of linguistic frameworks is serious business.

Carnap tells us that "the philosophical statement of the reality of properties – a special case of the thesis of the reality of universals – is devoid of cognitive content." He does *not* claim the statement to be devoid of meaning *simpliciter:* cognitive content is not the only kind of meaning there is. He seeks to illuminate the role of the ontological question "Do universals exist?" in a new light – one which involves

> a practical, rather than a theoretical question; it is the question of whether or not to accept the new linguistic forms. The acceptance cannot be judged as being either true or false because it is not an assertion. It can only be judged as being more or less expedient, fruitful, conducive to the aim for which the language is intended.
> *(Carnap 1950, 214)*

Whether any gain in understanding is achieved by explaining ontological practice in terms of pragmatic utility rather than representation of facts and accurate description of reality is not to be adjudicated here; the current task is to articulate Carnap's view, not to evaluate it. But it is clear that Carnap's conception of ontology does *not* portray it as shallow or easy: to portray ontological conflict as a clash of commitments rather than a clash of beliefs is not to patronize, minimize, or deflate it: it is simply to relocate it on the map of human endeavor, one likely to be friendlier to empiricist constraints.

This highlights the pragmatist aspect of Carnap's conception of ontological dispute. The expressivist aspect must now be discussed.

Not all philosophical illumination takes the form of reductive analysis, meaning equivalence, or translational paraphrase. In some cases it suffices to have an *explanation* of "what's going on" when we think or talk in certain ways. Consider emotivist theories of moral discourse: such theories do not purport to translate moral claims into claims about sentiments; the claim, rather, is that moral judgments serve to express sentiments, manifest attitudes, or articulate commitments. Emotivism is a paradigm case of an irrealist or nondescriptivist theory, according to which some region of commitment (in this case, moral-evaluative discourse) is portrayed as a non-fact-stating mechanism that serves some vital, nondescriptive purpose. Such a theory might or might not be satisfactory: it might not accommodate various aspects of moral discourse. More sophisticated variants of such an approach are provided by Allan Gibbard's norm expressivism, Simon Blackburn's projectivism, and other such "irrealist" theories.[5]

Carnap applies this explanatory strategy to ontological discourse. Thought and talk about the existence or nonexistence of various kinds of entities – propositions, numbers, mind-independent physical objects, etc. – are portrayed as mechanisms for articulating commitments, manifesting stances, or some other such nondescriptive enterprise. This is a fascinating strategy, whatever its ultimate prospects: one virtue is that it removes the burden of providing non-circular explications or meaning-preserving translations of "existence" talk into some other kind of talk. Talk of existence can be treated as irreducible, but explicable as a mechanism for achieving certain results. Talk about what exists, according to Carnap, is best understood as manifesting commitments to adopting certain linguistic forms.

This conception of ontological discourse captures the core of Carnap's suggestion that ontological questions are "external questions" about the advisability of adopting one or another "linguistic form." On this interpretation, Carnap's theory is precisely analogous to emotivism (with which he was familiar), insofar as it provides a nonreductive explanation – irrealist in form – of a puzzling region of discourse. The key idea is that ontological claims – e.g., "universals exist" – are expressions of commitment to the pragmatic advisability of adopting specific linguistic frameworks – e.g., one that permits quantification over predicate variables.

These are the rudiments of Carnap's theory. As with any irrealist explanation, profoundly difficult challenges arise – some endemic to nondescriptivist strategies generally, and some unique to the application of such strategies to ontological discourse. Having told us that ontological discourse is nondescriptive – that is, plays a non-fact-stating role – careful work must be done to specify precisely what role *is* played by such discourse. This Carnap has done, by invoking the idea of pragmatically motivated adoption of linguistic forms. But additional work is required: Carnap must explain the connections between the attitudes manifest in ontological discourse and the "thinner," non-ontological reality that prompts them. This might not be possible: circularities appear to loom (to be discussed shortly). And there's more: Carnap must explain how ontological discourse interacts with descriptive discourse: how the logic works and how Boolean combinations of truth conditional and non-truth-conditional indicatives are to be semantically evaluated. He must deal with Frege-Geach problems involving ontological claims occurring as antecedents of conditionals; he must show that 'Properties exist' and other ontological claims conform to standard Tarskian constraints on quotation and disquotation.

These are formidable challenges; had Carnap been aware of the costs incurred by any sort of expressivism, he might have softened his empiricist requirements and allowed ontology to be viewed as descriptive, fact-stating discourse. But hindsight is 20/20: after several decades of intensive systematic inquiry into the details and dynamics of expressivist theories, the costs incurred by Carnap's theory of ontology are more apparent to us than they were to him.[6]

Problems and complexities

Evaluation of Carnap's theory of ontology lies beyond the scope of this discussion; but full disclosure requires noting additional challenges it confronts.

Ontological relativity

Carnap's explanatory strategy requires that for each kind of entity about which ontological questions might arise, there exists a corresponding linguistic framework the adoption of which is at issue. Questions about the existence of properties, for example, manifest deliberations about the pragmatic advisability of adopting a linguistic framework which permits higher-order quantification; questions about the existence of propositions manifest deliberations about the advisability of adopting a framework that allows sentential variables and quantification over sentential positions. And so on.

A formidable challenge to Carnap's theory emerges in light of Quine's notorious claim that there is no fact of the matter about the objects that must be reckoned among values of variables so as to simultaneously satisfy a given set of sentences (Quine 1969). Given such relativity, there are no "real" ontological commitments of any theory, insofar as adoption of the theory is consistent with a variety of ontologies. Choice of linguistic framework is not, therefore, equivalent to choice of ontology. If, for example, sentences formulated in the language of material objects can be satisfied in a domain of natural numbers, then there is no basis for alleging that acceptance of those sentences – and the decision to use the language in which they are formulated – constitutes commitment to the existence of entities over and above natural numbers. Analogously: if the "linguistic framework" of Folk Psychology – a framework that contains such predicates as '____ believes that P' and '____ desires X' – permits formulation of sentences satisfiable in a domain of behavioral uniformities and/or neurochemical states and events, then acceptance of that framework is not, in itself, acceptance of an ontology of nonphysical states and events.

The upshot is that it is misguided to inquire into the specific ontological commitment undertaken in accepting a given linguistic form, for there is no unique such commitment. The most that can be said is how to interpret an accepted theory into some background theory. And Quine's doctrines of Indeterminacy of Translation and Ontological Relativity counsel that there is no unique such interpretation: if there is one way to construct it, then there are many ways.

All of this purports to work against Carnap. For he suggests that

(p) There are propositions.

is the expression of a commitment to the pragmatic advisability of adopting a linguistic framework which contains sentential variables and propositional quantification. But if adoption of such a framework is consistent with the *non*-existence of propositions – if sentences formulable within that framework admit of interpretations in domains that exclude propositions – then commitment to the language of propositional quantification is no sufficient condition for the truth of (p). Carnap has not, after all, captured the realities of ontological practice.

Carnap should not be singled out for bad publicity here. Even if Quine is right about (some instances of) ontological relativity, no fatal blow is thereby struck against Carnap's expressivist theory. For, as Quine insists, indeterminacy of translation begins at home, as does the relativity of ontology: even our own "background" discourse, in which we cavalierly formulate hypotheses about what does and does not exist, itself carries no unique ontological commitments. We might seek to block non-intended interpretations of our background language by stomping our feet and indignantly asserting "I am *not* talking about natural numbers or sets; I am talking about physical objects and mental events."[7] But if Quine's strategy is viable and our indignant assertion has a model in arithmetic, even our explicit repudiations of neo-Pythagorianism fail. There is – assuming certain constraints – no way to run.

Insofar as the concern here is whether Carnap's theory withstands Quine's thesis of Ontological Relativity, fairness requires noting that his critics are vulnerable to their own challenge: for if there is slack between ontology and commitment to the advisability of adopting a given linguistic framework, there is equal slack between ontology and existence claims construed in a non-Carnapian way.

Here is why: Carnap's "realist" critic invokes the idea of ontological relativity to establish that the existence of numbers (for example) is not uniquely determined by linguistic commitments, conventions, or decisions regarding explanatory ineliminability of linguistic forms; but that critic must confront precisely the same (alleged) relativity insofar as *it infects her own discourse about what exists*. Quine's argument applies to (p) above. If ontological relativity is correct, this critic's own remarks about the existence of certain kinds of objects fail to fix a unique ontology; thus Carnap's expressivist theory of ontological discourse encounters no *additional* problems here.

The doctrines of translational indeterminacy and ontological relativity might ultimately be incoherent: their very formulation appears to assume their own falsity. How can it coherently be claimed that there is no fact of the matter as to whether 'gavagai' refers to rabbits or rabbit stages, unless, as a matter of fact, 'rabbit' refers to rabbits rather than rabbit stages?[8] No matter. Intelligibility of ontological relativity notwithstanding, the present point is that no special challenge to Carnap's strategy is forthcoming on the basis of such concerns.

No exit

Irrealist explanations – like any explanations – require ingredients for their implementation. A thoroughgoing expressivism about moral discourse, for example, requires a well-defined set of noncognitive sentiments and/or stances – call it the *projective base* – elements of which are (according to the theory) expressed in moral indicatives. When Hume speaks of "gilding and staining natural objects with the colors borrowed from internal sentiment," he assumes the existence of psychological states sufficiently rich to do the job, and "colors" which – whatever their origin – are real enough to get smeared onto the world. Emotivism exploits a set of "boo/hooray" attitudes; Kripke's Wittgenstein, in explaining the role played by rule-following attributions, exploits feelings of confidence that an agent will continue a mathematical series in a certain way. (Kripke 1982). In each of these cases, a certain phenomenological base is presupposed: the emotivist's moral sentiments, the Wittgensteinian's feelings of confidence, etc. The irrealist, in suggesting that indicatives formulated with a given fragment of discourse serve to manifest stances, express commitments, or evince noncognitive attitudes, needs a rich story about such stances, commitments, and attitudes, a story which does not backhandedly advert to the discourse under analysis.

But the irrealist also needs a story about the objects and events that serve as prompting stimuli to the agent who manifests stances, projects attitudes, and/or undertakes commitments. Such objects and events must *exist* if they are to fulfill their explanatory role: how can nonexistent objects prompt speakers to undertake commitments? Thus the circularity: an adequate explanation of pragmatic decision and commitment requires deployment of ontological discourse. More starkly: an irrealist explanation of 'exists' and cognate expressions requires engagement in discourse about what does and does not exist. Thus talk of existence is deployed in theorizing about talk of existence; there is No Exit from the discourse of reality and existence, even when theorizing about the discourse of reality and existence.

It is unclear whether this predicament vitiates Carnap's enterprise. On the one hand, if the goal is to theorize about certain commitments – those, for example, allegedly manifest in ontological discourse – it is no surprise that such commitments have already been undertaken, and thus permeate the theoretical enterprise: talk of existence will likely force itself upon any effort to theorize about talk of existence. But it is not clear that – or how – a puzzling concept can be illuminated by a theory that deploys that very concept. Such discomfort might be defused by glibly invoking Neurath's ship and reminders about Quine's laudable (and fruitful) efforts "to ponder our talk of physical phenomena as a physical phenomenon, and our scientific imaginings as activities within the world that we imagine" (Quine 1960, 5). But Quine's scientific inquiry into science provides no helpful analogy: for unlike Carnap's inquiry, which stems from dismissive puzzlement about ontological discourse, Quine's inquiry stems from no such puzzlement about physical phenomena and scientific imaginings. Acquiescing in a discourse while simultaneously questioning its legitimacy is methodologically suspicious: if there is sufficient comfort with such discourse to deploy it in one's theorizing, why the initial fuss?

The basic point is that Carnap's pragmatist/expressivist theory appears to suffer from circularity. But there are many forms of circularity, not equally pernicious; the fact that Carnap is unable to exit ontological discourse while theorizing about it might or might not constitute a fatal flaw. (See Humberstone 1997; Keefe 2002).

No return

Carnap faces yet another puzzle – one that arises in connection with any implementation of expressivism, whatever the target discourse. It turns upon the fact that participants in a practice might feel undermined by the irrealist explanation of what they are up to and thus, once informed, be unable to return to their practice in good faith, going on as before. Some explanations destabilize that which they purport to explain. Thus: No Return.

Christine Korsgaard, reflecting upon Bernard Mandeville's ironic assertion "that virtue is just an invention of politicians, used to keep their human cattle in line," notes that

> The trouble with a view like Mandeville's is not that it is not a reasonable explanation of how moral practices came about, but rather that our commitment to these practices would not survive our belief that it was true. Why give up your heart's desire, just because some politician wants to keep you in line?
>
> *(Korsgaard 1996, 9)*

Korsgaard's concern is applicable to any irrealist explanation: participants in a practice might find certain explanations of what they are doing so disruptive that their willingness (or ability) to go on doing it is undermined. To this extent, irrealist explanation suffers from an essential instability. If the ontologist, having learned of Carnap's pragmatist/expressivist explanation of her practice, says "Well, if *that's* what I'm doing when I engage in ontology, I'm not sure I want to continue doing it," then it's not clear whether the irrealist explanation should be retained – even if it predicts and explains its own rejection.

Cases differ. Perhaps a zealous Freudian therapist, having informed his patient of what's really happening, stands prepared for militant resistance to his explanations, and stands equipped with a robust Freudian explanation of the patient's resistance to his Freudian explanation. An adequate theory of a practice should pack the resources to explain why the theory might be rejected by practitioners. But this might not be a helpful analogy to all cases.

The sting of this concern depends upon whether Carnap intends his theory of ontology to be *conservative*: that is, consistent with continued participation in the practice of ontology. But if so, it is not obvious that such conservativism can be sustained. We all know of reflective moralizers who, upon reading Ayer, Stevenson, or Hare, say "Well, if THAT'S what I'm doing when I moralize, then I had better stop, because I don't like doing that." Some noncognitivists regard this as no refutation of their theory. On the other hand, the informed moralizer might remain on track, accepting the noncognitivist explanation and saying "Very well; in moralizing I am gilding and staining with colors borrowed from my internal sentiments. I shall continue the practice as before; it's for a good cause." Responses – including disruptions – might vary. It is an empirical question whether a reflective participant in a given practice can or will, upon learning that theorists explain his activities in expressivist terms (for example), continue to participate as before. And it is not clear how this bears upon the plausibility of expressivism as a semantic theory.

Suppose a working ontologist, having learned of Carnap's account, responds by saying: "If *that* is what I am doing when I make claims about the existence of propositions, numbers, and/or mind-independent physical objects, I shall do so no more." Indeed, suppose this reaction to Carnap's theory to be fairly uniform among metaphysicians. Does the prevalence of such reaction disconfirm Carnap's theory? Not clear. Perhaps it is no requirement that a philosophical theory of a practice be such that coming to believe the theory is consistent with continued engagement (relatively unchanged) in the practice. Perhaps correct philosophical description and explanation of institutional practice need not be conservative.

These questions apply across a wide range of philosophical areas. Any philosophical project that involves theorizing about a fragment of discourse from an external explanatory perspective must sooner or later confront questions about the methodological constraints on such explanations.

Carnap paid insufficient attention to the possibly destabilizing impact of his irrealist explanatory strategy upon practitioners, and what such impact might show. In assessing Carnap's contribution, we might endorse this general constraint:

> Methodological Credo: if metaphysicians would have difficulty continuing their ontological practice upon learning of certain accounts of "what they are doing" when thus engaged, then those accounts should be rejected as incorrect.

But it is not clear what principled justification might be provided for such a Credo, and why destabilizing explanations should be deemed incorrect.

Conclusion

Expressivist explanations of a region of discourse are symptomatic of one kind of pragmatism: one that eschews *representation* as the basic notion in terms of which semantic content is to be explicated, focusing instead upon the *functional roles* occupied by locutions within a broader, normatively constrained social practice. Thus the familiar pragmatist question "What are we DOING when engaged in this discourse?" replaces the customary representationalist question "What is this discourse ABOUT, and what are its characteristics?"

On the current understanding of Carnap's strategy, an indicative assertion of the form "Natural numbers exist" is best understood not as the ascription of a property ("reality" or "existence") but rather as the expression of a commitment to the adoption of certain linguistic/conceptual resources. This commitment, in turn, is grounded in practical considerations concerning the advisability of adopting those resources, given one's interests, goals, and agendas, and given the way the world pushes back. The commitment might or might not be well advised; but it shows itself in a variety of ways, one of which is assertions formulated within ontological discourse. Thus the connection between pragmatism and expressivism.[9]

Here is Carnap's description of his agenda concerning ontology:

> it will be shown that using [a language referring to abstract entities] does not imply embracing a Platonic ontology but is perfectly compatible with empiricism and strictly scientific thinking.
>
> *(Carnap 1950, 206)*

This is not the best description of his enterprise. Carnap is better regarded not as having shown that adopting the framework of thing properties does not imply embracing a Platonic ontology, but rather that embracing such an ontology simply *is* the expression of a commitment to adopting that linguistic framework, and is thus less offensive to the tenets of "empiricism and scientific thinking" than one might initially have assumed.[10]

Notes

1 Similar perspectives are conveyed in Friedman 1991; O'Grady 1999; Eklund 2009; and elsewhere.
2 Transcribed from an interview with Murray Gell-Mann, "Working on quantum mechanics, the work of Everett," available at www.youtube.com/watch?v=bx2cUfPrbGU.
3 Assuming such paraphrases to be available; if not, a compelling argument for the existence of universals is thereby revealed. See Bealer 1993.
4 Explanatory and justificatory strategies deploying universals are further explored in Kraut 2010.
5 See, e.g., Blackburn 1984, 1993; Gibbard 1990; Schroeder 2010. Challenges faced by expressivist semantic strategies are forcefully articulated in Schroeder 2008. See also Rosen 1998.
6 Problems and prospects for Carnap's expressivist theory of ontology are explored in greater detail in Kraut 2016; see also Kraut and Scharp 2015.
7 Such sentiments occasionally emerge in Lewis' discussion of Putnam's model-theoretic arguments against Realism; see Lewis 1984.
8 The objection is familiar. See, for example, Hockney 1975; Searle 1987.
9 The connections are further explored in Kraut 1990. See also Williams 2013.
10 Portions of earlier versions of this chapter were presented at the University of Connecticut and the 3rd European Pragmatism Conference (University of Helsinki). I am grateful to Stewart Shapiro, Rasmus Jaksland, Ethan Brauer, Jamie Shaw, Huw Price, Delia Belleri, Simon Blackburn, Juliette Kennedy, Michael Lynch, Tristram McPherson, Dorit Bar-On, Bill Lycan, and an anonymous reader for helpful suggestions and/or critical discussion.

References

Bealer, George. 1993. "Universals," *Journal of Philosophy* 60: 5–32.
Blackburn, Simon. 1984. *Spreading the Word*. Oxford: Oxford University Press.
Blackburn, Simon. 1993. *Essays in Quasi-Realism*. Oxford: Oxford University Press.
Carnap, R. 1935. *Philosophy and Logical Syntax*. Bristol: Thoemmes Press.
Carnap, R. 1950. "Empiricism, Semantics, and Ontology." In *Meaning and Necessity*. Chicago: University of Chicago Press: 205–21.
Carnap, R. 1959. "The Elimination of Metaphysics Through Logical Analysis of Language," in A. J. Ayer (ed.), *Logical Positivism*. Glencoe, IL: The Free Press.
Einstein, A., Podolsky, B., and Rosen, N. 1935. "Can Quantum-Mechanical Description of Physical Reality Be Considered Complete?," *Physical Review* 47: 777–80.
Eklund, Matti. 2009. "Carnap and Ontological Pluralism," in Chalmers, Manley, and Wasserman (eds.), *Metametaphysics: New Essays on the Foundations of Ontology*. Oxford: Oxford University Press: 130–56.
Friedman, Michael. 1991. "The Re-evaluation of Logical Positivism," *Journal of Philosophy* 88: 505–19.
Gell-Mann, Murray. No Date. Transcribed from an interview with Murray Gell-Mann, "Working on quantum mechanics, the work of Everett," available at www.youtube.com/watch?v=bx2cUfPrbGU.
Gibbard, Allan. 1990. *Wise Choices, Apt Feelings*. Cambridge, MA: Harvard University Press.
Hockney, Donald. J. 1975. "The Bifurcation of Scientific Theories and Indeterminacy of Translation," *Philosophy of Science* 42: 411–27.
Humberstone, I. L. 1997. "Two Types of Circularity," *Philosophy and Phenomenological Research* 57: 249–80.
Jaksland, Rasmus. 2017. "A Dilemma for Empirical Realism: Metaphysical Realism or Instrumentalism," *Philosophia* 45: 1195–1205.
Keefe, Rosanna. 2002. "When Does Circularity Matter?," *Proceedings of the Aristotelian Society* 102: 253–70.
Korsgaard, Christine M. 1996. *The Sources of Normativity*. Cambridge: Cambridge University Press.
Kraut, Robert. 1990. "Varieties of Pragmatism," *Mind* 99: 157–83.
Kraut, Robert. 2010. "Universals, Metaphysical Explanations, and Pragmatism," *Journal of Philosophy* CVII: 590–609.

Kraut, Robert. 2016. "Three Carnaps on Ontology," in S. Blatti and S. Lapointe (eds.), *Ontology After Carnap*. Oxford: Oxford University Press: 31–58.

Kraut, Robert and Scharp, Kevin. 2015. "Pragmatism Without Idealism," in Christopher Daly (ed.), *The Palgrave Handbook of Philosophical Methods*. London and New York: Palgrave Macmillan: 331–60.

Kripke, Saul A. 1982. *Wittgenstein on Rules and Private Language*. Cambridge, MA: Harvard University Press.

Lewis, David. 1984. "Putnam's Paradox," *Australasian Journal of Philosophy* 62: 221–36.

O'Grady, Paul. 1999. "Carnap and Two Dogmas of Empiricism," *Philosophy and Phenomenological Research* 59: 1015–27.

Quine, W. V. O. 1960. *Word and Object*. Cambridge, MA: MIT Press.

Quine, W. V. O. 1969. "Ontological Relativity," in *Ontological Relativity and Other Essays*. New York: Columbia University Press: 26–68.

Rosen, Gideon. 1998. Review of Simon Blackburn's *Essays in Quasi-Realism*. *Noûs* 32: 386–405.

Schroeder, Mark. 2008. *Being For: Evaluating the Semantic Program of Expressivism*. Oxford: Oxford University Press.

Schroeder, Mark. 2010. *Noncognitivism in Ethics*. London and New York: Routledge.

Searle, John. 1987. "Indeterminacy, Empiricism, and the First Person," *Journal of Philosophy* LXXXIV: 123–46.

Shapiro, Stewart. 1998. "Philosophical Issues in Second-order Logic," *Routledge Encyclopedia of Philosophy*. Taylor & Francis. Available online at www.rep.routledge.com.

Thomasson, Amie. 2016. "Carnap and the Prospects for Easy Ontology," in S. Blatti and S. Lapointe (eds.), *Ontology After Carnap*. Oxford: Oxford University Press: 122–44.

Williams, Michael. 2013. "How Pragmatists Can Be Local Expressivists," in Huw Price (ed.), *Expressivism, Pragmatism and Representationalism*. Cambridge: Cambridge University Press: 128–44.

Woodin, W. Hugh. 2001. "The Continuum Hypothesis, Part I," *Notices of the AMS*: 567–76.

4

QUINE'S METAMETAPHYSICS

Karl Egerton

W. V. Quine stands out as one of the foremost figures of twentieth-century analytic philosophy. This chapter aims to show that a significant part of his work's enduring value lies in its contribution to metametaphysics, which will include showing how some more contentious aspects of Quine's thought can be seen as indispensable to it; we will problematise the widespread belief that one can isolate basic elements of Quine's metametaphysics without eroding their warrant.

§1 introduces the broad context. §2 examines Quine's most clearly metametaphysical work (and the desired backdrop for many analytic philosophers): 'On what there is'. Finding the story incomplete here, we explore other elements of Quine's corpus in turn. §3 analyses the nascent naturalism evident in 'Two dogmas of empiricism', §4 explores how the principle of charity becomes significant in *Word & Object*, and §5 shows how the eponymous principle of 'Ontological relativity' aims to defuse the puzzles of indeterminacy. In the process we will see how Quine's concerns stemming from naturalism in general, and from the problems of indeterminacy in particular, make it hard to separate the basic picture from his more controversial full-blown approach – hard, that is, to avoid ontological relativity. This is bad news for those wishing to use Quine as a neutral backdrop to analytic metaphysical debate, but good news for those who value the distinctive philosophical tradition within which Quine's work is a key development.

1 The view from a distance

There is widespread agreement that Quine's ideas, for better or worse, had a substantial impact on metaphysics, in particular on that part of the discipline called *meta*metaphysics.[1] The term seems to postdate the period of Quine's greatest influence, though not Quine's life as a whole;[2] metametaphysics has, however, come into greater focus in the twenty-first century, prompted especially by the 2009 collection *Metametaphysics*. That collection in turn was partially inspired by Peter van Inwagen's 'Meta-ontology' (1998), which purported to articulate Quine's methodology for ontology and thereby to expose the foundations of a popular – by some lights, the *dominant* – tradition in analytic philosophy.

Quine's contribution, however, is often reduced to one paper – 'On what there is' [OWTI] (1948) – treating it as the locus of critical consideration of method in analytic metaphysics. This view is not restricted to those merely acquainted with Quine, having been encouraged even by those who are intimately familiar with his work. For instance Hilary Putnam, who studied under and engaged extensively with Quine, writes that

> when Quine published a famous paper titled 'On what there is' ... [he] single-handedly made Ontology a respectable subject.
>
> *(2004: 78–9)*[3]

This focus is understandable – without doubt the paper is significant – but can lead to the assumption that the wider body of Quine's work is irrelevant. It can also slip into a yet narrower focus which further reduces Quine's contribution to metametaphysics to a few easily misinterpreted sound-bites. Analytic metaphysicians are familiar with the adage '[t]o be is to be the value of a variable' (Quine 1948: 34), typically called the Quinean criterion of ontological commitment, but this brief claim may conceal much, and may even lead to trouble if we rely on the radically incomplete picture it provides. So I will claim.

Here, therefore, is our strategy: unpack that sound-bite along with other elements of Quine's metametaphysics from OWTI (§2), before discussing (§§3–5) the metametaphysical theses that feature elsewhere in the 52(!) years of work that Quine published *after* this seminal paper. This helps us to evaluate the viability of adopting elements of Quine's metametaphysics piecemeal. It is hoped by many analytic metaphysicians that this can be done easily; I contend that it's far from easy.

Two brief warnings before proceeding: first, for those familiar with the debate, we will be exploring *Quine's* metametaphysics, not *Quinean* metametaphysics (whose locus is van Inwagen (1998, 2009), but see also Eklund 2006, Jenkins 2010 and Berto and Plebani 2015). The task of comparing these must be left for elsewhere (on this subject see Price 2009; see also my 2016). I will therefore take a more historical focus than has been the norm. By approaching the question of the methodology for metaphysics via Quine's developing body of work, we will be able to focus more heavily on Quine himself than on self-proclaimed Quineans.

Second, as we explore later, metaphysics and metametaphysics aren't strictly separate enterprises for Quine as they might be for other thinkers, primarily because of his holism, but insofar as is possible without going astray, we will here put aside Quine's specific metaphysical views. More in-depth exploration would involve considering Quine's attempts to dispense with properties, *possibilia*, etc., but we'll have to leave that for elsewhere.

2 'On what there is'

> A curious thing about the ontological problem is its simplicity. It can be put in three Anglo-Saxon monosyllables: 'What is there?' It can be answered, moreover, in a word – 'Everything'.
>
> *(Quine 1948: 21)*

Thus begins OWTI – with a deceptively simple statement about how to understand the ontological/metaphysical project. We want to know what there is, or what exists. And we can say what there is without engaging in any difficult work, as long as we don't mind trite (facetious?) answers – *everything* exists. The problem, as Quine acknowledges, is that we don't know what that 'everything' comprises, so we must put in more hours.

Of course it's not that simple anyway – objecting to the one-word answer, someone might insist that not *everything* exists – not, say, Pegasus, or round squares. At first Quine seems stuck should he deny that some kind K of entity exists, because even saying 'Ks do not exist' seems to admit that Ks are *something*. Were this the case, Quine's only way to resolve ontological disputes in the negative would be to refuse to say anything whatsoever, but this looks unsatisfying, indeed implausible. Surely it's coherent to deny that Ks exist! To see why non-existent objects *don't* disturb Quine's easy answer, why he can insist on this simple statement, we must delve further.

The solution is to invoke quantification. Quine follows Russell (1905): there's nothing strange in saying 'The current King of France does not exist', provided we treat this as disguised quantification. We don't intend to say 'You know the current King of France? It turns out he doesn't exist', but rather 'There is no such entity as the current King of France'. To speak more formally, what we mean to say can be represented as '$\neg\exists x(KoF(x))$' or equivalently '$\forall x(\neg KoF(x))$'. We can conclude our ontological task negatively by using a description and stating either that there is no such thing, or that each thing there is fails to satisfy the description. This is the significance of the aforementioned phrase 'to be is to be the value of a variable' – to say that something of kind K exists just is for it to be a replacement-instance of a bound variable in a statement that says, or entails, '$\exists x(Kx)$'.

This is a basic, but crucial, component of Quine's metametaphysics. It forms the beginnings of a methodology, accepting which cuts off a range of putative ways of investigating ontology and renders the questions 'Are there Ks?', 'Do Ks exist?' and 'Is "$\exists x(Kx)$" true?' equivalent. Other notions in the vicinity, like 'subsistence', are either subsumed within this equivalence or disallowed as having no clear sense. This is the core of how Quine 'made Ontology a respectable subject' – by formulating a proposal about how to interpret the ontological question and thereby limiting the threat of disputants talking past one another. In order to disagree with Quine on ontology, one has to either lay out their differences in terms of entities quantified over (which requires that they regiment their language enough to enable an answer to the quantificational question) or reject Quine's conception of ontology outright.[4]

But this treatment of ontological debate is far from complete. For all the above version of the criterion says, mere use of the term 'K' might be enough to make the statement '$\exists x(Kx)$' true, and thereby to make the statement 'Ks do not exist' a contradiction. Perhaps mere ability to meaningfully use 'K' demonstrates that Ks are something. This is not an option Quine wants available, so we need a second, stronger statement of the Quinean criterion for ontological commitment:

> a theory is committed to those and only those entities to which the bound variables of the theory *must be capable of referring* in order that the affirmations made in the theory be true.
>
> *(1948: 33, my emphasis)*

This adds a further condition – if you don't *need* to speak of Ks, you are not committed to Ks' existence. We should accept only the minimum number of entities (or kinds) to allow our theory to be true. To maintain my overall theory I don't need to be capable of referring to anything fitting the description 'current King of France' because I need the description only in the context of disputes with those who are confused about history and/or politics, and '$\exists x(KoF(x))$' is by no means a consequence of my claims in such disputes because I simply provide the condition *King of France* (or *entity identical with Pegasus*) and charge my opponent to find the entity meeting the condition in the list of things 'to which the bound variables of the theory must be capable of referring'.

We find ourselves, then, in a position to represent ontological debate. Two interlocutors start with a term 'K', and one claims to hold a theory which must accept (explicitly or implicitly) statements like '$\exists x(Kx)$' in order to be true, while the other does not. This also permits some understanding of how to *resolve* ontological debates: for when we investigate what sentences require the truth of '$\exists x(Kx)$', we may find that those sentences have different consequences from those first suspected, or that we don't wish to accept them after all. In the simplest cases we can avoid foolish debate about the existence of, e.g., sakes, or heebie-jeebies. Nevertheless much remains unanswered about how we resolve ontological debates.

OWTI closes with some relevant remarks that are seldom considered with care – Quine asks how we adjudicate between rival ontologies, and after initially saying that adopting an ontology is 'similar in principle to our acceptance of a scientific theory' (ibid.: 35) he goes on to say that we may pursue many options *in tandem*, and in the absence of strong reasons to prefer one ontology, 'the obvious counsel is tolerance and an experimental spirit' (ibid.: 38). Unless we encounter trouble with one or the other, we can allow them to sit side-by-side as alternative theories. This result might look unwelcome: for in the debate over the existence of Pegasus, I'm not satisfied with the result that I can now maintain its non-existence if this just means that my opponent and I retreat to our respective theories and mind our own business. After all, we are *opponents*; each thinks that the other ought to *come around to their way of thinking*. Furthermore, the mention of tolerance naturally evokes Carnap's 'principle of tolerance' (see, e.g., Carnap 1937: §17), a pluralist position which has been generally understood as seriously in tension with Quine.

In OWTI, Quine says little by way of clarification. Alongside his gestures toward scientific theory, he also makes what looks a merely dogmatic claim: he says that an 'overpopulated universe ... offends the aesthetic sense of us who have a taste for desert landscapes' (ibid.: 23). Relative to its actual significance, this quote has played a greater role in establishing Quine's reputation than any other. While Quine's vague remarks about tolerance are forgotten, his vague remarks about sparseness are taken with utmost seriousness – hence the widespread view that '[t]he Quinean method is eliminativist by design' (Schaffer 2009: 372), inherently biased in favour of casting out entities. If we work only on the basis of what is given here, though, it's equally consistent with Quine's metametaphysics to exercise tolerance!

The degree of openness found in OWTI *may* be virtuous (to form the foundation of a broad school, one might think, it would have to be consistent with various developments). Nevertheless Quine himself certainly had a more developed programme that manifested elsewhere in his work, so to treat his remarks about a 'desert landscape' as crucial is as short-sighted as it would be to take on wholesale his remarks about tolerance. Instead we should broaden our horizons and consider Quine's more mature work.

3 'Two dogmas of empiricism'

In order to develop Quine's metametaphysics beyond OWTI, we will first consider perhaps his only work that rivals it for fame: 'Two dogmas of empiricism' [TD] (1951a). Here Quine famously critiques empiricism through its 'dogmas' of analyticity and reductionism, and introduces the epistemological holism that is now well-known under the label of the 'web of belief'. Why, though, insist that this work is relevant to metametaphysics?

First, it develops OWTI's merely gestured-at 'scientific' attitude to our metaphysical (among other) theories, so if we want to clarify this, TD is a good place to look. This might seem outweighed by the fact that the dogmas introduce new philosophical territory: we're now engaging in philosophy of language (analyticity) and epistemology (reductionism), so why bring in metaphysics? But aside from the fact that the holism here introduced shows that precise subject divisions are not Quine's style, it should also be clear that it's part of investigating the nature of metaphysical questions to ask how one comes to know their answers, and what it means to ask/answer one. Interestingly, Quine later describes himself as doing 'the epistemology of ontology' (1983: 500). This is natural, for we can't properly understand a research programme without some notion of what would constitute a significant result within it. Let us, then, examine what TD introduces to Quine's metametaphysics.

Here, as elsewhere, Quine demonstrates a complex relationship with the empiricist tradition. On one hand, Quine is widely held to undermine at least one strand of the empiricist project – logical positivism – and TD substantially contributes to that. On the other, Quine clearly has extensive sympathies with their project: in the closing section, titled 'Empiricism without the dogmas', he stresses '[a]s an empiricist I continue to think of the conceptual scheme of science as a tool, ultimately, for predicting future experience' (41). Later he is even more committal:

> I haven't thought of myself as destroying [logical empiricism, but] as contributing to what it seemed to me needed further development ... What I was taking issue with was pretty much, I think, in the domain of logical rigour, and also of being more completely empiricistic. And certainly I felt that I was insisting on the ideals of the Vienna Circle more than they, and saying what I thought they ought to be saying.
> (Quine and Fara 1994)

To see how the tension can be resolved, we must examine TD's move from empiricism to naturalism.

In OWTI Quine already demonstrated commitment to empiricist ideas: when speaking of deciding between ontologies, he sees the 'phenomenalist' option as important because it takes 'epistemological priority' (1948: 38). Quine's conflicted comparison of phenomenalism (privileging sense data) and physicalism (privileging physical objects) showcases his empiricist scruples: the former has better epistemological credentials because its most basic ingredients – sense data[5] – are immediate to experience, but the physicalist conceptual scheme inherits the virtue of association with the successes of physics despite relying heavily on unobservable theoretical posits whose very positing requires extensive assumptions. This conflict continued to bother Quine and in TD the reason emerges. In critiquing the 'dogmas' of analyticity and reductionism, Quine crystallises a growing loss of faith in the idea that our words, concepts or sentences can be understood as being traceable back to their ultimate implications for our experience. We'll briefly consider why.

Quine starts by examining the notion of analyticity, or truth in virtue of meaning. The idea that some sentences are known to be true simply in virtue of the meanings of their constituent words, like 'Vixens are female foxes' or 'Bachelors are unmarried', has a long history, and played an important role for the logical positivists. For them it validated a distinction between empirical knowledge, whose source is clear (the senses), and apparently *a priori* knowledge. The latter had been an empiricist stumbling block because it was mysterious how such knowledge was acquired if not through the senses – but if based purely on meaning, analyticity would involve no special content, being founded merely on linguistic competence. However Quine finds truth in virtue of meaning impossible to unpack properly, since (i) we lack reason to believe in special discoverable entities, *meanings*, appeal to which would be a marker of truth, and (ii) the notion of sameness of meaning, which is needed to isolate *meanings*, is hard to clarify. We cannot make the notion more manageable by applying epistemic standards, e.g. universal willingness to assent to the sentence under the same conditions, since this doesn't exclude empirical claims that inspire universal assent (compare 'Vixens are female foxes' and 'There are dogs'). After discussing several other candidates for making sense of analyticity, e.g. semantic rules, Quine concludes that the notion cannot do the work the logical positivists require of it.

The way this feeds into Quine's metametaphysics becomes clearer with the second dogma: reductionism. Quine insists that we must abandon the idea that there corresponds to each statement a selection of possible empirical data that count as the evidence for or against it. If this were true one would be able to definitively give each statement's implications, but it doesn't survive scrutiny. Even for apparently purely observational statements, I might reject parts of my theory to hold those observational statements constant, or vice versa. If I make an observation of a neutrino that implies that it travelled beyond light-speed, I must choose between abandoning an important, deep-seated principle of my theory (that faster-than-light travel is impossible) and claiming that my observation was faulty somehow. I will choose based on the circumstances and wider implications, and for Quine, the rational choice is the path of least disturbance. The manoeuvres needed to settle the disturbance of rejecting what one has seen will often be less drastic than those needed to revise a central principle, though the balance may shift if enough observations accumulate.

This is a statement of *epistemological holism*: we cannot rationally accept or reject any claim without reference to our wider theory, so the siloed enquiry reductionists require is a distortion. Quine regards reductionism and analyticity as really the same dogma viewed from different angles: just as observing that matters of meaning and of fact both contribute to a statement's truth can trick us into thinking we can isolate that sentence's meaning-giving and fact-stating aspects, awareness that observations bear on some statements more than others can trick us into thinking we can pinpoint what's necessary to finally confirm a statement and conclusively settle its truth value. Rather everything faces the ongoing test of coherence with our best theory, which is overall science – with this step empiricism develops into naturalism. This importantly applies to all statements, including apparently metaphysical ones: we cannot perform any study into the nature of things that floats free of the enquiries comprising our broad science, but just as there can be no isolated metaphysical enquiry, there can be no wholesale ruling-out of metaphysics. This aspect of Quine's work was taken to scuttle the logical positivists' project, since their verification principle had ruled out metaphysical statements as meaningless because neither verifiable nor meaning-giving. Now, on Quine's picture, one cannot rule out a category as meaningless in advance since a lack of relevant implications for the remainder of our theory cannot be guaranteed. Yet importantly, a version of the verification principle survives, applied to *whole theories*: later, Quine asks of a hypothetical, completely empirically confirmed, theory, 'In what sense could the world then be said to deviate from what the theory claims?' and his answer is simple: 'Clearly in none' (1981: 22).

What, then, should we now add to our stock of ideas from Quine on the nature, or methodology, of metaphysics? I think it is clear: In TD, epistemological holism becomes inevitable because *naturalism* becomes a key part of Quine's metametaphysics. In saying this I go against van Inwagen especially, who says that Quine's attitude to science was 'a consequence of certain of his epistemological commitments and not of his metaontology' (2009: 506, n. 53). As I've already mentioned, that this is epistemological is no reason to suppose that it's not metametaphysical.

And we need it in order to understand two points briefly introduced above. Without naturalism we get neither the result, important for Quine, that metaphysical statements *are* (within certain restrictions) candidates for inclusion in our theory, nor the result that metaphysical statements are meaningful only insofar as they bear genuine connections to the rest of our theory. For the holism required by Quine's naturalism tells us that all respectable knowledge is continuous with science, which is governed by what is often called the Quine-Duhem thesis.

We've already uncovered a significant additional principle in Quine's metametaphysics by looking beyond the core of OWTI. Now we'll move on to an idea that's developed more extensively in Quine's great constructive project, *Word & Object*.

4 Word & Object

In *Word & Object* [WO] (1960), Quine develops the insights of TD with a more constructive focus. The book centres on a project of rational reconstruction – through the thought experiment of radical translation, Quine considers what a stranded linguist could learn about a wholly alien language, with this developing into an account of the theory that could be regarded as underpinning *our own* language. The lack of fixity in this theory is the notorious finding of the indeterminacy of translation.

We'll postpone exploring this, though, as on the way we find another idea of metametaphysical importance introduced. So far we've just seen that the ontological question requires us to establish what we must quantify over if our statements are to be true, and that our statements are to be taken as a corporate body rather than divided into independent clusters. But who's to say that we shouldn't take large quantities of the statements we habitually utter to be false? Our corporate body of truths might be very small, very distant from what's typically taken to be true, or both.

However, in a further development of Quine's naturalism, we see in WO an attempt to preclude that. Our construction of an overall theory needs a guiding principle, otherwise we can make sense of nothing at all – and that is that by and large people are getting things right. This guiding principle is the *principle of charity*.

Some would be surprised to see this treated as central to Quine's ideas, since it has been more often associated with the work of Donald Davidson (see his 2001). However it is indeed important for Quine. When exploring the creative revisions required in the course of radical translation Quine says that '[t]he maxim of translation underlying all this is that assertions startlingly false on the face of them are likely to turn on hidden differences of language' (1960: 59), and in an accompanying footnote identifies this as the principle of charity. He goes on to claim that 'the more absurd or exotic the beliefs imputed to a people, the more suspicious we are entitled to be of the translations' (ibid., 69). The idea is that since interpreting someone requires me to attribute beliefs to them, I must impute some degree of coherency and rationality to them in order for the attributions themselves to be coherent or rational. Otherwise I will have no reason to hypothesise even that two occurrences of a symbol are more likely to signify the same than are two occurrences of different symbols, and without such hypotheses I can't even start interpreting.

As Davidson helpfully puts it, there are two directions of pull to charity. We need to assign sense to utterances, and that means assuming some sharedness in what is believed true – 'we must maximize agreement, or risk not making sense of what the alien is talking about' (2001: 27), but we also need to understand why certain sentences and not others are assented to, and this means imputing rationality – 'we must maximize the self-consistency we attribute to him, on pain of not understanding *him*' (ibid.). So if I take there to be a distinction between hawks and handsaws, but on my interpretation my interlocutor recognises no such distinction, then *ceteris paribus* my interpretation is probably wrong. On the other hand if my interlocutor seems to routinely display some attitude but my interpretation attributes to them something radically at odds with that attitude, that suggests that *ceteris paribus* my interpretation is likely to be wrong *even if* I think the first, more internally coherent, attitude is obviously false.

Again, one might ask, why is this relevant to Quine's metametaphysics? The reason is that it advises us how widely our responsibilities range during metaphysical enquiry. We can't draw the limits of our theory within a restricted area, for instance saying that fundamental physics is the most respectable picture of reality so metaphysicians can safely ignore the question of how to reconcile it with talk of the macroscopic world. Within Quine's approach we *may* find talk of macroscopic objects to be misguided, but only with sufficient reasons to overturn the weight of our useful discourse about macroscopic objects. This also identifies more clearly what we're interested in when doing metaphysics: truth. We're not especially interested in entities, or essences, or fundamentality, but in *what is the case*.

One of the most significant aspects of the work Quine does in WO we have so far neglected – indeterminacy – but this is because it is more comprehensively explored elsewhere. While indeterminacy plays a significant role in WO, the *response* to indeterminacy is clearer in Quine's infamous paper 'Ontological relativity'.

5 'Ontological relativity'

The thesis of the indeterminacy of translation is well-known in analytic philosophy, as is the thesis of the inscrutability of reference, and these ideas, introduced in WO, become the central concern in 'Ontological relativity' [OR] (1968). In order to see their role in Quine's metametaphysics, however, we must first clear up a terminological issue. It's easy to get the mistaken impression that the indeterminacy and inscrutability theses are fundamentally different, but while they should be distinguished, they are at root similar. Indeed Quine himself later indicates that his choice of words was unfortunate and that 'indeterminacy of reference' would have better stated what he wished to convey (1992: 50). We'll now see why.

The thesis of the *indeterminacy of translation* is that two translation manuals for a language might agree on how we would expect language-users to behave on the basis of their translations despite those two manuals being incompatible with one another. We're invited to imagine two radical translators, operating independently, who each generate hypotheses about how to translate the utterances of an entirely alien community and amend them based on observation, refining hypotheses until they build up a vocabulary they match with their own to render them capable of communicating with that community. Given the many ways that one could consistently interpret and systematise a community's behaviours, Quine sees it as implausible that there wouldn't be multiple translations that successfully tracked all behaviour yet differed somewhere in what they took statements to signify.

The thesis of the *inscrutability/indeterminacy of reference* is that two translations of a language might agree even on the above and yet diverge regarding reference. That is, what entities the translations take the language to be picking out can differ *without* one assigning any statement the value 'true' where the other assigns that statement 'false'. A simple example of this idea is complement-based interpretation. Take the sentence 'My cat is an animal': one can ask why this should be analysed as speaking about my cat rather than *everything but* my cat. The obvious answer, that the statement would turn out false if understood as speaking of the *my-cat-complement* (which is not an animal but a vast aggregate of physical space), goes nowhere because by stipulation the complement-based interpretation also takes predicates like 'is an animal' to range over complements. So when I say 'My cat is an animal', I can consistently be interpreted as saying 'The my-cat-complement complement-is an animal.' Both sentences are true, but on the latter interpretation what I am talking *about* is not my cat, and what I am attributing is not an intrinsic property. Rather I am talking about everything but my cat, and attributing a clearly extrinsic property.[6]

One view of the indeterminacy theses is that they have sceptical results: I cannot know what someone else means, nor even what they refer to! This would suggest that whenever I try to understand anyone I'm taking a leap, performing a new radical translation. But this negative interpretation, on which I might be succeeding at every task that seems constitutive of communicative success yet still be getting my translation wrong, sits poorly with Quine. Since Quine holds that language is inherently public – '[i]n psychology one may or may not be a behaviorist, but in linguistics one has no choice' (1992: 37–8) – the standards for communicative success we set are all that's required to be right. Rather than saying that either translator may be failing despite apparent success, Quine wants to say that both can succeed despite their different interpretations. Hence in both cases, we are not dealing with mere inscrutability (inability to tell), we are dealing with *indeterminacy*.

Furthermore, the indeterminacy of translation may be based on a controversial claim that linguists have felt they can challenge, but the indeterminacy of reference has a much sparser base. The former is a hypothesis about what remains unfixed by observations yet is implied by an interpretation. It may therefore be too restrictive, too permissive, or both. It may be too restrictive since the resources Quine allows his radical translators aren't all that rich, and good translators might do more to pin down which statements should come out true; it may be too permissive since it allows good translations that differ on remote parts of theory to pronounce on things toward which the community might bear no attitudes whatsoever. However, the indeterminacy of reference is secured on simpler, technical grounds. We need only appeal to the basic representative machinery of model theory.[7] If we can offer any model on which a theory comes out true, we can offer multiple models, some of which will assign to the expressions of that theory different referents. One such model would be mathematical; we can preserve truth while taking our ordinary-object terms to range over not medium-sized dry goods but the natural numbers, as long as we reinterpret systematically.

This might ring alarm-bells – does Quine's metametaphysics now collapse, bringing the whole notion of his methodology to nothing? Quine is aware that the situation looks dire:

> We seem to be maneuvering ourselves into the absurd position that there is no difference on any terms, interlinguistic or intralinguistic, objective or subjective, between referring to rabbits and referring to rabbit parts or stages; or between referring to formulas and referring to their Gödel numbers.
>
> *(1968: 200)*

This would be absurd if nothing else on Moorean grounds – we couldn't be justified believing this with anything like as strong a conviction as that rabbits are not rabbit-complements. Accordingly Quine does *not* think this the end of the story, just a twist.

This is where the last of Quine's key metaphysical principles comes in: *ontological relativity*: '[s]pecifying the universe of a theory makes sense only relative to some background theory, and only relative to some choice of a manual of translation of the one theory into the other' (1968: 205). We make sense of reference relative to a background theory, in effect by accepting that theory as the metalanguage to the object language under scrutiny.

For Quine an absolute reference relation is untenable because any attempt to give it would simply set up more truth-preserving reinterpretations, but instead of conceding that reference is nonsense, Quine treats it as theory-relative. When speaking of interpretations of one theory in another, we treat the interpreting theory as fixed and understood; when speaking of interpretations of our own theory, we treat the theory through which we clarify the original as understood. There is then a difference between referring to rabbits and referring to rabbit-complements, *because there is a difference from the perspective of our theory*: as Quine playfully puts it, '"rabbit" denotes rabbits, whatever *they* are' (1992, 52: emphasis in original). Whether the theory is indeed understood will then depend on our success in satisfying the demands of (charitable) naturalism.

The scale of the metametaphysical implications should immediately be clear. We must hold our metaphysics relative to the theory in which we situate it. In one sense this isn't worrying

– after all, it was *our* theory that we wanted to better understand. In another, though, it's highly disruptive. Much traditional metaphysics trades on the notion of investigating the deepest nature of things, with the assumption that this will lead to a result that is not parochial but all-encompassing. Certain notions are effectively incompatible with ontological relativity – *fundamentality* as featured in much metaphysics, for instance, since this is supposed to identify what is most basic in a special metaphysical sense intended to float free of standard theorising. Strictly speaking we might allow fundamentality a smaller role, as a theoretical term of the theory under discussion, but that would substantially change the notion in play.

But for Quine we have just recognised something that's obvious elsewhere:

> to ask what reality is really like ... apart from human categories, is self-stultifying. It is like asking how long the Nile really is, apart from parochial matters of miles or meters.
>
> *(1992: 9)*

Furthermore, to return to our starting point, while we have moved far afield of the sparse beginnings of OWTI, OR really works from that base – by accepting that our interest is in quantification, and taking on the representative machinery, we furnish the tools for recognising that countless theories could systematise our language, provided those theories retained the right structure. The concerns motivating the principle of ontological relativity are not local to a later, special project of Quine's, but extend right back to the base that has been seen by some as uncontroversial.

Quine was aware throughout that his view re-imagined metaphysics – as indicated earlier, he usually avoided the term. Tellingly, when Carnap criticises Quine for using the term 'ontology' to describe his project, protesting that the term is meaningless, Quine responds that 'meaningless words ... are precisely the words which I feel freest to specify meanings for' (1951b: 66). As is gradually being recognised through re-evaluation of the history of logical positivism and the Quine-Carnap debate by historians of analytic philosophy, Quine develops the logical positivists' project, abandoning the untenable goal of ruling out metaphysics wholesale and instead finding a niche for something legitimately describable as metaphysics. That niche is helping to clarify – picking up a term that held great traction for both Carnap (1947: 8) and Quine (1960: §56), to *explicate* – elements of our theory. Quine was aware of this trajectory from early on: he states in TD his intention to oversee both 'a blurring of the supposed boundary between speculative metaphysics and natural science' and 'a shift toward pragmatism' (1951a: 20). Quine's introduction of his criterion for ontological commitment, supported by charitable naturalism and underpinned by ontological relativity, transforms the subject-matter of metaphysics: it moves from the pursuit of deep truths underlying all theories to the project of clarifying our theories without stepping outside them.

Conclusion

Far more could be said about what we've explored above: it has been possible only to trace the shape of Quine's metametaphysics across several stages of articulation, without delving into either the opportunities it presents or the serious challenges it of course faces. However this first step has at least three dividends. First, by drawing out the implications of Quine's metametaphysics we illuminate it, allowing comparative work that is blocked if the theory remains a shadowy background presence. Second, we broaden our understanding of what metaphysics can be by acknowledging a metametaphysics that adopts a serious methodology despite circumscribing the subject's ambitions, opening up a space between 'fundamental' metaphysics and wholeheartedly deflationary projects. Finally, we introduce a challenge for those availing themselves of parts of Quine's machinery. If they embrace the more radical underpinnings, this is in itself interesting (and, I hazard, for the better!), but if they refuse, it raises the question whether they can construct a coherent alternative that retains the appeal of Quine's version of naturalism.

Notes

1 For explicit instances of this claim, see Manley 2009 and Berto and Plebani 2015.
2 The term goes back at least to 1988, though this occurs in a discussion of Derrida (see Silverman 1988: 206). Given Quine's documented dismissal of Derrida (he was a signatory to an open letter protesting Cambridge University awarding him an honorary doctorate), this instance at least is likely to have passed him by.
3 Let us resolve one point early regarding ontology and metaphysics. On my assumptions here, ontology is part of metaphysics, so those who speak of Quine's contribution specifically to meta*ontology*, I take it, thereby speak of his contribution to metametaphysics. For reasons that emerge later, Quine rarely used the term 'metaphysics' (he didn't even use 'ontology' all that much), and the terms 'metaontology' and 'metametaphysics' both postdate the majority of Quine's work. I use 'metametaphysics' throughout, though little hangs on this. My own view, noted here primarily for reference, is that Quine preferred 'ontology' because it held fewer associations with approaches he rejected, resting content to let his opponents keep the term 'metaphysics'. Nevertheless ontology for Quine is not self-contained, as his approach relies on cooperation (and conflict) between ontology and *ideology*. If the project these two comprise is appropriately called 'metaphysics', then Quine's conception of how to pursue that project is appropriately called 'metametaphysics'.
4 Of course plenty *do so* – some insist on separating being and existence, others on distinguishing between different modes of existence, and yet others that these notions should not interest us in ontology whatsoever because the *real* ontological question is 'What is fundamental?'. More on how fundamentality fits (or doesn't) into Quine's conception of metaphysics later.
5 Naturalism would lead Quine to reject sense data, as he saw them as posits rather than theory-free building blocks – ultimately, not fruitful posits. Nevertheless he saw the attraction of their apparent immediacy.
6 This is merely an illustrative way of distinguishing between theory and complement-theory, not Quine's gloss on it – ways of making the intrinsic/extrinsic distinction are orthogonal to what we're discussing.
7 Putnam focuses heavily on model theory when developing these ideas from Quine, which play an important role in his earlier work (e.g. Putnam 1977); for analysis of Putnam's model-theoretic arguments see Button 2013.

References

Berto, F. and M. Plebani. (2015). *Ontology and Metaontology: A Contemporary Guide*. London: Bloomsbury.
Button, T. (2013). *The Limits of Realism*. Oxford: Oxford University Press.
Carnap, R. (1947). *Logical Syntax of Language*. Trans. A. Smeaton. London: Kegan Paul, Trench & Trubner.
Chalmers, D., D. Manley and R. Wasserman. (2009). *Metametaphysics: New Essays on the Foundations of Ontology*. Oxford: Oxford University Press.
Davidson, D. (2001). *Inquiries into Truth and Interpretation*, 2nd edn. Oxford: Oxford University Press.
Egerton, K. (2016). Getting off the Inwagen: a critique of Quinean metaontology. *Journal for the History of Analytical Philosophy* 4(6): 1–22.
Eklund, M. (2006). Metaontology. *Philosophy Compass* 1(3): 317–34.
van Inwagen, P. (1998). Meta-ontology. *Erkenntnis* 48(2–3): 233–50.
van Inwagen, P. (2009). Being, existence, and ontological commitment. In Chalmers *et al.*
Jenkins, C. (2010). What is ontological realism? *Philosophy Compass* 5(10): 880–90.
Manley, D. (2009). Introduction: a guided tour of metametaphysics. In Chalmers *et al.*

Price, H. (2009). Metaphysics after Carnap: the ghost who walks? In Chalmers *et al.*
Putnam, H. (1977). Realism and reason. *Proceedings and Addresses of the American Philosophical Association* 50(6): 483–98.
Putnam, H. (2004). *Ethics Without Ontology*. Cambridge, MA: Harvard University Press.
Quine, W. V. (1948). On what there is. *Review of Metaphysics* 2: 21–38.
Quine, W. V. (1951a). Two dogmas of empiricism. *The Philosophical Review* 60(1): 20–43.
Quine, W. V. (1951b). On Carnap's views on ontology. *Philosophical Studies* 2(5): 65–72.
Quine, W. V. (1960). *Word & Object*. Cambridge, MA: MIT Press.
Quine, W. V. (1968). Ontological relativity. *Journal of Philosophy* 65(7): 185–212.
Quine, W. V. (1981). *Theories and Things*. Cambridge, MA: Harvard University Press.
Quine, W. V. (1983). Ontology and ideology revisited. *The Journal of Philosophy* 80(9): 499–502.
Quine, W. V. (1992). *Pursuit of Truth*, 2nd edn. Cambridge, MA: Harvard University Press.
Quine, W. V. and R. Fara. (1994). In conversation: W. V. Quine (video interview).
Russell, B. (1905). On denoting. *Mind* 14: 479–93.
Schaffer, J. (2009). On what grounds what. In Chalmers *et al.*
Silverman, H. (1988). *Philosophy and Non-Philosophy since Merleau-Ponty*. Evanston, IL: Northwestern University Press.

5
METAPHYSICAL REALISM AND ANTI-REALISM

Jussi Haukioja

What is metaphysical realism?

Discussions of metaphysical realism have for the most part been motivated by *objections* to it: in particular, by Hilary Putnam's arguments against metaphysical realism from the 1980s, and metaphysical realists' answers to them. This chapter is no exception: my main focus will be on challenges to metaphysical realism. But first, let us try to get clearer on exactly what metaphysical realism is.

Metaphysical realism, is obviously, a form of realism. Realist views in philosophy come in many forms, but in general, realism about a subject matter (a range of objects, properties, facts, and so on) is characterized by two aspects (cf. Miller 2014): *existence* and *independence*. A realist about Xs holds, roughly, that Xs exist, and that Xs exist independently of what way say or think about Xs. *Anti*-realist views can then challenge one of the two aspects of the realist view, either by denying that Xs exist at all (e.g. Error theory, expressivism), or by accepting that Xs exist, but denying that they are mind-independent (e.g. Idealism, social constructivism).

It is important to recognize that most realism/anti-realism debates in philosophy concern *local* realisms: they are concerned with the ontological status of a given class of entities, such as numbers, values, unobservables in science, and so on. Just as importantly, most of us are, at least pre-theoretically, realists about some entities (ordinary physical objects, for example), and anti-realists about others (say, humour or fashion). In the sense of these local debates, it makes little sense to say that one is a realist or an anti-realist, without specifying a subject matter.

Metaphysical realism differs from local realisms in making an overarching claim about the entire reality, or, as it is often put, 'the world'. However, metaphysical realism and anti-realism should not be understood simply as views that are either realist or anti-realist in the above sense, about absolutely *everything*.[1] For one thing, and as just noted, most of us are anti-realists about *some* areas, and a commitment to local anti-realism about, say, humour or morals, does not alone disqualify one as a metaphysical realist. For another, as we will see below, critics of metaphysical realism typically want to embrace common-sense realism about ordinary material objects, claiming that such entities exist, and are in *some* important sense independent of our thought and talk of them: a metaphysical anti-realist does not typically hold that we *causally* bring about tables, mountains, and apples by thinking about them. Rather, metaphysical realism is committed to some more substantial claim about reality, and the relationships between our thought and talk on the one hand, and the world on the other.

This is quite clearly visible in how two philosophers who will figure prominently below, Hilary Putnam and Tim Button, understand metaphysical realism.[2] Putnam's characterization directly involves a claim, not merely about the entities which comprise the world, but about our descriptions of them, and of truth:

> The world consists of some fixed totality of mind-independent objects. There is exactly one true and complete description of 'the way the world is'. Truth involves some sort of correspondence relation between words or thought-signs and external things and sets of things.
>
> *(Putnam 1981, 49)*

In a similar vein, Button (2013, 7–11) understands metaphysical realism as the combination of three principles (or 'Credos'): *Independence, Correspondence,* and *Cartesianism.*

Independence: The world is (largely) made up of objects that are mind-, language-, and theory-independent.

Correspondence: Truth involves some sort of correspondence relation between words or thought-signs and external things and sets of things.

Cartesianism: Even an ideal theory might be radically false.

Metaphysical realism paints, roughly, the following basic picture of the world and our place in it. Even though some things and properties in the world are, in one sense or another, of our making (say, houses and fashionableness), most things and properties in the world do not owe their existence, or their nature, to us. As cognitive agents, we do our best to make our conception of the world fit the way the world is, in the sense of constructing theories and systems of belief that consist of singular terms aiming to denote objects, general terms aiming to denote properties instantiated by those objects, and so on. Our theories are true in so far as there *are* objects denoted by the singular terms we use, and in so far as those objects in fact *do* possess the properties denoted by our general terms, in accordance with how the theories ascribe such properties to the objects (and similarly for relations and so on). However, it is conceivable that the world is not at all like our best theories and descriptions claim it to be; we might be radically mistaken.

It should be obvious that metaphysical realism, so described, is a familiar position from the history of philosophy. Arguably, it is also pre-theoretically appealing, at least judging from the ease with which generation after generation of philosophy students are led into taking Cartesian doubts seriously, without much intuitive resistance concerning the *coherence* of sceptical scenarios which appear to rely on metaphysically realist background assumptions. Metaphysical realism is, because of its pre-theoretical appeal, often accepted as a default position, and detailed arguments for it considered unnecessary: the debates concerning metaphysical realism have practically without exception focused on arguments against the view, and realist counter-arguments. When something like metaphysical realism *is* argued for, the focus is typically on the Independence Credo (e.g. Devitt 1991, 39–82). However, as we will see below, critics of metaphysical realism may well accept Independence: accordingly, a defence of that principle does not yet amount to a defence of metaphysical realism, as understood here.

Relationship to semantics and epistemology

Since realism, in general, is committed to existence and independence, one may wonder why the above characterizations of metaphysical realism include semantic and epistemological assumptions. Indeed, Michael Devitt (1991, 2010, and elsewhere) has consistently warned us of 'putting the semantic/epistemological cart before the metaphysical horse': according to Devitt, metaphysics is primary, and it is only against a background of a metaphysical view that it makes sense to start formulating semantic and epistemological theories. In other words, we should first settle the metaphysical issue concerning existence and independence, and move on to semantics and epistemology only thereafter. Realism (with a capital 'R'), for Devitt, is the claim that "Tokens of most current common-sense and scientific physical types objectively exist independently of the mental" (Devitt 1991, 23 and elsewhere). This claim, Devitt insists, does not (even implicitly) say anything about truth or knowledge, or about other semantic and epistemological matters.

The initial plausibility of Devitt's comments notwithstanding, one may wonder whether such separation between metaphysics on the one hand, and semantics and epistemology on the other, is in practice feasible. The problem is that critics of metaphysical realism, such as Putnam and Button, would *accept* Devitt's characterization of Realism, at least under some plausible interpretation of "independence" (cf. Button 2013, 2; Putnam 1990, 30). And the issue is not merely a terminological disagreement about what deserves to be called "realism", Putnam and Button insisting to include semantic and epistemological ingredients in what they choose to call "realism", and Devitt refusing to do so. The key to understanding this is to note something that is especially clear in Button's presentation, but that was also clearly visible in Putnam: the arguments against metaphysical realism are put forward as a *reductio*: the point is to show that the whole metaphysical realist package is somehow incoherent, or unstable. Especially the model-theoretic argument against metaphysical realism, to which we will turn next, aims to establish that the metaphysical realist "cannot say what he or she wants to say" (Button 2013, 2). As we will see below, Button claims (and reads Putnam as claiming) that the independence and existence dimensions of metaphysical realism cannot be read in a philosophically substantive (and interesting) sense, as doing so would make it impossible for the metaphysical realist to explain how we can represent the world in thought and talk.

Arguments against metaphysical realism

The model-theoretic argument

The most central argument against metaphysical realism is Putnam's so-called model-theoretic argument. Putnam presented the argument in various different forms, and in various degrees of technicality (cf. Putnam 1978, 1981, 1983); we will here present the argument informally (for recent discussions and restatements of the argument, not overlooking its technical aspects, see Taylor 2006, 49–85; Button 2013, 14–31 and 225–240; and Khlentzos 2016, 18–26).

The basic idea of the model-theoretic argument is (deceptively) simple. According to metaphysical realism, as we have seen, the world consists of some mind-independent totality of objects, and truth of a correspondence between a theory and the world. Suppose now that we have a theory, T, the terms of which can be mapped onto the world in such a way that the theory comes out true: there is a correspondence relation between the sentences of the theory and the way the world is. Now, it can be shown that whenever there is *one* way of mapping the terms of the theory onto the world, making the theory come out true, there will be *many* ways of doing so. This can be illustrated by way of a permutation argument: if there is an interpretation of the theory that makes the theory true, we can always come up with alternative interpretations, shuffling the interpretations of both the names and the predicates in the theory in a way which makes exactly the same sentences true.[3] It can moreover be shown that similar permutations can be performed such that even the truth-*conditions* of all the sentences in the theory (conceived of as sets of possible worlds) remain the same. It follows that what the theory says about the world is not sufficient for singling out an *interpretation* for its terms, that is, for determining what the extensions of the terms in the theory are. Or, in other words, if you are given a theory and told that all sentences of the theory are true, that is not sufficient for saying what objects there are in the world, and what properties they possess.

This is already problematic for the metaphysical realist: if truth is to consist in correspondence between the theory and the world, this would seem to require that there is a *single* interpretation of the terms of the theory. But things are in fact even worse, according to Putnam: it follows that an ideal theory *cannot be false*, directly contradicting the metaphysical realist's epistemic assumption (the Cartesian Credo). Suppose we have an ideal theory, satisfying all our 'operational and theoretical constraints' (roughly: corresponding ideally well with all observational data and satisfying all requirements of simplicity, economy, etc.). Such an ideal theory will be consistent, and any consistent theory has a model that makes it true; moreover, if it has one model, it will have many. So the ideal theory has a model, and there is an interpretation of the terms of the theory onto the model which makes the theory true.

The metaphysical realist will, of course, object that it is not enough that there is *an* interpretation of the ideal theory which makes it true. What matters is whether or not the theory is true on the *intended* interpretation. But we have already seen that what the theory says about the world is not enough to single out one, intended, interpretation for its terms. So how is the intended interpretation determined, if not by operational and theoretical constraints? Here, many metaphysical realists today would appeal to causation: the intended interpretation is determined by some appropriate causal connections between us and the world (the details would need to be worked out in a complete causal theory of reference). Here, Putnam responds with his much-discussed 'Just More Theory' manoeuvre (Putnam 1981, 45–48; Putnam 1983, 17–18; and elsewhere). According to Putnam, an appeal to a causal theory of reference, or any other naturalistic reduction of the reference relation, is merely an addition to the ideal theory, 'just more theory'. Including such a theory of reference in our ideal overall theory will not help with the permutation argument: given that our theory has one interpretation that makes it true, it will have many, and these different interpretations will assign different extensions to the terms used in the theory (including 'causes' and 'refers').

The 'just more theory' manoeuvre has generally been considered as the weak point in Putnam's argument. Many commentators have held that it is question begging (cf. Devitt 1983; Lewis 1984; Wright and Hale 2017). It is one thing to say that, for example, causation singles out the intended interpretation for our theories, and another to say that *causation-talk* does that. The metaphysical realist who appeals to causation intends the former, not the latter: it is causation, not the causal theory, that determines how our terms refer. Therefore, to deny the realist the possibility of appealing to causal constraints in explaining word-world relations – including the word-world relations for terms such as "causation" – is to beg the question.

More recently, however, some commentators have expressed doubts about this (near) consensus. Douven (1999) and Button (2013) have sought to reconstruct the model-theoretic argument in ways that define the intended target of the argument more clearly, as a certain sort of metasemantic naturalism. According to metasemantic naturalism, our theory of what singles out an intended interpretation for our words is an empirical theory, completely on a par with other scientific theories. Putnam is assuming that any metaphysical realist will be committed to metasemantic naturalism, and it is *this* assumption that makes metaphysical realist vulnerable to the 'just more theory' manoeuvre. According to metaphysical realism, an epistemically ideal theory may be false. Given metasemantic naturalism, our metasemantic theories are simply a part of our overall theory of the world: consequently, the metasemantic constraints are 'just more theory' *by the realist's own lights* (cf. Douven 1999, 488–490). Button (2013, 27–31) presents the just more theory manoeuvre as a dilemma posed to the metaphysical realist. Either the realist's metasemantic theory has empirical content, or it doesn't. If it does, it is 'just more theory'. If it doesn't, then the realist is not living up to his or her own metaphilosophical ideals, giving up on the idea that a naturalistic reduction of reference is possible.[4]

Other arguments: brains in vats and response-dependence

In this section I will briefly review two other arguments that are naturally seen as arguments against metaphysical realism, as understood above: Putnam's 'brains in a vat' argument and Pettit's argument for global response-dependence. Both arguments point in the same direction: our best explanations of how cognitive agents like us can represent the world in thought and talk are in tension with the metaphysical realists' Cartesianism.

The 'brains in a vat' argument (Putnam 1981, 1–21) attacks the epistemological component of metaphysical realism directly. As we have seen, according to metaphysical realism we might be radically mistaken in our beliefs about the world, even when it comes to our best theories of it. A common way of illustrating this is to appeal to radical sceptical scenarios, such as the 'brains in a vat' scenario: for all we know, we might be, and always have been, brains in a vat floating in a pool of nutrients, with our nerve endings connected to a supercomputer which is feeding us sense experiences as of an external world with tables, mountains, electrons, and so on. But in reality, according to the sceptical story, none of those things exist. No matter how far fetched the scenario may seem, we cannot rationally rule it out, and thereby even our best theories of the world might be mistaken: the world might not be at all as our best theories describe it.

Against this, Putnam argued, strikingly, that we *cannot* be brains in a vat of the kind just described – that the scenario is, in a way, self-refuting. Even though the thought experiment describes a logically possible state of affairs, we can know that *we* are not brains in a vat (of the kind described). His argument is based on a general assumption about reference: that to refer to a thing, or a kind of things, there needs to be some kind of a causal connection between the subject and the thing or (instances of the) kind of thing referred to. Now, imagine a brain in a vat having sense experiences as of seeing a tree, and thinking "There is a tree over there". By hypothesis, the brain in a vat has never causally interacted with tree. It follows, by the above assumption about reference, that the brain's tokens of "tree" cannot refer to trees! What do they refer to, then? Here, Putnam is not committed to a determinate answer, but suggests that they might refer to "'trees in the image', or to the electrical impulses that cause tree experiences, or the features of the [computer] program that are responsible for those electrical impulses" (Putnam 1981, 14). But, crucially, they do not refer to real trees. Similarly, the brains' tokens of "brain" and "vat" do not refer to brains and vats, but to something else (if they refer to anything at all). It follows that a brain in a vat cannot think or say that it is a brain in a vat: when a brain in a vat thinks "I am a brain in a vat", she is not thinking of brains and vats, but something else. By hypothesis, she is thinking something *false:* she is not having sense experiences as of being a brain in a vat (whatever they might be). This takes Putnam to his striking conclusion: "if we are brains in a vat, then the sentence 'We are brains in a vat' says something false (if it says anything).

In short, if we are brains in a vat, then 'We are brains in a vat' is false. So it is (necessarily) false" (*ibid.*, 15).

Putnam's argument provoked a massive discussion, in which various more precise formulations of the argument were proposed, and their soundness debated.[5] For our purposes here, it is important to recognize that Putnam's argument was not put forward as an anti-*sceptical* argument in epistemology: taken as such, its scope would be all too narrow. The argument is aimed at a very specific kind of a sceptical hypothesis: for example, it would be toothless against variants of the 'brains in a vat' scenario where we have previously had causal contact with trees, brains, vats, and so on, but been recently 'envatted' by (say) an evil scientist. Moreover, even if we accept the argument as establishing that we are not brains in a vat of the Putnamian kind, one may feel that it does not really answer the sceptical worries: one may worry that one is, in some sense, in a similar predicament as the brains in vats, and that all Putnam's argument shows is that we cannot express our predicament in words.[6] But the main point of the argument is not epistemological, but metaphysical: to cast doubt on the metaphysical realist's standard illustration of Cartesianism.

Another argument to the effect that an explanation of our ability to represent things, properties, and states of affairs in thought requires us to give up, or at least moderate, some assumptions traditionally associated with metaphysical realism, is found in Philip Pettit's thesis of 'global response-dependence' of concepts (Pettit 1991). This argument has not received nearly as much attention as Putnam's model-theoretic argument or the 'brains in a vat' argument, but I believe it deserves to be noted, especially since (as I will suggest in the final section) Pettit's positive view may represent a way forward with the problem of how to describe or formulate metaphysical *anti*-realism (or, more generally, views that do not fully accept metaphysical realism, whether they deserve to be called anti-realist or not.)

Like the model-theoretic argument, Pettit's argument is based on a problem concerning referential indeterminacy, and his solution to it. On Pettit's view, we have to acknowledge global response-dependence of concepts to make sense of our ability to acquire concepts from examples: response-dependence is a product of his solution to the Kripkensteinian problem of rule-following, as presented in Kripke (1982). Pettit is concerned with the question of how an agent could pick up a determinate rule – in particular, acquire a determinate concept – from a finite array of examples. According to him, one precondition for this to be so much as possible is that the examples should appear similar to each other to the subject, in such a way that the subject can develop a disposition to classify new things as being relevantly similar to the exemplars or not. That is, having been exposed to a range of objects which are F, the subject begins to classify new cases as F or non-F, depending on whether they *seem* F or non-F to her.

However, such a classifying agent will also make mistakes. Suppose that, on a given occasion, a new case x seems non-F to the agent, and is classified as non-F by her, but x is, in fact, F. How do we explain this? In particular, what is it about the agent that makes it the case that her concept of F-ness correctly applies to x, even though she just categorized x as non-F? Pettit, and many others, will here appeal to 'normal' or 'favourable' conditions: the subject's concept of F-ness applies to x, because in *normal* conditions, x *would* seem F to the subject. It follows that the extensions of our basic concepts – concepts which we can acquire by being exposed to examples are determined by the relevant entities' dispositions to *seem* a certain way to us, in normal conditions: for a basic concept F, it will be *a priori*[7] that: x is F iff x would seem F to normal subjects in normal or favourable conditions. But such *a priori* biconditionals are precisely what characterize response-dependent concepts: thus, if Pettit is right, a solution to the rule-following problem entails that all our basic concepts are response-dependent: *global* response-dependence of concepts follows.[8]

There is a tension between global response-dependence and metaphysical realism, because the *a priori* biconditional entails that subjects are, in normal conditions, immune from making errors in their application of a basic concept F: in normal conditions, if an entity *seems* F to a normal subject, the entity *is* F. This, in turn, entails that we are immune from radical error in our application of basic concepts: although any particular application of such a concept can be mistaken, because of disturbing internal or external factors, *by and large* we are correct, because the extension of a given basic concept is determined by our inclinations to classify things as falling or not falling under the concept in question in normal conditions. This is in direct tension with the metaphysical realist's epistemological assumptions. The immunity from error is limited, in that a subject making a judgement using a basic concept will never be in a position to know whether she is operating in normal conditions or not. Nonetheless, *radical* error in our judgements concerning the extensions of our basic concepts is ruled out.[9]

Metaphysical anti-realism?

We have above looked at various arguments against metaphysical realism. What does the critic of metaphysical realism suggest in its place? What would metaphysical *anti*-realism look like? This may seem like an easy question to answer: a metaphysical anti-realist is simply someone who rejects at least one of the three Credos of metaphysical realism: Independence, Correspondence, or Cartesianism. In practice, things are much more complicated and murky, as evidenced by the fact that both of the critics of metaphysical realism that have figured centrally in the above – Hilary Putnam and Tim Button – characterize themselves as *realists*, just not *metaphysical* realists. For them, the denial of metaphysical realism is the denial of the availability of a 'God's-Eye' point of view, from which we could compare our theories and belief about the world to the world itself, as it is independently of our conceptual systems. But can we give positive content to the alternative view which denies the availability of such a point of view?

Putnam, in his 'internal' realist period, insisted that his view is not relativist[10] or idealist.[11] However, when he tried to give a positive characterization of internal realism, he quickly had to resort to metaphors which are not straightforward to interpret, and which can easily be seen as giving expression to a relativist or idealist view, after all; for example, when Putnam (1981) writes that "'objects' are as much made as discovered" (p. 54), or "If one must use metaphorical language, then let the metaphor be this: the mind and the world jointly make up the mind and the world" (p. Xi). But such remarks do not really help us in getting a precise picture of what the alternative view is committed to.[12]

Button, on the other hand, makes it clear that if we accept the arguments against metaphysical realism, it is the 'Cartesian Credo' – the idea that an epistemically ideal theory might be radically false – that has to go. The Independence and Correspondence Credos are simply part of common sense, and the model-theoretic argument and the 'brains in a vat' argument do not force us to abandon them[13] (Button 2013, 65–73). Still, Button's alternative positive view does not allow itself to be stated in simple terms, either. He accepts both the model-theoretic argument and the 'brains in a vat' argument, and concludes that radical sceptical scenarios can be resisted with the latter kind of argument. Still, more local sceptical scenarios remain unanswered, and it is not possible to draw a clear line between the sensible and the non-sensible kinds of scepticism. In a similar vein, Button concludes that sweeping claims of realism and anti-realism do not really have clear content: we should reject both metaphysical realism and internal realism, and simply be realists ("the *realist* sort of realists", *ibid.*, 181). The resulting picture is really one of metaphysical *quietism*, rather than anti-realism.

Conclusion

The realism/anti-realism debate was at its most active in the 1980s, following Putnam's controversial work. During the 1990s and 2000s, interest in metaphysical realism, and arguments against it, has gradually waned. However, Button (2013) breathes new life into the debates, giving a modern reconstruction of Putnam's arguments: maybe the time has come to reconsider them, bringing more recent theories of concepts, semantics, and metasemantics to bear on the issues. Button's conclusion may seem disappointing: we are really being told that common-sense realism – a view which both metaphysical realists and their critics have wanted to defend – is the only kind of realism that we can sensibly assert. Still, even if one accepts this, philosophical work remains to be done. If we reject metaphysical realism, and its Cartesian Credo in particular, more work is needed to explain *why* radical sceptical scenarios can be ruled out. The 'brains in a vat' argument, if successful, establishes that radical sceptical scenarios are not actual, but what does this tell us, more generally, about our concepts, and about meaning and reference? One possible way of trying to make progress in this area could be further development of Pettit's global response-dependence view.[14] If it is a precondition on our possessing a concept of a property that instances of the property seem primitively similar to us; moreover, if this connection between our responses on the one hand, and the property referred to on the other, entails immunity from radical error, maybe we have here the beginnings of an *explanation* of why radical sceptical scenarios can be ruled out. To me, this seems like a promising direction. But a more detailed exploration of this will have to wait for another occasion.

Notes

1 One does sometimes find characterizations of metaphysical realism that might seem to point in this direction. For example Drew Khlentzos (2016, 1) initially describes metaphysical realism as the view that "the world is as it is independently of how humans or other inquiring agents take it to be". However, Khlentzos goes on almost immediately to characterize metaphysical realism partly in semantic and epistemological terms, bringing his understanding of the view in line with that portrayed here.

2 Button prefers the label *external* realism; for the sake of uniformity, I will use "metaphysical realism" also when discussing his arguments.

3 For concrete examples, see e.g. Putnam 1981, 33–35; Button 2013, 14–17; Khlentzos 2016, 18–26.

4 Read this way, there are interesting parallels between Putnam's model-theoretic argument and Huw Price's arguments against representationalism, and for his 'global expressivism' (cf. Price 2011): Price argues against the possibility of substantial explanations of the reference relation, and maintains that a consistent deflationist attitude towards semantic notions leads to a kind of metaphysical quietism, dismissing traditional metaphysical issues such as the question of metaphysical realism.

5 For an excellent recent collection of essays relating to many aspects of the debate, see Goldberg 2016. Also noteworthy is Chalmers 2005, where the 'brains in a vat' argument is given a limited defence, and put to metaphilosophical use.

6 But see Button (2013, 130–140) for a response to the latter worry.

7 The reason that the biconditional will be knowable a priori is that, on the Pettit's view, our responses regarding F-ness in favourable conditions *fix the reference* of F. In other words, F expresses that property, whichever it happens to be, that fits our inclinations to judge things as F in normal conditions. The apriority of the biconditional follows for familiar Kripkean reasons (see Pettit 1991 and Haukioja 2001 for details).

8 Because of limitations of space, I am here overlooking major issues concerning the rule-following problem, especially the issue of how normal or favourable conditions are to be defined, as well as issues having to do with different conceptions of response-dependence. For details, see Pettit 1990, 1991, 1999; Haukioja 2007, 2013.

9 For more discussion of global response-dependence and realism, see Pettit 1998; Smith and Stoljar 1998; Devitt 2006.

10 "Denying that it makes sense to ask whether our concepts 'match' something totally uncontaminated by conceptualization is one thing; but to hold that every conceptual system is therefore just as good as every other would be something else" (Putnam 1981, 54).
11 "Human minds did not create the stars or the mountains, but this 'flat' remark is hardly enough to settle the philosophical question of realism versus antirealism" (Putnam 1990, 30).
12 I find Simon Blackburn's expression of the problem, from 35 years ago, as relevant and eloquent today as it was then:

> Idealists always face the problem of finding an acceptable way of putting what they want to say about the involvement of the mind in the world. Some fudge it: it is quite common to find people writing that 'objects' do not exist outside of our conceptual schemes, or that we 'create' objects (values, numbers) rather than discover them. This is not a good way to put anything. With the inverted commas off, such remarks are false. (We do not create trees and galaxies, nor the wrongness of cruelty, nor the evenness of the number 2. Nor can we destroy them either, except perhaps for the trees.) But what can they mean otherwise: what is meant by saying that 'trees' are mind-dependent, if trees are not? Perhaps just the platitude that if we did not have minds of a certain kind, we could not possess the concept of a tree. The problem for the idealist, or the anti-realist in general, is to steer a course between the platitude and the paradox.
>
> *(Blackburn 1984, 218–219)*

13 We do, however, have to read them simply as commonplace – and not very surprising – assertions about the world, and about the relationships between language and the world, rather than as making *philosophical* claims (*ibid.*, 65).
14 Pettit (1991, 588) suggests, in passing, that Putnam's internal realism could be interpreted as a commitment to global response-dependence.

References

Blackburn, S. (1984). *Spreading the Word*. Oxford: Clarendon Press.
Button, T. (2013). *The Limits of Realism*. Oxford: Oxford University Press.
Chalmers, D. (2005). "The Matrix as Metaphysics". In C. Grau (ed.), *Philosophers Explore the Matrix*, 132–176. Oxford: Oxford University Press.
Devitt, M. (1983). "Realism and the Renegade Putnam: A Critical Study of *Meaning and the Moral Sciences*". *Noûs* 17, 291–301.
Devitt, M. (1991). *Realism and Truth*, 2nd ed. Cambridge: Basil Blackwell.
Devitt, M. (2006). "Worldmaking Made Hard: Rejecting Global Response Dependency". *Croatian Journal of Philosophy* 6, 3–25. Reprinted in Devitt (2010).
Devitt, M. (2010). *Putting Metaphysics First: Essays on Metaphysics and Epistemology*. Oxford: Oxford University Press.
Douven, I. (1999). "Putnam's Model-Theoretic Argument Reconstructed". *The Journal of Philosophy* 96, 479–490.
Goldberg, S., ed. (2016). *The Brain in a Vat*. Cambridge: Cambridge University Press.
Haukioja, J. (2001). "The Modal Status of Basic Equations". *Philosophical Studies* 104, 115–122.
Haukioja, J. (2007). "How (Not) to Define Normal Conditions for Response-Dependent Concepts". *Australasian Journal of Philosophy* 85, 325–331.
Haukioja, J. (2013). "Different Notions of Response-Dependence", in M. Hoeltje, B. Schnieder, and A. Steinberg (eds.), *Varieties of Dependence*, 167–190. Munich: Philosophia Verlag.
Khlentzos, D. (2016). "Challenges to Metaphysical Realism". *Stanford Encyclopedia of Philosophy*. Available at https://plato.stanford.edu/entries/realism-sem-challenge/ [Accessed 6 March 2019].
Kripke, S. (1982). *Wittgenstein on Rules and Private Language*. Oxford: Basil Blackwell.
Lewis, D. (1984). "Putnam's Paradox". *Australasian Journal of Philosophy* 62, 221–236.
Miller, A. (2014). "Realism". *Stanford Encyclopedia of Philosophy*. Available at https://plato.stanford.edu/entries/realism/ [Accessed 5 March 2019].
Pettit, P. (1991). "Realism and Response-Dependence". *Mind* 100, 587–626.
Pettit, P. (1998). "Noumenalism and Response-Dependence". *The Monist* 81, 112–132.
Pettit, P. (1999). "A Theory of Normal and Ideal Conditions". *Philosophical Studies* 96, 21–44.
Price, H. (2011). *Naturalism without Mirrors*. Oxford: Oxford University Press.

Putnam, H. (1978). *Meaning and the Moral Sciences*. London: Routledge & Kegan Paul.
Putnam, H. (1981). *Reason, Truth and History*. Cambridge: Cambridge University Press.
Putnam, H. (1983). "Models and Reality". In *Philosophical Papers, Vol 3: Realism and Reason*, 1–25. Cambridge: Cambridge University Press.
Putnam, H. (1990). "A Defense of Internal Realism". In *Realism with a Human Face*, 30–42. London: Harvard University Press.
Smith, M. and Stoljar, D. (1998). "Global Response-Dependence and Noumenal Realism". *The Monist* 81, 85–111.
Taylor, B. (2006). *Models, Truth, and Realism*. Oxford: Clarendon Press.
Wright, C. and Hale, B. (2017). "Putnam's Model-Theoretic Argument against Metaphysical Realism". In C. Wright, B. Hale, and A. Miller (eds.), *A Companion to the Philosophy of Language*, 2nd ed. (Blackwell Companions to Philosophy), 703–730. Chichester: John Wiley & Sons.

6
FROM MODAL TO POST-MODAL METAPHYSICS

Nathan Wildman

From the early 1960s until the turn of the twenty-first century, metaphysics was dominated by broadly modal issues. This manifested itself in two inter-related ways. First, the metaphysics of modality was a favourite topic of discussion, with numerous metaphysicians attempting to limn the structure of modal reality. And second, metaphysicians used broadly modal tools to illuminate a range of philosophical notions in metaphysics, philosophy of language, mind and science, logic, epistemology, and metaethics (to name a few). In fact, during this period there was hardly any area of philosophical inquiry where modality did not play a major role. In light of this, it was natural to think that *the* chief role of metaphysics was to explore and map out the nature or applications of modality.

But times have changed. Metaphysics has outgrown its modal myopia. In its place, a number of philosophers have argued that various hyperintensional notions, including grounding, metaphysical explanation, structure, and fundamentality, are (or should be) the real focus of metaphysical inquiry.

The aim of this chapter is to detail the rise and fall of modal metaphysics. In particular, I hope to offer an introduction to some key figures and debates that represent the modal approach, highlight the subsequent critiques of it, and mark the shift towards post-modal metaphysics. More specifically, I contend that the argument modal metaphysicians used to motivate the general shift of focus from the metaphysics of what *is* to the metaphysics of what *must be* equally motivates a new revolution, centring upon a variety of post-modal, hyperintensional notions. This is not to say that there are no significant metaphysical questions remaining about modality, nor that modality/modal notions are metaphysically useless. Rather, even modal enthusiasts should agree that modality is not *the* central metaphysical notion.

Before doing so, it is useful to settle a bit of terminology. Let us say that a position in a sentence is *extensional* when other expressions with the same extension can be substituted into that position *salva veritate*. In contrast, a position is *intensional* just in case expressions that are necessarily co-extensive can be freely substituted without change in the sentence's truth value. Finally, a position is *hyperintensional* provided even substitution of necessary co-extensive expressions is not guaranteed to preserve truth-value. In the sentence 'Jobke is a dog', both 'Jobke' and 'is a dog' are extensional: other terms that co-designate Jobke (e.g. 'the oldest of Nathan's current pets') and predicates that have the same extension as 'is a dog' can be freely substituted in without changing the truth-value of the sentence. However, in the sentence, 'Necessarily,

Jobke is a dog', the place occupied of 'Jobke' is intensional: substituting another co-referring term can change the truth-value of the sentence. For example, while 'Necessarily, Jobke is a dog' is (plausibly) true, 'Necessarily, the oldest of Nathan's current pets is a dog' is false, since it is possible that I also have a pet, a cat, who is older than Jobke. Finally, the place of 'Jobke is a dog' in 'Nathan believes that Jobke is a dog' is hyperintensional, as substituting necessarily co-extensive phrases need not ensure the same truth value of the overall sentence. For example, suppose that, unbeknownst to Nathan, Jobke has been given the nickname 'Sweet-Pea' by Amanda. In that case, substituting the necessarily co-referring name 'Sweet-Pea' for 'Jobke' in the true sentence, 'Nathan believes that Jobke is a dog', results in a false sentence, 'Nathan believes that Sweet-Pea is a dog'.

We can, following Nolan (2014: 151–2), extend this treatment to non-linguistic entities by considering what constructions are (or should be) used to characterize them. The general idea is that an entity/phenomena is hyperintensional if hyperintensional language is needed to properly capture it, intensional if intensional but not hyperintensional language is required, and extensional if neither intensional nor hyperintensional language is needed.

The modal revolution

In the 1940s and 50s, metaphysics was dominated by a broadly extensionalist approach, according to which all viable philosophical notions could be understood in terms of a language of analysis wherein all co-refereeing names expressions could be substituted *salve veritate*: predicates with the same extension, like co-refereeing names, could be swapped for one another without a change in truth value of the relevant sentences. And while people like Carnap (1946, 1947), C.I. Lewis (1914, 1918; Lewis and Langford 1932), and Prior (1956, 1957) were developing theories of modality, it was clear that extensionalism, most famously championed by Quine, though also advocated for by Tarski, Goodman, and Davidson, was the received view.[1]

However, the beginning of the 1960s saw key developments in modal logic. In particular, while Kanger (1957), Montague (1960), Hintikka (1961), and Prior (1957) had all been treating modality in terms of worlds, the early work of Kripke (1959a, 1959b, 1963) fully fleshed out possible worlds semantics.[2] In so doing, these logicians offered resources to rebut extensionalist objections that modal resources were unclear, if not incomprehensible, the most famous example of this perhaps being Quine's (1953) 'animadversions' against quantified modal logic. The tools developed by these logicians fixed a solid logical foundation for modality, rendering it a topic worthy of consideration.

The literature that subsequently emerged was largely divided between two loose topics. The first concerned the metaphysics of modality itself. Two prominent issues here were how we should understand the nature of possible worlds – as abstract objects as defended by Plantinga (1974, 1976), Adams (1974), Fine (1977), and Chisholm (1981), as combinatorialist constructions, per Skyrms (1981) and Armstrong (1989, 1993, 1997), or as robust concrete objects, as infamously argued for by Lewis (1973, 1986) – and whether we should be committed actualists or posit the existence of possibilia. Developments in modal logic were also used to precisify long-standing notions like the *de re/de dicto* distinction.

The second, larger topic was that of applied modality. Here, metaphysicians employed broadly modal notions to make sense of other philosophical topics, including causation (Lewis 1973), mental content (Hintikka 1962), and physicalism (Lewis 1983; Jackson 1998).

The centrality of modal notions to metaphysics during this period is nicely encapsulated by two metaphysical works produced therein. First, Kripke's *Naming and Necessity*, first published in 1972, laid bare the significant impact modal thinking and tools can have not only in metaphysics, but also in philosophy of language, mind, and epistemology. It helped set the stage for much of the modal metaphysics that was to come. Second, Lewis' *On the Plurality of Worlds* is arguably the most influential work in metaphysics published in the twentieth century. Here, Lewis set out to show how his modal realism – the thesis that this world is just one among many like it, and that 'we who inhabit this world are only a few out of all the inhabitants of all the worlds' (1986: vii) – can be used to provide analyses of a variety of other notions. And while one of Lewis' central aims was to provide a reductive account of modality in terms of concrete possible worlds, this book epitomizes one of the themes in metaphysics during this period: that of using the tools broadly associated with modality (in this case, possible worlds) to make sense of other philosophical notions. That these two remain some of the most highly cited works in philosophy is a testament both to their import, and to the influence of the modality-based metaphysics they embody.

Further, Lewis and Kripke make excellent figure-heads for the shift from Quine-led extensionalism to broadly modal metaphysics. Quine rejected necessity and analyticity. In contrast, Kripke and Lewis fully embraced (in radically different ways) modality and intensionality.[3]

So why did this modal revolution occur? Why did modality suddenly take off, becoming the centre of metaphysical discussion for this extended period of time?

One reason was that many extensionalist explications of various notions were inadequate and clumsy, often entailing counter-intuitive results. A simple case of this concerns the identity conditions for properties. When is one property P the same as property Q? A purely extensionalist answer is that P and Q are identical iff they have the same extension – provided that all and only the same entities that have P also have Q, then P and Q are identical; otherwise, the properties are distinct.

However, consider the properties *being a cordate* (i.e. being a creature with a heart) and *being a renate* (i.e. being a creature with a kidney). These properties in fact have the same extension, as everything that in fact is a cordate is a renate, and everything that is in fact a renate is also a cordate. So, per the extensionalist account, they are identical. But this identification looks *wrong*. For one, it seems plausible that, had things been different – if evolutionary paths had varied ever so slightly – there could have been a cordate that is not a renate (similarly, there could have been a renate that is not a cordate). In other words, while they actually entirely overlap, there are possible scenarios where the two properties have different extensions. And the fact that they can have different extensions suggests that the two properties are two after all.

This leads to the second, closely linked reason for the modal revolution: it promised better tools, tools that philosophers could use to explicate various distinctions and concepts in new, superior ways. For example, in light of the problem facing extensionalist accounts of property identity, the natural move is to incorporate modality into our analysis, and say that properties P and Q are identical iff they have the same extension in every possible scenario – i.e. P = Q iff *necessarily*, the extension of P is the same as the extension of Q. This modalized account nicely circumvents the problem facing the extensionalist story, since it entails that being a renate and being a cordate are different properties.

A similar pattern repeated itself across a variety of contexts: a phenomenon would prove recalcitrant to analysis in extensionalist terms, either because no obvious analysis was possible or because the account on offer was unpalatable. Faced with this issue, metaphysicians would then wheel in their modal toolkit and produce something better.

This highlights the third and perhaps strongest reason behind the modal revolution: the broad *utility* of modal notions. Approaching metaphysics via the lens of modality was fruitful, offering benefits in theoretical unity – we can do a lot with just a commitment to possible worlds! – ideological economy – we can reduce/analyse a variety of notions in modal terms!

– and, due to the efforts of modal logicians, logical clarity. Commitment to broadly modal notions was, to borrow a phrase from Lewis, 'an offer you can't refuse' (1986: 4). If you wanted to do metaphysics, then doing so via modal notions looked like the way forward.[4]

Problems with modal metaphysics

In this way, the modal metaphysicians' case against extensionalism turned upon the fact that adopting the modal toolkit allows us to clarify new theoretical options, shed new light upon a range of metaphysical questions, and, most importantly, provide modal analyses of other metaphysical notions/concepts which were superior to any extensionalist alternative.

Of course, the success of this case hinges upon the modal metaphysicians' ability to deliver the goods. That is, the offer is only un-refusable provided that broadly modal notions can in fact be used to provide analyses of relevant metaphysical concepts/notions. The problem is that, for all the promise it offered, it is unclear just how ultimately illuminating the modal toolkit turns out to be. For many of the purported modality-based success stories turn out to be clunky, awkward, and inadequate, too.

We can already see this if we consider the example of property identity. While the modal account is definitely *better* than the extensionalist story, it certainly isn't perfect. For one consequence of the modal story is that all necessarily co-extensive properties are identical. But consider *triangularity* and *trilaterality*. These two properties necessarily apply to all and only the same objects, yet they seem to be distinct properties. Consequently, a proper account of property identity looks like it requires hyperintensional tools – modality alone is not going to cut it.

A similar pattern has repeated itself across a variety of contexts: a phenomenon appears to be successfully accounted for in broadly modal terms, but, after closer scrutiny, problems emerge. These problems indicate that the proposed modal story fails to adequately capture the relevant data. Further, no modal variant can be created to avoid the highlighted difficulties. To draw this out, I would like to consider three particular cases.

The first case concerns *essence*. One of the central questions in metaphysics is what things are, in the metaphysically significant sense of the phrase. And the metaphysical notion of essence is meant to provide (or be) the answer to this question: saying what x *is*, is to explicate x's essence. Yet much about essence remains mysterious. An analysis of the notion would help to clarify its nature.

Modality appears to offer a way to make sense of essence. Specifically, the modal analysis of essence – 'modalism' for short – is an attempt to characterize the metaphysically significant concept of essence in terms of metaphysical necessity. The standard modalist account defines essence via a necessitated existence-conditional:

ME   x is essentially F iff necessarily, if x exists, then x is F.[5]

Here the sense of necessity was understood to be metaphysical, rather than e.g., logical, conceptual, or nomic. One consequence of ME, and modalism in general, is that essentialist truths are in fact just a special case of modal truth, and essence is simply a sub-species of (metaphysical) necessity.

For most of the twentieth century, modalism was the dominant account of essence, being 'so wide-spread that it would be pointless to give references' (Correia 2005: 26).[6] However, the 1990s saw a series of problems, raised by Dunn (1990) and Fine (1994a, 1994b), which demonstrated that, 'the notion of essence which is of central importance to the metaphysics of identity is not to be understood in modal terms or even to be regarded as extensionally equivalent to a modal notion' (Fine 1994a: 3).

The first problem concerns properties that are necessary. Specifically, properties derived from necessary facts (e.g. the property of being such that 2 + 2 = 4, the property of being such that there are infinitely many prime numbers) and meta-essential properties (e.g. the property of being such that Socrates is essentially human) will, given ME, be essential to every object. But this is strongly counter-intuitive: it simply isn't part of what Socrates is that there are infinitely many prime numbers. Nor does the Eiffel Tower's nature say anything about Socrates' essence.

The more troublesome problem is that modalism is unable to account for what Dunn (1990: 90) calls 'asymmetric essential predicates'. Take Socrates and his singleton, {Socrates}. According to basic principles of modal set theory, necessarily, if Socrates exists, then he is a member of {Socrates}. Given ME, this means that both Socrates is essentially a member of {Socrates}. But,

> intuitively, this is not so. It is no part of the essence of Socrates to belong to [{Socrates}].
> ... There is nothing in the nature of a person ... which demands that he belongs to this or that set or which demands, given that the person exists, that there even be any sets.
>
> (Fine 1994a: 5)

Moreover, it *is* intuitive that {Socrates} is such that it essentially has Socrates as a member. Yet there is no way to use modal resources alone to capture this asymmetry (for further discussion, see Torza 2015). So no purely modal account can in fact provide a suitable analysis of essence.

In light of these problems, metaphysicians have largely abandoned the prospect of defining essence in purely modal terms. Some have offered accounts that incorporate non-modal tools, so as to circumvent the problem cases (e.g. Brogaard and Salerno 2013; Cowling 2013; Denby 2014; Livingstone-Banks 2017; and Wildman 2013, 2016). A more popular alternative, primarily developed in the work of Fine (1994a, 1994b, 1995a, 1995b, 2000, 2015) and Correia (2006, 2012, 2013) has been to claim that modalism gets the story backwards: instead of defining essence as a sub-species of modality, we should define modality as a sub-species of essence.[7] Regardless, consensus is that a straightforward modalist account of essence is simply untenable.

The second case concerns *ontological dependence*. Roughly, ontological dependence is the relation that holds between two entities when one depends upon the other for its existence. Understood in this manner, ontological dependence plays a prominent role in a wide range of philosophical theorizing. For example, a broadly Aristotelian account of the metaphysical notion of *substance* takes a substance to be a concrete, ontologically independent entity – i.e. an entity that does *not* ontologically depend upon anything (Lowe 1998, ch. 6). Further, the Aristotelian conception of *in re* universals holds that universals ontologically depend upon their exemplifiers, in contrast to the Platonic *ante re* view, which denies this dependence. Similarly, in philosophy of mind, some claim that the mind is distinct from, but ontologically dependent upon, the body. Finally, events have been said to depend upon their participants, non-empty sets on their members, holes upon their hosts, and boundaries upon the corresponding extended objects.

In fact, there is a range of ontological dependence relations. Perhaps the simplest is *rigid existential dependence*, which is where the existence of a particular individual x depends upon the existence of a specific individual y. This contrasts with *generic existential dependence*, where the existence of a particular individual x depends upon the existence of another individual of a certain sort. For example, the dependence between {Socrates} and Socrates is plausibly one of rigid existential dependence – the set depends upon *this* specific individual – while the dependence between a mereologically complex object and its parts is generic existential dependence – the complex object needs parts to exist, but which particular objects those parts are is not important.

Like with essence, it is natural to think that modal tools could be leveraged to provide suitable accounts of rigid and generic existential dependence. The standard way of doing so went along the following lines:

MRD   x rigidly ontologically depends upon y iff$_{df}$ necessarily, if x exists, then y exists

MGD   x generically ontologically depends upon an F iff$_{df}$ necessarily, if x exists, then some F exists

As with the modalist accounts of the essence, the sense of necessity here was understood to be metaphysical, rather than e.g. logical, conceptual, or nomic.

This modal account of ontological dependence is *prima facie* appealing. For one, it offers a radically simplified ideology: instead of taking them as primitive, we can cash out notions of rigid and generic existential dependence in terms of metaphysical necessity and some basic logical concepts. For another, one natural way to express x's existential depending on y is by saying that, 'x cannot exist unless y exists'; this makes explicit use of a modal notion, strongly suggesting that there is an intimate connection – a connection that MRD and MGD make explicit.

For all of this, both modal accounts are flawed, delivering the wrong results. MRD, for example, is unable to accommodate asymmetries in dependence between necessarily co-existing entities. Consider again {Socrates} and Socrates. Necessarily, {Socrates} exists iff Socrates does. Consequently, according to MRD, the two ontologically depend upon each other. Yet, suggests Fine (1995a: 271), Socrates does *not* depend upon the set, though the set does depend upon Socrates. Similarly, in every world where Socrates exists, there also exists a temporally extended event that is his life. Per MRD, Socrates and his life ontologically depend upon each other. But, according to Lowe, this is wrong: 'we want to say that Socrates' life only exists because Socrates does, whereas it would be putting the cart before the horse to say that Socrates exists because his life does' (1998: 145).

Another major problem is that both MRD and MGD entail that every entity ontologically depends upon every necessarily existing entity.[8] And it seems strange to say that Socrates rigidly depends upon the number 4735, or that the event of your having breakfast this morning generically depends upon the existence of pure sets. In both cases the depended-upon entities look irrelevant to the dependent, but modal tools alone do not let us filter out these irrelevancies.

In response to these issues, metaphysicians have largely given up on modal definitions of rigid and generic dependence, and instead have developed post-modal alternatives. Two particularly fruitful avenues here have been to use non-modalist conceptions of essence (see e.g. Fine 1995) and metaphysical explanation (see e.g. Correia 2005 and Schnieder 2006a), since both notions support more fine-grained distinctions than purely modal accounts while also seeming to explain the intuitive link between dependence and modality.[9]

The third and final case is about counterfactuals. The 'closest world' treatment of counterfactual conditions developed by Stalnaker (1968) and Lewis (1973) is often put forward as a smashing success story for modal metaphysics. The rough-and-ready idea is to use a notion of *comparative similarity* between worlds to provide truth-conditions for counterfactual conditionals, where one world is taken to be closer to actuality than another if the first resembles (i.e. is more similar to) the actual world more than the second does. Using this similarity relation, we can define the truth conditions for counterfactuals in the following manner:

MCF   'A □→C' is true in the actual world iff
  (i) there are no possible A-worlds; or
  (ii) some A-world where C obtains is closer to the actual world than any A-world where C does not obtain

This modal account of counterfactuals has received a large amount of push-back in recent years. The key point of contention has concerned counterpossibles – that is, counterfactuals with impossible antecedents. Since counterpossibles feature necessarily false antecedents, there are no worlds where the antecedent is true; hence, by clause (i), they are all trivially true. But this leads to counter-intuitive results. Consider the following pair of counterpossibles:

a   If Hobbes had succeed in squaring the circle, then all geometricians would have been amazed
b   If Hobbes had secretly squared the circle, then sick children in the mountains of South America at the time would have cared[10]

Intuitively, (a) is true – if, per impossible, Hobbes had squared the circle, then geometricians really would have been amazed! – but (b) is false. This suggests, contra MCF, that not every counterpossible is (vacuously) true. Consequently, it seems that we need to go beyond (standard) modal notions to make sense of counterpossibles and, more generally, counterfactuals; instead, we need to invoke other, hyperintensional notions (for further discussion, see e.g. Beall and van Fraassen 2003; Brogaard and Salerno 2013; Jago 2014; Nolan 1997; and Restall 1997).

The failures of the modal analyses of essence, ontological dependence, and counterfactuals are indicative of the fact that, in many cases, while modal tools certainly are finer grained than those available to the extensionalist, they often prove to not be fine grained *enough*. Like their extensionalist predecessors, purely modal approaches for many phenomena turn out to be inadequate, providing awkward, counter-intuitive results. The modal revolution had come, but it had not delivered on all of its promises.

Onwards to the post-modal future

At the dawn of the twenty-first century, metaphysics found itself in a position similar to the one it was in the early 1960s: an older paradigm is slowly playing itself out. New notions and tools were needed. To that end, metaphysicians have, in recent years, begun developing and refining new hyperintensional notions, which promise to shed new light on a diverse range of phenomena. The emergence of these, the use of which is rapidly expanding, marks the shift to a new, *post-modal* paradigm for metaphysics.

Perhaps the most significant of these new notions – or at least the one most discussed in the literature – is *metaphysical grounding*. The past two decades have seen an absolute explosion of literature on grounding. And while there is significant disagreement about the nature of grounding – see e.g. Audi (2012), Bliss and Trogdon (2016), Correia and Schnieder (2012), Fine (2012), Schaffer (2009), Skiles and Trogdon (this volume), and Wilson (2014) – all grounding-advocates agree that grounding is a hyperintensional notion. And it seems like grounding can make sense of cases that purely modal approaches cannot. For example, take {Socrates}-Socrates again: the latter's existence is said to *ground* the former's existence and not vice versa, though the two necessarily exist in all and only the same worlds.

A second, closely related, hyperintensional notion is *metaphysical explanation*. The nature of metaphysical explanation, and especially the relationship between grounding and metaphysical explanation is complex – for more, see e.g. Maurin (2018) and Thompson (this volume). But, like with grounding – metaphysical explanation is universally agreed to be a hyperintensional notion. And it also lends itself to certain applications that are not feasible with purely modal tools. For example, according to Schnieder (2011: 445), a hyperintensional notion of explanation allows us to make sense of the Aristotelian insight that p is true because p (but not vice versa). This is because 'p' and 'that p is true' are intensionally equivalent: necessarily, p is true iff p. Further, a hyperintensional notion of explanation allows us to capture the intuitive idea that some necessities explain other necessities – for example, the disjunctive fact that Socrates is wise or 2 + 2 = 4 is necessary because 2 + 2 = 4 is necessary. This wouldn't be possible given a purely modal conception of explanation, since we could not distinguish explanans from explanandum.[11]

A third hyperintensional notion is Sider's (2011) *structure*. A bit of background: Lewis (1983) noted that properties are cheap and abundant, but that this abundance of properties is in fact unable to perform many of the central tasks that properties are postulated to do. In particular, as every pair of objects shares infinitely many abundant properties, judgements about (dis)similarity become effectively impossible. As a solution, Lewis proposed we distinguish the special, *natural* properties, which demarcate the fundamental, objective ways that things are (dis)similar. Naturalness, then, was a property of properties, with the natural properties serving as the elite, fixing the qualitative joints of nature.[12]

Sider's notion of *structure* is a generalization and extension of naturalness. Like naturalness, structure concerns carving nature at its joints. However, unlike naturalness, structure applies to any grammatical category, rather than merely to predicates. This allows Sider to claim that objects, logical connectives, and even quantifiers are structural.

Sider places structure at the very heart of metaphysics:

> Metaphysics, at bottom, is about the fundamental structure of reality. Not about what's necessarily true. Not about what properties are essential. Not about conceptual analysis. Not about what there is. Structure.
>
> *(2011: 1)*

And while it is not clear that Sider is correct – see e.g. Barnes (2014) for some criticism – structure certainly seems to be a fruitful postulate for future metaphysical theorizing.

Finally, there is the notion of *fundamentality*. Recently, metaphysicians have begun exploring questions about the nature of fundamentality, as well as putting this notion to use to try and make sense of other philosophical concepts (for more discussion, see the papers in Bliss and Priest (2018); for criticism concerning fundamentality-driven metaphysical approaches, see Lipman (2018)).

The same general argument that undergirded the modal revolution seems to apply to these notions as well: the old tools have proven to be ineffective, while these new tools offer the promise of explicating various distinctions and concepts in new, superior ways. Further, metaphysicians have begun putting these notions to use in a variety of philosophical contexts. As before, their utility motivates the adoption of the tools. To paraphrase Lewis (1986: 4), the price is right; the benefits in theoretical unity and economy are worth the commitment.

Conclusion

The modal revolution is, for better or for worse, over. Modality is no longer queen of metaphysics, no longer *the* tool within the metaphysician's toolkit. And the general case for the modal revolution seems to equally motivate a post-modal, hyperintensional revolution, one centred upon concepts like grounding, metaphysical explanation, structure, and fundamentality. This is not to say that modality is no longer of metaphysical interest – we can, for example, still ask deep and important questions about the relationship between modality and the new hyperintensional notions[13] – and debates about the metaphysics of modality are still raging. However, it is clear that we have moved into the *post*-modal era of metaphysics. Only time will tell whether the post-modal revolution will be as successful as the modal one.

Notes

1 In fact, the title of one of Quine's posthumous publications is *Confessions of a Confirmed Extensionalist and Other Essays*. Further, as Nolan (2014: 153) points out, the roots of this movement stretch much further back, at least through the logical positivists as well as Russell and Whitehead's work on logical analysis.
2 In particular, possible world semantics allows the modal operators to be treated as quantifiers over possible worlds. Specifically, 'Possibly, P' is true iff P is true at some possible world, and 'Necessarily, P' is true iff P is true at every possible world. In this way, possible world semantics offered an extensional semantic theory for modal logic.
3 It is worth noting that the extent to which Lewis moved away from Quine is a topic of some debate. In particular, the cornerstone of Lewis' philosophical views – his modal realism – was, arguably, broadly *extensional*, since it was designed to provide a reductive characterization of modality in purely non-modal terms. Whether Lewis succeeded in his reductive ambitions is another topic of much debate (see e.g. Cameron 2012). Regardless, for present purposes, we can treat Lewis as part of the broad modal revolution. For more on Quine's (meta)metaphysics, see Egerton (this volume), and for more on Lewis, see the various chapters in Loewer and Schaffer (2015).
4 It is also likely that there were powerful sociological factors that had a hand in modality's success, though I set these aside here.
5 There was also an alternative, simple modalist account:

  MSE   x is essentially F iff$_{df}$ necessarily, x is F

 However, this simple account seems to entail that contingent existents have no essential properties. Take some contingently existing object like Cicero. In worlds where Cicero does not exist, he has no properties – he simply isn't there to have them. But then there are no properties such that, necessarily, Cicero possesses them. So, given MSE, it follows Cicero has no essential properties.
6 That said, see e.g. Cartwright (1968), Kripke (1971), Marcus (1967, 1971), McKay (1975), Parsons (1971), and Plantinga (1974).
7 For more on the nature of essence, see Leech (this volume).
8 This also means that nothing is an Aristotelian substance, since every entity ontologically depends upon something (and hence is not ontologically *in*dependent).
9 For more on ontological dependence, see Correia (2005, 2008), Koslicki (2012, 2013), and Tahko and Lowe (2016).
10 These examples are based upon ones used by Nolan (1997) and Williamson (2017).
11 A subset of these cases will be mathematical explanations: plausibly, some mathematical truths explain others and not vice versa. But, assuming that mathematical truths are necessary, all of them are necessarily equivalent.
12 In fact, depending upon one's metaphysics of properties, Lewisian naturalness is itself hyperintensional.
13 One such debate concerns whether full grounding entails metaphysical necessitation. Some (Audi 2012; Dasgupta 2014; Trogdon 2013) argue that it does, while others (Leuenberger 2014; Schnieder 2006b; Skiles 2015) claim it does not. A second is about whether fundamentalia are necessarily or merely contingently fundamental (Wildman 2018).

References

Adams, R. M. (1974). Theories of Actuality. *Noûs* 8: 211–231.
Armstrong, D. M. (1989). *A Combinatorial Theory of Possibility*. New York: Cambridge University Press.
Armstrong, D. M. (1993). A World of States of Affairs. *Philosophical Perspectives* 7: 429–440.
Armstrong, D. M. (1997). *A World of States of Affairs*. New York: Cambridge University Press.
Audi, P. (2012). Grounding: Toward a Theory of the In-Virtue-of Relation. *Journal of Philosophy* 109: 685–711.
Barnes, E. (2014). Going Beyond the Fundamental: Feminism: *An Introduction to Modal and Many-Valued Logic*. Oxford: Oxford University Press.
Bliss, R. and Priest, G., eds. (2018). *Reality and its Structure: Essays in Fundamentality*. Oxford: Oxford University Press.
Bliss, R. and Trogdon, K. (2016). Metaphysical Grounding. In Zalta, E., ed., *The Stanford Encyclopaedia of Philosophy*. URL = https://plato.stanford.edu/archives/win2016/entries/grounding/.
Brogaard, B., and Salerno, J. (2013). Remarks on Counterpossibles. *Synthese* 190(4): 639–660.
Cameron, R. P. (2012). Why Lewis's Analysis of Modality Succeeds in its Reductive Ambitions. *Philosophers' Imprint* 12(8): 1–21.
Carnap, R. (1946). Modalities and Quantification. *Journal of Symbolic Logic* 11(2): 33–64.
Carnap, R. (1947). *Meaning and Necessity*. Chicago: University of Chicago Press.
Cartwright, R. (1968). Some Remarks on Essentialism. *Journal of Philosophy* 65: 615–626.
Chisholm, R. (1981). *The First Person*. Minneapolis: University of Minnesota Press.
Correia, F. (2005). *Existential Dependence and Cognate Notions*. Munich: Philosophica.
Correia, F. (2006). Generic Essence, Objectual Essence, and Modality. *Noûs* 40(4): 753–767.
Correia, F. (2008). Ontological Dependence. *Philosophy Compass* 3(5): 1013–1032.
Correia, F. (2012). On the Reduction of Necessity to Essence. *Philosophy and Phenomenological Research* 84(3): 639–653.
Correia, F. (2013). Metaphysical Grounds and Essence. In Hoeltje, M., Schnieder, B., and Steinberg, A., eds., *Varieties of Dependence*. Munich: Philosophia Verlag, pp. 271–291.
Correia, F. and Schnieder, B., eds. (2012). *Metaphysical Grounding: Understanding the Structure of Reality*. Cambridge: Cambridge University Press.
Cowling, S. (2013). The Modal View of Essence. *Canadian Journal of Philosophy* 43(2): 248–266.
Dasgupta, S. (2014). On the Plurality of Grounds. *Philosophers' Imprint* 14: 1–28.
Denby, D. A. (2014). Essence and Intrinsicality. In Francescotti, R., ed., *Companion to Intrinsic Properties*. Berlin: De Gruyter, pp. 87–109.
Dunn, J. M. (1990). Relevant Predication III: Essential Properties. In Dunn, J. M. and Gupta, A., eds., *Truth or Consequences: Essays in Honor of Nuel Belnap*. Dordrecht: Kluwer Academic Publishers, pp. 77–95.
Fine, K. (1977). Prior on the Construction of Possible Worlds and Instants. Postscript to Prior, A. N. and Fine, K., *Worlds, Times and Selves*. Amherst, MA: The University of Massachusetts Press, pp. 116–161.
Fine, K. (1994a). Essence and Modality. *Philosophical Perspectives* 8: 1–16.
Fine, K. (1994b). Senses of Essence. In Sinnott-Armstrong, W., ed., *Modality, Morality, and Belief*. New York: Cambridge University Press, pp. 53–73.
Fine, K. (1995a). Ontological Dependence. *Proceedings of the Aristotelian Society* 95(3): 269–290.
Fine, K. (1995b). The Logic of Essence. *Journal of Philosophical Logic* 24(3): 241–273.
Fine, K. (2000). Semantics for the Logic of Essence. *Journal of Philosophical Logic* 29(6): 543–584.
Fine, K. (2012). Guide to Ground. In Correia, F. and Schnieder, B. eds., *Metaphysical Grounding*. Cambridge: Cambridge University Press, pp. 37–80.
Fine, K. (2015). Unified Foundations for Essence and Ground. *Journal of the American Philosophical Association* 1: 296–311.
Hintikka, J. (1961). Modalities and Quantification. *Theoria* 27(3): 119–128.
Hintikka, J. (1962). *Knowledge and Belief*. Ithaca, NY: Cornell University Press.
Jackson, F. (1998). *From Metaphysics to Ethics*. Oxford: Oxford University Press.
Jago, M. (2014). *The Impossible: An Essay on Hyperintensionality*. Oxford: Oxford University Press.
Kanger, S. (1957). *Provability in Logic*. Acta Universitatis Stockholmiensis, Stockholm Studies in Philosophy, Vol. 1, Stockholm: Almqvist and Wiksell.
Koslicki, K. (2012). Varieties of Ontological Dependence. In Correia, F. and Schnieder, B., eds., *Metaphysical Grounding*. Cambridge: Cambridge University Press, pp. 186–213.

Koslicki, K. (2013). Ontological Dependence: An Opinionated Survey. In Schnieder, B., Hoeltje, M., and Steinberg, A., eds. *Varieties of Dependence: Ontological Dependence, Grounding, Supervenience, Response-Dependence*. Munich: Philosophia Verlag, pp. 31–64.
Kripke, S. A. (1959a). A Completeness Theorem in Modal Logic. *Journal of Symbolic Logic* 24(1): 1–14.
Kripke, S. A. (1959b). Semantical Analysis of Modal Logic. *Journal of Symbolic Logic* 24(4): 323–324.
Kripke, S. A. (1963). Semantical Considerations on Modal Logic. *Acta Philosophica Fennica* 16: 83–94.
Kripke, S. A. (1971). Identity and Necessity. In Munitz, M., ed., *Identity and Individuation*. New York: New York University Press.
Kripke, S. A. (1972). *Naming and Necessity*. Cambridge, MA: Harvard University Press.
Leuenberger, S. (2014). Grounding and Necessity. *Inquiry: An Interdisciplinary Journal of Philosophy* 57(2): 151–174.
Lewis, C. I. (1914). The Calculus of Strict Implication. *Mind* 23(1): 240–247.
Lewis, C. I. (1918). *A Survey of Symbolic Logic*. Berkeley: University of California Press.
Lewis, C. I. and Langford, C. H. (1932). *Symbolic Logic*. London: Century. 2nd edition 1959, New York: Dover.
Lewis, D. (1973). *Counterfactuals*. Cambridge, MA: Harvard University Press.
Lewis, D. (1983). New Work for a Theory of Universals. *Australasian Journal of Philosophy* 61: 343–377.
Lewis, D. (1986). *On the Plurality of Worlds*. Oxford: Blackwell.
Lipman, M. A. (2018). Against Fundamentality-Based Metaphysics. *Noûs* 52(3): 587–610.
Livingstone-Banks, J. (2017). In Defence of Modal Essentialism. *Inquiry: An Interdisciplinary Journal of Philosophy* 60(8): 1–27.
Loewer, B. and Schaffer, J., eds. (2015). *A Companion to David Lewis*. London: Wiley-Blackwell.
Lowe, E. J. (1998). *The Possibility of Metaphysics*. Oxford: Clarendon Press.
Marcus, R. B. (1967). Essentialism in Modal Logic. *Noûs* 1: 91–96.
Marcus, R. B. (1971). Essential Attribution. *Journal of Philosophy* 68: 187–202.
Maurin, A. (2018). Grounding and Metaphysical Explanation: It's Complicated. *Philosophical Studies* 176: 1573–1594.
McKay, T. (1975). Essentialism in Quantified Modal Logic. *Journal of Philosophical Logic* 4: 423–438.
Montague, R. (1960). Logical Necessity, Physical Necessity, Ethics, and Quantifiers. *Inquiry: An Interdisciplinary Journal of Philosophy* 3(1–4): 259–269.
Nolan, D. (1997). Impossible Worlds: A Modest Approach. *Notre Dame Journal of Formal Logic* 38: 535–572.
Nolan, D. (2014). Hyperintensional Metaphysics. *Philosophical Studies* 171(1): 149–160.
Parsons, T. (1971). Essentialism and Quantified Modal Logic. Reprinted in Linsky, L., ed., *Reference and Modality*. Oxford: Oxford University Press, pp. 73–87.
Plantinga, A. (1974). *The Nature of Necessity*. Oxford: Oxford University Press.
Plantinga, A. (1976). Actualism and Possible Worlds. *Theoria* 42: 139–160.
Prior, A. N. (1956). Modality and Quantification in S5. *Journal of Symbolic Logic* 21(1): 60–62.
Prior, A. N. (1957). *Time and Modality*. Oxford: Clarendon Press.
Quine, W. V. O. (1953). Three Grades of Modal Involvement. Proceedings of the XIth International Congress of Philosophy, 14: 65–81, reprinted in *The Ways of Paradox and Other Essays*. New York: Random House, pp. 156–174.
Restall, G. (1997). Ways Things Can't Be. *Notre Dame Journal of Formal Logic* 38: 583–596.
Schaffer, J. (2009). On What Grounds What. In Chalmers, D., Manley, D., and Wasserman, R., eds., *Metametaphysics: New Essays on the Foundations of Ontology*. Oxford: Oxford University Press, pp. 347–383.
Schnieder, B. (2006a). A Certain Kind of Trinity: Dependence, Substance, Explanation. *Philosophical Studies* 129: 393–419.
Schnieder, B. (2006b). Truth-Making without Truth-Makers. *Synthese* 152: 21–46.
Schnieder, B. (2011). A Logic for 'Because'. *The Review of Symbolic Logic* 4(3): 445–465.
Sider, T. (2011). *Writing the Book of the World*. Oxford: Oxford University Press.
Skiles, A. (2015). Against Grounding Necessitarianism. *Erkenntnis* 80(4): 717–751.
Skyrms, B. (1981). Tractarian Nominalism. *Philosophical Studies* 40(2): 199–206.
Stalnaker, R. (1968). A Theory of Conditionals. In *Studies in Logical Theory*, American Philosophical Quarterly Monograph Series, 2. Oxford: Blackwell, pp. 98–112.
Tahko, T. E. and Lowe, E. J. (2016). Ontological Dependence. In Zalta, E., ed., *The Stanford Encyclopedia of Philosophy* (Winter 2016 Edition), URL = https://plato.stanford.edu/archives/win2016/entries/dependence-ontological/.

Torza, A. (2015). Speaking of Essence. *Philosophical Quarterly* 65(261): 754–771.
Trogdon, K. (2013). Grounding: Necessary or Contingent? *Pacific Philosophical Quarterly* 94: 465–485.
Wildman, N. (2013). Modality, Sparsity, and Essence. *Philosophical Quarterly* 63(253): 760–782.
Wildman, N. (2016). How to Be a Modalist about Essence. In Jago, M., ed., *Reality Making*. Oxford: Oxford University Press, pp. 177–195.
Wildman, N. (2018). On Shaky Ground? In Bliss, R. and Priest, G., eds., *Reality and its Structure: Essays in Fundamentality*. Oxford: Oxford University Press, pp. 275–290.
Wilson, J. M. (2014). No Work for a Theory of Grounding. *Inquiry: An Interdisciplinary Journal of Philosophy* 57(5–6): 535–579.
Williamson, T. (2017). Counterpossibles in Semantics and Metaphysics. *Argumenta* 2(2): 195–226.

PART II

Neo-Quineanism (and its objectors)

7
ONTOLOGICAL COMMITMENT AND QUANTIFIERS

Ted Parent

When is a speaker committed to the existence of a thing? Or in the philosophical jargon, when is your statement *ontologically committing*?[1]

At first, this may seem straightforward. For example, if a person sincerely asserts the following, a commitment to God's existence is indicated:

(1) God loves us and wants us to be happy.

Notoriously, however, natural language can be misleading. The term 'God' can be used in statements that do *not* commit one to God's existence, as with the following idiomatic expression:

(2) Some acts of God are covered under this policy.

When an insurance salesperson sincerely asserts (2), she does not mean to introduce theology into the discussion. And of course, a person can insincerely assert a God-statement. Sometimes, in the face of adversity, a non-religious person can be heard saying:

(3) God hates me.

But the most important case of an "ontologically idle" term occurs in a literal reading of a negative existential, such as the following:

(~P) Pegasus does not exist.

This use of the name 'Pegasus' clearly should not make (~P) ontologically committing. After all, the name is used precisely to *deny* the existence of Pegasus.

And here arises one of the oldest philosophical conundrums, going back to Parmenides – the problem of Non-Being. If one assumes the truth of 'Pegasus lacks being,' then it follows (does it not?) that there is nothing to which the subject-term refers. So it does not refer to Pegasus in particular. In which case, the statement fails to say anything in reference to Pegasus. But of course, it *is* saying something in reference to Pegasus – that Pegasus is not. But if you can refer to Pegasus, it seems that Pegasus must in some sense "be." Legions of responses to this problem have ensued.

What interest does this have for metametaphysics? The Problem of Non-Being is an issue in first-order metaphysics, specifically ontology (the study of what exists).[2] Yet the Problem makes vivid that we cannot directly "read off" ontological commitments from the names that a speaker uses. So this introduces a question about *methodology in ontology*: By what criterion can we identify the ontological commitments incurred by a statement? If the use of a name does not tip us off to an ontological commitment, what does? The present chapter is a slightly opinionated review of the three most prominent factions on such metaontological[3] questions: Quineans, Carnapians, and Meinongians. At the end, I also offer some related considerations about ontology, touching on ideas I have developed in various publications.

1 Quine and his successors

In his seminal work, Quine (1948/1961; 1960, etc.) approaches matters, first, by "regimenting" our statements – by paraphrasing them into the precise language of first-order quantificational logic. This effectively dispenses with idioms and metaphors like that in (2) and (3), along with other vagaries and infelicities of natural language. (Thus, "acts of God" might be re-framed as talk of "unforeseen accidents" or the like.) It still leaves us with negative existentials like (~P) however.

On this matter, Quine (1948/1961) first brushes aside two proposals, attributed to two fictional philosophers "McX" and "Wyman," although the latter is thought to be Meinong in a thin disguise.[4] (For more on Meinong, see section 3.) McX holds that 'Pegasus' refers to an idea in our minds, whereas Wyman claims that 'Pegasus' refers to a "unactualized possible object." But against McX, Quine observes that when one asserts (~P), one is not trying to deny the existence of an *idea*. Rather, one is denying the existence of a specific *animal*, a winged horse. Wyman's view, on the other hand, is criticized in several ways. Quine's most influential point here is that it is difficult to *individuate* non-actual objects. So to illustrate: I am presently thinking of a non-actual fat man standing in the doorway. And now ... I am thinking of a non-actual bald man standing in the doorway. Question: Is this the same man on both occasions? As it stands, there seems to be no fact of the matter. And Quine asks rhetorically "what sense can be found in talking of entities which cannot meaningfully be said to be identical with themselves and distinct from one another?" (p. 4).[5]

For his part, Quine endorses the analysis of (~P) from Bertrand Russell (1905; 1919, etc.). Generally, Russell held the view that the meaning of an ordinary proper name[6] should be analyzed in terms of a *definite description*, a description that is uniquely satisfied by the denotation of the name – if such there be. To illustrate, suppose a Russellian regards the name 'Pegasus' as equivalent to the definite description 'the winged horse captured by Bellerophon.' Then, (~P) will be seen as equivalent to:

(~P★) The winged horse captured by Bellerophon does not exist.[7]

This in turn can be symbolized into first-order quantificational logic as follows (where 'Wxb' translates 'x is a winged horse captured by Bellerophon'):

(4) $\sim\exists x\,(Wxb\ \&\ \forall y\,(Wyb \supset y = x))$

This says: Nothing is a winged horse captured by Bellerophon (which is identical to any such horse).[8] And crucially, its truth does *not* require Pegasus to exist as a referent. It is enough if everything in existence fails to satisfy 'Wxb,' i.e., fails to be described as a "winged horse captured by Bellerophon."

This Russellian analysis would explain, moreover, why the use of the name 'Pegasus' is not ontologically committing. The name is seen as equivalent a definite description, and the compound descriptor is meaningful (thanks to its constituent predicates like 'horse'), even if nothing actually satisfies it. What's more, Quine recognizes that Russell's analysis suggests a *different* criterion of ontological commitment.[9] As (4) makes clear, the truth of the statement depends on each object failing to satisfy the quantified-formula, when the object is assigned as the value of the variable 'x.' Thus, a commitment to the statement amounts to an ontological commitment against such a satisfier. *Mutatis mutandis* for 'Pegasus exists,' where it is analyzed as:

(5) $\exists x\, (Wxb\, \&\, \forall y\, (Wyb \supset y = x))$

In this case, truth requires that some object indeed satisfies the quantified-formula, when the object is assigned as the value of 'x.' And thus a commitment to (5) is an ontological commitment to such an object. Generalizing, we thus arrive at Quine's criterion of ontological commitment:

> (QC) An object o is an ontological commitment of a regimented (set of/) statement(s) iff o is required to make the statement(s) true (where o is assumed to be in the range of the bound variable(s)).

Quine (1948/1961) puts it this way: "To be assumed as an entity is … to be reckoned as the value of a variable" (p. 13).[10] Concordantly, Quine adds that "the *only* way we can involve ourselves in ontological commitments [is] by our use of bound variables. The use of alleged names is no criterion" (p. 12, his italics).[11]

It is fair to say that (QC) has been the most influential part of Quine's philosophy on contemporary writers. Indeed, some credit Quine with "reviving metaphysics" from the slumber induced by his positivist predecessors. However, Quine himself is clear that his interest is not metaphysics as much as metaontology – or more specifically, a criterion of ontological commitment: "We look to bound variables in connection with ontology not in order to know what there is, but in order to know what a given remark or doctrine … *says* there is" (ibid.).[12]

Van Inwagen (1998; 2014, ch. 3) is one metaontologist who carries the Quinean tradition. Yet there are enough surface differences that it is useful to say something particular to van Inwagen's view. He sums up his position in a series of five theses:

1. Being is not an activity.
2. Being is the same as existence.
3. Existence is univocal.
4. Existence is expressed by the existential quantifier.
5. (QC) is a procedural norm for ontological disputes.

One apparent departure from Quine is that van Inwagen's theses 1–4 seem concerned with the metaphysics of existence rather than a linguistic criterion of ontological commitment. But (assuming standard disquotational principles) 1–4 can be seen as having implications for the meaning of the words 'being,' 'exist,' and the like. And as we saw at the outset, the key issue is to decide when the use of natural language is ontologically committing.

Thesis 5 is not given a one-sentence formulation in van Inwagen; he says that it is really a "family of theses" (2014, p. 85). But the above formulation seems to capture the core of it. The thought is that (QC) lays down one of the "rules for engagement" for ontological disputes.

Van Inwagen illustrates this using the Platonism vs. nominalism debate about numbers. Consider that, assuming (QC), our best scientific theories are committed to the existence of numbers, as when physics tells us:

(6) The mass in grams of a homogeneous object is the product of its density in g/cm^3 and its volume in cm^3.

This would naturally be regimented as quantifying over numbers, i.e., as having numbers in the range of bound variables. Going by (QC), then, the nominalist is obligated to regiment (6) in a different manner. Quine himself did not believe that suitable nominalist paraphrases for mathematical physics were available, although see Field (1980) for an impressive attempt. But the present point is that (QC) defines an essential task for nominalism, if nominalism is to remain a viable option in the philosophy of mathematics. And it is this task-setting role which (QC) has for ontological disputes, per thesis 5.

Another contemporary Quinean is Sider (2009; 2011). Sider is an interesting case since, while he endorses something like (QC) (see 2011, p. 12), he also rejects descriptivism, given his affinity for Lewis' (1984) semantic doctrine of "reference magnetism." In fact, this combination of (QC) with anti-descriptivism seems common after Kripke's (1972/1980). Yet the continued popularity of (QC) is odd in one respect, since Quine's argument for (QC) assumed descriptivism. Granted, (QC) itself may still be defensible by some non-descriptivist means. Even so, (QC) itself may remain incompatible with externalist views such as "reference magnetism." (See Parent 2017a.)

2 Carnap and his successors

The most prominent opponent of Quine's metaontology is Carnap. Carnap's disagreement is best known via Carnap (1950/1956), although the Quine-Carnap debate extends well beyond that. Indeed, Quine's famous "Two Dogmas of Empiricism" (1951a) is in large measure a reply to Carnap. See also Quine (1951b), Carnap's (1955) replies, and the follow-up in Quine (1960, ch. 2), and Quine (1969, ch. 2).

Carnap's (1950/1956) basic idea is that existence-statements, even on their literal reading, often do *not* incur any absolute ontological commitments. Take for example a statement about number:

(N) There is an even prime.

Quine would regiment this as an existentially quantified formula, whose truth would require the number 2 in the range of the variable – whence it is ontologically committing. Carnap, in contrast, begins by reflecting on the linguistic rules by which number-terms were introduced in the first place. He says: "If someone wishes to speak in his language about a new kind of entities, [s/]he has to introduce a system of new ways of speaking, subject to new rules; we shall call this procedure the construction of a linguistic *framework* for the new entities in question" (p. 206). With the natural numbers specifically:

> The framework for this system is constructed by introducing into the language new expressions with suitable rules: (1) numerals like 'five' and sentence forms like 'there are five books on the table'; (2) the general term 'number' for the new entities, and sentence forms like 'five is a number'; (3) expressions for properties of numbers (e.g. 'odd,' 'prime'), [and so on].

(p. 208)

Once this framework is in place, (N) is straightforwardly a deductive consequence. Or, since the rules can be seen as *defining* the number-theoretic vocabulary, it turns out that (N) is *analytically true* with respect to those definitions (plus a few other axioms).

The metaphysical urge, however, is to insist on asking "But is (N) *really* true? Does an even prime *really* exist?" The question can seem odd: We just noted that within the framework for number-talk, it is uncontroversial that an even prime exists. Thus, Carnap says, the metaphysician's question must not be a question raised *internal* to the linguistic framework. Instead it must be an *external* question; it is a question about what exists "outside" the framework or independently of what the framework states. Yet here, the metaphysician's question remains odd, for the linguistic framework lays down the rules for the use of number-terms. Thus, if those rules are set aside, you get linguistic anarchy – number-theoretic talk is undefined. For this reason, Carnap concludes that the external question, the distinctly metaphysical question raised outside the framework, is *meaningless*.

Nonetheless, Carnap adds that those who ask metaphysical questions may be indirectly asking "a practical question, a matter of ... whether or not to accept and use the forms of expression in the framework" (p. 207). Yet the practical question is not adjudicated so much by evidence, but more by the utility of the framework, the "efficiency, fruitfulness, and simplicity" of the framework (p. 208). Carnap is clear, moreover, that utility is not evidence for the (external) truth of the framework.

So against Quine, points of ontology (when meaningful, and not indirectly practical) are uniformly uncontentious. Again, the question "Is (N) *really* true?" is patently affirmative within the number-theoretic framework, and unintelligible beyond it – except to debate theoretical utility in a roundabout way. N.B., internal ontological questions are not always affirmative: In the framework of evolutionary biology, the answer is negative if we ask "Does Pegasus exist?" The mistake for Carnap, however, is in thinking that there is a *framework-independent* answer to such questions, when really, such questions are incomprehensible outside of any framework. For Carnap, existence is a pluralistic affair, and whether x "exists" is always relativized to a framework.[13]

Several contemporary writers follow Carnap by adopting an internal/external distinction for existence-questions, and judging the legitimacy of such questions by this distinction. However, the details sometimes diverge from Carnap significantly. Consider first Thomasson (2009; 2015). In many respects, she is entirely with Carnap; for instance, she understands existence-terminology as governed by conventional "rules of use," and that under these rules, existence-questions are to be answered either empirically or analytically. And she rejects metaphysicians' attempts to answer existence-questions by *other* means, for this requires ignoring the established rules for use. Granted, rules can be stipulated ad hoc in a conversational context; however, this means that any metaphysical disagreement is merely verbal; this has disputants simply using existence-terms with different definitions.

Thomasson diverges from Carnap, however, in rejecting his pluralism about existence, where a question about the truth of (N) receives different answers relative to different frameworks. Thomasson's attitude is rather that there is only one set of linguistic conventions, and existence-questions should be addressed uniformly with respect to that set. In this respect, Thomasson seems more Quinean. One is reminded of Price (2009), where Quine is portrayed as asking Carnap rhetorically:

> [W]hat is to stop us treating all ontological issues as internal questions within a single grand framework? Why shouldn't we introduce a single existential quantifier, allowed

to range over anything at all, and treat the question of the existence of numbers as on a par with that of the existence of dragons?

(2009, p. 328)

(The interested reader is strongly encouraged to consult Thomasson's own chapter in this volume.)

Another contemporary Carnapian is Hofweber (2009; 2016). Hofweber begins by observing that 'There is' in (N) is polysemous, admitting of both an "internal" and "external" reading. But as a disanalogy to Carnap, Hofweber's "external" reading has 'there is' expressing an objectual or Quinean existential quantifier. (Though like Carnap, Hofweber's internal reading has it expressing an inferential-role or substitutional "some"-quantifier.[14]) The leads to a further disanalogy, namely, that statements about existence *are* meaningful on Hofweber's "external" reading. Nonetheless, this should not be an encouragement to metaphysicians, since Hofweber thinks that their truth should be settled by empirical science, not armchair speculation. In this, he subscribes to what he calls metaphysical "modesty." (For more on modesty, see Hofweber's chapter "Is Metaphysics Special?" in this volume.)

Nonetheless, he simultaneously decries "unambitious" metaphysics, which merely "works out the consequences" of scientific theory (2009, p. 264). Metaphysics should be "ambitious"; it should have its own set of questions and its own discoveries about the world to offer. I interpret Hofweber as criticizing Quinean "regimentation," where we translate scientific theory into quantificational logic, and then read off its ontological commitments as per (QC). My own view, however, is that regimentation is not simply a *translation* of scientific theory, but a *refinement* of it, subject to various desiderata. In which case, regimentation is not just the stale task of logical symbolization (see Parent 2015a for details).

There are several other prominent Carnapians on the contemporary scene, such as Yablo (1998; 2009), Chalmers (2009; 2012), Hirsch (2011), etc. See also Putnam (1987) for an earlier but influential Carnapian view. In addition, while Carnapians seem to be in ascendency, new criticisms are also emerging; see, e.g., Eklund (2016). However, it is not possible to cover all these ideas here. But see Sud and Manley's chapter on "Quantifier Variance," as well as Balcerak Jackson's chapter on "Verbal Disputes and Metaphysics" in this volume.

3 Meinong and his successors

Quine, recall, spoke of "Wyman" as one of his foils. Wyman holds that (~P) is true in virtue of a non-existent object which nonetheless has "being," thus clearly aligning him with Meinong (1904/1960). But thanks to Quine's critique (and Russell's 1905 scorn for those lacking a "robust sense of reality") Meinongianism remains a fringe view on the Problem of Non-Being. Yet Meinongianism has been strikingly persistent, and it creates some intriguing metaontological issues. First, however, let us take a look at what Meinong himself said.

Generally, Meinong (1904/1960) was focused less on *linguistic* representations like (~P), and more on *knowledge* of non-existents. But like Quine and Carnap, Meinong paid particular attention to mathematical objects, which for him, were undoubtedly objects of knowledge. He writes:

> the totality of what exists ... is infinitely small in comparison with the totality of the Objects of knowledge ... [This] is supported by the testimony of a very highly developed science – indeed the most highly developed one: mathematics. [Yet]

We would surely not want to speak of mathematics as alien to reality, as though it had nothing to do with what exists.

(pp. 79–80)

Meinong thus suggests that mathematical objects indeed have a kind of being called "subsistence," despite their non-existence. Some contemporary Meinongians, the "Noneists," reject subsistence; they prefer to say that mathematical objects do not exist in any sense at all (e.g., Routley 1980; Priest 2005/2016).[15] Now even Meinong rejected subsistence for fictional objects like the golden mountain and for impossibilia like the round square.[16] Yet Meinong himself did not wish to demote numbers and the like in quite the same way; thus, they are bestowed with subsistence.

But even apart from subsistence, some critics (e.g., van Inwagen 1977; Lycan 1979) protest that the very idea of a "nonexistent object" is unintelligible, even to the point of being "literally gibberish" (Lycan 1979, p. 290). However, some Meinongian ways of speaking are familiar to ordinary speakers. It is not esoteric philosophy-talk to say "Some things are the stuff of myth." Granted, Meinong did not help his cause with the notorious pronouncement: "Those who like paradoxical modes of expression could very well say: *'There are objects of which it is true that there are no such objects'*" (1904/1960, p. 83, italics mine). Later, I shall refer to the italicized claim as "Meinong's shocker." But as made clear by the clause prefixed to it, Meinong himself took the claim with a grain of salt.

Besides such defensive maneuvers, Meinongians have positive reasons in their favor. A striking case has been made by Brock (2004), showing that the widely respected Kripkean anti-descriptivist arguments work best on referentially "empty" names like 'Pegasus' – even better than on Kripke's own examples of non-empty names like 'Aristotle.' Such considerations might lead one to think of 'Pegasus' as *directly referential*, i.e., as a rigid designator for an unreal object.

So again, for Meinong, some things do not exist, like the golden mountain and round squares. But n.b., although a round square neither exists nor subsists, Meinong still admits that it is *an object of knowledge*: "Any particular thing that isn't real must at least be capable of serving as the Object for those judgments which grasp its *Nichtsein*" (1904/1960, p. 82). For Meinong, then, objects fall into *three* ontological categories: Those that exist, those that subsist but do not exist, and those that do not exist in any sense at all (cf. Chisholm 1973).[17] But thus far, all this concerns the Problem of Non-Being, a problem in first-order ontology. It does not yet directly address metaontology, and writers sometimes do not clearly separate Meinongian ontology from Meinongian metaontology.[18] Yet, as with van Inwagen in section 1, we may first observe that Meinong's metaphysics of existence naturally suggests a view about existence-terminology. This, in turn, is important to regimentation and to a criterion of ontological commitment. Thus, regarding (N) ("There is an even prime") Meinong would interpret the range of 'There is' as including objects that do *not* exist, contra the Quinean. But Meinong can still allow that 'There is' sometimes concerns only existing objects. Thus unlike Quine, we have a variability in what such terminology means. (And unlike Carnap, none of these uses are deemed meaningless.)

The variability in existence-terms is made explicit in the Meinongianism of Priest (2005/2016). Priest's regimentation includes not only the Quinean existential quantifier '$\exists$,' but also the distinctively Meinongian quantifier '$\mathscr{S}$,' standing for "some" in the sense of "at least one." The range of the latter includes the range of '$\exists$,' but includes more as well. (In Lewis' 1986 terms, '$\mathscr{S}$' is Priest's unrestricted quantifier, and '$\exists$' is restricted to existing members of that domain.) Berto's (2013) regimentation is similar, except '$\mathscr{S}$' is used as the Meinongian quantifier. Thus, the Priest-Berto regimentation of (~P) would be something like '$\sim\exists x\ x = p$,' where 'p' names

Pegasus, a non-existent in the range of the Meinongian quantifier. Whereas Meinong's shocker, in Berto's notation, could be regimented as:

(7) $\mathscr{S}y \sim\exists x\, x = y$

Yet (7) is quite consistent, for it just expresses that some (non-existent) object in the unrestricted domain of '$\mathscr{S}$' is not in the restricted domain of '$\exists$.'

Other Meinongians also portray existence-terminology as variable in meaning, although Parsons (1980) and Zalta (1983; 1988, etc.) deploy only one "particular" or "some"-quantifier, namely '$\exists$.' And they interpret '$\exists$' as the *Meinongian* quantifier. To express existence, Parsons and Zalta instead use an existence-*predicate* 'Ex..' Thus on the Parsons-Zalta approach, (~P) would be regimented as '~Ep,' and from this, Meinong's shocker follows by $\exists$-generalization:

(8) $\exists x \sim Ex$

But like (7), this is logically consistent as well, and for an analogous reason.

Berto and Priest each note that a Meinongian could use an existence-predicate in lieu of having two "some"-quantifiers. But when illustrating this, Berto-Priest continue to use '$\exists$' as the *Quinean* quantifier, and use the existence-predicate to define it. Whereas again, Parsons-Zalta use '$\exists$' as the *Meinongian* quantifier. This may seem to be just a disagreement about notation, but it brings to the fore a key metaontological question: Is quantifying over an object using '$\exists$' criterial for a commitment to the object? Since different Meinongians interpret '$\exists$' differently, it seems they should not give the same answer.

Things are further complicated by the fact that, for a Meinongian, an *ontological* commitment is apparently not the same as an *existential* commitment. (This is made clear especially by Berto 2013.) Consider again Meinong's shocker. It does not express an *existential* commitment, since the commitment is directed at a non-existent object. Nevertheless, since it states that *there is* a non-existent object, one might naturally think it is ontologically committing.

This distinction between existential vs. ontological commitment might help adjudicate what a Meinongian criterion should be. The Meinongian could say that her "ontological commitments" are incurred by quantifying over objects using the Meinongian quantifier, and that "existential commitments" are incurred by quantifying over objects using the Quinean quantifier (or if preferred, by having the objects in the extension of an 'exist'-predicate). As for the interpretation of '$\exists$,' the dispute indeed starts to look merely terminological. Since '$\exists$' is part of an artificial language, we are free just to *legislate* that '$\exists$' express whichever quantifier we want (cf. Crane 2013, ch. 2).

Yet in the same way, it starts to look merely terminological whether the regimented language should express existence using a predicate vs. a quantifier. Again, since we are talking about an artificial language, we appear at liberty to designate existing objects using either type of symbol. But in this instance, it is more contentious to dismiss the issue this way. There is a long tradition (which some trace back to Kant) on whether existence is a predicate. Or more perspicuously, it is a debate on whether existence-terms ascribe a genuine *property* to an object. It is not obvious, however, why this "existence-as-property" debate should bear on which symbolization we should use in our regimentation. But the matter is involved, and there are considerations on both sides which we are not able to cover here. (See Moltmann 2013, Berto op. cit., and Crane, op. cit., for recent discussions.)

In any event, can Meinongians at least agree that a criterion of *ontological* commitment is membership in the range of a *Meinongian* quantifier? It seems not, for two reasons.

One is that Meinongians seem equally well-placed to revive a *name*-based criterion of ontological commitment, of the sort that Russell and Quine opposed. After all, a Meinongian would see any named object as in the range of the Meinongian quantifier. In which case, having a name would be sufficient for a Meinongian ontological commitment. However, it is dubious whether having a name is *necessary* for such a commitment. (The real numbers are uncountable, and in a regimented language, one can introduce only countably many names.[19]) Nevertheless, if having a name is enough for a Meinongian ontological commitment, then such a commitment would be apparent from a sentence like 'Pegasus has wings.' There would be no need to fuss with the meaning of quantifiers and about how quantifier-rules should interface with the sentence.

The second reason to hesitate over a Meinongian criterion is that a Noneist like Routley (1980) explicitly refuses that his Meinongian quantifier indicates his ontological commitments (see, e.g., p. 424). On his view, round squares have no being in any sense at all – so it would be awkward for him to accept an "ontological commitment" to them. After all, even if one tolerates "there are shapes which don't exist," it is a further step to tolerate "there are shapes which are not." Be that as it may, Routley still claims the capacity to say "*some* shapes are not." In this, his Meinongian quantifier is really just a *quantifier*, used to designate a *quantity*, and there is no built-in assumption that the quantity of stuff has any sort of being. Yet if an ontologically neutral quantifier strikes you as a contradiction in terms, you are not alone. It is Routley's "neutral" quantifier which prompted Lewis (1990) to ask whether Routley was really a *None*ist. Perhaps he is better characterized as an "*All*ist," given that he countenances *everything* as an object.

On the other hand, if Routley's neutral quantification is jarring, it may be that you have been "tainted by philosophy." As Azzouni (2007) argues, there is good reason to think that quantifiers in natural language are ontologically neutral. (See also Priest 2008.) Further, Azzouni thinks the semantics for a neutral quantifier is unproblematic. Just take the neutral quantifier in your native tongue as part of the metalanguage, and use it to define a neutral quantifier into your regimented object language! (See Azzouni 2004; 2007, etc.) (Azzouni also thinks such a definition obviates the need for "Meinongian objects," but whether he is right about that is another matter.)

4 Closing remarks

I have devoted more space to Meinong, given that the metaontological aspects of his view remain underappreciated. (In the present volume, his name likely occurs at a fraction of the rate of 'Quine' or 'Carnap.') I also confess more sympathy for Meinong, although the importance of Quine's method of regimentation cannot be overstated. I also applaud the pragmatism of Quine and Carnap. "Pragmatism" here does not imply a controversial instrumentalist thesis about the nature of truth. Rather, it is a recognition that we are always working "internal" to a theory – and our choice of theory is rationally guided only by (theory-laden) observation and so called "pragmatic" constraints like conservativeness, simplicity, scope, etc. (Cf. Parent 2017b, ch. 1, section 3.) This need not be at odds with Meinong, though he tended to speak as if we had direct access to reality, or unreality as the case may be.

Importantly, pragmatism in Quine and Carnap leads to a *deflationist* view of ontology, making them surprisingly similar metaontologically.[20] Price (2009) expresses this well:

> Carnap's internal issues were of no use to traditional metaphysics ... And Quine's move certainly does not restore the non-pragmatic external perspective required by metaphysics ... Quine himself has sunk the metaphysician's traditional boat, and left all

of us, scientists and ontologists, clinging to Neurath's Raft ... [T]he force of Quine's remarks is not that metaphysics is like science as traditionally (i.e. non-pragmatically) conceived, but that science ... is like metaphysics as pragmatically conceived.

(pp. 326–327)

Neither this brand of pragmatism nor deflationary metaontology is wildly popular today. But in what remains, let me say something in support of the view. I am able to provide here only a sketch of what has been elaborated elsewhere. Yet I hope I might draw some attention to these issues.

Much of ontology looks like semantic theory. Talk about the "furniture of the world" soon turns into talk about our talk, and in particular, about what our terms denote. Thus, the reference of 'Pegasus' immediately became the focus in the Problem of Non-Being; similarly, the range of 'There is' in (N) is what draws scrutiny. The method regimentation only encourages this. But there is something odd in interpreting our (linguistic or mental) representations, using those very representations. The limit case of this is a homophonic interpretation. Consider here Carnap's own example (1950/1956, p. 217):

(f) 'five' designates five.

Since the term 'five' is interpreted by the self-same term, (f) is uninformative.[21] At least, it is hard to see how it could advance our understanding of ontology. But matters do not improve much if, during regimentation, we use the ordinary term to define a formal name:

(f★) 'f' designates five.

Our concern is usually with the ontology underlying English, yet (f★) primarily informs us about a formal symbol. However, things may look better if we instead use a heterophonic interpretation like:

(9) 'five' designates the successor of four.

Yet in a key sense, this only "pushes back" the question onto the definite description on the right-hand side. What sort of object, if any, does 'the successor of four' denote (assuming the question is meaningful[22])? Indeed, since the ontology of numbers is at issue – and since the right-hand side of (9) helps itself to a number-term – it effectively ignores the real question.

The example of (9) suggests a broader lesson. If the aim is to specify the ontology of a language L, an interpretation cannot answer what object an expression of L denotes, if the interpretations are themselves L-expressions. Briefly, that's because the ontology of such interpretations would naturally be in question as much as the expressions they interpret. So in order to settle the question of ontology, the interpretations themselves would need to be interpreted, and thus a regress.[23] For the purposes of specifying the ontology of L, there is no escape from the "circle of language."

Again, I cannot defend this line in detail here, but see Parent (2015b). Hopefully, it is at least suggestive of why one might be attracted to a "pragmatism" like that in Carnap and Quine. Before closing, however, let me quickly consider one important objection.

It is clear that the following addendum would not distinguish a Carnapian view from a Platonic one:

(10) There is a successor of four.

After all, Carnap might embrace this as an analytic consequence of the number framework, just like his acceptance of (N). But – why can't we introduce a regimented quantifier to express *bona fide* ontological commitments, a quantifier that unequivocally concerns what exists *external* to our theory? For instance, we could regiment (10) as '$\exists x\, S(4) = x$,' and make explicit the interpretation of the quantifier as follows:

(★) "$\exists x\, \Phi x$" is true iff *there really is* an object that is Φ.

This seems to be how the Neo-Carnapian Hofweber (op. cits.) escapes the "circle of language." But the problem is this. An unequivocally committal quantifier would need to be defined by an unequivocal English expression. Unfortunately, however, that *all* existence-terminology is equivocal between the "internal" and "external" readings – or between the ontologically neutral and committal readings. (It is not as if we are forbidden from using some of these terms in writing a novel.) Context can always "defang" existence-terminology so that it is ontologically non-committal. That is so, even if the existence-terms are italicized, iterated, put in caps, etc.

To illustrate, suppose we are discussing the play-within-a-play in Act V.i of *A Midsummer Night's Dream*. (This example comes from Parent 2014). Suppose you insist that there is no lion in the nested play. Then, I might reply emphatically by saying:

(11) I'm *not making this up!* THERE REALLY AND TRULY IS a lion named 'Snug'!

My utterance would be true, even though we all know that that Snug is fictional. Indeed, he is fictional even within the fiction of the play.[24] The point, again, is that if all existence-talk has both the neutral and committal readings, then any definition like (★) will be similarly equivocal. Whence the regimented language cannot contain a strictly unequivocal quantifier to function in a criterion of ontological commitment.

Notes

1 Typically in the literature, one speaks of the ontological commitments of a *statement* or a *theory* (a collection of statements), rather than the ontological commitments of a *person*. However, I vacillate between these two modes of expression, harmlessly I assume, since the ontological commitments of a person can be parlayed into ontological commitments of the theory the person believes. An anonymous reviewer points out that the notion of a "commitment" remains somewhat unclear, although this often is not noticed in the literature. For instance, if I see an orange and sincerely assert "there is an orange," I seem ontologically committed to at least one orange. But am I ontologically committed to *that particular* orange which I am seeing? It may seem so, although writers often speak as if I have just a commitment to oranges, and not any specific oranges. I detail similar complexities with "ontological commitment" in Parent (2017a), section 1. See also Berto (2013) for further discussion along these lines.
2 Recently, some have argued that "ontology" should instead be understood as the study of "what grounds what." We shall ignore this here, but the interested reader should consult Skiles and Trogdon's chapter on "Grounding" in this volume. Also relevant is Mormodoro's chapter on "Hylomorphic Unity" and Bliss' chapter on "Fundamentality."
3 Following Tahko (2015), we can distinguish metaontology and metametaphysics with help from the more customary distinction between ontology and metaphysics. Ontology is roughly the study of what exists, and is seen as a sub-discipline of metaphysics, which is concerned more broadly with the nature of reality, and with especially puzzling bits of reality (time, universals, freewill, etc.). Thus understood, metaontology can then be seen as the study of ontology and metametaphysics as the study of metaphysics.
4 When asked, Quine stated that 'McX' and 'Wyman' were not pseudonyms for any specific philosophers – they simply represent two views that were in the air at the time. (See Boynton Quine 2017,

#22 under Email Updates.) Regardless, the resemblance between Wyman and Meinong is undeniable. And to my mind, McX seems inspired by Frege's view that a name in intensional discourse refers not to its ordinary referent, but to its "sense" (which is basically a mental content for Frege). Though in deference to Quine, we should not insist on identifying McX or Wyman with any actual philosopher.

5 This exemplifies the famous Quinean dictum "no entity without identity." See Quine (1969, p. 23). By the way, this dictum also explains why Quine's regimentation was limited to *first*-order quantificational logic. In higher-order logics, one quantifies over properties, yet properties are individuated intensionally, thus frustrating attempts to give straightforward identity-conditions. E.g., the property of being triangular and the property of being trilateral are different, yet they are extensionally individuated by the same set of polygons.

6 "Ordinary" proper names is meant to contrast with Russell's notion of a "genuine" proper name. The latter sort of name was *not* analyzed into a definite description; its referent was instead "known by acquaintance" rather than "known by description." In this vein, Russell held that the only objects that had genuine proper names were sense-data, since only these could be "known by acquaintance." However, for our purposes, I shall ignore the case of genuine proper names in the main text. Indeed, Quine himself did not distinguish between names in this way, as all names were paraphrased into descriptive material. (This was in keeping with his strong stance that the *only* way to incur ontological commitment was by bound variables.)

7 In the end, the descriptivist should also analyze away the name 'Bellerophon' into a definite description, but we may ignore this for simplicity's sake.

8 The clause in parentheses can be put more colloquially as "there is at most one such horse," which is necessary to secure the "uniqueness" implied by the definite article 'The' in (~P). Also, Russell's view suggests adding a third conjunct '$\exists z\ z = x$' to (4). But this is logically redundant in classical logic, and so I have omitted it to reduce clutter.

9 In *Word and Object*, Quine comes to prefer 'ontic commitment' over the term 'ontological commitment'; see p. 120n. But 'ontological commitment' is the more common term by far.

10 In an earlier version of Quine's paper, the sentence here reads "To be is to be the value of a bound variable." This formulation is a bit more catchy, and is better known among philosophers. But I have quoted the revised version of the statement since it captures better Quine's intent. His point is not that existence is somehow *constituted* by the fact that an object lies in the range of a bound variable. Instead, this determines whether the theory *says* that the object exists.

11 One difficulty for (QC) concerns a non-existent without an associated definite description, e.g., if we use 'Tom' as a name for "some unicorn or other" without specifying which one. Quine's fix is to introduce a predicate 'x Tom-izes' so that the relevant negative existential is rendered as "there is no unique x that Tom-izes." However, there are many other rough spots in (QC) which would need further sharpening. For helpful discussion, see Rayo (2007).

12 Unfortunately, this section presents only a portion of Quine's systematic and fascinating views on ontology. But for further details, see Egerton's chapter on Quine, this volume.

13 As with the section on Quine, this section on Carnap is quite minimal. See Kraut's chapter on Carnap, this volume, for more details.

14 Briefly, a substitutional "some"-quantifier "Ex" can be defined as: "E$x\ \Phi x$" is true iff "Φa" is true for at least one name a (where a can be an empty name or not).

15 Routley's Nonism is especially clear that mathematical objects, while qualifying as "objects," do not even have *being*. Commentators have of course wondered about the intelligibility of this view.

16 Technically, a "round square" is a geometrical object, hence, a mathematical object. But in the main text, talk of "mathematical objects" shall be restricted to consistent mathematical objects only.

17 The round square is a key moment in Meinong (1904/1960) for other reasons. For instance, he holds that we know the "essence" (*Sosein*) of the round square, and that this essence subsists – even though, again, the round square does not subsist (pp. 82, 84–85). Indeed, the *Sosein* of the round square is one key reason he is led to his well-known "independence principle" viz., that the subsistence of a *Sosein* does not require the (definitive) being of the object. N.B., some of the most interesting debates among Meinongians concern the *Sosein* of non-existents. Do Meinongians really want to say Pegasus is a winged *horse*, a flesh-and-blood mammal? That seems to make him too real. Accordingly, the way in which Pegasus is a "horse" has been reinterpreted in various ways. (See Parsons 1980 on "watered-down" properties, Zalta 1988 on "encoding," Parent ms. on actual + yet nonactual property-instances, etc.). Also, Priest (2005/2016) has shown that restrictions must be placed on the "characterization

principle" for *Sosein*s, the principle that any characteristic defines a *Sosein*. Otherwise, the principle allows us to prove anything whatsoever.
18 Two recent introductory works that are occasionally unclear on Meinongian metaontology vs. ontology are Tahko (2015) and Berto and Plebani (2015). Though in general, I highly recommend these books, especially given their generous coverage of Meinong.
19 This was a sore spot for Barcan Marcus (1961/1993), who needed all objects to have names in order to adequately define her substitutional quantifiers. She humbly calls attention to the "awful simplicity" of this assumption, given the uncountability of the reals (p. 12). Although remarkably, when the issue comes up later (in the discussion with Quine and Kripke, added as an appendix), she instead expresses misgivings about the uncountability of the reals (p. 27)! What is also remarkable is that Quine responds by sympathizing with such misgivings (ibid.). At any rate, if any posit of a theory has a name, then the appearance of a name would be both necessary and sufficient for an ontological commitment. See Janssen-Lauret (2015) who interprets Barcan Marcus as holding such a criterion for ontological commitment, at least in relation to those names which Barcan Marcus calls "tags."
20 Soames (2009) also stresses the metaontological similarity of Quine and Carnap. But he traces it back not to a shared pragmatism (although that may be implicated), but rather to a shared verificationism about meaning. One surprising consequence is that for both Quine and Carnap, theories which are observationally equivalent are ipso facto equivalent in ontology. (Any apparent ontological disparities would be merely verbal.) Soames memorably calls this equivalence-thesis shared between Quine and Carnap their "stunning counterintuitive bedrock of ontological agreement" (p. 441).
21 Carnap himself calls the sentence "analytic," but I would rather just say it is uninformative. Both judgments assume, of course, that 'x designates y' is not ontologically committing on the value for y, except perhaps relative to a framework.
22 For my part, such an "external" metaphysical question is intelligible in that it is not gibberish; its interest and its force can be felt by first-year undergraduates. This aspect of my view marks a contrast with Carnap's own "pragmatism."
23 The regress argument here bears some kinship with the one in Quine (1969, ch. 2). However, Quine's regress arises in the context of his semantic indeterminacy, and no such assumption is made here.
24 The example also makes clear how the argument bears on Yablo (1998). Yablo hopes to do Carnapian ontology by defining internal vs. external in terms of the figurative/literal distinction, rather than the analytic/synthetic distinction. My example, however, reveals that existence-terminology can always be given a "non-literal" spin, even when all the signs are that they should be taken "literally." (The scare quotes here indicate that I also do not really believe that the fiction-internal uses of existence-terms are non-literal uses. I sympathize with Azzouni's view, noted at the end of section 3, that English language quantification in its "natural" state is ontologically non-committal.)

References

Azzouni, J. (2004) *Deflating Existential Consequence: A Case for Nominalism*, Oxford: Oxford University Press.
Azzouni, J. (2007) "Ontological Commitment in the Vernacular," *Noûs* 41(2): 204–226.
Berto, F. (2013) *Existence as a Real Property: The Ontology of Meinongianism*, Dordrecht: Springer.
Berto, F. and Plebani, M. (2015) *Ontology and Metaontology: A Contemporary Guide*, London: Bloomsbury.
Boynton Quine, D. (2017) *Willard Van Orman Quine Guest Book Volume 1*, available at: www.wvquine.org/guestwq1.html. Accessed July 28, 2018.
Brock, S. (2004) "The Ubiquitous Problem of Empty Names," *Journal of Philosophy* 101(6): 277–298.
Carnap, R. (1950/1956) "Empiricism, Semantics, and Ontology," *Revue Internationale de Philosophie* 4: 20–40. Pagination is from the reprint in his (1956).
Carnap, R. (1955) "Meaning and Synonymy in Natural Languages," *Philosophical Studies* 7: 33–47. Reprinted in his (1956).
Carnap, R. (1956) *Meaning and Necessity*, 2nd edition, Chicago: University of Chicago Press.
Chalmers, D. (2009) "Ontological Anti-Realism," in Chalmers et al. (eds.), pp. 77–129.
Chalmers, D. (2012) *Constructing the World*, Oxford: Oxford University Press.
Chalmers, D., Manley, D., and Wasserman, R. (eds.) (2009) *Metametaphysics: New Essays on the Foundations of Ontology*, Oxford: Oxford University Press.
Chisholm, R. (1973) "Beyond Being and Nonbeing," *Philosophical Studies* 24: 245–257.
Crane, T. (2013) *The Objects of Thought*, Oxford: Oxford University Press.

Eklund, M. (2016) "Carnap's Legacy for Contemporary Metaontological Debate," in S. Blatti and S. Lapointe (eds.) *Ontology after Carnap*, Oxford: Oxford University Press, pp. 165–189.

Field, H. (1980) *Science Without Numbers*, Princeton, NJ: Princeton University Press.

Hirsch, E. (2011) *Quantifier Variance and Realism: Essays in Metaontology*, Oxford: Oxford University Press.

Hofweber, T. (2009) "Ambitious yet Modest Metaphysics," in Chalmers et al. (eds.), pp. 260–289.

Hofweber, T. (2016) *Ontology and the Ambitions of Metaphysics*, Oxford: Oxford University Press.

Janssen-Lauret, F. (2015) "Meta-ontology, Naturalism, and the Quine-Barcan Marcus Debate," in F. Janssen-Lauret and G. Kemp (eds.) *Quine and His Place in History*, Basingstoke: Palgrave Macmillan, pp. 46–167.

Kripke, S. (1972/1980) *Naming and Necessity*, Cambridge, MA: Harvard University Press.

Lewis, D. (1984) "Putnam's Paradox," *Australasian Journal of Philosophy* 62: 221–236.

Lewis, D. (1986) *On the Plurality of Worlds*, Oxford: Blackwell.

Lewis, D. (1990) "Allism or Noneism?," *Mind* 99: 23–31.

Lycan, W. (1979) "The Trouble with Possible Worlds," in M. Loux (ed.) *The Possible and the Actual*. Ithaca, NY: Cornell University Press, pp. 274–316.

Marcus, R. (1961/1993) "Modalities and Intensional Languages," *Synthese* 1(4): 303–322. Pagination is from the reprint in her *Modalities: Philosophical Essays*, Oxford: Oxford University Press, pp. 3–23.

Meinong, A. (1904/1960) "The Theory of Objects," in I. Levi, D. B. Terrell, and R. Chisholm (trans.), *Realism and the Background of Phenomenology*, Glencoe, IL: Free Press, pp. 76–115.

Moltmann, F. (2013) "The Semantics of Existence," *Linguistics and Philosophy* 36: 31–63.

Parent, T. (2014) "Ontic Terms and Metaontology – or, On What There Actually Is," *Philosophical Studies* 170(2): 199–214.

Parent, T. (2015a) "Theory Dualism and the Metalogic of Mind-Body Problems," in C. Daly (ed.) *The Palgrave Handbook to Philosophical Methods*, New York: Palgrave, pp. 497–526.

Parent, T. (2015b) "Rule Following and Metaontology," *Journal of Philosophy* 112(5): 247–265.

Parent, T. (2017a) "Content Externalism and Quine's Criterion Are Incompatible," *Erkenntnis* 82(3): 625–639.

Parent, T. (2017b) *Self-Reflection for the Opaque Mind: An Essay in Neo-Sellarsian Philosophy*, New York: Routledge.

Parent, T. (ms.) "Conservative Meinongianism: An Actualist + Meta/Ontology," available at http://tparent.net/ConservativeMeinong.pdf.

Parsons, T. (1980) *Nonexistent Objects*, New Haven, CT: Yale University Press.

Price, H. (2009) "Metaphysics after Carnap: The Ghost Who Walks?," in Chalmers et al. (eds.), pp. 320–346.

Priest, G. (2005/2016) *Towards Non-Being: The Logic and Metaphysics of Intentionality*, 2nd edition, Oxford: Oxford University Press.

Priest, G. (2008) "The Closing of the Mind: How the Particular Quantifier Became Existentially Loaded Behind Our Backs," *Review of Symbolic Logic* 1: 42–55.

Putnam, H. (1987) "Truth and Convention: On Davidson's Refutation of Conceptual Relativism," *Dialectica* 41: 69–77.

Quine, W. (1948/1961) "On What There Is," *Review of Metaphysics* 2: 21–38. Pagination is from the reprint in his (1961).

Quine, W. (1951a) "Two Dogmas of Empiricism," *Philosophical Review* 60: 20–43. Reprinted in his (1961).

Quine, W. (1951b) "On Carnap's Views of Ontology," *Philosophical Studies* 2: 65–72.

Quine, W. (1960) *Word and Object*, Cambridge, MA: Harvard University Press.

Quine, W. (1961) *From a Logical Point of View*, 2nd edition, Cambridge, MA: Harvard University Press.

Quine, W. (1969) *Ontological Relativity and Other Essays*, Cambridge, MA: Harvard University Press.

Rayo, A. (2007) "Ontological Commitment," *Philosophy Compass* 2(3): 428–444.

Routley [Sylvan], R. (1980) *Exploring Meinong's Jungle and Beyond: An Investigation of Noneism and the Theory of Items*, Canberra, Australia: Central Printery at ANU.

Russell, B. (1905) "On Denoting," *Mind* 14: 479–493.

Russell, B. (1919) *Introduction to Mathematical Philosophy*, London: George Allen & Unwin.

Sider, T. (2009) "Ontological Realism," in D. Chalmers et al. (eds.), pp. 384–424.

Sider, T. (2011) *Writing the Book of the World*, Oxford: Oxford University Press.

Soames, S. (2009) "Ontology, Analyticity, and Meaning: The Quine-Carnap Dispute," in Chalmers et al. (eds.), pp. 424–443.

Tahko, T. (2015) *Introduction to Metametaphysics*, Cambridge: Cambridge University Press.
Thomasson, A. (2009) "Answerable and Unanswerable Questions," in Chalmers et al. (eds.), pp. 444–471.
Thomasson, A. (2015) *Ontology Made Easy*, Oxford: Oxford University Press.
van Inwagen, P. (1977) "Creatures of Fiction," *American Philosophical Quarterly* 14(4): 299–308.
van Inwagen, P. (1998) "Meta-ontology," *Erkenntnis* 48: 233–250.
van Inwagen, P. (2014) *Existence: Essays in Ontology*, Cambridge: Cambridge University Press.
Yablo, S. (1998) "Does Ontology Rest on a Mistake?," *Proceedings of the Aristotelian Society* 72: 229–261.
Yablo, S. (2009) "Must Existence-Questions Have Answers?," in Chalmers et al. (eds.), pp. 507–526.
Zalta, E. (1983) *Abstract Objects*, Dordrecht: Kluwer Academic Publishers.
Zalta, E. (1988) *Intensional Logic and the Metaphysics of Intentionality*, Cambridge, MA: MIT Press.

8
QUANTIFIER VARIANCE

Rohan Sud and David Manley

Imagine three tribes.[1] The first tribe speaks *Nihilese* – their linguistic dispositions are the same as those of English speakers, except that in their ordinary interactions they speak as though the theory of mereological nihilism is unproblematically true.[2] If we were to ask them, "What kinds of things are there? Tables? People? Things entirely composed of tables and people?" they'd say things like:

1   Although there are simples arranged table-wise and people-wise, there are no tables or people, or things composed of tables and people.

In the second tribe, people speak *Universalese*. Their linguistic dispositions are also the same as those of English speakers, except that in their ordinary interactions they speak as though the theory of mereological universalism is unproblematically true.[3] If we were to ask the homophonous question in their language, they'd say things like:

2   There are tables and people, as well as things entirely composed of tables and people.

The third tribe's members are speakers of *Shmenglish*, and they differ linguistically from English speakers only (if this is a difference) insofar as they speak as though a common-sensical theory of mereological composition is unproblematically true. In response to the homophonous question in their language, they'd say things like:

3   There are tables and people, but nothing composed of just a table and a person.

According to proponents of *quantifier variance*, the three tribes assert truths in their respective languages when they make these utterances.[4] The reason they don't contradict each other is (at least in part) that their uses of the phonetic strings 'something' and 'there are' have different truth-conditional contributions, even though the resulting expressions function in their respective languages similarly to our unrestricted quantifier. To put the claim more concisely: the unrestricted-quantifier-like expressions in these languages express different unrestricted-existence-like notions. Moreover, none of the tribes has a way of describing reality that is more metaphysically distinguished or perspicuous compared with the rest.

Of course, none of this entails that what exists depends on what language we speak – for example, that there wouldn't be any tables if we chose to speak Nihilese instead of English (Hirsch 2002b). That inference would be a simple use-mention error. If we were to speak Nihilese, the sentence homophonous with 'There are no tables' would be true. And if we were to speak Universalese, the sentence homophonous with 'There are no tables' would be false. But adopting another language would have no effect at all on whether there are tables.

However, quantifier variance about a debate may have some other dramatic consequences. For one thing, it seems to *deflate the significance* of the relevant debate.[5] Take a quintessential dispute about mereology between three contemporary metaphysicians: a nihilist, a universalist, and a common-senser. They treat each other as having a substantial disagreement and attempt to resolve it by subjecting their theses to a range of quasi-scientific criteria, such as parsimony considerations. Now, if quantifier variance is true, then there is a sense in which each participant in the debate has an equally good way of talking about the world. This is not to say that they are all speaking truly – since arguably they must all be treated as speaking the same language. But regardless of who is speaking truly – or even whether it's determinate who is speaking truly – there is no deep victory to be had by any of the disputants. Maybe the universalist is technically wrong when she says *in English* that there are things composed of people and tables; but if we were all to adopt her way of talking, her utterances would all be *true* in our new shared language. And we'd be no worse off, metaphysically speaking, for deciding to talk that way.

If this is right, not only does it deflate the significance of the original debate, but it also seems to have implications for how we should go about pursuing whatever's left of that debate. The idea is that, insofar as we still care which of the three ontologists is speaking truly in English, standard methods in ontology are ill-conceived. Instead, we should be asking, in effect, which of the various possible languages we're actually speaking. And the answer to *that* question is properly in the domain of semantics (and meta-semantics).[6]

So much for broad brush-strokes. In this chapter, we'll start by stating the thesis itself with more care, and then sketch the main arguments that support it (§2). We'll then explore some challenges to the two main components of quantifier variance (§3 and §4). And finally, we'll ask whether, if we accept quantifier variance and its allegedly deflationary consequences about an ontological debate, we should just cede the remaining shallow victory to common-sense ontology (§5).

1 The claim

The thesis of quantifier variance has two separable components:[7]

Quantifier variance about a debate (rough)
(i) Corresponding to each answer in the debate, there is an unrestricted-existence-like notion interpreted in terms of which that answer is true, and (ii) none of these is more metaphysically distinguished than any other.

Call the first claim *pluralism* and the second claim *egalitarianism*. Both could use sharpening, especially around the phrases 'corresponding to', 'unrestricted-existence-like notion', and 'metaphysically distinguished'.

Let's start with pluralism. Letting the tribe thought experiment be our guide, and closely following Dorr (2014), we can flesh out pluralism about a debate as follows:

For each answer in the debate, there are possible isolated communities such that: (i) they speak languages superficially like ours (and are like us in other respects) except that they treat the relevant uninterpreted sentences as unproblematically true; (ii) the sentences are in fact true in their languages; and (iii) the counterparts to our unrestricted quantifier in those languages have different meanings.

Two main questions of clarification arise.

a   What exactly does it mean to say that a tribe treats the relevant sentences as "unproblematically true" (Dorr 2014)? This could mean that the tribe merely treats the sentences as true but doesn't worry about their truth. On stronger interpretations, the tribe considers the sentences to be *obvious*, or even as quasi-analytic, and treats those who reject the sentences as if they are not competently speaking the same language. As we'll see, as we vary the strength of 'unproblematic truth', we vary the plausibility of the corresponding interpretation of quantifier variance and its consequences.
b   What exactly does it take for an expression in some possible language to be considered its counterpart to our unrestricted quantifier? Can we specify a syntactic or semantic feature in virtue of which a term counts as an unrestricted-quantifier-like expression (or 'UQE') in the counterfactual languages? We consider this question in detail in §3.1. In brief: such a feature is at least very tricky to specify directly, but that may not matter much for the purposes of quantifier variantists.

The second component of quantifier variance is *egalitarianism*: the claim that none of the relevant UQEs is more metaphysically distinguished than any other.

But what does it mean for a UQE to be metaphysically distinguished? We might say that a UQE is distinguished to the degree that its semantic value is *fundamental* or *natural*.[8] Or we might focus on the *expressive power* of its language, where this is a measure of the range of propositions, facts, or truth-conditions it can express.[9] Alternatively, we might say that a UQE is distinguished to the degree that there are special "metaphysical reasons" to use it.[10] In what follows, we'll try to avoid treating any particular variation of egalitarianism as canonical.

2 Arguments for quantifier variance

A direct style of argument for quantifier variance appeals directly to our ability to competently make judgments about meanings and patterns of entailment in the kinds of scenarios sketched in the introduction. On this line of reasoning, if we judge that the relevant sentences as uttered in those scenarios are true, and that the UQEs in the respective languages have different meanings, we have reason to accept quantifier pluralism. And if we then intuit that none of the languages in our tribe scenario are better off, metaphysically speaking, we have reason to accept egalitarianism as well.[11]

An alternative form of argument for pluralism appeals to a *principle of interpretive charity* that creates pressure to interpret our tribes as speaking truly and inferring correctly in their languages.[12] The idea is that it is more charitable to interpret the tribes as meaning different things with counterpart sentences than to interpret them as making mistakes about what exists. In particular, Hirsch argues that we can't get away with ascribing only subtle and reasonable mistakes to the speakers: instead, we'd have to ascribe error to their perceptual, a priori, and necessary judgments. And these are precisely the sorts of judgments for which considerations of charity are particularly weighty. (This claim is most plausible for counterfactual scenarios where speakers treat the relevant claims as quasi-analytic, partially constituting linguistic competence, even if such scenarios depart somewhat from actual patterns of use.)

A charity-based argument along these lines requires three additional things:

1. First, it requires that there actually *are* multiple candidate semantic values available to relieve the *prima facie* pressure to assign different meanings to UQEs. We discuss this assumption in §3.2.
2. Second, it requires the relevant differences in meanings between tribes to stem from differences in the meanings of their UQEs, rather than from differences in the meanings of their *other* expressions (such as their predicates). Arguably, positing variation in the UQE-meaning provides the most economical explanation (Dorr 2005). Moreover, as Sider (2009b, sec. 4) points out, the tribes will also diverge in which *counting sentences* they accept, where these state how many things there are using only UQEs, truth-functional connectives, the identity predicate, and perhaps the predicate 'concrete'.
3. Finally, it requires that charity considerations are not swamped by other factors that determine meaning. For example, on some meta-semantic views, metaphysical naturalness acts as a "reference magnet," creating pressure to interpret expressions as having more natural candidate meanings (see Lewis 1983; 1984). So, if one candidate meaning for UQEs is particularly metaphysically natural, this could potentially outweigh contrary pressure from charity.[13] On the other hand, tribes that treat the relevant claims as quasi-analytic are plausibly exhibiting dispositions that can resist even the pull of a reference magnet.[14]

As we'll see, considerations of naturalness are relevant to both components of quantifier variance. One could use considerations of naturalness either to block quantifier pluralism by appealing to reference magnets, or one could grant pluralism while still holding that one of the tribes has a more natural UQE than the rest. Accepting pluralism while rejecting egalitarianism still arguably avoids the deflationary consequences of quantifier variance, especially if we hold that ontologists are using a specialized quantifier intended to have the most metaphysically distinguished meaning (see §4.2).

3 Challenges for quantifier pluralism

Let's now turn to three of the most influential challenges facing the first component of the thesis of quantifier variance: quantifier pluralism.

3.1 *The demand for a direct specification of UQEs*

We noted earlier that it is difficult to specify what it takes for an expression in some possible language to be a counterpart to our unrestricted quantifier – for the expression to be functioning in its language like an unrestricted quantifier. But if we can't do this, it might be argued that the thesis of quantifier variance can't properly be stated, let alone adequately motivated.

We have our doubts about this complaint. Suppose that in fact the tribes in our counterfactual scenarios utter the relevant sentences truly in their own languages, and that those languages are metaphysically on a par. For the purposes of raising deflationary concerns about ontological disputes, it doesn't seem to matter whether we can independently specify the semantic or syntactic feature in virtue of which the tribes can be said to be using "unrestricted-quantifier-like expressions." If there were such a feature, it should perhaps not be very surprising if it's beyond our ability to directly specify that feature in English – just as it appears to be beyond our ability to specify what knowledge amounts to without using the term 'knowledge'. (We should be mindful of the expressive limitations of our home language, and shun the fantasy that we can use it to directly specify every thinkable intension.) If there is no such feature, then perhaps the term "quantifier variance" is a misnomer, but the deflationist concerns about ontological debates seem just as pressing. As long the tribes in our thought experiments all speak truly, and their languages are on a par, the deflationist concerns about ontological debates seem just as pressing – even if the term "quantifier variance" is a misnomer it seems we could have chosen to speak in any of these ways without missing out on anything, metaphysically speaking.

With these dialectical qualifications in mind, let's see how far we can get with a more direct method of specifying what it takes to count as a UQE. At a first pass, the semantic role shared by the natural language word 'something' and the existential quantifier of first-order logic is this: there is a domain of objects such that, when these terms are conjoined with a predicate, the resulting sentence is true just in case something in that domain satisfies the predicate. (We'll be ignoring, along with the bulk of the literature, the fact that most English quantifiers are binary and syntactically require restriction.)[15] So it's tempting to delineate the UQEs as those that, in this sense, range over that domain.

Unfortunately, this doesn't capture the idea of a UQE as intended by the quantifier variantist. Since our own unrestricted quantifier by definition ranges over everything, we can say that every domain is a subset of the domain of our unrestricted quantifier. But this seems[16] to imply that we can only give a quantifier variantist account of our tribes' languages if universal mereological composition is true. Otherwise, what domain does the UQE of Universalese range over? Quantifier variantists should allow that there are UQEs in more plenitudinous languages than ours – languages whose speakers would describe our quantifier as one that 'doesn't have everything in its domain'.

Another strategy is to treat the UQEs in a language as the expressions that share, in that language, the core inference rules associated with unrestricted quantification.[17] We could start, for example, with some natural-language analog of the classic introduction and elimination rules of the formal quantifier ($\exists$):

Classical $\exists$-Introduction: $\varphi\left(\frac{\tau}{x}\right)$ entails $\exists x \varphi(x)$, where $\varphi\left(\frac{\tau}{x}\right)$ is the result of replacing a name-like expression τ for all free occurrences of x in $\varphi(x)$.

Classical $\exists$-Elimination: If Γ together with $\varphi\left(\frac{\tau}{x}\right)$ entails ψ, and τ doesn't occur in Γ, $\varphi(x)$ or ψ, then Γ together with $\exists x \varphi(x)$ entail ψ.

One immediate concern for this approach is that it appeals to the category of 'name-like' expressions – but what does this amount to? For the same reason that we cannot (in English) delimit UQEs as those that range over some domain, we cannot delimit name-like expressions as those that refer to an item in a domain (Sider 2007, pp. 217–218). A potential solution is to characterize UQEs and name-like expressions *simultaneously* – they are the categories of expressions in the language which *together* play the characteristic inferential and grammatical roles of quantifiers and names. (Similarly for predicate-like and variable-like expressions.)[18]

However, the relevant inference rules may not be rich enough to delimit the intended class of expressions. For example, there are expressions that play the roles specified by these rules but are too foreign from the ordinary idea of existence to count as 'quantifier-like'. For example, imagine a language that maps each sentence φ to the proposition that *according to Bob, φ*, where

Bob is "logically perfect, maximally opinionated, and totally nuts" in the sense that his beliefs are completely unhinged from reality (Sider 2009b, pp. 391–392). In this language, the expression 'there is' will obey the inference rules above, but its meaning seems too distant from our notion of existence for it to count as "functioning like an unrestricted quantifier" in the language.[19]

In fact, it is not clear that we are all on the same page about what ought to count as "unrestricted-quantifier-like" even when it comes to actual expressions of natural language. For example, suppose we take plural quantification in English as primitive rather than as singular quantification over pluralities. In that case, should we treat the unrestricted plural existential quantifier as a UQE, despite its syntactic and semantic differences with the singular quantifier?

Looking beyond English, there's an even richer array of words that resemble our toy formal paradigm for unrestricted quantifier expressions in some ways but not others. For example, languages have obligatory classifier systems for quantification, making it unclear whether quantifiers are ever truly unrestricted. Or consider the particles *ka* and *mo* in Japanese, which combine with interrogative phrases (e.g. "who is happy," "why did you do that," and "where did you go") to yield quantification within the corresponding categories (e.g. people, reasons, and locations, respectively). In many languages, "the same particles [that] build quantifier words [also] function as connectives, additive and scalar particles, question markers, existential verbs, and so on" (Szabolcsi 2013).[20] This makes the quantifier words semantically compositional, and complicates the task of delineating the inferential roles of the relevant particles.

In short, those who wish to nail down exactly what it takes to count as a UQE face a bewildering array of choice points and complications when it comes to actual languages, let alone merely possible ones. Of course, as we stressed earlier, it's far from clear that we can saddle the quantifier variantist with the burden of providing a sharp definition of a UQE.

3.2 The demand for inter-linguistic semantics

Much of the debate around quantifier variance has focused on the ability to provide semantic theories of the requisite languages. Schematically, consider a quantifier variantist who posits two languages – Biglish and Smallish – with different UQEs, where a plenitudinous ontological theory comes out true in Biglish and a sparse theory comes out true in Smallish. Under this supposition, we can ask whether speakers of each language can give a semantic theory for the other. Assuming speakers of Biglish can give a semantic theory for their own language, it's plausible that they can also give a semantic theory for Smallish, by interpreting speakers of Smallish as using a quantifier with a smaller domain. But it's less clear whether speakers of Smallish can give a semantic theory for Biglish. And one might take the absence of such a theory as a basis for undercutting some dialectical motivations for quantifier variance.

First, let's ask: how important is it to the quantifier variantist that speakers of Smallish be able to formulate a semantic theory for Biglish?

It will be helpful to distinguish a couple of things we might mean by "a semantic theory" for Biglish.[21] Here are two candidates: (1) a *minimal semantic theory* for Biglish is a theory that specifies a *translation function* from every sentence in Biglish to a sentence in Smallish with the same truth-conditions and (2) a *compositional semantic theory* for Biglish is a theory that specifies a *compositional function* from each word in Biglish to a semantic value, together with combination rules for determining the truth-conditions of sentences in Biglish.

The availability of a Smallish minimal semantic theory of Biglish might help convince a sparse ontologist that there is in fact an interpretation of speakers of Biglish on which they are speaking truly (supporting pluralism) and that every intension expressible in Biglish can be expressed in Smallish (supporting egalitarianism).[22] And, if that semantic theory is compositional, it might help convince the sparse ontologist that the truth-preserving interpretation is psychologically realistic.

However, arguably a sparse ontologist shouldn't need a Smallish compositional semantic theory of Biglish to be convinced that there is a psychologically realistic truth-preserving interpretation of speakers of Biglish (Hirsch 2005, p. 158). After all, the sparse ontologist herself is often in the business of providing paraphrases of plenitudinous language that are supposed to be intentionally equivalent to the originals. For example, for 'there is a table', nihilists offer 'there are parts arranged tablewise', and so on. If she can do this, she herself instantiates a translation function from Biglish to Smallish that provides an interpretation of speakers of Biglish on which they are speaking truly. She needn't be able to articulate exactly how that translation function works! And given that she can actually produce this pairing, the interpretation not only exists but is apparently psychologically realistic.

Alternatively, one might justify the demand for a Smallish semantic theory of Biglish on the grounds of expressiveness equivalence. One might claim that without a Smallish compositional semantic theory of Biglish, Biglish would be more expressive than Smallish – viz. in that it can provide a semantic theory for Biglish.[23] Perhaps this expressive superiority might then be used to undermine egalitarianism and thus quantifier variance.

We will discuss significance of expressive superiority further in §4.1. However, note that the issue turns on subtle questions about how to understand expressive power. For instance: semantic theories are arguably necessary, so if expressiveness is understood in terms of which intensions are expressible, Smallish can trivially express a sentence that is intensionally equivalent to the Biglish semantic theory of Smallish.

A related basis for demanding an inter-linguistic semantics focuses on reference and its relationship with truth. Eklund (2009) and Hawthorne (2006) have pointed to a tension between quantifier variance and the following Tarskian principle:

T-Strong For any sentence of *any* language of the form "*F(a)*" to be true, the singular term must refer.

T-Strong (and similar principles for predicates) encodes a demand for a compositional semantic theory *given in terms of reference* (a 'referential semantic theory'). But it's hard to see how speakers of Smallish could give a referential semantic theory for Biglish. If no such theory is possible, then perhaps a sparse ontologist who accepts T-Strong should hesitate to accept that speakers of Biglish are speaking truly.

However, it's not clear why we should accept T-Strong as opposed to:

T-Weak For any sentence in *this* language of the form "*F(a)*" to be true, the singular term must refer.

which does not underwrite a demand for a Smallish referential semantic theory of Biglish (Hirsch 2009, pp. 240–243). One might try to justify T-Strong over T-Weak on the basis of highly controversial principles about the nature of truth, such as the principle that truth must be explained by reference.

Such principles can be denied (Hirsch 2009, pp. 238–239). Moreover, applying such principles requires some subtlety in the context of quantifier variance. After all, as Sider (2007) points out, speakers of Smallish plausibly shouldn't interpret Biglish sentences as even coming in predicate-name form. As he puts it: "names and quantifiers are connected. [Speakers of

Smallish] should deny that [the Biglish] expression 'a' is a name (i.e. deny that it is a name$_{small}$" (p. 218). In other words, speakers of Smallish can claim: what appear to be names in Biglish are in fact merely name-*like* expressions. That is, they play, in Biglish, the characteristic grammatical and inferential role of a name, but are not genuine names. Genuine names refer; but name-like expressions need not. (Compare the case with quantifier-like expressions and domains.) Of course, speakers of Biglish will still truly utter the sentence homophonous to 'Names refer'. So it seems chauvinistic to demand that truth be explained only by the relation picked out by 'reference' in Smallish rather than the relation picked out in Biglish. Indeed, arguably the Biglish sentence 'truth is explained by reference' is true, just as the homophonous sentence in Smallish is true. In this case, perhaps the quantifier variantist can supplicate intuitions about referential semantics by adopting the principle that truth in each language is explained by whatever plays the reference-role in that language.

So even if the variantist cannot articulate a Smallish semantic theory of Biglish, this doesn't obviously undercut the variantist position. But in fact the variantist arguably *can* provide such a theory. Dorr (2005), for example, suggests a highly general strategy for translating sentences of Biglish into Smallish that exploits the semantics of fictionalist or counter-possible conditionals: each sentence 'S' in Biglish will be mapped to the (apparently sensical) Smallish sentence 'In the fiction of such-and-such metaphysical theory, S' or 'If such-and-such metaphysical theory were the case, then it would be the case that S'. Modulo concerns about impossible antecedents and fictionalist operators, such a translation scheme seems relatively straightforward, at least for simple fragments of Biglish. Alternatively, we might pursue more specific translation schemes for particular ontological disputes.[24] For example, speakers of Nihilese can help themselves to plural quantification ('there are some Xs'), plural terms ('the As'), and a stock of plural predicates ('are arranged cupwise') to translate a significant fragment of Universalese. Universalese sentences like 'there is a cup' and '*a* is a cup' are translated as 'there are simples arranged cup-wise' and 'the As (taken plurally) are arranged cup-wise'.

These translational schemes can even be extended to a compositional semantics. Dorr, for example, sketches such a compositional semantics by letting the semantic value of predicate-like expressions be properties and the semantic value of quantifier-like expressions be second-order properties, so that a sentence like '$\exists x Fx$' is true iff the semantic value of the predicate-like term instantiates the semantic value of the quantifier-like expression. According to speakers of Smallish, the semantic value of 'cup' in both Biglish and Smallish is the property *being a cup*. The semantic value of the Smallish quantifier is the property of properties *being instantiated* while the semantic value of the Biglish quantifier-like expression is the property of properties *would be instantiated if such-and-such metaphysical theory were the case*. Similarly with more specific translation schemes: for instance the Nihilist minimal semantics of Universalese that deploys plural quantification can in principle be extended to a compositional semantics.[25]

To be sure, there are various complications and difficulties with these suggested semantic theories – especially when we consider attitude and modal operators. But giving a semantics for any language is a complicated and difficult business – at the very least, it's not obvious that the demand for a Smallish semantics of Biglish cannot be met by the quantifier variantist, even assuming she has this dialectical burden.[26]

3.3 Collapse arguments

According to pluralism, there is a plurality of UQEs with different meanings contained in various possible languages. So-called collapse arguments object to this claim.[27] These arguments are inspired by the following observation: any two expressions in the same first-order language which grammatically combine with all variables and formulas in the way typical of quantifiers and which obey **Classical ∃-Introduction** and **Classical ∃-Elimination** are logically equivalent – their inferential roles 'collapse' (Harris 1982). To see this, suppose we have a language with two such quantifier-like expressions, $\exists_1$ and $\exists_2$. Consider an arbitrary formula $\varphi(x)$ free in x, letting $\varphi\left(\frac{\tau}{x}\right)$ be the result of replacing all free occurrences of x with a name τ that doesn't appear elsewhere in $\varphi(x)$. By **Classical ∃-Introduction** applied to $\exists_2$, we have: $\varphi\left(\frac{\tau}{x}\right) \vdash \exists_2 x\varphi(x)$. And by **Classical ∃-Elimination** applied to $\exists_1$, we have: $\exists_1 x\varphi(x) \vdash \exists_2 x\varphi(x)$. *Mutatis mutandis* in the other direction. Call this result 'the collapse result'.

If we thought that the inference rules that delineate UQEs include **Classical ∃-Introduction** and **Classical ∃-Elimination** (and we suppose logically equivalent expressions have the same meaning), then pluralism posits two inequivalent expressions both of which obey these rules. Does this contradict the collapse result? Not obviously. The collapse result shows that any two expressions that obey these rules are equivalent *if they are in the same language*. But pluralism posits unrestricted-quantifier-like expressions with different meanings *in different languages* (Warren 2015).

Bridging this gap seems to require a 'combined' language that (i) includes two UQE terms with the same meanings of the UQE terms in the separate languages and (ii) includes a common stock of variable-like, name-like, predicate-like, etc. terms that play the characteristic grammatical and classical-inferential role with respect to both quantifier-like terms. (That is, both quantifier-like terms can combine with the same formulas and obey the classical inferential rules with respect to those formulas.) Given the existence of such a language, the collapse result should lead us to conclude that the two UQEs are logically equivalent, and thus arguably have the same meaning. And since each has the same meaning as one of the UQEs in the separate languages, the latter expressions also share a meaning, contradicting quantifier pluralism.

But is there such a combined language? It's not clear that we can stipulate *both* that the quantifier-like expressions in the combined language have the same meanings as the quantifier-like expressions in the separate languages *and* that they obey the requisite classical inferential roles *relative to a common stock of expressions* (see Sider 2007, p. 218; Warren 2015).

Suppose we attempt to construct such a language by taking one of the separate languages, say Schmenglish, and adding terms that have the same meanings as the terms in the other language, say Universalese. From the perspective of Schmenglish, the terms in Universalese are not (unrestricted) quantifiers, names, or predicates – they are merely (unrestricted) quantifier-like, name-like, and predicate-like, playing the corresponding grammatical and inferential roles in Universalese. (Some name-like terms in Universalese don't refer to anything, so aren't names.) The inferential rules of the Schmenglish quantifier governs how it interacts with sentences with names – not sentences with name-like terms. Insofar as a quantifier-like term interacts with these merely name-like terms in the way given by the classical inference rules, that quantifier-like term appears to no longer have the same meaning as our quantifier. (Indeed, it appears to have the meaning of the Universalese UQE!) And, insofar as we can insist that there is a term in the combined language with the same meaning as the Schmenglish quantifier (which can sensically combine with formulas in requisite way), we wouldn't expect it to obey classical, as opposed to, say, free, inference rules.[28]

Summing up: the collapse arguments for the equivalence of UQEs within a language don't straightforwardly apply to the equivalence of UQEs across languages. That's not to say that this gap cannot be bridged, but doing so is not straightforward.[29]

4 Challenges for quantifier egalitarianism

As we saw at the end of §2, those who think some potential quantifier meanings are more metaphysically distinguished than others can reject quantifier pluralism by holding that all of the tribal UQEs in our thought experiment should be interpreted as having the most metaphysically distinguished meaning. (On one version, this is because ordinary speakers harbor some deep linguistic intention to use the quantifier in a distinguished way, and this intention weighs more heavily even than charity to the truth of claims they find unproblematic; on another version, there is "semantic magnetism" that operates despite a lack of such intentions.) On this view, the thesis of egalitarianism, as we've stated it, becomes moot because all the tribes have the same quantifier meaning.

In this section, we'll discuss an alternative view, which acknowledges that the tribes all have different quantifier meanings – thus granting quantifier pluralism – but denies that all of these meanings are on a par, metaphysically speaking. On this kind of view, one might also claim that, while English speakers may be using a sub-par quantifier, contemporary ontologists are best interpreted as making their assertions in a language that has the most metaphysically distinguished meaning for its UQE.

Some reject egalitarianism on the grounds that one of the languages is more expressive than the others; others on the grounds that one is more natural than the others. We'll consider each strategy in turn.

4.1 Expressiveness

Assume a coarse-grained conception of content.[30] Can languages with existence-like notions that are more restrictive than ours express all of the coarse-grained contents that we can express with our more generous existence-like notions (see Hawthorne 2009, §2)?

Suppose we are speaking Shmenglish, and that some simples are first arranged tablewise to fuse a table *a* and later arranged tablewise to fuse a distinct table *b*. And suppose further that we think there is a possible world in which the simples composes *a* without *b* existing at all, and also a world in which the simples compose *b* without *a* existing at all – and that these worlds are indistinguishable with respect to the simples, differing only in the haecceitistic identities of *a* and *b*. The (Schm)english sentence '*a* exists' is true at the first world and false at the second. But it's hard to see how speakers of Nihilese can express this intension.

Spotting the modal assumptions required for this argument, does it undermine quantifier variance? Of course, the expressive differences conflict with certain regimentations of egalitarianism and not others.[31] The important question is how the expressiveness of the languages interacts with the purported deflationary consequences of quantifier variance (see Manley 2009a, §7).

It might be argued that ontological disputes are taking place in – or should be resurrected in – the expressively superior language.[32] It's unclear, however, how far this line of thought extends. For one thing, even if some languages are expressively impoverished, there may be multiple languages corresponding to participants in an ontological dispute for which pluralism and egalitarianism are true. Furthermore, even if there were a unique most expressive language, it's not entirely clear how moving to this language would make the ensuing ontological debate more significant or how it would vindicate the quasi-scientific methodology traditionally used to adjudicate such debates.

However, a much deeper problem lies in the vicinity. What the above discussion shows is that which propositions and possibilities one countenances will depend on one's other ontological views. So, in the sorts of ontological disputes that the variantist seeks to deflate, ontologists will not only be asserting contrary claims about, say, composition. They will also be asserting contrary claims about which propositions (and distinctions in modal space) there are. In the above example, the common-senser will assert, in addition to (2), sentences like:

4   For some simples, it's possible that they are arranged in exactly the same way, there are no other simples, and yet there is something that doesn't actually exist.

Meanwhile, the Nihilist will deny this sentence.

The problem for quantifier variance is that, when we include these claims among those under dispute, it's less clear that we should consider the corresponding communities for both sides of the debate to be speaking truly. Deflating debates over the existence of material objects is one thing; deflating debates over distinctions in modal space is another.

In replying to this challenge, the quantifier variantists face the question of how substantive or deflationary they take disputes about *de re* modality to be. If one treats them as substantive, one can deny that (4) is true even expressed in Shmenglish – though it seems surprising that quantifier variance would require a particular set of *de re* modal claims. Alternatively, given that quantifier variantists are deflationist about whether a particular object exists, it should hardly be surprising if they treat disputes about the trans-world identity of that object in the same way. In particular, one could claim that in asserting and denying (4) respectively, both communities are speaking truly after all, because the difference in quantifier meaning induces a difference in the meaning of modal expressions.[33]

4.2 Naturalness of the quantifier

Another strategy for rebutting quantifier variance is to extend Lewis's (1983) influential notion of natural properties to quantifier-like expressions. This directly undercuts strong forms of egalitarianism if the UQEs in some languages pick out more natural semantic values, assuming naturalness guarantees metaphysical distinction. If one candidate semantic value of UQEs is especially distinguished, perhaps ontologists can mimic traditional ontological debates by treating ontological theories as expressed in "Ontologese," a language whose UQE is stipulated to have that semantic value.

This kind of approach can be found fleshed out in different ways in Sider (2001, 2009b, 2011). The rough idea, setting aside some complexities of Sider's considered view, comes in three steps.[34]

Step 1. First, extend the theoretical role of naturalness to UQEs. Typically, natural properties are taken to be those that play the following theoretical role: they make for similarity, are easier to refer to, appear in more explanatory laws, etc.[35] At least some of these roles can be extended straightforwardly to UQEs: for instance, the more natural candidate semantic values of UQEs are those that appear in more explanatory laws, are easier to refer to, and perhaps make for greater similarity of facts (see Sider 2009b, pp. 404–405).

Step 2. Assert that something in the world answers to this theoretical role. This could be backed up by a variant of Lewis's argument that positing naturalness helps explain the various phenomena connected by its theoretical role (lawhood, similarity, reference, etc.) – it offers us a unifying theory of the otherwise disparate phenomena.

Step 3. Next, argue that a single UQE expresses a particularly natural semantic value relative to all the rest. Again, we can extend Lewis's methodology by holding that the terms that appear in our best scientific and/or metaphysical theories will express natural semantic values. If we have independent access to the goodness of theories formulated in the various languages, we can exploit the connection between goodness of theories and naturalness. So, if we judge theories stated using a particular UQE as best, this gives us reason to treat that term's semantic value as particularly natural.[36]

So, how might a quantifier variantist respond to this idea? They could reject the extension of naturalness to the semantic value of UQEs, by denying an independent grip on the notion's supposed theoretical role, at least beyond its application to predicates. Alternatively, if we can explain phenomena like similarity and the explanatoriness of laws by appealing solely to the naturalness of properties, we might worry that the extended notion of naturalness is theoretically unnecessary.[37]

Quantifier variantists can also resist Step 3 in various ways. One option is to deny that we have independent epistemic access to the goodness of theories formulated in the various languages; this will at least leave us unsure about egalitarianism (see Dorr 2013 and Warren 2016). Another option is to hold that the theories expressed in the relevant languages are equally good. Even if some UQEs are more natural than others, there may be an equivalence class of equally natural UQEs, and as long as that contains all the UQEs in the tribes corresponding to positions in the relevant debate, egalitarianism with respect to that debate is preserved. Finally, one might ask why we should think that the best theory will include a quantifier-like term at all.[38]

5 Quantifier variance and common-sense ontology

Let's now suppose that we grant quantifier variance about some debate, and perhaps even accept that there is some sense in which this deflates the significance of that debate. We are still left with a question about which, if any, of the ontologists was *right* – if only superficially. In other words, which ontologist, if any, was speaking truly in the shared language of the debate?

A natural line of reasoning suggests that (in this sense) it's the common-senser who was right all along:[39]

i Without overwhelming meta-semantic pressure to the contrary, English speakers are speaking whichever possible language makes their linguistic dispositions most reasonable. (Supported by the principle of charity)
ii There isn't overwhelming pressure to the contrary. (Supported by egalitarianism)
iii English speakers are disposed to treat the common-senser's claims as unproblematically true and the common-senser's inferences as unproblematic.
iv The language that makes these dispositions most reasonable is one on which those claims *are true* and the inferences *are entailments*, if such a language is possible. (Supported in part by egalitarianism)
v Such a language *is* possible. (Supported by quantifier pluralism)
vi So: English speakers are speaking a language on which the common-senser's claims are true.
vii The metaphysicians are making their assertions in English.
viii So: the common-senser's claims are true, and conflicting claims are false. Furthermore, {speaking English} there are tables and people, but nothing composed of a table and a person.

Of course, a number of these steps can be challenged. We'll focus on two key points.

1. First, there's a tension between premise (iii) and premise (iv), where depending on how strongly we interpret 'unproblematically true', one or the other premise becomes vulnerable.

Suppose that it's easy to count as treating the common-senser's claims as "unproblematically true" – it just requires finding them pretty obvious on their face, but not as quasi-analytic or constitutive of linguistic competence. Then it's hard to deny that English speakers do find the common-senser's claims unproblematically true.[40] But at the same time, it may not be so uncharitable to treat them as making some reasonable mistakes about what exists.

After all, perhaps they are disposed to retract their claims when presented with *other* considerations they find unproblematically true, such as the principle that no two objects can occupy exactly the same space at the same time.[41] How to weigh this conflict will depend on the specifics of the conflicting claims and how they are treated by English speakers (as analytic? obvious? potentially subject to revision?)[42] We may just face a meta-semantic challenge of weighing conflicting dispositions to use the expressions, all of which are relevant to interpretation. One potential upshot is that the meaning of the completely unrestricted English quantifier is indeterminate.

On the other hand, as we strengthen our reading of 'unproblematically true', it would be harder to charitably interpret people as making mistakes with claims they find unproblematically true. But it also becomes easier to reject the premise that English speakers really find the common-senser's claims unproblematically true in the relevant sense. Do ordinary English speakers really treat rival ontological claims as evincing a kind of linguistic incompetence? As Sider has stressed, adequately testing this kind of question on subjects arguably requires that they are properly prompted: for example, made to understand that the quantifiers are 'wide-open' and that certain paraphrases are not under dispute (2004, p. 680).

2. A second option is to reject (vii). As we've seen, some inegalitarians about quantifier meanings hold that metaphysicians are making their claims not in English but in the language of "Ontologese," in which the UQE has the unique perfectly natural candidate semantic value. But even granting egalitarianism, it might be that, the meaning of the quantifier in the ontology room is not simply settled by use facts in ordinary talk. After all, ontologists are trying to do something different with their words – unlike ordinary speakers, for example, they don't treat the questions they're asking as having straightforward answers. And in some cases they are explicitly intending to be using the quantifier to express a metaphysically distinguished notion.[43]

So what happens if there is no good candidate semantic value for the quantifier that answers to these intentions? In that case, perhaps the meaning of their quantifier reverts to that of the ordinary English one. But if our meta-semantics give their distinctive intentions enough weight, perhaps their quantifier just fails to express anything at all. Or if there is a tie among several most distinguished notions, it might be indeterminate which one gets expressed by the quantifier in the ontology room. And finally, we might decide that what is meant by ontologists' quantifier varies from one dispute to another, and hinges on the particular intentions involved in each one. (For example, some ontologists explicitly proclaim that they are using the ordinary English quantifier, while others view themselves as employing a specialized meaning.) In short, there are a variety of meta-semantic pictures that place enough weight on ontologists' dispositions to block the conclusion that the common-senser is technically right – even if we grant both components of quantifier variance.

6 Conclusion

The debate around quantifier variance is flourishing, and we have only been able to offer a brief overview. As we have seen, accepting the combined theses of quantifier pluralism and egalitarianism about an ontological debate arguably has major implications for that debate, both by deflating its significance and by motivating a new methodology. The stakes are high for those who pursue first-order ontological questions.

Notes

1 This story follows the opening thought experiments from Dorr (2005; 2014).
2 Mereological nihilism is the view that there are no proper parts.
3 Mereological universalism is the view that any two concrete objects compose a third.
4 See especially Hirsch (2011). Predecessors include the views of Carnap and Putnam.
5 For a prominent deflationary view about ontology that's not committed to quantifier variance, see Thomasson (2015).
6 Quantifier variance matters for other reasons too: for its connection to neo-Fregeanism, see Hale (2007), Hawley (2007), and Sider (2007); for its connection to indefinite extensibility, see Warren (2017); for its connection to vague existence see Liebesman and Eklund (2007), Sider (2009a), and Sud (2017; 2018).
7 Sometimes 'quantifier variance' is used to refer only to what we here call 'quantifier pluralism' (e.g. Manley 2009a, which uses 'egalitarian quantifier variance' for the stronger claim).
8 Sider is notably concerned with how natural or (in his terms) 'structural' the meanings of UQEs in various languages are (2007; 2009b; 2011).
9 See e.g. Eklund (2009). See also Manley (2009a; 2009b), who doubts that expressiveness confers metaphysical distinction.
10 As Hirsch put it, the language is preferable "on purely metaphysical (rather than pragmatic) grounds" (Hirsch 2011, p. xv); but he doubts that having natural meanings confers such grounds (Hirsch 1993).
11 A more oblique case for egalitarianism appeals to the judgment that the meanings of UQEs are highly sensitive to use, which on some meta-semantic theories will clash with the claim that any of them has a candidate meaning that is especially metaphysically natural. (For more on linguistic judgments and metaphysical distinction, see Manley 2017.)
12 We're leaving it open whether charity is an epistemic principle that guides meaning-judgments, or a metaphysical principle that is somehow *constitutive* of meanings. See Warren (2017, sec. 2) for a more specific meta-semantic argument for quantifier pluralism.
13 Elements of this reply can be found in Sider (2001, Introduction), Sider (2009b, p. 410), Sider (2011).
14 For related discussion, see Hirsch (2004; 2005, p. 176; 2008a; 2008b, p. 193 and fn 21). Hirsch suggests that naturalness at best tips the scales among equally charitable candidates.
15 Moreover, in treating *something* and *there are* as semantic atoms akin to the existential quantifier of first-order logic, we're ignoring some subtleties of natural language. Arguably, *something* combines the meanings of *some* and *thing*, where *some* is a binary quantifier whose first argument serves to restrict the domain at issue; but even on such a view, *something* can be unrestricted in the sense at issue in the paper as long as *thing* functions only as a "dummy" predicate that applies to the whole domain. Meanwhile, *there are* looks to be composed of two atoms, and also, when conjoined with a plural noun phrase *Fs*, results in a true sentence only if there is *more* than one F.
16 This implication is not obvious. For example, a nihilist might interpret the Universalese quantifier as ranging over sets of mereological simples (cf. Hirsch and Warren 2017).
17 See e.g. Hirsch (2011, Introduction) and Sider (2007; 2009b; 2011). Hirsch delineates the expressions by their 'formal-syntactic inferential role' in the language. Note that we're using a single notion of entailment to characterize the UQEs. However, if the meaning of 'entails' differs across the languages, we should arguably use each language's own notion of entailment to characterize its UQE (see Dorr 2014).
18 This strategy also needs some account of what it means for an expression to 'obey an inference rule'. For example, inference rules are often taken to give patterns of entailment; but is entailment in the relevant sense a matter of necessary truth preservation, conceptual or a priori deducibility, or what?
19 Another worry is that, even if these inferential rules characterize what it is for an expression to function like a quantifier, they don't seem limited to quantifiers that are *unrestricted*. After all, a restricted quantifier in a language with few or no names can obey these rules. A minimum requirement for quantifier-like expression '∃' to be unrestricted is that there is no *other* quantifier-like expression '∃o' in the language such that '∃o $x\phi(x)$' entails '∃$x\phi(x)$' but not vice versa (see Turner 2010). This requirement, however, is too easy to satisfy for languages that have only one quantifier-like expression. We might require the truth of explicit pronouncements in the language that the expression is supposed to be interpreted unrestrictedly; but it's not clear how one might make such a pronouncement non-trivially.

We could also try to exploit the open-endedness of natural languages by requiring that the expressions continue to obey the inference rules across natural ways the language may change, including the addition of new name-like and quantifier-like expressions to the language (Warren 2017, p. 90).

20 See also Shimoyama (2006) on Japanese *ka* and *mo* and Seibt (2015) on various classifier systems and their relationship to ontological categories.
21 We'll assume that a semantic theory is a theory of the truth-conditions of sentences, rather than a dimension of meaning other than truth-conditions (e.g. cognitive significance, assertibility-conditions, conversational effect, etc.).
22 Some arguments (which we don't appeal to here) for the deflationary consequences of quantifier variance (e.g. Hirsch 2005; 2009) include the premise that *both* sides of an ontological dispute can charitably interpret the other side as speaking truly.
23 Eklund (2009) presses for a Smallish *referential* semantic theory (discussed below) on these grounds.
24 See, e.g., Hirsch (2009, sec. 4) for a discussion of translation schemes in the context of disputes over persistence. Hirsch thinks that these translation schemes are conceptually more basic than Dorr's fictionalist and counter-possible schemas.
25 Indeed, depending on how exactly one understands "referential semantics," a Smallish "referential semantics" of Biglish is perhaps in the offing, at least for certain disputes. Hirsch and Warren (2017), for example, offer speakers of Nihilese a "Tarksian" semantics for Universalese, on which Universalese names for material objects are mapped to sets of simples (similarly for predicates of material objects). Although, they acknowledge this does not constitute a "referential semantics" on more stringent understandings of the expression.
26 See also Sider (2007) for an 'algebraic' characterization of the meanings of quantifier-like expressions in terms of the structure these meanings are meant to play in the theory of quantifier variance.
27 See, for example, Hale and Wright (2009). For a related argument, see McSweeney (2016).
28 As a final reply, the quantifier variantist might choose to use non-classical inference rules to define the class of quantifier-like expressions. This form of quantifier variance isn't susceptible to an analogous collapse result. (See Turner 2010 for a related discussion.) And given the use of empty names in English, perhaps the English notion of existence is best characterized by the free rules. (It's not clear, however, whether this move can support quantifier variantists' deflationary ambitions. Perhaps contemporary ontologists can simply introduce a new notion of existence, using the classical rules, and conduct their debates using this classical notion.)
29 For an advanced discussion on bridging this gap see Dorr (2014) who explores various collapse arguments that apply directly to the semantic values of the UQEs instead of going via a combined language.
30 On a structured conception of propositions, things get tricky. Consider the following sentences:

   (i) There are simples arranged tablewise.
   (ii) There is something composed of simples arranged tablewise.

Call the structured propositions that (i) and (ii) express in Schmenglish '*a*' and '*b*' respectively. One might hold that, even if the sentence (i) in Nihilese expresses *a*, no sentence in Nihilese expresses *b*, giving Schmenglish an expressive advantage. (See McGrath 2008.)

Here are two replies due to (Hirsch 2008a). First, a similar argument gives Nihilese a corresponding expressive advantage. Consider:

(iii) There is nothing composed of simples arranged tablewise.

Is there a sentence in Schmenglish that expresses what (iii) expresses in Nihilese? If not, then arguably neither language is strictly more expressive than the other. Second, one might insist that (ii) can be expressed by a speaker of Nihilese after all: perhaps she can express it by saying:

(iv) If it were the case that composition were common-sensical, then it would be the case that there is something composed of the simples arranged table-wise.

Of course, speakers of Nihilese and speakers of Shmenglish will differ on their acceptance of sentences like 'the proposition expressed in Nihilese by (iv) has n constituents'; but this could be because the two languages differ on what is meant by 'proposition', 'expressed', or 'constituents' (see Hirsch 2008a, fn.18 for a related discussion). For example, if structured contents are sets, even the axioms of impure set theory will be expressed in quantificational terms and thus, arguably, what's meant by 'set' will differ between the two languages.

31 As we've seen, one could use expressive superiority to target pluralism by treating superior expressiveness as an interpretive pressure that competes with the presumption that the tribes are speaking truthfully in their respective languages. But the pressure from expressive superiority would have to be quite strong. See McGrath (2008).
32 See McGrath (2008) for more discussion of how considerations of expressive power can be used to undercut some arguments deflating ontology.
33 This response fits more naturally with some theories of *de re* modality than others: for example, some counterpart-theoretic or conventionalist views may have the flexibility to identify suitable semantic values for modal expressions in both languages.
34 In fact, Sider's variant of *natural* (viz. *S* for *structural*) isn't predicated of semantic values: it's an operator that attaches to terms of any grammatical category. So, for instance, where we say *The semantic value of 'N' is structural*, Sider would prefer *S(N)*.
35 See Dorr and Hawthorne (2013) for more on construing Lewis's (1983) theory of naturalness as a term-introducing theory that defines the term 'natural' in the spirit of Lewis (1970).
36 There are various ways one could cash out the notion of goodness for a metaphysical theory: some candidate virtues include being more explanatory; having more natural expressions; describing a simpler world; having a simpler language; having a language whose simplicity matches the simplicity of the world. See Manley (2016).
37 For example, Hirsch (2008b, p. 195) argues that the reference magnetic influence of naturalness for properties derives from its similarity-conferring power; but the naturalness of quantifier-like expressions doesn't make for similarity, so it's not a factor in fixing meaning.
38 For more on this strategy see Dasgupta (2009; 2017); Sider (2011, especially sec. 9.6.4); Turner (2011; 2017); Donaldson (2015).
39 The argument given below is most similar to "the argument from charity" in Hirsch (2002a), although there are various differences. An alternative argument in Hirsch's later (2005) work goes via claims about the verbalness of ontological disputes (see Jackson 2012 for a reply).
40 However one might also reject (iii) by holding that, although they *appear* to be treating the claims of the common-senser as true, ordinary English speakers are using a contextually restricted quantifier or are speaking loosely. See Hirsch (2002a, pp. 104–107) for a convincing reply.
41 Hirsch (2002a) calls these "conflicts of charity."
42 Hirsch thinks (roughly) that such general principles should not command much meta-semantic weight, relative to specific claims that conflict with them (Hirsch 2002a, sec. 3).
43 See Dorr (2005), Sider (2009b, pp. 411–413), and Sider (2011, sec. 9) for more on speaking Ontologese. See also our §4.2.

References

Dasgupta, S. (2009) 'Individuals: An essay in revisionary metaphysics', *Philosophical Studies*, 145, pp. 35–67.
Dasgupta, S. (2017) 'Can we do without fundamental individuals? Yes', in Barnes, E. (ed.) *Current controversies in metaphysics*. Routledge, pp. 7–23.
Donaldson, T. (2015) 'Reading the book of the world', *Philosophical Studies*, 172(4), pp. 1051–1077.
Dorr, C. (2005) 'What we disagree about when we disagree about ontology', in Kalderon, M. E. (ed.) *Fictionalism in metaphysics*. Oxford University Press, pp. 234–286.
Dorr, C. (2013) 'Reading *Writing the Book of the World*', *Philosophy and Phenomenological Research*, 87(3), pp. 717–724.
Dorr, C. (2014) 'Quantifier variance and the collapse theorems', *The Monist*, 97(4), pp. 503–570.
Dorr, C. and Hawthorne, J. (2013) 'Naturalness', *Oxford Studies in Metaphysics*, 8, pp. 3–77.
Eklund, M. (2009) 'Carnap and ontological pluralism', in Chalmers, D. J., Manley, D., and Wasserman, R. (eds) *Metametaphysics*. Oxford University Press, pp. 130–156.
Hale, B. (2007) 'Neo-Fregeanism and quantifier variance', *Proceedings of the Aristotelian Society*, 107, pp. 375–385.
Hale, B. and Wright, C. (2009) 'The metaontology of abstraction', in Chalmers, D. J., Manley, D., and Wasserman, R. (eds) *Metametaphysics*. Oxford University Press, pp. 178–212.
Harris, J. H. (1982) 'What's so logical about the "logical" axioms', *Studia Logica*, 41(2–3), pp. 159–171.
Hawley, K. (2007) 'Neo-Fregeanism and quantifier variance', *Proceedings of the Aristotelian Society Supplementary Volumes*, 81, pp. 233–249.

Hawthorne, J. (2006) 'Plentitude, convention, and ontology', in *Metaphysical essays*. Oxford University Press.
Hawthorne, J. (2009) 'Superficialism in ontology', in Chalmers, D. J., Manley, D., and Wasserman, R. (eds) *Metametaphysics*. Oxford University Press, pp. 213–230.
Hirsch, E. (1993) *Dividing reality*. Oxford University Press.
Hirsch, E. (2002a) 'Against revisionary ontology', *Philosophical Topics*, 30, pp. 103–127. Reprinted in Hirsch (2011).
Hirsch, E. (2002b) 'Quantifier variance and realism', *Philosophical Issues*, 12, pp. 51–73. Reprinted in Hirsch (2011).
Hirsch, E. (2004) 'Comments on Theodore Sider's *Four Dimensionalism*', *Philosophy and Phenomenological Research*, 68(3). Reprinted in Hirsch (2011).
Hirsch, E. (2005) 'Physical-object ontology, verbal disputes, and common sense', *Philosophy and Phenomenological Research*, 70, pp. 67–97. Reprinted in Hirsch (2011).
Hirsch, E. (2008a) 'Language, ontology, and structure', *Noûs*, 42(3), pp. 529–538. Reprinted in Hirsch (2011).
Hirsch, E. (2008b) 'Ontological arguments: Interpretive charity and quantifier variance', in Sider, T., Hawthorne, J., and Zimmerman, D. (eds) *Contemporary debates in metaphysics*. Oxford University Press. Reprinted in Hirsch (2011).
Hirsch, E. (2009) 'Ontology and alternative languages', in Chalmers, D. J., Manley, D., and Wasserman, R. (eds) *Metametaphysics*. Oxford University Press. Reprinted in Hirsch (2011).
Hirsch, E. (2011) *Quantifier variance and realism: Essays in metaontology*. Oxford University Press.
Hirsch, E. and Warren, J. (2017) 'Quantifier variance and the demand for a semantics', *Philosophy and Phenomenological Research*. doi: 10.1111/phpr.12442.
Jackson, B. B. (2012) 'Metaphysics, verbal disputes, and the limits of charity', *Philosophy and Phenomenological Research*, 86(2), pp. 412–434.
Lewis, D. (1970) 'How to define theoretical terms', *The Journal of Philosophy*, 67, pp. 427–446.
Lewis, D. (1983) 'New work for a theory of universals', *Australasian Journal of Philosophy*, 61(4), pp. 343–377.
Lewis, D. (1984) 'Putnam's paradox', *Australasian Journal of Philosophy*, 62(3), pp. 221–236.
Liebesman, D. and Eklund, M. (2007) 'Sider on existence', *Noûs*, 41(3), pp. 519–528.
Manley, D. (2009a) 'Introduction: A guided tour of metametaphysics', in Chalmers, D. J., Manley, D., and Wasserman, R. (eds) *Metametaphysics*. Oxford University Press, pp. 1–37.
Manley, D. (2009b) 'When best theories go bad', *Philosophy and Phenomenological Research*, 78(2), pp. 392–405.
Manley, D. (2016) 'Keeping up appearances: a reducer's guide'. MS.
Manley, D. (2017) 'Moral realism and semantic plasticity'. MS.
McGrath, M. (2008) 'Conciliatory metaontology and the vindication of common sense', *Noûs*, 42(3), pp. 482–508.
McSweeney, M. (2016) 'An epistemic account of metaphysical equivalence', *Philosophical Perspectives*, 30, pp. 270–293.
Seibt, J., (2015) 'Ontological scope and linguistic diversity: Are there universal categories?', *The Monist*, 98(3), pp. 318–343.
Shimoyama, J., 2006. 'Indeterminate phrase quantification in Japanese', *Natural Language Semantics*, 14(2), pp. 139–173.
Sider, T. (2001) *Four-dimensionalism*. Oxford University Press.
Sider, T. (2004) 'Replies to Gallois, Hirsch and Markosian', *Philosophy and Phenomenological Research*, 68(3), pp. 674–687.
Sider, T. (2007) 'NeoFregeanism and quantifier variance', *Aristotelian Society*, Supplementary Volume 81, pp. 201–232.
Sider, T. (2009a) 'Against vague and unnatural existence: Reply to Liebesman and Eklund', *Noûs*, 43(3), pp. 557–567.
Sider, T. (2009b) 'Ontological realism', in Chalmers, D. J., Manley, D., and Wasserman, R. (eds) *Metametaphysics*. Oxford University Press, pp. 384–423.
Sider, T. (2011) *Writing the book of the world*. Oxford University Press.
Sud, R. (2017) 'Ontological deflationism, vague existence, and metaphysical vagueness'. MS.
Sud, R. (2018) 'Vague naturalness as ersatz metaphysical vagueness', in Bennett, K. and Zimmerman, D. W. (eds) *Oxford studies in metaphysics*. Oxford University Press, pp. 243–277.

Szabolcsi, A. (2013) 'Quantifier particles and compositionality', Proceedings of the 19th Amsterdam Colloquium.

Thomasson, A. (2015) *Ontology made easy*. Oxford University Press.

Turner, J. (2010) 'Ontological pluralism', *The Journal of Philosophy*, 107(1), pp. 5–34.

Turner, J. (2011) 'Ontological nihilism', in Bennett, K. and Zimmerman, D. W. (eds) *Oxford studies in metaphysics*. Oxford University Press, pp. 1–50.

Turner, J. (2017) 'Can we do without fundamental individuals? No', in Barnes, E. (ed.) *Current controversies in metaphysics*. Routledge, pp. 24–34.

Warren, J. (2015) 'Quantifier variance and the collapse argument', *The Philosophical Quarterly*, 65(259), pp. 241–253.

Warren, J. (2016) 'Sider on the epistemology of structure', *Philosophical Studies*, 173(9), pp. 2417–2435.

Warren, J. (2017) 'Quantifier variance and indefinite extensibility', *Philosophical Review*, 126(1), pp. 81–122.

9
VERBAL DISPUTES AND METAPHYSICS

Brendan Balcerak Jackson

UNIVERSALIST: There is an object composed of my nose and your ear.
ANTI-UNIVERSALIST: No, there is no such object!

Many philosophers are tempted to diagnose disputes like this one as merely verbal: the parties don't really genuinely disagree, but are merely talking past each other. Importantly, the diagnosis of mere verbalness is thought to have deflationary consequences. If a dispute in ontology or metaphysics is merely verbal, this is taken to show that there isn't really anything substantive at stake in the dispute; it is really just a disagreement about how to use language. Or it is taken to show that the dispute should not be regarded as being about a genuinely metaphysical matter, but instead about what is most interesting or what we should care about more. The aim of this chapter is to elucidate these issues. What is it for an apparent dispute to be merely verbal? If a dispute in a certain area of metaphysics turns out to be merely verbal, what (if anything) does this imply about the possibility of substantive metaphysical theorizing in that area?

1 Introduction

Some disagreements tend to strike us as being somehow defective. We might try to diagnose that defectiveness in different ways, depending on the case. For example, consider the following dispute about art:

A: Goya's *Saturn Devouring His Son* is a beautiful painting.
B: No, it isn't. It's hideous!

One fairly common reaction to a disagreement like this is to dismiss it as confused: beauty is in the eye of the beholder, and all that can be said about the painting is that it is beautiful-for-me or not beautiful-for-you; there is no further objective fact about beauty for A and B to disagree about (see e.g. Kölbel 2004; MacFarlane 2007). Sometimes a more epistemic diagnosis of the defectiveness seems appropriate. Observing an argument about whether there is intelligent life on other planets, we might be tempted to dismiss the dispute as futile: we know so little about what conditions in the rest of the universe are like, and about what conditions would be suitable for intelligent life, that there is very little justification for taking either side of this debate.

Certain disagreements in metaphysics and ontology are also sometimes thought to be defective. One oft-cited type of example has to do with van Inwagen's (1995) special composition question, the question of when some things compose a thing.

UNIVERSALIST: There is an object composed of my nose and your ear.
ANTI-UNIVERSALIST: No, there is no such object!

We could try to diagnose this dispute in the same way as the Goya dispute above. But few would be eager to accept the kind of relativism about composition (or existence) that such a diagnosis would entail (although see Chalmers 2009). An epistemic diagnosis along the lines of that of the alien intelligence dispute is also available (see Bennett 2009). But some philosophers who regard certain disputes in ontology or metaphysics as defective have tended to prefer a different diagnosis: the apparent disagreements are *merely verbal*; the parties don't really genuinely disagree, but are merely talking past each other (Hirsch 2005, 2009; Sidelle 1999, 2002, 2007).

Importantly, the diagnosis of mere verbalness is thought to have deflationary consequences. According to Eli Hirsch, the mere verbalness of debates about the special composition question and others shows that, "[n]othing is substantively at stake in these questions beyond the correct use of language," and that "[t]he only real question at issue is which language is (closest to) plain English. ... That's the sense – the only sense – in which" any party to the dispute can claim to be right (Hirsch 2005: 67, 70). Similarly, Alan Sidelle argues that the debate in Melville's *Moby Dick* about whether or not whales are fish is merely verbal, and that this shows that there are no substantive biological questions at stake in it; there are only questions about what 'fish' means in ordinary English, or about which meaning for 'fish' is most interesting or useful or important (Sidelle 2007). By the same token, according to Sidelle, if a disagreement like the one about composition above turns out to be merely verbal, it should be relocated "away from essentialist metaphysics ... and into semantics or a recognizably philosophical (or otherwise) issue over what is interesting or important" (Sidelle 2007: 97).[1]

The aim of this chapter is to elucidate these issues. What is it for an apparent disagreement to be merely verbal (Sections 2–4)? If a dispute in a certain area of metaphysics turns out to be merely verbal, what – if anything – does this imply about the possibility of substantive metaphysical theorizing in that area (Sections 5–6)?

2 Verbal disputes

The metametaphysical literature provides many putative examples of merely verbal disputes, such as apparent disagreements about whether Pluto is a planet (Sidelle 2007); about whether a glass is a cup (Hirsch 2005); or about whether an appletini is a martini (Bennett 2009). Let us begin with an example along the same lines. Suppose that the pedantic scientist Bunsen has the following exchange with Bocuse, an accomplished chef:

BUNSEN: You might be interested to know that tomatoes are a type of fruit.
BOCUSE: Nonsense! Tomatoes are not a fruit, they are a vegetable.

Even though this exchange has the form of a disagreement, it is natural to suspect that Bunsen and Bocuse are really just talking past each other. What Bunsen intends to convey is that tomatoes fit the biological characterization of a fruit – that is, that they are seed-bearing bodies that develop from the ovary at the base of the flower. What Bocuse intends to deny is not this, but rather the claim that tomatoes play the sorts of culinary roles typical of apples, cherries, and mangoes. Their apparent disagreement is merely verbal: *merely* verbal because they do not (we may suppose) disagree about whether tomatoes are fruits in either the biological or the culinary sense, and merely *verbal* because the appearance of disagreement somehow arises out of their divergent uses of the word 'fruit.'

Notice that here we are talking about disagreement or dispute as an *activity* rather than as a *state* (Cappelen and Hawthorne 2009: 60). Two thinkers are in a state of disagreement when they have conflicting attitudes. This can be the case even if they are not engaged in any conversational exchange, and indeed even if they are entirely unaware of each other. (For this reason, the question of whether an apparent disagreement, in the state sense, is merely verbal makes little sense.) But for A and B to disagree, in the activity sense, they must somehow interact – paradigmatically, via some sort of conversational exchange. A merely verbal dispute is also an activity, a conversational exchange that in some way resembles an activity of genuinely disagreeing while failing to be one.[2]

What makes the Bunsen/Bocuse dispute merely verbal? One possibility is that the apparent dispute arises because the word 'fruit' means something different when Bocuse utters it than when Bunsen does. Perhaps 'fruit' is homophonous in English, or perhaps Bocuse and Bunsen speak slightly different English dialects (or idiolects). According to this suggestion, what Bocuse says or asserts when he utters 'Tomatoes are not a fruit' is in fact compatible with what Bunsen says or asserts when he utters 'Tomatoes are a fruit'; and perhaps each party in fact accepts the proposition asserted by the other.

However, even if this is a correct description of the Bunsen/Bocuse case (which it may not be, as we will see shortly), verbal disputes can also arise in cases where we would surely not want to posit any differences in meaning. Suppose that A and B are discussing actors. A is thinking of Ryan Gosling's performance in the film *La La Land*, but gets confused about whether Gosling has the name 'Ryan Gosling' or 'Ryan Reynolds':

A: Ryan Reynolds is a mediocre singer, but a pretty good dancer.
B: What? Ryan Reynolds is not a good dancer.

This is a merely verbal dispute much like the Bunsen/Bocuse case. Is this because A's utterance of 'Ryan Reynolds' refers to Gosling while's B's refers to Reynolds? Probably not: orthodox views about the semantics of proper names imply that A's use of 'Ryan Reynolds' refers to Reynolds, just as B's does, even if the mental images, descriptive characteristics and so on that A associates with the name pick out Gosling rather than Reynolds (Evans 1973; Kripke 1980). So, what B says genuinely contradicts what A says; still, there is intuitively no genuine disagreement between A and B (Balcerak Jackson 2014; Sidelle 2007). In fact, an analogous objection might be raised to the initial description of the Bunsen/Bocuse case. Perhaps 'fruit' is a natural kind term that picks out the same biological category whether it is used by Bunsen or Bocuse (Putnam 1975). Even so, this does nothing to remedy our sense that their dispute is defective (Chalmers 2011).

Here is a different sort of example. Suppose that A is speaking hyperbolically, but that B fails to recognize this:

A: I have a million exams to grade today.
B: That's impossible! You don't have a million exams to grade today.

This too is a merely verbal dispute: A and B presumably agree that the actual number of exams that A has to grade is well below one million, and we may suppose that they also agree that A has a lot of exams to grade. The appearance of disagreement in this case arises because of a mismatch between the way A and B are using the phrase 'a million exams.' But the mismatch is not semantic. Both attach the same meaning to the phrase, and what B says genuinely contradicts what A says (which, in this case, is not what A means).

What these cases illustrate is that we cannot understand merely verbal dispute entirely in purely semantic terms. An apparent dispute's being merely verbal is not just a matter of the parties unwittingly attaching different meanings to some expression, or of their sentences expressing compatible propositions (in context) without realizing it. The confusion or mismatch concerns the purposes which certain expressions are being used to serve in the exchange; this is fundamentally a matter of pragmatics, not semantics (Balcerak Jackson 2014; Vermeulen 2018).[3] This is important to keep in mind when we turn to attempts to provide general characterizations of mere verbalness.

3 Accounts of mere verbalness

There are at least three broad types of general accounts that can be found in the literature.[4] The following characterization from Balaguer (2017: 3) is representative of the first:

> Let Smith and Jones be people and S be a sentence, and suppose that Smith says that S is true and Jones says that it's false. Then the dispute between Smith and Jones is merely verbal iff (a) Smith and Jones mean different things by S – or S expresses different propositions in the languages of Smith and Jones, or some such thing; and (b) Smith and Jones would agree that S is true in Smith's language and false in Jones language, or that the proposition that Smith takes S to express is true and the proposition that Jones takes S to express is false, or something along these lines.

In a similar vein, Bennett (2009: 40) says that an apparent dispute involving a sentence of the form 'There are Fs' is merely verbal when, "[t]he disputants assign different meanings to either the existential quantifier, the predicate '*F*,' or the negation operator, and are consequently talking past each other." Call these *semantic* accounts, since they trace mere verbalness to differences in the meanings the parties attach to expressions, to the propositions expressed by their sentences and the like.

As we have just seen in Section 2, semantic accounts are inadequate to capture the full range of merely verbal disputes. Two parties can be having a merely verbal dispute even when there are no relevant differences in the meanings they attach to expressions, and even when the sentences they utter express genuinely contradictory propositions.[5]

A second type of account is what we might call *interpretational*. Hirsch (2009: 239) suggests the following (see also Sidelle 2007):

> I would … define a verbal dispute as follows: It is a dispute in which, given the correct view of linguistic interpretation, each party will agree that the other party speaks the truth in its own language. This can be put more briefly by saying that in a verbal dispute each party ought to agree that the other party speaks the truth in its own language.

As Hirsch makes clear, "the correct view of linguistic interpretation" is one that gives great weight to considerations of charity; when we interpret a speaker's language we should do so in a way that avoids attributing to her obvious, irrational, or inexplicable mistakes insofar as possible. This might just seem like another version of a semantic account – and it is, if we assume that the facts about charitable interpretation determine the actual meanings of the expressions in each party's language. But Hirsch does not make this assumption. He allows, for example, that social externalist considerations might imply that a word like 'arthritis' in a speaker's language does not have the meaning that would be assigned to it by the most charitable interpretation of the speaker, but rather has the meaning that best corresponds to the way 'arthritis' is used by experts and others in the community (Burge 1979; Hirsch 2005). To put it another way, a speaker's "own language" – the one she is most charitably interpreted as speaking – may not be the language she actually speaks; and it is the parties' "own languages" that determine whether an apparent dispute between them is merely verbal.

For this reason, interpretational accounts might seem immune to the problem that was just seen to face semantic accounts. However, interpretational accounts have difficulty accommodating cases like the 'million exams' dispute above. As Vermeulen (2018) observes, considerations of interpretive charity apply to a speaker's language as a whole, and to her use of it across a wide range of actual and counterfactual circumstances. It is highly unlikely that an interpretation that treats 'a million' as meaning *many* in A's own language would satisfy the demands of charity.[6] Indeed, any plausible view of interpretation *ought* to have the result that A and B are to be interpreted as having the same language.[7] The difference between A and B is not in their languages (either their "own languages" or the languages they actually speak) but a difference in how they are using their common language in the current conversation. Interpretational accounts cannot accommodate this, and so they mis-classify the 'million exams' dispute as genuine.[8]

Observations like this motivate a third family of *pragmatic* accounts. On Vermeulen's own view, for example, the 'million exams' dispute comes out as merely verbal because what A herself means – regardless of what her words mean – is that she has many exams to grade, and what B himself means is that A does not have 1,000,000 exams to grade; these are compatible claims that (we may suppose) A and B both accept. In general, on Vermeulen's view, the status of an apparent disagreement as genuine or merely verbal depends on the speaker-meanings of the parties, which often diverge from the meanings of their sentences in context.

I have argued elsewhere (Balcerak Jackson 2014) for a different sort of pragmatic account, one that makes use of the discourse-theoretic notion of a *question under discussion* (Roberts 2012). On this view, what underlies a merely verbal dispute is a mismatch between the parties about which question they take to be under discussion, a mismatch that somehow arises because of the way the parties are using language. In the 'million exams' case, A intends to address the question of whether she has many exams to grade, while B intends to address the question of whether A has 1,000,000 exams to grade; they don't disagree in their answers to either of these questions.

According to either of these pragmatic accounts, mere verbalness is not just a matter of what the parties' words mean, but of the way the parties are using their words in the current exchange. Thus, they offer a more promising avenue than semantic and interpretational accounts for capturing the full range of merely verbal disputes.[9]

4 The Hirsch/Sider debate

One prominent recent strand in the metametaphysical literature is the debate between Sider and Hirsch about whether certain debates in metaphysics and ontology, such as the dispute between the mereological universalist and the anti-universalist from Section 1, are merely verbal. The foregoing discussion of mere verbalness helps shed light on that debate.

According to Hirsch, the dispute between the universalist and the anti-universalist is merely verbal, because the parties ought to be interpreted as speaking different languages, such that each should agree that what the other says is true in her own language (Hirsch 2005, 2009). In particular, on Hirsch's view, the difference in languages has to do with the meaning assigned to the existential quantifier phrase 'there is.' The universalist ought to interpret the anti-universalist as speaking the truth in his language when he says 'It is not the case that there is an object composed of my nose and your ear'; and the anti-universalist ought to interpret the universalist as speaking the truth in her language when she says 'There is an object composed of my nose and your ear.' According to Hirsch, these interpretations are the most faithful to the demands of charity.

In response, Sider argues that interpretation is constrained by more than just considerations of charity (Sider 2009, 2011). All else being equal, an interpretation that assigns more natural meanings, or meanings that do a better job of "carving reality at the joints" is to be preferred to one that assigns less natural or joint-carving meanings. Moreover, according to Sider, there is a most natural candidate meaning for 'there is.' Since this meaning at least approximately fits both the universalist's and the anti-universalist's uses of 'there is,' it is this meaning that both of them ought to be interpreted as having. Contra Hirsch, the universalist and the anti-universalist are *not* best interpreted as having different languages, and at most one of them can be speaking the truth in their language. The dispute is not merely verbal after all.

However, Sider's claim that the parties ought to be interpreted as speaking the same language – even if it is correct – does not suffice to establish that the dispute is genuine rather than merely verbal. As we saw in Section 3, an apparent dispute can be merely verbal even if the parties are best interpreted as speaking the same language; this is exactly the situation we observed in the 'million exams' case. Or consider the Bunsen/Bocuse dispute from Section 2. The biological category of fruit is quite plausibly more natural than the culinary category, so if Sider is correct then 'fruit' might pick out the biological category in both of their languages. But this is not enough to make their dispute genuine. What *Bocuse* means when he utters 'Tomatoes are not a fruit' is that they are not to be handled in such-and-such ways by chefs; this is compatible with what *Bunsen* means, which is that tomatoes are seed-bearing bodies of a certain sort. Someone who is convinced that the universalism dispute is merely verbal can insist that the same sort of thing is happening here: the parties are talking past each other, regardless of what their words mean or what the best interpretation of their language is.[10]

But the news is not all bad for Sider. The disagreement between Hirsch and Sider as presented so far concerns the question of how best to interpret the universalist and the anti-universalist. But even if Hirsch is correct that the parties ought to be interpreted as speaking different languages, Sider also has a fallback reply (Sider 2009, 2014). Sider argues that nothing prevents both sides from simply *deciding* to start speaking a language in which 'there is' picks out the most natural existential quantifier meaning. This would allow them to have a genuine, non-verbal disagreement about whether there are arbitrary mereological sums: do they exist in the sense picked out by the distinguished meaning of 'there is'? Hirsch (2008b) is skeptical that the parties to the dispute *can* simply decide to speak such a language; he doubts that merely stipulating that we are speaking a certain language can be counted on to secure the result that we really do so. But it should be clear from the discussion in Section 3 that this is not really essential for securing genuine disagreement. What matters is that the parties are somehow able to converge on a common question of whether arbitrary sums exist in the most natural or joint-carving sense of existence, and/or that they are able to make it clear that what *they* mean – regardless of what their words mean – are claims about the existence of arbitrary sums in the most natural or joint-carving sense. Assuming that such a concept of existence is at least coherent, it is hard to see why they should be unable to do so.[11]

5 Merely verbal deflationism

If an apparent dispute in some area of metaphysics turns out to be merely verbal, what does this imply about the possibility of substantive metaphysical theorizing in that area? In particular, should we conclude that the only sense in which either party can claim to be correct is in the sense that their way of using language corresponds more closely to ordinary English, as Hirsch would have it? Or should we conclude, with Sidelle, that the only worthwhile questions are not metaphysical ones, but rather questions about what is important or interesting? In previous work (Balcerak Jackson 2013, 2014) I argue that we should not.

To see why, imagine that A and B are discussing a new corporate regulation bill introduced by Senator Warren.

A: Have you read the Warren bill? I can't believe what a socialist policy that is!
B: No, the Warren bill is not a socialist policy.

Suppose that A and B are using the term 'socialist' differently, so that this is a merely verbal dispute (by whatever standard – pragmatic, semantic, or interpretationalist). If so, what A conveys is that the bill is socialist in a certain sense, call it socialist$_A$. What B conveys is that the bill is not socialist in a different sense, socialist$_B$. Ex hypothesi, A and B agree that the bill is socialist$_A$ and not socialist$_B$, so there is no real disagreement between A and B.

Does it follow that the only sense in which A can claim to be correct is in the sense that socialism$_A$ more closely matches the meaning of 'socialism' in ordinary English? Surely not. A can also claim to be correct (rightly or wrongly) in the sense that the bill really is a socialist$_A$ policy. Even though it turns out that B does not disagree with this, others might. And this might be an entirely genuine, non-verbal disagreement, whether about the features and effects of the bill, or about the nature of socialism$_A$, or both. Analogous remarks apply to B and socialism$_B$. Likewise, from the mere fact that the exchange between A and B is a merely verbal dispute, we cannot automatically conclude that the only worthwhile question that remains is whether socialism$_A$ or socialism$_B$ is more important or useful or interesting. The point is not that such questions are *not* worthwhile, but that we have no basis for saying that these are the *only* worthwhile questions.

The same goes for apparent disputes in metaphysics. Consider the following example from Balaguer (2017):

ETERNALIST: Dinosaurs exist.
PRESENTIST: No, dinosaurs do not exist.

Suppose that this is a merely verbal dispute arising from a difference in the way the parties use 'exist,' and that the presentist and the eternalist agree that dinosaurs exist$_E$, and that dinosaurs do not exist$_P$. Does it follow that the only remaining question is whether exist$_E$ or exist$_P$ is closer to the meaning of ordinary English 'exist'? Not at all. The eternalist's claim that dinosaurs exist$_E$ might be a fully substantive claim about which there can be genuine, non-verbal metaphysical disagreement; likewise for the presentist's claim that dinosaurs do not exist$_P$.[12] Nor does it follow that the only worthwhile questions that remain are about the relative importance or interest of existence$_E$ versus existence$_P$. The mere fact that A and B turn out to agree on the answer to a certain question tells us very little, on its own, about the nature of the question itself.

This point is gradually becoming more widely appreciated. For example, Sidelle (2016) acknowledges that merely showing that the parties to a metaphysical dispute are talking past each other is not enough to establish a genuine form of deflationism: "it needs to be the case that they *cannot be reformulated so as to constitute a substantive dispute*" (Sidelle 2016: 66; emphasis in original). And Balaguer (2017) distinguishes between "actual-literature verbalism," according to which some actual exchange between real metaphysicians happens to be a merely verbal dispute, and true "metametaphysical verbalism," according to which "there is no non-verbal (metaphysical) dispute to be had" about some metaphysical question.[13]

6 Triviality and semantic under-specification

Nevertheless, for many there remains a powerful temptation to move automatically from the premise that a given dispute in metaphysics is merely verbal to the conclusion that nothing is substantively at stake in it. Why is this?

I suspect that there are at least two main reasons. The first is a tendency to conflate mere verbalness with the idea that both parties' claims follow trivially (e.g. analytically, or as a matter of conceptual entailment) from neutral claims that both parties already accept. For example, I noted above that the mere verbalness of the eternalist/presentist dispute is compatible with the eternalist's claim that dinosaurs exist$_E$ being fully substantive. But one might wonder how this *could* be a substantive claim. Existence$_E$ is the eternalist's concept of existence – isn't it "built in" to that concept that if it was true in the past that there were dinosaurs then it follows that dinosaurs exist$_E$? If so, then anyone who accepts that it was true in the past that there were dinosaurs, and yet disagrees with the eternalist about whether dinosaurs exist$_E$, is making some kind of conceptual error. Trying to have substantive debate about whether dinosaurs exist$_E$ is like trying to have a debate about whether Kant was a bachelor; once we agree that he was an unmarried male, there's nothing left to discuss.

There is some evidence that this is how Hirsch thinks of the claims made by the parties to metaphysical disputes such as the eternalist/presentist one. Part of his argument for the mere verbalness of such disputes is that interpretations on which the parties genuinely disagree run afoul of a principle he calls *charity to understanding*:

> there must be the strong presumption that typical speakers of a language have a sufficiently adequate grasp of their linguistic and conceptual resources that they don't generally make a priori (conceptually) false assertions, especially when these assertions seem to be relatively simple, not involving any complicated calculations or computations.
>
> *(Hirsch 2005: 71–72)*

The suggestion, evidently, is that the eternalist would have to regard the presentist as making a simple conceptual error if she were to interpret the latter's utterance as meaning that dinosaurs do not exist$_E$, just as we would regard someone as making a conceptual error if he claims that Kant was not a bachelor after having agreed that he was an unmarried male.[14]

Perhaps it really is the case that the claim that dinosaurs exist$_E$ follows trivially from the fact that there have been dinosaurs. Nothing has been said here to challenge this.[15] Such a view is plausibly a form of deflationism – in fact, at least as applied to ontological disputes, it closely resembles Amie Thomasson's (2015) deflationary "easy ontology" view.[16] Of course, anti-deflationists about temporal ontology reject this picture. For them, the eternalist's ontology of dinosaurs is a consequence of a substantive metaphysical theory, not something that is built in to her concept of existence (whether or not this is the same as the presentist's concept). The important point, for present purposes, is that *this* disagreement is not resolved merely by arguing that the dispute about dinosaurs is merely verbal. The mere fact that A and B are talking past each other tells us nothing in one way or the other about whether the parties' claims follow trivially from agreed-upon facts. Just recall the Bunsen/Bocuse dispute, or the dispute about Ryan Gosling's talent as a dancer.

A second reason it is tempting to slide from the claim of mere verbalness to the claim that nothing substantive is at stake has to do with what we might call *semantic under-specification*. Returning to the debate about the Warren bill, one might worry that 'socialist' is such an imprecise or ill-defined term that there is simply no fact of the matter about whether the bill counts as 'socialist,' even after we pin down the precise features of the bill, its likely effects, and so on. This would give us an explanation of what is defective about the dispute: even after the parties reach agreement on all the relevant facts about the bill, the meaning of 'socialist' still leaves it open whether or not the term applies, and so there is simply no way for either party to show that she is speaking the truth. Applied to the eternalist/presentist dispute, the suggestion would be that 'there is' is semantically under-specified in such a way that even after we fix all the neutrally characterized temporal facts, there is no fact of the matter about whether 'There are dinosaurs' is true.[17]

Notice, however, that this is a different diagnosis of the dispute, one that is actually *in tension* with the diagnosis of mere verbalness. Rather than seeing the eternalist and the presentist as talking past each other, it sees them as both using 'there is' to convey claims that employ the *same* under-specified concept of existence.

One might instead reason as follows. If the meaning of 'there is' is under-specified, then there are multiple distinct ways that we could specify its meaning more fully. One such specification might have the result that 'There are dinosaurs' comes out true, while another might have the result that the sentence comes out false. Both meanings are compatible with the core (under-specified) meaning of the expression, and there is no substantive question about which of them is the "right" one.[18] This is plausibly a form of deflationism.[19] And one might even argue that the situation it describes is one that is ripe for merely verbal dispute: perhaps the eternalist employs a more fully specified meaning that makes 'There are dinosaurs' come out true, while the presentist employs a distinct one that makes 'There are no dinosaurs' come out true. This line of thought, or something like it, can be discerned in Sidelle (2007). As with the claims about triviality above, nothing has been said here to directly challenge this kind of deflationism. But it is important to see that mere verbalness, as such, is an afterthought; it is not doing any of the deflationary work.[20]

7 Conclusion

As a discourse phenomenon, merely verbal dispute is certainly worth investigating; we learn something about how communication ordinarily functions by examining ways it can malfunction. But as concerns its metametaphysical significance, the general thrust of the remarks here has been negative. If our apparent dispute over putatively metaphysical sentence M turns out to be merely verbal, then it is true that there isn't any point in you and I continuing to debate the truth value of M. But there may still be entirely substantive debates to be had. It doesn't follow that the truth value of M, as either you or I understand it, is an analytic or conceptual consequence of truths we both accept. Nor does it follow that M's meaning is under-specified in a way that makes it impossible to determine a truth value. Metametaphysicians who are drawn to – or worried about – these flavors of linguistic deflationism would be advised to approach them directly, and avoid the detour through mere verbalism.

Notes

1 The remaining philosophical issues might include normative debates over how we *ought* to use the terms in question, as advocates of "metalinguistic negotiation" have emphasized (see Belleri 2017; Plunkett and Sundell 2013; Thomasson 2017).
2 Vermeulen (2018: 333) helpfully describes merely verbal disputes as cases in which the parties seem to be engaged in an activity of disagreeing, although they are not in fact in a state of disagreement with each other.
3 The cases here are all ones in which an apparent dispute is merely verbal even though the participants assert contradictory propositions. There are also cases that illustrate that parties can genuinely disagree even when the propositions they assert are compatible (Balcerak Jackson 2019).
4 Due to limitations of space, I omit the account developed by Chalmers (2011) that does not easily fit into the tri-partite division given here. See Vermeulen (2018) for critical discussion.
5 This is not a substantial complaint against Balaguer or Bennett. As is clear from the quoted passage from Balaguer, his aim is not to advance an account that captures the precise contours of the category of merely verbal disputes. And Bennett is primarily interested in articulating an *epistemicist* view of ontological debates that contrasts with the view that they are verbal in the sense quoted.
6 Vermeulen (2018) bolsters this point with further ingenious cases in which merely verbal dispute arises because of scope ambiguity and other forms of structural ambiguity (such as 'I saw him with my telescope').
7 See also Hawthorne (2006).
8 For further criticism of Hirsch's interpretational account, and its application to metaphysical disputes, see Horden (2014).
9 A third pragmatic account is developed by C.S.I. Jenkins, who maintains that a verbal dispute is a case in which the parties have no (state) disagreement about the subject matter being discussed, but "present the appearance of doing so owing to their divergent uses of language" (Jenkins 2014: 11). All three accounts yield the same verdict on a wide range of cases, because there are very close connections between what speakers mean by their utterances in an exchange, the subject matter of their exchange, and the question(s) under discussion that they intend to address. However, I think there are cases where attending only to speaker meaning yields the wrong result, and that in these cases we need to look to the question under discussion – which, in my view, is what fixes the subject matter of the exchange (Balcerak Jackson 2019; Szabó 2017; Yablo 2014).
10 This point cuts both ways: it is also possible for a dispute to be genuine even though the parties' utterances express compatible propositions (see note 3). So, even if Hirsch is correct about how best to interpret the languages of the universalist and the anti-universalist, their dispute may not be merely verbal.
11 In fact, Parent (2006) argues, in effect, that Hirsch *must* allow that they can do so, even if we insist on conducting ontological debates in ordinary English. This is because Parent sees 'there is' and related vocabulary in ordinary English as having ontologically non-committing uses as well as ontologically committing ones; and it is up to speakers themselves to make sure that it is the commissive use that is in play when we ask 'What exists?' in ordinary English.
12 In reply to this point (first made in Balcerak Jackson 2013), Hirsch states that he only claims that certain *questions* debated by metaphysicians are non-substantive, not that any *statements* they advance are non-substantive (Hirsch 2013). But if A's statement that p is substantive (whatever exactly that amounts to) then surely there is a substantive question of whether or not it is the case that p.
13 Balaguer's metametaphysical verbalism corresponds closely to the notion of *unrevisability* that I define in Balcerak Jackson (2013: 424).
14 I should note, however, that Hirsch (2013) seems intent to deny that the parties' claims have this sort of status.
15 Of course, many philosophers are skeptical about the notions of analyticity and conceptual truth on which this form of deflationism seems to rely (see, for example, Williamson 2007). However, it may actually be possible to spell out a notion of analyticity *in terms* of mere verbalness: perhaps an analytic truth is, roughly, one about which any apparent disagreement must be merely verbal (see Balcerak Jackson 2014; Chalmers 2011).
16 See also the *trivialism* of Balaguer (2016), and the *easy ontology* entry in this volume.
17 Analogous suggestions have been made about various other disputes in metaphysics, such as disputes about personal identity (Sidelle 1999; Sider 2001), the metaphysics of material objects (Sidelle 2002), temporal ontology (Balaguer 2016), and the metaphysics of modality (Rosen 2006).

18 This is one way of making sense of Hirsch's doctrine of quantifier variance (Hirsch 2002, 2008a; see also the entry *quantifier variance* in this volume).
19 Although perhaps it is better seen as a form of *pluralism*, analogous to the logical pluralism of Beall and Restall (2005).
20 For more on the relationship between mere verbalness, triviality, and semantic under-specification, see Balcerak Jackson (2014).

References

Balaguer, M. (2016). Anti-metaphysicalism, necessity, and temporal ontology. *Philosophy and Phenomenological Research*, v. 92, pp. 145–167.
Balaguer, M. (2017). Why metaphysical debates are not merely verbal (or how to have a non-verbal metaphysical debate. *Synthese*.
Balcerak Jackson, B. (2013). Metaphysics, verbal disputes, and the limits of charity. *Philosophy and Phenomenological Research*, v. 86 (2), pp. 412–434.
Balcerak Jackson, B. (2014). Verbal disputes and substantiveness. *Erkenntnis*, v. 79 (S1), pp. 31–54.
Balcerak Jackson, B. (2019). Essentially practical questions. *Analytic Philosophy*, v. 60(1), pp. 1–26.
Beal, J. C. and Restall, G. (2005). *Logical Pluralism*. Oxford: Oxford University Press.
Belleri, D. (2017). Verbalism and metalinguistic negotiation in ontological disputes. *Philosophical Studies*, v. 174, pp. 2211–2226.
Bennett, K. (2009). Composition, colocation, and metaontology. In: D. Chalmers, D. Manley, and R. Wasserman, eds., *Metametaphysics: New Essays on the Foundations of Ontology*. Oxford: Oxford University Press.
Burge, T. (1979). Individualism and the mental. *Midwest Studies in Philosophy*, v. 4 (1), pp. 73–122.
Cappelen, H. and Hawthorne, J. (2009). *Relativism and Monadic Truth*. Oxford: Oxford University Press.
Chalmers, D. (2009). Ontological anti-realism. In: D. Chalmers, D. Manley, and R. Wasserman, eds., *Metametaphysics: New Essays on the Foundations of Ontology*. Oxford: Oxford University Press.
Chalmers, D. (2011). Verbal disputes. *Philosophical Review*, v. 120 (4), pp. 515–566.
Evans, G. (1973). The causal theory of names. *Proceedings of the Aristotelian Society*, v. 47 (1), pp. 187–208.
Hawthorne, J. (2006). Plenitude, convention, and ontology. In *Metaphysical Essays*. Oxford: Oxford University Press.
Hirsch, E. (2002). Quantifier variance and realism. *Philosophical Issues*, v. 12, pp. 51–73.
Hirsch, E. (2005). Physical-object ontology, verbal disputes, and common sense. *Philosophy and Phenomenological Research*, v. 70 (1), pp. 67–97.
Hirsch, E. (2008a). Ontological arguments: interpretive charity and quantifier variance. In: T. Sider, J. Hawthorne, and D. Zimmerman (eds.), *Contemporary Debates in Metaphysics*. Malden: Blackwell Publishing.
Hirsch, E. (2008b). Language, ontology, and structure. *Noûs*, v. 42, pp. 509–528.
Hirsch, E. (2009). Ontology and alternative languages. In: D. Chalmers, D. Manley, and R. Wasserman, eds., *Metametaphysics: New Essays on the Foundations of Ontology*. Oxford: Oxford University Press, pp. 231–259.
Hirsch. E. (2013). Charity to charity. *Philosophy and Phenomenological Research*, v. 86 (2), pp. 435–442.
Horden, J. (2014). Ontology in plain English. *Philosophical Quarterly*, v. 64 (255), pp. 225–242.
Jenkins, C. S. I. (2014). Merely verbal disputes. *Erkenntnis*, v. 79 (S1), pp. 11–30.
Kölbel, M. (2004). Faultless disagreement. *Proceedings of the Aristotelian Society*, v. 104 (1), pp. 53–73.
Kripke, S. (1980). *Naming and Necessity*. Cambridge: Harvard University Press.
MacFarlane, J. (2007). Relativism and disagreement. *Philosophical Studies*, v. 132 (1), pp. 17–31.
Parent, T. (2006). Ontic terms and metaontology: or, what there actually is. *Philosophical Studies*, v. 170 (2), pp. 199–214.
Plunkett, D. and Sundell, T. (2013). Disagreement and the semantics of normative and evaluative terms. *Philosopher's Imprint*, v. 13 (23), pp. 1–37.
Putnam, H. (1975). The meaning of 'meaning.' *Minnesota Studies in the Philosophy of Science*, v. 7, pp. 131–193.
Roberts, Craige. (2012). Information structure in discourse: towards an integrated formal theory of pragmatics. *Semantics and Pragmatics*, v. 5, pp. 1–69.
Rosen, G. (2006). The limits of contingency. In: F. MacBride, ed., *Identity and Modality*. Oxford: Oxford University Press, pp. 13–39.

Sidelle, A. (1999). On the prospects for a theory of personal identity. *Philosophical Topics*, v. 25 (1/2), pp. 351–372.

Sidelle, A. (2002). Is there a true metaphysics of material objects? *Philosophical Issues: Realism and Relativism*, v. 12, pp. 118–145.

Sidelle, A. (2007). The method of verbal dispute. *Philosophical Topics*, v. 35 (1/2), pp. 83–113.

Sidelle, A. (2016). Frameworks and deflation in "Empiricism, Semantics, and Ontology" and recent metametaphysics. In: S. Blatti and S. Lapointe, eds., *Ontology after Carnap*. Oxford: Oxford University Press, pp. 59–80.

Sider, T. (2001). Criteria of personal identity and the limits of conceptual analysis. *Philosophical Perspectives Volume 15: Metaphysics*, pp. 189–209.

Sider, T. (2009). Ontological realism. In: D. Chalmers, D. Manley, and R. Wasserman, eds., *Metametaphysics: New Essays on the Foundations of Ontology*. Oxford: Oxford University Press.

Sider, T. (2011). *Writing the Book of the World*. Oxford: Oxford University Press.

Sider, T. (2014). Hirsch's attack on ontologese. *Noûs*, v. 48 (3), pp. 565–572.

Szabó, Z. G. (2017). Finding the question. *Philosophical Studies*, v. 174 (3), pp. 779–786.

Thomasson, A. L. (2015). *Ontology Made Easy*. Oxford: Oxford University Press.

Thomasson, A. L. (2017). Metaphysical disputes and metalinguistic negotiation. *Analytic Philosophy*, v. 58 (1), pp. 1–28.

Van Inwagen, P. (1995). *Material Beings*. Ithaca: Cornell University Press.

Vermeulen, I. (2018). Verbal disputes and the varieties of verbalness. *Erkenntnis*, v. 83, pp. 331–348.

Williamson, T. (2007). *The Philosophy of Philosophy*. Malden: Blackwell Wiley Publishing.

Yablo, S. (2014). *Aboutness*. Princeton: Princeton University Press.

10
ABSOLUTE GENERALITY

Agustín Rayo

Absolutism, as I will understand it here, is the thesis that there is sense to be made of absolutely general quantification: quantification over absolutely everything there is. The aim of this chapter is to clarify the metaphysical debate between absolutists and some of their detractors.[1]

1 The role of absolutism

Absolutism plays an important role in contemporary philosophy. Here are some examples:

1.1 Ontology

Certain ontological claims are naturally seen as presupposing Absolutism (Williamson, 2003). Take Mathematical Nominalism: the claim that there are no mathematical objects. If Mathematical Nominalism is to have its intended meaning, it can't be understood as the claim that there are no mathematical objects within the range of a restricted quantifier. A natural way of ruling out this unintended interpretation is to take Mathematical Nominalism to presuppose Absolutism and read it as the claim that absolutely everything is non-mathematical.

1.2 Meta-metaphysics

Absolutism can be used to guard against the complaint that a metaphysical dispute is purely verbal.

Consider, for example, the dispute between a mereological nihilist, who thinks that everything is a mereological atom, and a mereological universalist, who thinks that any individuals have a fusion (van Inwagen, 1990). Now imagine a critic who takes the debate to be purely verbal. She thinks that the nihilist and universalist accept the same propositions but that they use different sentences to express those propositions because they use their quantifiers in different ways.

For instance, when the nihilist asserts "$\sim\exists x\,\text{Table}(x)$" she expresses the same proposition that the universalist expresses when she asserts "$\sim\exists x(\text{Atom}(x) \wedge \text{Table}(x))$". And when the universalist asserts "$\exists x\,\text{Table}(x)$" she expresses the same proposition that the nihilist expresses when she asserts "∃xxTableishly(xx)" (read: "some things are arranged table-ishly"). Notice, however, that the critic's proposal cannot get off the ground when the nihilist and the universalist are interpreted as absolutist positions, whose quantifiers range over absolutely everything.

1.3 Philosophy of logic

Absolutism is an important tool in the debate between singularists and pluralists in the philosophy of logic.[2]

The pluralist believes that plural terms like "them" or "they" might refer to several objects at once. Consider, for example, "There are four workers and they carried the piano together." The pluralist thinks that "they" in this sentence should be taken to refer, collectively, to the four workers. The singularist disagrees. She thinks that "they" must be taken to refer to a single object, though perhaps a single object of a special sort (e.g. the set that has each of the four workers as members). Pluralists and singularists have a corresponding disagreement about plural quantification. The pluralist thinks that a plural quantifier like "some things are such that ..." ranges over ordinary individuals in a special way (i.e. plurally). The singularist thinks that the only way to make sense of plural quantifiers is to see them as ranging over special objects (e.g. sets) in the ordinary way (i.e. singularly).

Absolutism can be used to generate a powerful family of arguments against singularism. Consider, for example, the sentence:

1   There are some things that include all and only the individuals that are not members of themselves.

It is natural to think that (1) is true. But when the plural quantifier is read as ranging singularly over, e.g. sets, it is equivalent to:

2   There is a set that includes all and only the individuals that are not members of themselves.

which can be used to prove a contradiction by asking whether the relevant set is a member of itself. The singularist might point out that the contradiction would be avoided if the relevant set is outside the range of "all" in (2). But the pluralist can deploy Absolutism to block this move by stipulating that "all" is to be read as ranging over absolutely everything.

1.4 Philosophy of mathematics

Vann McGee (1997) has shown that the absolutist is in a position to give a categorical axiomatization of set theory (i.e. an axiomatization that can only be satisfied by different mathematical structures when the structures are isomorphic to one another). McGee's starting point is an axiom system that Ernst Zermelo showed to be *quasi-categorical* (i.e. it can only be satisfied by different mathematical structures *of the same size* when the structures are isomorphic to one another). McGee then enriches Zermelo's system with a further axiom, whose quantifiers are to be understood as absolutely general:

Urelement Set Axiom For any non-sets, there is a set that has exactly those individuals as members.

With the new axiom in place, McGee is able to prove that there must be just as many sets in the absolutely general domain as there are individuals, and therefore that any two structures satisfying the extended axiom system must be of the same size. So Zermelo's quasi-categoricity result becomes a categoricity result.

We have just considered one respect in which Absolutism is a boon to the philosophy of mathematics. But it is worth noting that Absolutism can also lead to trouble. For instance, Absolutism entails that there is good sense to be made of the question of how many individuals there are. But if there is an answer to this question, there must be consistent mathematical theories that are, in fact, false. For however many individuals are in the absolutely general domain, there will be consistent mathematical theories that can only be true if there are more.[3] This generates a tension between Absolutism and standard mathematical practice, since mathematicians are, by and large, happy to work with any consistent theory they find interesting.[4]

2 Anti-absolutism

Why reject Absolutism? According to Timothy Williamson, "The most serious concern about [unrestricted uses of quantifiers] is their close association with the paradoxes of set theory ... However, one cannot generate [the relevant] contradiction just by using unrestricted quantifiers. The contradiction always depends on auxiliary assumptions ... [and we] can reject those assumptions without rejecting unrestricted quantification" (Williamson, 2013, p. 15).

I would like to suggest that this underplays the case against Absolutism. Although it is certainly true that the set-theoretic paradoxes have led some philosophers to question Absolutism,[5] I think the most important challenge to Absolutism is a particular conception of the relationship between our language and the world it represents. I shall refer to it here as *Recarving Anti-Absolutism*.

Recarving Anti-Absolutism is something of a household view. It has been a visible presence in analytic philosophy since Frege, and has been defended, in different forms, by many influential philosophers since.[6] And yet it has often been ignored by absolutists.

I think it is hard to appreciate why Absolutism is such a substantial claim, and why it might be regarded as controversial, without having an alternative along the lines of Recarving Anti-Absolutism firmly in view. For this reason, and because other aspects of the Absolutism debate have been adequately surveyed elsewhere,[7] I will devote much of the remainder of this chapter to articulating Recarving Anti-Absolutism and clarifying its relationship to Absolutism.

3 Recarving Anti-Absolutism

In this section, I will describe an especially straightforward version of the view:

Recarving Anti-Absolutism
A   To interpret a language is to assign a (coarse-grained) proposition to each sentence.[8] Any such assignment counts as an admissible interpretation of the language, as long as it respects logical entailments (and thereby respects the semantic structure that gives rise to those entailments).[9]
B   There is a definite fact of the matter about how the world is. (Recarving Anti-Absolutism is therefore a form of Realism, in at least a limited sense.)
C   The way the world is determines whether each sentence of an interpreted language is true by determining whether the (coarse-grained) proposition assigned to that sentence is true.
D   For a singular term of the language I am now speaking to refer is for it to occur in an atomic sentence that expresses a (coarse-grained) proposition other than the absurd proposition $\bot$.[10]

The best way to get clear about Recarving Anti-Absolutism is to contrast it with an alternative picture:

The metaphysical conception of language

A There is a metaphysically privileged articulation of the world into constituents. Some of these constituents are *objects* and make up the world's metaphysically privileged *ontology*; others are *properties* (or, more generally, relations) and make up the world's metaphysically privileged *ideology*.

B A *metaphysical assignment of reference* for a given language is a pairing of each singular term in the language with an object in the world's metaphysically privileged ontology, and each *n*-place atomic predicate in the language with an *n*-place relation in the world's ideology.

C Let *I* be a function that assigns a proposition to each sentence of the language. In order for *I* to count as an admissible interpretation of the language, there must be some metaphysical assignment of reference ϱ such that, for each atomic sentence $\ulcorner P(t_1,\ldots,t_n)\urcorner$ of L, $I(\ulcorner P(t_1,\ldots,t_n)\urcorner)$ is the set of worlds w such that the individuals $\varrho(t_1),\ldots,\varrho(t_n)$ are each part of w's metaphysically privileged ontology and are related by $\varrho(P)$ at w.

As an example, consider the language of first-order arithmetic, interpreted by the function A:

- A assigns the trivial proposition, $\top$ (i.e. the set of all worlds), to each arithmetical sentence that is standardly taken to be true;
- A assigns the absurd proposition, $\bot$ (i.e. the empty set of worlds), to each arithmetical sentence that is standardly taken to be false.

An immediate consequence of the Metaphysical Conception is that A only counts as an admissible interpretation of the language if the world has the right metaphysically privileged constituents. Suppose, for example, that the world's metaphysically privileged ontology consists entirely of contingently existing individuals. Then "0 = 0" cannot be assigned to the necessary proposition, $\top$, since any individual that might be selected as the referent of "0" will fail to be part of some world's metaphysically privileged ontology.

According to Recarving Anti-Absolutism, in contrast, all it takes for an assignment of propositions to sentences to count as an admissible interpretation of the language is for the assignment to preserve logical entailments. And A certainly satisfies this condition. Notice, moreover, that A can be specified without having to worry about whether the world's metaphysically privileged ontology contains numbers. It can be specified syntactically, for instance:[11]

$$A(\ulcorner t=s\urcorner) = \begin{cases} \top, \text{ if } \ulcorner t=s\urcorner \text{ can be syntactically reduced to } "0 = 0" \\ \bot, \text{ otherwise} \end{cases}$$

$$A(\ulcorner \sim\phi\urcorner) = \top - A(\phi)$$

$$A(\ulcorner \phi\wedge\psi\urcorner) = A(\phi) \cap A(\psi)$$

$$A(\ulcorner \exists x\, \phi\urcorner) = \begin{cases} \top, \text{ if } A(\phi\,[t/x]) = \top \text{ for some closed term } t \\ \bot, \text{ otherwise} \end{cases}$$

Notice, finally, that Recarving Anti-Absolutism entails that one can establish the existence of numbers on the basis of purely linguistic considerations. For suppose that first-order arithmetic is a sub-language of the language we speak. We can then use the following argument to show that the number 0 exists:

Since $A(\text{``}0=0\text{''}) = \top$, "$0=0$" is an atomic sentence of our language that expresses a non-absurd proposition. So, by Recarving Anti-Absolutism, "0" refers. But if "0" refers, it refers to 0. So 0 exists.

Analogous arguments can be used to establish the existence of any other natural number, on the basis of linguistic considerations. How can we make sense of this?

It is useful to start with a slightly different example. Let p be the (coarse-grained) proposition that Socrates died and consider an interpretation on which "Socrates died" and "Socrates's death occurred" are both assigned p. The Recarver has no reason to doubt the admissibility of this interpretation.[12] So following Frege (1980a, §64), she might claim that p can be "carved up" in different ways. When we assert "Socrates died" we carve p up into the object Socrates and the property of dying; when we assert "Socrates's death occurred" we carve it up into the event of Socrates's death and the property of occurring.

The case of first-order arithmetic is similar. Since, on interpretation A, every true sentence of first-order arithmetic expresses the necessary proposition $\top$, the Recarver will think that each such sentence corresponds to a carving of $\top$. When we assert "$0=0$", for example, we carve $\top$ into 0 and the identity relation; when we assert "$\sim(0=1)$" we carve it into 0, 1, the identity relation and the negation operation.

It is worth emphasizing that although Recarving Anti-Absolutism entails that one shouldn't be too concerned about ontological issues, it doesn't entail that anything goes. The Recarver believes that one could introduce an object *called* "Vulcan", and do so cost-free. (For instance, one could expand one's language to talk about modulo-one arithmetic and use "Vulcan" to refer to the relevant mathematical object.) But Recarving Anti-Absolutism does *not* entail that one could use "Vulcan" to refer to the planet Vulcan. For that would require an assignment of truth-conditions according to which "$\exists x(x=\text{Vulcan})$" is only true at the actual world if a particular astronomical condition obtains. Which condition? I can use my own language to identify the relevant coarse-grained proposition by uttering "there is a planet orbiting the Sun within Mercury's orbit". A speaker of a different language might identify the same coarse-grained proposition differently (perhaps by uttering a version of "there are planitizing simples in orbit around the sunish simples and within the orbit of the mercurish simples). Regardless of the linguistic resources one uses to identify it, the condition itself will fail to be satisfied at the actual world. So "$\exists x(x=\text{Vulcan})$" won't be true. And, for reasons discussed in Section 3.3, this ensures that "Vulcan" won't refer.

3.1 Why accept Recarving Anti-Absolutism?

In ordinary speech, we take a decidedly carefree attitude towards expansions of our domain of discourse. We do not hesitate to go from an event-free statement like "they started arguing" to an event-loaded counterpart, like "an argument broke out". We routinely expand our domain to keep track of our social practices: we talk about commitments, memes, internet accounts, votes and cabinet positions. We routinely expand our domain to streamline our descriptions of the natural world – as when we talk about orbits or tides – or to abstract away from irrelevant differences between objects – as when we talk about particle-types or letter-types. We do not hesitate to introduce talk of mathematical objects when it is expedient to do so. And so forth.

Recarving Anti-Absolutism captures this important phenomenon, since it entails that the project of expanding one's domain of discourse is not hostage to a certain kind of metaphysical fortune. The Recarver agrees that "an argument broke out" won't be true unless a certain worldly condition obtains – unless someone started arguing – but she insists there is no need to worry about whether a further ontological condition is satisfied: the condition that the world's metaphysically privileged ontology include *arguments* alongside people who argue.

The Metaphysical Conception of Language, in contrast, entails that "an argument broke out" cannot be true unless this additional condition obtains. Whatever the merits of such a position form a metaphysical point of view, it strikes me as bizarre as an account of our linguistic practice. An ordinary speaker might choose to assert "an argument broke out" rather than "a few people started to argue" on the basis stylistic considerations, or in order to achieve the right emphasis, or in order to create the right context for a future assertion. But it would be preposterous to suggest that her usage of such sentences, in marketplace contexts, turns on her views about whether arguments figure amongst the world's ultimate furniture.

I certainly do not mean to suggest that this is enough to settle the debate in favour of the Recarver. But I hope it suffices to highlight some of the reasons philosophers have found Recarving Anti-Absolutism attractive.[13]

3.2 A language-relative conception of objecthood

The Recarver thinks that we can carve a fact into objects by using singular-term-containing sentences to describe it. But she thinks there is no sense to be made of a *language-transcendent domain*, such as the one presupposed by the Metaphysical Conception of Language. In other words, she thinks there is no sense to be made of an articulation of the world into constituents that is significant from a purely metaphysical point of view and therefore significant independently of how the world happens to be represented.

As a result, the Recarver espouses a *language-relative* conception of object. In slogan form: an object (relative to the language I am now speaking) is the referent of one of my singular terms, or the value of one of my singular variables.[14] And what are the values of my variables? I can accurately answer this question by *using* my variables. I can assert "Any x is such that x is the value of a variable." I can go on to summarize the language-relative conception of object, as it applies to my language, by saying "Any x is such that x is an object" – or, indeed, "*absolutely everything* is an object". I can then give an account of the quantifiers of my language by saying "unrestricted quantifiers range over a domain consisting of absolutely everything; restricted quantifiers range over a subdomain of the absolutely general domain". In saying this, however, I do not mean that every individual in a language-transcendent domain is an object, or that my quantifiers range over the individuals in such a domain. There is, after all, no sense to be made of a language-transcendent domain. All I mean is that every individual into which the world gets carved up by my language is an object.

Consider an example. In "Empiricism, Semantics and Ontology", Carnap considers the "thing language", which is concerned with "the spatio-temporally ordered system of observable things and events". The thing language carves the world into objects with spatiotemporal locations. So a speaker of the thing language can use "everything has a spatiotemporal location" – or, indeed, "absolutely everything has a spatiotemporal location" – to express a true proposition. But now suppose she extends her language with arithmetical vocabulary. The enriched language delivers a carving of the world that yields both numbers and spatiotemporal objects. And since the new language subsumes the old, our speaker can now use "not everything has a spatiotemporal location" to express a true proposition.

One could, if one wanted, describe the situation by saying that the subject has "extended" the range of her quantifiers by moving to a richer language. But it is important to be clear that such a claim would have to be made from the point of view of a language that subsumes both the thing language and the language of arithmetic. The claim can be made from the point of view of the language that the subject has moved to, or from the point of view of our own language. But there is no sense to be made of the "thing ontology" from the perspective of a language that does not subsume the thing language. And, crucially, there is no such thing as a "neutral" point of view, from which one can compare the ontologies delivered by these different languages. Since the only intelligible conception of object is language-relative, ontological questions can only be assessed from within some language or other.

A critic might use this point as an objection against Recarving Anti-Absolutism, by claiming that it underscores the Recarver's expressive limitations. She might complain that the Recarver is unable to state a suitably general version of her own view. This is an important point, which I will address in Section 3.5.

3.3 Empty names

Recarving Anti-Absolutism allows for empty names. But it imposes the constraint that any atomic sentence containing an empty name must express the absurd (coarse-grained) proposition, $\bot$.

The Recarver's view of singular terms is based on two theses, both of them sensible. The first is that an atomic sentence $\ulcorner P(a_1,\ldots,a_n) \urcorner$ can only be true when its singular terms refer,[15] and, more generally, that the following is a logical truth:

$$P(a_1,\ldots,a_n) \rightarrow \exists x_1 \ldots \exists x_n (x_1 = a_1 \wedge \ldots \wedge x_n = a_n)$$

The second is the Kripkean thesis that "granted that there is no Sherlock Holmes, one cannot say of any possible person that he would have been Sherlock Holmes, had he existed", (Kripke, 1980, 158) and, more generally, that $\ulcorner \Diamond \exists x(x=n) \urcorner$ fails to be true whenever n is an empty name.

When the two theses are combined, they deliver the following:[16]

Name Principle: $\Diamond P(a_1,\ldots,a_n) \rightarrow \exists x_1 \ldots \exists x_n (x_1 = a_1 \wedge \ldots \wedge x_n = a_n)$

The Name Principle ensures that a singular term of my language will refer whenever it is part of an atomic sentence that expresses a coarse-grained proposition other than $\bot$. To get from here to the Recarver's view of reference, all we need is the claim that satisfying such a condition is *constitutive* of reference: *what it is* for a singular term of my language to refer is for it to be part of an atomic sentence that expresses a coarse-grained proposition other than $\bot$.

3.4 Question-begging

A critic might complain that by adopting a language-relative conception of object, the Recarver is simply changing the subject. The critic might claim, in particular, that the Recarver isn't *really* talking about objects in the sense that is relevant to metaphysicians. Here is an argument to that effect:

> The Recarver's conception of reference is a poor substitute for genuine reference: just because a singular term has been shown to refer by the Recarver's lights, there is no reason to think that it *really* refers. As a result, her conception of object, which is tied up with her conception of reference, is a poor tool for genuine ontological inquiry: just because one speaks a language that contains the syntactic string "0 = 0" and is interpreted in such a way that the string is paired with a consistent proposition, there is no reason to think that the world's ontology really contains numbers.

Our critic would be begging the question. According to Recarving Anti-Absolutism, what it is for a singular term of one's language to *really* refer is for it to figure in an atomic sentence that expresses a non-absurd proposition. And all it takes for me to *really* speak truly when I say that 0 is an object is for my language to be interpreted in such a way that "0 = 0" expresses a consistent proposition.

It is useful to consider an analogy. Suppose that there is no sense to be made of the claim that ascots are "objectively" fashionable, as opposed to fashionable by the lights of some community or other. Then the notion of objective fashionability is irreparably misguided. The only notion of fashionability that deserves a place in our cognitive lives is the community-relative notion: to be fashionable *just is* to be fashionable relative to the relevant community. The Recarver sees a language-transcendent conception of object as similarly misguided. So she thinks that the only notion of object that deserves a place in our cognitive lives is the language-relative notion: to be an object *just is* to be an object relative to the relevant language. From the point of view of the Recarver, the only way of *really* talking about objects is to talk about objects in the language-relative sense.

The critic might try a different line of attack. She might claim that the Recarver doesn't really avoid commitment to a language-transcendent conception of object. For she is committed to the conception of object that results from considering the referents of singular terms of *every possible language*.

Once more, our critic would be begging the question. For just like the anti-absolutist thinks that there is no sense to be made of a language-transcendent notion of object, she thinks there is no sense to be made of a language-transcendent notion of reference. So even if she were to grant that there is sense to be made of "all possible languages",[17] she would think there is no neutral perspective from which one can make sense of the combined ontology that all possible languages would deliver. Let me explain.

Suppose that Recarving Anti-Absolutism is true and suppose that I want to make sense of the claim that the singular term t refers. If t is a term in my language, there is no problem. For if t does indeed refer, some atomic sentence $\ulcorner P(t, a_1, \ldots, a_n) \urcorner$ expresses a non-absurd (coarse-grained) proposition. So, by the Name Principle of the preceding section, $\ulcorner \exists x(x = t) \urcorner$ is true. So I can conclude that t refers to something, and go on to assert $\ulcorner \text{'}t\text{'} \text{ refers to } t \urcorner$.

Now suppose that t is not a term in my language. If t refers, it must refer to something – and therefore to some object. But the only conception of object I am able to understand is the one that is linked to the terms and variables of my own language. In other words: I can only make sense of an object insofar as it is one of the constituents into which my own language carves up the world.[18] So there is no guarantee that I will be able to find a suitable referent for t, even if the user of some other language would be justified in treating it as referential. (Notice, in contrast, that from the point of view of the Metaphysical Conception of Language, I can make good sense of the claim that t refers. For as long as I am able to make sense of a metaphysically privileged articulation of the world into constituents, I can understand the claim that t refers as the claim that it is suitably paired with a member of the world's metaphysically privileged ontology.)

3.5 Anti-Absolutism beyond one's language

The Recarver's views about reference are restricted to "the language I am now speaking". Recall Part D of the definition of Recarving Anti-Absolutism in Section 2:

> **Reference Thesis** For a singular term *of the language I am now speaking* to refer is for it to occur in an atomic sentence that expresses a non-absurd (coarse-grained) proposition.

The importance of this restriction is brought out by a point I made earlier. Let *t* be an alien singular term. In order to make sense of the claim that *t* refers, I must be able to make sense of the claim that it refers to something – and therefore to some object. But there is no guarantee that I will be able to find a suitable object amongst the constituents into which my own language carves up the world. So there is no guarantee that I will be able to make sense of the claim that *t* refers.

This raises the question of whether the Recarver is in a position to state her view in a way that is general enough to deserve being described as an account of the relationship between language and the world it represents, as opposed to simply an account of one particular language.

The answer, I think, is that she can get part-way there. The Recarver is committed to thinking that one cannot make sense of reference for arbitrary languages. But she takes herself to have a recipe for explaining what reference consists in, and she thinks that this recipe can be used by speakers of alien languages to make sense of a language-relative conception of language centred on their own language. (In the case of our language, the recipe yields the Reference Thesis; in the case of an alien language, it yields an analogue of the Reference Thesis for that language.)

Consider an analogy. I am not able to make sense of a way for the world to be with true contradictions. I nonetheless have a good enough understanding of paraconsistent logic to describe a formalism that could be used by a dialetheist to make sense of a scenario in which, e.g. the Liar Sentence is both true and false. So although I cannot make it all the way to understanding a true contradiction, I can get part-way there. For I am in a position to articulate a recipe that could be used by dialetheists to make sense of a situation in which the Liar is both true and false. The Recarver takes herself to be in an analogous position with respect to reference. Even if she can't make sense of a notion of reference for a sufficiently alien language, she is able to articulate a recipe that could be used by speakers of that language to make sense of the relevant notion.[19]

4 Strong Absolutism

At the beginning of the chapter, I characterized Absolutism as the thesis that there is sense to be made of absolutely general quantification: quantification over absolutely everything there is.

Thus stated, Absolutism is not as restrictive a thesis as it first appears. Notice, in particular, that the Recarver has no objection to quantification that is absolute, in the sense of being unrestricted. Recall, for example, our speaker of Carnap's thing language. When she uses "absolutely everything has a spatiotemporal location" to express a true proposition, she is quantifying over absolutely every object without restriction, in the only sense of object that is well-defined from her point of view. So quantifier restriction is not really what's at issue.[20]

At the same time, it is clear that Recarving Anti-Absolutism is in tension with Absolutism, as Absolutism tends to be used in the literature. One way to see this is to think back to the examples I gave in Section 1 to illustrate the role of absolute generality in contemporary philosophy. Consider, for instance, an absolutist statement of Mathematical Nominalism, according to which absolutely everything is non-mathematical. It would be perverse to suggest that the speaker of Carnap's "thing language" is a nominalist in the intended sense, even though she accepts "absolutely everything has a spatiotemporal location". For if she later enriched her language with arithmetical vocabulary and accepted "there are numbers" she would not take herself to have abandoned a metaphysical position she previously held. She will merely take herself to have adopted a different representational system with which to describe the world.

I would like to suggest that what is really at issue between Recarving Anti-Absolutism, on the one hand, and the forms of Absolutism that tend to be defended in the literature, on the other, is not the question of whether quantification can be absolutely unrestricted, but the question of whether there is sense to be made of a language-transcendent domain: an articulation of the world into constituents that is significant from a purely metaphysical point of view, and therefore significant independently of how the world happens to be represented. Accordingly, I suspect that contemporary metaphysicians who think of themselves as absolutists are often presupposing the following:

> **Strong Absolutism** There is sense to be made of a language-transcendent domain, which consists of all objects. For a quantifier to be absolutely general is for it to range over every entity in this domain.

Strong Absolutism creates a clear dividing line. Views like Recarving Anti-Absolutism, which deny the intelligibility of a language-transcendent domain, are on one side of the line; views that support applications of absolutely general quantification of the sort we discussed in Section 1 are on the other side of the line.

5 Quantifying over everything

Neither Absolutism nor Strong Absolutism entails that absolute generality is, in fact, attainable. Absolutism is the claim that there is sense to be made of absolutely general quantification and Strong Absolutism is the claim that there is sense to be made of absolutely general quantification over a language-transcendent domain. Each of these views allows for the concept of absolutely general quantification to be intelligible but unrealized in practice.

On the other hand, it is unlikely that anyone would want to accept Absolutism, on either of its forms, while denying that absolutely generality is, in fact, attainable. Notice, in particular, that it is not clear that the view in question could be endorsed without creating an instability. For it is hard to see how one might explain what one means by absolute generality without purporting to quantify absolutely generally. (For instance, the statement of Absolutism we have been using here glosses absolutely general quantification as "quantification over absolutely everything there is".) Notice, in contrast, that there is nothing unstable about *denying* Absolutism, on either of its forms: the claim is simply that there is no sense to be made of expressions like "absolutely everything", as they are intended by one's opponent. So, *pace* Williamson (2003, §V), I do not think that anti-absolutists face a special difficulty in stating their view.

When anti-absolutist views of the kind we have considered in this chapter are ignored, discussions of Absolutism are open to a certain kind of pitfall. They run the risk of taking it for granted that there is sense to be made of absolute generality and proceed on the assumption that the only open question is whether absolute generality is, in fact, attainable in practice. David Lewis (1991), for example, wonders whether a foe of absolutism would need to resort to the idea that "some mystical censor stops us from quantifying over absolutely everything without restriction" (p. 68). This suggests that Lewis is thinking of the anti-absolutist as someone who agrees that there is sense to be made of an absolutely general domain and is worried only about whether our quantifiers can be stretched wide enough to encompass all of the domain at once. I think a more interesting target is an anti-absolutist who denies that there is sense to be made of an absolutely general domain.[21]

Notes

1. There is a large literature on Absolutism, which has been surveyed elsewhere. (See Florio, 2014 and the introduction to Rayo and Uzquiano, 2006.) I will not duplicate such efforts here. I will also not engage with the debate within linguistics on how quantifier domain restriction is best understood. (See, for instance, Stanley and Szabó, 2000.)
2. The seminal text is Boolos, 1984. For more recent discussion, see Oliver and Smiley, 2013; Florio and Linnebo, typescript.
3. Here is a simple example. Let the individuals aa consist of every individual that in fact exists. Then the following higher-order theory is consistent, but it is also unsatisfiable (for Cantorian reasons):

    $\sim\exists R(\forall \alpha \exists a \prec aa(R\alpha a) \wedge \forall \alpha \forall \beta \forall a(R\alpha a \wedge R\beta a \rightarrow \alpha = \beta))$.

4. Here is David Hilbert: "if the arbitrarily given axioms do not contradict each other with all their consequences, then they are true and the things defined by them exist. This is for me the criterion of truth and existence" (Letter to Frege, December 29, 1899; in Frege, 1980b). And here is Georg Cantor:

    > Mathematics is in its development entirely free and only bound in the self-evident respect that its concepts must both be consistent with each other and also stand in exact relationships, ordered by definitions, to those concepts which have previously been introduced and are already at hand and established.
    >
    > (Cantor, 1883)

    Both quotations are drawn from Linnebo, 2018, §1.3.
5. Michael Dummett, in particular. See, for instance, Dummett, 1963 and Dummett, 1991, 316–319. See also Parsons, 1974, 1977; Glanzberg, 2004.
6. Here is a partial list: Carnap, 1950; Dummett, 1981; Wright, 1983; Rosen, 1993; Stalnaker, 1996; Sidelle, 2002; Glanzberg, 2004; Burgess, 2005; Hellman, 2006; Parsons, 2006; Chalmers, 2009; Eklund, 2009; Hirsch, 2010; Thomasson, 2015; Linnebo, 2018.
7. See, for instance, Williamson, 2003 and Florio, 2014.
8. A coarse-grained proposition is a way for the world to be. Here it will be modelled as a set of metaphysically possible worlds.
9. In particular: if ψ is a logical consequence of φ, then the proposition expressed by φ must entail the proposition expressed by ψ. For a more detailed specification of the needed condition, see Rayo, 2013, §1.3.
10. The absurd proposition is just the empty set of worlds. Why insist on a non-absurd proposition? See Section 3.3. Why restrict attention to the singular terms of one's own language? See Section 3.5.
11. I assume that t and s are closed terms, built from "0", the successor function " ′ ", the addition function "+" and the multiplication function "×". A syntactic reduction from $\ulcorner t = s \urcorner$ to "0 = 0" is a two-step transformation. The first step is to eliminate every occurrence of "+" and "×" from t and s by repeated application of the following transformations:

    $n + m' \rightarrow (n + m)'$ $n + 0 \rightarrow n$
    $n \times m' \rightarrow (n \times m) + n$ $n \times 0 \rightarrow 0$

    The resulting terms, $t^\star$ and $s^\star$, are each of the form "$0'^{\cdots'}$". The second step of the transformation is to simultaneously eliminate an occurrence of " ′ " from each of $t^\star$ and $s^\star$ in $\ulcorner t^\star = s^\star \urcorner$ until one of the terms is "0". If the result is $\ulcorner 0 = 0 \urcorner$, we say that $\ulcorner t = s \urcorner$ is syntactically reducible to "0 = 0".
12. Note that I am not claiming that "Socrates died" and "Socrates's death occurred" express the same proposition as they are actually used in English. The claim is only that the Recarver has no reason to question the admissibility of a regimentation of English on which the two sentences express the same proposition.
13. For a sustained defence of Recarving Anti-Absolutism, see Rayo, 2013.
14. Does this mean that there couldn't be objects without language? No. The objects there could have been (relative to the language I am now speaking) are the values of my actual singular variables at different worlds, regardless of the linguistic resources available at those worlds. (Compare Linnebo, 2018, §2.2.)
15. Wait! Aren't there contexts in which one can correctly assert e.g. "Sherlock Holmes is a detective"? The issue is controversial. But here I will follow Williamson, 2013, §4.1 in presupposing that there are no contexts in which it is both the case that "Sherlock Holmes is a detective" is true and "Sherlock Homes" is an empty name. For example, when "Sherlock Homes is a detective" is used in Conan

Doyle's writings (assuming he actually used that sentence), it used as part of a story. So there is no presumption that the sentence is true (as opposed to true according to the story). Maybe you think that "Sherlock Holmes is a detective" can be true as used outside the story, perhaps in the context of a lecture on British crime fiction. In such a context, however, "Sherlock Homes" is not an empty name: it refers to a fictional character (as opposed to a human, which is what Holmes is according to the stories). The same idea applies to more complex cases. Consider, for example, "James Bond isn't the only example of a famous nonexistent murdering spy", which was helpfully suggested by a referee for this volume. The most natural way of getting a true reading from this sentence is to interpret it as roughly equivalent to the following: "James Bond (the fictional character) isn't the only example of something that is: (a) famous, (b) a fictional character and (c) according to the relevant fiction, a murdering spy." I have no idea how one might generate such a reading semantically, with no need for pragmatic repair. If the reading cannot be generated semantically, it is not a threat. If it can, we'll have a true sentence without an empty name, since "James Bond" will refer to a fictional character.

16 *Proof:* Suppose $\ulcorner \Diamond(P(a_1,\ldots,a_n))\urcorner$ is true. By the first thesis, $\ulcorner \Diamond(\exists x_1 \ldots \exists x_n(x_1 = a_1 \wedge \ldots \wedge x_n = a_n))\urcorner$ is true. So, by the second thesis, $a_1,\ldots,a_n$ are non-empty names. So $\ulcorner \exists x_1 \ldots x_n(x_1 = a_1 \wedge \ldots \wedge x_n = a_n)\urcorner$ is true.

17 For relevant discussion, see Eklund, 2006 and the debate between Eklund, 2014 and Rayo, 2014.

18 Could we assess reference claims from the point of view of the "maximal" language that results from combining the linguistic resources of "all possible" languages? According to Recarving Anti-Absolutism, there can be no such thing as a maximal language. For if you give me a supposedly maximal language, I can use its resources to dream up a non-trivial extension. For example, I can introduce the notion of a "hyper-set", with the stipulation that for any objects countenanced by the given language, there is a hyper-set with exactly those objects as hyper-members.

19 Some anti-absolutists might hope to go beyond the part-way strategy I am offering here by developing theoretical tools that can do some of the work that absolutists think can be played by absolutely general quantifiers. One strategy is to use schemas (Parsons, 1974, 2006; Glanzberg, 2004; Lavine, 2006). Another strategy is to use a modal construction. Fine, 2006, for instance, argues that necessarily, for any range of objects, there could be a set-like entity not among them whose members are exactly those objects. (See also Hellman, 2006).

20 For related discussion, see Fine, 2006; Hellman, 2006; Parsons, 2006.

21 I was lucky enough to have Ari Koslow as my research assistant on this project and am grateful for her excellent work. I am also grateful to Salvatore Florio, Nick Jones and an anonymous referee, for their many helpful comments.

References

Benacerraf, P., Putnam, H., eds 1983, *Philosophy of Mathematics*, second edn, Cambridge University Press, Cambridge.

Boolos, G. 1984, "To be is to be a value of a variable (or to be some values of some variables)", *The Journal of Philosophy* **81**, 430–449. Reprinted in Boolos (1998), 54–72.

Boolos, G. 1998, *Logic, Logic and Logic*, Harvard University Press, Cambridge, MA.

Burgess, J. 2005, "Being explained away", *Harvard Review of Philosophy* **13**, 41–56.

Butts, R. E., Hintikka, J., eds 1977, *Logic, Foundations of Mathematics, and Computability Theory*, Reidel, Dordrecht.

Cantor, G. 1883, *Grundlagen einer allgemeinen Mannigfaltigkeitslehre. Ein matematisch-philosophischer Versuch in der Lehre des Unendlichen*, Teubner, Leipzig. English translation in Ewald (2005).

Carnap, R. 1950, "Empiricism, semantics and ontology", *Analysis* **4**, 20–40. Reprinted in Benacerraf and Putnam (1983), 241–257.

Chalmers, D. J. 2009, "Ontological anti-realism", in Chalmers et al., pp. 77–129.

Chalmers, D. J., Manley, D., Wasserman, R. eds 2009, *Metametaphysics: New Essays on the Foundations of Ontology*, Oxford University Press, Oxford.

Dummett, M. 1963, "The philosophical significance of Godel's theorem", *Ratio* **5**, 140–155. Reprinted in Dummett (1978).

Dummett, M. 1978, *Truth and Other Enigmas*, Duckworth, London.

Dummett, M. 1981, *Frege: Philosophy of Language*, second edn, Harvard University Press, Cambridge, MA.

Dummett, M. 1991, *Frege: Philosophy of Mathematics*, Duckworth, London.

Eklund, M. 2006, "Neo-Fregean ontology", *Philosophical Perspectives* **20**(1), 95–121.

Eklund, M. 2009, "Carnap and ontological pluralism", *in* Chalmers *et al.*, pp. 130–156.
Eklund, M. 2014, "Rayo's metametaphysics", *Inquiry* **57**(4), 483–497.
Ewald, W. B. 2005, *From Kant to Hilbert, Volume 2: A Source Book in the Foundations of Mathematics*, Oxford University Press, Oxford.
Fine, K. 2006, "Relatively unrestricted quantification", *in* Rayo and Uzquiano, pp. 20–44.
Florio, S. 2014, "Unrestricted quantification", *Philosophy Compass* **9**(7), 441–454.
Florio, S., Linnebo, Ø. typescript, *The Many and The One: A Philosophical Study*.
Frege, G. 1884, *Die Grundlagen der Arithmetik*. English Translation by J. L. Austin, *The Foundations of Arithmetic*, Northwestern University Press, Evanston, IL, 1980.
Frege, G. 1980a, *The Foundations of Arithmetic*, Northwestern University Press, Evanston, IL. English Translation by J. L. Austin.
Frege, G. 1980b, *Philosophical and Mathematical Correspondence*, University of Chicago Press, Chicago, IL. Edited by G. Gabriel, H. Hermes, F. Kambartel, C. Thiel and A. Veraart.
Glanzberg, M. 2004, "Quantification and realism", *Philosophy and Phenomenological Research* **69**, 541–572.
Hellman, G. 2006, "Against 'absolutely everything'!", *in* Rayo and Uzquiano, pp. 75–97.
Hirsch, E. 2010, *Quantifier Variance and Realism: Essays in Metaontology*, Oxford University Press, Oxford.
Kripke, S. A. 1980, *Naming and Necessity*, Harvard University Press, Cambridge, MA.
Lavine, S. 2006, "Something about everything: Universal quantification in the universal sense of universal quantification" *in* Rayo and Uzquiano, pp. 98–148.
Lewis, D. 1991, *Parts of Classes*, Blackwell, Oxford.
Linnebo, Ø. 2018, *Thin Objects*, Oxford University Press, Oxford.
McGee, V. 1997, "How we learn mathematical language", *Philosophical Review* **106**, 35–68.
Oliver, A., Smiley, T. 2013, *Plural Logic*, Oxford University Press, Oxford.
Parsons, C. 1974, "Sets and classes", *Noûs* **8**, 1–12. Reprinted in Parsons (1983).
Parsons, C. 1977, "What is the iterative conception of set". In Butts and Hintikka (1977). Reprinted in Parsons (1983).
Parsons, C. 1983, *Mathematics in Philosophy*, Cornell University Press, Ithaca, NY.
Parsons, C. 2006, "The problem of absolute universality", *in* Rayo and Uzquiano, pp. 203–219.
Rayo, A. 2013, *The Construction of Logical Space*, Oxford University Press, Oxford.
Rayo, A. 2014, "Reply to critics", *Inquiry* **57**(4), 498–534.
Rayo, A., Uzquiano, G., eds 2006, *Absolute Generality*, Oxford University Press, Oxford.
Rosen, G. 1993, "The refutation of nominalism (?)", *Philosophical Topics* **21**, 149–186.
Sidelle, A. 2002, "Is there a true metaphysics of material objects?", *Philosophical Issues* **12**(1), 118–145.
Stalnaker, R. C. 1996, "On what possible worlds could not be", *in* S. Stich, A. Morton, eds, *Benacerraf and His Critics*, Blackwell, Oxford, pp. 103–119.
Stanley, J., Szabó, Z. G. 2000, "On quantifier domain restriction", *Mind and Language* **15**(2,3), 219–261.
Thomasson, A. L. 2015, *Ontology Made Easy*, Oxford University Press, New York.
van Inwagen, P. 1990, *Material Beings*, Cornell University Press, Ithaca, NY.
Williamson, T. 2003, "Everything", *in* J. Hawthorne, D. Zimmerman, eds, *Philosophical Perspectives 17: Language and Philosophical Linguistics*, Blackwell, Oxford, pp. 415–465.
Williamson, T. 2013, *Modal Logic as Metaphysics*, Oxford University Press, Oxford.
Wright, C. 1983, *Frege's Conception of Numbers as Objects*, Aberdeen University Press, Aberdeen.

11
THE METAMETAPHYSICS OF NEO-FREGEANISM

Matti Eklund

1 Introduction

Fregean approaches to branches of mathematics characteristically combine platonism (these branches of mathematics are about mind-independent abstract objects) and logicism (these branches reduce to logic). Today's neo-Fregeanism has mostly focused on arithmetic, and has centrally done so via appeal to

Hume's principle (HP): the number of Fs is identical to the number of Gs if and only if the Fs and the Gs are equinumerous,

where the Fs and the Gs are equinumerous if and only if there are exactly as many Fs as Gs. As the name suggests, neo-Fregeanism derives inspiration from Frege's ideas. The main proponents are the theorists who put this view on the contemporary map, Crispin Wright and Bob Hale, but the view also owes a lot to earlier writings by Michael Dummett.

2 Background

Frege's logicism is the view that arithmetic reduces to logic. Frege's own attempt at reducing arithmetic to logic famously failed. What Frege did was to devise his own logical system, and show how arithmetic could be deduced from axioms of logic within that system. Crucial here was the axiom Basic Law V, saying, roughly, that the "value-range" of one "concept" is identical to that of another just in case the concepts are coextensive. Frege's "concepts" are, roughly, properties and relations; and "value-ranges" can be thought of as extensions. Russell's paradox showed that Basic Law V rendered Frege's logic inconsistent. Frege found no suitable way of restricting or modifying Basic Law V in light of what Russell pointed out.[1]

Here is a way of explaining the issue. The value-range of a property can be thought of as the set of things that have the property and Frege's theory then says that for every property there is a set of exactly the things that has the property. Among all the properties there is is the property of not being a member of oneself. Frege is committed to there being a set, or value-range, of exactly the things that have this property. Call this set R. Is R a member of itself or not? Whatever we say, we land in contradiction. If it is, it isn't. If it isn't, it is.

There is another way to defend something like Frege's view on arithmetic. In *Die Grundlagen der Arithmetik* (1884/1950), Frege had considered another way to reduce arithmetic to logic plus definitions, relying not on Basic Law V but instead on HP. Frege himself rejected this idea because of what has come to be known as the Julius Caesar problem: HP is insufficient to show that Julius Caesar is not a number.[2] But HP has other nice features. Charles Parsons (1965) showed that (in second-order logic) the Peano axioms can be derived from Hume's principle plus what uncontroversially are definitions. George Boolos (1986) showed that Frege arithmetic – the theory obtained by adding HP to second-order logic – is consistent if classical analysis is. So HP is strong enough to found arithmetic on, while the specter of inconsistency is avoided. Neo-Fregeanism can be characterized in different ways, but one core idea is to revive something much like Frege's view, but through relying on HP instead of Basic Law V.

While Parsons had done the necessary formal groundwork, and ideas about ontology congenial to neo-Fregeanism had been defended by Dummett, e.g., in (1973), the first systematic defense of neo-Fregeanism was Crispin Wright's *Frege's Conception of Numbers as Objects* (1983). Neo-Fregeanism has since been defended in Bob Hale's *Abstract Objects* (1987), and by many papers authored by Wright and Hale, separately and jointly. Many of these papers are collected in Hale and Wright's *The Reason's Proper Study* (2001). While neo-Fregeanism is chiefly associated with Wright and Hale, some other authors have defended views in much the same spirit.[3] In Oystein Linnebo (2018), basic ideas of neo-Fregeanism, including reliance on principles similar to HP, are employed not only in philosophy of mathematics, but also in the case of some issues in the ontology of material objects. Jeffrey Sanford Russell (2017) gives an account of mereological sums using the same strategy as the neo-Fregean uses in the case of numbers. Stewart Shapiro and Geoffrey Hellman (2017) give a neo-Fregean account of geometrical points. Neo-Fregeanism has generated a vast literature, both literature criticizing neo-Fregean ideas, and literature exploring these ideas further.[4]

Neo-Fregeans seek to found arithmetic on (second-order) logic plus HP. Like Frege himself, they take number-talk to be referring talk, and the take expressions like "the number of Fs" to be singular terms purporting to refer to numbers. Notice a prima facie tension – one I will be coming back to – between reducing arithmetic to logic and taking number terms to be referring terms: given the latter tenet, some arithmetical truths require for their truth that there be certain objects, numbers. How can it be a matter of logic plus definitions that such-and-such objects exist? One could defend the view that arithmetic is reducible to logic plus definitions by saying that arithmetical truths don't really concern objects – the surface structure of the language we use to do arithmetic is misleading – or perhaps by saying that truths of arithmetic really should be thought to have conditional form "if the natural numbers exist, then ...."[5] But that is not the way of Frege and the neo-Fregeans.

My focus here will be on the platonism the neo-Fregeans defend, and on what general views on metaphysics underlie this platonism and the defense of it. Of special concern is the neo-Fregean's characteristic marriage of logicism (mathematics, or branches thereof, can be reduced to logic plus definitions) and platonism (these branches of mathematics are about mind-independent abstract objects). How can it be true both that mathematics is about mind-independent abstract objects and that it amounts to nothing more than logic plus definitions? There are other important questions about neo-Fregeanism and ontology. Some concern the reliance on higher-order logic, and how higher-order quantification is best understood. Some concern the above-mentioned Julius Caesar problem. But those other questions will not be the focus here.

Apparently crucial to the neo-Fregean program is the question of the status of HP. It is supposed to have the status of a definition, while being ontologically committing. Even before getting into what, in the context, it is to have the "status of a definition," the question seems pressing: how could HP pull off the feat of having both these features? The simplest suggestion is to simply advert to its being an *abstraction principle*: a principle of the form

$§α = §β$ if and only if $α \sim β$

where $α$ and $β$ are variables of some type, $§$ is a function from what these variables range over to objects, and $\sim$ is an equivalence relation on what the variables range over.[6] Another well-known and seemingly plausible abstraction principle is the direction principle,

The direction of a = the direction of b iff a and b are parallel.

Maybe there is something about abstraction principles generally that allows for the combination of being ontologically committing and being true by definition. Maybe, to prefigure a later theme, the sentences flanking an instance of an abstraction principle correspond to *recarvings* of one and the same content, and that is what is special about abstraction principles. However, simply saying that HP has these features because of being an abstraction principle won't really do, regardless of the details. For Basic Law V is also an abstraction principle, but Basic Law V is inconsistent. One project for the neo-Fregeans is that of delineating the class of acceptable abstraction principles. The simplest suggestion would be just to appeal to consistency: an abstraction principle is acceptable if it is consistent. But Boolos (1990) showed that some consistent abstraction principles are inconsistent with each other. The general version of this problem is known as the bad company problem. HP keeps *bad company* – abstraction principles which should not be accepted as having the virtues that HP has according to the neo-Fregeans. How can one separate the good abstraction principles from the bad? The bad company objection to neo-Fregeanism is the objection that there is no principled way to do so.[7] However, even given an extensionally satisfactory response to the bad company objection – a proposal that classifies the abstraction principles we want to accept as good and classifies the ones we don't want to accept as bad – there still remains the issue of how the good abstraction principles could pull off the trick of being analytic while still having the status of definitions.

Thus far I have spoken of Frege's, and the neo-Fregeans', project as being to the effect of reducing arithmetic to logic plus definitions. But this talk of reduction stands in need of some discussion and unpacking. It would be possible for a theorist to hold that arithmetic is "reducible to logic and definitions" but not see this as being of great importance, for example on the ground that logical truth is as philosophically problematic as arithmetical truth to begin with. A reduction of arithmetic to logic plus definitions is philosophically important if, for example, logical truths and truths by definition are assumed to have certain epistemic and/or metaphysical features – they can be known on the basis of semantic or conceptual competence alone, and they are true purely by virtue of meaning. If so then the reduction of arithmetic to logic plus definition can show that arithmetical truths have these features too. Arithmetical truths maybe seemed worrisome before; the reduction shows that they have the same features. For example, maybe logical truths and truths by definition are *analytic*, and the reduction shows that also arithmetical truths are analytic. The talk of analyticity only helps given certain conceptions of analyticity: ones where analytic truths are guaranteed to have certain features. For example, given an understanding of analyticity where a sentence is analytic if understanding the expressions involved and the mode of composition is sufficient for knowledge of the truth of the sentence, analyticity has epistemic significance. Showing mathematical truths to be analytic in this sense is showing them to be knowable by virtue of semantic competence alone. I will understand the neo-Fregean to intend the reduction to have philosophically significant implications.

3 Platonism and logicism

Given the above discussion we can characterize neo-Fregeanism in the philosophy of mathematics somewhat more carefully. We can see it as the conjunction of the following tenets. *Platonism*: there are mind-independent abstract objects. *Neo-logicism*: mathematics is analytic (for example in the epistemic sense characterized above). *Abstraction principles*: key to the justification of the first two tenets is the reliance on abstraction principles, such as HP. The idea is that abstraction principles, or some of them, are analytic, while at the same time the truth of the left-hand side of an abstraction principle is ontologically committing, to mind-independent abstract objects.

The above paragraph characterizes neo-Fregeanism as a general philosophy of mathematics. But the focus is often specifically on arithmetic; and HP is specifically in the service of a philosophy of arithmetic. One can defend neo-Fregeanism solely about arithmetic, and give a different account of other parts of mathematics. In many of Hale and Wright's writings the focus is on arithmetic. But applications of neo-Fregeanism to set theory and analysis have also been discussed.[8] Note that there is at least a potential problem here. It may be thought that the philosophical problems regarding mathematics arise in similar ways for many or all branches of mathematics, in such a way that all of these branches of mathematics should receive similar philosophical treatment. Then if, for example, there are no suitable abstraction principles on which to base analysis or set theory, neo-Fregeanism fails: for it cannot then be extended to all branches of mathematics.

Sometimes Hale and Wright present neo-Fregeanism as primarily an epistemological project.[9] Thus conceived, its main aim is to account for our knowledge of mathematics. However, even if the project is epistemological, it can still have distinctive ontological implications. And some aspects of it are overtly ontological, like the commitment to platonism.

Before turning to discussions in the literature of the neo-Fregean approach to ontology, it may be useful to first consider in the abstract how one can seek to make platonism and logicism compatible, and while making abstraction principles central in this project.

One could have particular views on analyticity and on knowledge such that the analyticity of mathematical claims is compatible with all sorts of metaphysical views on mathematics. Paul Boghossian (1996) introduced the celebrated distinction between epistemic and metaphysical analyticity. As Boghossian presents the distinction, a sentence is *metaphysically analytic* if it is true purely by virtue of meaning; it is *epistemically analytic* if one is justified in believing it purely by virtue of semantic competence. Having introduced the distinction, Boghossian argued that while no sentence can be metaphysically analytic, some sentences can still be epistemically analytic. The argument to the effect that no sentence is metaphysically analytic runs as follows. In general for a sentence S to be true, the following two conditions must be satisfied: (i) S means that p, and (ii) p. Since the second condition must always be satisfied, the sentence is not true in virtue of meaning alone.[10]

The reason Boghossian's stance is relevant in the context is that it is metaphysical analyticity that is most obviously in tension with platonism: how can ontologically committing claims be true purely by virtue of meaning? To be sure, one can also wonder how the epistemic analyticity of a sentence is compatible with its being ontologically committing. But here is a model of how this can be so: the concept's semantic value is what makes true the constitutive principles, when there is a possible semantic value that does so. The determination of semantic value can work in this way even when it is a substantive claim that there in fact is something that makes true the constitutive principles.

Against this general background, it can be suggested that HP can be epistemically but not metaphysically analytic. The idea is that even if HP it is not metaphysically analytic, it can still be epistemically analytic – and the world can be such that it is in fact true, even if laying it down as a stipulation does not itself guarantee that it is in fact true. I do not know of anyone who has actually taken this route, but it seems a reasonable route to explore.

Some other ways of rendering platonism and neo-logicism compatible are more metaphysical in nature, and involve defending the metaphysical analyticity of mathematics, or specific branches thereof. Before I turn to the specific suggestions, some remarks on Boghossian's argument against metaphysical analyticity. The argument is quick. It overlooks that one idea related to that of truth in virtue of meaning is that some propositions are *vacuously true*; they don't demand anything of the world.[11] If sentence S means that p and the proposition that p is such a proposition is such a proposition, then condition (ii) is vacuously satisfied, and S can be said to be true by virtue of meaning alone. Needless to say, the notion of vacuous truth can itself be questioned, but that is another matter.

Given this as background, suppose that one takes number facts to be grounded in facts about equinumerosity, in such a way that HP is true. And suppose further that grounding is to be so understood that if the fact that A is fully grounded in the fact that B, then the fact that A is *nothing over and above* the fact that B. Then fact A does not require anything more of the world than that B obtains – even if the fact that A is a fact to the effect that such-and-such entities exist and fact B does not concern such entities. HP or its instances can then be metaphysically analytic even if the left hand side requires numbers and the right hand side does not concern such entities. They are so, if fact A's requiring nothing of the world over and above what fact B requires means that the fact expressed by "A iff B" requires nothing of the world.[12] While this latter claim of course can be denied, it is easy to see how it can be found attractive.

This kind of metaphysical route has been explored by some authors. Gideon Rosen (2010), Robert Schwartzkopff (2011), Tom Donaldson (2017), Linnebo (2018), and Ciro de Florio and Luca Zanetti (2020) have explored what is sometimes called *the Rosen-Schwartzkopff principle*. In Donaldson's rough formulation:

> For any properties F and G, if the number of things that have the property F is identical to the number of things that have the property G, then this fact is grounded by the fact that the things that have the property F and the things that have the property G can be paired one-to-one.[13]

While this principle is more complex, it is intended to be a grounding-theoretic counterpart of HP.

A different way to marry neo-Fregeanism and metaphysical analyticity is suggested by recent work due to Agustín Rayo (2013). Rayo's discussion is centered on an extended critical argument against what Rayo calls *metaphysicalism*. Rayo's metaphysicalist holds that for an atomic sentence of a given form to be true, there needs to be the right "kind of correspondence" between "the logical form of a sentence and the metaphysical structure of reality." More specifically, for an atomic sentence to be true, the singular terms and the predicate must refer to objects and a property such that these objects and this property are carved up by the world's metaphysical structure. Rayo rejects metaphysicalism. Given this rejection of metaphysicalism, Rayo thinks that there is nothing that blocks maintaining, e.g., that for the number of dinosaurs to be zero *just is* for there to be no dinosaurs. The thought is this. Were metaphysicalism to be true the differently structured constructions flanking the "just is" would demand different things of the world; but now we can instead say that they demand the very same thing. Given Rayo's view it can be held that for the left hand side of an instance of HP to hold just is for the right hand side to hold. This is a different route to metaphysical analyticity of the instances of HP.

Again this demands that if for A to hold just is for B to hold, then the biconditional that A iff B is metaphysically analytic.

Ross Cameron (2008) holds that the neo-Fregean would be wise to relate (HP) to truth-making: more specifically, to say that "the number of Fs = the number of Gs" and "the Fs and the Gs are equinumerous" have the same truth-maker, that the Fs and the Gs are equinumerous. Given this, then again all that is demanded of the world for it to be true that the number of Fs = the number of Gs, and hence for there to be numbers, is that the Fs and the Gs are equinumerous. Again this is naturally put in terms of metaphysical analyticity. If the sentences flanking an instance of HP have the same truthmaker, then if one is true, nothing more is demanded of the world for the other to be true as well.

4 The neo-Fregean as quietist

Let me now start inching towards what Hale and Wright themselves have actually said about the metaontology of neo-Fregeanism. But before attending directly to what they write, let me discuss how they have been understood by commentators. The reason for waiting before introducing the stars of the show is that it is somewhat simpler to discuss Hale and Wright once we have more of the theoretical map clearly in view.

Several authors commenting on neo-Fregeans' writings have, in different ways, presented suggestions regarding the metaontology of neo-Fregeanism – regarding the stance with respect to ontological questions which underlies neo-Fregeanism. This project is to some extent one of interpretation (what do the neo-Fregeans have in mind?) and to some extent one of rational reconstruction (what is the best kind of view on ontology for someone who wishes to embrace the neo-Fregeans' theses?)

Much of the earliest commentary on neo-Fregeanism concerned other aspects of the neo-Fregean project. Some commentators focused heavily on the bad company objections and on technical aspects of the neo-Fregean program. Hale, Wright, and Dummett had an important debate over whether number terms introduced by HP refer in the same sense as that in which ordinary referring terms do.[14] The more general critical discussion of the metaontology of neo-Fregeanism took off fairly late – somewhat paradoxically so, since general pronouncements on the nature of ontology were more prominent in 1980s work by Wright and Hale, such as Wright (1983) and Hale (1987), than they have tended to be in later work.

A relatively early discussion of the metaontology of neo-Fregeanism is found in a part of Fraser MacBride's overview article (2003). In his more recent (2016), MacBride revisits the relevant theme from that article, presenting his point in terms of a supposed dilemma for the neo-Fregean:

> If reality is crystalline then their view that reality contains a sufficient plenitude of objects, properties and relations arranged thus-and-so to make their periphrases for established truths true is left hostage to cosmological fortune. Whereas if reality is plastic then it becomes dubiously coherent to conceive of our ordinary, scientific and mathematical claims about a diversity of objects as being genuinely true or false of an independent reality.[15]

(If reality is "crystalline" then it "consists of states of affairs – objects, properties and relations arranged thus-and-so – whose structure is fixed quite independently of language".[16])

The thought is this. Either the existence of mathematical objects is mind-dependent or it is not. If it is not, then our reliance on HP seems to be on shaky ground. From where does the assurance come that there mind-independently are the objects there need to be in order for HP to be true? If, by contrast, it is mind-dependent what mathematical objects there are, this epistemological mystery may be averted, but now we have abandoned the realist view that mathematical claims are made true or false by an independent reality. And this – and perhaps this ought even to be added to the above characterization of neo-Fregeanism – is a realist view that the neo-Fregean seeks to accept.

MacBride has a suggested way out for the neo-Fregean. It is that we can deny "that there is an intelligible question to be raised" regarding "how language hooks onto reality."[17] If we can deny this, then we can agree that a sentence S is true "without thereby becoming embroiled in the uncomfortable consequences of having to say what *makes* S true."[18] On the view suggested, there is no perspective we "can adopt whereby language-as-a-whole can be significantly compared to something else, *viz.* reality." Summing up the suggestion, MacBride says,

> We are no more capable of adopting such a perspective outside of language from which we can measure language against reality and pronounce upon its representational efficacy than we can step outside our own skins. Consequently there is no "ontological gap" to be bridged between what is truly said and what the world is like.[19]

I see two substantive problems with MacBride's discussion. One is that many key formulations are epistemic rather than metaphysical. For example, in the above, the material before the "consequently" is only epistemic: it only comments on our limitations as *knowers*, and on what we can measure against what. That is compatible with all sorts of *metaphysical* views. But what follows the "consequently" is about metaphysics and not epistemology. What is the connection between the epistemic assumptions and the metaphysical conclusions supposed to be? Second, more importantly, note that MacBride describes the ontological gap as between what is "truly said" and what the world is like. MacBride describes quietism as a distinctive and radical doctrine but there is a sense in which, on most views, there is no ontological gap of the kind described: if it is truly said that p, then here's one way the world is like: the world is such that p. I am not merely harping on a bad formulation. Just below, MacBride speaks of falling prey to the temptation "[t]o demand supernumerary assurance of word-world co-operation before being willing to acknowledge the existence of objects that correspond to the occurrence of syntactically singular expressions in true sentences."[20] Again the truth of the sentences is presupposed, and the question concerns what the truth of the sentence entails. This may be thought innocuous: if the truth of p entails that q, then a requirement for it to be the case that p is that q. But it is a problem that the truth of the sentences is just assumed, since many would find it exceedingly plausible that given that the sentences are indeed true, then the corresponding objects exist. It seems more reasonable to seriously question whether the relevant sentences are true.

MacBride does talk about submitting "to the norms of our discourse," but the connection between this and what his quietism officially concerns remains unexplored.[21] One reason why I bring up this issue regarding MacBride is that we will see similar issues arise regarding Wright and Hale.

5 Easy existence

Two other proposed interpretations of the metaontology of neo-Fregeanism instead take the neo-Fregeans to rely on an assumption to the effect that existence is *easy*. Speaking intuitively:

the worry that numbers may simply fail to exist is due to thinking of existence as a significant achievement, but existence is in fact easily had. But this way of thinking is mistaken. The point is usefully explained in terms of MacBride's dilemma. What if facts about what objects there exist are mind-independent – in accordance with the second horn – but existence is easily had? If so, then the worry that, say, numbers may fail to exist can anyway be set aside as misguided.

One of these two easiness interpretations is the *maximalist* interpretation defended by myself (2006, 2016) and by Katherine Hawley (2007).[22] Maximalism is here the view that anything which satisfies already very weak conditions also does exist. Maximalism is itself just a view on what exists. It is not itself a metaontological view. But the idea behind the maximalist understanding of neo-Fregeanism is that there is some metaontological view – some view on the nature of existence – that neo-Fregeanism relies on, and this metaontological view has maximalism as a consequence.[23] For example, I argue in my (2006) that Wright holds reference to be secondary to truth in a particular way and that is what drives the whole thing. In arguing what I do I compare arguments given in a different context in Wright (1992). Given a metaontological view of the kind described, there is a guarantee that so long as some newly introduced terms satisfy some very weak conditions – in the case of pure abstracta, something like that the stipulations introducing them are coherent – those terms refer.

The other easiness interpretation, due to Sider (2007), is the *quantifier variance* interpretation. The doctrine of quantifier variance – introduced, under that name, by Eli Hirsch is, at a first stab, the doctrine that there are different possible existential quantifier meanings, and no one quantifier meaning is privileged over all the others.[24] If the neo-Fregean relies on quantifier variance, then she can say that what guarantees the success of the stipulation of an abstraction principle (satisfying weak conditions) is that there will be some meaning or other for the existential quantifier such that when the quantifier expression has that meaning, the abstraction principle is true. The guarantee isn't that the newly introduced terms refer to entities that exist in the ordinary sense of "exists" but that the terms refer to entities in some sense of "exists," and what the stipulation of HP effects is that we come to use "exists" in such a way that the sentence employed to state HP comes out true.

Neither easiness interpretation immediately addresses epistemological concerns. Under both interpretations, the metaontological claim emphasized must be supplemented by other assumptions in order to properly explain knowledge of the existence of mathematical objects.

6 Early Wright and Hale

Having discussed the metaontology of neo-Fregeanism in general terms and having discussed commentators, let me now finally turn to what Wright and Hale themselves have said about the matter. They have said somewhat different things over the years. I will discuss some different themes that have come up.

Much of Wright (1983) was devoted to discussion of the notion of an object which his neo-Fregeanism is said to rely upon. The focus is on Frege's context principle, and on the syntactic priority thesis – "the thesis of the priority of syntactic over ontological categories"[25] – which Wright says follows from it. What has come to be known as the context principle receives some different formulations in Frege's *Grundlagen* (1884), but in one famous formulation, it is the principle that "it is only in the context of a sentence that words have any meaning."[26] Wright spells out the syntactic priority thesis as follows:

> According to this thesis, the question of whether a particular expression is a candidate to refer to an object is entirely a matter of the sort of syntactic role which it plays on

whole sentences. If it plays that sort of role, then the truth of appropriate sentences in which it so features will be sufficient to confer on it an objectual reference; and questions concerning the character of its reference should then be addressed by philosophical reflection on the truth-conditions of sentences of the appropriate kind.[27]

Needless to say, it is not immediate or uncontroversial that the syntactic priority thesis follows from, or even is very closely related to, the context principle as intended by Frege.

Note that it is apparently fully consistent with the syntactic priority thesis so understood to deny that number terms refer – even if they syntactically are singular terms. What follows from the syntactic priority thesis is only that given that they syntactically are singular terms then they are candidates to refer to objects, and so if appropriate sentences containing them are true there are objects to which they refer. But nothing is said which addresses whether any appropriate sentences are true, or how to determine that.

What I am complaining about is the same sort of thing I earlier complained about regarding MacBride. The truth of the relevant sentences is just taken for granted; what is problematized is only what follows from the truth of these sentences.

Here is one more illustration. Hale (1987) lays out what he calls the "Fregean argument" as follows:

1   If a range of expressions function as singular terms in true statements, then there are objects denoted by expressions belonging to that range
2   Numerals and many other numerical expressions besides, do so function in many true statements (of both pure and applied mathematics)

Hence

3   There exist objects denoted by these numerical expressions (i.e. there are numbers)[28]

Hale then launches straight into a thorough discussion of the criteria for being a singular term, and the case that can be made for thinking that many numerical expressions. He discusses a number of objections to the platonism he defends, including Benacerraf's arguments concerning mathematical knowledge, and concerning indeterminacy.[29] But he never pauses on what positive case can be made for thinking that the relevant kinds of statements are *true*.

In a somewhat later piece, the introduction to their joint (2001), Hale and Wright describe neo-Fregean platonism as relying on two ideas. One is that objects just are whatever singular terms refer to. The other is that

> no more is to be required, in order for there to be a strong prima-facie case that a class of apparent singular terms have reference, than that they occur in true statements free of all epistemic, modal, quotational, and other forms of vocabulary standardly recognized to compromise straightforward referential function.[30]

Striking, again, is that they do not pause on the question of the truth of the statements.

All this said, I think it is clear, *intuitively*, what picture Wright and Hale are operating with. If the category of object is explained via that of the category of (referring) singular term, then there is nothing that stands in the way of the truth-conditions of appropriate sentences containing singular terms purporting to refer to Ks being the same as the truth-conditions of sentences without such K-terms. For given some sentences of the latter kind one can introduce sentences with what syntactically are singular terms with the stipulation that the newly introduced sentences are to have the same truth-conditions as the old ones; and if the category of object is simply understood in terms of that of singular term, then there is no real possibility that such a stipulation will fail simply on the ground that the world happens to fail to contain the objects in question. But it still remains that the syntactic priority thesis itself does not say anything about this. No general principle has been produced which would justify this broad claim.

It is natural to think that in the background of Hale and Wright's neo-Fregeanism lies something like Rayo's anti-metaphysicalism and not what Wright calls the syntactic priority thesis. If metaphysicalism is rejected, then the path lies open to reasonably holding that for it to be the case that the number of Fs = the number of Gs just is for it to be the case that the concepts F and G are equinumerous. Rejecting metaphysicalism relates to what truth demands, by saying something about what truth does not demand: the truth of a sentence does not require there to be a structural correspondence between the sentence and reality.

Later in (1983), Wright turns directly to HP and lays out the impressive formal case that can be made for reliance on HP: in second-order logic, all of second-order arithmetic follows from HP plus what uncontroversially are definitions, and the system is equiconsistent with ordinary second-order Peano arithmetic. Moreover, regardless of what to say about theoretical matters, HP does have an air of being a definition. There is what may be called a possible "particularist" line – particularist in the sense that it does not seek to motivate HP by appeal to general considerations – which focuses on HP because it has these good-making features but which eschews any attempt at a general theoretical scaffolding. Another possible reason why Wright and Hale do not pause on the truth of the relevant mathematical sentences might be thought to be that they can justifiably simply rely on this as an assumption. Given both the success of mathematics and what common sense suggests, there is prima facie reason to take these sentences to be true. Then so long as arguments to the contrary can be rebutted we can indeed reasonably take these sentences to be true. No more direct argument for their truth is required. But for what it is worth, Wright and Hale do seem to wish to provide more by way of theoretical scaffolding than a thoroughgoing particularism would allow; and they do seem to seek to provide direct arguments for the claim that some ontologically committing mathematical claims are true.

7 Content recarving and implicit definition

In some of their writings, the neo-Fregeans relied on a notion of *content recarving*.[31] The same content can be recarved in different ways, and this is what is going on with acceptable abstraction principles: the sentences flanking an instance of such a principle in some sense have the same content, only carved up in different ways. The question is what criterion of individuation of content should be employed here. If the criterion is very coarse-grained (e.g., metaphysically necessary equivalence), then the claim of sameness of content is plausible, but it is correspondingly implausible that sameness of content of two sentences implies that the sentences thereby have the same epistemic and metaphysical status. If the criterion is more fine-grained (e.g., the content of a sentence is the structured proposition it expresses, where propositional structure is isomorphic to linguistic structure), it is plausible that sentences with the same content thereby have the same epistemic and metaphysical status, but it is implausible that sentences flanking an instance of HP indeed have the same content. Linnebo (2018) calls this the Goldilocks problem: "Is there a notion of content with granularity that is just right, neither too coarse nor too fine?"[32]

The neo-Fregeans' attempts to solve the Goldilocks problem have met with serious criticism.[33] Maybe as a result, the neo-Fregeans have more recently tended to emphasize content recarving less – although the notion is again appealed to in Hale and Wright (2009b), which I will get to in the next section. What Hale and Wright have turned to instead is implicit definition, and criteria for something to be an implicit definition.

In his (2009), John MacFarlane asks the question: what good reason might the neo-Fregean have for focusing on abstraction principles rather than directly on the axioms of Peano arithmetic? He focuses specifically on the theory of implicit definition in Hale and Wright (2000). His answer is negative: the neo-Fregean has no good reason for focusing on abstraction principles. If HP satisfies their criteria for being an acceptable implicit definition, then so do the Peano axioms. In their (2009a), Hale and Wright attempt to reply to MacFarlane.

Whether or not MacFarlane's question is an embarrassing one for Hale and Wright it can certainly seem embarrassing for the offered interpretations. Quietism: couldn't the neo-Fregean equally well lay down the Peano axioms by stipulation, and answer doubts about their truth by appeal to quietism? Maximalism: if existence is easy in the way the maximalist interpretation holds, then it is also easy for the Peano axioms to come out true. Quantifier variance: if stipulating that the sentence used to state HP comes out true manages to affect the meaning of the existential quantifier in such a way that this sentence comes out true, why can't we achieve the same by stipulating that the sentences used to express the Peano axioms come out true? MacFarlane's question arises also for the suggestion that the neo-Fregean simply appeal to Boghossian's distinction between different kinds of analyticity: if HP can be epistemically analytic despite making a substantive claim about the world, can't this go also for the Peano axioms? The ideas that might fare better are the ones I introduced in connection with metaphysical analyticity. Focus first on the Rosen-Schwartzkopff idea: HP can be used as an account of wherein the number facts may be grounded (equinumerosity relations between concepts), but the Peano axioms do not in the same way suggest a similarly attractive story about grounding. Similarly, Rayo's strategy relies on "just is"-statements, and while HP can obviously be recast as a "just is"-statement, there appears to be no natural way of so using the axioms of PA. Turning to Cameron, the point he makes is that the sentences flanking an instance of HP can have the same truthmaker, and there is no straightforward way of developing that idea in the case of the axioms of PA. All this said, MacFarlane brings up suggestions regarding how the PA axioms could be relevantly analogous here too. If one can stipulate HP, can one not equally well stipulate

"$x(x=x) \equiv PA$?[34]

Elaborating on what MacFarlane says, couldn't one say that the PA axioms are ungrounded, zero-grounded (grounded but not in anything), or grounded in The True? Couldn't one say that for there to be numbers just is for The True to obtain? Couldn't one say that the PA axioms don't need truthmakers, or have the same truthmaker as The True?

8 The metaontology of abstraction

A more recent article by Hale and Wright, their (2009b), focuses on metaontology, and contains criticisms of the maximalist and quantifier variance interpretations, as well as Hale and Wright's own statement of their positive view. The following passage sums up their main message:

> The kind of justification which we acknowledge *is* called for is precisely justification for the thought that no such [metaphysical] collateral assistance is necessary. There is no hostage to redeem. A (good) abstraction *itself* has the resources to close off the alleged (epistemic metaphysical) possibility. The justification needed is to enable – clear the

obstacles away from – the recognition that the truth of the right-hand side of an instance of a good abstraction is *conceptually* sufficient for the truth of the left. There is no gap for metaphysics to plug, and in that sense no 'metaontology' to supply. This view of the matter is of course implicit in the very metaphor of *content* recarving.[35]

Hale and Wright understand the friends of the maximalist and quantifier variance interpretations to take the neo-Fregean claims to stand in need of metaphysical "collateral assistance": to take the neo-Fregean to need to rely on some underlying metaphysical assumptions.[36] They reject this supposed demand, as seems reasonable. Whatever the underlying metaphysical assumptions may be, one can ask what justifies them in turn. However, seen as a criticism of the two other interpretations offered, this seems unfair. What the interpretations fasten on is just that the neo-Fregean defense of HP does not take place in a theoretical vacuum, but the neo-Fregeans adduce reasons for thinking HP has the features ascribed to it – and these reasons adduced seem like they generalize beyond the case at hand.

More interestingly, Wright and Hale describe their own positive account. Central to their discussion is the distinction between so-called sparse and abundant conceptions of properties. They take the abundant conception as a model for what they want to say about objects:

> On one way of taking it, the relevant notion of genuine property is akin to that in play when we conceive it as a nontrivial question whether any pair of things which both exemplify a certain set of surface qualities – think, for example, of a list of the reference-fixers for "gold" given in a way independent of any understanding of that term or an equivalent – have *a property in common*. When the question is so conceived, the answer may be unobvious and negative ... However this conception stands in contrast with that of the more "abundant" theorist, for whom the good standing, in that sense, of a predicate is *already* trivially sufficient to ensure the existence of an associated property, a (perhaps complex) *way of being* which the predicate serves to express. For a theorist of the latter spirit, predicate sense will suffice, more or less, for predicate reference.[37]

In general terms, the abstractionist metaphysics of abstract objects, and of reference to them – sometimes called *minimalism* – stands to the conception of the matter that underwrites the reference-fixing model as an abundant conception of properties stands to a sparse one.[38]

They discuss the obvious disanalogy that on the abundant conception of properties, any predicate with a sense also ascribes a property whereas the analogous claim about objects seems absurd.

> On the abundant view of properties, predicate sense suffices for reference. But it is not the abstractionist view of singular terms that sense suffices for reference – the view is that the truth of atomic contexts suffices for reference. However everyone agrees with that. The controversial point is what it takes to be in position reasonably to take such contexts to be true. The point of analogy with the abundant view is that this is not, by minimalism, conceived as a matter of hitting off, Locke-style, some "further" range of objects. We can perfect the analogy if we consider not simple abundance but the view that results from a marriage of abundance with Aristotelianism. Now the possession of sense by a predicate no longer suffices, more or less, for reference. There is the additional requirement that the predicate be true of something, and hence that some atomic statement in which it occurs predicatively is true.[39]

But what they say here is problematic, and the problems are of a kind we have seen before. They say first that the view is not simply that the truth of atomic contexts suffices for reference – "everyone agrees with that."[40] The more central issue is: what does it take for atomic contexts to be true? Their only answer to this is negative: it is not a matter of "hitting off ... some 'further' range of objects." This may not be fully clear – but more importantly, it does not address what exactly *is* needed. This is the same thing I raised complaints about earlier, when discussing early Wright and Hale. My complaint there was that the issue of truth is not explicitly addressed. Here the issue of truth is explicitly addressed, and it is at least clear that Hale and Wright take themselves to have a general point about what truth demands. However, their only remark is negative.

9 Concluding remarks

I have here provided an overview of issues regarding the metaontology of neo-Fregeanism. I have discussed various things Hale and Wright have said, various interpretations of their views found in the literature, as well as some other possible views the neo-Fregean might in principle appeal to.

As mentioned, I have myself elsewhere (2006, 2016) defended a particular view on what the neo-Fregeans commit to, the maximalist interpretation. I have not tried to again make the case here for this view.

I have also not tried to issue a general verdict on the neo-Fregean ideas, or even on those ideas that have to do with metaontology specifically. Instead I have merely tried to provide a helpful map of the territory. For what it is worth, I find the neo-Fregean project extremely attractive. But my aim here has not been to provide a brief for neo-Fregeanism, but to discuss different questions pertaining to the metaontological issues that come up in connection with the view.

Notes

1 Discovered by Russell in 1901, communicated to Frege in the letter Russell (1902).
2 It is a nice question why Frege did not think that the problem does arise equally for Basic Law V: Basic Law V seems insufficient to show that Julius Caesar is not a value-range.
3 See, e.g., Cook (2009b), Linnebo (2018), and Rayo (2016).
4 See perhaps especially Fine (2002), Heck (2011), Linnebo (2018), the essays on neo-Fregeanism in Boolos (1998), the essays in Linnebo (2009), and the essays in Ebert and Rossberg (2016), in addition to other material referred to in this overview.
5 See here Field (1984).
6 Characterization adapted from Linnebo (2018), p. xii.
7 See, among other things, Boolos (1990), Cook (2009a), Cook and Linnebo (2018), Linnebo (2018), Studd (2016), Weir (2003), and Wright (1999).
8 See, e.g., Boolos (1986) and (1993), Hale (2000a, 2000b), Linnebo (2018), Shapiro (2000), Studd (2016), and Wright (2000). Logan (2015, 2017) has applied neo-Fregean ideas to category theory.
9 Wright (2016), p. 161.
10 Boghossian (1996), p. 364.
11 Boghossian only considers another response: that S's meaning that p somehow makes it the case that p (1996, p. 364f).
12 Needless to say, some steps here can be questioned; and there are conceptions of grounding given which the connection between grounding and "nothing over and above" does not hold.
13 Donaldson (2017), p. 775. Much of Donaldson's article is then devoted to finding the best formulation of the principle.
14 See Dummett (1991, 1998), Hale (1994), and Wright (1998a, 1998b).
15 MacBride (2016), p. 96.

16 MacBride (2016), p. 94.
17 MacBride (2016), p. 97.
18 MacBride (2016), p. 97.
19 MacBride (2016), p. 97.
20 MacBride (2016), p. 97.
21 MacBride (2016), p. 106. Note that the talk of submitting to the norms of our discourse is again epistemic, as it seems to address the issue when we should accept the sentences as true. That issue is different from the issue of the conditions under which the sentences are true. The relevant metaphysical view would be that the sentences are true exactly when they are true according to the norms of our discourse. Another view still would profess agnosticism about the metaphysics, and simply say that the best we can do is to submit to the norms of our discourse.
22 Some qualifications are needed. Hawley only compares the maximalist interpretation with the quantifier variance interpretation (see just below) and argues that the former is preferable. And in my (2006) I only took myself to be discussing a strand of thought in earlier neo-Fregean writings.
23 Compare the liberal ontological views defended in Schiffer (2003) (regarding abstracta) and Thomasson (2015) (more generally).
24 See the essays collected in Hirsch (2011). In Eklund (forthcoming), I discuss and compare some different views that have been discussed under the heading quantifier variance.
25 Wright (1983), p. 51.
26 Frege (1884), §53.
27 Wright (1983), p. 51f. There are very similar passages found in earlier writings by Dummett, see e.g. his (1956), p. 40f.
28 Hale (1987), p. 11.
29 See Benacerraf (1965) and (1970).
30 Hale and Wright (2001), p. 8.
31 See e.g. Hale (1997) and Wright (1997).
32 Linnebo (2018), p. 78.
33 For Hale's proposal, see Hale (1997) (and the postscript to it in Hale and Wright (2001)). For critical discussion, see e.g. Fine (2002), pp. 39–41, Potter and Smiley (2001) and (2002), and chapter 4 of Linnebo (2018).
34 MacFarlane (2009), p. 455.
35 Hale and Wright (2009b), p. 193. I mentioned earlier that in later writings, Wright and Hale have tended to focus less on content recarving – but note that content recarving is prominently mentioned here.
36 There is no discussion of MacBride's quietist interpretation in Hale and Wright (2009b).
37 Hale and Wright (2009b), p. 197f.
38 Hale and Wright (2009b), p. 207.
39 Hale and Wright (2009b), p. 208.
40 There is more controversy here than Hale and Wright let on. What Hale and Wright think is that if "F(a)" is true, then "a" refers and hence a exists. So from the truth of atomic contexts they think it can be concluded that there exists something which the singular term or terms in the sentence refer to. This is something some theorists would deny.

References

Benacerraf, Paul: 1965, "What Numbers Could Not Be," *Philosophical Review* 74: 47–73.
Benacerraf, Paul: 1970, "Mathematical Truth," *Journal of Philosophy* 70: 661–79.
Boghossian, Paul: 1996, "Analyticity Reconsidered," *Noûs* 30: 331–68.
Boolos, George: 1986, "Saving Frege from Contradiction," *Proceedings of the Aristotelian Society* 87: 137–51; reprinted in Boolos (1998).
Boolos, George: 1990, "The Standard of Equality of Numbers," in Boolos (ed.) *Meaning and Method: Essays in Honor of Hilary Putnam*, Cambridge University Press, Cambridge, pp. 261–77. Reprinted in Boolos (1998).
Boolos, George: 1993, "Whence the Contradiction?," *Proceedings of the Aristotelian Society Suppl. Vol.* 67: 213–34. Reprinted in Boolos (1998).
Boolos, George: 1998, *Logic, Logic and Logic*, Harvard University Press, Cambridge, Massachusetts.

Cameron, Ross: 2008, "Truthmakers and Ontological Commitment: or How to Deal With Complex Objects and Mathematical Ontology Without Getting Into Trouble," *Philosophical Studies* 140: 1–18.

Cook, Roy: 2009a, "Hume's Big Brother: Counting Concepts and the Bad Company Objection," *Synthese* 170: 349–69.

Cook, Roy: 2009b, "New Waves on an Old Beach: Fregean Philosophy of Mathematics Today," in Oystein Linnebo and Otavio Bueno (eds.), *New Waves in Philosophy of Mathematics*, Palgrave Macmillan, London, pp. 13–34.

Cook, Roy T. and Oystein Linnebo: 2018, "Cardinality and Acceptable Abstraction," *Notre Dame Journal of Formal Logic* 59: 61–74.

De Florio, Ciro and Luca Zanetti: 2020, "On the Schwartzkopff-Rosen Principle," *Philosophia* 48:405-19.

Donaldson, Tom: 2017, "The (Metaphysical) Foundations of Arithmetic," *Noûs* 51: 775–801.

Dummett, Michael: 1956, "Nominalism," *Philosophical Review* 65: 491–505.

Dummett, Michael: 1973, *Frege: Philosophy of Language*, Duckworth, London.

Dummett. Michael: 1991, *Frege: Philosophy of Mathematics*, Duckworth, London.

Dummett, Michael: 1998, "Neo-Fregeans: In Bad Company?," in Schirn (1998), pp. 369–88.

Ebert, Philip and Marcus Rossberg (eds.): 2016, *Abstractionism*, Oxford University Press, Oxford.

Eklund, Matti: 2006, "Neo-Fregean Ontology," *Philosophical Perspectives* 20: 95–121.

Eklund, Matti: 2016, "Hale and Wright on the Metaontology of Neo-Fregeanism," in Ebert and Rossberg (2016), pp. 79–93.

Eklund, Matti: forthcoming, "Collapse and the Varieties of Quantifier Variance," in James Miller (ed.), *The Language of Ontology*, Oxford University Press, Oxford.

Field, Hartry: 1984, "Critical Notice of Crispin Wright's *Frege's Conception of Numbers as Objects*," *Canadian Journal of Philosophy* 14: 637–62.

Fine, Kit: 2002, *The Limits of Abstraction*, Oxford University Press, Oxford.

Frege, Gottlob: 1884, *Die Grundlagen der Arithmetik*, Wilhelm Koebner, Breslau. English translation by J.L. Austin, *The Foundations of Arithmetic*, Blackwell, Oxford, 1950.

Hale, Bob: 1987, *Abstract Objects*, Blackwell, Oxford.

Hale, Bob: 1994, "Dummett's Critique of Wright's Attempt to Resuscitate Frege," *Philosophia Mathematica* 2: 122–47. Reprinted in Hale and Wright (2001).

Hale, Bob: 1997, "*Grundlagen* §64," *Proceedings of the Aristotelian Society* 97: 243–61. Reprinted with a postscript in Hale and Wright (2001).

Hale, Bob: 2000a, "Reals by Abstraction," *Philosophia Mathematica* 8: 100–23. Reprinted in Hale and Wright (2001).

Hale, Bob: 2000b, "Abstraction and Set Theory," *Notre Dame Journal of Formal Logic* 41: 379–98.

Hale, Bob and Crispin Wright: 2000, "Implicit Definition and the A Priori," in Paul Boghossian and Christopher Peacocke (eds.), *New Essays on the A Priori*, Oxford University Press, Oxford, pp. 286–319. Reprinted in Hale and Wright (2001).

Hale, Bob and Crispin Wright: 2001, *The Reason's Proper Study*, Oxford University Press, Oxford.

Hale, Bob and Crispin Wright: 2009a, "Focus Restored: Comments on John MacFarlane," *Synthese* 170: 457–82.

Hale, Bob and Crispin Wright: 2009b, "The Metaontology of Abstraction," in David Chalmers, David Manley, and Ryan Wasserman (eds.), *Metametaphysics*, Oxford University Press, Oxford, pp. 178–212.

Hawley, Katherine: 2007, "Neo-Fregeanism and Quantifier Variance," *Proceedings of the Aristotelian Society Suppl. Vol.* 81: 233–49.

Heck, Richard G.: 2011, *Frege's Theorem*, Oxford University Press, Oxford.

Hirsch, Eli: 2011, *Quantifier Variance and Realism*, Oxford University Press, New York.

Linnebo, Oystein (ed.): 2009, Special issue on the bad company objection, *Synthese* 170: 321–482.

Linnebo, Oystein: 2018, *Thin Objects*, Oxford University Press, Oxford.

Logan, Shay Allen: 2015, "Abstractionist Categories of Categories," *Review of Symbolic Logic* 8: 705–21.

Logan, Shay Allen: 2017, "Categories for the Neologicist," *Philosophica Mathematica* 25: 26–44.

MacBride, Fraser: 2003, "Speaking With Shadows: A Study of Neo-Logicism," *British Journal for the Philosophy of Science* 54:103–63.

MacBride, Fraser: 2016, "NeoFregean Metaontology," in Ebert and Rossberg (2016), pp. 94–112.

MacFarlane, John: 2009, "Double Vision: Two Questions About the Neo-Fregean Program," *Synthese* 170: 443–56.

Parsons, Charles: 1965, "Frege's Theory of Number," in Max Black (ed.), *Philosophy in America*, Cornell University Press, Ithaca, New York, pp. 180–203.

Potter, Michael and Timothy Smiley: 2001, "Abstraction by Recarving," *Proceedings of the Aristotelian Society* 101: 327–38.
Potter, Michael and Timothy Smiley: 2002, "Recarving Content: Hale's Final Proposal," *Proceedings of the Aristotelian Society* 102: 301–4.
Rayo, Agustín: 2013, *The Construction of Logical Space*, Oxford University Press, Oxford.
Rayo, Agustín: 2016, "Neo-Fregeanism Reconsidered," in Ebert and Rossberg (2016), pp. 203–21.
Rosen, Gideon: 2010, "Metaphysical Dependence: Grounding and Reduction," in Bob Hale and Aviv Hoffmann (eds.), *Modality: Metaphysics, Logic, and Epistemology*, Oxford University Press, Oxford, pp. 109–36.
Russell, Bertrand: 1902, "Letter to Frege," in Jean van Heijenoort (ed.), *From Frege to Gödel*, Harvard University Press, Cambridge, Massachusetts, 1967, pp. 124–5.
Russell, Jeffrey Sanford: 2017, "Composition as Abstraction," *Journal of Philosophy* 114: 453–70.
Schiffer, Stephen: 2003, *The Things We Mean*, Oxford University Press, Oxford.
Schirn, Matthias (ed.): 1998, *Philosophy of Mathematics Today*, Clarendon Press, Oxford.
Schwartzkopff, Robert: 2011, "Numbers as Ontologically Dependent Objects – Hume's Principle Revisited," *Grazer Philosophische Studien* 82: 353–73.
Shapiro, Stewart: 2000, "Frege Meets Dedekind: A Neologicist Treatment of Real Analysis," *Notre Dame Journal of Formal Logic* 41: 335–64.
Shapiro, Stewart and Geoffrey Hellman: 2017, "Frege Meets Aristotle: Points as Abstracts," *Philosophia Mathematica* 25: 73–90.
Sider, Theodore: 2007, "Neo-Fregeanism and Quantifier Variance," *Proceedings of the Aristotelian Society Suppl. Vol.* 81: 201–32.
Studd, James: 2016, "Abstraction Reconceived," *British Journal for the Philosophy of Science* 67: 579–615.
Thomasson, Amie: 2015, *Ontology Made Easy*, Oxford University Press, Oxford.
Weir, Alan: 2003, "Neo-Fregeanism: An Embarrassment of Riches," *Notre Dame Journal of Formal Logic* 44: 13–48.
Wright, Crispin: 1983, *Frege's Conception of Numbers as Objects*, Aberdeen University Press, Aberdeen.
Wright, Crispin: 1992, *Truth and Objectivity*, Harvard University Press, Cambridge, Massachusetts.
Wright, Crispin: 1997, "On the Philosophical Significance of Frege's Theorem," in Richard Heck (ed.), *Language, Thought, and Logic: Essays in Honour of Michael Dummett*, Oxford University Press, Oxford, pp. 201–44. Reprinted in Hale and Wright (2001).
Wright, Crispin: 1998a, "On the Harmless Impredicativity of N = (Hume's Principle)," in Schirn (1998), pp. 339–68. Reprinted in Hale and Wright (2001).
Wright, Crispin: 1998b, "Response to Dummett," in Schirn (1998), pp. 389–405. Reprinted in Hale and Wright (2001).
Wright, Crispin: 1999, "Is Hume's Principle Analytic?," *Notre Dame Journal of Formal Logic* 40: 6–30. Reprinted in Hale and Wright (2001).
Wright, Crispin: 2000, "Neo-Fregean Foundations for Analysis: Some Reflections," *Notre Dame Journal of Formal Logic* 41: 317–34.
Wright, Crispin: 2016, "Abstraction and Epistemic Entitlement," in Ebert and Rossberg (2016), pp. 161–85.

12
EASY ONTOLOGY

Amie L. Thomasson

If you look at metaphysics books and articles from the 1950s onwards, you will find a great many heated debates about what really exists: Do numbers exist? Do properties exist? Do mereological sums exist? Do social groups exist? Does consciousness exist? Do ordinary objects such as tables and chairs exist? Do persons exist?, and so on.

Yet it seems easy to answer such questions using simple arguments that apparently take us from an uncontroversial premise, via a conceptual truth, to a conclusion about what exists. So, for example:

1. (Uncontroversial truth): There are two cups on the table.
2. (Conceptual truth): If there are n Ks, then the number of Ks is n.
3. (Derived claim): The number of cups on the table is two.
4. (Ontological claim): There is a number.

1. (Uncontroversial truth): That barn is red.
2. (Conceptual truth): If x is red, then x has the property of redness.
3. (Derived claim): That barn has the property of redness.
4. (Ontological claim): There is a property.

In each case, the third premise sounds intuitively redundant with respect to the first (lending credence to the idea that they are linked by a conceptual truth), and the conclusion is just an existential generalization from the third premise. Those who favor 'easy ontology' accept such arguments, and so think that existence questions are easy to settle – generally in the positive.

Yet such arguments have often been dismissed or held to be controversial or even paradoxical. Stephen Yablo (2000) calls these 'overeasy existence proofs,' and writes, "such arguments carry with them a palpable sense of daring and a distinct feeling of pulling a rabbit out of a hat. Nobody supposes that there are *easy* proofs, from a priori or empirically obvious premises, of the existence of abstracta" (2000, 275).

But why should such straightforward arguments be thought to be paradoxical? Suspicions against *a priori* arguments for the existence of things trace back at least to Kant's criticisms of ontological arguments for the existence of God.[1] But the easy arguments at issue here are importantly different from the ontological arguments Kant criticized. For they do not treat 'exists' as a predicate, and adding reference to entities like numbers and properties (unlike adding reference to God) 'conservatively extends' the prior theory – roughly, not altering the theory's take on the causal order (see Schiffer 2003, 52–61).

In the contemporary context, as Yablo (2000) brings out, easy arguments are thought to be paradoxical largely because they appear to conflict with the widely accepted 'Neo-Quinean' methodology for addressing existence questions. The Neo-Quinean approach dominated metaphysics from the mid-twentieth century through the early twenty-first – so much so that the approach has also simply been labeled "mainstream metaphysics" (Manley 2009, 4).[2] Tracing their approach back to Quine's influential article, "On What there Is" (1948/1953), Neo-Quinean 'mainstream metaphysicians' hold that we are ontologically committed to, and only to, what we must quantify over in order to render the statements of our best theory true. On this approach, answering an existence question is difficult. It requires, first, establishing one's theory as the 'best total theory.' Since competing ontological theories are generally (according to all disputants) not distinguished empirically, and so don't differ in their empirical adequacy, this generally means establishing that the favored theory has other 'theoretical virtues,' such as explanatory power and parsimony of ontology or ideology. Second, it is thought to require establishing that the entities in dispute *must* be quantified over to render the statements of the theory true – and so that there is no viable *paraphrase* that would do the job as well. As a result, on this model, arguing that numbers or properties exist, say, is no easy matter to be settled via trivial inferences. Instead, it requires extensive debates about which is the best 'total theory,' and about the adequacy of various strategies to paraphrase away apparent reference to them.

Easy arguments, however, are in tension with the standard Neo-Quinean way of arguing about existence questions. As Thomas Hofweber puts it, "How can it be that substantial metaphysical questions have apparently trivial answers, answers that are immediately implied by ordinary everyday statements that we all accept and that apparently are not disputed among those who disagree about ontology?" (2005a, 260). Given the dominance of the neo-Quinean approach, it was once fairly common to assume that one must reject the easy arguments, 'show where they go wrong.' Yablo, for example, simply assumes that the easy arguments must go: "I am going to assume without argument that [accepting easy arguments] is out of the question. Our feeling of hocus-pocus about the 'easy' proof of numbers (etc.) is really very strong and has got to be respected" (2000, 278).

Easy arguments have been important, then, not merely because they seem to answer existence questions, but also because this appearance raises important methodological and metaontological questions that call into question a way of approaching existence questions that was so dominant as to be virtually unquestioned for decades. Those questions include: How are we to go about answering existence questions? Does answering them require extensive, hard to settle, debates about which 'total theory' has the greatest virtues, and whether the relevant talk can be successfully paraphrased? Or are such extended 'deep' debates in ontology somehow deeply misguided, since existence questions are answerable easily?

1 A brief history of easy arguments

Easy arguments, particularly in the philosophy of mathematics, are often traced back to Frege. Frege held that the ontological category of *object* is simply the correlate of the logical category of *proper name* (Dummett 1973/1981, 55–6), where proper names are taken to include all singular terms. Proper names, on Frege's view, must be associated with a criterion of identity, and for this he appeals to Hume's idea: "When two numbers are so combined as that the one has always an unity answering to every unity of the other, we pronounce them equal" (Frege

1884/1968, sec 62). Bob Hale suggests that the Fregean idea that in general our ontological categories are correlates of the "categorization of the types of expressions by means of which we refer to them" (2010, 403) leads to a deflationary approach to existence questions. For we can argue for the existence of entities of a certain kind just by arguing that there are true atomic statements (say, 'The number of cups equals the number of saucers') in which the relevant expressions function as singular terms (2010, 406). Hale puts it as follows, "Under the Fregean approach, questions about the existence of entities of this or that kind are transformed into questions about *truth* and *logical form* – are there true statements [of the right form] incorporating expressions of the appropriate logical type?" (2010, 406).

Another historical source of easy arguments is Carnap's work on existence claims taken in what he called the 'internal' sense – questions about the existence of certain entities asked *within* a given linguistic or conceptual framework. Carnap famously held that such internal existence questions could be answered straightforwardly, either by 'logical' (broadly: analytic) methods (for questions like "Is there a prime number greater than a hundred?" (1950/1956, 208–9)) or by empirical methods (for questions like, "Are unicorns and centaurs real or merely imaginary?" (1950/1956, 207)). Even a general claim such as "There are numbers," Carnap suggests, "follows from the analytic statement 'five is a number' and is therefore itself analytic" (1950/1956, 209). Similarly, "there are propositions" follows from "*That Chicago is large* is a proposition" and is "analytic ... and even trivial" (1950/1956, 210). Any further questions about the reality of entities of various kinds, Carnap held, are either – taken literally – ill-formed pseudo-questions, or can be reinterpreted as practical questions about whether or not to adopt a certain linguistic framework.

In the contemporary debate, easy arguments were first raised explicitly in the philosophy of mathematics, and were directly inspired by Frege. Bob Hale and Crispin Wright (Hale and Wright 2001, 2009; Wright 1983) develop and defend a 'neo-Fregean' or 'abstractionist' position in the philosophy of mathematics. Building on Frege's idea that objects are just the correlates of singular terms, Hale argues that one can argue for the existence of numbers as follows "If ... there are true statements incorporating expressions functioning as singular terms, then there are objects of some corresponding kind. If the singular terms are such that, if they have reference at all, they refer to numbers, there are numbers" (2010, 406). Hale and Wright (again following Frege 1884/1968, Section 63–5) make use of Hume's Principle, expressed more formally as: The number of ns = the number of ms iff the ns and ms are equinumerous. They take this to be a conceptual truth which enables one to argue as follows:

- The cups and saucers are equinumerous.
- The number of ns = the number of ms iff the ns and the ms are equinumerous.
- The number of cups = the number of saucers.

Since the conclusion is a true identity claim, they argue that the singular terms in it ('the number of cups,' 'the number of saucers') must refer, and so we can conclude that there are numbers. Oystein Linnebo (2018) develops a more recent 'abstractionist' approach in this tradition. He begins from "the Fregean idea that an object, in the most general sense of the word, is a possible referent of a singular term," so that the question 'what objects are there?' may be answered by addressing what forms of singular reference are possible (2018, xii). He goes on to argue that singular reference is easy to achieve, giving us an easy ontology of numbers, considered as 'thin' objects.

Stephen Schiffer (1994, 1996, 2003) broadens the applications of easy arguments substantially. Schiffer's driving concern is in the theory of meaning, and his central aim is to give a deflationary ontology of propositions considered as 'pleonastic entities.' But (in part to warm us up to the idea of pleonastic entities and provide a clear exposition of them) he also provides easy arguments for the existence of properties, events, states, and fictional characters. Unlike the neo-Fregeans, Schiffer does not require that we make use of an identity statement to infer that the relevant singular terms refer; we may simply make arguments like the following:

1   Lassie is a dog.
2   *That Lassie is a dog* is true (Schiffer 2003, 71)

Schiffer treats these as pleonastic 'something from nothing' inferences. They are pleonastic since (2) is intuitively redundant with respect to (1): Someone who said "Lassie is a dog and *that Lassie is a dog* is true" would seem to be just pretentiously repeating themselves. They are 'something from nothing' inferences since we begin in (1) with no terms that even aim to refer to propositions (only to a dog), and yet can derive a true sentence in (2) with a new singular term ('*That Lassie is a* dog') which, as Schiffer puts it, is apparently guaranteed to refer to a proposition – entitling us to infer that there are propositions. Yet these propositions, he insists, are merely ontologically 'lightweight' 'pleonastic' entities, entities "whose existence is secured by something-from-nothing transformations" (2003, 51), and which "have 'no hidden and substantial nature for a theory to uncover'" (2003, 63). Schiffer takes these inferences to be connected by a conceptual truth ("a truth knowable a priori via command of the concept" (2003, 52)); in this case the relevant conceptual truth would be: If P, then *that P* is true. It is worth noting that, while neo-Fregeans require that the conceptual truth be an equivalence principle (Hume's principle) and require that the derived claim take the form of an identity statement, Schiffer's pleonastic inferences have no such requirements, enabling us to broaden the sorts of entity for which we can apparently get pleonastic arguments (see Thomasson 2015, 138–9).

Amie Thomasson (2015) broadens the approach still further and aims to explicitly draw metaontological conclusions from it. Beyond accepting Schiffer's easy arguments for properties, propositions, events, states, and the like, Thomasson maintains that parallel arguments can lead us to conclusions about the existence of ordinary concrete objects. Consider disputes about the existence of composite inanimate material objects such as tables and chairs. While realists and eliminativists about tables disagree about whether there are tables, they will agree to the following statement (given in the eliminativist's 'language of refuge')[3]:

- There are particles arranged tablewise.

Eliminativists introduce these ways of speaking to account for the truth-value of our ordinary claims apparently about tables – so we can still allow that it is true (or 'nearly as good as true') that there is a table in my dining room, and false (or 'nearly as good as false') that there is a table on my head. These paraphrases also enable eliminativists to distinguish their view from the 'madman's' view that we are all massively deceived in our table experiences. To mimic the truth-conditions of ordinary table-talk, 'there are particles arranged tablewise' is supposed to build in all of the ordinary conditions for being a table *except for the existence of composite inanimate material objects* – so it is supposed to require not only certain spatial arrangements of 'particles,' but also their joint ability to fulfill the characteristic functions of tables, the presence of whatever intentions and/or cultural contexts are thought necessary for there to be tables, and so on. Given that all of those conditions are in place, Thomasson argues, a competent speaker who has mastered use of both 'table' and 'particles arranged tablewise' is entitled to infer from 'there are particles arranged tablewise' to 'there is a table,' thereby answering the ontological question about the existence of ordinary objects. The fact that we can get easy arguments for ordinary concrete objects (as well as events and states), Thomasson argues, shows that we should not think that the entities we can become committed to via easy arguments have some lightweight or reduced ontological status – and she accordingly drops Schiffer's term "pleonastic entity." Instead, she insists, these arguments lead us to conclusions that the disputed entities exist in the only sense that has sense (2015, Section 3.3). Thomasson also argues that in many cases we do not need to make such inferences to answer existence questions: often we can just make use of our conceptual competence (say, with 'table') and go have a look in the restaurant to know that there are tables. But whether the existence questions are answered via trivial inferences or ordinary observation, Thomasson follows Carnap (on internal questions) in insisting that well-formed, answerable existence questions can be answered by nothing more mysterious than straightforward empirical and conceptual work – this is another sense in which answering existence questions is 'easy.' Thomasson explicitly draws a metaontological conclusion from this: that the neo-Quinean debates about what (*really*) exists are misguided.

2 The importance of easy ontology

Easy ontological arguments are important both to first-order philosophical problems and, more broadly, to methodological and metaphilosophical concerns. At the first-order level, they are relevant to resolving old debates about such things as numbers, propositions, or properties. Their relevance here is not merely that they give us a straightforward answer ('yes') to the question of whether such things exist. Taking an easy ontological approach also is thought to bring clear benefits in solving problems about how we could come to *know about* such entities. Indeed in the philosophy of mathematics, the original motivations for a neo-Fregean position were largely epistemological (see Hale and Wright 2009). There have been longstanding problems in the philosophy of mathematics about how we could possibly come to know about numbers and other mathematical objects given that, as Hale and Wright put it, such entities "do not, seemingly, participate in the causal swim" (2009, 178). Neo-Fregeans aim to approach the problem of how we could come to *know* mathematical entities "by reference to an account of how meaning is conferred upon the ordinary statements that concern such objects" (2009, 178–9). For on the easy ontological model, we do not need to 'peer into' a Platonic heaven to discover whether certain numbers are identical – we need only be inducted into a way of speaking by adopting certain implicit definitions that entitle us to make inferences, say, from the fact that the cups and saucers are equinumerous, to conclude that the number of plates equals the number of saucers. As long as we can have a non-mysterious account of how we know, e.g., that the cups and saucers are equinumerous, we can have a non-mysterious account of how we can come to know 'the identities and distinctions' among numbers (Hale and Wright 2009, 179). The full neo-Fregean story indeed aims to go further in showing how to recover other mathematical knowledge as well (see Hale and Wright 2009, 180).[4]

In Schiffer's work, the most direct concerns are with defending a 'face value' theory of propositions as an important contender in theories of content. But epistemological concerns also play a role here. For once we have a pleonastic account of certain entities (propositions, properties, etc.),

> the account in these terms is used to show how our knowledge of something-from-nothing entailments (e.g. if Lassie is a dog, then Lassie has the property of being a dog) is a priori conceptual knowledge, thus explaining how we're able to have knowledge and reliable beliefs about these sorts of entity.
>
> *(2003, 3)*

Similarly, an important point Schiffer raises in favor of the pleonastic account of fictional characters is that it enables us to answer the question of how it is possible for us to have knowledge of fictional entities (2003, 52). Epistemological concerns are again at the forefront in Thomasson's development of easy ontology. For on her view, easy inferences enable us to acquire knowledge of such things as modal facts and properties – thereby resolving the long-standing problems of modal epistemology (Thomasson 2018 and 2020).

The most significant impact of easy arguments, however, has to do with their relevance to methodological and meta-philosophical issues. Most particularly, the availability of easy arguments raises questions about whether the existence questions are suitable topics for 'deep' philosophical debate. Here again, the underlying motivations are epistemological. For accepting easy arguments for the existence of various disputed entities – and accepting that we never need more than empirical and/or conceptual methods to answer those existence questions that are well-formed and answerable – demystifies the epistemology of metaphysics. Thomasson argues (2015, 2017) that the Neo-Quinean approach makes the epistemology of metaphysics completely mysterious. Its defenders typically insist that the answers to the relevant existence questions cannot be known through conceptual nor straightforward empirical means – they are (in Theodore Sider's phrase) "epistemically metaphysical" (2011, 187). Some aim to defend their views by appeal to general metaphysical principles, but it remains unclear how these are to be known. Others appeal to the 'theoretic virtues' exhibited by the favored theory. But Bennett (2009) and Kriegel (2013) present detailed analyses in which prominent competing theories simply trade off one theoretic virtue for another, leaving doubt that comparisons of theoretic virtues can settle many metaphysical disputes. Neo-Quinean metaphysicians tend to respond to concerns about epistemological mystery by insisting that there is only a difference of degree between ontological theories and scientific ones. As Sider puts it "as a general epistemology of metaphysics I prefer the vague, vaguely Quinean, thought that metaphysics is continuous with science," with metaphysics and science employing many of the same criteria for theory choice, though he admits that "metaphysical inquiry is by its nature comparatively speculative and uncertain" (2011, 12). Yet there are crucial differences between rival metaphysical versus scientific theories that should not be ignored. For it is rare for competing scientific theories to be equally empirically adequate; on the other hand, it is rare for rival ontological theories to *differ* in their empirical adequacy or predictions (Paul 2012, 12). Where the theoretic virtues other than empirical adequacy are concerned, it is very much open to doubt that these are truth-conducive (Bricker, forthcoming).[5]

In short, one important consequence of accepting easy ontological arguments is that it helps make it non-mysterious how we can come to have knowledge of certain disputed entities (including numbers, properties and other abstracta, modal facts and properties, etc.), and – more broadly – how we can come to have knowledge in ontology – knowledge of what exists. While easy arguments threaten the neo-Quinean approach, they also give us hope for a less mysterious epistemology for metaphysics.

3 Objections to easy arguments

Easy arguments are intuitively acceptable – even obvious. The moves that take us from the premise to the conclusion, as Schiffer nicely points out, seem intuitively redundant. Given their apparent obviousness and acceptability, why would anyone reject them?

As mentioned at the outset, one reason such arguments are often rejected, or simply assumed to be "out of the question" (Yablo 2000, 278) is that the idea that existence questions could be answered easily conflicts with the dominant neo-Quinean approach to addressing existence questions, which takes these to be "substantial metaphysical questions" (Hofweber 2005a). But despite its popularity, one should not be so quick to assume that the neo-Quinean approach is correct – or one we should embrace. For, as mentioned above, it leads to epistemological mysteries and an apparent rivalry with science (didn't we think it was the job of the natural sciences to tell us what exists?). Moreover, in practice the approach has led to nothing but an overwhelming and ever-diverging set of opinions about 'what really exists,' without the disputants even agreeing on a methodology that would tell us how such debates can be resolved. The move from: 'Neo-Quinean ontology is right' to 'Easy arguments must have gone wrong somewhere' could only even look acceptable in a context in which the neo-Quinean approach was dominant and unquestioned. But in the new era of metametaphysical debates, that is no longer the case, and the neo-Quinean approach has been subjected to sustained and important lines of criticism in recent years that cannot be merely ignored.[6] Assessing easy ontological arguments requires more of substance than just noting the conflict with neo-Quinean methodology.

Both Yablo and Hofweber take up this challenge, offering targeted assessments of where easy arguments go wrong – and why (despite appearances) they don't really give us ontological conclusions that could render neo-Quinean arguments out of place. Yablo's idea (developed in various ways in his 1998, 2000, 2001), is that the conclusions of easy arguments should not be taken seriously as assertions of existence – for they are implicitly in the scope of a fiction or game of make-believe (or are simulating, metaphorical, or otherwise non-literal). As Yablo puts it "the *a priori* approach to existence questions is undermined by doubts about literality" (2000, 276). Childhood games of make-believe (on Kendall Walton's 1990 influential analysis) often involve props (like pats of mud) and principles to generate make-believe truths (such as: where there is a pat of mud, pretend there is a pie). Fictionalists aim to give a similar analysis of easy arguments. What an easy ontologist treats as a conceptual truth ("if there are n Ks, then the number of Ks is n"), a fictionalist treats as a generative principle in a game of make-believe (Yablo 2001, 77). As a result, the fictionalist holds that the conclusions of easy arguments don't give us genuine ontological claims, but only a *pretense* – and so easy arguments don't answer ontological questions. Thomasson (2013 and 2015, Chapter 5) replies extensively to the fictionalist threat. She argues that the fictionalist only begs the question against the deflationist who accepts easy arguments. Moreover, she argues that fictionalists face the daunting challenge of contrasting the allegedly fictional uses of the terms in question ('number,' 'property,' etc.) with literal uses, and that deflationists can offer a better account of the discourse.

Another attempt to diagnose where easy arguments 'go wrong' is developed by Thomas Hofweber (2005a, 2005b, 2007). Hofweber appeals to work in linguistics to argue the new noun terms introduced in easy arguments (e.g., 'number') are merely introduced to bring in a 'focus effect.' That is, when we shift from saying 'There are two bagels on the table' to 'The number of bagels is two,' this focuses our attention on *quantity* (how many bagels) rather than on other aspects of content (such as: what's on the table). As a result, Hofweber argues, such terms do not 'have the function of referring,' and the quantifier used in the conclusion cannot be an ontologically committing one. Hofweber thus distinguishes two uses of the quantifier: an 'internal' or 'inferential role' use (which he claims is at work in the conclusions of easy arguments), and an 'external' or 'domain conditions' use, which is existentially committing. This enables him to diagnose the 'puzzle' of ontology by holding that easy arguments only give us conclusions that use the *internal* quantifier, whereas genuine existence questions must be answered using the *external* quantifier. For further discussion of and response to these arguments, see Brogaard (2007), Moltmann (2013), Balcerak Jackson (2013), Felka (2014), and Thomasson (2015, Chapter 9).

Beyond these targeted attempts to show where easy arguments 'go wrong,' there are more global sources of resistance. One of the most venerable sources of resistance to easy ontological arguments is the idea that the (supposed) implicit definitions or conceptual truths used in easy arguments may keep 'bad company' with principles that are clearly problematic. This form of argument was first raised against neo-Fregean arguments in the philosophy of mathematics, and has generated a large literature.[7] The original idea was that, while neo-Fregean arguments for numbers rely on Hume's Principle, other principles with a similar form notoriously lead to trouble.[8] More generally, the problem for easy ontological arguments is that existence-entailing conditionals that superficially look like the conceptual truths in easy arguments can lead to outright contradictions or conflicts with known facts. Completely general principles for introducing properties, for example, may lead to contradiction. If we accept a principle that says: "If there is a description, then there is a property corresponding to that description," then we run into trouble with the description 'being a property that doesn't apply to itself.' Or (as Schiffer argues (2003, 53)) suppose we introduce an implicit definition for a new term 'wishdate' as follows: "If x wishes for a date, then x gets a wishdate," where a 'wishdate' is understood as "a person whose existence supervenes on someone's wishing for a date, every such wish bringing into existence a person to date" (Schiffer 2003, 53). This implicit definition then seems to entail, falsely, that whenever someone wishes for a date, a person pops into existence.

To reply to bad company arguments, defenders of easy arguments must find a way to draw the line between (purported) conceptual truths that are acceptable and can be used in non-problematic something-from-nothing inferences, and those that are unacceptable – and to do so without being ad hoc. One prominent strategy (carefully developed in Schiffer 2003, Chapter 2) is to require that the concepts introduced be 'conservative extensions' of the prior theory, roughly in the sense that they do "nothing to disturb the pre-existing causal order," as described by the prior theory (Schiffer 2003, 55). Introducing proposition talk by allowing that, if Lassie is a dog, then the proposition that Lassie is a dog is true, does nothing to alter our take on the causal order. By contrast, it does alter our take on the causal order if we introduce 'wishdate' talk by allowing that, if Joanne wishes for a date, then some person pops into existence. Thomasson (2015, Chapter 8) develops a strategy for replying to bad company arguments that involves thinking about what it would take to (just) extend a language. She motivates requirements along these lines, requiring that legitimate existence-entailing conditionals can be seen as "object-language reflections of rules of use that could successfully and minimally introduce new terms or concepts to a (previously) more restricted language" (2015, 260) – that is, introduce them in ways that are learnable, and that ensure they don't add any new empirical commitments, only an enriched language (including sufficient coapplication conditions to enable us to make judgments of identity and distinctness for things of the kind).

More recently, some have objected to easy arguments by suggesting that even if such inferences look obvious, they may be subject to 'defeaters' (Eklund 2017; Korman 2019), so that serious metaphysical debate may be as relevant as ever in defeating the entitlements or assessing putative defeaters. Such claims raise interesting questions about whether (if the easy arguments are based on conceptual truths) they are really subject to 'defeat' in ways that empirical claims are (for discussion see Thomasson 2019). Other questions include whether (some or all of) the principles that the easy ontologist takes to be conceptual truths express *correctness rules* or mere *'rules of evidence'* (Yablo 2014, 495–6) that can be over-ridden (for discussion and response see Thomasson 2014).

Perhaps the most general source of resistance to easy arguments comes from resistance to the idea that there are the needed conceptual or analytic truths to take us from the uncontroversial truth to the ontological conclusion. For, given Quine's influential arguments in "Two Dogmas of Empiricism" (1951/1953), it has become commonplace to reject the idea that there are any analytic truths. Timothy Williamson (2007) has more recently raised other arguments against the idea that there are any analytic truths. Those who accept easy inferences then have two options. They may (with Jonathan Schaffer 2009, 360) take the easy claims of existence to be just 'obvious' or 'compelling,' or even to have a kind of Moorean certainty that trumps any philosophical claims to the contrary – without taking them to be inferences based on an analytic or conceptual truth. Or they may defend the idea that there are (something like) analytic or conceptual truths in the sense in which they are needed to make the easy arguments. Thomasson takes the latter route, responding to Quinean arguments against analyticity (in her 2007) and to Williamson's arguments (in her 2015). On Thomasson's view, all one needs to legitimate the easy arguments is a view on which "*mastery* of relevant linguistic/conceptual rules governing the terms/concepts employed in the inference, plus knowledge of an undisputed truth, *licenses or entitles* one to make the relevant inference using those terms, without the need for any further investigation" (2015, 233). And that, she argues, is not undermined by Williamson's arguments.[9]

One increasingly common line of response to easy arguments is to suggest that, even if easy arguments, as stated in English, are sound, there are deeper questions about existence suitable for debate by ontologists. We have already seen one version of this, in Hofweber's claim that there are two senses of the quantifier at work: a merely internal use in the conclusions of easy arguments, and an external use which could be used in properly answering questions about what 'exists in reality.' Cian Dorr (2007) suggests a similar idea that, in the conclusions of easy arguments for numbers and properties which say 'there are numbers/properties,' 'there are' is being used in a 'superficial' sense. On Dorr's view, this contrasts with a 'fundamental' sense of 'there are' we can use in asking 'substantive ontological questions' about whether there really are numbers or properties. Ross Cameron (2008) suggests that we distinguish questions about what *exists* (which may well be answered as the neo-Fregean suggests) from questions about what *really* exists, which "can only be settled (if it can be) by doing metaphysics" (2008, 13). The fullest development of an idea along these lines has come in the work of Theodore Sider, who allows that even those who are committed to the idea that ontological debates are 'deep' and 'substantive' and to be pursued (roughly) by neo-Quinean methods may allow that, "when applied to *English* quantification, the easy ontology picture might well be correct" (2011, 196–7). Yet deep ontological debates, he argues, might nonetheless remain sensible, since we could conduct them in a new language of Ontologese – a language "in which the quantifiers are stipulated to carve at the joints" (2011, 197). Thus, debates about the relevance of easy ontological arguments to the prospects for serious ontological debates (done in neo-Quinean fashion) also involve us in debates about whether or not one can make sense of a 'fundamental,' or 'Ontologese' quantifier and reconstruct debates about existence in those terms.

4 Conclusion

Twenty years ago, it was common to assume that easy arguments raised only the puzzle of how to pinpoint where they had gone wrong, given their apparent conflict with the dominant neo-Quinean approach to ontology. This assumption has changed. Now, an increasing number of metaphysicians accept that easy arguments, at least as expressed in English, should be accepted. In fact, growing acceptance of the idea that existence questions can be answered easily has led several non-deflationary metaphysicians to suggest that metaphysics should turn away from questions about existence to focus on some other project (Fine 2009, 158; Schaffer 2009, 357; Cameron 2010). The question of whether we should accept easy arguments for the existence of disputed entities raises deep questions not only about the acceptability of a neo-Quinean approach to existence questions, but also about what we should be doing in metaphysics. For if existence questions are so easily resolved, it is hard to see them as the appropriate subject for serious and deep inquiries. Discussions of easy ontology thus bring to center stage questions about what we legitimately can and should be doing in metaphysics, and what role answering existence questions should play in our work.

Notes

1 And so, immediately after introducing easy arguments like those above, Stephen Schiffer memorably quips, "Maybe you feel like reading Kant to me" (2003, 52). He goes on to say what distinguishes these arguments from traditional ontological arguments for the existence of God – namely that adding reference to such things as numbers and properties may *conservatively extend* a theory whereas adding reference to God does not (2003, 52–61).
2 While the approach is standardly labeled as 'neo-Quinean,' it is not at all clear that it is true to the historical Quine, who (in his own words), was "no champion of traditional metaphysics" (1951, 66). For discussion see Price (2009).
3 This language is developed by Peter van Inwagen (1990, Chapters 10 and 11) to paraphrase talk about tables and other composite inanimate material objects. For arguments for elimination of ordinary objects such as tables and chairs, see his (1990) and Trenton Merricks' (2001).
4 For further discussion of the neo-Fregean approach, see Eklund (2006) and Hawley (2007).
5 For doubts about whether theoretic virtues are truth-conducive for metaphysical debates, see Shalkowski (2010), Saatsi (2016), and Thomasson (2017).
6 For criticisms of the approach (from various different directions), see Eli Hirsch (2002a, 2002b, 2011), Karen Bennett (2009), Uriah Kriegel (2013), David Chalmers (2009), Yablo (2009), and Thomasson (2015, 2017).
7 The bad company problem was originally developed in Boolos (1990) and Heck (1992). For further discussions of it see Linnebo (2009a) and the other contributions to Linnebo (2009b).
8 That is to say: certain other abstraction principles that involve a bi-conditional connecting an identity statement to an equivalence relation lead to trouble. The classic case in point is Frege's own Basic Law V, introducing talk of extensions of sets as follows: that the extension of (set) F equals the extension of set G iff, for all x, Fx iff Gx. As Russell showed, this abstraction principle leads to contradiction if we combine it with Frege's comprehension principle (that there is a concept corresponding to every expressible condition on objects) and the principle that every concept has an extension.
9 Scharp (2013, 48–9) similarly argues that those who possess a concept are 'quasi-entitled' to the constitutive principles governing that concept, where that is to say that they would be entitled if there were no countervailing evidence. He does not do so, however, in defense of easy arguments, and it is not clear that quasi-entitlement would be strong enough for the needed defense of easy arguments.

References

Balcerak Jackson, Brendan (2013). "Defusing Easy Arguments for Numbers." *Linguistics and Philosophy* 36 (6): 447–61.
Bennett, Karen (2009). "Composition, Colocation and Metaontology." In Chalmers et al., 38–76.
Boolos, George (1990). "The Standard Equality of Numbers." In Boolos, ed., *Meaning and Method: Essays in Honor of Hilary Putnam*. Cambridge: Cambridge University Press, 261–77.
Bricker, Phillip (forthcoming). "Realism without Parochialism." In Bricker, *Modal Matters: Essays in Metaphysics*. Oxford: Oxford University Press.
Brogaard, Berit (2007). "Number Words and Ontological Commitment." *Philosophical Quarterly* 57 (226): 1–20.
Cameron, Ross (2008). "Truthmakers and Ontological Commitment: Or How to Deal with Complex Objects and Mathematical Ontology without Getting into Trouble." *Philosophical Studies* 140 (1): 1–18.
Cameron, Ross (2010). "How to Have a Radically Minimal Ontology." *Philosophical Studies* 151: 249–64.

Carnap, Rudolf (1950/1956). "Empiricism, Semantics and Ontology." In Carnap, *Meaning and Necessity*, Second Edition. Chicago: University of Chicago Press.
Chalmers, David (2009). "Ontological Anti-Realism." In Chalmers et al.
Chalmers, David, Ryan Wasserman, and David Manley, eds. (2009). *Metametaphysics*. Oxford: Oxford University Press.
Dorr, Cian (2007). "There Are No Abstract Objects." In Theodore Sider, John Hawthorne, and Dean Zimmerman, eds., *Contemporary Debates in Metaphysics*. Oxford: Blackwell.
Dummett, Michael (1973/1981). *Frege: Philosophy of Language*. Second Edition. Cambridge, Massachusetts: Harvard University Press.
Eklund, Matti (2006). "Neo-Fregean Ontology." *Philosophical Perspectives* 20: 95–121.
Eklund, Matti (2017). Review of *Ontology Made Easy*. *Notre Dame Philosophical Reviews*, February 16, 2017. https://ndpr.nd.edu/news/ontology-made-easy/.
Felka, Katharina (2014). "Number Words and Reference to Numbers." *Philosophical Studies* 168 (1): 261–82.
Fine, Kit (2009). "The Question of Ontology." In Chalmers et al.
Frege, Gottlob (1884/1968). *The Foundations of Arithmetic*. Translated by J. L. Austin. Evanston, Illinois: Northwestern University Press.
Hale, Bob (2010). "The Bearable Lightness of Being." *Axiomathes* 20: 399–422.
Hale, Bob and Crispin Wright (2001). *The Reason's Proper Study: Essays Towards a Neo-Fregean Philosophy of Mathematics*. Oxford: Clarendon.
Hale, Bob and Crispin Wright (2009). "The Metaontology of Abstraction." In Chalmers et al.
Hawley, Katherine (2007). "Neo-Fregeanism and Quantifier Variance." *Aristotelian Society Supplementary Volume* 81: 233–49.
Heck, Richard (1992). "On the Consistency of Second-Order Contextual Definitions." *Noûs* 26: 491–5.
Hirsch, Eli (2002a). "Quantifier Variance and Realism." In Ernest Sosa and Enrique Villanueva, eds., *Realism and Relativism. Philosophical Issues* 12. Oxford: Blackwell, 51–73.
Hirsch, Eli (2002b). "Against Revisionary Ontology." *Philosophical Topics* 30: 103–27.
Hirsch, Eli (2011). *Quantifier Variance and Realism*. New York: Oxford University Press.
Hofweber, Thomas (2005a). "A Puzzle about Ontology." *Noûs* 39 (2), 256–83.
Hofweber, Thomas (2005b). "Number Determiners, Numbers, and Arithmetic." *Philosophical Review* 114 (2): 179–225.
Hofweber, Thomas (2007). "Innocent Statements and their Metaphysically Loaded Counterparts." *Philosopher's Imprint* 7 (1): 1–33.
Korman, Daniel (2019). "Easy Ontology without Deflationary Metaontology." In Symposium on *Ontology Made Easy*. *Philosophy and Phenomenological Research* 99 (1): 236–43.
Kriegel, Uriah (2013). "The Epistemological Challenge of Revisionary Metaphysics." *Philosopher's Imprint* 13 (12): 1–30.
Linnebo, Oystein (2009a). "Introduction." *Synthese* 170 (3): 321–9.
Linnebo, Oystein, ed. (2009b). *Synthese Special Issue: The Bad Company Problem*. 170 (3).
Linnebo, Oystein (2018). *Thin Objects: An Abstractionist Account*. Oxford: Oxford University Press.
Manley, David (2009). "Introduction: A Guided Tour of Metametaphysics." In Chalmers et al.
Merricks, Trenton (2001). *Objects and Persons*. Oxford: Clarendon.
Moltmann, Friederike (2013). "Reference to Numbers in Natural Language." *Philosophical Studies* 162 (3): 499–536.
Paul, Laurie (2012) "Metaphysics as Modeling: The Handmaiden's Tale." *Philosophical Studies* 160: 1–29.
Price, Huw (2009). "Metaphysics after Carnap: the Ghost who Walks?" In Chalmers et al.
Quine, W. V. O. (1948/1953). "On What there Is." In Quine, *From a Logical Point of View*. Cambridge, Massachusetts: Harvard University Press.
Quine, W. V. O. (1951). "On Carnap's Views on Ontology." *Philosophical Studies* II (5): 65–71.
Quine, W. V. O. (1951/1953). "Two Dogmas of Empiricism." In Quine, *From a Logical Point of View*. Cambridge, Massachusetts: Harvard University Press.
Saatsi, Juha (2016). "Explanation and Explanationism in Science and Metaphysics." In Matthew Slater and Zanja Yudell, eds., *Metaphysics and the Philosophy of Science: New Essays*. Oxford: Oxford University Press, 163–91.
Schaffer, Jonathan (2009). "On What Grounds What." In Chalmers et al.
Scharp, Kevin (2013). *Replacing Truth*. Oxford: Oxford University Press.
Schiffer, Stephen (1994). "A Paradox of Meaning." *Noûs* 28: 279–324.

Schiffer, Stephen (1996). "Language-Created Language-Independent Entities." *Philosophical Topics* 24 (1): 149–67.
Schiffer, Stephen (2003). *The Things We Mean*. Oxford: Oxford University Press.
Shalkowski, Scott (2010). "IBE, GMR and Metaphysical Projects." In Bob Hale and Aviv Hoffmann, eds., *Modality: Metaphysics, Logic and Epistemology*. Oxford: Oxford University Press.
Sider, Theodore (2011). *Writing the Book of the World*. Oxford: Oxford University Press.
Thomasson, Amie L. (2013). "Fictionalism versus Deflationism." *Mind*, doi.10.1093/mind/fzt055.
Thomasson, Amie L. (2014). "Quizzical Ontology and Easy Ontology." *Journal of Philosophy* 111 (9–10): 502–28.
Thomasson, Amie L. (2015). *Ontology Made Easy*. New York: Oxford University Press.
Thomasson, Amie L. (2017). "Metaphysics and Conceptual Negotiation." *Philosophical Issues* 27 (2017) doi:10.1111/phis.12106.
Thomasson, Amie L. (2018). "How Can We Come to Know Metaphysical Modal Truths?" *Synthese* https://doi.org/10.1007/s11229-018-1841-5.
Thomasson, Amie L. (2019). Book Symposium on *Ontology Made Easy:* Replies to Comments on *Ontology Made Easy*. *Philosophy and Phenomenological Research* 99 (1): 251–64.
Thomasson, Amie L. (2020). *Norms and Necessity*. New York: Oxford University Press.
Van Inwagen, Peter (1990). *Material Beings*. Ithaca, New York: Cornell University Press.
Walton, Kendall (1990). *Mimesis as Make-Believe*. Cambridge, Massachusetts: Harvard University Press.
Williamson, Timothy (2007). *The Philosophy of Philosophy*. Oxford: Blackwell.
Wright, Crispin (1983). *Frege's Conception of Numbers as Objects*. Aberdeen: Aberdeen University Press.
Yablo, Stephen (1998). "Does Ontology Rest on a Mistake?" *Proceedings of the Aristotelian Society* Supplementary Volume 72: 229–61.
Yablo, Stephen (2000). "A Paradox of Existence." In Anthony Everett and Thomas Hofweber, eds., *Empty Names, Fiction, and the Puzzles of Non-Existence*. Palo Alto: CSLI Publications, 275–312.
Yablo, Stephen (2001). "Go Figure: A Path through Fictionalism." *Midwest Studies in Philosophy* 25: 72–102.
Yablo, Stephen (2009). "Must Existence-Questions Have Answers?" In Chalmers et al.
Yablo, Stephen (2014). "Carnap's Paradox and Easy Ontology." *Journal of Philosophy* 111 (9–10): 470–501.

13
DEFENDING THE IMPORTANCE OF ORDINARY EXISTENCE QUESTIONS AND DEBATES

Jody Azzouni

I Ordinary questions about existence

Well-trained philosophers easily lose touch with what continues to fascinate nonphilosophers. This isn't entirely condescension, although that plays an undeniable role. Another cause is unavoidable professionalization. The philosophical concerns that grip nonphilosophers become specialized subject areas (involving many things nonphilosophers can't see any motivation for). One effect of this, surely, is the apparent irrelevance of professional philosophy from the rest of intellectual/cultural life (Carus 2007, 5–6) – especially in the United States. Another, more important for this chapter, is the inability of philosophers to appreciate or engage with the still-living philosophical concerns of the general public.

Existence questions – Ontology (according to one description of it) – illustrates this well. Existence *questions* and *debates* are linked: if an existence question can be asked, then sides can be taken and arguments offered. There are *specific* existence questions that (most) nonphilosophers care (and debate) about more than anything else. God remains the best example (Aikin and Talisse 2011; Craig 2017); in the contemporary setting, people still willingly kill or maim one another because of their belief (or disbelief) in that particular supernatural entity. The cultish character of contemporary life, however, is suffused more generally with – this is the best way to put it – *numerous* ontological disagreements. There are the many disagreements over *facts*, to be sure. But intertwined with these, and as important, are beliefs and disbeliefs about *kinds* of entities – extra-terrestrials, fossils (items taken to have origins *millions* of years ago), as well as things like chakras, astrological energies (ones denied by the sciences), and run-of-the-mill supernatural items: angels, devils, heavens and hells, ghosts.

Most philosophers *can't* take seriously the existence questions the general public takes seriously; it's equally hard for nonphilosophers to take seriously (or even understand) the existence questions that are raised in contemporary metaphysics courses (Loux and Zimmerman 2003; Sider et al. 2008; Kim et al. 2011). Do numbers exist? Properties? The Civil War? Is anything an instantiated nounphrase covers ("sakes," "ways to fly," "days and weeks," etc.) a *thing*? When Yablo (1998, 230) writes that a "line of research aimed at determining whether Chicago, April, Spanish, etc. really exist strikes this cast of mind as naïve to the point of comicality," he isn't channeling a global attitude of risibility towards existence questions *of all sorts* by salt-of-the-earth sensible folk; he's channeling the nonphilosophical impression of *certain* existence questions – ones induced by professional concerns.

The failure by some philosophers to recognize that nonphilosophers are concerned with *ontological questions* has been going on for some time. It already appears in Carnap and Quine, when Carnap (1950, 207) writes that the "question [of the reality of the thing world itself] is raised neither by the man in the street nor by scientists, but only by philosophers," and when Quine (1981, 9) writes that ontological "concern is not a correction of a lay thought and practice; it is foreign to the lay culture, though an outgrowth of it." A similar view is expressed by those metaphysicians who claim (as many do) that one first learns about ontology in classrooms (or in the "philosophy room"), or that nonphilosophers don't mean by "exist" what philosophers mean by it (a lightweight "ordinary" use of "exist," versus a heavy "ontological" use: Chalmers 2009, 81). Meanwhile, chakras and extra-terrestrials continue to excite (much of) the general public; chemists and physicists (around 1900) debated the existence of atoms (Azzouni and Bueno 2016; Seth and Smith forthcoming for the history; van Fraassen 2009 for denials of this); until recently cosmologists wondered if extrasolar planets exist; and many still wonder about life on other planets.

There are more powerful factors behind the widespread failure of analytic philosophers to recognize the public's continuing concern with metaphysics – apart from the sociology just mentioned. Certain *interpretations*, as we'll see in III, IV, and V, of the logical formalisms used to model natural languages lead to characterizations of formal languages *and* natural languages as unsuitable for expressing existence questions and debate. The apparent corollary: existence questions and debates *themselves* make no sense or are a matter of language-relative application conditions for words. This is because there are taken to be no language frameworks (artificial *or* natural) within which serious versions of existence questioning and debates can occur. The result is *weird* because philosophers, in particular, continue to engage in serious ontological debate and raise existence questions. (*Somehow*, in *some* language medium, they're managing this. So are the rest of us.)

It largely remains unrecognized that the view that natural and artificial languages aren't possible frameworks for serious ontological discussion arises in large measure from assumptions *about* formal tools – specifically logical ones, and from how the properties of those formal tools are taken to show properties of natural languages; it doesn't arise from the intrinsic properties of those tools. Relatedly, most of the striking metaontological disagreements between philosophers arise from differing empirical presuppositions about specific locutions in natural languages as well as disagreements about what formalizations of those locutions are possible. This is the case I try to make in this chapter.

It's hard to *overstate* the importance of ordinary ontological questions and debates. That we can raise such questions, argue about them, and reasonably (perhaps) decide them one way or the other is central to our collective self-image as creatures aspiring to be rational. **We can find out what's in the world and what isn't; we can discover what's made up by us, or projected by us onto the world, and what, instead, is real.** It's *this* that fuels the general public's continuing fascination with ontological questions – and not (as a superficial glance might indicate) faddish, intellectually outdated, or primitive concerns with magic and the supernatural.

II Conditions for cogent/serious ontological questioning and debates

Consider the current debate about extra-terrestrial life, or early twentieth-century arguments among professional scientists about atoms. Or consider contemporary philosophical existence disagreements about numbers. These seem to have several properties in common (Azzouni 2017, 23).

1   They occur in the same language.
2   The disputants understand what (kind of) object the disagreement is over. Each opponent can express the disagreement, and the other's position, in her own terms.
3   Although ontological disagreements can be presented descriptively, they can be characterized purely existentially – purely in terms of what exists (and doesn't exist).
4   Disputants needn't agree on the descriptions of the entities in question. They can disagree over the existence of atoms, for example, or God, and *also* disagree on the properties atoms and God have (or would have) if they existed.

These, nearly enough, are truisms. This doesn't mean they don't need further elucidation; and it doesn't mean they're *true*. The question at hand is what existence questions and debates look like if we (have to) give any of these conditions up. Notice, though, this *platitude*: Two individuals can't *have* an ontological debate – any debate, really – unless they speak the same language. They need words *in common* to debate *with*. Thus, for different parties to raise questions about the possible existence of extra-terrestrials, or chakras, it seems they need *in common* the concepts of existence, chakras, and extra-terrestrials. Similarly, to entertain the existence or nonexistence of something (talking bears), one needs – so it seems – certain words, or concepts, to do this, "exist" and "talking bears" among them. The words "exist" and "talking bears," or the concepts corresponding to these words, anyway, also need certain properties if posing questions about the existence of talking bears is cogent or nontrivial. I'll illustrate this in later sections.

III How Carnap and application-conditions-neoCarnapians eliminate ontological debate

Carnap denies there are languages in which ontological questions can be raised, and real ontological debates occur. *Natural* languages are useless for serious discourse: they're useless for science (including semantics) and philosophy. He replaces them with a family of artificial languages, individuated by their logical and lexical vocabulary as well as by rules governing that vocabulary. In particular, the quantifiers of these languages ("$(\exists x)$," "(x)") differ from the quantifiers in other languages in their **discourse-domains** (Carnap 1950; Yablo 1998; Eklund 2009; Azzouni 2017; Kraut, this volume; see Carus 2007 for a historically nuanced discussion of Carnap's evolving positions).

The upshot is that ontological debate is ruled out for lack of a medium for it to take place in. Carnap takes the truth of "$(\exists x)Px$" within a language L as the satisfaction of a certain set of operations encapsulated by the rules (in an artificial language) governing "P." That there are numbers, according to a number language, follows directly from the rules of that language; that 2 is prime follows from a proof in the number language using logical and arithmetical rules; that there are rocks follows from the application rules of "rock" in a thing language – when, for example, something is examined and determined (by the rules for "rock") to be a rock; a far more difficult set of operations does the same for "electron." These are all "internal questions." These questions may be hard to answer, and they may engender debates; nevertheless, they're not genuine ontological questions like: "Are there *really* rocks or electrons?" There are Carnapian "cartoon languages." Here too questions can arise, e.g., "How many daughters did Mickey Mouse have?" This isn't an ontological question – but it *is* a Carnapian internal question. We *know* Mickey Mouse doesn't exist.

There are two kinds of questions that Carnap replaces purported ordinary existence questions with. Internal questions are one kind. On this reconstrual of real metaphysical questions, when chemists and physicists debated the existence of atoms, they were asking an internal question. Similarly, some are asking: Are there Tasmanian tigers (still) left? – this is answered by the language-internal rules of "Tasmanian tiger." Ontology thus is "easy ontology" (Thomasson 2015; also Fine 2009 and Schaffer 2009 for a different route to trivial existence inferences; Daly and Liggins 2014 for objections): These "ontological questions" are resolved by using the application conditions governing words. Thus, "how many daughters did Mickey Mouse have?," on this Carnapian reconstrual of ontological questions as internal ones *is* as much an ontological question as questions about electrons.

The other sort of admissible Carnapian question is an "external" one. On this reconstrual, our apparently ordinary metaphysical questions – "Are there atoms, *really*?" "Are there Tasmanian tigers, *really*?" are practical questions like whether we should be using a language in which "Tasmanian tiger" occurs, or whether we should adopt a new language which has different application conditions for the word "atoms." These "external questions" are about the pragmatic usefulness of languages having certain vocabulary with particular application conditions. External questions, thus, aren't about the references of words, although ontological questions seem to be. Anything that appears to be a genuine ontological question, on Carnap's view, is recalibrated as either an internal or an external question.

These interpretations, however, seem to misrepresent these debates. The problem is that there *were no* fixed application rules, conventions or linguistic rules governing the word "atom," or if there were, the rules seem to have been (and still are) *defeasible* ones mutating easily under pressure from *both* empirical results and conceptual considerations *without changing the identity of the words in question or what they refer to*. Perrin's atoms were governed by classical physics; those of most physicists shortly into the twentieth century were *not*. Everyone nevertheless took everyone else to be talking about the *same* atoms. That seems to be a crucial aspect of ordinary ontological debate – as condition 4 of section II indicates. Everyone *did not* think they were only arguing over the same *word*.

Claims that "Tasmanian tiger," and other "natural-kind" words haven't rules of application was an early objections to Carnap's picture – and to others like it. The point is perhaps first made by Wittgenstein (1958), and this rejoinder has only been deepened and sustained by contemporary lexical semantics and studies in concepts (Jackendoff 1983; Jackendoff 1990; Fodor 1998). Austin (1962, 120–121), noted that ordinary words ("pig," "tit") haven't such rules; the point was argued for by Quine (1953b) and Putnam (1962), and used by them to undermine the analytic/synthetic distinction; Kripke (1980) used it to support his early causal theory of meaning applied to natural-kind terms like "gold," "zebra," and the like.

The nonexistence of *correct* application conditions for "natural-kind" words in natural languages shouldn't be confused with stronger claims the above philosophers also rejected. At issue isn't that many ordinary words don't have definitions or necessary and sufficient conditions (although that's also claimed by these philosophers), and it isn't that there are no pure "observational" or "verificationist" application rules (although that was *stressed* by them). The point is only that there are no discernable rules we acquire in learning natural languages that can provide anything but temporary de facto (nonsemantic) application conditions for a very large class of words. These **de facto** rules ("stereotypes" – Putnam 1975) can't be described as application conditions that provide the "correct" applications of words – they can't be treated as giving the "conceptual content" for these words. If these philosophers are right about ordinary usage (if their thought experiments about the possibility of blue gold, robot-cats, etc., are correct), ordinary and scientific ontological questions can't be reconstrued internally in Carnap's sense

(Putnam 1962, 42–46; Putnam 1965); but see Thomasson 2015, and Thomasson, this volume, for the view that there are such rules of application in ordinary language, and Price 2009 for similar claims).[1]

Many natural-language words *do* have application conditions that their speakers recognize: legal words like "debt," "contract," "notorized document," "jury," "marriage"; calendar words like "Thursday," "public holiday," "day"; defined game-words, "touchdown," "strike"; and others too, e.g., "fictional character." Nonphilosophers recognize (consciously or subliminally) these special words to have **stipulated** application conditions: they never raise ontological questions about such things. As Yablo's remarks indicate, nonphilosophers find ontological debates about these weird and off-putting ("Does *chess* exist?" "Huh?"). This is no accident. It's words that *lack* such conditions, "atom," "zebra," "bird," "vegetable," "electricity," "God," gold," – and others, like "exist" and "object" – that *ordinary* ontological questioning and debate focuses on. (See Azzouni 2017, 38–46, for a distinction between criterion-transcendent and criterion-immanent expressions.)

IV Quine's attempted resurrection of ontological debate

Quine (1953a) denies Carnap's internal/external distinction. The result, he thinks, allows genuine ontological debate: McX and Wyman think Pegasus exists, Quine doesn't. Further, Quine gives an ecumenical criterion for when *any* first-order *discourse* is committed to something: That something falls within the "range of its variables." His criterion is still controversial: Is it semantic, as Quine seems to describe it, or syntactic? How does it apply to non-first-order languages? (Cartwright 1954; Church 1958; Schleffler and Chomsky 1958; Boolos 1984; Bricker 2016; Azzouni 2015; Egerton, this volume; Parent, this volume.)

Quine also rejects Carnap's characterization of artificial languages as various (alternative) mediums for cogent discourse, ones differing in their descriptions of "what there is" by differing in rules that govern their differing quantifier domains. Instead, Quine (1960, 1–3) argues that we're trapped in the language/conceptual scheme we inherit. Carnap's artificial languages are still valuable, but only as "regimentations": transcriptions of portions of natural language into formal structures that – given our expressive needs – exhibit required syntactic and semantic properties.

This means "what there is" is what falls under the variables of the one language we're (collectively) speaking. Despite this, Quine's ecumenical criterion apparently allows us to debate others who disagree with our ontology. We debate with, say, other speakers of our own discourse about whether there are talking bears; that is, do our variables range over talking bears? This isn't a debate over what our variables *should* range over because we don't get to choose *that* – although we do get to choose whether we *think* our variables range over talking bears or not.[2] This characterization of ontological debate is extended to speakers of alternative (first-order) discourses, or of non-first-order discourses that we've transcribed into first-order terms. They assert (∃x)Talking-bearsx; we deny (∃x)Talking-bearsx; we're debating with those speakers about what's in our (common) domain of discourse. Quine's characterization of existence debates changes the ordinary picture in both subtle and obvious ways.

The Santa Claus paradigm. Oliver thinks Santa Claus exists; Anna Ella thinks he doesn't. Both agree he's a jolly white-bearded man who flies around in sleds. They don't disagree about the description of Santa Claus (they agree on what he's *like*) – it's a disagreement over his *existence*. Similarly, Sherlock Holmes and Saint Patrick, we both agree, are depicted in stories. But although both sets of stories are false (so we think) Saint Patrick actually existed although Sherlock Holmes didn't. Moses is a similar case: we seem able to separate arguments about whether the stories about him are true from those about whether he existed. Different answers to these questions are possible (Kripke 1980).

For Quine, debates about truths don't separate from debates about ontology; (3) is wrong. Ontological disagreements, strictly speaking, are only over the truth of a set of sentences understood in common – and only some of which are existential. The believer in Pegasus thinks true some set of sentences that implies that some complex predicate is instantiated ("a winged horse born from the blood of the slain Medusa … who was believed in by ancient Greeks …"); her opponent thinks false some or all of these sentences.

To support Quine's view, it can be claimed that the impression that an existence disagreement is over *an entity* is an **aboutness illusion**: existence debaters are talking about a *specific something* they affirm and deny the existence of. This (Quine 1953a) is unfair to existence-deniers. If something doesn't exist, no one can talk about *it* (Parmenides 1984; Quine 1953a; Cartwright 1954; Reimer 2001; Azzouni 2014). Meinongians can retort that this is no illusion: some discourses about known fictions are true, the sentences in them distinguish between different fictions ("Hercules is strong," "Mickey Mouse isn't") and so the truth of those discourses **require** distinctive nonexistent (Meinongian) entities whose properties support truth conditions (Chisholm 1972; Parsons 1980; Routley 1980; Zalta 1988; Priest 2005).

Condition (4) is rejected by Quine as well – this rejection follows from his rejection of condition (3). There's no sense to a distinction between disputants who disagree on the properties of the same particle or who disagree about the existence of different particles. Quine (1960, 16) applauds this result in his "parable of the neutrino."

A widely accepted presupposition of the Quinean approach to replacing ontological debate about entities with ontological debate about sets of sentences is **charity**: Our "understanding" of words comes to the holding true sentences that those words appear in. To possess interpreted words in common, therefore, is to possess truths in common these words appear in. This is a corollary to a particular approach to the interpretation of other languages: formulate a Tarski-style theory of truth (Tarski 1983) of those languages in our own. This enables an understanding of foreign discourse by mapping the sentences they take true to ones we take true (Quine 1960; Davidson 1984; Hawthorne 2006; Eklund 2008; Hirsch 2011).

The charity model of understanding and interpretation, however, interferes with the view that ontological *opponents* understand each other. (Quine's approach puts pressure on condition (2).) Deniers of Pegasus' existence can't understand proponents of his existence by means of the sentences they and their opponents *agree on* (what's important are the sentences they *don't* agree on). But, on the charity model, words are understood in common by means of truths containing those words held in common. As far as the *truths* Anna Ella accepts are concerned, Santa Claus and the tooth fairy are the same. (It's the falsehoods believed by Oliver – falsehoods according to Anna Ella – that distinguish "Santa Claus" and "the tooth fairy.") Thus Anna Ella doesn't have the words "Santa Claus" and "the tooth fairy" in common with Oliver because they don't *share* the needed truths. According to the charity model, Anna Ella doesn't understand Oliver's *words* "Santa Claus" and "the tooth fairy."

If ontological debates are characterized – as Quine does – as disagreements over truths, ontological debaters must characterize both the truths *and* falsehoods of their opponents to make *sense* of their disagreements.[3] But charity, as Quine understands it, doesn't allow this. I've argued, therefore, that although Quine's approach to ontological debate leaves intact some aspects of ontological debates, it undercuts something significant that's part of our ordinary practice: Opponents *understand* one another.

V Equivocality ontology

A and *B*, let's say, have different ontological commitments. Perhaps *A* is committed to werewolves and *B* isn't. Perhaps *A* is committed to incars and outcars but not cars, and *B* is committed to cars, but not incars and outcars. Perhaps *A* is committed to ordinary objects, but *B* is committed in addition to the mereological sums of such objects.[4] And, perhaps, *A* is a nominalist (a believer in concrete objects but not mathematical abstracta) but *B* is a Platonist (a believer in *both* mathematical abstracta and concrete objects). Let us understand the words they use, "exist," "there is," "real," etc., along the lines of the formal "($\exists x$)": a discourse-domain interprets these words.

According to the **univocality** approach, if *A* and *B* *debate* the existence of something, they *share* the ontological notions – of "exist," "there is," and "real" – and share their accompanying discourse-domain. They disagree over what's in that domain. (Compare: two people with the same notion of "gold" disagree about whether *that ring* is gold.) Call this a *univocality disagreement* about existence. Using the same words with the same interpretation, there is a disagreement over whether those words apply to a particular something (or a particular kind of somethings).

A popular alternative (Putnam 1987; Hawthorne 2006; Bennett 2009; Eklund 2008; Eklund 2009; Cameron 2010; Hirsch 2011; Sider 2011) is to take *different* quantifier domains as indicating different meanings for "exist," "there is," "real," etc., in just the same way that the word "bank" means different things depending on whether its extension contains all and only financial institutions or instead all and only sides of rivers. (Sometimes this alone is called "quantifier variance," e.g., Manley 2009, 18, and sometimes additional conditions are added to it, e.g., Hirsch 2011, xiv.[5]) We can thus think of an *A* and *B* who differ in their existence commitments as having a **domain disagreement** when their words correspond to different domains. Call this an *equivocality disagreement* about ontology, and describe *A* and *B* as speaking different **ontological languages** (Hirsch 2011, xii).

Some domain disagreements, it has been argued, are substantial and some are only verbal. Given two different ontological languages, L_A and L_B, suppose there is a mapping between them taking the truths *B* holds (about werewolves, incars and outcars, ordinary objects and mereological sums of ordinary objects) to the truths *A* holds (about certain humans and/or dogs, cars, ordinary objects and groups of objects). Then one can say the apparent disagreement between *A* and *B* is only verbal. *A* understands *B*'s claims via *A*'s own sentences that *B*'s claims are mapped to (and vice versa). On the other hand, suppose no mapping exists between the two ontological languages L_A and L_B that takes *B*'s truths about concrete objects to *A*'s truths about concrete objects and abstracta. Then, that domain disagreement can be presumed substantial – not verbal (Bennett 2009; Hirsch 2011).

There is controversy over whether and what kinds of mappings are possible. For example ("the semantic objection") it's claimed that appropriate interpretations must be Tarskian, which means the mappings must take account of what the terms refer to (Hawthorne 2006; Eklund 2008; Eklund 2009). It may be countered that all that's needed for mutual interpretation is the preservation of the truths believed (along the lines of charity, described in the last section): "coarse-grained" sentence-to-sentence mappings are sufficient and exist, and these don't (or needn't) map the words used by apparent ontological opponents to one another (Hirsch 2011; Azzouni 2017 worries whether such coarse-grained mappings exist).

For purposes of evaluating whether equivocality characterizations of ontological debates and questions leave them intact, this issue can be set aside. Let's agree that there sometimes are such suitable mappings and that a distinction between substantial and verbal domain disagreements can be drawn in terms of whether they exist or not. What's the status of existence debates and questions?

Distinguish between existence debates, as they occur *between* A and B, speakers of alternative ontological languages, and existence debates between equivocalist *philosophers* using speakers like A and B as examples to study ontology. If A and B are *debating*, they must (so it seems) be doing so via translations between the (relevant) sentences of their respective languages. Consider first, *substantial* debates. Suppose no truth-to-truth mappings between the languages of Platonists and nominalists exist. Then Platonist A and nominalist B can't talk to one another because they don't speak one another's languages. No debate. If, on the other hand, *A and B* speak both languages, they can talk to one another, but they'll still not recognize an *existence* disagreement that's the grounds of a *debate*. True, there is a *domain-disagreement*, but all *that* means is that the *different* existence-words "plat-existence" and "nom-existence," respectively of each language, don't mean the same things. It's no contradiction – it's not even controversial – that plat-existing things needn't nom-exist. A and B can agree on that!

On the other hand, suppose there *are* truth-to-truth mappings that can be used by A and B to interpret one another regarding some domain-disagreement. The result is a translation of their respective remarks that involves no disagreements – ontological or otherwise. Again: No debate.

What about equivocalist *philosophers* – can *they* engage in debates? We have the same problem with *their* supposed debates that we've just seen ordinary debates to have. What, for example, follows about *ontology* from some domain disagreements being only apparent (because the appropriate truth-to-truth mappings exist)? One apparent position is realist about "what there is" because *one* such language matches the world's own natural quantificational structure (it "carves reality at the joints"); the others don't (Cameron 2010; Sider 2011). Call this language "Ontologese." One can deny Ontologese, however: by denying the world has any such natural quantificational structure that can match *any* such language; one can thus apparently be an anti-realist about "quantificational structure." One can also, more neutrally, claim that a large class of "truth-conditionally-equivalent ontological languages" – ones between which such truth-to-truth mappings exist – are equally metaphysically meritorious (Hirsch 2011, xiv).

One oddity about *this* debate is that – contrary to the equivocality assumption – philosophers aren't honoring the initial assumption that different *meanings* for quantifier expressions are expressed in these various languages. What's needed to make the philosophical debate a *debate* is reading the quantifiers of all these languages as if they indicate the *same notion* of existence. Similarly, a second-order debate about banks may occur by asking of financial institutions and sides of rivers: But which are the *real* banks? Talk of "carving reality at its joints," or "matching the world's 'quantificational structure' " that philosophers use in these debates are euphemisms for the straightforward, "what really exists" and the like. Recognizing this brings us back to earth, to the original domain disagreement between the nonphilosophical A and B. Equivocal approaches eliminate the possibility of ontological debate between them. Equivocalist philosophers restore sense to their own ontological debates only by holding *in common* "carve reality at the joints" or "ontological structure." But why is this consistent with the letter or spirit of equivocalism? Why aren't second-order words "carve reality at the joints" or "ontological structure" different in meanings (as used by these philosophers) just as the first-order words "exist" and the like are?

There is a closely related debate (among philosophers) about natural-language words that, if any do, indicate ontology: "exist," "there is," "real," etc. Are such words are polysemous or not? (Yes: Hofweber 2016; in some sense yes: Fine 2009; no: Quine 1953a, Schaffer 2009, van Inwagen 2009, Thomasson 2015, Azzouni 2017.) The only point to make about *this* debate is that, like all debates about "exist," "real," "there is," and other natural-language words, its resolution requires the tools of linguistics and usage evidence. (We've seen this already in section II with respect to supposed application-condition rules for natural-kind words.) That is, questions of ambiguity and polysemy must be resolved by means of the subtle empirical methods, among them, substitution tests and the Chomsky conjunction test (Hofweber 2007, 27; Azzouni 2017, 64–67, for philosophical applications – and disagreements – about what the conjunction test shows; Wierzbicka 1996, 95, and elsewhere, for illustrations of a substitution test). The metaphysical conclusions come afterwards.

A similar point holds about natural-language quantifier domain restriction ("There is no beer," contextually signaling, "there is no beer in this refrigerator") and whether that shows a maximalist, trivialist, or lightweight ontology at work in natural language (Yes: Lewis 1986, Schaffer 2009, Thomasson 2015; no: Daly and Liggins 2014, Azzouni 2017, 112 – in particular because domain restriction phenomena doesn't apply to "exist").

I return to the relationship between linguistics and metaphysics in section VII.

VI Coupling univocality with neutrality

Is there a way to satisfy the four conditions on ordinary ontological debates and questions *directly*? Yes. I'll illustrate this in a formal setting first. A technical move (not the only one possible) introduces *two* formal devices, a quantifier "$(\exists x)$" and a predicate expression, "E." Interpret the quantifier neutrally: it "quantifies" over what exists *and* what doesn't (Raley and Burnor 2011; Bueno 2014; Azzouni 2017, appendix to the general introduction).[6] "E" is an existence predicate. Its extension is all and only what exists.

Characterize ontological debates between disputants this way: speakers have in common a quantifier domain (associated with their shared quantifiers) as well as a shared extension for "E." The disagreement isn't about what's in the quantifier domain; it's about what's in the extension of the existence predicate. They don't disagree that werewolves, incars, outcars, cars, ordinary objects, mereological sums of such objects, abstracta, and much else, are in the shared domain – all that means is that they share a vocabulary they can argue with; they disagree about which of these *exist*. A thinks that werewolves exist; B doesn't; A thinks incars and outcars exist, but not cars; B thinks *vice versa*; A thinks ordinary objects exist, but not mereological sums of them; B thinks both exist. Finally, A thinks concreta are in the extension of E but not mathematical abstracta; B thinks both abstracta and concreta are in that extension.

In formal terms (regarding werewolves), A thinks $(\exists x)(Wx \;\&\; Ex)$ – because, after all, A doesn't think *all* werewolves exist; B thinks $(x)(Wx \rightarrow \neg Ex)$. And both think: $(\exists x)Wx$. (1) They're speaking the same language; (2) they share the word "werewolf," which they both understand; (3) their disagreement is solely over whether the extension of E includes werewolves; (4) they can disagree about what werewolves are like – A may think, regardless of whether werewolves exist, they can't be killed by silver bullets; B thinks they can.

Exactly how the natural-language words, "there is," "exist," "real" exhibit the above formal structure – and indeed, if they do at all – are subtle empirical matters. There many options here. One possibility is that the natural-language "there is" and "exist" correspond rather directly, respectively, to the formal "$(\exists x)$" and "E." Another is that natural language is supple – natural-language words play different roles with respect to one another so that speakers can flexibly indicate ontological disagreements and debates when they have them. If so, this is to be established on the basis of usage. (For disagreements on usage with respect to this issue, see Hofweber 2009; Schaffer 2009; Asay 2010; Azzouni 2017.)

VII Concluding homily about linguistic science and metametaphysics

There is a rumor that many contemporary philosophers think metaphysics, and metametaphysics too, must free themselves from the notorious "linguistic turn"; related, there has (recently) been an announcement of an "ontological turn" (Cameron 2010, 8). *Pace* such views, one lesson of this chapter is that almost every debate in contemporary ontology – even though it's about metaphysical matters – turns on (i) prior empirical assumptions about natural-language locutions, on (ii) prior assumptions about what kinds of (formal) properties artificial languages can have, or (iii) both. Some "easy ontologists" think that all natural-kind words come with application conditions; opponents disagree; others think there are easy analytic ontological inferences in natural language; opponents disagree. Some ontologists think the words "exist," "there is," and "real" are univocal and always ontologically committing; others think such words, although univocal, are ontological committing only in certain contexts – either because they function according to quantifier domain-restrictions or for other broader reasons involving pragmatic vectors; still others think the words are polysemous in ways relevant to ontology – on some readings they're committing but not on others. Among philosophers who think such words are sometimes or always committing, some think there is a criterion for what exists associated with the meaning of those words; others think there is such a criterion but it's not associated with the words by virtue of their meanings; and, finally, some think these words involve no such criterion. Some philosophers think these words are never ontologically committing. Many think that the properties of the natural-language words, "there is," "exist," "real," and others, make them useless for ontological research – for one of the above sorts of reasons just given, or for others. Either, in this case, a special language must be regimented in which the appropriate ontological commitments can be expressed, or this isn't possible (and perhaps, isn't even coherent). Ontologese proponents (or those who think the semantic properties of natural languages are easily changed) may couple metaphysical assumptions about natural kinds with "reference magnetism" claims about how words can be semantically sensitive to metaphysical structure.

The presuppositions dividing metaontologists aren't always transparent. I've not fully described all the disagreements among metaontologists, nor have I even illustrated in this chapter all the disagreements that I sketched in the last paragraph; but I've illustrated enough of them to support the claim of the last paragraph. In principle, therefore, adjudication of these issues ought to be a matter of the linguistic science of natural (and formal) languages. We should write on the gate above the entrance to metaphysics an alternative to what's rumored to have been Plato's dictum:

Let no one ignorant of linguistics enter here.

Notes

1 Thomasson (2015, 38–9) writes about such supposed application rules that "the point is the simple, almost trivial observation that for a question to be asked meaningfully the terms in it must be governed by rules of use: we must be using a linguistic framework to ask an (internal) existence question." See Thomasson (2015, 91) for conditions on such rules, especially that they induce "correct applications" of the words in question. See Azzouni (2017, 49–53) for how words like "coral," "berry," "bread," and many others *have* mutated, and how that shows there are no such rules for these words. Devitt and Sterelny (1999, 73) argue, however, that all such words must have some conceptual content – and so they must have some application conditions – because of the "qua problem" for how the references for "natural kind" words are grounded in samples. A sample item, that is, often belongs to more than one kind, "not only an echidna, but also a monotreme, a mammal, a vertebrate, and so on." For the reference of a word to be successfully grounded in samples, the kind has to be indicated.
2 Quine is a fallibilist. If "There are talking bears," is true, then talking bears fall within the range of our variables – whether we realize it or not.
3 Competently applied charity may need to make sense of the intentions of speakers too (see Horden 2014). If so, charity may require that we recognize the speaker (sometimes) intends something she knows we regard as false.

4 Incars (respectively, outcars) have persistence conditions so that they exist only within (outside) garages. An incar transforms into an outcar whenever it's driven out of a garage, and vice versa. The mereological sum of a napkin and a fork is the spatially scattered object composed of both napkin and fork.
5 "Quantifier variance" is *awful* terminology. First, it means more than one thing in the literature. More importantly, it's originally due to a characterization of the relevant locutional linchpins for ontology being the quantifiers ("(∃x)," "("x)") with associated quantifier domains – as they're characterized in formal languages (and when employing Tarskian semantics). But, none of the locutions typically used for ontological commitment ("there is," "exist," "real") *are* natural-language quantifiers. Genuine natural language quantifiers ("many," "some," "all," "a good number of") ironically, are rarely directly used by speakers to ontologically commit themselves (Azzouni 2017). I've thus instead adopted the words "univocal," and "equivocal," and their cognates, to distinguish positions here.
6 "Quantifies over," as well as "domain" are misleading metaphors for neutralists; there aren't collections of things (somewhere or other) that the quantifiers of a language operate in relation to. See Azzouni (2017, appendix to the general introduction).

References

Aikin, S. F. and Talisse, R. B. (2011). *Reasonable atheism: A moral case for respectful disbelief*. Amherst, NY: Prometheus Books.
Asay, J. (2010). How to express ontological commitment in the vernacular. *Philosophia Mathematica* 18(3), pp. 293–310.
Austin, J. L. (1962). *Sense and sensibilia*. Oxford: Oxford University Press.
Azzouni, J. (2014). Freeing talk of nothing from the cognitive illusion of aboutness. *Monist* 97(4), pp. 443–459.
Azzouni, J. (2015). The challenge of many logics: A new approach to evaluating the role of ideology in Quinean commitment. *Synthese* [online] Available at: doi:10.1007/s1129-015-0657-9.
Azzouni, J. (2017). *Ontology without borders*. Oxford: Oxford University Press.
Azzouni, J. and Bueno, O. (2016). True nominalism: Referring versus coding. *British Journal for the Philosophy of Science* 67(3), pp. 781–816.
Bennett, Karen. (2009). Composition, collocation, and metaontology. In D. J. Chalmers, D. Manley, and R. Wasserman, eds., *Metametaphysics: New essays on the foundations of ontology*, 1st ed. Oxford: Oxford University Press, pp. 38–76.
Boolos, G. (1984). To be is to be the value of a variable (or to be some values of some variable). *Journal of Philosophy* 81, pp. 430–449.
Bricker, P. (2016). Ontological commitment. [online] Available at: Edward N. Zalta, ed., *The Stanford Encyclopedia of Philosophy* (Winter 2016 Edition), URL = https://plato.stanford.edu/archives/win2016/entries/ontological-commitment/.
Bueno, O. (2014). Nominalism in the philosophy of mathematics. [online] Available at: Edward N. Zalta, ed., *Stanford Encyclopedia of Philosophy*, CSLI; URL = http://plato.stanford.edu/archives/spr2014/entries/nominalism-mathematics/.
Cameron, R. P. (2010). Quantification, naturalness, and ontology. In A. Hazlett, ed., *New waves in metaphysics*, 1st ed. Houndmills, Basingstoke, Hampshire, Great Britain: Palgrave Macmillan, pp. 8–26.
Carnap, R. (1950). Empiricism, semantics, and ontology. In *Meaning and necessity* (1956), 1st ed. Chicago: University of Chicago Press, pp. 205–221.
Cartwright, R. (1954). Ontology and the theory of meaning. *Philosophy of Science* 21(4), pp. 316–325.
Carus, A. W. (2007). *Carnap and twentieth-century thought: Explication as enlightenment*. Cambridge: Cambridge University Press.
Chalmers, D. (2009). Ontological anti-realism. In D. J. Chalmers, D. Manley, and R. Wasserman, eds., *Metametaphysics: New essays on the foundations of ontology*, 1st ed. Oxford: Oxford University Press, pp. 77–129.
Chisholm, R. (1972). Beyond being and nonbeing. In P. van Inwagen and D. W. Zimmerman, eds., *Metaphysics: The big questions* (2008), 1st ed. Oxford: Oxford University Press, pp. 40–50.
Church, A. (1958). Symposium: Ontological commitment. *Journal of Philosophy* 55, pp. 1008–1014.
Craig, W. L. (2017). *God and abstract objects: The coherence of theism: Aseity*. Cham, Switzerland: Spring.
Daly, C. and Liggins, D. (2014). In defense of existence questions. *The Monist* 97(4), pp. 460–478.
Davidson, D. (1984). *Inquiries into truth and interpretation*. Oxford: Oxford University Press.

Devitt, M. and Sterelny, K. (1999). *Language & reality: An introduction to the philosophy of language*. Cambridge, MA: MIT Press.
Eklund, M. (2008). The picture of reality as an amorphous lump. In T. Sider, J. Hawthorne, and D. W. Zimmerman, eds., *Contemporary debates in metaphysics*, 1st ed. Malden, MA: Blackwell, pp. 382–396.
Eklund, M. (2009). Carnap and ontological pluralism. In D. J. Chalmers, D. Manley, and R. Wasserman, eds., *Metametaphysics: New essays on the foundations of ontology*, 1st ed. Oxford: Oxford University Press, pp. 130–156.
Fine, K. (2009). The question of ontology. In D. J. Chalmers, D. Manley, and R. Wasserman, eds., *Metametaphysics: New essays on the foundations of ontology*, 1st ed. Oxford: Oxford University Press, pp. 157–177.
Fodor, J. A. (1998). *Concepts: Where cognitive science went wrong*. Oxford: Oxford University Press.
Hawthorne, J. (2006). Plenitude, convention, and ontology. In D. J. Chalmers, D. Manley, and R. Wasserman, eds., *Metametaphysics: New essays on the foundations of ontology*, 1st ed. Oxford: Oxford University Press, pp. 53–70.
Hirsch, E. (2011). *Quantifier variance and realism: Essays in metaontology*. Oxford: Oxford University Press.
Hofweber, T. (2007). Innocent statements and their metaphysically loaded counterparts. *Philosophers' Imprint* 7(1), pp. 1–33.
Hofweber, T. (2009). Ambitious, yet modest, metaphysics. In D. J. Chalmers, D. Manley, and R. Wasserman, eds., *Metametaphysics: New essays on the foundations of ontology*, 1st ed. Oxford: Oxford University Press, pp. 260–289.
Hofweber, T. (2016). *Ontology and the ambitions of metaphysics*. Oxford: Oxford University Press.
Horden, J. (2014). Ontology in plain English. *The Philosophical Quarterly* 64(255), pp. 225–242.
Jackendoff, R. (1983). *Semantics and cognition*. Cambridge, MA: MIT Press.
Jackendoff, R. (1990). *Semantic structures*. Cambridge, MA: MIT Press.
Kim, J., Korman, D. Z., and Sosa, E. (2011). *Metaphysics, an anthology*, 2nd ed. Malden, MA: Wiley-Blackwell.
Kripke, S. (1980). *Naming and necessity*. Cambridge, MA: Harvard University Press.
Lewis, D. (1986). *On the plurality of worlds*. London: Blackwell.
Loux, M. J. and Zimmerman, D. W. (2003). *The Oxford handbook of metaphysics*, 1st ed. Oxford: Oxford University Press.
Manley, D. (2009). Introduction: A guided tour of metametaphysics. In D. J. Chalmers, D. Manley, and R. Wasserman, eds., *Metametaphysics: New essays on the foundations of ontology*, 1st ed. Oxford: Oxford University Press, pp. 1–37.
Parmenides. (1984). *Parmenides of Elea*. Toronto, Canada: University of Toronto Press.
Parsons, T. (1980). *Nonexistent objects*. New Haven, CT: Yale University Press.
Price. H. (2009). Metaphysics after Carnap: The ghost who walks? In D. J. Chalmers, D. Manley, and R. Wasserman, eds., *Metametaphysics: New essays on the foundations of ontology*, 1st ed. Oxford: Oxford University Press, pp. 320–346.
Priest, G. (2005). *Towards non-being: The logic and metaphysics of intentionality*. Oxford: Oxford University Press.
Putnam, H. (1962). The analytic and the synthetic. Reprinted in *Mind, language and reality: Philosophical papers*, vol. 2, 1st ed. Cambridge: Cambridge University Press (1975), pp. 33–69.
Putnam, H. (1965). How not to talk about meaning. Reprinted in *Mind, language and reality: Philosophical papers*, vol. 2, 1st ed. Cambridge: Cambridge University Press (1975), pp. 117–131.
Putnam, H. (1975). The meaning of "meaning." In *Mind, language and reality: Philosophical papers*, vol. 2, 1st ed. Cambridge: Cambridge University Press, pp. 215–271.
Putnam, H. (1987). *The many faces of realism*. La Salle, IL: Open Court.
Quine, W. V. (1953a). On what there is. In *From a logical point of view* (1980), 3rd ed. Cambridge, MA: Harvard University Press, pp. 1–19.
Quine, W. V. (1953b). Two dogmas of empiricism. In *From a logical point of view* (1980), 3rd ed. Cambridge, MA: Harvard University Press, pp. 20–46.
Quine, W. V. (1960). *Word and object*. Cambridge, MA: MIT Press.
Quine, W. V. (1981). Things and their place in theories. In *Theories and things*. Cambridge, MA: Harvard University Press, pp. 96–99.
Raley, Y. and Burnor, R. (2011). The predicate approach to ontological commitment. *Logique et Analyse* 54, pp. 359–377.
Reimer, M. (2001). A "Meinongian" solution to a Millian problem. *American Philosophical Quarterly* 38, pp. 312–318.

Routley, R. (1980). *Exploring Meinong's jungle and beyond: An investigation of noneism and the theory of items*, interim ed. Philosophy department monograph 3. Canberra: Research School of Social Sciences, Australian National University.

Schaffer, J. (2009). On what grounds what. In D. J. Chalmers, D. Manley, and R. Wasserman, eds., *Metametaphysics: New essays on the foundations of ontology*, 1st ed. Oxford: Oxford University Press, pp. 347–383.

Scheffler, I. and Chomsky, N. (1958). What is said to be. *Proceedings of the Aristotelian Society* 59, pp. 71–82.

Seth, R. and Smith, G. (forthcoming). *Brownian motion and molecular reality*. Oxford: Oxford University Press.

Sider, T. (2011). *Writing the book of the world*. Oxford: Oxford University Press.

Sider, T., Hawthorne, J., and Zimmerman, D. W. (eds.) (2008). *Contemporary debates in metaphysics*, 1st ed. Malden, MA: Blackwell.

Tarski, A. (1983). The concept of truth in formalized languages. In J. Corcoran, ed., *Logic, semantics, metamathematics* (trans. J. H. Woodger), 2nd ed. Indianapolis. IN: Hackett Publishing Company, pp. 152–278.

Thomasson, A. (2015). *Ontology made easy*. Oxford: Oxford University Press.

Van Fraassen, B. C. (2009). The perils of Perrin, in the hands of philosophers. *Philosophical Studies* 143, pp. 5–24.

Van Inwagen, P. (2009). Being, existence, and ontological commitment. In *Existence: Essays in ontology* (2014). Cambridge: Cambridge University Press, pp. 50–86.

Wierzbicka, A. (1996). *Semantics: Primes and universals*. Oxford: Oxford University Press.

Wittgenstein, L. (1958). *Philosophical investigations*, 3rd ed. New York: Macmillan.

Yablo, S. (1998). Does ontology rest on a mistake? *Proceeding of the Aristotelian Society*, 72 (sup.), pp. 229–262.

Zalta, E. N. (1988) *Intensional logic and the metaphysics of intentionality*. Cambridge, MA: MIT Press.

14
ONTOLOGICAL PLURALISM

Jason Turner

Ontological Pluralism is said in many ways, at least if two counts as 'many.' On one disambiguation, to be an 'ontological pluralist' is to be accommodating about theories with different ontologies. This is the pluralism of Carnap (1950) and Putnam (1987) who grant that there are competing ontological visions of reality but deny that any has an objectively better claim to correctness. We are free to accept an ontology of numbers or not, and there's no good philosophical debate to be had about whether there really are any numbers. This is the pluralism of the pluralistic society, where opposing ontological visions need to learn to just get along.

We will discuss the other disambiguation of 'Ontological Pluralism' here. According to it, there are different *ways of being*. In *The Problems of Philosophy*, for instance, Bertrand Russell (1912: 90, 98) tells us that, while there are relations as well as people, relations exist in a deeply different way than people do.[1] The way in which the world grants being to relations is radically different, on this picture, than the way in which it grants it to people.

Despite its pedigree, during the twentieth century analytic philosophers grew suspicious of the notion. Presumably, when the logical positivists tried to kill metaphysics, the idea things could exist in different ways was supposed to die with it. Quine's resurrection of metaphysics linked ontology with quantifiers, making it hard to see what 'things exist in different ways' could mean. Zoltán Szabó puts it this way:

> The standard view nowadays is that we can adequately capture the meaning of sentences like 'There are *F*s,' 'Some things are *F*,' or '*F*s exist' through existential quantification. As a result, not much credence is given to the idea that we must distinguish between different kinds or degrees of existence.
>
> *(2003: 13)*

And this, I think it is fair to say, was the dominant view for quite some time.

But why think there is just one existential quantifier? If, as Russell tells us, the abstract exists in a different way than the concrete, perhaps this is because there are two different existential quantifiers – '$\exists_a$' and '$\exists_c$' – one of which ranges over *abstracta* and the other which ranges over *concreta*. Then we could make sense of multiple ways of existing while respecting the Quinean identification of existence with existential quantification. This idea – that there are multiple existential quantifiers we should use for metaphysics – fuels the contemporary revival of ontological pluralism.

This contemporary revival been carried out largely by Kris McDaniel (2009, 2010, 2017), who both puts the view in historical context and explores several of its potential applications. And there have, of course, been objections. A number of those objections have been addressed in McDaniel's work and my (2010), and I'm not going to rehash them all here.[2] Rather, I'll be looking at – and, to some extent, responding to – some more recent arguments against the view. Before doing that, though, we'll want to get a bit clearer about just what ontological pluralism is.

Elite quantifiers

Contemporary thinking about ontological pluralism links it with *quantificational pluralism*, which says that are multiple existential quantifiers. Left unadorned, though, quantificational pluralism is cheap. Start with any existential quantifier '$\exists$.' We can define a 'smaller' quantifier – that is, an expression that acts like a quantifier over a smaller domain – by restriction. We pick an expression ψ and define $\ulcorner \exists_{small} x \varphi(x) \urcorner$ by $\ulcorner \exists x(\psi(x) \wedge \varphi(x)) \urcorner$. And we can also define a 'bigger' quantifier – that is, an expression that acts like a quantifier over a larger domain. Suppose that P is a proposition that completely describes a possible world just like the actual one except for the addition of a single unicorn. Then $\ulcorner \exists_{big} x \varphi(x) \urcorner$ means $\ulcorner$if P had been true, $\exists x \varphi(x) \urcorner$.[3] So there are surely multiple quantifiers. Quantificational pluralism comes easy.

Intuitively, '$\exists_{small}$' seems to have little to do with whether there are non-ψs, and '$\exists_{big}$' has little to do with whether there are unicorns. Quantifiers like these are mere linguistic artifacts. We should ignore them when doing ontology. Only certain quantifiers – following Ben Caplan (2011) we'll call them *elite* quantifiers – matter for metaphysics. '$\exists_{small}$' and '$\exists_{big}$' aren't elite quantifiers, and so don't tell us a straight metaphysical story about what there is.

Metaphysics should only care about what elite quantifiers range over. We can use this to answer the initial complaint against ontological pluralism. Pluralism isn't just the (unremarkable) claim that there are multiple existential quantifiers. It is the (surprising) claim that there are multiple *elite* existential quantifiers.

What makes a quantifier elite? That's a hard question. It's complicated by the fact that eliteness may come in degrees: Some non-elite quantifiers may be more elite than others. But it is a question that everyone faces, whether pluralists or not. We all need to explain why '$\exists_{small}$' and '$\exists_{big}$' just aren't important for metaphysics.

Some accounts of eliteness may render pluralism incoherent. But at least one popular account does not. Ted Sider (2001, 2009, 2011) has argued that we should extend the Lewisian (1983) notion of *naturalness* – of 'carving nature at the joints' – to expressions of any syntactic category. Some quantifier expressions 'carve nature at the joints' better than others, and the more joint-carving they are, the more elite they are. Nothing this account of eliteness rules out multiple existential quantifiers having it, and in fact contemporary proponents of pluralism have gravitated towards such an account.[4,5]

Can pluralism be defined?

Contemporary ontological pluralists want to defend the idea that, for instance, the abstract and the concrete could exist in different ways. And they do it by saying that the abstract and the concrete could be ranged over by different elite existential quantifiers, say '$\exists_a$' and '$\exists_c$.' This suggests that we could *define* pluralism in terms of elite quantification. As a first pass, we might try:

Ontological Pluralism (first pass): There are multiple ways of being iff there are multiple elite existential quantifiers.

But this definition generates false positives. Consider George, an ontological monist. He is convinced that there are irreducibly *plural* quantifiers. Unlike the singular existential quantifier of first-order logic, which ranges over its domain one thing at a time, the plural quantifier ranges over its domain in groups. We use these plural quantifiers in claims such as 'Some tanks surrounded the fort.' These quantifiers can't be reduced to first-order existential quantifiers.[6] So George thinks there are two elite existential quantifiers: a singular one and a plural one.[7]

But George, it seems, is no pluralist. He doesn't come to believe that there is a different, plural *way of being* when he comes to accept an irreducibly plural quantifier. He comes to believe rather that there is a different, plural *way to quantify* over some things the being of which he had already accepted.

To sort this out, let's back up and ask: Why did our first-pass definition put things in terms of elite *existential* quantifiers rather than elite quantifiers *generally*?

Here's one reason: Someone might think that an existential quantifier '∃' is just as elite as its universal dual '∀.'[8] Such a theorist doesn't think that there are two ways of being, though. They simply think that there are two ways of quantifying over some things with the same mode of being. So as to not classify them as pluralists, we restricted ourselves to existential quantifiers.

A quantifier '∃' and its dual '∀' are part of a *family* of quantifiers. It's baked into their semantics that they quantify over the same things. They just do it in different ways: one at a time or all at a time. The family may have more members as well. It may have plural quantifiers, or generalized quantifiers such as 'most' or 'uncountably many.' These quantifiers are all part of the family because it's baked into their semantics that they range over the same domain.

We focused on existential quantifiers because they belonged to the same family as universal ones, and we wanted a way to pick one representative from the family. But there may be more than one existential quantifier in a given family. This suggests we should have started with a more general proposal. Call a family of quantifiers elite when at least one of its members is elite. Then we can try:

Ontological Pluralism (second pass): There are multiple ways of being iff there are multiple elite quantificational families.

This revised definition no longer misclassifies George.

In principle, different ways of being can overlap (depending on just what those ways of being are). So it could happen that there are two ways of being, and so two elite existential quantifiers, '$\exists_1$' and '$\exists_2$,' that range over precisely the same domain. We might worry that '$\exists_1$' and '$\exists_2$' would be equivalent and our second-pass definition misclassifies this as a monistic view. But notice that, even if this happens, it is not part of the *semantics* of these two quantifiers that they range over the same domain – just as it's not part of *semantics*, but rather part of *geometry*, that 'three-sided planar figure' and 'three-angled planar figure' are necessarily equivalent. '$\exists_1$' and '$\exists_2$' can count as part of different families even if they range over the same domains so long as it's not baked into their meanings that they share a domain.

Ross Cameron (2018) and Arturo Javier-Castellanos (2019) both raise a worry which applies to the second-pass definition just as much as it does to the first. Eli Hirsch (2002a, 2002b) thinks that many ontological disputes aren't 'deep,' because we could speak so as to make one party right or speak so as to make the other party right, and neither way of speaking is more metaphysically privileged than the other. For instance, when some particles are arranged in a tablewise fashion, we could use a 'compositionalist' quantifier '$\exists_{comp}$' to truly say '$\exists_{comp}x(x$ is a table),' or we could use a 'mereological nihilist' quantifier '$\exists_{nihl}$' to truly say 'it's not the case that $\exists_{nihl}x(x$ is a table).' According to Hirsch, neither way of speaking has metaphysical primacy; the dispute is instead verbal.

If Hirsch is right and neither way of speaking is metaphysically privileged, '$\exists_{comp}$' and '$\exists_{nihl}$' ought to be equally elite (and no less elite than any other contender). Since they don't range over the same domains, they must be in different, equally elite families. In that case, Hirsch would count, on the second-pass definition, as believing in different ways of being.

McDaniel (2017: 37–38) expresses some sympathy with the thought that Hirsch should be categorized as an ontological pluralist. But this seems wrong. Pluralism seems ontologically inflationary, whereas Hirsch wants to deflate.[9] Hirsch wants his multiple quantifiers to represent *alternative* ways of describing reality; saying '$\exists_{nihl}xs(xs$ are particles arranged tablewise)' is supposed to be an alternative, equally metaphysically good way of saying '$\exists_{comp}x(x$ is a table).' If we describe reality just using '$\exists_{nihl}$,' we haven't left anything out. But ontological pluralism shouldn't be like this. If we describe the world just using a quantifier for *concreta*, we leave out the *abstracta*. If there are multiple ways of being, then we must talk about all of them to say all that deserves saying.

This suggests a third attempt:

> **Ontological Pluralism (third pass):** There are multiple ways of being iff there are some (multiple) elite quantificational families, and if any language that uses only quantifiers from some of these families is expressively impoverished compared to some language that uses all of them.

Javier-Castellanos (2019: §IV) considers a suggestion like this and raises a concern about it, related to the first worry about the second-pass objection. Here is a modified form of his concern. Suppose Zed believes that there are two modes of being, but that those modes overlap completely, so that everything has both modes. Zed will only count as a pluralist if he thinks that each of the quantifiers corresponding to those modes of being are expressively ineliminable. Javier-Castellanos expresses a concern that this might be asking too much of Zed.

But I don't see why. When Zed says '$\exists_2 y(y=a)$,' he either says something more than when he says '$\exists_1 y(y=a)$' or he doesn't. If he doesn't, then it is very hard to see in what sense he thinks there are two modes of being. But if he does, then '$\exists_2$' is expressively ineliminable: There is something Zed can say with it that he can't say without it. This strikes me as the right result.

Even with these refinements, the definition faces some hard cases. One comes from higher-order quantification – quantification into positions other than name positions. Some have argued (e.g. Prior 1971: 35, Rayo and Yablo 2001, Williamson 2003: 458–460) that higher-order quantification isn't in the business of ranging over things. The variables don't *have* values, so there's no question about what kind of being their values have. But higher-order quantifiers may be elite and expressively indispensable. It doesn't seem that someone like Williamson, who eschews multiple first-order quantifiers, should be thought of as an ontological pluralist. (Cf. Turner 2010: 12–13, McDaniel 2017: 40–41).

Here's a second kind of case. (Cf. Cameron 2018: 792–793 and McDaniel 2017: 71–72.) Some philosophers think that the world is filled with some undifferentiated *stuff*, which is then further divided into the countable *things* that it makes up. Defenders of such a view may, as Ned Markosian (2015) suggests, think that we need one kind of quantifier to range over the stuff and another to range over the things. And since stuff and things are equally metaphysically important, both quantifiers (and hence, their families) will need to be elite.

Both cases raise several tricky issues. For what it's worth, it isn't obvious to me that friends of higher-order quantification or stuffy quantifiers aren't thereby dabbling in ontological pluralism. It seems strange to say that they are; but it also seems a bit strange to say that they aren't. In tough cases like this, we might do either of two things. First, we might defer to the theory: If our account makes good sense of the easy cases, we ought to trust it on the hard ones as well. Second, we might defer to our judgments: If stuff or the values of second-order variables seem relevantly disanalogous to paradigm ways-of-being cases, we might reject the definitions instead.

So far I've raised concerns about the right-to-left direction of the proposed definitions. We can raise concerns about the left-to-right direction, too. A metaphysician who thought that Wittgenstein basically had it right in the *Tractatus* and that ontology should be done in terms of names rather than quantifiers might still think that there are multiple modes of being. Perhaps she will mark this by using a different color or font for names, or something, to mark the 'way of being' under which their referents are being referred to. (Cf. Turner 2010: 10.) I don't know that this would be a particularly *good* theory; but I do not see why it would not just as well count as a form of ontological pluralism.

But why, we may wonder, were we trying to *define* ontological pluralism in the first place? We struggle to give unobjectionable definitions of many philosophical views. There is little consensus about precisely how to define 'deontology,' 'physicalism,' or 'epistemic internalism,' for instance. Yet most of us still think that there are coherent deontological, physicalist, or internalist views worth engaging with. Some of us might even think that some of these views are right.

We appealed to elite quantifiers to answer a charge of incoherence. That charge stemmed from the idea that existence is tied to existential quantification and that only one existential quantifier matters to metaphysics. We defuse the objection by noting that metaphysics might care about more than one existential quantifier. That only requires that some views (a) be recognizably pluralistic and (b) involve multiple elite existential quantifiers. This doesn't mean that every view which is (a) must also be (b), or vice versa. It is a difficult and interesting question as to whether we can give necessary and sufficient conditions for a view to be pluralistic, but the coherence of pluralism itself doesn't hinge on our ability to do this. (Cf. McDaniel 2017: 46.)

Pluralism and generic quantification

A 'generic' quantifier is a quantifier that ranges over things with any kind of being whatsoever. The pluralist ought to admit that we use such quantifiers, both at home and at work, for the pluralist herself says that some things have different kinds of being. If that is to be both true and non-trivial the 'some' in 'some things' must range over things with different kinds of being.

Trenton Merricks has recently (2017) argued that this causes problems for pluralism. Consider Annie, an ontological pluralist who thinks that there are two ways of being, represented by '$\exists_1$' and '$\exists_2$' and their duals '$\forall_1$' and '$\forall_2$.' Annie admits that there is a generic quantifier '$\exists$' and its dual, '$\forall$.' She can also define a pair of quantifiers which, by her lights, are satisfactory proxies for the generic quantifiers. She does it by:

(Df. $\exists\star$) $\ulcorner \exists\star x\varphi(x) \urcorner := \ulcorner \exists_1 x\varphi(x) \vee \exists_2 x\varphi(x) \urcorner$.

(Df. $\forall\star$) $\ulcorner \forall\star x\varphi(x) \urcorner := \ulcorner \forall_1 x\varphi(x) \wedge \forall_2 x\varphi(x) \urcorner$.

Ontological pluralism

Now, asks Merricks, what does Annie think about the relationship between the *generic* quantifiers and the defined ★-quantifiers? Is, as he puts it, '∀' just a *shorthand* for '∀★'? In other words, when Annie says 'everything is F,' is that supposed to be mere shorthand for 'everything$_1$ is F and everything$_2$ is F'? Or not? Merricks argues, in essence, that problems loom either way:

The 'shorthand' argument
1 If '∀' is just shorthand for '∀★,' then Annie has no way to express her disagreement with a pluralist who thinks that there are three ways of being. (This is a problem.)
2 If '∀' is not shorthand for '∀★,' then Annie believes that there is a third way of being enjoyed both by the things that exist$_1$ and the things that exist$_2$.
3 If Annie believes that there is a third way of being enjoyed both by the things that exist$_1$ and the things that exist$_2$, then ontological pluralism is unmotivated. (This is a problem.)
4 So either way, Annie has a problem.

Since Annie is a stand-in for any pluralist, if the argument is sound, pluralists have a problem.

But is it sound? First, why think (1)? Merricks argues as follows. Suppose that Boris is an ontological pluralist who thinks that there are three ways of being. Annie needs some way to express her disagreement with Boris. Presumably she will express this disagreement with

(Two) Everything either exists$_1$ or exists$_2$,

which she expects Boris to deny. In symbols, (Two) reads:

(Two′) $\forall x (\exists_1 y (y = x) \vee \exists_2 y (y = x))$.

In (Two) and (Two′), 'everything' and '∀' express the generic quantifier, which ranges over things no matter what kind of being they have.

But if '∀' is just shorthand for '∀★,' then (Two′) becomes

(Two★) $\forall_1 x (\exists_1 y (y = x) \vee \exists_2 y (y = x)) \wedge \forall_2 x (\exists_1 y (y = x) \vee \exists_2 y (y = x))$.

Unfortunately, (Two★) is trivially true. So Boris won't deny it. Thus, it can't capture his disagreement with Annie.

So much for (1). Now, why think (2)? Here, Merricks is less explicit, but as far as I understand it, the thought goes like this. If generic existence is more than just existence★, then it is a genuine something over-and-above the specific ways of being, and not to be understood in terms of them. And if it is not to be understood in terms of them, then (since no other options for understanding present themselves), we ought to take it to be an elite quantifier. Then generic existence is thus a third mode of being.

Third, why think (3)? Here Merricks takes us through a host of historical motivations for pluralism which all rely on a core thought: some things are so radically unlike each other that they must have different modes of being. The particulars of the motivations are different in each case, but they each echo Russell's insistence that one kind of thing is so different from another that "we cannot say that it exists *in the same way* in which" (1912: 90) the other does. If the two things share a third, generic mode of being, then we can indeed say this, and the motivation is scuppered.

I suspect it's possible to quibble with (3), but I won't do that here. Instead I'll focus on premises (1) and (2). These premises use the phrase 'is shorthand for,' and I am not entirely sure how I am supposed to understand it. As a result, I cannot tell which of (1) or (2) Annie should reject.

Suppose Annie says the following:

> By my lights, '∀' (and '∃') are *analyzed* as '∀⋆' (and '∃⋆'). My analysis isn't 'conceptual,' by which I mean that our cognitive architecture probably doesn't explicitly represent generic quantification as conjunctions or disjunctions of other kinds of quantification. Instead, my analysis is *metaphysical*. It says that *what it is* for something to generically exist is for it to either exist$_1$ or exist$_2$, regardless of how we represent it to ourselves. But while I think I am right, I also think my claim is substantive, and reasonable people can disagree with it.

Now that she has said all this, we can wonder: Does Annie think that '∀' is *shorthand* for '∀⋆'?

It depends on whether we use 'shorthand' to include analysis or not. Suppose we do. In this case, Annie should think that premise (1) is false and that the argument for it is just Moore's paradox of analysis, badly disguised. Boris will of course accept the trivial (Two⋆). But he would only thereby accept (Two′) if he also accepted Annie's analysis, which he doesn't.

We should know by now that disagreement about an analysis needn't entail disagreement about any particular analysandum. The utilitarian thinks that it's a substantive thesis – worth arguing for – that the right action is that which maximizes utility, while thinking it's a logical truth that the action which maximizes utility is that which maximizes utility. And the deontologist will happily admit that a given action maximizes utility without thereby agreeing that the action is *right*. But none of this should force the utilitarian to think that 'is right' doesn't mean the same as 'maximizes utility.' It's the *equivalence of meaning* which is substantive, and the parties can disagree about rightness without disagreeing about utility-maximization by disagreeing about whether 'right' and 'maximizes utility' mean the same thing.

Like the utilitarian, Annie thinks that (Two′) is substantive – worth arguing for – while (Two⋆) is trivial. And like the deontologist, Boris can happily agree to (Two⋆) without thereby having to agree to (Two′), so there's no reason (Two′) can't capture Annie's and Boris's disagreement. But none of this should force Annie to think that (Two′) doesn't mean the same as (Two⋆). It's the equivalence of meaning which is substantive, and Annie can disagree with Boris about (Two′) without disagreeing about (Two⋆) by disagreeing about whether these mean the same thing. So if one way of being shorthand for something is being an analysis of it, then, premise (1) is false.

But perhaps 'shorthand' is meant to *exclude* cases of analysis, applying only to cases of something like stipulative definition. In this case Annie ought to deny that '∀' is shorthand for '∀⋆' while continuing to insist that the former analyzes the latter. And then she should deny (2). For if '∀' is analyzed in terms of other quantifiers, it is not elite, and so no way of being.

Of course, '∀' will be a quantifier we can use to range over whatever exists$_1$ plus whatever exists$_2$. But Annie denies that this makes them share any sort of way of being – in the same way that '∃$_{big}$' 'ranging over' a unicorn fails to give unicorns any being at all. Annie grants that some linguistic expression covers both the existents$_1$ and the existents$_2$, but she denies that this implies the sort of similarity that Russell said those things couldn't have. According to Annie, saying that an existent$_1$ and an existent$_2$ both generically exist is a lot like saying that a green thing (observed before the set time) and a blue thing (observed after that time) are both grue. That's strictly speaking true, but we shouldn't get overly metaphysically excited about it. And this holds even if 'grue' is not *stipulatively* defined, but part of a community's native linguistic endowment, the way '∀' is part of our native endowment. As long as generic existence has an analysis, things that generically exist need not in any sense share a way of existing – just as things that are both grue need not in any sense share a color.

Notational variance

Sometimes, a single theory can be presented in different guises. And sometimes we can mistakenly think that what is one theory is in fact two. So far, I have acted as though ontological pluralism is a different theory than ontological monism. But we might worry that it isn't.

Suppose that Cynthia considers herself an ontological pluralist, and thinks there are just two modes of being: the abstract one ('$\exists_a$') and the concrete one ('$\exists_c$'). Furthermore, like Annie above, Cynthia thinks generic quantification is analyzed by these two. Dan considers himself an ontological monist, replacing these two ways of being with the predicates 'is abstract' and 'is concrete.' Aside from this disagreement, Dan and Cynthia agree about everything else.

We can translate everything Cynthia says into something Dan will accept, and vice versa. We translate Cynthia's claims for Dan by turning her pluralist quantifiers into a generic quantifier restricted by 'is abstract' or 'is concrete,' as appropriate. And we translate Dan's claims for Cynthia by trading generic quantification using the recipe in (Df. ∀★) and (Df. ∃★) above, and translating 'x is abstract' and 'x is concrete' as '$\exists_a y(y=x)$' and '$\exists_c y(y=x)$,' respectively.

The existence of these translations might make us suspect that Cynthia's and Dan's theories are really the same one in disguise after all. There are two versions of this worry. On the 'soft' version, we might simply observe Cynthia and Dan and wonder whether they take themselves to be disagreeing or to be simply expressing themselves using different symbols. On the 'hard' version, we assume that Cynthia and Dan *do* take themselves to be disagreeing, and then insist that they *can't* be, because their theories are really notational variants of each other.

I've argued elsewhere (2012) against this worry. Although I didn't make the distinction there, I had the hard version of the worry in mind, though what I said tells against the weak version as well. The idea is that, if two theories are notational variants of each other, the translations between them ought to preserve logical truth. And this doesn't happen with the proposed translations. When Dan says

(1) Everything is either abstract or concrete,

that is not a logical truth; but it gets translated as

(2) $\forall_a x(\exists_a y(y=x) \vee \exists_c y(y=x)) \wedge \forall_c x(\exists_a y(y=x) \vee \exists_c y(y=x))$,

which is. The only wrinkle is that Dan might define 'is abstract' as 'is not concrete,' in which case (1) would be logically true after all. But in this case Dan must also think that

(3) Nothing is both abstract and concrete,

is logically true. Cynthia's translation of this won't be, though; it's no part of pluralism that the different ways of being won't overlap.

Bruno Whittle (forthcoming) has recently argued against this response. He suggests that we ought to treat 'is abstract' and 'is concrete' as logical expressions too. This, by itself, doesn't quite secure the desired result; we also have to say something about the 'logic' that they obey. In particular, we have to assume that the logic of 'is abstract' and 'is concrete' make (1) logically true. Once we do that, the objection from logical equivalence is blocked.

Whittle is right. My argument relies on some assumptions about what counts as 'logical,' and if those assumptions are rejected the argument won't go through. Furthermore, I don't really know quite how to *justify* those assumptions. Nonetheless, I'm still inclined to think that the observations about (1) and (2) are suggestive, and can help us resist the thought that pluralism must be a variant of monism.

First, the appeal to logic can be seen as an attempt to get at something deep about how we structure reality with different kinds of expressions. The argument can then be seen as suggesting that theories are different when they structure reality differently.

On Cynthia's account, 'something has no way of being' seems ruled out *by the very nature of quantification*. It doesn't matter whether the quantifiers are for *abstracta* and *concreta* or anything else instead. The quantifier 'flavors' aren't really relevant. By contrast, 'something is neither abstract nor concrete' isn't ruled out just by the nature of predication. It matters what the predicates mean, and that we think the two are exhaustive. Treating 'is abstract' and 'is concrete' as logical terms doesn't defuse the argument by itself; we also have to suppose that (1) is logically true. Nothing about predication generally makes (1) logically true. It's rather something specific to abstractness and concreteness. On the other hand, the logical truth of (2) stems merely from general facts about quantification and identity. The fact that the quantifiers stand in for abstract and concrete ways of being is immaterial. This suggest that the theories aren't notational variants, because they represent reality's structure very differently – and that calling (1) a logical truth is merely *hiding* this fact.

Second, I worry that Whittle's response overgeneralizes. Consider a toy case. Ethel and Finnegan both think that space is infinite. They agree about almost everything, with one exception: Finnegan thinks, and Ethel denies, that one point of space is the objective 'center' of the universe.[10] Intuitively, Ethel and Finnegan have different theories. Finnegan believes that there is *more* structure in the world than Ethel does, because Finnegan believes it makes sense to ask which point is the objective center of the universe, and Ethel does not.

Now comes Gerald, who tries to argue that Ethel's and Finnegan's theories are notational variants of each other. Suppose that p is the point of space that Finnegan thinks is at the center. Finnegan thinks 'x is the objective center of the universe' is satisfied by p; Ethel claims this predicate makes no sense. Gerald points out that Finnegan's predicate could be translated into '$x = p$,' which of course Ethel can make sense of. Ethel will think that everything Gerald says is true under this translation.

Does this make Ethel's and Finnegan's theories notational variants after all? To argue otherwise, we might point out that

(4) p is the objective center of the universe,

is not a logical truth according to Finnegan, but turns into one (namely, '$p = p$') under Gerald's proposed translation. This is one way to emphasize that Finnegan recognizes more theoretical possibilities than Ethel does, which is a good reason to think their views distinct. But Gerald may insist that 'is the objective center of the universe' must also be treated as logical, and in a way that renders (4) logically true, rendering the objection toothless.

This seems misguided to me. Ethel and Finnegan's theories really are different, and the difference in status between (4) and '$p=p$' seems diagnostic of this fact. I'm leery of any argumentative move that would suggest that it isn't.

Note that these first two responses to Whittle's objection speak to both the strong and weak versions of our original worry. If neither of them are persuasive, though, a third consideration, which speaks only to the strong worry, might be. On the strong version of the worry, Cynthia and Dan *take themselves* to have different theories. Given this, why should any hostile third party get to insist that (1) be treated as a logical truth? It's Dan's theory, after all; he should be the one to decide what logic he accepts for his terms. If he doesn't want his theory to be a notational variant of Cynthia's, then he shouldn't treat (1) as logically true; and I don't see where we get the authority to tell him he's wrong about that. If (1), as it shows up in Dan's theory, is not logically true – and Dan is the one who should tell us whether it is or not – then the original response goes through.

If Whittle means only to be defending a weak form of our original worry, this fallback response won't speak to his objection: If Dan *wanted* his theory to be a notational variant of Cynthia's, he could perhaps make it so by treating (1) as logically true. Maybe so. For the record, while I think a logical-truth-preserving translation between two theories is a necessary condition for notational variance, I doubt it is sufficient.[11] Still, if Dan wants his theory to be Cynthia's in disguise, he can treat (1) as logically true to ensure that this necessary condition is met. But that will be a special case driven by Dan's particular decisions. There is still no argument that, *in general*, pluralist theories are always disguised versions of monist ones. Pluralists who want to be genuinely different from their monist counterparts can deny that certain truths are *logically* true.

Conclusion

Ontological pluralism is certainly in a much better position today than it was a decade ago. Its rehabilitation by appeal to elite quantifiers has helped resurrect it from the positivist's graveyard. But that's not to say it has come to dominate the metaphysical scene. The view remains niche, with detractors eager to argue against it. I have tried to evaluate a few of those arguments here. Surely, though, more arguments will be brought to bear before all is said and done.[12]

Notes

1 See also Moore 1903/1953: 110–113.
2 Spencer 2012 contains a nice introduction and overview of some of these arguments.
3 Compare Dorr 2005: 256–257, Sider 2009: 391–392, and Turner 2010: 15–16 for similar tricks.
4 See e.g. Turner 2010: 8–9 and McDaniel 2017: 35–37. This is not to say that friends of such accounts cannot argue against ontological pluralism on other grounds, perhaps by arguing that pluralistic theories should be rejected for broadly theoretical reasons. (Although he doesn't explicitly make such an argument, this seems to be the subtext of Sider 2011: 206–208.) But the naturalness account at least renders pluralism intelligible.
5 Theorists who prefer the grounding-theoretic framework (e.g. Schaffer 2009; Rosen 2010) could instead perhaps define eliteness along these lines: a quantifier '$\exists^*$' is elite iff, for every formula φ, if a truth of the form $\ulcorner \exists^* x \varphi(x) \urcorner$ is grounded in any truth of the form ψ, then ψ is itself of the form $\ulcorner \exists^* x \theta(x) \urcorner$ for some θ. More prosaically: '$\exists^*$' is elite iff it's never grounded in anything else. If (as is often insisted) existential quantifications must be grounded in their instances, fancier footwork is needed. Perhaps this will do: '$\exists^*$' is elite iff, for every formula φ, if a truth of the form $\ulcorner \exists^* x \varphi(x) \urcorner$ is grounded in any truth of the form ψ other than something of the form $\ulcorner \varphi(a) \urcorner$, then ψ is itself of the form $\ulcorner \exists^* x \theta(x) \urcorner$ for some θ. I leave it to those more attracted to the grounding-theoretic framework to work out further kinks in this suggestion.

6 Boolos 1984. McDaniel (2017: 39–40) seems surprisingly sympathetic to the view that, if both singular and plural existential quantifiers are elite, they correspond to different ways of being. I think this is too quick, as I hope this section will show.
7 Assuming, as is often done, that irreducible ideology is elite.
8 This follows from Sider's (2011: 217–222) views on the eliteness of logical expressions, for example.
9 McDaniel recognizes this and suggests it's what fools us into thinking Hirsch isn't a pluralist. I think he's right, except about the 'fools' bit.
10 Finnegan's view is pretty silly, and I know of nobody who endorses it. But the disagreement between Ethel and Finnegan is structurally similar to real debates, such as the debate between those who accept Newtonian 'absolute velocities' and the Galileans who don't, or the Lorentzians who accept 'absolute simultaneity' and the special relativists who don't. I could make the same point with respect to either of these debates; it would just require more setup.
11 See McSweeney 2017 for an argument for its necessity. On sufficiency, note that the advanced modal logic of Cresswell 1990 can be shown to be intertranslateable with a language involving quantification over possible worlds, and in a way that preserves logical truth. But I doubt that the world-involving theory and the world-free one are notational variants of each other.
12 Thanks to Ross Cameron, Kris McDaniel, Trenton Merricks, Bruno Whittle, and an anonymous referee for helpful comments.

References

Boolos, G. (1984), 'To be is to be a value of a variable (or to be some values of some variables),' *The Journal of Philosophy* **81**, 430–449.
Cameron, R. P. (2018), 'Critical study of Kris McDaniel's *The Fragmentation of Being*,' *Res Philosophica* **95**(4), 785–795.
Caplan, B. (2011), 'Ontological superpluralism,' *Philosophical Perspectives* **25**, 79–114.
Carnap, R. (1950), 'Empiricism, semantics, and ontology,' *Revue Internationale de Philosophie* **4**, 20–40. Reprinted in Carnap 1956, pp. 205–221.
Carnap, R. (1956), *Meaning and Necessity*, 2nd edn, University of Chicago Press, Chicago.
Chalmers, D., Manley, D., and Wasserman, R., eds (2009), *Metametaphysics*, Oxford University Press, Oxford.
Cresswell, M. J. (1990), *Entities and Indices*, Kluwer, Dordrecht.
Dorr, C. (2005), 'What we disagree about when we disagree about ontology,' in M. E. Kalderon, ed., *Fictionalism in Metaphysics*, Oxford University Press, Oxford, pp. 234–286.
Hirsch, E. (2002a), 'Against revisionary ontology,' *Philosophical Topics* **30**, 103–127.
Hirsch, E. (2002b), 'Quantifier variance and realism,' *Philosophical Issues* **12**, *Realism and Relativism*, 51–73.
Javier-Castellanos, A. (2019), 'Quantifier variance, ontological pluralism and ideal languages,' *Philosophical Quarterly* **69**(275), 277–293.
Lewis, D. (1983), 'New work for a theory of universals,' *The Australasian Journal of Philosophy* **61**, 343–377. Reprinted in Lewis 1999: 8–55.
Lewis, D. (1999), *Papers in Metaphysics and Epistemology*, Cambridge University Press, Cambridge.
Markosian, N. (2015), 'The right stuff,' *The Australasian Journal of Philosophy* **93**(4), 665–687.
McDaniel, K. (2009), 'Ways of being,' in Chalmers et al. (2009).
McDaniel, K. (2010), 'A return to the analogy of being,' *Philosophy and Phenomenological Research* **81**(3), 688–717.
McDaniel, K. (2017), *The Fragmentation of Being*, Oxford University Press, Oxford.
McSweeney, M. M. (2017), 'An epistemic account of metaphysical equivalence,' *Philosophical Perspectives* **30**(1), 270–293.
Merricks, T. (2017), 'The only way to be,' *Noûs* **53**(3).
Moore, G. E. (1903/1953), *Principia Ethica*, Cambridge University Press, Cambridge.
Prior, A. N. (1971), *Objects of Thought*, Clarendon Press, Oxford.
Putnam, H. (1987), *The Many Faces of Realism*, Open Court Press, La Salle, Ill.
Rayo, A. and Yablo, S. (2001), 'Nominalism through de-nominalization,' *Noûs* **35**(1), 74–92.
Rosen, G. (2010), 'Metaphysical dependence: Grounding and reduction,' in B. Hale and A. Hoffman, eds, *Modality: Metaphysics, Logic, and Epistemology*, Oxford University Press, Oxford, pp. 109–135.
Russell, B. (1912), *The Problems of Philosophy*, Home University Library. Reprinted by Oxford University Press, Oxford, 1959.

Schaffer, J. (2009), 'On what grounds what,' in Chalmers et al. (2009).
Sider, T. (2001), 'Criteria of personal identity and the limits of conceptual analysis,' *Philosophical Perspectives* **15**, 189–209.
Sider, T. (2009), 'Ontological realism,' in Chalmers et al. (2009).
Sider, T. (2011), *Writing the Book of the World*, Oxford University Press, Oxford.
Spencer, J. (2012), 'Ways of being', *Philosophy Compass* **7**(12), 910–918.
Szabó, Z. (2003), 'Nominalism,' in M. J. Loux and D. W. Zimmerman, eds, *The Oxford Handbook of Metaphysics*, Oxford University Press, Oxford, pp. 11–45.
Turner, J. (2010), 'Ontological pluralism,' *The Journal of Philosophy* **107**(1), 5–34.
Turner, J. (2012), 'Logic and ontological pluralism,' *The Journal of Philosophical Logic* **41**(2), 419–448.
Whittle, B. (forthcoming), 'Ontological pluralism and notational variance,' in D. W. Zimmerman and K. Bennett, eds, *Oxford Studies in Metaphysics*, Vol. 12, Oxford University Press, Oxford.
Williamson, T. (2003), 'Everything,' *Philosophical Perspectives* **17**, 415–465.

PART III

Alternative conceptions of metaphysics

15
GROUNDING

Alexander Skiles and Kelly Trogdon

Introduction

Metametaphysics concerns foundational metaphysics. Questions of foundational metaphysics include: What is the subject matter of metaphysics? What are its aims? What is the methodology of metaphysics? Are metaphysical questions coherent? If so, are they substantive or trivial in nature? Some have claimed that the notion of *grounding* is useful in addressing such questions. In this chapter, we introduce some core debates about whether – and, if so, how – grounding should play a role in metametaphysics.[1]

It is undeniable that *in fact* grounding plays at least *some* role in how *metaphysics* is conducted, and has perhaps done so since the beginning.[2] Two roles stand out in particular. First, the notion of grounding is routinely used to state, at least in an intuitive way, what is at issue in various metaphysical disputes. Examples include debates over what (if anything) is the ground of mentality and what (if anything) is the ground of modality. Second, the notion is also routinely used to state, at least in an intuitive way, what various other notions of metaphysical interest amount to. Examples include what it is for a property to be intrinsic rather than extrinsic and what it is for an entity to be a substance rather than a mode.

But what conclusion is to be drawn about *what if any* role grounding *should* play in *metametaphysics*, in answering foundational questions about the nature of metaphysics? Some claim that the notion should play an absolutely central role. For example, Jonathan Schaffer writes that "metaphysics as I understand it is about what grounds what" in part due to the alleged inadequacies of the 'Quinean' approach that dominated metaphysics throughout most of the twentieth century, which focuses on what there is rather than "what is fundamental, and what derives from it" (2009: 379). For another example, Kit Fine writes that although questions about what grounds what "are not without interest to naïve metaphysics" (which concerns "the nature of things without regard to whether they are real"), nonetheless "they are *central* to realist metaphysics" (which concerns whether reality does in fact contain things with that nature). "Indeed," Fine writes, "if considerations of ground were abolished, then very little of the subject would remain" (2012: 41, emphasis in original; cf. 2001: 28–29).

Nonetheless, the issue of what role, if any, grounding should play in metametaphysics remains controversial. Some are skeptical about the very coherence of the notion of grounding or suspect that talk about what grounds what isn't non-trivially truth-evaluable.[3] Some who find the notion coherent argue that its allegedly essential, central theoretical utility has been overblown.[4] And among even those who find the notion both coherent and useful, there is substantial disagreement regarding the nature of grounding itself, which will interact in certain ways about what use we ultimately put the notion to.[5] For the purposes of this chapter, we shall assume that there is at least something coherent to the notion, and that at least some grounding statements are true. As for the nature of grounding, we shall proceed as much as possible without taking on substantive commitments, although we shall also indicate when questions about the relationship between grounding and metametaphysics may turn on which commitments are ultimately taken up.

There is, however, one issue about the nature of grounding that we should address before proceeding. Consider the following representative grounding claim: the fact that the bowl is brittle is grounded by the fact that the chemical bonds of its atoms are covalent. Some stipulate that by 'grounding' they mean a distinctive form of determination, where to determine is, roughly speaking, to *produce* or *bring about*. In this case to say that the brittleness of the bowl is grounded by the covalent bonds of the bowl's constituent atoms is to say that the bonding of the atoms produces or brings about the brittleness of the bowl.[6]

Others stipulate that by 'grounding' they mean a distinctive form of explanation. In this case, to say that the brittleness of the bowl is grounded by the covalent bonds of the atoms is to say that the bowl is brittle *because* the bonding of the atoms is covalent.[7] Compare: many interpret causal claims (e.g. "The stone striking the glass caused the window to shatter") as targeting in the first instance causation understood as a distinctive form of determination rather than causal explanation; some, however, see ordinary causal claims as targeting in the first instance causal explanation.[8] Rather than plump for one view or the other – or cast the dispute aside as "largely verbal" (cf. Dasgupta 2017: fn. 8) – let us simply speak of the distinction between *determination*$_G$ and *explanation*$_G$ and use 'grounding' when potential differences between the two can be safely elided.

In what follows, we focus on three of the most interesting and widely discussed roles that have been assigned to grounding in metametaphysics. Specifically, we consider how grounding might be relevant to whether metaphysical questions are substantive (§1), how to choose between metaphysical theories (§2), and how to understand so-called 'location problems' (§3).

1 Substantive questions

A widely held view – both nowadays, and during much of the twentieth century – is that the subject matter of metaphysics chiefly concerns *existence*, and that its central task is to address questions like "Do numbers exist?," "Do properties exist?," "Do ordinary objects like tables and chairs exist?," and so on. However, some have argued that so-called 'existence questions' such as these are *trivial*, in the sense that they have obvious answers. For example, Thomasson (Chapter 12, this volume) argues that ontological disputes involving existence are typically resolvable in a straightforward manner by appealing to conceptual and empirical truths. Do tables exist? Of course! For it is a conceptual truth that if there are entities such as wood and screws arranged table-wise, then there is a table; and it is an empirical truth that there are, in fact, entities arranged table-wise. Similarly, Fine (2009), Hofweber (2005), and Schaffer (2009) consider arguments involving seemingly innocuous inferences that, if valid, would quickly trivialize classic existence questions. Do numbers exist? Well, yes: (i) Jupiter has four moons; so (ii) the number of moons of Jupiter is four; so (iii) there is a number which is the number of moons of Jupiter, namely four; so (iv) there are numbers, among them the number four. Seems pretty straightforward!

If the task of metaphysics is to answer existence questions, and if existence questions are trivial, then it would appear as though there is little of interest or value for metaphysics to do. So if you think that metaphysics *is* worth doing, how might you respond? One option, of course, is to argue that at least some existence questions of interest to metaphysics are substantive after all, despite what the arguments above suggest (Daly and Liggins 2014). But, supposing for the moment that these existence questions are indeed trivial, how else might you respond?

A response under active consideration in recent metametaphysics – sometimes associated with the broader, so-called 'neo-Aristotelian' approach to metaphysics re-emerging as of late – is that the subject matter of metaphysics chiefly concerns *grounding* rather than existence, and that its central task is instead to address questions like "What (if anything) grounds the existence of numbers?," "What (if anything) grounds the existence of properties?," "What (if anything) grounds the existence of ordinary objects?," and so on (Schaffer 2009). On the face of it, these questions have substantive answers even if the corresponding existence questions do not.[9]

As a case study, take the classic metaphysical debate between *monists* and *atomists*. Do tables exist? Historically, at least, on this question the monist and the atomist typically *agree*: yes, they do. Rather, they disagree about what grounds the existence of tables. Now, *how* to state their disagreement in grounding-theoretic terms will depend to some extent on what views about the nature of grounding hold in the background. But here is one grounding-theoretic disagreement (perhaps among many) that they might have.[10] Let a *concrete fact* be one that solely concerns the existence and/or features of concrete objects. And say that one fact is *ultimately* grounded in other facts only if none of the latter facts are themselves grounded in further facts.[11] Then which concrete facts ultimately ground the existence of tables? According to atomists, these are facts solely concerning the existence and/or features of their constituent mereological atoms (i.e. objects without proper parts). According to monists, instead these are facts solely concerning the existence and/or features of the entire cosmos (i.e. the fusion of all concrete objects).

Questions about what ultimately grounds what certainly seem substantive. At any rate, the literature on the grounding question above certainly *treats* it as substantive, as it has been thought to turn on highly non-trivial matters concerning the nature of composition, necessity and possibility, causation and lawhood, intrinsicality, the interpretation of quantum theory, and – of course – grounding.[12] This piece of evidence, however, is far from decisive, since even a cursory look at the literature on existence questions shows that these have been thought to turn on highly non-trivial matters too.[13]

Moving on, what about the substantivity of questions of *non-ultimate* grounding: i.e. questions about what grounds what such that the grounds themselves presumably have further grounds? Perhaps at least some of these grounding questions have trivial answers. Does the fact that there are entities arranged table-wise ground the fact that there is a table? You might think the answer is obviously 'yes,' as it's a conceptual truth that facts to the effect that there are entities arranged table-wise ground facts to the effect that there are tables, and it's an empirical truth that there are entities arranged table-wise. The claim that the former is a conceptual truth may be especially attractive if you're already on board with the idea that it's a conceptual truth that there are entities arranged tables-wise only if there are tables.

Putting aside the issue of whether there are such conceptual truths about grounding, the general idea that some grounding claims are trivial may be the most plausible if by 'grounding' we have in mind explanation$_G$ rather than determination$_G$. On the one hand, since questions of what determines$_G$ what ostensibly concern reality's metaphysical structure, it seems that such questions are substantive. On the other hand, some questions concerning what explains$_G$ what may be fairly trivial given certain broader accounts of what explanation amounts to. Where Δ is a collection of facts and [p] the fact that p, suppose that it's enough for Δ to explain$_G$ [p] that citing Δ is a relevant, truthful answer to the question "What makes [p] obtain?" in a context in which causal answers have been set to one side (cf. Thompson 2019). Whether or not there are conceptual connections between entities being arranged table-wise and the existence of tables, if empirical investigation reveals that there are some pieces of wood screwed together table-wise here, we seem to have a relevant and truthful answer to the question "What makes it the case that there is table here?" in contexts in which we aren't looking for causes. This line of reasoning – grounding being a form of explanation combined with a lightweight conception of explanation renders certain grounding claims non-substantive – displays not only how conceiving of grounding as explanation$_G$ versus determination$_G$ is relevant to the relationship between grounding and metametaphysics, but also how we conceive of explanation$_G$ in particular.[14]

An alternative take on the table example is this: it's an *essential* truth about tables rather than a conceptual truth that facts to the effect that there are entities arranged table-wise ground facts to the effect that there are tables. In this case, while it's *conceptually* possible for there to be pieces of wood screwed together table-wise but no tables – and thus conceptually possible for there to be pieces of wood screwed together table-wise that do not ground the existence of any tables – nonetheless this is *metaphysically* impossible. The thought is that *part of what it is to be* a table – part of the *real* (as opposed to 'nominal') *definition* of being a table – is that if there are some things arranged table-wise, then this grounds the fact that there is a table. Such essence claims are, it seems, substantive. This in turn suggests that our original grounding claim – the fact that there are entities arranged table-wise grounds the fact that there is a table – itself is substantive.[15]

What these considerations suggest is that if we're going to wheel in grounding to resolve the problem that the supposed triviality of existence claims poses for metaphysics, we need to get clear on the extent to which questions of non-ultimate grounding are substantive in nature. If it turns out that such questions normally are trivial, then appealing to grounding in this context isn't going to be as effective in showing that metaphysics is worth doing than it might have initially seemed.

2 Theoretical economy

Even if grounding claims like those discussed above are substantive, metaphysics does not consist *solely* in answering a hodgepodge of questions about the structure of reality. Another goal of metaphysics is to construct and rationally evaluate *theories* capable of answering such questions. Metaphysical theories, like scientific theories, can be judged in various ways: for instance, in terms of how comprehensive they are and the degree to which they unify what might seem to be disparate phenomena. Another potential application of grounding is to a third characteristic by which a theory can be judged: how *economical* it is.

Focus now on *ontological* economy. (Another type, *ideological* economy, will be discussed later.) According to Occam's Razor – henceforth *the Razor* – the ontological economy of a theory is measured in terms of the entities it posits, or as it is sometimes put, by its *ontological commitments*. Roughly speaking, the more existence-involving a theory is – the more entities it posits – the less ontologically economical it is. (Here and in what follows we put aside the distinction between *quantitative* and *qualitative* economy, which respectively concern whether we count how many individual entities a theory posits as opposed to how many *kinds* of entities it posits (cf. Lewis 1973: 87).

Although initially plausible, a number of philosophers reject the Razor because they claim that it leaves out a crucial notion, that of *fundamentality* (Bennett 2017: Ch. 8; Cameron 2010; Schaffer 2015, Sider 2013). Space prevents a full account of the many ways in which this idea has been implemented, not all of which make explicit (or any) appeal to grounding per se. But since this is an entry on grounding, here is a simple approach that makes use of the notion. Let us identify the fundamentality of a *fact* with the notion of *ungroundedness*: a fact is fundamental just when it both obtains and is not grounded in further facts. And let us say that an *entity* (e.g. an object or a property) is fundamental just when some fundamental fact concerns that entity, and derivative otherwise. Finally, let us suppose that properties and relations exist, and understand what it is for a fact to 'concern an entity' broadly enough so that, e.g., if it is a fundamental fact that Gothenburg is a city, then both the object Gothenburg and the property *being a city* count as fundamental entities.[16]

With entity fundamentality so understood, we can capture the idea that the Razor wrongly leaves out fundamentality thus: the ontological economy of a theory is measured not by the entities it posits per se, but instead by the *fundamental* entities it posits. Roughly speaking, the more fundamentality-involving a theory is – the more fundamental entities it posits – the less ontologically economical it is. On this view, one may draw a distinction between the ontological commitments of a theory and the *costs* of those commitments. While you have to pay for fundamental entities, derivative entities are on the house. To contrast this principle with the Razor, call it *the Laser* (Schaffer 2015).

Why prefer the Laser to the Razor? Adapting an example from Schaffer (2015), compare two toy scientific theories. The first theory posits 100 types of fundamental particle, while the second posits 10 types of fundamental strings, which in varying combinations make up the 100 types of particle described by the first theory. Suppose that these theories are tied with respect to all theoretical virtues, save for perhaps ontological economy. (Although for sake of simplicity, let us suppose that the fundamental facts posited by both theories contain exactly the same properties.) Now, if there is a good reason to choose between the two theories, the choice must turn on how they compare to one another with respect to ontological economy. In this case, the Razor says that the *first* theory is preferable to the second because the former posits fewer entities (100) than the latter (110). Yet the Laser seems to predict what seems clearly correct, namely that the *second* theory is preferable to the first. It predicts this because the former posits fewer fundamental entities (10) than the latter (100).

Of course, those who wield the Razor might claim that the situation above is not described correctly. Once the content of these theories is spelled out in more detail (they might claim), surely the two theories will differ with respect to theoretical virtues that do not directly involve ontological economy, such as their relative depth, unification, or elegance. (Indeed, in his discussion of this example, Schaffer himself describes the theories as differing along these lines.) And this (they might claim) is why the second theory is preferable to the first. Supposing that the second theory is, say, more elegant than the first, this is why the former is preferable to the latter, not because the former posits fewer fundamental entities than the latter (cf. Baron and Tallant 2018).

In response, proponents of the Laser might concede that the second theory is, say, more elegant the first, but insist that this is because the former posits fewer fundamental entities than the latter. While the theories do indeed differ with respect to various theoretical virtues, it's nevertheless the Laser that ultimately explains why the second theory is preferable to the first. Another option is to argue that there are methodological principles that underpin the Laser specifying how ontological economy interacts with other desirable features of theories, and these principles also suggest that the second theory is preferable to the first.[17]

Moving on, let us suppose for the remainder of the discussion that one ought to wield the Laser rather than the Razor when evaluating theories. What are some potential consequences for metaphysics? Here is one. *Mereological nihilists* reject the existence of complex concrete objects: they claim that all concrete objects are simple and fundamental. Given our account of fundamentality above, the atomists discussed in the previous section will agree with mereological nihilists that all fundamental objects are simple.[18] Yet atomists claim that there are complex objects as well, each of which decomposes into simple objects. Suppose that mereological nihilists and atomists agree on which entities are simple, and so agree on which objects are fundamental. Then the mereological nihilist claims that these are all the objects, while the atomist claims that there are many more: all the derivative, complex ones. In this case, some would suggest that economy considerations point in favor of mereological nihilism: as Williams puts this point, "[t]hose who love desert landscapes should applaud the elimination of mereological composition from the fundamental furniture of the world (along with much else)" (2006: 494). But if the Laser rather than Razor is true, this is not right – the two theories are *on a par* with respect to ontological economy. If so, then one would need to appeal to other considerations in choosing between them.

So far we have focused on one species of parsimony, ontological economy. But we should note that there is another species, *ideological* economy. The ideological economy of a theory is measured not by the notions it invokes per se, but instead by the *primitive* – 'undefined,' 'unanalyzed' – notions it invokes. Roughly speaking, the more primitive notions a theory invokes, the less ideologically economical it is. In some cases, ontological and ideological economy go hand-in-hand: two theories that differ only in whether they include ghosts in their fundamental ontology will likely differ in whether they include notions like being made of ectoplasm in their fundamental ideology as well. But ontological economy is no guarantee for ideological economy. (Consider two eternalist theories that agree about what fundamental entities there are across all time, yet disagree on whether 'A-theoretic' notions like that of being past can be analyzed in terms of 'B-theoretic' notions like that of being earlier than, where both theories take the latter notions as primitive.) Nor is ideological economy a guarantee for ontological economy. (Consider two theories that take the notion of parthood as primitive, yet disagree on whether some fundamental entities are parts of others.)

As with the Laser, while you have to pay for primitive notions, defined ones are on the house.[19] And as with ontological economy, ideological economy seems relevant to rational theory choice both in metaphysics and elsewhere. And here again as well, there are potential applications of grounding – two in particular.

First, if important notions in metaphysics can be defined in terms of grounding, then grounding will be relevant to the ideological economy of metaphysical theories. In addition to the notion of fundamentality discussed above, grounding-based definitions have been given of the notions of social construction (e.g. Epstein 2015; Griffith 2018), intrinsicality (e.g. Rosen 2010; Witmer et al. 2005), truthmaking (Correia 2014; Rodriguez-Pereyra 2005), identity criteria of various sorts (Fine 2016; McDaniel 2015), and many others.

Second, many have thought that what it is for a notion to be, or fail to be, analyzable in terms of other notions – and thus, how to understand what ideological economy amounts to – must be given in ground-theoretic terms. On the one hand, it does not seem sufficient for a notion to be definable that every fact about it be grounded. As a case in point, some have suggested that, even though grounding is a primitive notion, facts about it must be systematically grounded for there to be any derivative entities at all; otherwise, given the account of fundamentality above, if it were a fundamental fact that the table's being beige grounds its being brown, then the table, *being beige*, and *being brown* would all be fundamental entities (cf. Bennett 2011 and deRosset 2013). On the other hand, a number of accounts entail that it is necessary for one notion to be definable in terms of other notions that the facts about one be grounded in facts about the others, along with certain other conditions being met (cf. Correia 2017; Horvath 2018; Rosen 2015; Skiles 2014).

3 Location problems

Let's return to the disagreement between atomists and monists concerning tables. Putting aside grounding for the moment, we can think of them as offering different proposals about how tables 'fit into' the world. The problem that atomists and monists are addressing here is similar to what Jackson (1998: 2–4) calls a *location problem*. These problems concern how familiar, manifest facts (e.g. the existence of tables) 'fit into' a world that is assumed to be ultimately physical in nature. On the face of it, location problems do not concern what exists, but rather how facts that seem to have importantly different 'subject matters' relate to one another.

Location problems have played a prominent role in metaphysics for at least the past fifty years. Consider, for example, contemporary metaphysics of mind. One location problem in this context that has received an enormous amount of attention is this: provided that the actual world is ultimately physical (and no fundamental physical fact is a mental fact), how do manifest mental facts (e.g. Gomer is having a greenish experience) fit into the world? As Kim describes the problem:

> Through the 1970s and 1980s and down to this day, the mind-body problem – *our* mind-body problem – has been that of finding a place for the mind in a world that is fundamentally physical. The shared project of the majority of those who have worked on the mind-body problem over the past few decades has been to find a way of accommodating the mental within a principled physicalist scheme, while at the same time preserving it as something distinctive – that is, without losing what we value, or find special, in our nature as creatures with minds.
>
> *(1998: 2)*

Note how the problem is not framed in terms of existence questions about minds – mental realism is taken for granted. The issue instead is how the subject matter of psychology relates to the subject matter of fundamental physics.

In thinking about location problems, a question immediately arises. What is the sense of 'fitting into' or 'having a place for' at issue here? When philosophers first started to address location problems as such, they tended to formulate matters in purely *modal* terms. One way in which things were cast goes like this: to establish how a manifest fact 'fits into' a world that it is ultimately physical in nature is to find some physical fact (preferably a fundamental physical fact) that modally entails it, where the modality in question is metaphysical in nature.

It was recognized by Kim and others, however, fairly early on that purely modal accounts of 'fitting into' are too weak. To see why, return to the mind-body problem. Roughly speaking, the physicalist (about the mental) claims that (i) there are possible worlds that are ultimately physical yet have a place for the mental, and (ii) the actual world is such world. The dualist, by contrast, denies (i), and thus denies (ii) as well. Thus the dualist claims, while the physicalist denies, that the truth of mental realism requires that the world not be ultimately physical. Nonetheless, the dualist might still claim that at least some mental facts are modally entailed by fundamental physical facts. (A coherent view, for example, is that the fundamental physical facts modally entail that there are Cartesian souls.) So the modal entailment take on 'fitting into' has the consequence that dualism is compatible with (ii) from above, which is clearly the wrong result.

An attractive alternative proposal is this: for the mental to fit into the physical world is for fundamental physical facts to ground, and not just modally entail, the mental facts. Physicalists can either be *reductive* or *non-reductive*. For *reductive* physicalists, mental facts may be said to 'fit into' the actual world by dint of being identical to certain non-fundamental physical facts, which in turn are grounded in — and not merely modally entailed by — certain fundamental physical facts. *Non-reductive* physicalists deny that mental facts are identical to these non-fundamental physical facts; nonetheless, they may claim that mental facts are grounded in them all the same, and thus grounded in whatever fundamental physical facts these non-fundamental physical facts are grounded in. More generally speaking, we can understand physicalism as the thesis that (i) there are possible worlds in which the mental facts are ultimately grounded by fundamental physical facts, and (ii) the actual world is such a world. And we can understand dualists as denying (i) and thus (ii) as well.

Let *modal dualism* be dualism plus the claim that there are systematic modal connections between the mental and the fundamental physical. Some suggest that a key difference between modal dualists and physicalists is this: while the latter are committed to the idea that all such modal connections have physicalist-friendly explanations, the former deny that all such connections have these sorts of explanations (Horgan 1993). The grounding-theoretic characterization of physicalism set out above is useful in this context, as it suggests how the relevant explanations might proceed for physicalists. The thought is that physicalists might reasonably claim that the grounding connections between the mental and the fundamental physical themselves explain the relevant modal connections. If grounding is explanation$_G$, the idea is that (i) fundamental physical facts explain$_G$ the mental facts; and (ii) these facts about what explains$_G$ what themselves explain$_G$ why the mental and fundamental physical are modally yoked in the way that they are. If grounding is instead determination$_G$, the idea is that (i) fundamental physical facts determine$_G$ the mental facts; (ii) these facts about what determines$_G$ what themselves determine$_G$ why the mental and fundamental physical are modally yoked in the way that they are; and (iii) the determination$_G$ facts at issue in (ii) back or underwrite an explanation of why the mental and fundamental physical are modally yoked in the way that they are.

Are the proposed explanations, however, physicalist-friendly? Roughly speaking, an explanation is physicalist-friendly when the material that goes into its explanans doesn't undermine physicalism. For example, while causal/nomic explanations of mental content are typically considered to be physicalist-friendly, competing explanations cast in terms of fundamental phenomenal properties aren't. The explanans that interests us here consists of the grounding connections between the mental and the fundamental physical — does this explanans include material that undermines physicalism? You might think that the answer is 'yes,' claiming that these grounding connections themselves fall within the physicalist's explanatory ambit, yet they apparently lack physicalist-friendly explanations.

There are at least two things to say in response. First, Bennett (2011) and deRosset (2013) develop and defend a theory of iterated grounding relative to which grounding facts between the mental and fundamentally physical clearly have physicalist-friendly explanations. The basic idea is this: provided that thus-and-so fundamental physical facts, the Ps, ground mental fact M, the Ps also ground [the Ps ground M], [the Ps ground [the Ps grounds M]], and so on. In this case, the fundamental physical facts themselves explain the relevant grounding connections. And an explanans that consists entirely of fundamental physical facts obviously doesn't contain material that undermines physicalism.

Second, it may be improper to view physicalists as being committed to telling a story about what grounds facts to the effect that fundamental physical facts ground mental facts in the first place. This is so if the facts about what grounds what fall outside the explanatory order altogether (cf. Dasgupta 2014). But let's suppose for the sake of argument that these facts do have explanations. Still, there might not be a problem for the physicalist here. Of course, 'physicalism' and 'dualism' are terms of art, and there is at least some degree of freedom in how we choose to use them. One reasonable constraint, however, is that we use them in a way that is at least in the ballpark of what self-described physicalists and dualists have in mind. Returning to the mind-body problem understood as a location problem, physicalists as we have seen are interested in finding a place for the *manifestly mental* facts in the physical world, such as that Gomer is having a greenish experience. Perhaps the physicalist, then, ought not to be concerned with finding a place for *other* sorts of facts in the world, such as facts about what grounds what. This is so even though the relevant grounding facts *involve* manifestly mental facts, and even if there is an attenuated sense of 'mental' whereby such grounding facts count as being mental themselves.[20]

Here is a final thought on the matter. There are at least two senses of 'explanation' in terms of which we might describe the explanatory burden of physicalism with respect to grounding. First, there is the idea that physicalists need to *metaphysically explain* the fact that these fundamental physical facts ground those mental facts by citing what grounds this very fact about grounding. (Perhaps there are other ways for the physicalist to metaphysically explain a fact about grounding; but let us set this complication aside.) Second, there is the idea that physicalists must provide *well-justified reason to believe* that these fundamental physical facts ground those mental facts. These projects are quite different. Physicalists, of course, must do the latter – they need to provide at least some evidence for buying into their network of grounding claims. No surprise here – this is just to say that physicalists should have good arguments for their position! So the point is that if you don't properly distinguish between providing an epistemic reason and providing a metaphysical ground, it may seem that part of the physicalist's job is to provide grounds for certain facts about what grounds what.[21]

4 Conclusion

In this chapter we considered three points of contact between grounding and foundational metaphysics – how grounding might be relevant to whether metaphysical questions are substantive, how to choose between metaphysical theories, and how to understand so-called 'location problems.'

We draw things to a close by returning to a matter raised in the last section, the issue of providing reasons for thinking that this grounds that. Just how we should think about the epistemology of grounding is an interesting matter – this itself is a further question of foundational metaphysics. What are plausible diagnostics for grounding, principles that specify the conditions under which claims about what grounds what are plausible? Much of the literature on grounding so far has focused on clarifying the notion. These discussions, however, provide little guidance for formulating grounding diagnostics – it seems clear, for example, that nothing in these discussions could be operationalized into a discovery procedure that doesn't crucially depend on our already having knowledge about what grounds what.

The epistemology of grounding is perhaps an underexplored area of research. While we don't have the space to explore these issues in any detail here, here is one thought. Some implausible grounding claims seem implausible because we have no sense of *how* the grounding is supposed to work. This is the case, for example, with respect to the claim Gomer having a greenish experience grounds the fact that Socrates is a philosopher. This claim is implausible because there is no reasonable story to tell about how the connection is supposed to run between the Gomer fact and the Socrates fact. A comparison to causal mechanisms may be helpful here. Suppose you make a claim about what causes a neurochemical event such as the release of neurotransmitters. Since biochemistry is a subject matter in which causal relations have underlying causal mechanisms, if it's unclear what sort of underlying causal mechanism might be operative in this case, this counts against your causal claim.

So it seems that one way to justify a grounding claim (or at least show that it's not obviously implausible) is to tell a plausible story about how the connection runs between the relevant facts. How do you do that? Well, one option is to lean heavily on the comparison with causal claims and underlying causal mechanisms. To specify how the connection runs between the ground and grounded is to specify what we might call a "metaphysical mechanism" linking these facts together, either directly or via their constituents (Trogdon 2018). And another option is to develop substantive principles specifying relations of counterfactual dependence between the relevant facts (Schaffer 2016).[22]

Notes

1. For introductions to and surveys of recent literature on grounding, see Bliss and Trogdon (2014), Clark and Liggins (2012), Correia and Schnieder (2012), Raven (2015, 2019), and Trogdon (2013a).
2. Three figures who ostensibly theorized about and in terms of grounding are Aristotle, Spinoza, and Bolzono – for discussion, see Corkum (2016), Newlands (2018), and Schnieder (2014), respectively. For more on the role that grounding has played in the history of philosophy, see entries in Raven (2020).
3. See Daly (2012) and Hofweber (2016, Ch. 13).
4. See Koslicki (2015), Turner (2017), and Wilson (2014).
5. One disagreement concerns the formal features of grounding. For example, while many assume that it's transitive, some argue that grounding (understood as a binary relation) isn't – see Schaffer (2012).
6. See Audi (2012), Schaffer (2016), and Trogdon (2013a).
7. See Dasgupta (2017), Litland (2015), and Rosen (2010).
8. For more on the later view, see Strevens (2008).
9. See Marmadoro (Chapter 22, this volume) for further discussion of neo-Aristotelian metaphysics.
10. As an alternative to the disagreement described below, there might (also, or instead) be disagreements pertaining to the grounding relationships directly holding between certain types of objects themselves, rather than between certain facts about them (Schaffer 2010).
11. See Bliss (Chapter 16, this volume) for more on the relationship between ungroundedness and the closely related notion of fundamentality, as well as section 2 of the present entry.
12. See Trogdon (2017) for an overview of some of these considerations.
13. For a small sample, see the debate between Dorr (2008) and Swoyer (2008) on the existence of properties.
14. For more on the related notion of metaphysical explanation, see Thompson (Chapter 17, this volume).
15. See Fine (1994) for a canonical discussion of essence in the operative sense and Leech (Chapter 19, this volume) for further discussion. For further discussion of the grounding-essence interface, see Audi (2012), Correia and Skiles (2019), Dasgupta (2014), Rosen (2010), Trogdon (2013b), and Zylstra (2019).
16. What if there are no fundamental facts – does this imply that there are no fundamental entities? For a more sophisticated grounding-based account of fundamentality meant to deal with such concerns, see Raven (2016).
17. See Schaffer (2015) for discussion related to this last point. For further objections to the Laser, see Baron and Tallant (2018) and Fiddaman and Rodriguez-Pereyra (2018).
18. Or at least, all *concrete* ones. For sake of simplicity, let us ignore putative fundamental non-concrete entities, but see Trogdon and Cowling (2019) for discussion.
19. See Cowling (Chapter 29, this volume) for related discussion.
20. Crucial here is that by "physicalism" we mean physicalism about the (manifest) mental. Contrast this view with physicalism *tout court*, according to which any grounded fact is ultimately grounded by fundamental physical facts. Advocates of this thesis are in the business are showing what grounds facts about what grounds what (Dasgupta 2014).
21. See Trogdon (2015) for more on understanding location problems in terms of grounding.

22 We wish to thank Ricki Bliss, two anonymous referees, and the members of the Metaphysical Explanation research group (lead by Anna-Sofia Maurin and funded by Riksbankens Jubileumsfond) for their helpful comments. Thanks to the Department of Philosophy, Linguistics, and Theory of Science at the University of Gothenburg and the Swiss National Science Foundation (grant 10012_150289) for research funding. Finally, thanks to Café Publik in Gothenburg for putting up with us as we worked out the ideas in this chapter.

References

Audi, P. 2012. "Grounding: toward a theory of the in-virtue-of relation," *Journal of Philosophy* 109: 685–711.
Baron, S. and J. Tallant. 2018. "Do not revise Ockham's Razor without necessity," *Philosophy and Phenomenological Research* 96: 596–619.
Bennett, K. 2011. "By our bootstraps," *Philosophical Perspectives* 25: 27–41.
Bennett, K. 2017. *Making Things Up*. Oxford: Oxford University Press.
Bliss, R. and K. Trogdon. 2014. "Metaphysical grounding." In E. Zalta (ed.) *The Stanford Encyclopedia of Philosophy*.
Cameron, R. 2010. "How to have a radically minimal ontology," *Philosophical Studies* 151: 249–264.
Clark, M., and D. Liggins. 2012. "Recent work on grounding," *Analysis* 72: 812–823.
Corkum, P. 2016. "Ontological dependence and grounding in Aristotle," *Oxford Handbooks Online*. Oxford: Oxford University Press.
Correia, F. 2014. "From grounding to truth-making: some thoughts." In A. Reboul (ed.), *Mind, Values, and Metaphysics: Philosophical Papers Dedicated to Kevin Mulligan*, Vol. 1. Berlin: Springer.
Correia, F. 2017. "Real definitions," *Philosophical Issues* 27: 52–73.
Correia, F. and B. Schnieder. 2012. "Grounding: an opinionated introduction." In F. Correia and B. Schnieder (eds.), *Metaphysical Grounding: Understanding the Structure of Reality*. Cambridge: Cambridge University Press.
Correia, F. and A. Skiles. 2019. "Grounding, essence, and identity," *Philosophy and Phenomenological Research* 98: 642–670.
Daly, C. 2012. "Scepticism about grounding." In F. Correia and B. Schnieder (eds.), *Metaphysical Grounding: Understanding the Structure of Reality*. Cambridge: Cambridge University Press.
Daly, C. and D. Liggins. 2014. "In defense of existence questions," *Monist* 97: 460–478.
Dasgupta, S. 2014. "The possibility of physicalism," *The Journal of Philosophy* 111: 557–592.
Dasgupta, S. 2017. "Constitutive explanation," *Philosophical Issues* 27: 74–97.
deRosset, L. 2013. "Grounding explanations," *Philosophers' Imprint* 13: 1–26.
Dorr, C. 2008. "There are no abstract objects." In T. Sider, J. Hawthorne, and D. Zimmerman (eds.), *Contemporary Debates in Metaphysics*. Oxford: Blackwell.
Epstein, B. 2015. *The Ant Trap: Rebuilding the Foundations of the Social Sciences*. Oxford: Oxford University Press.
Fiddaman, M. and G. Rodriguez-Pereyra. 2018. "The razor and the laser," *Analytic Philosophy* 59: 341–358.
Fine, K. 1994. "Essence and modality," *Philosophical Perspectives* 8: 1–16.
Fine, K. 2001. "The question of realism," *Philosophers' Imprint* 1: 1–30.
Fine, K. 2009. "The question of ontology." In D. Chalmers, D. Manley, and R. Wasserman (eds.), *Metametaphysics: New Essays on the Foundations of Ontology*. Oxford: Oxford University Press.
Fine, K. 2012. "Guide to ground." In F. Correia and B. Schnieder (eds.), *Metaphysical Grounding: Understanding the Structure of Reality*. Cambridge: Cambridge University Press.
Fine, K. 2016. "Identity criteria and ground," *Philosophical Studies* 173: 1–19.
Griffith, A. 2018. "Social construction and grounding," *Philosophy and Phenomenological Research* 97: 393–409.
Hofweber, T. 2005. "A puzzle about ontology," *Noûs* 39: 256–283.
Hofweber, T. 2016. *Ontology and the Ambitions of Metaphysics*. Oxford: Oxford University Press.
Horgan, T. 1993. "From supervenience to superdupervenience: Meeting the demands of a material world," *Mind* 102: 555–586.
Horvath, J. 2018. "Philosophical analysis: the concept grounding view," *Philosophy and Phenomenological Research* 97: 724–750.
Jackson, F. 1998. *From Metaphysics to Ethics*. Oxford: Oxford University Press.

Kim, J. 1998. *Mind in a Physical World*. Cambridge, MA: MIT Press.
Koslicki, K. 2015. "The coarse-grainedness of grounding." In K. Bennett and D. Zimmerman (eds.), *Oxford Studies in Metaphysics* (Vol. 9). Oxford: Oxford University Press.
Lewis, D. 1973. *Counterfactuals*. Oxford: Blackwell.
Litland, J. 2015. "Grounding, explanation, and the limit of internality," *Philosophical Review* 124: 481–532.
McDaniel, K. 2015. "Propositions: individuation and invirtuation," *Australasian Journal of Philosophy* 93: 757–768.
Newlands, S. 2018. *Reconceiving Spinoza*. Oxford: Oxford University Press.
Raven, M. 2015. "Ground," *Philosophy Compass* 10: 322–333.
Raven, M. 2016. "Fundamentality without foundations," *Philosophy and Phenomenological Research* 93: 607–626.
Raven, M. 2019. "Metaphysical grounding." In D. Pritchard (ed.), *Oxford Bibliographies in Philosophy*. Oxford: Oxford University Press.
Raven, M. 2020. (ed.), *The Routledge Handbook of Metaphysical Grounding*. New York: Routledge.
Rodriguez-Pereyra, G. 2005. "Why truthmakers?" In H. Beebee and J. Dodd (eds.), *Truthmakers: The Contemporary Debate*. Oxford: Oxford University Press.
Rosen, G. 2010. "Metaphysical dependence: grounding and reduction." In R. Hale and A. Hoffman (eds.), *Modality: Metaphysics, Logic, and Epistemology*. Oxford: Oxford University Press.
Rosen, G. 2015. "Real definition," *Analytic Philosophy* 56: 189–209.
Schaffer, J. 2009. "On what grounds what." In D. Chalmers, D. Manley, and R. Wasserman (eds.), *Metametaphysics: New Essays on the Foundations of Ontology*. Oxford: Oxford University Press.
Schaffer, J. 2010. "Monism: the priority of the whole," *Philosophical Review* 119: 31–76.
Schaffer, J. 2012. "Grounding, transitivity, and contrastivity." In F. Correia and B. Schnieder (eds.), *Metaphysical Grounding: Understanding the Structure of Reality*. Cambridge: Cambridge University Press.
Schaffer, J. 2015. "What not to multiply without necessity," *Australasian Journal of Philosophy* 93: 644–664.
Schaffer, J. 2016. "Grounding in the image of causation," *Philosophical Studies* 173: 49–100.
Schnieder, B. 2014. "Bolzano on causation and grounding," *Journal of the History of Philosophy* 52: 309–337.
Sider, T. 2013. "Against parthood," *Oxford Studies in Metaphysics* 8: 237–293.
Skiles, A. 2014. "Primitivism about intrinsicality." In R. Francescotti (ed.), *Companion to Intrinsic Properties*. Berlin: De Gruyter, pp. 221–252.
Strevens, M. 2008. *Depth: An Account of Scientific Explanation*. Cambridge, MA: Harvard University Press.
Swoyer, C. 2008. "Abstract entities." In T. Sider, J. Hawthorne, and D. Zimmerman (eds.), *Contemporary Debates in Metaphysics*. Oxford: Blackwell.
Thompson, N. 2019. "Questions and answers: metaphysical explanation and the structure of reality," *Journal of the American Philosophical Association* 5: 98–116.
Trogdon, K. 2013a. "An introduction to grounding." In B. Schnieder, M. Hoeltje, and A. Steinberg (eds.), *Varieties of Dependence*. Munich: Philosophia Verlag.
Trogdon, K. 2013b. "Grounding: necessary or contingent?" *Pacific Philosophical Quarterly* 94: 465–485.
Trogdon, K. 2015. "Placement, grounding, and mental content." In C. Daly (ed.), *The Palgrave Handbook on Philosophical Methods*. London: Palgrave.
Trogdon, K. 2017. "Priority monism," *Philosophy Compass* 12: 1–10.
Trogdon, K. 2018. "Grounding-mechanical explanation," *Philosophical Studies* 175: 1289–1309.
Trogdon, K. and S. Cowling. 2019. "Prioritizing platonism," *Philosophical Studies* 176: 2029–2042.
Turner, J. 2017. "Curbing enthusiasm about grounding," *Philosophical Perspectives* 30: 366–396.
Williams, J.R.G. 2006. "Illusions of gunk," *Philosophical Perspectives* 20: 493–513.
Wilson, J. 2014. "No work for a theory of grounding," *Inquiry* 57: 535–579.
Witmer, D.G., B. Butchard, and K. Trogdon. 2005. "Intrinsicality without naturalness," *Philosophy and Phenomenological Research* 70: 326–350.
Zylstra, J. 2019. "The essence of grounding," *Synthese* 196: 5137–5152.

16

FUNDAMENTALITY

Ricki Bliss

It's not uncommon to hear philosophers speak about how things fundamentally are: that the world is fundamentally physical, or fundamentally just, or even that it exists fundamentally. Sometimes, the expression 'fundamental' is just synonymous with terms like 'basically', 'really' or 'objectively'. What we mean to say in claiming that the world is fundamentally just, for example, is that the world is just, in and of itself, and not, somehow, as the result of human imagination. Sometimes, though, the term 'fundamental' is actually used as a technical term, where although it is importantly conceptually connected to ideas about how things basically, objectively or in and of themselves actually are, it takes on a nuanced and sophisticated life of its own. This entry is about what metaphysicians of a certain stripe mean when they use 'fundamental' in this technical sense.

The view

Let us begin by considering how the notion of fundamentality has been understood in very recent terms. Although the idea is a very old and illustrious one, there is value in beginning with contemporary discussions, for it is there that fundamentality has been presented in terms more familiar.

Consider the relationship between a shadow and the willow tree that casts it. Or between the fact that a country is at war and the facts that its troops are massed at the border, its citizens are living under conditions of austerity and its leader has reinstated the draft. In both cases, some kind of dependence relation would seem to be in operation – indeed several of them. In the first case, the shadow would seem to depend in some deeply important way for both its existence and its identity on the tree that casts it. In the second case, the fact that a country is at war seems to be grounded in facts about its troops, austerity and reinstatement of the draft (amongst still other facts, surely). We say that it is *because* a nation has troops massed at the border and a reinstatement of the draft that that nation is in a state of war.

In both cases, it also looks like the dependence relation runs in one direction and not the other. The tree does not appear to depend on its shadow; just as being in a state of war does not look to be constitutive of what it is to have troops massed at the border, for example. Of course, being in a state of war might *cause* the massing of troops at a border, and the tree is causally involved with the existence of its shadow, but this is a different dependence relation.

The question of whether causation is a dependence relation, and how it ought to be classified if it is, is a thorny one.[1] Here it is enough to note that whatever particular relations are involved here, they are not only causal, and they look to run in one direction and not the other.

In addition to this, it seems reasonable to suppose that willow trees don't existentially depend upon themselves, and the fact that a nation is at war cannot be said to be the case as a result of the fact that a nation is at war. It also seems reasonable to suppose that just as willow trees depend upon the existence of their willow tree parts, and those parts depend upon the existence of those further parts, the willow tree also depends on the parts of the parts as well.

Let us call the general class of relations that we are concerned with here *metaphysical dependence relations*. There is much debate in the contemporary literature surrounding issues as to how we are supposed to understand these relations. Some folks are happy to understand (the full fleet of) metaphysical dependence relations as synonymous with *grounding*.[2] Advocates of this approach also tend to be of the view that metaphysical dependence or grounding relations are *the* primary structuring relations, where relations of, for example, parthood, membership and so on just *are* grounding relations. We can think of this approach as the big-D Dependence approach.[3]

The big-D Dependence view can be contrasted with the small-d dependence view. According to the small-d dependence account, there are a fleet of metaphysical dependence relations: grounding, identity, parthood, membership, realisation, composition, and still others (possibly causation). The term 'metaphysical dependence relation' stands to 'grounding', 'identity' and so on, as do genus terms to species. In other words, just as cats are always, in fact, tabby cats, lions, pumas and so on, instances of metaphysical dependence relations are always instances of identity, parthood, etc. Just as there are no cats running around in the world that aren't particular species of cats, there are no metaphysical dependence relations that aren't particular small-d dependence relations.

The issues here are many and largely peripheral to the interests of the current discussion.[4] They will, at various points, have a bearing on how we think about fundamentality, however, and I shall flag these along the way. Let us just say more broadly, for now, that reality is arranged by metaphysical dependence relations.

So far, I have encouraged the idea that these metaphysical dependence relations induce a hierarchy. Metaphysical dependence is asymmetric, transitive and irreflexive.[5] Let us suppose that the cosmos metaphysically depends on the galaxies that compose it. Let us suppose that those galaxies depend on their stars and gravitational fields. And that the stars depend on their star parts, and so on. At some point, one might think, this all needs to stop somewhere. One might think there needs to be a bottom or basic level below which there is nothing further; that the dependence chains that order reality need to terminate in some thing or some things that don't depend upon anything else.

Absolute fundamentality

The first and arguably central notion of fundamentality is that of *absolute fundamentality*.[6] What it is to be absolutely fundamental, by the lights of most contemporary accounts, is to be independent. A first criterion on fundamentality, then, is independence, where independence is understood as not metaphysically dependent upon anything else:

> Absolute Fundamentality: if something is fundamental, then it is does not metaphysically depend upon anything else.

As stated, independence serves as a necessary but not sufficient condition on fundamentality. According to many influential accounts of fundamentality, however, independence is both a necessary *and* sufficient condition on being fundamental. According to these accounts, then:

> Absolute Fundamentality₁: something is fundamental if and only if it does not metaphysically depend upon anything else.

A few interesting points of note regarding fundamentality so understood. First, fundamentality defined in terms of independence gives us a fairly thin concept – there's nothing more to fundamentality than being independent. Where such a thin conception of fundamentality might yield unintuitive or even undesirable consequences is when we consider the kinds of things that this might allow to count as fundamental. Let us suppose that sense data are the kinds of things that just don't appear to metaphysically depend upon anything else. Or let us suppose that there are metaphysical atoms. On our very thin conception of fundamentality both sense data and metaphysical atoms are fundamental. Why think this is odd? Well, you might think that whatever is fundamental also plays something like a metaphysically important role in the overarching structure of reality, and that sense data just don't seem like the kinds of things that are fit to play such a role. It does not follow from something's being independent that it is connected to reality in the right (or any!) kind of way. Where what it is to be fundamental is just to be independent, metaphysical atoms utterly disconnected from everything else are apt to be fundamental. In such a case, we could have 'island universes' of fundamentalia drifting freely from the rest of reality. Aside from the intuitive oddness of such views, reasons to believe this to be the case will be presented later in this discussion.

Of course, if we define fundamentality in terms of independence, then what it is to be fundamental just is to be independent. And if island universes turn out to be fundamental, then so be it. There is not necessarily anything particularly bad about employing such a thin concept. But historically at least, noteworthy accounts of fundamentality have seemed to be much richer or thicker. Consider God and His role in generating/causing/explaining absolutely everything else. Or consider the prime matter of the Aristotle of the *Metaphysics*, without which there would be nothing else.

I will circle back to a discussion of thicker conceptions of fundamentality in a moment, but for now I wish to address an additional issue for fundamental₁. As already mentioned, there is disagreement over whether there is a generic metaphysical dependence relation that orders reality – big-D dependence – or whether metaphysical dependence is more like a genus that collects together different species of small-d dependence relations: grounding, parthood and so on. On the first approach, to be fundamental is to be *categorically* independent. To understand what this might mean, consider God. God is absolutely and categorically independent of absolutely everything (except for, perhaps, Himself). Nothing else grounds God, causes God, is that of which God is a part or member, and so on.

On the second approach, fundamentality is *relative to a small-d dependence ordering*. This means that although some fundamentalium might be ungrounded, it might well be caused by or composed of something else, for example. Consider a fact about the existence of an electron, and suppose it is a fundamental fact. According to some very influential theories of facts, facts are (non-mereologically) composed of their parts. On the generic understanding of metaphysical dependence, this presents a problem as, on the one hand, fundamental facts are rock bottom, and, on the other hand they are not. Where big-D Dependence is the relation in operation, either we are forced to modify our metaphysics of facts, or to abandon the idea that facts are the kinds of things that can be fundamental after all (as it would be more accurate to say that some amongst their constituents are, provided that those constituents are simples).

On the second approach to metaphysical dependence, fundamentality is always relative to a small-d dependence ordering. In the case of the fact that an electron exists, we would then be able to say that *relative to the grounding ordering*, the fact is fundamental, but *relative to the ontological dependence ordering* it is not – because it further decomposes into parts. Facts that are fundamental relative to the grounding ordering might well be existentially, causally, mereologically, etc. dependent upon other things.

I am not quite sure which account of fundamentality is better – categorical absolute fundamentality or relativised absolute fundamentality. First, how these two accounts of metaphysical dependence interact with fundamentality surely bears consideration, but there are many additional and sophisticated problems that these approaches to metaphysical dependence relations face that need to be considered; problems that have nothing to do with interactions with fundamentality. Second, there seem to be competing factors pulling against both accounts. The demand for categorical absolute independence is arguably an impossibly high standard – are there any candidate entities that meet the criterion? On the other hand, relative absolute fundamentality looks to drain much of the significance from the notion of fundamentality. Perhaps a slightly less nebulous way of expressing the concern is that relativised absolute fundamentality renders problematic the idea of an overarching hierarchy. Where absolute fundamentality is relativised, there is no absolutely fundamental level, but, rather, fundamental entities interspersed throughout a structure whose hierarchical ordering – if such a structure it has – cannot be fixed by the dependence relations. That said, relativised absolute fundamentality tolerates domain-specific orderings. This allows us to talk about fundamental *beliefs* or *sets* as they are relevant to foundational epistemology and set theory, for example, without having the entire greater edifice of derivative reality bearing down upon them. I suspect the issues here are many, for now, though, let us move onto a second conception of fundamentality.

An additional, and to my mind much thicker, conception of fundamentality available in the literature is expressed in the language of *completeness*. In broad strokes, the idea is that the fundamental is complete insofar as it accounts for/generates, and/or explains everything else. In particular, what is fundamental forms the *minimally complete* basis for the rest of reality:

> Absolute Fundamentality$_C$: something is fundamental if and only if it is a member of a plurality of entities F, where F forms a minimally complete basis that determines everything else. The basis is minimally complete if no proper subset of the plurality of F is itself complete.[7]

In keeping with previous commitments, I leave it open what kinds of things can be members of the plurality – things, facts, etc. We are to assume that the relevant determination relation(s) are just those discussed above – our metaphysical dependence relations. The minimally complete basis, then, is the set of entities upon which everything else metaphysically depends.

How thick we understand this conception of fundamentality to be depends largely upon how we understand the metaphysical dependence relations at issue. One suggestion made in the literature says that absolute fundamentality$_C$ should be understood as providing a *complete minimal description of reality*. A stronger interpretation of absolute fundamentality$_C$ has the fundamentalia playing a much more active role. Such a role might express itself in the form of *explaining* or *generating* everything else.[8]

I believe it is reasonable to understand the dual conditions of independence and completeness as expressing the two faces of the notion of an unexplained explainer. For this reason, I believe it is interesting to wonder whether we can, or if we should, pull these two demands apart as the current thinking appears to. In general, philosophers seem to think that fundamentality is to be *defined* in terms of independence, with completeness merely serving as an auxiliary desideratum. I happen to think this is a mistake.

Well-foundedness

There is one further issue here that deserves our attention. I have used expressions such as 'dependence chains must terminate in something fundamental'. Expressions such as these are, at best, unclear, and at worst, misleading. Although a commitment to fundamentality involves a commitment to independent entities, it is, in fact, an open question how from any non-fundamental entity we are to arrive at what's fundamental.

In what is perhaps the most obvious or intuitive characterisation of fundamentality, fundamental entities sit at the bottoms of dependence chains. And, importantly, from any entity within a dependence chain, a fundamental entity can be arrived at in a *finite* number of steps. In other words, between any node on a dependence chain and the fundamental entity(s) that ultimately grounds that chain, there is a finite number of steps that must be taken in order to arrive there. There are also a variety of different ways in which those steps might be able to be made: this will depend partly on commitments regarding full and partial dependence, amongst other things. To claim that dependence chains terminate in something fundamental, where those chains are *downwardly finite* is to claim that dependence chains are *well-founded*.

We can contrast this approach to fundamentality with a view according to which it is *not necessarily* the case that dependence chains are downwardly finite. On this second approach, a dependence chain can be downwardly infinite where the members of the chain (and possibly the chain itself) are nonetheless grounded in something fundamental. To illustrate this way of understanding fundamentality, consider Euclidian space which is comprised of points and regions. Each region of space on such a view is divided into sub-regions. Facts about the existence of regions, then, are grounded in facts about the existence of their sub-regions, and so on *ad infinitum*. It is also the case, however, that facts about the existence of regions are also grounded in facts about the existence of the points between which the regions occur. Facts about regions (and sub-regions) are members of downwardly infinite grounding chains where those members are nonetheless grounded in something fundamental.[9]

Relative fundamentality

Not everyone would agree that the correct starting place in a discussion of fundamentality is with absolute fundamentality. One might believe that the central notion of fundamentality, at least contemporarily, is *relative* as opposed to absolute fundamentality. Recall from the previous section that reality is said to be structured by metaphysical dependence relations that induce a hierarchical ordering: galaxies depend on their stars, shadows upon their hosts and so on. What is at issue with establishing this ordering is having the means to elucidate the picture according to which things are more or less fundamental than other things. In other words, what it is to say that the shadow depends upon its host is just to say that the shadow is *less fundamental than* its host, and the host is *more fundamental than* the shadow that it casts. We might also think that shadows reside at the same level of the ordering as each other and are, thus, equifundamental. Relative fundamentality is, then, expressed in terms of relations of less fundamental than, more fundamental than and equifundamental with.

On the face of it, the notion of relative fundamentality looks easy to capture: the relation(s) doing the ordering just need to induce the right kind of structure. But as is often the case in philosophy, things are not so simple here as they might at first appear to be. The formal hierarchy-inducing properties are asymmetry, irreflexivity and transitivity. According to what we can think of as the orthodox view of grounding, relations of ground have these very properties.[10] According to heterodox accounts of grounding, however, they do not.[11] According to heterodox accounts of grounding, grounding might well be non-symmetric, non-reflexive and non-transitive. It has been suggested that one reason to prefer the orthodox account of grounding is that it provides the means to vindicate a picture of reality according to which it has a hierarchical structure – the account gives us the means to capture relative fundamentality. This, of course, is circular, as we cannot appeal to reality's being hierarchically structured in order to justify the view that reality is hierarchically structured. *If* reality is hierarchically arranged – *if* things are more or less fundamental than other things – then the orthodox account of grounding might be useful in capturing that, but separate argumentation is required to establish this in the first place.

Notice that the point about formal properties centred on the notion of grounding. A further issue for capturing relative fundamentality is that not all of the purported metaphysical dependence relations share the right formal features. Identity, for example, is reflexive and set membership is non-reflexive. Not all the relations that look like candidate metaphysical dependence relations share the right formal properties to guarantee the kind of hierarchy that talk of relative fundamentality is supposed to capture.

Debates over what ought to count as a metaphysical dependence relation and what the formal properties of said relations are aside, there are some additional problems with how best to capture relations of relative fundamentality. Let us suppose that I am less fundamental than my parts. The relative fundamentality ordering between me and my parts is, in this case, secured by us entering into parthood (or metaphysical dependence) relations with one another. But how are we supposed to capture what seems like an obvious relative fundamentality ordering between things that do *not* enter into metaphysical dependence relations with one another at all?[12] There is at least one electron in Brisbane, Australia that I do not appear to depend upon, but that we would surely want to say is more fundamental than me. There are a host of reasons for thinking that extant accounts of metaphysical dependence are not adequate for straightforwardly capturing relative fundamentality.

It is interesting to note that in spite of the centrality of the thought that reality has a hierarchical structure to theorising about reality and its over-arching structure, little attention has been paid in the literature to the finer details of relative fundamentality.[13] For this reason, in combination with the very natural thought that fundamentality talk is first and foremost concerned with absolute fundamentality, in what remains of this discussion we shall place our focus there.

What work the fundamental?

It is one thing to understand what it is to be fundamental, and quite another to understand why there needs to be any such kind of thing in the first place. So why suppose there is, or must be, something fundamental? It is important to recognise that our two disjuncts might lead us in the direction of different answers: reasons to believe there *is* something fundamental might not be reasons to believe that there has to be. As there needing to be something fundamental entails that there is something fundamental, I shall begin the discussion there. At the end of this section, I will circle back and address reasons to believe it is at least true of the actual world that it contains something fundamental.

Although discussions in the literature on *why* there should be anything fundamental are not as commonplace as discussions which assume that there is, there are enough remarks and arguments available to be able to construct the case in its favour.[14]

Perhaps the most powerful reasons available in the literature for supposing that there must be something fundamental are variations on arguments from vicious infinite regress. Consider the fact that you exist. What explains that fact? Let's suppose that facts about your parents, functioning vital organs, DNA, amongst still others metaphysically explain the fact that you exist. But what explains those facts? Presumably facts about your grandparents, your parents' functioning vital organs, DNA and so on. And what explains those facts? It appears as though we are off to the races. Why suppose the chains need to bottom out in something fundamental?

One reason for thinking as much is that if the chains do not terminate, then we have not actually explained each of the facts that we encounter along the way. If I can go on *ad infinitum*, I haven't really explained the fact that I exist after all. Schaffer, I take it, is expressing something like this concern when he says, 'being is infinitely deferred and never achieved'.[15]

On a variation of the problem, Dasgupta claims that fundamentality is, at least plausibly, motivated by the thought that we would like to answer the question 'Good grief, how come it all turned out like *this*?' where 'being told that the mountain is here because some particles are arranged thus and so does not (even if true) answer the question'.[16] What Dasgupta is suggesting is that the regress is vicious not because ordinary things are left without explanation – contra Schaffer – but because no appeal to ordinary entities/explanations can help us answer the question 'how come it all turned out like this?'.

Kit Fine, on the other hand, claims that

> there is still a plausible demand on ground or explanation that we are unable to evade. For given a truth that stands in need of explanation, one naturally supposes that it should have a 'completely satisfactory' explanation, one that does not involve cycles and terminates in truths that do not stand in need of explanation.[17]

Unlike both Schaffer and Dasgupta, Fine's remarks seem to suggest that the reason to believe there is something fundamental is involved in a different game altogether; that of explanatory satisfaction.

I do not happen to agree with the arguments presented or suggested by Schaffer, Dasgupta and Fine. For reasons of economy, I shall not elaborate upon what I believe to be the problems with Dasgupta and Fine's formulations of the problem.[18] Instead, I shall focus upon an additional interpretation of the regress problem as a means of motivating fundamentality that invariably involves rejecting Schaffer's understanding of the problem as well.

Let us return, then, to the Schaffer-style formulation of the problem as presented above. I have suggested that we are to understand the situation as one in which the regress is vicious because where the chain does not terminate, we have not explained the fact that you exist. I believe, however, that we can seriously wonder if this is the case. Although the story of the fact of your existence is a fairly complex one, we know how to tell it. Exactly what the fact that you exist depends upon are facts about the existence of your parents, your DNA and so on. In the normal course of things, a perfectly adequate explanation of the fact that you exist involves appeal to your ancestry and genetic material. In the normal course of things, no one would find an explanation of the fact that you exist wanting because it does not make appeal to the origins of the universe. The regress, as presented, would appear to be dripping with perfectly adequate explanations.

The regress as formulated is not vicious, then, as it is not the case that we have not explained the fact that you exist. Indeed, we appear to have done a whole lot of explaining about a number of things. Is there a way of understanding the regress such that it is vicious? One would hope so, as an argument from vicious infinite regress is about the most powerful argument we have for thinking there has to be something fundamental.

Again to the fact that you exist. What explains this fact? Facts about your genetic inheritance, functioning vital organs and so on. What explains those facts? At each stage of the regress, we explain the facts at the level above. What we appear not to have explained, however, is how the whole lot of it got to be in the first place. In other words, the regress is vicious when we switch from the *local* perspective – this fact here and that fact there – to the *global* perspective – a totality of facts, a kind of fact, or something sufficiently general.[19]

Let us take stock. What the proposed account of the regress problem as it is employed to justify fundamentality suggests is that fundamentality is closely tied to explanation. And I believe this to be true at least twice over. First, as I mentioned above, terminating the regress in something fundamental – something independent – is justified on the grounds that it is only by doing so that a certain kind of explanation can be achieved. Second to this, the account of viciousness I have presented indicates that whether or not a regress is vicious is tied to what we take the explanatory project to be. If we are looking for explanations for local matters of fact – such as the fact that you exist – then the generated regress is not a problem. If we are after global explanations, then the generated regress is unacceptable and, therewith, vicious.

So where, then, does this leave us? One reason to suppose that there must be something fundamental is that we need an explanation for a certain kind of fact or thing. What *kind* of fact or thing might this be? A natural suggestion would seem to be that we need an explanation for the collection of non-fundamental things. How exactly we ought to understand what this means is something of a thorny issue, though. If we suppose that wholes metaphysically depend on their parts, the totality of dependent things would depend on its parts, the dependent things. How the metaphysics of totalities interacts with the metaphysics of metaphysical dependence would seem to suggest that whatever we are seeking an explanation for, it cannot be the collection of dependent things.[20]

A more compelling formulation of the regress problem in defence of fundamentality takes as its target the thought that we need an explanation or ultimate ground for why there are any dependent things whatsoever.[21] Taking this as our explanatory target avoids any conflict with the metaphysics of totalities and looks to deliver a non-question-begging reason to suppose there must be something fundamental. Although there is much more to be said about this formulation of the regress problem, let us move along.

A second cluster of arguments seem to target the thought that we ought to commit to the idea that there *is* something fundamental; that it is at least true of the actual world that it contains something fundamental. These arguments are arguments from theoretical virtue. Perhaps the most developed of such arguments is put forward by Ross Cameron in the literature. According to Cameron, we ought to think that it is at least true of the actual world that it contains something fundamental as a theory of reality that posits fundamentalia is more *unified* than one that does not.[22] The broad idea seems to be that explaining chairs, stars, peonies and people in terms of fundamentalia gives us a more unified theory of reality than one that explains chairs, stars, peonies and people in terms of legs, seats, hydrogen, electrons, petals, stamen, DNA, great-grandparents and so on *ad infinitum*. The intricacies, and indeed correctness, of Cameron's view are well beyond the ken of the current discussion. For now, it is enough to recognise the outline of at least one argument from theoretical virtue.

A further argument based on theoretical considerations is put forward in the literature by Bliss (2019). According to this argument, the commitment to fundamentalia – independent, unexplained entities – is a theoretical cost. If this cost is to be worth paying, then the fundamentalia need to be of theoretical utility or value. A valuable role to be played by the fundamentalia is if they can explain something that non-fundamental things cannot. A good candidate for something that non-fundamental entities cannot explain is why there are any non-fundamental entities in the first place. We have, thus, an additional reason for thinking the formulation of the regress problem as presented above is a good one.

It is also reasonable to suppose that there will be other arguments in defence of fundamentality motivated by theoretical considerations. One immediate thought is that a picture of reality committed to the existence of something fundamental might be more parsimonious or perhaps more elegant than one that posits an infinite cascade of ever more fundamental entities. The details of such arguments are yet to be worked out.

Open issues

We have ridden roughshod over many deep, sophisticated and important issues. Unfortunately, constraints have meant we have not been able to do justice to most of the above-mentioned topics. In this final section, I would like to highlight what are surely interesting and valuable future directions of research.

Although intrinsically interesting, debates over whether we should accept big-D or small-d accounts of metaphysical dependence are significant to many of the issues that metaphysical dependence intersects with. How we are to understand both absolute and relative fundamentality is impacted by what account of metaphysical dependence we hold. As I mentioned above, both what I called categorical absolute independence and relativised absolute independence have issues that speak against them. They also have features that speak in their favour. In either case, these accounts look to give as remarkably different pictures of the way the world is, and remarkably different pictures of fundamentality in particular. Further investigation of how these pictures might look, and what the reasons are for holding them, are central to pushing debates over reality and its structure forward.

I have suggested that the reason to believe there is something fundamental is that we need an explanation for why there are any dependent entities whatsoever, or something of the like. The first thing to note is that, as it stands, our explanatory target looks incomplete. The question is, in fact, a contrastive question. It is interesting to note that not only must we supply a contrast, but different contrasts generate importantly different questions. 'Why are there any dependent entities whatsoever, rather than no dependent entities?' is a different question to 'why are there any dependent entities whatsoever, rather than only fundamental entities?'. Exploring the contrastive nature of the questions that might be motivating fundamentality is an important part of placing the view on firm epistemic footing.

The second point of note as regards arguments in defence of fundamentality that centre around the regress problem is that they very often look to appeal to some version or other of the Principle of Sufficient Reason (PSR). One might worry that a principle which says that, say, every fact has an explanation is incompatible with a view that posits the existence of something fundamental. But we need only look to the history of cosmological arguments to the existence of God to see that the PSR has a long history of being used to motivate foundationalist views. Contemporary debates would benefit much from returning to older, very well-developed discussions of the PSR and cosmological arguments.

Finally, although by no means exhaustively, there are interesting and important questions surrounding what we might think of as the two substantial alternatives to the view described here: *metaphysical infinitism* and *metaphysical coherentism*. Metaphysical infinitism is a view according to which reality is hierarchically arranged, with chains of entities ordered by metaphysical dependence relations that are asymmetric, transitive and irreflexive, but where there happens to be nothing independent.[23] Metaphysical coherentism, on the other hand, is a view that minimally tolerates dependence loops. I am somewhat circumspect in my presentation of coherentism as there are a variety of ways in which coherentism could be.[24] Indeed, although one can define coherentism as a species of anti-foundationalism, it need not be. A view on which the dependence relations were anti-symmetric and anti-reflexive could tolerate loops whilst at the same time allowing there to be something fundamental. That species of metaphysical coherentism might be viable seems to be garnering some attention in the contemporary literature.[25] In broad strokes, metaphysical coherentism is an interesting view. Particularly interesting for our purposes is how absolute fundamentality would look on a coherentist picture.

Much remains to be explored and understood about both absolute and relative fundamentality. The purpose of this entry has been to offer the reader an overview of what philosophers are talking about when they talk about fundamentality and why such a view, or perhaps better, research programme, is enjoying so much contemporary attention.

Notes

1 See Bennett 2017 esp. ch. 4 for a discussion of causation and what she calls 'building' – of which grounding and existential/ontological dependence are types. See Schaffer forthcoming for a rejection of the thought that causation is a grounding relation.
2 Schaffer 2009 can be read as endorsing such a view.
3 I adopt this terminology from Wilson 2014 in which she distinguishes between big-G Grounding and small-g grounding. Here, I prefer to speak in the language of 'dependence' for the reason that one needs not buy into the grounding framework whatsoever to make use of notions of fundamentality.
4 For excellent discussions of some of the relevant issue, see Koslicki 2015 and Wilson 2014.
5 For a discussion of the formal properties that metaphysical dependence relations could have, see Bliss and Priest 2018.
6 See Bennett 2017, ch. 5.
7 Bennett 2017, ch. 5, esp. sections 5.3 and 5.4.
8 See, again, Bennett 2017, ch. 5 and Tahko 2018, esp. section 1.3.
9 Please see Bliss and Trogdon 2014, esp. section 6.2. Please see also Dixon 2016 and Rabin and Rabern 2016 for excellent and thorough discussions of the many nuanced ways of understanding what it is for a dependence chain to be well-founded.
10 For a defence of the orthodox conception of grounding see Raven 2013.
11 For a discussion of how one might go about defending a heterodox account of grounding please see the Introduction in Bliss and Priest 2018.
12 See Bennett 2017, ch. 6 for a discussion of this and other related problems.
13 The most extensive and illuminating discussion of relative fundamentality available in the current literature is to be found in Bennett 2017.
14 See for example, Bliss 2013 and 2019, Dasgupta 2016, Raven 2016 and Schaffer 2009 and 2010.
15 Schaffer 2010, p. 62. Elsewhere, he states 'there must be a ground of being. If one thing exists only *in virtue of* another, then there must be something from which the reality of the derivative entities ultimately derives', Schaffer 2010, p. 37.
16 Dasgupta 2016, p. 5.
17 Fine 2010, p. 105. I am aware that Fine may well not have intended this remark to serve as a reason to suppose that there is something fundamental. Whether Fine intended it as such, one *could* still use this remark as a means of arguing in defence of fundamentality.
18 See Bliss 2019 for a more substantive discussion.
19 For a detailed discussion of these issues, see Bliss 2019.
20 See Bliss 2019 for discussion.
21 See Bliss 2013 and 2019.
22 See Cameron 2008. See also Paseau 2010 for a critique of Cameron's account.
23 For discussions of features of views that could be classified as species of infinitism see Bliss 2013, Bliss and Priest 2017 and Bohn 2018.
24 See Bliss and Priest 2018 for a more detailed discussion of the possibilities regarding coherentism.
25 For discussions of features of views that could be classified as species of coherentism see Barnes 2018, Jenkins 2011, Morganti 2018, Priest 2013 and Thompson 2016.

References

Barnes E., 'Symmetric Dependence', in *Reality and its Structure: Essays in Fundamentality*, Ricki Bliss and Graham Priest (eds). Oxford University Press (2018).
Bennett K., *Making Things Up*, Oxford University Press (2017).
Bliss R.L., 'Viciousness and the Structure of Reality', *Philosophical Studies*, vol. 166, no. 2 (2013), pp. 399–418.
Bliss R., 'What Work the Fundamental', *Erkenntnis*, vol. 84, no. 2 (2019), pp. 359–379.
Bliss R. and Priest G., 'Metaphysical Dependence and Reality: East and West', in *Buddhist Philosophy: A Comparative Survey*, Steven Emmanuel (ed.). Basil Blackwell (2017).
Bliss R. and Priest G., 'Fundamentality: A Geography', in *Reality and its Structure: Essays in Fundamentality*, Ricki Bliss and Graham Priest (eds). Oxford University Press (2018).
Bliss R. and Trogdon K., 'Metaphysical Grounding', *Stanford Encyclopaedia of Philosophy* (2014).
Bohn E.D., 'Indefinitely Descending Ground', in *Reality and its Structure: Essays in Fundamentality*, Ricki Bliss and Graham Priest (eds). Oxford University Press (2018).
Cameron R., 'Turtles All the Way Down', *Philosophical Quarterly*, vol. 58, no. 230 (2008), pp. 1–14.
Dasgupta S., 'Metaphysical Rationalism', *Noûs*, vol. 48 (2016), pp. 1–40.
Dixon S., 'What Is the Well-Foundedness of Grounding?', *Mind*, vol. 125 (2016), pp. 439–468.
Fine K., 'Some Puzzles of Ground', *Notre Dame Journal of Formal Logic*, vol. 51, no. 1 (2010), pp. 97–118.
Fine K., 'Guide to Ground', in *Metaphysical Grounding: Understanding the Structure of Reality*, Fabrice Correia and Benjamin Schnieder (eds). Cambridge University Press (2013).
Jenkins C., 'Is Metaphysical Dependence Irreflexive', *The Monist*, vol. 94 (2011), pp. 267–276.
Koslicki K., 'The Coarse-Grainedness of Grounding', in *Oxford Studies in Metaphysics*, Karen Bennett and Dean Zimmerman (eds). Vol 9. Oxford University Press (2015).
Morganti M., 'The Structure of Reality: Beyond Foundationalism', in *Reality and its Structure: Essays in Fundamentality*, Ricki Bliss and Graham Priest (eds). Oxford University Press (2018).
Paseau A., 'Defining Ultimate Ontological Basis and the Fundamental Layer', *Philosophical Quarterly* vol. 60, no. 238 (2010) pp. 169–175.
Priest G., *One: Being an Investigation into the Unity of Reality and its Parts, Including the Singular Object Which Is Nothingness*, Oxford University Press (2013).
Rabin G.O. and Rabern B., 'Well Founding Grounding Grounding', *Journal of Philosophical Logic*, vol. 45, no. 4 (2016), pp. 349–379.
Raven M., 'Is Ground a Strict Partial Order', *American Philosophical Quarterly*, vol. 50 (2013), pp. 191–199.
Raven M., 'Fundamentality Without Foundations', *Philosophy and Phenomenological Research*, vol. 93, no. 3 (2016), pp. 607–626.
Schaffer J., 'On What Grounds What', in D Chalmers, D Manley and R Wasserman (eds). *Metametaphysics: New essays in the foundations of ontology*, Oxford University Press (2009).
Schaffer J., 'Monism: The Priority of the Whole', *Philosophical Review*, vol. 119, no. 1 (2010), pp. 31–76.
Tahko T., 'Fundamentality', The Stanford Encyclopedia of Philosophy (Fall 2018 Edition), Edward N. Zalta (ed.), https://plato.stanford.edu/archives/fall2018/entries/fundamentality/.
Thompson N., 'Metaphysical Interdependence', in *Reality Making*, Mark Jago (ed.). Oxford University Press (2016).
Wilson J., 'No Work for a Theory of Grounding', *Inquiry*, vol. 75, no. 5 (2014), pp. 535–579.

17
METAPHYSICAL EXPLANATION

Naomi Thompson

Explanation has been studied in philosophy mostly under the guise of *scientific explanation*. We usually think of scientific explanation (and hence of explanation in general) as telling us *why something happened*. So, we might explain why the toaster exploded in terms of how hot it got and what it's made of, why a child is crying in terms of what his brother just said to him, and why two nations are at war in terms of their recent history. All of these explanations seem to be of the same type. But recently philosophical interest has begun to turn towards other sorts of explanations, not of why something happened, but of *what makes something the case*. What makes it the case that the child is crying is that there are tears streaming down his cheeks and a wailing sound coming out of his mouth. What makes it the case that two nations are at war is that they are bombing one another and that their soldiers are shooting at each other. These kinds of *metaphysical* explanations are usually thought to be synchronic rather than diachronic, and constitutive rather than causal.[1]

Where we might think of the notion of causal explanation as relating things on a horizontal plane, metaphysical explanation relates things vertically; across levels of fundamentality. When one fact or thing metaphysically explains another, the former is usually considered more fundamental than the latter. Metaphysical explanations thus describe reality's structure, and as such are a key metametaphysical notion. In this chapter we'll explore this notion of metaphysical explanation, but first some general remarks about the expression 'explanation'.

1 'Explanation'

The term 'explanation' is unfortunately somewhat vague, and can be used to express a few different (though related) things. First, we might take 'explanation' to refer to a communicative act, as when Rab provides an explanation to his mother as to the state of his clothes. Here an explanation might be something like 'my shirt is muddy because a bus drove through a puddle and splashed me'. But we also use the term 'explanation' to refer not to the communicative act, but to what is expressed in that communicative act (this is a second way in which we might use the term). We might think of what is expressed as a representational entity such as a proposition, in which case we might also use the term 'explanation' in a third way – for whatever worldly thing (such as worldly facts or states of affairs and the relation between them) is represented. In this case, that the shirt's being muddy is accounted for by the bus driving through a puddle and splashing its wearer.

That we use the term 'explanation' in different ways muddies the waters when we consider the extent to which explanation is sensitive to the context in which explanations are presented, the questions asked in order to elicit the relevant explanatory information, and how explanation is related to notions such as understanding. We might think of these sorts of things as epistemic constraints on explanation, and consider the extent to which we can make sense of a notion of explanation free of epistemic constraints.

Explanation in the first sense of a communicative act is plausibly heavily epistemically constrained. An explanation in this sense will likely only be successful if the explanation-seeker is able to understand the answer to her question and to assimilate it with what she already knows, and if she takes it to be a relevant answer to her question. (Suppose Rab is playing a video game where the object is to avoid being splashed by passing vehicles, and his own shirt is bloody from an earlier nosebleed. In that case, the explanation offered would not satisfy his mother.)

It is plausible too that there are some epistemic constraints on explanation in the second sense, though this is more controversial. The proposition <the Morning Star is bright> plausibly does not explain why Venus is easy to spot in the night sky.[2] One might argue that only <Venus is bright> is explanatory in the relevant way, even though 'the Morning Star' and 'Venus' refer to the same object. Again, it is tempting to think that a reason <the Morning Star is bright> fails to explain why Venus is easy to spot in the night sky is that understanding an explanation expressing the relevant proposition would require some extra knowledge.

Very few would be willing to accept that there are epistemic constraints on explanation in the third sense. At that level, explanation seems to have nothing to do with agents or with representation, it is just a matter of how the world fits together.[3] We can thus make sense of three different conceptions of explanation: an *epistemic* conception according to which explanations are merely informative representations; an *ontic* conception according to which explanations are simply portions of the world's structure; and a *hybrid* conception according to which explanations have both epistemic and ontic components.[4]

Most often in discussions of metaphysical explanation, participants seem to have a conception of explanation with at least some ontic component in mind. But note that the sense of 'explanation' most familiar from our folk discourse is plausibly epistemic. We usually think of explanation as what is offered in response to a request for information in a certain context, often with the aim of enhancing understanding. We must therefore be willing to set aside some of our usual assumptions about explanation in order to give due consideration to an ontic conception of explanation, but in doing so, may find we lose some of our grip on the notion.

2 Why believe in metaphysical explanation?

So far we have seen a couple of purported examples of metaphysical explanation, and noted that discussions in the literature most often have in mind explanation in an ontic sense. To further support the claim that there is some distinctive notion of explanation in play here, it is worth looking at a few more examples:

a   Black has won the game of chess because the white king is in checkmate
b   To be kind is pious because it is loved by the gods
c   The painting is beautiful in virtue of the arrangement of colours and lines
d   It is that Jack and Jill underwent a ceremony, signed a book and were issued with a certificate that explains why they are married
e   Since the ball is crimson, it is red

Whilst each of these examples are examples of what we earlier called *what-makes-it-the-case-that* explanations, note that none of them employ that terminology. The expression 'because' can be used in the presentation of any kind of explanation. Confusingly, we sometimes even come across terminology usually associated with causal explanation in the presentation of metaphysical explanations, as in (d). (Contrast (d) with the kind of explanation we would usually expect to receive in answer to the question 'why are Jack and Jill married?' Such a question would seem to require an answer detailing the genesis of their relationship, their shared interests and their love for each other.) We should not therefore expect to be able to recognise a metaphysical explanation by the terminology employed (though a helpful test is to see if the statement could be rephrased as a what-makes-it-the-case-that statement). Instead, we might recognise such claims by their 'metaphysical flavour': they are non-causal, and concern the constitutive generation of an outcome (Schaffer 2017: 303).[5]

Dasgupta (2017) argues that this notion of 'constitutive explanation' (as he calls it) is 'utterly familiar', not only in philosophy but also in other academic fields and in everyday life. Dasgupta takes metaphysical explanation to play an important role in philosophy, namely that of 'limning many issues of intellectual interest' (2017: 76). Deep philosophical debates very often centre around what it is that metaphysically explains some phenomenon or phenomena we care about (our mental states, for example). This notion of metaphysical explanation is important and importantly different from the notion of causal explanation which also divides logical space along important joints. Many debates in science and some in philosophy concern causal rather than metaphysical explanation, but it would be a mistake not to recognise both as essential to our theorising.

3 Grounding and metaphysical explanation

The recent literature in metaphysics has seen something of an explosion of work on *grounding* (see Skiles and Trogdon, this volume). It is generally claimed that grounding is an explanatory relation, where the sort of explanation at issue is the metaphysical explanation that is the topic of this chapter. One way to think of metaphysical explanation then is by analogy with the relationship between causation and casual explanation: worldly causal relations and worldly grounding relations back causal explanations and metaphysical explanations respectively.[6]

Raven (2015: 326) distinguishes between positions he calls *separatism* and *unionism*. The separatist position is roughly that described above; grounding relations are distinct from the metaphysical explanations that track them.[7] The unionist position holds that grounding *just is* a kind of metaphysical explanation.[8] Adopting a unionist position usually requires that we think of metaphysical explanation in the ontic sense outlined above. Failure to do so would constitute a challenge to the objectivity and mind-independence of grounding, because if metaphysical explanation is to some extent an epistemic phenomenon and grounding *just is* metaphysical explanation, then grounding too must be to some extent an epistemic phenomenon (see Thompson 2016). The separatist by contrast has the resources to be more flexible, as metaphysical explanations might communicate epistemically constrained information about worldly grounding relations.

There are, however, additional complications that the separatist must deal with. The view must be fleshed out in such a way that the relationship between grounding and explanation is itself explained. One option is to say that that relationship is one of grounding; grounding relations *ground* metaphysical explanations. This is fine so long as we have an independent handle on the notion of grounding, but there is a threat of epistemic circularity if we are to appeal to the notion of metaphysical explanation in order to elucidate that of grounding (as in e.g. Audi 2012; Correia and Schnieder 2012).

There are other ways in which the relationship between grounding and explanation might be accounted for. For example, we might take metaphysical explanations to provide information about the grounding relations that give rise to them, or to represent those relations, or we might think that those relations are themselves *parts* of the relevant metaphysical explanation. In what follows, we will consider the notion of metaphysical explanation without presupposing any particular connection between metaphysical explanation and grounding. We'll talk instead of *production* relations – relations that *produce* metaphysical explanations, of which grounding will often be considered one.[9]

4 Models of metaphysical explanation

Because metaphysical explanation is a species of explanation, and explanation has most often been studied in the context of scientific explanation, extant models of metaphysical explanation generally proceed by adapting models of causal explanation in various ways. This has produced a number of different ways of thinking about metaphysical explanation, with little consensus in the literature as to which (if any) is correct. There is perhaps also scope for a more ambitious approach to thinking about metaphysical explanation unconstrained by previous work on scientific explanation, but this project has not yet been widely undertaken and might perhaps seem unattractive, as there is intuitive appeal to thinking of the different notions of explanation as reasonably unified.

Backing model

The separatist view described above is an instance of what is sometimes called the 'backing model' of metaphysical explanation, since it takes metaphysical explanations to be backed by metaphysical production relations. In the above discussion, the relevant backing relations are all relations of grounding, but an alternate approach to metaphysical explanation might instead take some other single metaphysical production relation to back all metaphysical explanations. Any such relation would have to exhibit a high degree of generality, given the diversity of the situations in which we seem to encounter metaphysical explanations. (This is one reason why grounding as conceived as a relation between facts is a good candidate for the relevant production relation; grounding relations can relate facts about any number of different kinds of things.) Alternatively, one might think that a plurality of metaphysical production relations could serve as metaphysical explanation-backers. Set-membership relations, determinate-determinable relations, property realisation, and composition all seem like good candidates to back metaphysical explanations involving their relata.

The backing model is a quite general account of metaphysical explanation, and fleshing out the proposal requires further work. Some more detailed accounts of metaphysical explanation can be thought of as instances of the backing model (e.g. the mechanical model discussed below). That the backing model builds in the existence of a worldly production relation might be seen as a benefit or a cost, depending on how seriously we take the claim that metaphysical explanation is an ontic phenomenon. The backing model has unsurprisingly been popular, since an ontic approach to metaphysical explanation is by far the most common (though this model is not the only way to defend an ontic conception of metaphysical explanation).

Mechanical model

The mechanical model of metaphysical explanation derives from the causal-mechanical model of scientific explanation associated with Salmon (1984). The basic idea is that a causal explanation explains *how* one thing causes another; it gives an account of the relevant processes that produce the effect, representing the causal relation as being an instance of a causal mechanism. Mechanical models of metaphysical explanation give an account of the process of metaphysical production which produces the explanation. For example, Trogdon (2018) takes (some) metaphysical explanations to represent grounding relations as being instances of grounding mechanisms. Trogdon takes grounding to be a relation between facts, and a grounding mechanism is the metaphysical determination relations that obtain between the constituents of those facts. A metaphysical explanation thus tells us *how* the grounding connection between two facts operates, by giving us a specification of the determination relations involved. Here's a simple example. The fact that Socrates exists grounds the fact that {Socrates} (the set with Socrates as its sole member) exists. A mechanical explanation of this fact will say that a set-membership relation obtains between Socrates and {Socrates}, which are the constituents of the relevant facts.[10]

Trogdon is clear that his view only applies to some metaphysical explanations, but we can imagine a stronger view whereby *all* metaphysical explanation proceeds in this way. The stronger view is appealing in that it offers a unified account of metaphysical explanation (and of explanation in general, if we think of causal explanation along the same lines). However, it seems that we can sometimes give a metaphysical explanation in the absence of a relevant mechanism. For example, that an object has four sides of equal length seems to metaphysically explain that the object is a square, but there does not seem to be any metaphysical mechanism underlying that explanation.

Subsumption model

One of the most well-known models of scientific explanation is the Deductive-Nomological (DN) model, according to which explanations take the form of a sound deductive argument. That is, in a successful explanation the *explanandum* (what is to be explained) must be a logical consequence of the *explanans* (that which purports to explain the explanandum), and the explanans must be true. In addition, the explanans must contain at least one law of nature, without which the argument would be invalid (see Hempel 1965).

The subsumption model is so-called because the explanation is subsumed under a law. In the case of scientific explanation the relevant law is a law of nature. A natural extension then takes successful metaphysical explanation to be a sound deductive argument where at least one essential premise in the explanans is a *metaphysical* law. Kment (2014) defends such an account of metaphysical explanation. Wilsch (2016) also defends a version of a DN account, though he denies that we need think of the law as *included* in the explanans (as opposed to simply underlying the explanation). A metaphysical law (according to Wilsch) is a quantified material conditional which has its direction built in; it takes us from more fundamental to less fundamental truths. So, if we take it to be a metaphysical law that whenever an object exists, the singleton set of that object exists, we can move from the premise that Socrates exists and that law to the conclusion that {Socrates} exists.

A common objection to the DN model is that a deductive-nomological inference is neither necessary nor sufficient for scientific explanation. Against necessity, some scientific explanations seems to involve statistical or probabilistic laws, according to which future events cannot be deduced from past events. But no analogue of this worry arises for metaphysical explanation because there are no metaphysical analogues of statistical or probabilistic explanations (Wilsch 2016: 15). Against sufficiency, it has been argued that some law-like regularities do not make for successful explanations. It is a law-like regularity that objects fall to the ground when dropped by a woman, but that an object was dropped by a woman is not a good explanation of its falling to the ground.[11] This kind of worry is not avoided when the subsumption model is extended to cover metaphysical explanation. It is a general lesson that in adopting subsumption models of explanation we must be careful about how a law is defined in order to prevent bogus laws or arbitrary generalisations from counting, and this also holds in the metaphysical case.

A further objection to the DN model in general is that inferences are symmetric: we can't only infer and thus explain the length of the shadow from the height of the flagpole and the laws of the propagation of light, but we can also deduce and thus explain the height of the flagpole from the laws and the length of the shadow it casts (Bromberger 1966). This problem does not arise if, as per Wilsch's view, the laws take the form of material conditionals and thus have their directionality built in. This, however, is a substantial commitment of Wilsch's view. Other extensions of the model with different accounts of the laws might have to give a different answer to this objection.

Unificationism

Unificationist accounts of explanation take the goal of explanation to be that of providing a unified account of a range of different phenomena. In the literature on scientific explanation, unificationism is defended by Kitcher (1981; 1989), who identifies scientific progress as involving increasing unification. We judge that explanatory progress has been made when our theorising becomes more unified. This is why, for example, it is more explanatory to posit a general law in place of the special laws it subsumes, rather than positing a number of basic special laws. As applied to metaphysical explanation, it is far less obvious that philosophical progress involves increasing unification (though unification is considered a virtue in philosophical theorising, and it is arguably the case that general laws are more explanatory than the special laws they subsume). One way for the unificationist to motivate her position is to establish that progress in metaphysics involves increasing unification, but this might be a difficult task.

Rather than linking unification to metaphysical progress, the friend of unification for metaphysical explanation might instead appeal to the connection between explanation and understanding to motivate her position.[12] Roughly speaking, unificationists think of explanations as corresponding to deductively valid arguments, and take the more explanatory theories to be those that use a small number of stringent argument patterns with a small number of premises to derive a broad range of conclusions. (Argument patterns are sequences of schematic sentences with variables in place of some non-logical vocabulary, filling instructions telling us what the variables are to be substituted for, and a classification into premises, conclusion and inference rules).[13] Baron and Norton (2019) adapt Kitcher's account in order to present a theory of metaphysical explanation. The aim of metaphysical explanation is to unify beliefs implicated in metaphysical explanations. Roughly speaking, a derivation is a metaphysical explanation iff it is an instance of an argument pattern that unifies metaphysical beliefs. Baron and Norton claim that their account is able to ensure that metaphysical explanations are irreflexive, asymmetric and non-monotonic, though (as they note) it remains to be seen whether our intuitions about what metaphysically explains what generally line up with what is predicted by the account.

An advantage of the unificationist account is that it corresponds to an intuitive way of thinking about understanding: we understand better when we can derive large amounts of information from a fairly small number of premises and a small number of argument patterns. The ability to do so suggests an ability to identify connections between things; to fit things together. This is perhaps a more promising avenue for the unificationist to pursue in motivating her account than an appeal to the notion of metaphysical progress. However, for those who maintain that metaphysical explanation is an entirely ontic affair, this will not be very attractive. On the other hand, one might claim that an advantage of the unificationist account motivated in this way is that it allows us to respect and to utilise the intuitive connection between explanation and understanding, and thus renders metaphysical explanation less mysterious than it may seem on some other accounts of the notion.[14]

Pragmatic account

The pragmatic account of explanation takes us still further towards thinking of explanation as an epistemic rather than an ontic notion. According to this view, all explanations are answers to questions, and as such are sensitive to the context in which the relevant questions are asked and answered. The first systematic pragmatic account of scientific explanation is developed in Van Fraassen (1980), who diagnoses the failure (as he sees it) of other accounts of explanation in their failure to recognise or to pay proper attention to pragmatic or contextual factors in explanation (Woodward 2017). Van Fraassen's work on explanation comes in the context of his constructive empiricism: the anti-realist view according to which empirical adequacy rather than truth is the aim of science. This background is not, however, essential to adopting a version of the pragmatic account of explanation, and for its application to metaphysical explanation.

According to van Fraassen, requests for explanation ask why the topic of the question occurred as opposed to some other member of the contrast class. As applied to metaphysical explanation, the relevant questions are plausibly what-makes-it-the-case-that questions, so we might ask 'what makes it the case that the ball is red?' The context in which our question is asked determines whether we are interested in what makes it the case that the ball is red rather than some other colour, or red rather than covered in writing, or red now when it was previously green, and so on. The context also determines what kind of answer will be relevant, depending on (for example) what the questioner is interested in. One kind of answer to our question is that the ball is red because it is scarlet. A different kind of answer is that the ball is red because it reflects light with a wavelength in the red spectrum. Both of these answers are answers to the question 'what makes it the case that the ball is red rather than some other colour?', and each would be appropriate in a particular context (see Thompson 2019).

A concern about pragmatic accounts of explanation, at least as developed in the way described above, is that relevance is not subject to any constraints, and so given an appropriate context, anything might be said to explain anything else. (Kitcher and Salmon 1987). In the scientific case the pragmatist can respond by taking current scientific knowledge (or consensus) as part of the relevant context, thus ruling out as irrelevant answers in conflict with that body of knowledge. The absence of consensus in the metaphysical case makes this response less attractive. One option then is to bite the bullet, and to maintain that there really are far fewer constraints on what might count as explanatory than has generally been supposed. Another is to insist (for example) that both explanans and explanandum must be true.

Note that pragmatic accounts are not the only accounts of explanation where questions play an important role. As suggested above, it is very often the case that a request for explanation is elicited by a question, and it is fairly standard to think that which answers are appropriate will be constrained by the context of the request, and the background of the questioner and respondent (as in the first sense of 'explanation' discussed in section two above). What is distinctive about the pragmatic account is that there is no underlying explanation in an ontic sense which also helps to constrain the appropriateness of an answer to the question.

5 Realism and anti-realism

We started out with the idea that most prefer to think of metaphysical explanation as (at least partially) an ontic phenomenon. In other words, they are *realists* about metaphysical explanation. The first two models discussed above are realist accounts of metaphysical explanation because they involve worldly determination relations underwriting the relevant explanations. Wilsch (2016: 2) is explicit that his is also to be considered a realist account of metaphysical explanation, but we can imagine a version of the subsumption model according to which what is key is that the explanandum is to be *expected* given or *understood in terms of* the explanans.[15] On this kind of reading, it is possible to dispense with worldly determination altogether in giving an account of explanation, and instead adopt an anti-realist account of explanation whereby for A to be explanatory with respect to B is simply for B to be expected or understood, given A. This is not the only way to be an anti-realist about metaphysical explanation. The unificationist and pragmatic models above represent a second and third possible form of anti-realism, because they too don't (or at least need not) involve worldly determination relations.

The assumption of realism about metaphysical explanation has been used in various ways. For example, Kim (1990: 24) argues that supervenience is not an explanatory relation because it is not in itself a determination relation. The view that supervenience is not explanatory is common, but adopting an anti-realist account of metaphysical explanation would allow supervenience to count as explanatory in the right circumstances.

Realism about metaphysical explanation is also used in arguments for the existence of a grounding relation. Audi (2012: 105) is a particularly clear example. His argument is as follows:

1 If one fact explains another, then the one plays some role in determining the other.
2 There are explanations in which the explaining fact plays no causal role with respect to the explained fact.
3 Therefore, there is a non-causal relation of determination.[16]

Note that the first premise is only true on a realist conception of explanation. Audi's defence of the premise is somewhat cursory:

> Some will object to (1) on the ground that it makes explanation too ontologically robust, when in fact explanation is a merely pragmatic and heavily interest-relative affair. Here, I will assume that this is false. The correctness or incorrectness of an explanation, I assume, is at least in part a matter of its matching up with the structure of the world, structure that is conferred by the determination relations that hold among the world's inhabitants.
>
> *(Audi 2012: 105)*

Given the availability and attractiveness of some anti-realist accounts of metaphysical explanation, the dismissal of such accounts by Audi and some others might be seen as an oversight.

Taylor (2018) argues against explanatory realism. Her argument focuses on a few classes of seemingly problematic cases for the realist. One such example involves what she calls 'rule-based explanation'.[17] Suppose we want to explain what makes it the case that Izzy has cheated at Scrabble. One cheats in virtue of breaking one or more of the rules of the game, so the fact that Izzy cheated is metaphysically explained by her breaking one or more rules of Scrabble. But there is arguably no metaphysical determination relation involved here, and so the realist about metaphysical explanation might struggle to offer a suitable explanation.

Another class of seemingly problematic explanations for the realist are reductio-ad-absurdum explanations. The fact that the Kelvin scale has an absolute zero is explained by the ideal gas laws. The laws say that the volume of a gas is directly proportional to its temperature. Since the volume of a gas cannot be less than zero, neither can its temperature, and so the temperature scale must stop at that point (Taylor 2018: 211). It is hard to come up with any determination relation that could be playing a role in this explanation, since it doesn't seem to be the case that the ideal gas laws *ground* the fact that the Kelvin scale has an absolute zero (they merely imply it).

A unionist about metaphysical explanation who claims that metaphysical production or determination relations are by their nature explanatory will either deny that there is a genuine explanation in these cases, or insist that there is some determination going on. We don't have the space here to consider the relevant cases in more detail, but it should at least be clear that there are reasons to be suspicious of a wholesale endorsement of realism about metaphysical explanation.

6 Concluding remarks

Increased interest in a distinctive notion of metaphysical explanation is likely to continue to be philosophically beneficial; there are good reasons to believe in such a notion, and it can be put to interesting use. There is more work to be done to establish the best way to think about metaphysical explanation, and to explore connections between metaphysical explanation and related notions such as grounding and determination.

Notes

1 There might be many examples of constitutive explanations given in science (and causal explanations in metaphysics), and so the distinction is not supposed to be between explanations in two different domains of discourse. It is also possible to elicit a metaphysical explanation with a why-question, and sometimes even a causal explanation with a what-makes-it-the-case-that-question. Nevertheless, thinking in terms of this distinction between types of question provides a useful heuristic.
2 We are assuming here a fairly fine-grained, non-Russellian conception of propositions.
3 It is open to us to deny that there is any such sense of explanation. Van Fraassen (1980) famously defends such a view.
4 The distinction between epistemic and ontic conceptions of explanation is made by Salmon (1984; 1989). The hybrid conception is defended in Kim (1994), Ruben (2012) and Trogdon (2018).
5 It is worth noting too that the distinction between causal and metaphysical explanation is not universally recognised; some deny that there is any such distinction to be made.
6 One might also take the class of relations playing a role in metaphysical explanations to include more than just grounding. Other plausible candidates are other synchronic non-causal determination relations such as constitution, set-membership, the determinate-determinable relation and mereological relations. Some take grounding to be just one relation amongst many in this category (e.g. Bennett 2017) whilst others take grounding-talk to eclipse talk of these more fine-grained relations (e.g. Raven 2012; Rosen 2010).
7 Defenders of this sort of view include Audi (2012); Maurin (2019); Schaffer (2012); and Trogdon (2013).
8 Dasgupta (2014), Fine (2012) and Raven (2012) all prefer this way of thinking about the relationship between grounding and metaphysical explanation.
9 Kovacs (2017) introduces the language of production into the debate.
10 In more complex cases, the explanation will have more components. See Trogdon (2018) for examples.
11 The reason for this is that explanations are *non-monotonic*. If some fact A explains a further fact C, it is not necessarily the case that A plus some arbitrary extra information B jointly explain C.

12 Kovacs (2019) develops a theory of unificationism for metaphysical explanation in which he appeals to this motivation.
13 Stringency is a matter of the strictness of the filling instructions and the classification, and the condition prevents argument patterns being so loosely constrained as to spuriously cover maximal numbers of derivations.
14 Other accounts of metaphysical explanation, particularly those which exemplify a hybrid conception of explanation, might also make use of this connection as part of their motivation.
15 Some of Hempel's own writing on the topic suggests such a reading. See, for example Hempel (1965: 337).
16 Whilst Audi takes (3) to indicate the presence of a grounding relation, we don't have to follow him in this.
17 Taylor doesn't take *all* rule-based explanation to pose a threat to explanatory realism, but considers a large set of such explanations which do appear problematic.

References

Audi, P., 2012. A Clarification and Defense of the Notion of Ground. In: F. Correia and B. Schnieder (eds), *Metaphysical Grounding: Understanding the Structure of Reality*. Cambridge: Cambridge University Press, pp. 101–121.
Baron, S. and Norton, J., 2019. Metaphysical Explanation: The Kitcher Picture. *Erkenntnis*, online first.
Bennett, K., 2017. *Constructing the World*. Oxford: Oxford University Press.
Bromberger, S., 1966. Why Questions. In: R. Colodny (ed.), *Mind and Cosmos: Essays in Contemporary Science and Philosophy*. Pittsburgh: University of Pittsburgh Press, pp. 86–111.
Correa, F. and Schnieder, B., 2012. Grounding: An Opinionated Introduction. In: F. Correia and B. Schnieder (eds), *Metaphysical Grounding: Understanding the Structure of Reality*. Cambridge: Cambridge University Press, pp. 1–36.
Dasgupta, S., 2014. The Possibility of Physicalism. *The Journal of Philosophy*, 111, pp. 557–592.
Dasgupta, S., 2017. Constitutive Explanation. *Philosophical Issues*, 27(1), pp. 74–97.
Fine, K., 2012. A Guide to Ground. In: F. Correia and B. Schnieder (eds), *Metaphysical Grounding: Understanding the Structure of Reality*. Cambridge: Cambridge University Press, pp. 37–80.
Hempel, C., 1965. *Aspects of Scientific Explanation and Other Essays in Philosophy of Science*. New York: Free Press.
Kim, J., 1990. Supervenience as a Metaphysical Concept. *Metaphilosophy*, 21(1–2), pp. 1–27.
Kim, J., 1994. Explanatory Realism, Causal Realism, and Explanatory Exclusion. In: J. Kim, *Essays in the Metaphysics of Mind*. Oxford: Oxford University Press.
Kitcher, P., 1981. Explanatory Unification. *Philosophy of Science*, 48, pp. 507–531.
Kitcher, P., 1989. Explanatory Unification and the Causal Structure of the World. In: P. Kitcher and W. Salmon (eds), *Scientific Explanation*. Minneapolis: University of Minnesota Press.
Kitcher, P and Salmon, W., 1987. Van Fraassen on Explanation. *Journal of Philosophy*, 84, pp. 315–330.
Kment, B., 2014. *Modality and Explanatory Reasoning*. Oxford: Oxford University Press.
Kovacs, D., 2017. Grounding and the Argument from Explanatoriness. *Philosophical Studies*, 174(12), pp. 2927–2952.
Kovacs, D., 2019. Metaphysically Explanatory Unification. *Philosophical Studies*, online first.
Maurin, A.-S., 2019. Grounding and Metaphysical Explanation: It's Complicated. *Philosophical Studies*, 176, 1576–1594.
Raven, M., 2012. In Defence of Ground. *Australasian Journal of Philosophy*, 90, pp. 687–701.
Raven, M., 2015. Ground. *Philosophy Compass*, 10(5), pp. 322–333.
Rosen, G., 2010. Metaphysical Dependence: Grounding and Reduction. In: R. Hale and A. Hoffman (eds), *Modality: Metaphysics, Logic, and Epistemology*. Oxford: Oxford University Press, pp. 109–136.
Ruben, D., 2012. *Explaining Explanation*. 2nd Edition. London: Routledge.
Salmon, W., 1984. *Scientific Explanation and the Causal Structure of the World*. Princeton: Princeton University Press.
Salmon, W., 1989. *Four Decades of Scientific Explanation*. Pittsburgh: Pittsburgh University Press.
Schaffer, J., 2012. Grounding, Transitivity, and Contrastivity. In: F. Correia and B. Schnieder (eds), *Metaphysical Grounding: Understanding the Structure of Reality*. Cambridge: Cambridge University Press, pp. 122–138.
Schaffer, J., 2017. Laws for Metaphysical Explanation. *Philosophical Issues*, 27, pp. 302–321.

Taylor, E., 2018. Against Explanatory Realism. *Philosophical Studies*, 175, pp. 197–219.

Thompson, N., 2016. Grounding and Metaphysical Explanation. *Proceedings of the Aristotelian Society*, 116(3), pp. 395–402.

Thompson, N., 2019. Questions and Answers: Metaphysical Explanations and the Structure of Reality. *Journal of the American Philosophical Association*, 5(1), pp. 98–116.

Trogdon, K., 2013. An Introduction to Grounding. In: M. Hoeltje, B. Schnieder and A. Steinberg (eds), *Varieties of Dependence*. Munich: Philosophia Verlag, pp. 97–122.

Trogdon, K., 2018. Grounding-Mechanical Explanation. *Philosophical Studies*, 175, pp. 1289–1309.

van Fraassen, B., 1980. *The Scientific Image*. Oxford: Oxford University Press.

Wilsch, T., 2016. The Deductiuve-Nomological Account of Metaphysical Explanation. *Australasian Journal of Philosophy*, 94(1), pp. 1–23.

Woodward, J., 2017. *Scientific Explanation*, s.l.: The Stanford Encyclopedia of Philosophy.

18
TRUTHMAKING AND METAMETAPHYSICS

Ross P. Cameron

Truthmaking is a relationship between things in the world, and truths about how the world is. A truthmaker, A, for a truth, p, is an object whose existence makes it true that p. It is true that p in virtue of A existing. According to truthmaker theory, what is true is ultimately determined by what exists.

There are many questions we can ask about the nature and behavior of both truthmakers and the truthmaking relation, but this is not the purpose of this chapter. (For an introduction to such issues, see Cameron 2018.) In this chapter, we look at how truthmaker theory interacts with certain metametaphysical concerns. In section 1, we will look at how truthmaker theory offers an account of ontological commitment very different from the orthodox view defended by W.V. Quine. In section 2, we will look at how truthmaker theory can be used to defend the idea that there is a hard, and distinctively metaphysical, ontological project to be undertaken, in light of challenges from philosophers such as Amie Thomasson who think that ontological questions are easily answered.

1 Ontological commitment: Quine versus truthmaking

Everyone is interested in the extent of ontology: whether reality's contents are exhausted by the beings studied by the natural sciences, such as protons, protozoa, or planets, or whether it includes supernatural entities such as gods, ghosts, or goblins; whether reality's contents are exhausted by purely physical entities or whether it includes irreducibly mental beings; whether reality's contents are exhausted by atoms in the void or whether it includes complex phenomena like tables, people, universities, or galaxies.

How are we to settle the extent of ontology? Of course, we have many ways of investigating reality, which deliver to us theories of what the world is like. However, it is not always straightforward to take a theory of what the world is like and draw from this a conclusion about the extent of ontology. We might observe living beings and arrive at a biological theory that tells us, e.g., that wolves and dogs are both of the genus Canis. So now we know that that is true … what is the extent of ontology? Does it include wolves and dogs and the genus Canis, or just wolves and dogs, or some other answer? Doing more biology won't tell you that. We might do some physics and discover that the speed of light is finite. Okay … so what is the extent of ontology? Does it include light, or both light and speeds, or light and speeds and numbers?

Doing more physics won't tell you that. We need an account of ontological commitment to tell us what, given a particular claim or theory about the world, the extent of ontology is according to that claim or theory.

It is common to treat the question of the extent of ontology as equivalent to the question of what there is. Quine (1948, p21) said that the ontological problem can be asked in three words: "What is there?," and answered with one: "Everything," and many have been happy to agree with him. However, in saying this Quine is taking a substantive metaontological stance. Others think that the ontological problem is not what there is but rather what *exists*, or what *really exists*, or what has *being*, or etc. Quine, of course, would see all of these as equivalent – what exists is what there is, which is what has being, which is what is real. But other theorists disagree and hold that these claims come apart. Meinong (1904), for example, saw a distinction both between what there is and what has being, and between the things that have being and exist and the things that have being but do not exist (but merely subsist). Cameron (2008) and Fine (2009) see a distinction between what exists and what really exists, or is real, and argue that the ontological problem should be cast in terms of the latter notion, not the former. Sider (2011) sees a distinction between what we can truly say exists while speaking English, and what we could say exists if we were speaking a language where all our terms carved nature at its joints; the ontological question, then, is what exists when we use 'exists' with this special meaning that limns reality's ontological structure. These theorists are going to reject Quine's characterization of the ontological problem as asking (in English) what there is.

To avoid taking a stance at the outset on this particular metaontological issue, let's simply talk about solving the ontological problem. When a metaphysician proposes an answer to the ontological problem, they propose a list of things that there are/that exist/that really exist/that have being (delete as appropriate). For want of a better term, let's call such a list an ontology, and say that when a metaphysician proposes putting something on such a list, they are adding it to their ontology. An account of ontological commitment is an account of how to determine an ontology, given your theory of the world, or a claim you have made.

Quine's account of ontological commitment, which is orthodoxy in contemporary analytic philosophy, is as follows.[1] First, take your best theory of the world as delivered by the empirical sciences. Second, regiment that theory in a first-order language. Finally, ask what must belong to the domain of the quantifier in order for that theory, thus regimented, to be true. The domain of quantification yields the ontological commitments of the theory.

Quine's account has substantial implications for the extent of ontology. Suppose our theory tells us that there are distinct things that are alike in some particular way: they share a feature, F-ness. We want to know whether we should admit just the things to our ontology, or whether we should admit both them and the property of F-ness. Quine's account of ontological commitment settles the issue: admit the things, but not the property. That is because when we regiment the claim in first-order logic we get $\exists x \exists y (x \neq y \,\&\, Fx \,\&\, Fy)$, and in order for that to be true, the domain of quantification has to include two things, each of which are F, but it does *not* have to include the property of F-ness. Thus, Quine uses his account of ontological commitment to argue for nominalism, a particular theory of the extent of ontology: the ontology of the world includes individuals that are various ways, but it does not include ways that those individuals are (i.e., properties).

While Quine's approach is very popular, it is not without its dissenters. One cause for dissatisfaction is Quine's focus on the language of first-order logic as the privileged language of regimentation. Many philosophers think other languages are just as legitimate. So for example we might use a second-order language to say that *a* and *b* are alike in some way:

20) $\exists F(Fa \,\&\, Fb)$

(Read as: There is some way that a is, and b is also that way.)

Or we might use a modal language to say that, despite being an only child, I could have had a brother who could have been an astronaut:

M) $\Diamond \exists x(Bxr \,\&\, \Diamond Ax)$

(Read as: It is possible that there is some thing that is the brother of Ross and [in that possibility] it is possible that that thing be an astronaut.)

Or we might use a plural language to say that some critics only admire other critics:

GK) $\exists xx[\forall u(u < xx \rightarrow Cu) \,\&\, \forall y \forall z(((y < xx \,\&\, A(y,z)) \rightarrow (z < xx \,\&\, z \neq y)))]$

(Read as: There are some things, and everything that is one of those things is a critic, and for any two things you like, if the first is one of those things and admires the second, then the second is also one of those critics, and is distinct from the first.)

It seems like there are good questions to be asked concerning the ontological commitments of (2O), (M), or (GK). Does (2O) commit us to properties, or just to the individuals a and b? Does (M) commit us to merely possible brothers? Does (GK) commit us to a new kind of thing, a *plurality*? But Quine thinks these are not good questions, for ontological commitment is only defined over sentences of a first-order language. In order to even ask what our ontological commitments are when we try to make claims like the above, we have to first translate them into a first-order language, thinks Quine. So while Quine himself wouldn't want to say anything like (2O), a Quinean could say a first-order paraphrase:

2O★) $\exists x(Px \,\&\, Iax \,\&\, Ibx)$

(Read as: There is something that is a property, and a instantiates it, and so does b.)

And while Quine himself eschewed modal talk like (M), his student Lewis gave us a way of regimenting modal claims in a first-order language, the language of counterpart theory (Lewis 1968), and would say instead of (M):

M★): $\exists w_1 \exists w_2 \exists x \exists y \exists z(Ww_1 \,\&\, Ww_2 \,\&\, Ixw_1 \,\&\, Iyw_1 \,\&\, Cxr \,\&\, Bxy \,\&\, Izw_2 \,\&\, Czy \,\&\, Az)$

(Read as: There is a world at which there is a counterpart of Ross that has a brother, and that brother of Ross's counterpart has a counterpart at a world that is an astronaut. 'Wx' is to be read as 'x is a world,' 'Ixw' as 'x belongs to world w,' and 'Cxy' as 'y is a counterpart of x.')

And we can eliminate plural quantification over critics in favor of regular first-order quantification over the set of critics:

GK★): $\exists x(Sx \,\&\, \exists y Myx \,\&\, \forall y(Myx \rightarrow Cy) \,\&\, \forall r \forall z((Mrx \,\&\, Arz) \rightarrow (Mzx \,\&\, r \neq z)))$

(Read as: There is a set, and there is something that is a member of that set, and every member of that set is a critic, and for any two things, if the first is a member of that set and admires the second, the second is a distinct member of that set.)

Once regimented, we can use Quine's criterion to discover the ontological commitments of these claims, and we can see that (by that measure) (2O★) commits us to properties as well as individuals, that (M★) commits us to possible worlds and to possible individuals at those worlds, and that (GK★) commits us to sets.

To those of us who do not feel the need to limit our descriptive resources to those of first-order logic, this may seem like the wrong result. Some people (see *inter alia* Prior 1971, Rayo and Yablo 2001, Wright 2007, Cameron 2019) have argued that second-order quantification is ontologically innocent – in particular that it does not bring a commitment to properties or sets – and would object to being forced to render 'a and b are alike in some way' as (2O★), which quantifies over properties, rather than (2O) which (according to those who think second-order quantification is ontologically innocent) does not. Similarly, some have argued that what we say within the scope of a possibility operator doesn't itself bring any ontological commitment, so would object to being forced to render 'Ross could have had a brother who could have been an astronaut' as (M★), which quantifies over possible worlds and possible individuals, rather than (M) which does not. (See Sider 1999 for relevant discussion.) And some have argued that plural quantification is not quantification over some new kind of entity – a plurality – but is instead a new way of quantifying over the same ordinary individuals that the first-order quantifier ranges over, and hence that 'Some critics admire only other critics' should be rendered as (GK) – which quantifies only over what this sentence appears to be about: critics! – rather than as (GK★), which quantifies over sets. (For an entry into the debate over the 'Geach-Kaplan' sentence, as it's known, see Boolos 1984; Quine 1986, Ch.5; and Rayo 2002.)

Quine's demand that we regiment our claims and theories in a first-order language before we assess them for ontological commitment is a substantive demand, then, and can result in ontological commitments different from what we might have expected. Now, one could of course argue about whether or not, for each of the above claims, the first-order regimentation gets the ontological commitments right. But I want instead to push another thought: there is something uncomfortable about this paradigmatically metaphysical question – what the extent of ontology is – being dependent on our choice of what language we select as the language of regimentation. Quine tells us to regiment our theories in a first-order language, and the ontological commitments of our theory of the world will be those things that must be in the domain of the quantifier in order for the first-order regimentation of that theory to be true. Someone who allowed a language with modal operators, or second-order quantifiers, of plural quantification, or deontic operators, or tense operators, etc., might see different ontological commitments from the same theory. But then how are we to determine the *real* ontological commitments? Of course, some people might think there is no such thing as the real ontological commitments: Quine (1968) himself was happy with the idea that what exists is relative to the way we conceptualize the world. But most contemporary metaphysicians are not ontological relativists: they believe that there are objective, mind-independent ontological facts. How then are they to determine the ontological commitments of a theory, when regimenting that theory in different languages yields different *apparent* ontological commitments?[2]

The truthmaker account of ontological commitment[3] is a thoroughly metaphysical account of ontological commitment, that eschews anything like a privileged language of regimentation. According to the truthmaker account, the ontological commitments of a claim or theory are those things that must exist to *make true* that claim/theory. Truthmaking is a relation between things in the world and propositions that describe a way for the world to be, and whether or not some thing(s) make true some proposition is not sensitive to how we happen to linguistically convey that proposition, so the truthmaker theorist will not end up in the embarrassing position above, where we get different ontological commitments depending on what language we choose to convey the claims we make. What a theory is ontologically committed to is determined by metaphysical facts – what makes what true – not by our choice of a language of regimentation.

For illustration, suppose I have a very simple theory that says three things: that a is red, that b is blue, and that a and b are alike somehow. Quineans think that the first claim commits one to a but not to redness, and the second to b but not to blueness, because when I predicate something of an individual I only need to quantify over the individual, not a property. But they have trouble resisting commitment to properties if they say the third claim, for that appears to demand that we say that there is some property that a and b share. (Van Inwagen 2004 concludes, on Quinean grounds, that nominalism about properties is untenable.) Someone who, unlike Quine, allowed second-order quantification, and who thought it was ontologically innocent, might instead think that none of these claims commit you to properties. The truthmaker theorist says that the issue of whether this theory commits you to properties shouldn't depend on whether or not we allow second-order quantification. Rather, it depends on a metaphysical issue: are properties needed as truthmakers for true predications concerning how ordinary individuals are?

Likewise, if I say that I could have had a brother who could have been an astronaut: the truthmaker theorist says the issue isn't whether we allow a language with modal operators, but rather what must exist to *make true* this modal claim – must reality contain possibilia to make true claims like this, or can they be made true by actual things? And for the Geach-Kaplan sentence, while logicians might argue over its logical form, the truthmaker theorist says that as far as its ontological commitments are concerned, the issue is solely what must exist to make it true: just critics (and their properties and relations), or sets of critics? Those are hard questions to answer, of course, and metaphysicians fight about what is needed to make true predicational claims, modal claims, etc. But it places the issue where it should be: as a hard metaphysical question, rather than a question about what language to use to state our theories.

The truthmaker account can both overgenerate and undergenerate ontological commitments relative to the Quinean criterion. While Quine thinks that an ordinary predication such as 'Electron e is charged' commits you to the electron, but not to the property of being charged, the truthmaker theorist might think that you need the property in addition, or even a complex state of affairs built out of the electron and the property, to account for this truth. (See Armstrong 2004, Ch.4.) And thus the truthmaker theorist sees ontological commitments that the Quinean does not. On the other hand, Quine (1948) believes in mathematical objects because he thinks quantification over them is indispensable to our best science. The truthmaker theorist can agree that such quantification is indispensable, but might nonetheless resist the claim that we are ontologically committed to mathematical entities, on the grounds that the fact that those mathematical entities exist is made true by the non-mathematical world in some way (Cameron 2008). And thus the truthmaker theorist eschews ontological commitments that the Quinean sees.

2 Truthmaking, easy ontology, and (very) hard ontology

There are many advantages one might claim for the truthmaking account of ontological commitment. Here I will focus on one particularly pleasing metametaphysical feature of the truthmaking account: it allows us to recognize that existential questions can oftentimes be easy – sometimes to the point of triviality – to answer, while still allowing that there is a deep, difficult, and interesting ontological project to be undertaken.

Let's start by considering a challenge to the idea that there is a deep and distinctively philosophical ontological question. We can all agree that something is happening table-wise in the room. While some rooms are lacking with respect to all table-ish phenomena, this room is not. One can sit down to dinner. But many metaphysicians think there is still an interesting metaphysical question to be asked: is there a table in this room? They accept that there are particles in the room that are arranged in a table-wise manner, but think that there is a further question that this leaves open: whether, *in addition*, there is a table. Philosophers who embrace a commonsense ontology (Korman 2016) and universalists about composition (e.g. Lewis 1986, pp. 212–213 and Sider 2001) say that there *is* a table in addition to the things arranged table-wise; those who believe that complex objects must meet some stringent condition like being alive (van Inwagen 1990) or being conscious (Merricks 2001), or those who hold that there are no complex objects (Dorr 2002; Sider 2007; Cameron 2010),[4] say that there is not – there are just the particles arranged table-wise. They all agree that it is a substantive metaphysical question whether, in addition to particles arranged in a certain way, there is also a table.

To others, there seems to be something wrong with this debate. Some (e.g. Hirsch 1993) see it as a merely verbal dispute. Some (Bennett 2009) see it as a genuine, but epistemically intractable, debate. But the most plausible grounds for dismissal of the debate, in my view, is the idea that while it is indeed a genuine question whether there is a table, it is one that can be answered *easily*. *Of course* there is a table, given that there are particles arranged table-wise – that's just what it *takes* for there to be a table.[5]

Amie Thomasson, e.g., argues that there is no distinctively metaphysical question to be asked concerning the existence of tables. (See Thomasson 2007 and, especially, 2015, forthcoming a, and Ch.12 this volume.) There is the question of what conditions must be met in order to appropriately deploy the concept *table* – and this is to be answered by engaging in conceptual analysis – and then there is a straightforward empirical question as to whether those conditions are met, which anyone can answer by looking in the room. Once we verify that the conditions are or are not met, we know whether the concept is appropriately deployed, and there is no further ontological question remaining, because for Ks to exist just is for appropriate conditions for the deployment of the concept of Ks to obtain. There is no work for the metaphysician here. And of course, there's nothing special about tables. According to Thomasson, for every kind of thing, the Ks, there is the question of when to deploy the concept K – a question to be resolved by conceptual analysis – and there is the question of whether those conditions are met – a question that can be answered by straightforward empirical means. Applying Thomasson's view beyond the realm of ordinary objects, we might ask whether, e.g., there is some thing that numbers both the branches of the US government and the books composing *The Lord of the Rings*. Well, says Thomasson, first we should do some conceptual analysis and discover the application conditions for number terms. Perhaps we will arrive at the following: the conditions required for there being a thing that numbers both the branches of the US government and the books composing *The Lord of the Rings* are just that there is a one-one correspondence between the branches of government and the books. (Cf. the neo-Fregean project in Wright and Hale 2001.) That there *is* such a one-one correspondence is a straightforward empirical fact about the world that is agreed to on all sides. So again, we might conclude, there is no deep metaphysical question here: anyone competent with the concept *number* is in a position to see that it is appropriate to deploy the concept in such a manner under such conditions, and that is all it takes for there to be numbers.

Thomasson (forthcoming a) calls this easy ontology approach the biggest threat to the project of 'deep' ontology. I agree. The threat is that the question of whether things of a certain kind exist simply reduces to the questions of when the concept of that kind is appropriately deployed and whether the world meets those conditions. The latter question is often easily answered (Is *table* appropriately deployed? Well, can you put your dinner down?); and even when it is not easily answered (Is the concept *silicon-based life-form* appropriately deployed?) it is not a *philosophical* question.

To my mind, however, all this just shows us that the project of ontology should not be understood in terms of cataloging what exists. I agree with Thomasson that existential questions are often easy to answer (Of course there are tables – look, my dinner's on one.), and that for many traditional ontological debates the existence of the entities in question follows trivially from obvious truths (Of course there are numbers: there are eight planets. Of course there are properties: the fire truck and the postbox are both red.). But there is still a deep, hard to answer, ontological question: what makes what true? Do tables, numbers, properties, etc., figure amongst the *truthmakers* for the way the world is? This, I contend, cannot be answered by the combination of conceptual analysis and empirical observation: it is a substantive metaphysical question. (Cf. Schaffer 2009b.)

So, there is an inconsistent triad:

A Ontology is about what exists
B Claims about what exists can be settled easily/trivially
C Ontological claims cannot be settled easily/trivially

Thomasson accepts (A) and (B) and thus rejects (C), thereby rejecting ontology as a deep metaphysical issue. The traditional Quinean ontologist accepts (A) and (C), thereby rejecting (B), and so she has to argue that there is a gap between, e.g., there being things arranged table-wise and there being a table, or between there being a one-one correspondence between the Fs and the Gs and there being a number that numbers both the Fs and the Gs, etc. I accept (B) and (C) and reject (A): existential questions can (often) be easily or trivially settled, but ontology is not about what exists but rather about what makes the true claims about reality true, and that is not easily or trivially settled.

Thomasson is unconvinced that the Truthmaker view offers a rehabilitation of deep ontology. The Truthmaker approach tells us that many *apparent* ontological commitments of the claims we make about the world are not genuine commitments. So for example, while Thomasson takes the apparent ontological commitments of 'There is a table in this room' and 'There is a number that numbers the Beatles' at face value – i.e., that they are committal to tables and numbers respectively – and concludes that the ontological questions as to whether there are tables and numbers is easily settled in the affirmative, the truthmaker theorist will accept that tables and numbers *exist* but will say that there is an additional hard metaphysical question as to whether those claims bring genuine ontological commitment to tables and numbers.

For this approach to work, says Thomasson, we need a criterion that tells us when an apparent ontological commitment is not a genuine commitment. She says (Thomasson forthcoming b):

> [I]t seems that even if truthmaker theorists don't give us a recipe for determining what we *are* committed to, they must at least provide a criterion that will tell us when we *aren't committed to* entities of a kind apparently referred to in a given sentence or theory. Under what conditions can we say that an entity *isn't needed* as a truthmaker and so deny the apparent commitment?

She offers the following suggestion for such a principle (ibid.):

> [W]e can avoid ontological commitments … in cases where a sentence 'A' is non-vacuously entailed by an ontologically alternative sentence 'B,' enabling us to look to the truthmakers for 'B' to serve also as the truthmakers for 'A' and thereby reduce our commitments.

The idea here is that because 'There are tables' is (non-vacuously[6]) entailed by 'There are particles arranged table-wise' and 'There is a number that numbers both the books in *The Lord of the Rings* and the branches of the US government' entailed by 'There is a one-one correspondence between books in LotR and government branches' we can conclude that tables and numbers are not needed as truthmakers for those claims – rather, they are made true simply by whatever makes true the sentences from which they are entailed. So what makes it true that there are tables is just whatever makes it true that there are particles arranged table-wise. And, one might naturally conclude, that is merely the particles, or perhaps the state of affairs of the particles being so arranged, but no need for a table.

But Thomasson argues, against the truthmaker proposal, that this won't work. First, note that if we follow her principle, we should not end up committed to even the particles arranged tablewise, for 'There are particles arranged tablewise' is itself non-vacuously entailed by a sentence that does not refer to particles, namely 'It's particling around here (in a tablish manner).' The idea here is that whenever we have a sentence that talks about some things, we can always instead use a feature-placing language to describe the same feature of reality without talking about those things. (See O'Leary-Hawthorne and Cortens 1995 and Turner 2011.) Just as we can say 'It is raining' and the 'it' there does not aim to refer to some object that is undergoing rain, so can we say 'It is tabling over there' instead of 'There is a table over there,' or 'It is peopling' instead of 'There are people,' or 'It is particling in a tablish manner' instead of 'There are particles arranged tablewise.' In this manner, we could if we wish speak a language in which all reference to objects is eliminated. So by the principle Thomasson offers the truthmaker theorist, we would never end up ontologically committed to anything.

Now of course, the truthmaker theorist could conclude that this is an excellent argument for ontological nihilism: the view that ontology contains no things. However, there is something very suspicious about this apparently radical ontological view following from the apparently innocuous linguistic fact that we can speak a feature-placing language that avoids reference to objects. But also, why should we conclude that there are no tables just because 'There are tables' is non-vacuously entailed by 'It is tabling'? 'It is tabling' is equally non-vacuously entailed by 'There are tables' after all. If each entails the other, why focus on one as revealing what there really is? Thomasson's objection is that given how flexible language is, for any sentence we start with that contains terms that purportedly refer to Xs, we will always be able to find, or invent, a sentence that non-vacuously entails it that does not contain any term that purportedly refers to Xs. As she says, "language seems to be ontologically flexible: it seems that there are quite typically entailments to a target sentence from an ontologically alternative one (whether one we already have in the language or can invent)." And she concludes from this that the idea of chasing back these chains of entailments to find the *real* ontological commitments is a doomed project.

I think the truthmaker theorist has to concede that it cannot simply be because 'There are tables' is non-vacuously entailed by 'There are particles arranged table-wise' that tables are not needed as truthmakers for 'There are tables.' By that reasoning, as Thomasson says, *no* truthmakers would be needed for anything, since for any claim at all we can find a 'nihilist friendly' claim in a feature-placing language that will entail the target claim without talking about *anything* objectual.

Thankfully, I do not think this concession hurts the truthmaker theorist. The truthmaker theorist should not claim that 'There are tables' does not require the existence of tables to make it true because it is a trivial consequence of 'There are particles arranged table-wise.' Rather, they should claim that *either* these sentences both require the existence of tables to make them true, or neither do. It is then a further theoretical question whether or not tables are necessary to make either claim true. Maybe we have *pro tanto* reason for believing the more minimal ontological theory simply on parsimony grounds (see Cameron 2008, p12), but we don't have a reason for the minimal ontology simply in virtue of the fact that the more expansive ontological claims are entailed by the more minimal ones.

This echoes an old point of Alston's (1963) concerning paraphrase. Consider the Quinean strategy of avoiding ontological commitment by paraphrasing away the offending sentences. Suppose, for example, that for every sentence that apparently posits the existence of a property or relation, we have a paraphrase into a nominalist-friendly sentence. Why, asks Alston, should we think of this as showing that we are not committed to properties and relations rather than as revealing a hidden commitment to such things in the allegedly nominalist-friendly claims? Similarly, if 'There is a number that numbers both the Fs and Gs' is equivalent, in some sense, to 'There is a one-one correspondence between the Fs and the Gs,' why take that to show that we can do without numbers, rather than revealing that numbers were required all along for the truth of claims about equinumerosity. No reason! The equivalence tells you only that the *same* things are needed in each case, and it does not tell you *what* is needed. To work that out, we have to do some metaphysics.

Simply looking at the pattern of entailment between 'There are tables,' 'There are particles arranged tablewise,' 'It is tabling,' etc., won't tell you which of these sentences, if any, perspicuously describes the ontology of the world: whether our ontology is one containing complex material objects, or one containing merely particles scattered in the void, or whether it contains no objects at all. There is no *syntactic* or *semantic* reason to privilege one of these theories over the other. But that doesn't mean there is *no* reason to privilege one theory over the other. It simply means that any such reason will be resolutely metaphysical. It is a deep metaphysical question whether the world is, at bottom, a world of *things* or whether ontological nihilism really does have the right account of the fundamental nature of reality, e.g., There may indeed be mutual entailment between a feature-placing description of reality and an objectual one, but that is not in conflict with the claim that one of these descriptions gets the fundamental nature of the world *right*. That the descriptions mutually entail one another suggests, at most, that they describe the same portion of reality. Thus they will be made true, ultimately, by the same features of the way things are. But there is still the metaphysical question of *what* those features are. Looking at what entails what won't tell us what those features are, but that doesn't mean there's not an answer.

So, I claim that the ontological commitments of a theory are those things that must exist to make it true. Thomasson (forthcoming b) objects:

> there is typically *no* unique statement of what entities *must* be in the ontology of the world to account for its truth: we can generally get an entailment up to the truth of the target sentence from an ontologically alternative one, which would apparently require a different ontology to make it true.

But I say that there *is* a unique statement of what entities must be in the ontology of the world to account for its truth. It is just that we cannot work out *what* that statement is simply by looking at what entails what. It is a mistake to think that the true description of reality at its most fundamental level must be one that entails any true derivative description, but is not itself entailed by any such derivative description. Derivative truths can entail fundamental ones. Patterns of entailment do not track patterns of truthmaking.

Thus I reject Thomasson's offered answer for when we aren't committed to certain entities that are apparent commitments of a claim. What, then, do I put in its place? Well, I've already given the answer: you are not committed to certain entities when those entities are not needed to make true what you are claiming. Thomasson wants something more illuminating. But I don't see why there need be anything more illuminating to say, at least in general terms. Why should the truthmaker theorist expect some general principle about what is or is not needed to make some claim true? We can only expect to answer on a case by case basis, by engaging in the highly speculative and highly fallible project of metaphysics. Is 'Ball is red' made true merely by Ball, or is it made true by the particular redness of Ball, or by the state of affairs of Ball being red, or is reality ultimately not even a reality of objects? It would be nice if there was some general principle concerning what makes what true that we could apply to this claim and churn out the answer. But there is not. We have to get our hands dirty and engage with all the arguments metaphysicians give concerning the benefits of an ontology of tropes versus an ontology of states of affairs, or of the costs and benefits of ontological nihilism. Ontology is not just not easy, it is very, very hard on this view. And no doubt the multitude of rival answers to this question about what makes this simple intrinsic predication true should give us cause to not be too confident in our own answer. But all this means is that we should have a healthy epistemic humility regarding what we take the ontological commitments of our theories to be. We are ontologically committed to what must exist to make true the claims we make about the world, and there *is* a unique answer to the question of what those truthmakers are,[7] even if all we can do in answering it is make our best educated guess.[8]

Notes

1 For useful discussion see Rayo (2007) and Bricker (2014).
2 See Cameron (2019) for a much more detailed elaboration of this issue.
3 For defenses of which see Heil (2003), Armstrong (2004), Cameron (2008), and Rettler (2016). Cf. Schaffer (2008), who rejects the claim that the ontological commitments of a theory are what must exist to make it true, but accepts instead the claim that the things which a theory commits to being *fundamental* are what must exist to make it true. However, he also holds (Schaffer 2015) that when judging theories with respect to ontological parsimony we should look not to their respective ontological commitments, but rather at what they claim to be fundamental: the more parsimonious theory is the one that posits fewer fundamentalia. Given this, it seems to me that there is merely a terminological difference between the truthmaker account and Schaffer's view: we both agree that what's important when weighing up the ontology a theory commits us to is what must exist to make true that theory, not what can be truly said to exist if that theory is true.
4 Although note that both Sider and Cameron (and Dorr 2005) want to allow for the strict truth of the claim that there is a table, thus making the view more like the kind of view about to be discussed.
5 Thomasson (2015, forthcoming a). Cf. Rayo (2013), although note that it is no part of Rayo's view that ontological claims are *easy* to answer: while Rayo holds that *what it is* for there to be a table is for there to be things arranged table-wise, *what it is* statements are themselves substantive theoretical claims which require substantive defense. Also cf. Schaffer (2008, 2009a) who agrees with Thomasson that many existence questions are easy to the point of triviality to answer, but concludes from this that the deep ontological question is not about what exists, but about what grounds what. If we see truthmaking as a species of grounding (as in Cameron 2018), Schaffer's approach and the truthmaking approach are very much alike.
6 Classically, everything follows logically from a set of premises that are contradictory. The requirement on non-vacuous entailment is to restrict attention to entailments where the premises are genuinely relevant to the conclusion.
7 I've been making a simplifying assumption throughout: that there's always a unique collection of things that must exist to make any particular truth true. But really, that's not plausible: That there are humans could be made true by any humans, and no *particular* collection of humans is needed to make it true; that there is jade could be made true by some jadeite but also by some nephrite, and no particular substance is needed to make it true. What should we say about the ontological commitments when there is no particular (collection of) thing(s) that *must* exist to make our claim true? In those cases, I think

our commitment is simply disjunctive: we are ontologically committed to one or other of the potential collections of truthmakers. See Cameron (2019) for more detail, and cf. Rettler (2016) for discussion of alternate proposals.
8 Thanks to Elizabeth Barnes, Joshua Spencer, Amie Thomasson, Jason Turner, Robbie Williams, and an anonymous referee, for helpful discussion.

References

Alston, William, (1963), 'Ontological Commitments,' *Philosophical Studies* 14, pp. 1–8.
Armstrong, D. M., (2004), *Truth and Truthmakers*, Cambridge: Cambridge University Press.
Bennett, Karen, (2009), 'Composition, Colocation, and Metaontology,' in David Chalmers, David Manley, and Ryan Wasserman (eds) *Metametaphysics*, Oxford: Oxford University Press, pp. 38–76.
Boolos, George, (1984), 'To Be Is to Be the Value of a Variable (or to Be Some Values of Some Variables),' *Journal of Philosophy* 81 pp. 430–449.
Bricker, Phillip, (2014), 'Ontological Commitment,' *The Stanford Encyclopedia of Philosophy* (Winter 2014 Edition), Edward N. Zalta (ed.), URL = http://plato.stanford.edu/archives/win2014/entries/ontological-commitment/.
Cameron, Ross P., (2008), 'Truthmakers and Ontological Commitment: or, How to Deal with Complex Objects and Mathematical Ontology Without Getting into Trouble,' *Philosophical Studies* 140(1), pp. 1–18.
Cameron, Ross P., (2010), 'How to Have a Radically Minimal Ontology,' *Philosophical Studies* 151(2), pp. 249–264.
Cameron, Ross P., (2018), 'Truthmakers,' in Michael Glanzberg (ed.) *The Oxford Handbook of Truth*, Oxford: Oxford University Press, pp. 333–354.
Cameron, Ross P., (2019), 'Truthmaking, Second-Order Quantification, and Ontological Commitment,' *Analytic Philosophy* 60(4), pp. 336–360.
Dorr, Cian, (2002), *The Simplicity of Everything*, PhD dissertation, Princeton University.
Dorr, Cian, (2005), 'What We Disagree About When We Disagree About Ontology,' in Mark Kalderon (ed.) *Fictionalist Approaches to Metaphysics*, Oxford: Oxford University Press, pp. 234–286.
Fine, Kit, (2009), 'The Question of Ontology,' in David Chalmers, David Manley, and Ryan Wasserman (eds.) *Metametaphysics: New Essays on the Foundations of Ontology*, Oxford: Oxford University Press, pp. 157–177.
Heil, John, (2003), *From an Ontological Point of View*, Oxford: Oxford University Press.
Hirsch, Eli, (1993), *Dividing Reality*, Oxford: Oxford University Press.
Korman, Dan, (2016), *Objects: Nothing Out of the Ordinary*, Oxford: Oxford University Press.
Lewis, David, (1968), 'Counterpart Theory and Quantified Modal Logic,' *The Journal of Philosophy* 65(5), pp. 113–126.
Lewis, David, (1986), *On the Plurality of Worlds*, Oxford: Blackwell.
Meinong, Alexius, (1904), 'On the Theory of Objects,' reprinted in Roderick Chisholm (ed.) *Realism and the Background of Phenomenology*, Free Press, (1960), pp. 76–117.
Merricks, Trenton, (2001), *Objects and Person*, Oxford: Clarendon Press.
O'Leary-Hawthorne, John and Cortens, Andrew, (1995), 'Towards Ontological Nihilism,' *Philosophical Studies* 79(2), pp. 143–165.
Prior, A. N., (1971), 'Platonism and Quantification,' in his *Objects of Thought*, Oxford: Oxford University Press, pp. 31–47.
Quine, W. V., (1948), 'On What There Is,' *The Review of Metaphysics* 2(1), pp. 21–38.
Quine, W. V., (1968), 'Ontological Relativity,' *The Journal of Philosophy* 65(7), pp. 185–212.
Quine, W. V., (1986), *Philosophy of Logic*, Second Edition, Cambridge, MA: Harvard University Press.
Rayo, Agustin, (2002), 'Word and Objects,' *Noûs* 36, pp. 436–464.
Rayo, Agustin, (2007), 'Ontological Commitment,' *Philosophy Compass* 2(3), pp. 428–444.
Rayo, Agustin, (2013), *The Construction of Logical Space*, Oxford: Oxford University Press.
Rayo, Agustin and Yablo, Stephen, (2001), 'Nominalism through De-Nominalization,' *Noûs* 35(1), pp. 74–92.
Rettler, Bradley, (2016), 'The General Truthmaker View of Ontological Commitment,' *Philosophical Studies* 173(5), pp. 1405–1425.
Schaffer, Jonathan, (2008), 'Truthmaker Commitments,' *Philosophical Studies* 141(1), pp. 7–19.

Schaffer, Jonathan, (2009a), 'On What Grounds What,' in David Chalmers, David Manley, and Ryan Wasserman (eds) *Metametaphysics*, Oxford: Oxford University Press, pp. 347–383.

Schaffer, Jonathan, (2009b), 'The Deflationary Metaontology of Thomasson's *Ordinary Objects*,' *Philosophical Books* 50(3), pp. 142–157.

Schaffer, Jonathan, (2015), 'What Not to Multiply Without Necessity,' *Australasian Journal of Philosophy* 93, pp. 644–664.

Sider, Theodore (1999), 'Presentism and Ontological Commitment,' *The Journal of Philosophy* 96, pp. 325–347.

Sider, Theodore (2001), *Four-Dimensionalism*, Oxford: Oxford University Press.

Sider, Theodore (2007), 'Against Parthood,' *Philosophical Review* 116, pp. 51–91.

Sider, Theodore, (2011), *Writing the Book of the World*, Oxford: Oxford University Press.

Thomasson, Amie L., (2007), *Ordinary Objects*, Oxford: Oxford University Press.

Thomasson, Amie L., (2015), *Ontology Made Easy*, Oxford: Oxford University Press.

Thomasson, Amie L., (forthcoming a), 'Easy Ontology and its Consequences,' in Gary Ostertag (ed.) *Meanings and Other Things (Essays on the Work of Stephen Schiffer)*, Oxford: Oxford University Press.

Thomasson, Amie L., (forthcoming b), 'Truthmakers and the Project of Ontology,' *Oxford Studies in Metaphysics*.

Turner, Jason, (2011), 'Ontological Nihilism,' in Dean W. Zimmerman and Karen Bennett (eds.) *Oxford Studies in Metaphysics* Volume 6, Oxford: Oxford University Press, pp. 1–50.

van Inwagen, Peter, (1990), *Material Beings*, Ithaca, NY: Cornell University Press.

van Inwagen, Peter, (2004), 'A Theory of Properties,' in Dean W. Zimmerman (ed.) *Oxford Studies in Metaphysics* Volume 1, pp. 107–138.

Wright, Crispin, (2007), 'On Quantifying into Predicate Position: Steps Towards a New(trialist) Perspective,' in Mary Leng, Alexander Paseau, and Michael Potter (eds.) *Mathematical Knowledge*, Oxford: Oxford University Press, pp. 151–174.

Wright, Crispin and Hale, Bob, (2001), *The Reason's Proper Study*, Oxford: Oxford University Press.

19
ESSENCE

Jessica Leech

1 Introducing essence

What is essence? In rough terms, the essence of a thing is *what it is to be* that thing. So, perhaps *what it is to be* Socrates is, in part, to be human. Or *what it is to be* the number 2 is to be the successor of 1. Or *what it is to be* the Mona Lisa is, in part, to have been painted by da Vinci. Such a phrase has its origins in Aristotle (e.g. *Metaphysics* 1029b; *Topics*: 101b38–102a1).[1] Indeed, the notion of essence is one that can be found throughout the history of philosophy, although in this chapter we will focus on the contemporary context of this notion.[2] Sometimes the essence of a thing is also called its *nature*.[3] As we will see, essence is sometimes connected to the notion of *truth in virtue of the nature of* something.

The notion of essence is at work in the distinction between *essential properties* and *accidental properties*. Roughly, the essential properties of something *belong to* or *flow from* or *are required* by its essence; its accidental properties do not. One might take the essence of a thing to be identical to its essential properties (e.g. Hale 2013: 152–153, note 17), although notably Lowe has argued that essences are in fact no kind of entity at all.[4] In this chapter, we will not be concerned with the ontological question about essence: what kind of entity an essence is (property, nothing, something else?). Rather, we will be primarily concerned with how to define the essence of a thing, regardless of its ontological type, and some important applications and refinements of the notion of essence.

There are at least two importantly different ways one might begin to think about essence via the essential properties of a thing (regardless of whether one takes the essence to *be* those properties or to otherwise determine them). One might take essential properties to be *necessary* or *both necessary and sufficient* for the existence of a thing. Thus: first, one might claim that Socrates is essentially human, meaning that *necessarily*, if Socrates exists, he is human. Of course, being human is not *sufficient* for being Socrates, for there are many other humans (such as you and I) that are not Socrates. Second, then, one might claim that Socrates essentially originated from the particular gametes from which he actually originated, meaning that such an origin is *necessary and sufficient* for being Socrates: necessarily, if Socrates exists, he originates from his actual original gametes, *and* necessarily, anything that originates from those gametes is Socrates. If one holds these necessity and sufficiency claims together, one posits an *individual essence* of Socrates.[5]

245

Of these options, the necessity claim is almost always intended. The sufficiency claim is made variously, depending upon the specific thesis under discussion. There are, indeed, a range of first-order metaphysical claims[6] under debate, concerning what properties are essential in various of the three senses just outlined. In particular:

1. Essentiality of kind: if a is of kind K, then a is essentially of kind K.
2. Essentiality of origin: if a has origin O, then a essentially has origin O.
3. Essentiality of constitution: if a is composed of material M, then a is essentially composed of M.[7]

(1) is primarily a necessity claim; (2) and (3) can be considered as mere necessity claims, but often also incorporate a sufficiency claim. Such essentialist theses have received considerable philosophical attention,[8] but our focus in this chapter will be elsewhere – with a theoretical understanding of essence and its relations to other (meta)metaphysical notions, rather than specific claims involving essence.

The necessity implicated in discussions of essence is typically *de re metaphysical necessity*. *De re* necessity is distinguished from *de dicto* necessity. Informally, *de re* necessities pertain to a thing, no matter how we refer to it; *de dicto* necessities concern statements about things, where it matters how one refers to those things in that statement.[9] For example,

> 8 is necessarily greater than 5

is a statement of *de re* necessity. It says of the numbers 8 and 5 that the former is greater than the latter, no matter what, no matter how we refer to them. This is plausibly true. By contrast,

> Necessarily, the number of planets in our solar system is greater than 5

is a statement of *de dicto* necessity, and it is plausibly false. It matters that the number 8 is referred to in this statement by the description 'the number of planets in our solar system' rather than the numeral '8'. For whilst it is plausible that the statement '8 is greater than 5' is necessarily true, it is not plausible that the statement 'the number of planets in our solar system is greater than 5' is necessarily true. For it seems perfectly possible that when the solar system formed, the cosmic detritus orbiting the sun could have combined to form only 3 planets, rather than 8. The essences of things are intended to concern those things independently of how we talk about or describe them, hence they concern *de re* necessity. Quine famously mounted an attack against the kinds of *de re* necessary, essentialist claims that are the topic of this chapter; he was concerned with how to adequately formalize such statements, and with the metaphysical commitments that a well-formed statement of *de re* necessity might engender. This is not the place to defend *de re* modalities against Quine, but it is important to note that this kind of challenge looms in the background.[10]

Rosen (2006) sketches metaphysical necessity as 'the strictest real necessity'.

> Let's call any modality that is alethic,[11] non-epistemic, and sometimes substantive or synthetic a *real* modality. ... Metaphysical necessity is the strictest real necessity and metaphysical possibility is the least restrictive sort of real possibility in the following sense: If P is metaphysically necessary, it is necessary in every real sense: If P is really possible in any sense, then it's possible in the metaphysical sense.
>
> *(Rosen 2006: 16)*

Metaphysical necessity is also typically distinguished from logical necessity on the one hand, and natural necessity on the other. Whilst all logical necessities are metaphysically necessary, there are some metaphysical necessities that are not logically necessary. The examples given are often connected to essentialism, for example, that it is metaphysically necessary that Socrates is human (since he is essentially human), but it is no logical contradiction to suppose that he might have been a boiled egg. Likewise, whilst all metaphysical necessities are naturally necessary, not every natural necessity is metaphysically necessary. For example, one might think that the laws of nature are naturally necessary, but 'one has the palpable sense – though philosophy might correct it – that some of the laws of nature might have been otherwise' (Rosen 2006: 16). It is a genuine challenge to make better sense of metaphysical necessity by adding to this sketch. We will see below that some essentialists have offered an account of the nature of metaphysical necessity as having its source in the essences of things.

2 Modalism about essence

2.1 First wave modalism

Claims of essence almost always come along with necessity claims. One might, then, simply define the essence of a thing in terms of its necessary properties. Hence, we have what I will call *Modalism* about essence. In its most simple form:

Categorical Modalism: a is essentially F if, and only if, necessarily, a is F.

However, this threatens to rule out contingent entities, such as Socrates, from having an essence. For, if necessarily, a is F, then 'a is F' is true in all possible worlds. But one might think that if 'a is F' is true, then a exists.[12] Hence, a exists in all possible worlds, that is, a exists necessarily. Due to such considerations, modalism about essence usually takes a conditional form.

Conditional Modalism: a is essentially F if, and only if, necessarily, if a exists, a is F.

Arguably, much of the discussion of essence throughout the latter half of the twentieth century assumed some version of modalism. Much of that work was inspired by Kripke on essential properties in *Naming and Necessity*, and one can plausibly interpret Kripke there as taking essential properties just to be *de re* necessary properties (see Kripke 1981: 110). However, philosophical discussion about essence changed significantly towards the end of the twentieth century and into the twenty-first, largely ignited by Fine's 1994 paper 'Essence and Modality'. In that paper, Fine presents a series of counterexamples to both Categorical and Conditional Modalism and urges us to reconsider modalism about essence in general. We will consider Fine's positive view in due course. But first, let us review his purported counterexamples to modalism, and potential responses from the committed modalist.

All of the following examples are intended to challenge Conditional Modalism; the first four also act against Categorical Modalism. They challenge the right-to-left directions of modalism, not left-to-right. That is, Fine, and others, are happy to accept that essence implies necessity, that is,

[Essence to Necessity]$_{CON}$: if a is essentially F, then, necessarily, if a exists, a is F.

[Essence to Necessity]$_{CAT}$: if a is essentially F, then, necessarily, a is F.

But they want to challenge the claim that necessity is sufficient for essence, that is,

[Necessity to Essence]$_{CON}$: if necessarily, if a exists, a is F, then a is essentially F.

[Necessity to Essence]$_{CAT}$: if necessarily, a is F, then a is essentially F.

As Conditional Modalism is already the more plausible account, I will explicitly outline the counterexamples as against this view.

[Singleton Socrates] Consider the set that contains only Socrates; the singleton set of Socrates. Given standard views in modal set theory, necessarily, if Socrates exists, then he is a member of singleton Socrates. According to the usual account of the identity of sets, it is part of what it is to be singleton Socrates that Socrates is a member of it, but it is no part of our usual understanding of what it is to be Socrates that Socrates be a member of his singleton set. 'Strange as the literature on personal identity may be, it has never been suggested that in order to understand the nature of a person one must know to which sets he belongs' (Fine 1994: 5).

[Distinctness] Socrates and the Eiffel Tower are distinct. Hence, necessarily they are distinct. But even though necessarily, if Socrates exists, he is distinct from the Eiffel Tower, it is no part of what it is to be Socrates that he is distinct, or otherwise related to, the Eiffel Tower, 'for there is nothing in his nature which connects him in any special way to it' (Fine 1994: 5).

[Necessary Truth] Take any necessary truth, e.g. $2 + 2 = 4$. Then, necessarily, if Socrates exists, he is such that $2 + 2 = 4$. But again, it is no part of what it is to be Socrates that $2 + 2 = 4$, or indeed, any other necessary truth that has no obvious connection to Socrates.

[Essential Truth] Take any essential truth unconnected to Socrates, for example, that the Eiffel Tower is essentially iron. This, Fine argues, must itself be necessarily true, for it implies a necessary truth, and necessary truths are themselves necessarily true. Hence, as in the previous example, it follows that necessarily, if Socrates exists, then he is such that the Eiffel Tower is essentially iron. But again, why should we think that essential truths about things unconnected to Socrates are part of his essence? 'O happy metaphysician! For in discovering the nature of one thing, he thereby discovers the nature of all things' (Fine 1994: 6).

[Existence] It is trivially true that necessarily, if Socrates exists, then Socrates exists. Hence, a straightforward consequence of Conditional Modalism is that Socrates essentially exists. But surely we don't want to attribute essential existence to all things, including contingent beings that might not have existed?

In all cases, we have a plausibly true claim about a necessary property of Socrates, but the claim that this property is essential to Socrates is plausibly false. Fine concludes that modalism about essence is wrong: we should not define essence in terms of necessity.

2.2 *Modalism bites back*

Many philosophers have attempted to respond to Fine's attack on modalism, whilst acknowledging the intuitive force of the proposed counterexamples.

One variety of response is to accept Fine's rejection of Categorical and Conditional Modalism, but argue that a refined version of modalism can be defended. Yes, not *all* of Socrates's necessary properties are essential to him, but perhaps we can simply add to the definition to delineate a subset of them to define his essence. The modalist account of essence has thus been amended in various different ways – restricting essential properties to those that are *non-trivial* (not had by all things just in virtue of being a thing, or following from such properties) (Della Rocca 1996), or *intrinsic* (Denby 2014), or *sparse* (Wildman 2013). I won't review in detail how these proposals are intended to respond to Fine, but to give a couple of examples: *existence* is a property (if it is a property at all) shared by all things whatsoever, and hence is a trivial property, so is ruled out from being an essential property; *being such that 2 + 2 = 4* is not a sparse property, and so is ruled out from being an essential property.

In response, the non-modalist can usually invent some further counterexamples, or otherwise object. For example, Skiles (2015) presents a series of putative essentialist claims that involve necessary but abundant properties (challenging the claim that necessity + sparseness provides a necessary condition for essence).

> Take, for instance, human artifacts such as the Eiffel Tower, which essentially exemplifies various abundant properties (e.g., *being a tower*) and essentially stand in abundant relations (e.g., the relation *was designed and constructed to perform such-and-such function by*, which it bears to some engineer or other, or perhaps to Gustave Eiffel in particular). Similarly goes for entities such as smiles (the essential nature of which include facts about faces), holes (the essential nature of which include facts about perforated surfaces), tropes (the essential nature of which include facts about the particular things they 'inhere' in), and events (the essential nature of which include facts about the objects, properties and times that 'participate' in them), among others.
>
> *(Skiles 2015: 106)*[13]

An alternative modalist response is to refine the modal notion involved. Correia (2007) proposes a modalist account of essence in terms of 'a non-standard, independently motivated conception of the metaphysical modalities' (Correia 2007: 67). Morvarid (2018) provides a helpful summary of the (rather complex) view.

> Correia's modal account exploits a distinction between two notions of necessity … which he dubs *local* and *global*, respectively (Correia 2007, p. 68). Roughly speaking, a state of affairs is locally necessary iff it is necessary for *intrinsic* reasons, and a state of affairs is globally necessary iff it is necessary either for intrinsic reasons or for *extrinsic* reasons (Ibid, p. 67).

> The main idea of Correia's account is to analyze essence in terms of local necessity … roughly … an object o is essentially F iff it is a local necessity that o is F. In this way, (global) necessities which follow from irrelevant external sources are banned from falling in o's essence. Since Fine's counter-examples are (global) necessities which follow from irrelevant external sources, the account is immune to them.
>
> *(Morvarid 2018: 4998)*

For example, it won't count as essential to Socrates that he be distinct from the Eiffel Tower, because it is not locally necessary – there could be worlds with facts about Socrates that lack facts about the Eiffel Tower, although it will still be globally necessary that Socrates is distinct from the Eiffel Tower.

Setting aside some potential objections to Correia's proposal,[14] one can question the extent to which it provides a genuine defence of modalism, insofar as the modal notion implicated has been altered. Indeed, Fine responds that

> In claiming that the notion of essence was not to be understood in modal terms I had in mind the familiar metaphysical modalities, i.e. the familiar notions of metaphysical necessity, metaphysical possibility and the like. But in proposing an account of essence

in terms of metaphysical modality, Correia has in mind a certain refinement of the familiar metaphysical modalities – which he calls 'local' as opposed to 'global' (p. 68) – and yet it was not my intention to argue against an account of essence in terms of any modal notions whatever. Indeed, the conceptual leap that is required to recognize the intelligibility of the local modalities is exactly the kind of conceptual leap for which I was arguing in the paper.

(Fine 2007: 85)

Modalism proper, then, concerns the 'familiar notion' of metaphysical necessity. Appeal to a new modal notion to explain essence shows, for Fine, that a conceptual leap is required to move from metaphysical necessity to essence. That said, even if Fine is right that Correia makes this conceptual leap, there is still an open question which set of concepts, and which developed theory, shed best light on essence and necessity.

3 Beyond modalism

Those who are persuaded by the arguments against any form of modalism about essence need to provide an alternative account. Even if one takes *essence* to be a primitive notion, one can still recognize the value of providing further elucidation. In this section, we will consider first various ways in which essence has been taken to be related to the *real definition* of a thing. We will further examine a distinction made by Fine, between *constitutive essence* and *consequential essence*. Finally, we will consider the attempt to give an account of necessity in terms of essence (the opposite of the modalist's attempt to give an account of essence in terms of necessity).

3.1 Essence and definition

In his 1994 paper, Fine proposes that we understand essence, not modally, but on the model of *real definition*.

> It has been supposed that the notion of definition has application to both words and objects – that just as we may define a word, or say what it means, so we may define an object, or say what it is. The concept of essence has then taken to reside in the 'real' or objectual cases of definition, as opposed to the 'nominal' or verbal cases.

(Fine 1994: 2)

Aristotle wrote,

> A definition is a phrase signifying a thing's essence.

(Topics: 101b36–37)

> Definition is the formula of the essence.

(Metaphysics: 1031a12)[15]

Such a view has also been taken up and developed by Hale (2013).

> The nature (or essence or identity) of a thing is simply what it is to be that thing. It is what distinguishes that thing from every other thing. ... A thing's nature or essence is what is given by its definition ... one might think of a full dress, or canonical,

definition as specifying what type of thing it is and what distinguishes it from everything else within its type.

(Hale 2013: 151)

Thinking of the essence of a thing as what is given by its real definition naturally suggests not only necessary, but also sufficient, properties for being a thing. Indeed, Hale explicitly notes that the definition of a thing not only tells us what type it is (e.g. Socrates is *human*), but also what distinguishes it from other things (e.g. Socrates alone originates from gametes g_1 and g_2).

Proponents of the notion of real definition point out that philosophy often concerns giving real definitions; whenever we want to say what something is, as opposed to giving the meaning of some word or concept, we are giving a real definition of it.

> The case can be made that contemporary analytic philosophy is up to its ears in idioms of definition, analysis, reduction and constitution that are best understood in a ... metaphysical key – as demands for *real definition* rather than linguistic or conceptual analysis. On this view, when we ask what it is for a thing to be a person or for a creature to be conscious or for a fact to be a law of nature or for two expressions to be synonymous or for an object to be colored or for an action to be free or for an artifact to be an artwork, we are best understood as seeking real definitions of the properties, kinds, and relations that figure in our questions, rather than semantic or conceptual equivalents ...
>
> *(Rosen 2015: 189)*

Even assuming that this is correct, one might hope for a more precise account of real definition itself. And indeed, in recent years the debate has moved on from taking essences to be straightforwardly expressed by real definitions (as in Fine 1994), to giving a more precise account of real definition, and a more nuanced account of its relation to essence and other (meta)metaphysical notions. So, for example, Rosen (2015) gives an account of real definition in terms of grounding;[16] Fine (2015) gives accounts of *essence* and *ground* in terms of constitutively necessary and sufficient conditions respectively, and although he appears still to endorse the view that statements of essence correspond to real definitions, he introduces an important variety of real definition corresponding to constitutively necessary-and-sufficient conditions;[17] Correia (2017) gives an account of real definition in terms of *generalized identity* (of which more below); Gorman (2005, 2014) proposes a non-modalist account of essence in terms of *explanation*, and an account of real definition as a statement of essence.[18] At least, many parties seem to endorse a close relationship between essence, grounding and definition, but there is room for disagreement on how they fit together.

3.2 Constitutive and consequential essence

However one gives a fuller non-modalist account of the nature of essence, further questions arise. Fine has also introduced a distinction between *constitutive essence* and *consequential essence*. His preliminary definition is,

> An essential property of an object is a constitutive part of the essence of that object if it is not had in virtue of being a consequence of some more basic essential properties of the object; and otherwise it is a consequential part of the essence.
>
> *(Fine 1995: 57)*

However, such a definition allows logical truths to be part of the consequential essence of all things. But some such properties are intuitively not essential, in the same spirit as Fine's original set of counterexamples to modalism. Just as it seemed wrong to say that Socrates is essentially such that $2 + 2 = 4$, it seems similarly out of step to say that Socrates is essentially such that $2 = 2$. There are thus a series of proposals for refining the definition of consequentialist essence.

Fine's own proposal further specifies that any objects involved in the essence of a thing must be 'pertinent' to that thing's essence, which he explains in terms of them not being able to be 'generalized away'.

> It is characteristic of the extraneous objects that they can be generalized away. Thus not only is it true in virtue of the identity of Socrates that $2 = 2$ but also that, for any object x, $x = x$. The objects pertinent to a thing's essence can therefore be taken to be ones which cannot in this manner be generalized away.
>
> *(Fine 1995: 59)*

For example, supposing that Socrates is essentially human, being human is part of his constitutive essence, because this can't be generalized away (it's not the case that, for any object x, x is human). It will also be part of Socrates's essence to be *human or clay*, because Socrates being human or clay is a consequence of his being human, and again, being human or clay can't be generalized away (not everything is human or clay). But it is no part of Socrates's essence to be such that $2 = 2$, for the reasons just given.

Correia (2012) broadens essence to concern not only a single thing, e.g. Socrates, but also pluralities of things, e.g. Socrates and the Eiffel Tower. He then restricts the logical consequence relation to draw only on rules of inference for those logical concepts that are included in the plurality whose essence is under consideration.

> α is derivatively essential to X just in case α does not belong to the basic nature of X, but is a logical consequence, relative to the set of all logical concepts in X, of the basic nature of X.
>
> *(Correia 2012: 648)*

Where *derivative essence* corresponds to *consequential essence*, and *basic nature* to *constitutive essence*. So, for example, if X is just Socrates, and no logical concepts, in particular not the concept of identity, then it will not be part of Socrates's derivative or consequential essence that he be such that $2 = 2$ (or such that for all x, $x = x$). But if X is, say, *disjunction* and Socrates, then if it is of Socrates's basic nature to be human, it will be derivatively essential to Socrates that he is human or not human.

Finally, the 'Fine-Rosen Proposal',[19] instead of defining consequential essence in terms of (what follows from) constitutive or basic essence, turns things around, and defines constitutive essence in terms of consequential essence and grounding.

> The *constitutive* claims of essence can [then] be taken to be those consequentialist statements of essence that are not partly grounded in other such claims.
>
> *(Fine 2012: 79)*

> p belongs to the constitutive essence of x iff p belongs to the consequential essence of x, and there are no propositions Γ such that p belongs to the consequential essence of x in virtue of the fact that Γ belongs to the consequential essence of x.
>
> *(Rosen 2015: 196)*

Rather than consequential *following from* constitutive essence, instead, we start with a wider understanding of the consequential essence of something and pick out from it those features or claims that are not grounded in other essential features or claims. Such a proposal draws yet another candidate link between *essence* and *ground*.[20] Note that the Fine-Rosen proposal for defining *constitutive essence* is similar to Gorman's proposal for defining *essence tout court*. There appears to be some measure of agreement *that there is a distinction here to be made*, but less agreement on which side of the distinction *essence* falls.

3.3 Essentialist theories of modality

The non-modalist about essence is often not content to take *essence* as a notion independent of *necessity*, but rather, their ambition is, *contra* the modalist, to give an account of necessity in terms of essence.

There are, most would agree, some necessary truths. But what is their source? Why are these truths necessary? Many answers to these questions have been offered.[21] Some proponents of a non-modalist account of essence also propose that we go further and give an account of necessity in terms of essence.

> An essentialist theory of modality should be distinguished from essentialism, as it is commonly understood – that is, as the doctrine, which goes back in one form to Aristotle, that things have some of their properties essentially, but others merely accidentally. ... An essentialist theory of modality involves a commitment to essentialism, but goes beyond it, by seeking to explain necessity and possibility – or at least metaphysical necessity and possibility – in terms of essence: what is metaphysically necessary is what is true in virtue of the nature (or essence) of things, and what is metaphysically possible is what is not ruled out by the natures of things.
>
> (Hale 2017: 835)[22]

(See also Fine 1994: 9.) The standard proposal is thus:

> It is metaphysically necessary that $p =_{df}$ it is true in virtue of the nature (essence) of all things that p.[23]

That is, take together all things and all of their essences; they together will generate the metaphysical necessities. Moreover, other kinds of necessity can be defined in terms of restricted kinds of things. For example, conceptual necessity becomes truth in virtue of the nature of all concepts.

The proposed Essentialist Theory of Modality faces a number of questions, which we will not have space to explore here.[24] Instead, let us briefly consider one quite general issue that any version of the theory will likely face. Given that the notion of essence is intended not to be defined in terms of metaphysical necessity, one might reasonably ask why one should expect essential properties, and whatever follows from them, to be metaphysically necessary. After all, necessity is no longer built into the very notion of what essence is, so what is it about essence that ushers in necessity? If the essence of a thing is *what it is to be* that thing, why think that it also tells us what that thing *must be*, or *couldn't have been*? What is it about essence that makes it suitable to provide a source for necessity?[25]

Hale presents an essentialist response to such questions.[26]

> Once it is granted (*vide infra*) that we can intelligibly speak of a thing's nature, or identity, it must be agreed that truths about it are necessary. For the supposition that a thing might have had a different nature immediately raises an obvious problem. Let α be the thing in question, and let Φ be its nature – that is, $\Phi\alpha$ says what it is for α to be the thing it is. Then the supposition that α might have had a different nature is the supposition that it might not have been the case that $\Phi\alpha$, and might have been that $\Phi'\alpha$ instead. Now there is, we may assume, no difficulty in the supposition that something *else*, and perhaps even something of the same type as α, lacks the property Φ and has the property Φ'. But our supposition has to be that α *itself* might have lacked Φ and been Φ' instead. This is equivalent to the suggestion that for some β, it might have been the case that $\beta = \alpha \wedge \neg\Phi\beta \wedge \Phi'\beta$. But how could this possibly be true? Given that $\Phi\alpha$ tells us *what it is for α to be the thing it is*, and that $\neg\Phi\beta$, β *lacks* what it takes to be that thing, it must be that $\beta \neq \alpha$. In short, the supposition that a thing's nature might have been different breaks down because it is indistinguishable from the supposition that something *else* lacks that nature.
>
> (Hale 2013: 133)

Hale's point is that if the essence of a thing is what it is to be that thing, then if we try to suppose that the thing could exist without that essence, there's nothing in that supposition to tell whether we've succeeded in thinking of *that thing* without its essential properties, or just some different thing altogether. Nevertheless, one could press the concern that the argument still assumes that what it *actually* takes to be a thing must carry over to also be how the thing *must* be. In Leech (2018) I attempt to develop this response. I argue that, for a number of plausible theoretical roles that essence might be introduced to play – such as providing a principle of individuation – such roles could just as well be filled by contingent as necessary properties.[27] Hence, at least according to these motivations for a notion of essence, essence need not imply necessity.

4 Extending essence

Few, if any, proponents of theories of essence intend only *objects* to have essences. Thus,

> Here I am using the word 'thing' in the widest possible sense, to include entities of every ontological type – not only objects, but properties, relations and functions (of each level), and whatever other kinds of entities there may be. Everything has a nature, no matter what kind of thing it is.
>
> (Hale 2013: 151)

To give another example, *dispositional essentialists* (e.g. Bird 2007) argue that properties have (dispositional) essences. So far in this chapter, we have been thinking largely in terms of *objectual* essentialist statements that predicate an essential property of an object, but there are other kinds of essentialist statements. For example,

[Objectual essence]: It's essential to Socrates to be a human.

[Generic essence]: It's essential to being a human to be a rational animal.

[Factual essence]: It's essential to Socrates's being a human that he be a rational animal.

(See *Correia and Skiles* 2019: 649)

Indeed, there is an important class of statements that are often classed as essentialist that I haven't even mentioned yet: theoretical identities involving kinds, such as 'Water=H$_2$O', or 'Gold is the element with atomic number 79'. By calling these 'essentialist', philosophers may sometimes only mean that they are necessarily true. But sometimes a claim stronger than mere necessity is intended. For example, one might claim that *what it is to be water* is to be H$_2$O, whereas *what it is to be H$_2$O* goes deeper into the chemistry and physics of the substance.[28] Such claims, if genuinely essentialist, and not merely about necessity, are plausibly to be classed as generic essence claims.

A further challenge in the theory of essence is to give an account of how to cash out such statements. For example, one might be tempted to reduce statements of generic essence and factual essence to statements of objectual essence involving a special kind of entity: properties, kinds, facts, and so on. For example, 'It's essential to being a human to be a rational animal' could become 'It's essential to *the property of being human* to be such that all humans have it.' But such a view would commit the essentialist to an ontology including whatever entities were introduced.[29] We might expect to be able to give an account of the source of necessity in terms of essence without thereby having to take a stance on other substantive metaphysical and ontological debates. Correia and Skiles (2019) propose an alternative account of essentialist statements in general in terms of a generalized notion of identity, which brings us nicely to our final section.

5 Essence and identity

Essence is often explained in terms of the *identity* of a thing. The sources we have been considering here, such as Fine (1994), Fine (1995) and Hale (2013), have been littered with mentions of identity in tandem with or in place of essence. Sometimes the essence of something is supposed to be what distinguishes that thing from everything else (Hale 2013), or provides it with a principle of individuation (Wiggins 2001). Kripke's renaissance of *de re* necessity and essentialist claims in *Naming and Necessity* centrally involves an examination of identity statements, particularly the theoretical identities mentioned in section 4. More work remains to be done on how precisely to understand the relation between essence and identity. But to close, it is worth noting the development of a new account of essence based on a notion of *generalized identity* (Correia and Skiles 2019).

A generalized or generic notion of identity is one that has been recently explored in the philosophical literature, e.g. Correia (2017), Dorr (2016). We normally think of identity statements as involving two names (or variables) standing for an object or objects flanking the identity sign, as in '$a = b$' – *a is b*. But there are other kinds of identification that we make, statements of the form 'To be F is to be G' or 'For it to be the case that *p* is for it to be the case that *q*.' For example,

1   George Orwell is Eric Blair.
2   To be a vixen is to be a female fox.
3   For it to be the case that you have 2 + 3 apples is for it to be the case that you have 3 + 2 apples.

Correia and Skiles introduce a more general kind of identity statement, using the operator '$\equiv$' rather than '$=$', provide a theory of how it works and use it to give accounts of essence and ground. Correia (2017) also gives an account of real definition in terms of generalized identity. Details aside, such a movement towards providing a unifying account of essence, real definition and grounding in terms of identity has the potential for significant advantages. If successful, this approach could (a) provide a non-modalist account of essence, but in terms of something more familiar than essence – *identity*; (b) make good on the existing links that are assumed to hold between essence and identity; and (c) provide a response to the worry raised in Section 3.3, about whether essence as understood by the non-modalist can be assumed to generate necessity. For if essence is to be understood in terms of a notion of identity, then its capacity to provide a source of necessity could perhaps be derived from the necessity of identity.

Notes

1 The phrase 'τὸ τί ἦν εἶναι' is usually translated into English directly as 'essence', but it has a literal meaning more like 'what it is to be'.
2 Perhaps most notably, Locke distinguished between *real* and *nominal essence*. Real essence is 'the very being of anything, whereby it is what it is'. Nominal essence concerns the general concept that we associate with kind terms: 'that abstract idea which the general, or sortal … name stands for' (Locke's Essay: III.iii.15).
3 Although one can challenge the connection between *essence* and *nature*. See, e.g., Cowling (2013).
4 'Although all entities *have* essences, essences themselves should never be thought of as *further entities*, somehow specially related to the entities whose essences they are' (Lowe 2008: 23).
5 See Roca-Royes (2011) for more on essential properties and individual essences.
6 Metaphysical claims, rather than meta-metaphysical claims.
7 The origin of these debates, at least in recent decades, is Kripke's *Naming and Necessity* (1981), in particular pp. 110–115.
8 See Roca-Royes (2011) for an instructive overview of these debates, including detailed recommendations for further reading.
9 One can give a formal account of the distinction in terms of scope: *De dicto* statements have quantifiers within the scope of the modal operator, e.g. '$\Box \exists x Fx$' – 'Necessarily, something is F', and *de re* statements do not, e.g. '$\exists x \Box Fx$' – 'Something is such that it is necessarily F', or '$\Box Fa$' – 'a is necessarily F'.
10 See, for example, Quine (1976), where he warns us against returning to 'the metaphysical jungle of Aristotelian essentialism'. See also Chapter 4 of this volume on Quine's Metametaphysics.
11 I.e. factive: if it is necessary that p, then p.
12 But see Fine (2005).
13 Skiles also argues against the sufficiency of Sparse Modalism.
14 See, for example, Morvarid (2018).
15 See Charles (2000) and Witt (1989) for more detail on Aristotle's own view.
16 For more on grounding, see Chapter 15 of this volume.
17 One might reasonably wonder how exactly to understand 'constitutively necessary' and 'constitutively sufficient'. Indeed, the difficulty in understanding these notions is part of Correia (2017)'s reason for not adopting Fine's account of real definition.
18 See Chapter 17 of this volume for more on metaphysical explanation.
19 As named by Nutting *et al.* (2018), based on Fine (2012) and Rosen (2015).
20 See Nutting *et al.* (2018) for a challenge to the Fine-Rosen view.
21 See Lewis (1986) for a classic presentation of a possible worlds approach, Sidelle (1989, 2009) for a defence of conventionalism and Vetter (2015) for an approach that finds the source of modality in the potentialities of things.
22 Although, note that Hale does not intend to *reduce* necessity to essence (Hale 2013: 150).
23 Fine introduced the notation '$\Box_x p$' to express 'it is true in virtue of the nature of x that p', although we shall not need to use it here.
24 For example, Wildman (2018) presents some purported counterexamples. Koslicki (2012) considers, among other things, how necessities follow from essences: Do statements of necessity follow logically from essentialist statements, or is a non-logical notion of consequence required? She argues that logical consequence alone is insufficient, and proposes something closer to Aristotle's own view: derivations of necessities from essences must go via *definitions*, where these definitions are explanatory of the conclusion. This draws yet further links between *essence*, *definition* and *explanation*. See Chapter 17 of this volume for more on metaphysical explanation.

25 See also Mackie (2020) for a discussion of this worry. Gorman (2014) acknowledges that his account of essence in terms of explanation has no consequence for the modal status of essential properties.
26 See also Wiggins (2001), particularly his discussion of 'the anchor constraint'.
27 See Mackie (2006) for comprehensive arguments against the move from a property *F* actually providing a thing *a* with a principle of individuation, to *a* being necessarily *F*. See also Mackie (2020).
28 See Bird (2009) and Ellis (2001) for more on essences and natural kinds.
29 See Correia (2006) for further objections to this kind of proposal, and further discussion of generic essence statements.

References

Aristotle, *Metaphysics*, in J. Barnes (ed.), *The Complete Works of Aristotle, Volume Two* (Princeton, NJ: Princeton University Press, 1984).

Aristotle, *Topics*, in J. Barnes (ed.), *The Complete Works of Aristotle, Volume One* (Princeton, NJ: Princeton University Press, 1984).

Bird, A. (2007) *Nature's Metaphysics: Laws and Properties* (Oxford: Oxford University Press).

Bird, A. (2009) 'Essences and Natural Kinds', in R. Le Poidevin *et al.* (eds), *The Routledge Companion to Metaphysics* (London: Routledge).

Charles, D. (2000) *Aristotle on Meaning and Essence* (Oxford: Oxford University Press).

Correia, F. (2006) 'Generic Essence, Objectual Essence, and Modality', *Noûs*, 40:4, 753–767.

Correia, F. (2007) '(Finean) Essence and (Priorean) Modality', *Dialectica*, 61:1, 63–84.

Correia, F. (2012) 'On the Reduction of Necessity to Essence', *Philosophy and Phenomenological Research*, 84:3, 639–653.

Correia, F. (2017) 'Real Definitions', *Philosophical Issues*, 27, Metaphysics.

Correia, F. and Skiles, A. (2019) 'Grounding, Essence, and Identity', *Philosophy and Phenomenological Research*, 98:3, 642–670, first published online 2017.

Cowling, S. (2013) 'The Modal View of Essence', *Canadian Journal of Philosophy*, 43:2, 248–266.

Della Rocca, M. (1996) 'Recent Work on Essentialism: Part 1', *Philosophical Books* 37, 1–13.

Denby, D. A. (2014) 'Essence and Intrinsicality', in R. Francescotti (ed.), *Companion to Intrinsic Properties* (Berlin: De Gruyter).

Dorr, C. (2016) 'To Be F Is to Be G', *Philosophical Perspectives*, 30, Metaphysics.

Ellis, B. (2001) *Scientific Essentialism* (Cambridge: Cambridge University Press).

Fine, K. (1994) 'Essence and Modality', *Philosophical Perspectives* 8, 1–16.

Fine, K. (1995) 'Senses of Essence', in W. Sinnott-Armstrong, D. Raffman and N. Asher (eds.) *Modality, Morality, and Belief: Essays in Honor of Ruth Barcan Marcus* (Cambridge: Cambridge University Press).

Fine, K. (2005) 'Necessity and Non-Existence', in his *Modality and Tense: Philosophical Papers* (Oxford: Oxford University Press).

Fine, K. (2007) 'Response to Fabrice Correia', *Dialectica*, 61:1, 85–88.

Fine, K. (2012) 'Guide to Ground', in F. Correia and B. Schnieder (eds.), *Metaphysical Grounding: Understanding the Structure of Reality* (Cambridge: Cambridge University Press).

Fine, K. (2015) 'Unified Foundations for Essence and Ground', *Journal of the American Philosophical Association*, 1:2, 296–311.

Gorman, M. (2005) 'The Essential and the Accidental', *Ratio*, 18, 276–289.

Gorman, M. (2014) 'Essence as Foundationality', in D. D. Novotný and L. Novák (eds.), *Neo-Aristotelian Perspectives in Metaphysics* (London: Routledge).

Hale, B. (2013) *Necessary Beings: An Essay on Ontology, Modality, and the Relations Between Them* (Oxford: Oxford University Press).

Hale, B. (2017) 'Modality: Postscript', in B. Hale, C. Wright and A. Miller (eds), *A Companion to the Philosophy of Language, Volume II*, Second Edition (Oxford: Wiley Blackwell).

Koslicki, K. (2012) 'Essence, Necessity, and Explanation', in T. E. Tahko (ed.), *Contemporary Aristotelian Metaphysics* (Cambridge: Cambridge University Press).

Kripke, S. (1981) *Naming and Necessity* (Oxford: Blackwell).

Leech, J. (2018) 'Essence and Mere Necessity', *Royal Institute of Philosophy Supplement*, 82, 309–332.

Lewis, D. (1986) *On the Plurality of Worlds* (Oxford: Blackwell).

Locke, J. (1689) *An Essay Concerning Human Understanding*, P. H. Nidditch (ed.) (Oxford: Clarendon Press, 1975).

Lowe, E. J. (2008) 'Two Notions of Being: Entity and Essence', *Royal Institute of Philosophy*, 83.4, 23–48.

Mackie, P. (2006) *How Things Might Have Been: Individuals, Kinds, and Essential Properties* (Oxford: Clarendon Press).

Mackie, P. (2020) 'Can Metaphysical Modality Be Based on Essence?', in Mircea Dumitru (ed.), *Metaphysics, Modality, and Meaning: Themes from Kit Fine* (Oxford: Oxford University Press).

Morvarid, H. (2018) 'Finean Essence, Local Necessity, and Pure Logical Properties', *Synthese*, 195, 4997–5005, first published online 2017.

Nutting, E. S., Caplan, B. and Tillman, C. (2018) 'Constitutive Essence and Partial Grounding', *Inquiry*, 61:2, 137–161.

Quine, W. V. O. (1976) 'Three Grades of Modal Involvement' in his *The Ways of Paradox and Other Essays*, rev. edn. (Cambridge, MA: Harvard University Press).

Roca-Royes, S. (2011) 'Essential Properties and Individual Essences', *Philosophy Compass*, 6:1, 65–77.

Rosen, G. (2006) 'The Limits of Contingency', in F. MacBride (ed.), *Identity and Modality*, Mind Association Occasional Series (Oxford: Oxford University Press).

Rosen, G. (2015) 'Real Definition', *Analytic Philosophy*, 56:3, 189–209.

Sidelle, A. (1989) *Necessity, Essence, and Individuation: A Defense of Conventionalism* (Ithaca, NY: Cornell University Press).

Sidelle, A. (2009) 'Conventionalism and the Contingency of Conventions', *Noûs*, 43:2, 224–241.

Skiles, A. (2015) 'Essence in Abundance', *Canadian Journal of Philosophy*, 45:1, 100–112.

Vetter, B. (2015) *Potentiality: From Dispositions to Modality* (Oxford: Oxford University Press).

Wiggins, D. (2001) *Sameness and Substance Renewed* (Cambridge: Cambridge University Press).

Wildman, N. (2013) 'Modality, Sparsity, and Essence', *The Philosophical Quarterly*, 63, 760–782.

Wildman, N. (2018) 'Against the Reduction of Modality to Essence', *Synthese*, online early.

Witt, C. (1989) 'Aristotelian Essentialism Revisited', *Journal of the History of Philosophy*, 27:2, 285–298.

20
FICTIONALIST STRATEGIES IN METAPHYSICS

Lukas Skiba and Richard Woodward

1 Introduction

We often find ourselves in a predicament. On the one hand, we find it congenial to talk about things of various sorts: the *hole* in the cheese, the *number* of planets, the *property* shared by all spiders, the *possible worlds* at which things are different, and so on and so forth. But on the other hand, we find it difficult to accept that there really are such things. For these things are troublesome: they offend our taste for desert landscapes and raise difficult questions about how we could know about them. So we are presented with a dilemma: either we give up our ontological scruples and embrace the existence of troublesome entities, or we reject their existence and revise our linguistic practices.

Fictionalists hope that we can have our cake and eat it, that our linguistic practices can be reconciled with our ontological scruples. After all, the things we say in the context of our engagement with fiction provide a paradigm case where we do not find it so problematic to talk about entities in whose existence we do not believe: we often say things like "some elves are nimble" or "Holmes is a detective" even though we do not believe that there really are elves or a brilliant detective living at 221b Baker Street. So perhaps our talk of fictional things can be used as a model for understanding our talk of troubling entities, and thereby a way of talking with the vulgar but thinking with the wise. Such fictionalist proposals are common, having been offered in the case of mathematical objects (Field 1980, 1989), unobservable entities (van Fraassen 1980), possible worlds (Rosen 1990), composite objects (Rosen and Dorr 2002), fictional characters (Brock 2002), scientific models (Frigg 2010), propositions (Armour-Garb and Woodbridge 2010), colours (Gatzia 2010) and beyond.

This conception of the fictionalist's project is broad and irenic, picking out a genus of which there are various species, and the more specific content of a fictionalist proposal will vary depending on exactly how the basic idea is fleshed out. Here, we provide an overview of what we take to be the core choice points facing the fictionalist, as well as a survey of some of the main issues facing the viability of fictionalist strategies.

2 The analogy with fiction

The term "fictionalism" and its cognates are used widely and wildly in the literature, and one would be forgiven for thinking that "fictionalism" is nothing more than a term of philosophical fashion, with nothing unifying the various proposals that have been given the label. For our part, we think that the most minimal and inclusive conception of fictionalism takes fictionalists about a given discourse – talk of numbers, or talk of properties, etc. – to accept something like the following (for alternative attempts to characterize fictionalism see, e.g., Kroon (2011: §2), Armour-Garb and Woodbridge (2015: Ch.1)):

The Analogy
The target discourse can usefully be interpreted by analogy with a natural way of interpreting paradigmatically fictional discourse, and the ensuing account of the target discourse supports an anti-realist account of its apparent subject matter.

To see the import of **The Analogy**, consider its application to a particular case. So, e.g., mathematical fictionalists emerge as holding that the sentences of mathematical talk should be interpreted by analogy with a natural way of interpreting sentences like 'most elves are nimble' and that the ensuing account of mathematics supports an account of the apparent subject matter of mathematical talk – numbers, sets, functions, etc. – that is anti-realistic in character. A number of points merit attention.

First, the fictionalist tells us only how mathematical talk *can usefully* be understood. Her claim is thus distinct both from the more straightforwardly descriptive claim that we *do in fact* understand mathematics along fictionalist lines as well as from the more straightforwardly normative claim that we *should* do so. Accordingly, the fictionalist's project is neither 'hermeneutic' nor 'revolutionary' to use some jargon that has become popular in the literature (see Stanley 2001; for the origins of the distinction see Burgess 1983 and Burgess and Rosen 1997). Both the hermeneut and the revolutionary agree that we can and should be fictionalists; they disagree on whether we are. **The Analogy** is endorsed by both hermeneuts and revolutionaries, and the core elements of a fictionalist proposal can be enriched in either direction. Indeed, fictionalism could be developed in neither direction, for the distinction between hermeneutic and revolutionary ficitonalisms is not an exhaustive one: for instance, van Fraassen's claim in the case of defending his fictionalist account of unobservables is not that scientists *are* fictionalists, nor that scientists *should* be fictionalists, but rather that scientists *can* be fictionalists (see Van Fraassen 1980, 1994; Rosen 1994). It is for this reason that we chose not to build a stronger descriptive or normative element into our formulation of **The Analogy**.

Second, what's important is *paradigmatically* fictional discourse. It's familiar that our engagement with fiction is multi-faceted. We have *authors* who create, *audiences* who consume, *critics* who analyse, and so on. The distinctions here are not clean cut, but different things still seem to be going on in each case. What's relevant to fictionalism, at least as we understand it, is the type of fictional discourse that is exemplified when someone is talking to you about *The Hobbit* and tells you that some elves are nimble. In this setting, it is consumers of fiction rather than authors or critics who serve as the paradigm. Our use of 'talk of fiction' and 'fictional discourse' should be understood in this light.

Third, the proposed analogy isn't between the target discourse and fictional discourse, but the target discourse and a *natural interpretation* of fictional discourse. Not only does this leave it open whether the natural interpretation is the right one, it's also consistent with there being various equally natural interpretations of fictional discourse and two fictionalists might disagree about which of these is relevant to the proposed analogy. In any case, we stress that it isn't built into fictionalism that what's going on when we say things about numbers (e.g.) is continuous with what's going on when we say things about elves. It might be odd to be a fictionalist (and hence an anti-realist) about, say, numbers while being a realist about fictional characters (by regarding them as, say, abstract artefacts) but such combinations of views shouldn't be ruled out by fiat, and the analogy is formulated to avoid doing just that: you can still regard an anti-realistic interpretation of talk about elves as natural, even if you don't think it is the view to be adopted all-things-considered. Moreover, even once the fictionalist has settled upon a particular basis for the analogy, we've yet to be told how tight it is meant to be. For instance, is the analogy with fiction just meant to remind us that we can use sentences without committing ourselves to the entities we seem to be talking about? Or is some richer analogy with fiction intended? If so, what is it and what purposes does it serve?

Finally, even once we've settled both the nature and tightness of the analogy, the account of mathematical talk that the fictionalist hopes to build on these foundations is only meant to support an anti-realist account of the *apparent* subject matter of mathematics, and that's consistent with the claim that the real subject matter of mathematics is number free. Indeed, some fictionalists have argued that the *real* subject matter of the target discourse is not the subject matter that one might initially identify (see Yablo 2001, who suggests that the real subject matter of a claim like "the number of planets is eight" only concerns planets rather than numbers too).

At least in these ways, then, endorsing **The Analogy** is the start of the story rather than the end of it. That's to say that the fictionalist has choices, and that more specific proposals can be delineated in terms of how these choices are made. One upshot is that the commitments of fictionalism, as well as its benefits and problems, will vary depending upon the exact nature of the proposal at hand. For the purposes of this survey, however, we choose to focus on issues that we think arise for a great many (if not all) fictionalist strategies.

3 Fictionalist paraphrases

Given that the core element of a fictionalist proposal is captured by a claim as weak as **The Analogy** it's unsurprising to learn that fictionalists disagree about how the approach should be cashed out and thereby about how fictionalism is best developed. Be that as it may, we can identify one core idea that is often found in textbook presentations of fictionalism: the importance of what we will call *fictionalist paraphrases*. A fictionalist paraphrase in the intended sense is simply a mapping from sentences that concern troubling entities to sentences that concern the content of a fiction. Some examples: Field (1980) paraphrases the claim that there are prime numbers as: *it is true according to standard arithmetic that prime numbers exist*; Rosen (1990) paraphrases the claim that there are possible worlds as: *it is true according to modal realism that possible worlds exist*; Brock (2002) paraphrases the claim that fictional characters exist as: *it is true according to fictional realism that fictional characters exist*; and Dorr (2005) paraphrases the claim that tables exist as: *it is true according to universalism that tables exist*.

What we have in each case, then, is a mapping from sentences which seemingly can only be true if reality contains certain things to sentences which seemingly can be true even if reality lacks those very things. For even if reality lets us down and fails to contain things like numbers or tables or fictional characters, it can still be true that such things exist according to certain theories. For it is well known that story-operators are *non-factive* in the sense that *p*'s being true according to some fiction or theory doesn't entail that *p* is true tout court. It would be madness to think we can infer that there really are elves from the fact that it is true according to *The Hobbit* that elves exist. Moreover, insofar as these kinds of 'theory-shadowing' paraphrases are distinctive of fictionalist proposals, they help to see the appropriateness of the label 'fictionalism' – after all, it is natural to think everyday utterances of sentences like "there are elves" are acceptable (perhaps even true) because in the relevant contexts those sentences are best interpreted as concerning what is true according to some salient work of fiction.

But even though the appeal to theory-shadowing paraphrases is distinctive of fictionalist proposals, the more interesting issue concerns not their presence but their purpose and significance. Indeed, the species of fictionalism can be demarcated, at least in part, precisely in terms of the respective theoretical roles that each associates with fictionalist paraphrases.

The main choice point facing the fictionalist can be illustrated nicely by analogy with the things we say about fictional characters, events and places. Suppose that Alice and Billy are discussing *The Lord of the Rings* and Alice claims that some elves are nimble. Despite the fact that there are no such creatures as elves, there is a natural sense in which Alice's claim is correct because it is true according to Tolkien's story that there are such creatures and that some of them are nimble. Indeed, given the topic of their conversation, it would seemingly be incorrect for Alice to claim that no elves are nimble. For despite the fact that there are no such creatures as elves (and hence, that none of them are nimble), such a claim seems incorrect because it is false according to Tolkien's story that no elves are nimble. The facts about what is and isn't true according to some relevant story thus often seem to determine whether or not the things we say about fictional characters are appropriate (correct) or inappropriate (incorrect). But whilst this observation is common ground in the debate, the crucial question concerns whether the relevant standard of correctness is distinct from truth. And here we are drawn in two competing directions. On the one hand, the fact that there are no such creatures as elves strongly suggests that Alice spoke falsely when she claimed that some elves are nimble. But on the other hand, the fact that she and Billy were discussing the goings-on in *The Lord of the Rings* tempts us to hold that Alice spoke truly because what she meant was that it is fictional that some elves are nimble. Hence, on the first proposal, Alice spoke falsely but correctly, whereas on the second proposal, she spoke correctly because she spoke truly.

We can accordingly distinguish two fictionalist strategies, based on two alternative conceptions of the role of fictionalist paraphrases. On the first proposal, the fictionalist about unobservable entities thinks that sentences like "there are electrons" are false but nonetheless appropriate or correct because it is true according to standard physics that there are such things, and the mathematical fictionalist thinks that sentences like "there are functions" are false but nonetheless appropriate because it is true according to standard maths that there are such things. We call this view *committal* fictionalism because the fictionalist who pursues this option thinks of sentences like "there are electrons" or "there are functions" as being *ontologically committed* to electrons or functions insofar as their truth requires the existence of such things. Indeed, it's precisely the fact that this requirement is not met – or so the fictionalist thinks – that explains why the committal fictionalist does not accept that these sentences are true.

On the second proposal, by contrast, the fictionalist thinks that the things we ordinarily think and say are not only correct but true because what is really meant by sentences like "there are electrons" or "there are functions" is that it is true according to standard physics that there are electrons or true according to standard maths that there are functions. We call this view *non-committal* fictionalism because the fictionalist who pursues this option thinks of sentences like "there are electrons" or "there are functions" as being *ontologically innocent* insofar as their truth does not require the existence of such things and instead merely requires that it is fictional that there are such things. Indeed, it's precisely the fact that this requirement is met – it is fictional that there are such things – that explains why the non-committal fictionalist accepts that these sentences are true.

Moreover, though we have distinguished these two proposals in broadly semantic terms, i.e. in terms of whether or not the things we ordinarily think and say are true, the committal and non-committal fictionalist will have correspondingly different accounts of other aspects of our linguistic practices. For instance, the non-committal fictionalist can straightforwardly accept that the speech acts that we perform when we say things about electrons and functions are assertions that are judged to be correct or incorrect depending on whether or not what is asserted is true: it's just that such a fictionalist appeals to fictionalist paraphrases to specify the *content* of what is asserted. And similarly, the non-committal fictionalist can also allow that we express beliefs when we say that there are electrons or functions: it's just that she appeals to her paraphrases to specify the *content* of what is believed. By contrast, the committal fictionalist cannot straightforwardly accept that we are making assertions or expressing beliefs, since on her view we speak falsely when we say things like "there are electrons" or "there are functions". Accordingly, the committal fictionalist will instead see us as performing a speech act that is distinct from assertion (typically called *quasi-assertion* or *pretend assertion*) and as expressing a mental attitude that is distinct from belief (typically called *acceptance*).

4 Objections to fictionalism

There is no shortage of objections to fictionalist approaches in metaphysics. But many of these objections only arise for more precise *species* of fictionalism and do not thereby threaten to establish that the *genus* is somehow problematic in a more global way. For instance, some have thought that mathematical fictionalism is empirically wrong given that young children who suffer from autism generally find it difficult to engage in pretence but do not generally find it difficult to learn mathematics (see Stanley 2001; for discussion see Liggins 2010 and Kim 2014). But obviously this worry only arises for fictionalists who both make pretence a central aspect of their account and also claim their account to be descriptively adequate. Similarly, some have thought that the fictionalists face a problem accounting for the ontology of fictions (see Nolan 1997): but even if we grant that fictions are abstract objects of some kind (sets of propositions, say), this worry will only arise with respect to fictionalist strategies focused on avoiding commitment to abstracta: no immediate worry arises with respect to fictionalist strategies focused on avoiding commitment to, for instance, composite objects or concrete possible worlds.

Given our focus on the features of fictionalism in general, then, we focus instead on objections that we think are more global, affecting many (if not all) fictionalist strategies. In particular, and continuing our emphasis on the role of fictionalist paraphrases, we will focus on two objections that arise due to specific fictionalist paraphrases that the fictionalist seems forced to accept. Both of these objections first arose with respect to the fictionalist account of possible worlds bruited by Rosen (1990), and we will present them in that context.

4.1 The incompleteness objection

To begin, it is worth noting that though Rosen calls his account "modal fictionalism", this is a misnomer: his goal is to develop a fictionalist account of possible worlds rather than a fictionalist account of modality itself. Rosen's fictionalist does not deny that some things are really possible – just that there really are other worlds at which these possibilities are realized. Hence, whereas other philosophers endorse biconditionals such as *it is possible for there to be blue swans just in case there is a possible world where swans are blue*, Rosen's fictionalist instead endorses biconditionals like *it is possible for there to be blue swans just in case according to the fiction of possible worlds, there is a possible world where swans are blue*. The distinctively fictionalist element is then that sentences such as "there is a possible world where swans are blue" can nonetheless be regarded as correct (and perhaps even true) because they can be understood as being elliptical for their fictionalist paraphrases.

However, ordinary fictions are incomplete. While we learn that Patrick Bateman has a worrying obsession with the aesthetic qualities of business cards, we do not learn what his favourite colour is. Thus, it would be wrong to say that, according to *American Psycho*, Bateman's favourite colour is blue. But it'd be equally wrong to say that, according to *American Psycho*, it's not the case that Bateman's favourite colour is blue. The story is simply silent on the issue. Now, if the fictionalist's chosen story is incomplete in the way that ordinary fictions are, trouble arises. For suppose that there is some claim Q⋆ about possible worlds such that neither Q⋆ nor its negation is true according to the fictionalist's story. (Rosen's specific example of such a claim is *there is a world containing k-many objects*, where *k* is some suitably large infinite cardinal.) Calling that story *Modal Realism*, we thus have

1 It is not the case that according to *Modal Realism*, Q⋆
2 It is not the case that according to *Modal Realism*, not-Q⋆

But remember that the modal fictionalist thinks that claims about the content of *Modal Realism* are systematically linked to underlying facts about possibility and necessity, as illustrated by her endorsement of biconditionals like: it is possible for there to be blue swans just in case according to the fiction of possible worlds, there is a possible world where swans are blue. But now let Q be the modal claim corresponding to the claim about Modal Realism negated in (1) and not-Q be the modal claim corresponding to the claim about Modal Realism negated in (2). (Rosen's specific examples of such claims are *it is possible for there to be k-many objects* and *it is not possible for there to be k-many objects*.) And recall that the fictionalist seems committed to endorsing the following two claims:

3 Q just in case according to *Modal Realism*, Q⋆
4 not-Q just in case according to *Modal Realism*, not-Q⋆

But now disaster follows since the fictionalist is committed to a flat-out contradiction: (1) and (3) commit the fictionalist to endorsing that it is not the case that Q whereas (2) and (4) commit the fictionalist to endorsing the negation of that very claim. Moreover, though the objection first arose with respect to modal fictionalism, it is not confined to that case. For example, mathematical fictionalists will face it too, when basing their fictionalism on a mathematical theory that is silent on certain relevant mathematical claims like the axiom of choice (Woodward 2012). Likewise, compositional fictionalists will face it when basing their fictionalism on a composition principle that allows for a certain kind of incompleteness (Skiba 2017).

We distinguish two kinds of response to the incompleteness objection. On the one hand, the fictionalist might grant that her story is incomplete – that is, grant the conjunction of (1) and (2) – but deny that contradiction follows. One strategy for doing so, initially suggested by Rosen but developed in more detail by Nolan (2011), does so by rejecting the application of the fictionalist's schemata (3) and (4) in cases where the fiction is incomplete, the idea being that when the fiction is incomplete with respect to some claim about worlds, the corresponding modal claims are neither true nor false. The most obvious problem with this suggestion is that it is ad hoc: the only reason that the fictionalist has for thinking that the relevant modal claims are truth-valueless is to fix a structural problem in her own theory. Absent independent motivation, the response smacks of desperation (see Rosen 1990; Woodward 2012). An alternative version of this strategy, suggested by Skiba (2017), accepts the incompleteness of the fiction but denies contradiction follows by holding that the apparently contradictory commitment is not actually a contradiction because, in context, *Q and not-Q* expresses the consistent claim more perspicuously captured by the conjunction of (1) and (2).

On the other hand, the fictionalist might apply her schemata across the board but deny that contradiction follows by denying that her fiction is incomplete (see Fine 2003; Brogaard 2006; Nolan 2011). For remember that the fictionalist is not a fictionalist about modality itself – her fictionalism concerns the existence of possible worlds, not the facts of possibility and necessity. So, for any modal claim, either that claim is true or it is not; in particular, either the modal claim *Q* is the case or its negation is. The subsequent idea is to use the modal facts themselves to generate the content of the fiction: if *Q* holds, then let *Q★* be fictionally true, and if not-*Q* holds, let not-*Q★* be fictionally true. One cost of this strategy is that any ambition to provide a reductive analysis of modality has to be given up, but this was unlikely to work anyway (see Nolan 1997 and section 5 below). Moreover, it is not clear how this solution can be extended to fictionalisms other than modal fictionalism (see Skiba 2017). An alternative version of this strategy, suggested by Woodward 2012, holds that the conjunction of (1) and (2) should be rejected because of how truth according to fiction works within the context of fictionalist strategies: given that *according to the fiction*, *P* is analysed in terms of the counterfactual conditional, *P would have been true had the fiction been true*, the fictionalist can motivate rejecting the conjunction of (1) and (2) on the grounds that it is a general structural feature of counterfactuals that, for any antecedent *A* and any consequent *C*, either the counterfactual *A would C* is true or the counterfactual *A would not-C* is true.

4.2 *The Brock-Rosen objection*

Just like the Incompleteness Problem, the Brock-Rosen objection first arose in the context of modal fictionalism (Brock 1993; Rosen 1993). It begins by noting that, since the modal realist conceives of the existence of many worlds as necessary rather than contingent, the fictionalist seems committed to endorsing the following claim:

(BR1) According to Modal Realism, at every world, there are many worlds.

But again, the fictionalist's proposal is that the facts about the content of *Modal Realism* are systematically linked to underlying modal facts. That is, we have:

(BR2) Necessarily, there are many worlds just in case according to *Modal Realism*, at every world, there are many worlds.

But now disaster follows. For (BR1) and (BR2) together entail that it is necessary that there are many worlds. Moreover, since the fictionalist is not a fictionalist about modality, she seems forced to accept that it is strictly and literally true that it is necessary that there are many worlds. But given that necessity implies truth, the fictionalist seems committed to endorsing that it is strictly and literally true that there are many worlds – which is rather unfortunate given that the entire point of modal fictionalism was to avoid this commitment. Again, the Brock-Rosen objection is not confined to modal fictionalism. Nolan and O'Leary-Hawthorne (1996) observe that, just like modal operators can be applied to statements about worlds, so numerical operators can be applied to statements about mathematical entities. By reasoning similar to the above, they show how mathematical fictionalists are forced to accept that it is strictly and literally true that there is at least one number (see also Yablo 2001).

As before, we distinguish two strategies of response available to the fictionalist. On the one hand, she might reject the initial premise of the Brock-Rosen objection, (BR1). The idea here is to be careful about exactly what it means to say, within the context of modal realism, that there are many worlds 'at' every world: for instance, swans exist 'at' our world because they are *part* of our world, but one might think that other worlds don't exist 'at' our world because they are not part of our world (see Noonan 1994 and Rosen 1995, building on Lewis 1968, for an alternative see Kim 2002 building on Bricker 2001). The most obvious problem here is that there is surely *some* sense in which the modal realist thinks that the existence of many worlds is necessary rather than contingent, which is what (BR1) was meant to express: the architect of modal realism, David Lewis, was after all explicit on the point (see Lewis 1986, p. 80; compare Divers 1999a). Whatever that sense is, then, there seems to be *some* sense in which the fictionalist is committed to endorsing (BR1) and thereby some sense in which she is committed to embracing the existence of many worlds.

On the other hand, the fictionalist might grant the initial premise of the Brock-Rosen objection, but deny the apparently ensuing commitment to the existence of other worlds. One version of this strategy has it that the relevant application of the fictionalist's general account, i.e. (BR2), fails in this case since it only applies to modal claims about ordinary objects rather than modal claims about possible worlds (see Nolan and O'Leary-Hawthorne 1996). An alternative strategy, suggested by Liggins (2008) and Woodward (2008), grants the premises of the Brock-Rosen objection but denies that the conclusion that follows is fatal to the fictionalist's project. The idea here is that the apparently fatal commitment is not actually a problem because, in context, *there are many worlds* expresses the innocent claim more perspicuously captured by (BR1).

5 The benefits of fictionalism

Suppose that all of the structural and technical problems that the fictionalist faces can be addressed. This would seem to put the fictionalist in a strong position to argue that her approach is preferable to alternative proposals since she could now argue that she has earnt the right to speak about troublesome things like possible worlds without thereby being forced to accept that there really are any such things. Appearances, however, can be misleading. For even if the fictionalist's theory is more parsimonious than its rivals, it remains to be seen whether or not fictionalism is the best overall approach. Put otherwise: even if fictionalism is significantly *cheaper* than its rivals, those other theories may have *benefits* that outweigh the extra commitments that they enforce upon us. Indeed, the kind of fictionalist account of talk of possible worlds that we have been considering is directly conceived as an answer to Lewis's (1986) challenge to deliver "paradise on the cheap" – an account of possible worlds that delivers the benefits of his own theory in a theoretically less costly manner.

Whilst the fact that the subsequent literature has focused largely on the "costs" side of the equation is understandable, it has encouraged a somewhat *laissez faire* attitude to the benefits associated with fictionalism. Rosen, for instance, tells us that fictionalists "can have all the benefits of talking about possible worlds without the ontological costs" – but tells us neither which benefits he has in mind nor why exactly the fictionalist is in a position to enjoy them (see also Sider 2002). Even putting aside the controversies surrounding the question of whether Lewis's theory really does deliver the benefits he claims, this is particularly problematical since some benefits that Lewis claims his theory delivers can quite obviously not be enjoyed by the fictionalist. For instance, one of the explanatory benefits Lewis claims of his theory is its ability to provide identifications of various kinds of entities with constructions out of possible worlds: thus *propositions* are identified with sets of possible worlds and *properties* are identified with sets of possible individuals (1986, §§ 1.4, 1.5). But all hands agree that it is a requirement on the success of these identifications that there is a plurality of possible worlds beyond the actual world and a plurality of possible individuals beyond the actual ones. And since the fictionalist cannot accept these commitments, at least some of the benefits of Lewis's theory are quite clearly off-limits to the fictionalist (compare Divers 2002).

Matters are more vexed with respect to the other benefits which Lewis associates with his theory, however, and here we will focus on two such benefits: the *conceptual* benefit of providing a reduction of modal concepts to non-modal concepts, and the *inferential* benefit of providing a first-order method of assessing the validity of modal arguments. We focus on these benefits not only because of their prominence in the relevant debate but also because they illustrate the ways in which different issues facing the fictionalist interact with each other. In particular, there is reason to think that she cannot simultaneously enjoy the conceptual and inferential benefits associated with Lewis's theory.

Lewis claims that his theory provides a distinctive explanatory benefit: a reduction of modal concepts to non-modal ones. For instance, the concept of possibility is reduced to the concept of truth at some world, which Lewis argues can be understood non-modally within the context of his theory. Lewis claims that other modal concepts can be understood non-modally: necessity reduces to truth at all worlds, contingency to truth at some but not all worlds, impossibility to truth at no world, and that even more complex modal concepts like counterfactual dependence and supervenience can also be analysed non-modally (see Lewis 1986 and Divers 2002 for discussion).

Now, recall that whereas Lewis's account is built around schemata like *Possibly P iff P is true at some world*, the fictionalist's account is built around schemata like *Possibly P iff it is fictional that P is true at some world*. Accordingly, even if we grant that the concepts that figure in Lewis's analysis are non-modal in character, an extra concept appears in the fictionalist's analysis, viz. the concept of *fictionality* or *truth according to a theory*. The question of whether the fictionalist's analysis is genuinely non-modal, then, turns on the question of whether the concept of fictionality is non-modal. And as Rosen himself notes, the prospects of providing a non-modal analysis of fictionality do not seem to be very good since there is an intuitive connection between a proposition p being true according to a fiction and the truth of the various modal claims like "the fiction necessitates p" and "p would have been true had the fiction been true" (cf. Divers 1999b; Dorr 2005; Woodward 2010). But the fictionalist cannot understand fictionality in either of these ways without compromising any ambition she has to deliver a non-modal analysis of modal concepts. Rosen's official reply is to reject any analysis of fictionality in modal terms and indeed take the notion of truth according a theory as a primitive. The most immediate problem here is that the concept of fictionality doesn't look like a particularly good one to take as primitive: it is a concept that stands in need of explication rather than one in terms of which other concepts are explicated. A less immediate, but in our view equally pressing problem, is that there are reasons to think that the fictionalist *should* understand the concept of fictionality in modal terms. Illustrating this point, however, requires us to look at a different kind of benefit associated with talking about possible worlds.

Talking about possible worlds is often thought to be beneficial insofar as doing so provides a first-order method of assessing the validity of modal arguments: on a possible worlds analysis, the validity of the argument from *necessarily p* to *possibly p* is explained since it is understood as the argument from *p is true at all worlds* to *p is true at some world* holds in virtue of the standard rules for quantifiers. There are obvious reasons why making these transitions is often called the practice of 'doing modal logic by proxy'. For rather than assessing the validity of modal arguments by relying on the inference rules of a particular modal logic, we rely instead on the inference rules of (non-modal) first-order logic. The language of first-order logic thus provides a proxy language for the language of modal logic. And the main motivation for doing modal logic by proxy – indeed, the main reason why the practice is considered to be beneficial – is that the practice is *inferentially economical* in the sense that the set of inference rules in the proxy language is smaller than the set of inference rules in the modal language. (Compare the discussion of the inferential benefits of mathematics in Field (1980), whereby the main benefit associated with talk of numbers is its ability to make our inferential lives easier.)

But, as Divers (1999b) observes, it appears that the fictionalist can justify the practice of doing modal logic by proxy if and only if she can justify her acceptance of the following *Safety Result* (SR):

(SR) Necessarily, if B★ is a consequence of A★, then B is a consequence of A

(where A and B are modal claims and their starred counterparts are their respective possible-world paraphrases). Divers goes on to argue that the fictionalist can establish the safety result – and hence can deliver the inferential benefits associated with talking about possible worlds – but in doing so analyses the concept of fictionality in modal terms. And though the specific analysis Divers uses, whereby *according to the fiction, p* becomes *the fiction necessitates p*, is not needed – Woodward (2010) shows how the safety result can be established given a counterfactual analysis of fictionality – the point emerges that whilst the fictionalist can enjoy the inferential benefits associated with talk of possible worlds, doing so seemingly requires her to understand the concept of fictionality in modal terms, and thereby admit that she cannot provide a thoroughly non-modal analysis of modal concepts. (A further potential cost of establishing (SR) with the help of the modal analysis of the fiction-operator is that she has to regard her fiction as merely contingently false, see Skiba 2019 for an attempt to avoid this.)

The case of the modal fictionalist, then, provides a nice case study not only of the structural and technical difficulties that fictionalist strategies in metaphysics must overcome, but also of the problems that proponents of such strategies face when it comes to delivering the same range of explanatory benefits as are offered by alternative approaches. Moreover, it illustrates the interplay between the choices that the fictionalist makes in constructing her account and the overall assessment of a specific fictionalist proposal in terms of not only the theoretical costs it incurs but also in terms of the explanatory benefits it offers.

References

Armour-Garb, B. and J. Woodbridge. 2010. "The Story about Propositions", *Noûs* 46: 635–674.
Armour-Garb, B. and J. Woodbridge. 2015. *Pretense and Pathology*. Cambridge: Cambridge University Press.
Bricker, P. 2001. "Island Universes and the Analysis of Modality", in G. Preyer and F. Siebelt (eds.), *Reality and Humean Supervenience: Essays on the Philosophy of David Lewis*, 27–55. Lanham, MD: Rowman & Littlefield.
Brock, S. 1993. "Modal Fictionalism: A Response to Rosen", *Mind* 102: 147–150.
Brock, S. 2002. "Fictionalism about Fictional Characters", *Noûs* 36: 1–21.
Brogaard, B. 2006. "Two Modal-isms: Fictionalism and Ersatzism", *Philosophical Perspectives* 20: 77–94.
Burgess, J. 1983. "Why I Am Not a Nominalist", *Notre Dame Journal of Formal Logic* 24: 93–105.
Burgess, J. and G. Rosen. 1997. *A Subject with No Object*. Oxford: Clarendon Press.
Divers, J. 1999a. "A Genuine Realist Theory of Advanced Modalising", *Mind* 108: 217–239.
Divers, J. 1999b. "A Modal Fictionalist Result", *Noûs* 33: 317–346.
Divers, J. 2002. *Possible Worlds*. London: Routledge.
Dorr, C. 2005. "What We Disagree about When We Disagree about Ontology", in Mark Kalderon (ed.), *Fictionalism in Metaphysics*, 234–286. Oxford: Oxford University Press.
Field, H. 1980. *Science without Numbers*. Princeton, NJ: Princeton University Press.

Field, H. 1989. *Realism, Mathematics and Modality*. Oxford: Blackwell.
Fine, K. 2003. "The Problem of Possibila", in M. J. Loux and D. W. Zimmerman (eds.), *Oxford Handbook of Metaphysics*, 161–179. Oxford: Oxford University Press.
Frigg, R. 2010. "Models and Fiction", *Synthese* 172: 251–268.
Gatzia, D. 2010. "Colour Fictionalism", *Rivista di Estetica* 43: 109–123.
Kim, S. 2002. "Modal Fictionalism Generalized and Defended", *Philosophical Studies* 111: 121–146.
Kim, S. 2014. "A Defence of Semantic Pretence Hermeneutic Fictionalism against the Autism Objection", *Australasian Journal of Philosophy* 92: 321–333.
Kroon, F. 2011. "Fictionalism in Metaphysics", *Philosophy Compass* 6/11: 786–803.
Lewis, D. 1968. "Counterpart Theory and Quantified Modal Logic", *The Journal of Philosophy* 65: 113–126.
Lewis, D. 1986. *On the Plurality of Worlds*. Oxford: Blackwell.
Liggins, D. 2008. "Modal Fictionalism and Possible-Worlds Discourse", *Philosophical Studies* 138: 151–160.
Liggins, D. 2010. "The Autism Objection to Pretence Theories", *Philosophical Quarterly* 60: 764–782.
Nolan, D. 1997. "Three Problems for Strong Modal Fictionalism", *Philosophical Studies* 87: 259–275.
Nolan, D. 2011. "Modal Fictionalism", in E. Zalta (ed.), *The Stanford Encyclopaedia of Philosophy* (Winter 2011 Edition), from http://plato.stanford.edu/archives/win2011/entries/fictionalism-modal/.
Nolan, D., and O'Leary-Hawthorne, J., 1996. "Reflexive Fictionalisms", *Analysis* 56: 26–32.
Noonan, H. 1994. "In Defence of the Letter of Fictionalism", *Analysis* 54: 133–139.
Rosen, G. 1990. "Modal Fictionalism", *Mind* 99: 327–354.
Rosen, G. 1993. "A Problem for Fictionalism about Possible Worlds", *Analysis* 53: 71–81.
Rosen, G. 1994. "What Is Constructive Empiricism?", *Philosophical Studies* 74: 143–178.
Rosen, G. 1995. "Modal Fictionalism Fixed", *Analysis* 55: 67–73.
Rosen, G. and C. Dorr. 2002. "Composition as a Fiction", in R. Gale (ed.), *The Blackwell Guide to Metaphysics*, 151–174. Oxford: Blackwell.
Sider, T. 2002. "The Ersatz Pluriverse", *Journal of Philosophy* 99: 279–315.
Skiba, L. 2017. "Fictionalism and the Incompleteness Problem", *Synthese* 194: 1349–1362.
Skiba, L. 2019. "Fictionalism, the Safety Result, and Counterpossibles", *Analysis* 79: 647–658.
Stanley, J. 2001. "Hermeneutic Fictionalism", *Midwest Studies in Philosophy* 25: 36–71.
Van Fraassen, B. 1980. *The Scientific Image*. Oxford: Oxford University Press.
Van Fraassen, B. 1994. "Gideon Rosen on Constructive Empiricism", *Philosophical Studies* 74: 179–192.
Woodward, R. 2008. "Why Modal Fictionalism Is Not Self-Defeating", *Philosophical Studies* 139: 273–288.
Woodward, R. 2010. "Fictionalism and Inferential Safety", *Analysis* 70: 409–417.
Woodward, R. 2012. "Fictionalism and Incompleteness", *Noûs* 46: 781–790.
Yablo, S. 2001. "Go Figure: A Path Through Fictionalism", *Midwest Studies in Philosophy* 25: 72–102.

21
GLOBAL EXPRESSIVISM

Stephen Barker

Expressivism is a doctrine that challenges the idea that representation is central to any explanation of linguistic meaning. Expressivism is usually applied to discourse about value. A standard non-expressivist approach to discourse about value holds that in assertions of sentences like 'truth-telling is good', 'murder is wrong', etc., speakers are representing states of affairs about things being good or wrong, etc. Expressivists deny that in such assertions speakers are, at the level of analysis, representing how things are. Rather speakers are *expressing* affective-states, such as approval or disapproval focused on objects or conditions – Blackburn (1984) and Schroeder (2007). Expressivists claim that when value-sentences appear, unasserted, in logical compounds speakers are not directly expressing attitudes, but are committing themselves to attitude-expression under certain conditions (see Geach 1965; Blackburn 1984, 1993; Barker 2004; and Schroeder 2007).

There are a number of reasons to be attracted to value-expressivism, but the chief one is that expressivism renders talk of value metaphysically unproblematic. The usual understanding of this is that value-expressivism implies there are no values. So, although nihilistic about facts and properties of value, expressivism, on this conception, nevertheless explains how there can be meaningful talk *about value*.

My concern in what follows is whether expressivism can be extended to all domains of discourse and, if it can, to understand what metaphysical conception of reality it implies. Below I set out several conceptions of global expressivism and evaluate their prospects. I then arrive at my favoured conception, according to which *(i)* all statements express non-cognitive states (which are affective-states in the case of value-sentences, otherwise not) but that *(ii)* the cognitive features of assertions – their truth-conditions and inferential features – are recovered not from expressed states but from other features of sentences. I then address the question of metaphysics. You might be worried that if value-expressivism is nihilistic about value, global expressivism should imply a (self-defeating) metaphysical nihilism about everything. I show, however, that global expressivism does not imply metaphysical nihilism, but rather a *metametaphysical nihilism*. Global expressivism implies that there is no ultimate structure or metaphysical nature to reality. That means global expressivism implies that metaphysics, the enterprise that 'limns the true and ultimate structure of reality' (Quine 1960: 221) is deeply misguided.

1 Globalizing expressivism

The term *global expressivism* seems to suggest that the kind of treatment that value-expressivists give of value-sentences is extended to all domains of discourse, viz., about value, modality, justification, causation, ordinary physical objects and their properties – colour, shape, etc. – and so on. The question here is what is the essence of the explanatory approach instanced by value-expressivism that we can extend to all domains? One answer is that we treat all assertions in all domains as expressing motivational states. For example, Schroeder (2007) contends that value-expressivists ought to propose that all atomic sentences involve expression of an affective-state, *being-for*. Being-for is approval of people acting as if p is the case. So in asserting 'the cat is fluffy', U expresses approval of people acting as if the cat is fluffy is the case. Schroeder's theory is not very promising as a route to global expressivism. For one, it presupposes a separate non-expressivist treatment of the contents p that are the focus of *being-for* states. So it's not really global after all. If we are going to globalize expressivism we need some other way of extending value-expressivism.

A better approach proceeds with the following ideas. Value-sentences express affective-states. These are non-cognitive states in the sense that *(a)* they are not beliefs; *(b)* they are not truth-assessable (or truth-apt) mental states; and *(c)* they are not states comprising a commitment to how things are. So we might propose that global expressivism is the idea that all utterances, at the level of analysis, express non-cognitive mental states (non-doxastic, non-truth-apt, and non-representational) of which affective-states are one kind. However, this proposal immediately faces an objection. Suppose we accept the orthodox thesis below – see Schroeder (2007):

> **Inheritance**: Sentences get their content (truth-conditions, inferential properties, etc.) from the mental-states they express.

Given **Inheritance** and the thesis that sentences always express non-cognitive states, then all sentences will lack cognitive-content, which is absurd. Global expressivism must then deny **Inheritance**, and hold instead that sentences gain their cognitive-content from another source, which though connected to the non-cognitive states expressed, is not reducible to those states. (See Barker 2007). This denial of **Inheritance** is crucial to the prospects of global expressivism. We shall see what form this alternative concept of cognitive-content takes below.

The programme for globalizing expressivism has then two components, summed up in *Global-E*:

> *Global-E*: *(i)* All domains of discourse are explained in terms of speakers expressing mental-states, Π, that are non-cognitive. *(ii)* The cognitive-contents of sentences are not derived from the underling states Π expressed by those sentences – **Inheritance** is false – but through some other set of facts about such sentences.

I now develop these two features, *(i)* and *(ii)*, in turn.

The value-expressivist's thesis that value-utterances express affective-states is an empirical hypothesis about the mental machinery underlying value-assertion. (It's not conceptual analysis.) *Global-E (i)* is such a hypothesis extended to all utterances. To articulate this hypothesis, I take the uncontroversial view that the mind-brain is a network of interconnected systems and sub-systems, characterized by input and output patterns, etc. The states expressed by utterances, Π-states, are sub-systems of the great network that each speaker U's mind embodies. Functional-states and systems, including Π-states, are compositional, but not semantically since they are not representational. The Π-state of an atomic sentence 'O is G' is composed of *functional-antecedents*, that is, sub-systems that causally condition production of words. The functional-antecedent of a term 'O' is a file, φ. The functional-antecedent of the predicate 'G' is a relatively-abstract functional-component ξ, which determines the kind of utterance 'O is G' is, viz., value-sentence, sentence about causation, and so on. The Π-state $\varphi + \xi$ is a state-type that through sub-doxastic processes can be tokened in various sub-systems of the speaker's language-functional system and interact with the functional-antecedents of logical-constants. (See Barker 2004, 2007.) Π-states have canonical inputs and non-canonical inputs and their outputs include utterances. Let's give a basic idea of how this works.

We suppose a sentence 'O is good' expresses a state, $\varphi + \gamma$, where φ is 'O''s file and γ is the functional-antecedent of 'good'. The canonical inputs for Π-states of 'O is good' are affect-states. We suppose the mind has a sub-system, AFFECT, that is a disposition to react to features of the world, either with positive-affective response – as in liking something, say a taste – or negative affective-response, as in finding something distasteful – as in finding a taste disgusting. AFFECT can be activated by perceived objects x – for which we have files φ – manifesting positive affective-response to x. In which case, the speaker's functional system enters a state $\varphi + \gamma$, whose main output is production of sentences like 'O is good'. (A negative output from AFFECT induces a Π-state $N[\varphi+\gamma]$, where N is the functional-antecedent of negation. The verbal output of $N[\varphi+\gamma]$ is utterance of 'O is not good'.) That's the basic picture of value-sentence production. But there is another important aspect we should mention. Not all utterances of 'O is good' (or 'O is not good') are based on affective-response. Sometimes we say something is good because someone tells us it's good. I may be told that Frida is a good (or bad) person. That means information pathways underlying testimony are another way through which $\varphi+\gamma$, or $N[\varphi+\gamma]$, can be tokened in U's system. So the Π-state, $\varphi+\gamma$ is not an affective-state as such, but a state whose canonical cause is affect, but which can get tokened in the system through other informational pathways.

The goal of *Global-E (i)* is to generalize this picture to all predicates. Take 'table'. Applying my toy-model, we suppose that use of 'table' by a given speaker is determined in part by a functional-antecedent: TAB. TAB is just a functional sub-system that, through visual, mainly, information, reacts in various ways – on, off, or undecided – in the presence of material things. So if an object, O, for which we have file φ, induces TAB's on-state in a perceptual situation, then this is the canonical cause of a state $\varphi+\tau$, which can induce production of 'O is a table', where τ is the functional-antecedent of 'table'. (Activation of TAB's off-state is the canonical input for tokening of $N[\varphi+\tau]$.) Activation of TAB is not the only way $\varphi+\tau$, or $N[\varphi+\tau]$, can be tokened. Information pathways underpinning reasoning or testimony can lead to the mental tokening of $\varphi+\tau$, or $N[\varphi+\tau]$, which can then generate 'O is (not) a table'. And so on

Take general-terms like 'causes' and 'depends on'. The Π-states of 'O_1 depends on O_2' have as their canonical input output of a functional-system, call it *MANIP*, which is linked to the agent's manipulation of the environment. You get to know about dependency through tweaking things. You manipulate pixels to get face-images to appear on screens. In such cases, MANIP's output is a state involving two files φ_1 and φ_2. This state corresponds to the information that one event is a *recipe* for getting the other. This output from MANIP is the input for tokening of a Π-state $\varphi_1+\chi+\varphi_2$, that has as verbal output utterances like 'the face-image depends on the pixel-pattern'. (See Barker 2012.) Again, non-canonical input to tokening $\varphi_1+\chi+\varphi_2$, can be pathways underlying reasoning and testimony.

That's my very simplified conception of *Global-E (i)* (see Barker 2007, 2011, 2015, 2018). We now have to put this brief sketch within a broader context of understanding. Expressivism is not a *response-dependent* treatment of value-sentences or any other class of sentences. Response-dependent approaches propose that the mental-antecedents of predicates have the core role of *fixing* the extension of the predicate. (See Jackson and Pettit 2002). So, in the case of 'good', U's disposition to affective response, AFFECT, would fix a class of objects – the good things from U's point of view – and so U's assertions of 'O is good' would manifest a representational-state with the content that 'good''s extension contains O. Thus $\varphi+\gamma$ would be a cognitive-state, *contra Global-E (i)*. If we accepted that all predicates work as the response-dependence theorist suggests, then all ∏-states would be cognitive-states, and so global expressivism would become a form of global-response dependence theory.

If we are to retain *Global-E (i)* – ∏-states are non-cognitive – then we must reject the response-dependent conception of predicates. Response-dependence theories of representational states are questionable anyway – see Gauker (2002). We ought to see expressivism as a way of avoiding the downsides of response-dependence. But to do that, and thus maintain that ∏-states are non-cognitive, we have to deny that the functional-antecedents of predicates, such as AFFECT, TAB, MANIP, etc., fix extensions of predicates. But if such functional-antecedents lack this meaning-fixing role, we have to reject **Inheritance**. Rather, as per *Global-E (ii)*, sentences gain their cognitive content – their truth-conditions and inferential features – through some other path. I now examine three versions of global-expressivism that give three different ideas of how cognitive content is acquired.

2 Global-expressivism I

One approach to global expressivism we might adopt is to deploy *semantic minimalism*. Horwich (1990) introduces the concept of truth-minimalism. This is the idea that truth is a metaphysically unproblematic property since it's implicitly defined through platitudes, viz., the conjunction of all instances of the T-schema – *S is true iff P*. Wright (1992) argues for minimalism about *truth-aptness* – the property that a sentence has when it is a fit object of truth-assessment. On this conception, sentences are truth-apt just in case they have disciplined use, are embeddable in most contexts, and enter into the scope of negation. Thus we propose that sentences expressing ∏-states are truth-apt, and being true is just a matter of disquotation. Furthermore, we might claim that utterances of truth-apt sentences expressing ∏-states are assertions. In which case, it is a short step to contending that belief-states are *minimally* dispositions to utter truth-apt sentences S sincerely – see Divers and Miller (1995). To these minimalist proposals about truth-aptness, assertion, and belief, we can also add minimalist treatments of fact and reference. For example, the infinite list of instances of the fact-schema, *There is a fact that P iff P* provide a real-definition of fact, or for reference: *'N' refers to O iff N is O*.

Assuming semantic minimalism, we can derive our first formulation of global expressivism, call it *GE-I*:

GE-I:

(i) All sentences express ∏-states – as per *Global-E (i)*.
(ii) Declarative sentences systematically used by speakers are minimally belief-expressing, truth-apt, assertive, and representational.

Sentences then don't get their truth-aptness from underling states expressed – we deny **Inheritance** – but rather by superficial features of syntax and use.

Does this conception of *GE-I* work? Here are some problems. Part of our idea of the objective world is that there are objective facts of similarity (Lewis 1983). If we accept *GE-I (ii)* then the objective representational properties of sentences are not inherited from ∏-states and predicate functional-antecedents. They must be inherited from minimal features of syntax. But it is completely mysterious how this is meant to happen.

Another concern is that using *GE-I* we have no basis for any claim that speakers mean the same by their predicates. We cannot say that to mean the same is to have exactly the same dispositions to response to reality, since, speakers meaning the same don't always have the same dispositions to response. Defining meaning-the-same in terms of an ideal version of the speaker brings us back to the response-dependent picture.

We might modify *GE-I* by proposing that some Π-states are representational, viz., we accept some form of response-dependence conception of content – see Blackburn (1993) and Sinclair (2012). If we adopt this idea, we have renounced global expressivism. We have global minimalism but local expressivism and local response-dependence.

3 Global-expressivism II (inferentialism)

Price (2007, 2011, 2013) puts forward an alternative picture of global expressivism which shares some features with *GE-I*. Instead of using minimalism he deploys Brandomian (1994) inferentialism. Brandom proposes that the speech-act of assertion is defined as a move in the language-game of giving and taking reasons, that is, of making inferences:

> **Assertion**: U asserts that S iff *(i)* U undertakes to justify S, if asked to, and *(ii)* permits speakers to use S as a premise in arguments.

Brandom sees inference as somehow prior to propositional content (something that's open to question), and defines propositional content in terms of inferential role. Price thinks that all assertions in all domains conform to **Assertion**. Price grafts this general inferentialist model of assertion onto a version of the *Global-E* conception of expressivism, in which, some Π-states are non-representational and some are representational. By representational here we mean tracking or response-dependent fixing of extensions. Price calls this *e-representation*. At the same time, Price thinks that all declarative sentences and predicates are representational in a deflationary, insubstantial sense – he calls this *i-representation*. This is based on Brandom's inferentialist (anaphoric) treatment of truth- and reference-attributions. Price's global expressivism is then:

GE-II:

(i) Sentences divide into those that express non-e-representational Π-states and those that express e-representational Π-states.
(ii) Declarative sentences are systematically defined by (Brandomian) inferential practice, and all exhibit i-representationality.

Is this global expressivism? *GE-II* does not respect *Global-E (i)*. But if we count Brandomian inferentialism as expressivism, then in virtue of being inferentialist it is globally expressivist. We have global-inferentialism mixed with partial expressivism in the sense of *Global-E*.

GE-II is problematic, at least for the following reasons:

a   Introducing e-representational Π-states as judgements about how things are conflicts with inferentialism. Suppose that utterance of 'the cat is on the mat' involves e-representation of a condition in the world. Assertion of this sentence comprises manifesting a mental state of commitment-to-how-things-are, partly constituted by tracking. Assertion so understood does not look like the display of a disposition to inferential activity. So Price's claim that all assertion conforms to **Assertion** is open to question.

b Price assumes that e-representation is a naturalistically definable relation – as in the response-dependent conception. So Price's programme inherits the issues afflicting response-dependence (see Gauker 2002).

c What is the metaphysical status of merely i-represented as opposed to e-represented reality? Price proposes quietism: with the folk we say values exist, and with the philosophers we are silent. But is silence a stable position? We know that if we treat values, epistemic facts, etc., as real, then we get placement problems (Jackson 1998). We cannot fit values, epistemic facts, etc., along with atoms, into one coherent whole, *the real world*. If so, why are we not committed, by *reductio*, to denying that values, etc., exist and thus moving from silence to denial?

4 Expressivism about meaning

GE-I and *GE-II* fail as conceptions of global expressivism, because they cannot explain how we can successfully deny **Inheritance**. I now sketch a third account that, unlike *GE-I* and *GE-II*, allows us to successfully deny **Inheritance**. It does so because it is thoroughgoing in its expressivism. *Global-E* tells us that all domains of sentences are treated expressively. We ought then to be expressivist about meaning-attribution. Meaning-expressivism is, I suggest, the key to global expressivism. Let's see how.

Suppose you understand a language perfectly. Then your understanding is almost automatic. Its automatic nature suggests strongly that a relatively autonomous, sub-doxastic, functional sub-system is in operation, which processes speech. Let's call it *MEAN*. MEAN underlies speakers' spontaneous comprehension of speech (and writing) of known languages. Your assertion of *U's term means table* involves an expression of a Π-state that is ultimately the output of MEAN.

MEAN takes as input perceived patterns of speech (or writing, or symbol production in general), often in the context of a physical environment and non-verbal behaviour. MEAN produces as output *activation* of the mental antecedents for the interpreter H's own meaningful symbol productions. Say H hears '*table*'. H's MEAN activates the functional antecedent, in H's system, for 'table'. By *activation* we mean that in processing U's speech there is a selection of the functional-antecedents that underlie H's potential speech production. These parts of H's functional system *light up*! There is evidence that perception by primates of actions is cognitively underpinned by activation of *mirror neurons*. When one agent perceives another doing *F*, neurological activity that would go on were the agent to perform *F* herself, goes on – see Stamenov and Gallese (2002). The basic conjecture here is that MEAN works analogously to sub-systems in primates governing action perception.

If we are meaning-expressivists then we have to be expressivists about illocutionary acts – like assertion – since assertions are contentful. We might suppose there is a functional-antecedent, call it *PRAG*, whose activation is expressed by claims of the form: *U is asserting that P*. Again, there is an analogy with mirror neurons. We may suppose that PRAG takes as input overt behaviours of speaker U and, as output, activates functional-pathways in H that underpin H's assertions. That means the relevant sectors of H's function-system that would be initiated were H to produce *S* manifesting a Π-state – which is the functional basis for H's asserting that *P* – *light up*. It's this activation that is expressed by H's assertion, *U is asserting that P*, when the assertion is based on perceiving U's behaviour. In this expressivist treatment we are not *analysing* assertion. Rather, we are providing a functional basis of attribution of assertive-status – describing what's manifested (expressed) in such attributions. We are also supposing that assertions are occurrences that *depend* on producing of strings *S* expressing Π-states.

Let's note three features about the meaning-expressivism programme.

a   Meaning is normative (Kripke 1982). If I say U means *table* by her term 'table', then I judge she ought to apply 'table' to this thing before me now, which in my judgement is a table. This disposition to infer that U ought to call something a table is explained through expressivism about meaning. If my functional-system MEAN-processes U's predicate '*table*' in terms of the functional-antecedent τ, underpinning my 'table' then I am disposed to say, *correct* or *incorrect*, to U's use of 'table', based on outputs of *my* functional-antecedent for 'table', which is governed by my functional-system TAB. My judgement that *U is correct to call this table* simply expresses my disposition to call this table, that is, to express output from TAB. Given I am so disposed, and I naturally approve of my own dispositions, I also express my approval of U's calling something table. Hence, I assert: *U is doing something correct*.

b   Meaning-attribution is holistic. In the meaning-attributor H's functional-system, MEAN and PRAG need to interact with other functional-systems that underpin H's general intelligence and theory of the world and persons. Suppose speaker U suffers various kinds of deficits and is prone to mistakes with her use of 'table'. U's disposition to use 'table' and even the kinds of inferences she accepts, don't match those of H, the attributor. Nevertheless, H's system, MEAN, given its internal interaction with more general functional-antecedents, can generate activation of the functional antecedents of H's 'table' as output in processing U's tokenings of 'table'. U's meaning table is not fixed by any single, isolatable disposition (such as Price's e-representational causal-co-variation model or Horwich's (2005) laws of use).

c   Just as meaning something is irreducible, so is meaning-the-same. Suppose U utters a sentence *S* and H interprets and judges that U means that *P* by *S*, which H expresses with her sentence *R*. This activity of interpretation on the present conjecture depends on the mirroring process:

> U utters *S* expressing Π^u and H's system MEAN-processes U's act in terms of her state Π^h, that can be manifested by utterance of *R*.

Our judgement that U and H mean the same is an utterance whose content depend on the fact that our systems MEAN-process both, U's *S* and H's *R* in terms of the same Π-state, Π^{us}. This is not a theory of what meaning-the-same consists in, but what goes on when we assert *U means the same as H*. Instead of offering a theory of what constitutes meaning the same, we offer a theory of what constrains meaning-attribution, that is, of how our postulated functional-systems MEAN and PRAG operate. These functional-sub-systems – and related systems that underpin interpretation – function in the same way across speakers and are non-arbitrary in their operation. It's up to cognitive-science to discover how they function.

Adopting expressivism about meaning means there is no reductive account of meaning, meaning-the-same, or analysis of the normativity of meaning. So, what is global expressivism up to? It cannot be thought of as a semantic theory (*pace* Schroeder 2007). It's no more a theory of meaning than value-expressivism is a theory of value. My suggestion is that it is, as Price would say, a genealogical theory, but one but with a more systematic cognitive-science twist (see Barker 2007, 2018). This is our third version of global expressivism, *GE-III*:

GE-III: From within our interpretative stance – ultimately underpinned by MEAN, PRAG, etc. – we pre-theoretically identify assertions, referrings, dispositions to inference, and so forth. We provide a theoretic characterization of the functional systems upon which this (intentional) activity depends – modelled on *Global-E*. Such a programme must provide a compositional theory of underlying functional states and systems. It's not a compositional semantics but rather a composition metasemantics – a compositional account of the functional-states and mental features that enable talk about everything.

Expressivism, taken to its global limit, becomes an anti-representationalist form of cognitive-science. But it is not eliminativist or reductionist about meaning. Rather, it merely describes – in a compositionally rich way using posited inner functional-systems – the functional machinery upon which such intentional states, capacities to refer or think about, and activities, such as assertion, *depend*.

This anti-reductionism about meaning, representation, and intentional states in general can be summed up:

Intentional-Dependency: Intentional states – O's capacity to think about/refer to x, to assert that *P*, etc. – are irreducible, but obtain in virtue of a complex functional architecture of sub-systems with functional descriptions, embedded in a larger environment.

The obvious objection to **Intentional-Dependency** is that this requires a duality of irreducible intentional-states dependent on non-intentional conditions. Is this not a pernicious metaphysical dualism? How is it that intentional states, which, on the *GE-III* view, are irreducible non-natural states, get to depend on natural (causal-dispositionally defined) functional states? What *is* meaning and representation in the end?

These are deeply metaphysical questions, whereas *GE-III* is a speculative proposal about how a programme in cognitive science could go and not per se metaphysics. The metaphysical concerns I have just voiced all reflect intuitions about the placement of things (in the sense of Jackson 1998), with questions like what is intentionality and how does it arise in the physical world? These are questions and concerns of metaphysics. It's with respect to this enterprise of metaphysics that the radical potentiality of *GE-III* reveals itself. By being expressivists in the way that *GE-III* proposes, we have in our hands a metasemantic critique of the very kind of metaphysical thinking that lies behind placement problems and metaphysical objections to **Intentional-Dependency**.

5 Metasemantics meets metametaphysics

Expressivism is typically presented as having a metaphysical payoff. As we noted above, standard value-expressivism embraces metaphysical nihilism about subject matter. Values don't really exist. They have, at best, a (fictional) *quasi-real*, as-if status (Blackburn 1993). Price thinks the payoff of his global expressivism, *GE-II*, is quietism about values, modal-facts, etc. That means that with the folk we say values exist, and with the philosophers we are silent. We deflect metaphysical inquiry. What is the metaphysical payoff for *GE-III*? The payoff is a *metametaphysical* nihilism. We don't, as in standard expressivism, deny that values exist, nor do we remain quiet. Rather, we claim there are values but deny that values have an ultimate metaphysical nature. Moreover, this denial of ultimate nature applies to everything. There is no ultimate nature to reality in which all things have a place. That's why Jacksonian placement problems can never get started.

What is this no-ultimate-nature conception and why does it follow if we accept *GE-III*'s programme? First, I describe the metaphysical view of reality that generates placement problems, and then argue that accepting this metaphysical view requires a metasemantic concept of language/thought, which is rejected in *GE-III*.

Positive metaphysics

A reluctance to allow values into *one's ontology* is not due to the mere recognition that we seem be committed to *their being*, but from trying to form a conception of what values *really* are. Values concern us because we think, in order to accept their reality, we have to identify what constitutes them – what it is to be a value, or in other words, what value consists in or what its ultimate nature is. To say what values ultimately are is to point to their *real definition*. Real definition is ontological analysis. It's what something is in terms of ultimate building blocks assigned to reality. So just as science affirms *water is H_2O*, in metaphysics, we make claims like properties are ultimately sets, and so on. Real-definitions are identities but with an explanatory order:

O is ultimately $X =_{df}$ *(i)* $O = X$ and *(ii)* X's features along with those of other things explain all of O's manifest characteristics.

So if we think water is H_2O, then features of water – boiling point, transparency, indeed, the reference of the term 'water' etc., are explained by features of H_2O, given laws and other atomic features. Similarly, in metaphysics, we assert identities like:

I_1: The property, F-ness, is the set of F-things.

I_2: The proposition that P is the set of worlds in which P.

I_3: The state of affairs that A is F is the non-mereological fusion of A with F-ness.

If we think properties are ultimately sets, then we think features of sets can explain manifest features of properties. If we think propositions are ultimately sets of worlds, then sets of worlds between explain all manifest features of propositions. And so on.

Positive metaphysics is an explanatory project. That's why, in analytic metaphysics, we are deeply concerned with applying Occam's Razor to questions of reality. Occam's Razor is an explanatory principle. The explanatory project afoot is summed up in **Ultimate-Success**:

Ultimate-Success: A completed ontological system is a selection of metaphysically simple things, which allow us to define, using metaphysically simple relations, everything.

You only have to look at the ambitions of Lewis's (1996) framework, a paradigm of positive metaphysics, to see total explanation is the goal. Quine (1960) clearly falls into this pattern as well. If you think reality must conform to **Ultimate-success,** then you are accepting **Real**:

Real: If we quantify over xs, then we need to provide a real definition of what xs are.

The real definition better be in terms of ultimate building blocks, the primitives of whatever system is chosen.

Different metaphysical systems choose different primitives through which to define everything. Primitives are metaphysically simple things – they depend on nothing else but themselves for their natures.

Realizing **Ultimate-Success** is fraught with difficulty. Hence Jackson's (1998) *placement problems*, which concern how entities are meant to *fit* into the physical world. Entities that don't fit are characterized as *queer* (Mackie 1977) or *ontological danglers* (Smart 1959). If entities cannot be defined, then we must exclude them from the domain of quantification, and save the language that apparently talks of them. Hence traditional forms of expressivism have been applied to talk of values for just this reason. Indeed, minimalism is just the idea that real definitions should use platitudes. Minimalists don't give up **Real**. Rather, they seek instead to limit the impact of things by making real definitions as low-key as possible.

Mind-body dualism is a serious worry if we accept **Ultimate-Success**. Given **Real**, dualism is bad because the real-definition of the mental and that of the physical are entirely separate from each other. If so, why are the mental and the physical connected? Why does the mental arise at all? Following **Real**, we can only find **Intentional-Dependency** highly objectionable.

Metasemantic assumptions

Fortunately, once we accept *GE-III* we can dispute the metaphysical argumentation, based on **Real**, that threatens **Intentional-Dependency**. Here's how. At the heart of the enterprise of positive metaphysics is affirming real-definitions. So, to *think* any positive metaphysical view as a truth a theorist has to be able to think, as truths, the real-definitions that characterize the position – such as I_1-I_3 above. I claim, however, there is a metasemantic condition on our ability to think a true real-definition.

Say U accepts 'O is ultimately X' expressing the truth that O is ultimately X. If this sentence is true, it must be a necessary truth. But the only way that it can be a necessary truth is that 'O''s reference to O is essentially connected to U's reference to X. That requires, given reference depends on information flow from the world, that X is part of the explanation of how 'O' in U's mouth gets to mean what it means. There two basic ways in which X can have this explanatory role:

   (i) 'O' is introduced by a description D and X uniquely satisfies D.
   (ii) U's referring to O holds in virtue of U's mind being causally or informationally impacted on by X.

Let's say that if either *(i)* and *(ii)* hold, then U's mind *latches onto* X, in which case:

   Identity: For U to think/assert/judge truly a real-definition, 'O is ultimately X', then U's capacity to use 'O' to refer to X is grounded in U's mind's *latching* onto X, which means some descriptive, causal, etc., mechanism securing the referent.

So, what's required for us to think (as truths) any theoretical identities, like I_1-I_3 above, is that the left-hand terms involved conform to **Identity**. That means, the terms, 'F-ness', 'the proposition that P', and 'the state of affairs that P' must all work by latching onto referents.

If this is correct then positive metaphysics, which seeks analysis of all things in accordance with **Ultimate-Success**, conforming to **Real**, is committed to:

   Mirror: The fact that U can think about/refer to x depends on U's mind *latching* onto x.

In effect, positive metaphysics is committed to the idea that our capacity to use language quantifying over entities depends on our minds latching onto – either descriptively/causally, etc. – these entities. Only then can we accept **Real**, as a binding constraint on our conceptions of reality.

It's precisely **Mirror**, however, that is rejected in *GE-III*. This rejection is not universal. Many terms conform to **Mirror**, that is, by some form of causal-descriptivism. So, for example, the following are non-analytic true identities – though only the second is real-definition:

I_v: Hersperus is Phosphorus.

I_w: Water is the stuff constituted by H_2O.

Your capacity to use 'Hesperus' and 'Phosphorus' to refer to x = Venus depends on functional-antecedents being impacted on by Venus itself. Similarly, your capacity to use 'water' and 'stuff constituted by H_2O' to refer to x depends on functional-antecedents being impacted on by the referents of these terms.

GE-III implies a partial but significant denial of **Mirror**, which is dictated by acceptance of *Global-E*. Your capacity to use a sentence with propositional content that P or use a predicate to mean F-ness, does not depend on your mind *latching onto* the content that P (or state of affairs) or property of F-ness:

> **No-Mirror:** O's capacity to mean/refer to/think about F-ness and her capacity to think/say/judge that P depends on functional-antecedents and formation of Π-states and their interaction with an environment, but none of this involves U's mind latching onto (in the sense specified above) properties – F-ness – or the state of affairs/proposition that P.

If we accept **No-Mirror** we cannot maintain a commitment to the project of positive metaphysics. That's because we know that we cannot affirm truly any thesis about what properties, propositions, or states of affairs, etc., are, as in I_1-I_3 above. So, all metaphysical hypotheses (plural theoretical identities) like those below are false:

H_1: Properties are ultimately sets.

H_2: Propositions are ultimately sets of worlds.

H_3: States of affairs are ultimately non-mereological fusions.

What's the right view about properties, propositions, states of affairs, etc.? No view is right, since to have the right view we would have to be able to think a true theoretical-identity claim, but the metasemantic conditions that have to be in place in order for us to be able to think truly any such claim are absent. So no view is the correct view. So, to say 'there is no true theoretical-identity for x' is the formal mode of saying 'there is no metaphysical nature to x'. The pursuit of seeking the ultimate nature of the general categories of reality must be a misguided enterprise.

So the reason why philosophers have failed to uncover what properties ultimately are is because properties have no ultimate nature. The reason why theorists have failed to explain what property-instantiation is – think of the third-man argument – is because there is no ultimate nature to property-instantiation. What philosophers are seeking – ultimate nature – is not there.

If this is correct, then we have to accept the idea of reality without ultimate real definition, which I sum up thus:

> **Emptiness**: There is no ultimate set of natures/entities that explain all things. For each x that we quantify over, it's false that x is ultimately X, for whatever X qua (proposed) ultimate nature.

The emptiness view is the complete negation of positive metaphysics summed up in **Ultimate-Success**. It's a negative metaphysical conception of reality.

Emptiness is not nominalism – we are not denying that properties, propositions, states of affairs, etc., exist. Nor is it deflationism. Deflationism is the thesis that the real definition of properties, propositions, states of affairs, facts, etc., are platitudinous. But the current view rejects even the idea that these things have platitudinous real-definitions. **Emptiness** is not quietism. To the question *Do values or properties, etc., really exist?* the answer is *yes*. We just deny they have any ultimate nature. On the emptiness view, reality is structured – science, common sense, folk psychology, etc. all describe features of the world that are really there. It's just this world of structured beings lacks any ultimate nature. Although **Emptiness** rules out the theoretical identities of metaphysics, it does not rule out those of science. As suggested above, identities like I_w, *water* = H_2O, are not ruled out. Water itself enters into the explanation of how you think about water (**Mirror** applies). (I submit that this holds generally for scientific theoretical-identities, though this claim cannot be established here.) But, of course, pursuing the question, *What is H_2O ultimately?*, bottoms out in reality for which there is no true-theoretical identity. All reality lacks an ultimate building-block nature.

We can now return to our question about intentionality grounded by functional states, viz., **Intentionality-Dependency** (§4). Mental states, given that they involve representation/reference, etc., are immediately metaphysically empty. There is no theory about what referring to tables, thinking of Vienna, or feeling sad about Brexit is. Similarly, there is no theory about what *depending* is. We are accepting in **Intentionality-Dependency** that the mental depends or supervenes on the mental, but the resulting duality is not a (positive) metaphysical one, which would require distinct ultimate natures mysteriously connected. The dependency of **Intentionality-Dependency** is a (hopefully) verifiable, empirical hypothesis about reality, but there is no metaphysical constitution to it – as with all things.

We have finally assembled an understanding of how *GE-III* enables us to deny **Inheritance** and thus avoid response-dependence (see §1). Doing so, I suggested, is the key to global expressivism. Expressivism about meaning given the metametaphysical thesis of emptiness enables us to hold that cognitive-content of utterances is not inherited from ∏-states. Still we have explained how cognitive-content, though irreducible, is dependent on the underlying system with its patterns of ∏-states.

Some think that global expressivism might operate as a metasemantics that can co-exist with the formal, science of semantics – the enterprise of assigning means, conceptualized in terms of the syntactic and semantic forms of logic, to provide compositional theories of truth-conditions for languages. See MacFarlane (2014). But *GE-III* is not consistent with semantics as a science – Barker (2007). Semantics is (positively) metaphysical insofar as it requires theoretical-identification of meanings with *formal entities*, for example, predicate meanings with extensions or propositions with sets of worlds, etc. *GE-III* with its emptiness conception of meaning won't allow that. Instead of compositional semantics it allows only a compositional metasemantics.

The emptiness view, the idea of reality as lacking ultimate nature, is not an established position in Western philosophy, and particularly not in contemporary analytic metaphysics. It's not eliminativism, fictionalism, minimalism, or internal realism. To find any precedent for this kind of view we need to leave Western philosophy and enter Eastern traditions. There we shall find, in the Buddhist canon, some echo of *GE-III*'s emptiness. This is Nāgārjuna's conception of *śūnyatā*, the idea that all things are empty of *svabhava* or *own nature*, which is a central doctrine of madhyamaka Buddhism. This seems to be a form of nihilism about ultimate reality. Commentators are divided, however, on how exactly to understand Nāgārjuna's *śūnyatā*. (See, for example, Garfield 1995; Westerhoff 2009; Priest 2013; Siderits 2013; and Bliss 2015.) My suggestion is that *GE-III*'s emptiness might be another understanding of what *śūnyatā* could be. We cannot assess that conjecture here.

References

Barker, S. J. 2004. *Renewing Meaning*. Oxford: Oxford University Press.
Barker, S. J. 2007. *Global Expressivism*. Book (Online Preprint Publication).
Barker, S. J. 2011. 'Faultless Disagreement, Cognitive Expressivism, and Absolute, but Non-Objective Truth', *Proceedings of the Aristotelian Society* 110: 183–199.
Barker, S. J. 2012. 'Expressivism about Making and Truth-Making', in B. Schnieder and F. Correia (eds.), *Grounding and Explanation* (Cambridge: Cambridge University Press): 272–293.
Barker, S. J. 2015. 'Expressivism about Reference and Quantification over Non-existent Entities without Meinongian Metaphysics', *Erkenntnis* 80: 215–234.
Barker, S. J. 2018. *Metaphysical Emptiness and the Mind*. Ms.
Blackburn, S. 1984. *Spreading the Word*. New York: Oxford University Press.
Blackburn, S. 1993. *Essays in Quasi-realism*. New York: Oxford University Press.
Bliss, R. 2015. 'On Being Humean about the Emptiness of Causation', in Y. Deguchi, J. Garfield, and G. Priest (eds.), *The Moon Points Back* (Oxford: Oxford University Press): 67–95.
Brandom, R. 1994. *Making it Explicit*. Cambridge, MA: Harvard University Press.
Divers, J. and Miller, A. 1995, 'Platitudes and Attitudes: A Minimalist Conception of Belief', *Analyst* 55: 37–44.
Garfield, J. 1995. *Fundamental Wisdom and the Middle Way: Nāgārjuna's Mūlamadhyamakakārikā*. New York: Oxford University Press.
Gauker, C. 2002. *Words without Meaning*. Cambridge, MA: MIT Press.
Geach, P. 1965. 'Assertion', *Philosophical Review* 74: 449–465.
Horwich, P. 1990. *Truth*. London: Blackwell.
Horwich, P. 2005. *Reflections on Meaning*. Oxford: Oxford University Press.
Jackson, F. 1998. *From Metaphysics to Ethics*. Oxford: Oxford University Press.
Jackson, F. and Pettit, P. 2002. 'Response-Dependence without Tears'. *Noûs* 36. *Philosophical Issues 12 Realism and Relativism*: 97–117.
Kripke, S. 1982. *Wittgenstein on Rules and Private Language*. Cambridge, MA: Harvard University Press.
Lewis, D. 1983. 'New Work for a Theory of Universals', *Australasian Journal of Philosophy* 61: 343–377.
Lewis, D. 1996. *On the Plurality of Worlds*. Oxford: Blackwell.
MacFarlane, J. 2014. 'Review of Huw Price's *Expressivism, Pragmatism, and Representationalism* (Cambridge University Press)', *Notre Dame Reviews*, University of Notre Dame.
Mackie, J. L. 1977. *Inventing Right and Wrong*. New York: Penguin.
Price, H. 2007. *Realism without Mirrors*. Oxford: Oxford University Press.
Price, H. 2011. 'Expressivism for Two Voices', in J. Knowles and H. Rydenfelt (eds.), *Pragmatism, Science and Naturalism* (Zürich: Peter Lang): 87–113.
Price, H. 2013. *Expressivism, Pragmatism, and Representationalism*. Cambridge: Cambridge University Press.
Priest, G. 2013. 'Between the Horns of Idealism and Realism: The Middle Way of Madhyamaka', in S. Emmanuel (ed.), *A Companion to Buddhist Philosophy* (Malden, MA: Wiley-Blackwell): 214–222.
Quine, W. V. 1960. *Word and Object*. Cambridge, MA: MIT Press.
Schroeder, M. 2007. *Being For*. Oxford: Oxford University Press.
Siderits, M. 2013. *Nāgārjuna's Middle Way: The Mulamadhyamakakarika*. Boston: Wisdom Publications.
Sinclair, N. 2012. 'Expressivism and the Value of Truth', *Philosophia* 40: 877–883.

Smart, J. J. C. 1959. 'Sensations and Brain Processes', *Philosophical Review* 68: 141–156.
Stamenov, M. and Gallese, V. (eds.). 2002. *Mirror Neurons and the Evolution of Brain and Language (Advances in Consciousness Research 42)*. Amsterdam: John Benjamins Publishing Co.
Westerhoff, J. 2009. *Nāgārjuna's Madhyamaka: A Philosophical Introduction*. Oxford: Oxford University Press.
Wright, C. 1992. *Truth and Objectivity*. Cambridge, MA; London: Harvard University Press.

22
HYLOMORPHIC UNITY[1]

Anna Marmodoro

How many unities?

Aristotle's hylomorphism has been discussed – more intensely than ever in recent decades – as a viable option in metaphysics, and 'hailed' as generating solutions in a variety of domains of philosophical inquiry, in particular in the field of the philosophy of mind. Yet, I submit that the division and unity of hylomorphic compounds, as Aristotle conceived of them, require further critical investigation.

One of the greatest metaphysical insights that Aristotle contributed to the history of philosophy is that objects may be partitioned in two ways: into *parts* and into *abstracta*.[2] The latter kind of division has not received due attention among contemporary extensional mereologists (who advocate division into parts only), and even in neo-Aristotelian quarters. Which of the types of division one applies to objects determines which type of explanation one offers for the unity of such objects – it is one thing to explain the unity of parts and quite another to explain the unity of abstracta. Further, there are many ways one can divide an object into abstracta; for instance, by abstracting from the object any of the forms Aristotle classifies in his categories, e.g. by abstracting away a quality from an object, or by abstracting away the substantial form, leaving the material substratum. In all cases, abstracting away an abstractum leaves behind an abstractum. Thus, e.g. abstracting 'being musical' from a musical subject leaves behind the abstractum 'a-musical-subject' (and not a non-musical-subject, which is not an abstractum but a concrete individual, such as Socrates).

Contemporary hylomorphists who recognise only division of objects into parts consider hylomorphism as such type of division; namely, they consider matter and form as parts of the hylomorphic compound. However, Aristotle explicitly denies that matter and form are parts; he writes in the *Categories*: "By 'in a subject' I mean what is in something, not as a part, and cannot exist separately from what it is in" (1a-24–25). To understand Aristotle's position, we need to realise that for him parthood presupposes discreteness and separateness. Discreteness separates, and separateness divides. Thus continuous entities, such as substances, do not have parts. Every substance is one and unified. As an illustration of this point, let us consider a colour spectrum and assume that is continuous; does it have parts? For Aristotle it doesn't, because it has no divisions and hence nothing separates any discrete parts. But it has differences (in colour); can we turn its differences into divisions? My answer is that yes, we can; we can divide a spectrum into its colours. If this is a physical division of the spectrum into coloured parts, then we will not have a spectrum any more, but only parts that are derived from it, and hence we could not show these parts to be parts of the spectrum. On the other hand, if the division of the spectrum is by abstraction, on the basis of its colours, then we are, in a sense, assuming the spectrum's continuity and also denying it; we assume that there is continuity but also discrete parts which are not physically there in the spectrum as parts.

As we will see below, there is some confusion in the use of the term 'part' in the literature. I will call the parts divided by abstraction *metaphysical parts*, which are abstracta but not discrete parts since they presuppose the continuity of the substance. By contrast, what I will call *physical parts* are discrete, and can be derived by physical division of a substance, but they are not present, discretely, in the substance. Abstracta, including metaphysical parts, are dependent on the whole they are abstracted from, but discrete physical parts that can be derived from a 'continuous' substance are not.[3]

In addition to clarifying which type of part is relevant for understanding Aristotle's hylomorphism, I urge that we need to critically re-examine certain assumptions we make in our study of Aristotle's theory. I will briefly outline here below which of such assumptions I will probe in this chapter.

A primary matter-to-form relation? I will argue that, contrary to how Aristotle presents his view to us and how it has been received in the scholarly tradition, an Aristotelian substance is *not* a hylomorphic composite of matter and form. Rather, I will show that each substance is a cluster of overlapping hylomorphic composites of matters and forms. I will argue that i) there are different types of form that Aristotle individuates by abstraction in a substance, which serve different metaphysical purposes, in relation to the organisation of the substance; ii) there are also different types of matter that remain, after abstraction, in a substance, and they, too, serve different purposes; and finally iii) there are various composites of forms and matters that we find by abstraction in a substance, each of which is hylomorphically composed and articulates significant structural features of the substance. I will conclude from this analysis that the fact that there are different types of form to be abstracted from a substance, leaving behind different types of matter, undermines in various ways Aristotle's attempts to account for the unity of a substance along a hylomorphic model, whereby a matter M and a form F are somehow 'bound up' into one by a primary matter-to-form relation. Aristotle suggested there is such a relation, which is responsible for the unity and oneness of a composite substance. However, his explanation introduces what I call *double hylomorphism*, which I show to be a circular account of the oneness of a substance. The hylomorphic composite is unified into one, by the form of the matter; this presupposes the oneness of the form; but Aristotle assumes that the form is unified into one hylomorphically (genus to species as matter to form) in its definition. This account is circular; for, if the hylomorphic unity of a substance derives from the unity of its form, the unity of the form cannot be hylomorphic, lest a regress ensues. What is Aristotle's hylomorphism then, if not a metaphysical account of the unity of substances? I submit that we should understand hylomorphism as a *schema* of substantial division. I will show that we can think of such a schema as applicable to a substance multiple times, in the explanation of different structural metaphysical features of the substance, such as essence *versus* substratum; function *versus* process; subject versus property; soul *versus* body; intelligible *versus* indefinite; etc., which are all individuated by abstraction.

As potentiality to actuality? Aristotle claims that the 'relation' between matter and form in a hylomorphic compound is analogous to that of potentiality to actuality. I will show that a close investigation of the notion of potentiality in Aristotle's metaphysics reveals that there are different types of potentiality at work in his system (where each is more fundamentally different from the other than Aristotle suggests); and that there are in consequence different types of 'relations' between potentiality and actuality. I call the two general types of potentiality *causal potentiality* (C-Pot) and *metaphysical potentiality* (M-Pot) respectively. C-Pot is the potentiality of the physical powers (such as e.g. heat, magnetism, etc.), which account for causal interactions within a substance and among substances; whilst M-Pot is the potentiality of the material substratum to be en-formed. I will show that, in accounting for the unity of the hylomorphic compound, Aristotle assumes that M-Pot is both *directed* (towards the form) and is *incomplete* (without the form). *Contra* Aristotle, I will argue that, when directed, potentiality is not incomplete (for instance magnetic powers attracting nothing are not incomplete; they are simply unactivated); and when incomplete, the potentiality of matter without form is not directed (for instance the marble of the statue is not directed towards being a statue). Therefore, M-Pot cannot do the unification work it is called upon to do by Aristotle, as if it were potential, as incomplete and directed.

The challenge: Aristotle is firmly committed to the stance that substances are *par excellence* unities;[4] I claim that his theory of substance, as it stands, does not deliver a sound account of substantial unity. I will supply such an account, which derives from principles within Aristotle's metaphysics, but differs from the account given to us by Aristotle. To understand the unity of a substance, I submit that we need distinguish between its *being united, being whole* and *being one*. I will argue that *causation* unites its relata into a cluster of many inter-connected physical parts; *holism* unifies the many, under a principle of re-identification, into a whole of many *metaphysical parts*;[5] and finally *individuating concepts* fuse the metaphysical parts by converting them to properties of a single substance. I will argue that Aristotle came very close to having a full account of the oneness of a substance, but fell short of it. He unified matter and form in a substance definitionally, but did not explain and justify the oneness of the definition.

Hylomorphic divisions

I will start by briefly outlining my understanding of the ontological status of matter and form in an Aristotelian substance, following Scaltsas (1994) (but with significant differences regarding how to interpret substantial unity, which I will explain in the penultimate section of the chapter). I interpret Aristotle as holding that given a substance, we discover its hylomorphic compositeness *by abstracting form from it*, thereby dividing the substance into two, leaving matter behind (as per the procedure that Aristotle describes in *Metaphysics* VII.3).[6] Different types of abstract entities (a form, a part, a type of matter, etc.) may be individuated from the same substance by means of abstraction. Aristotle generates multiple hylomorphic divisions, all by abstraction, which introduce *many different types of form and of matter*[7] in the make-up of a substance.[8] These different types of form and matter do different kinds of metaphysical explanatory work in a substance. I will give here some examples of form and matter of such different types in a substance, starting with forms:

i *Instantiated particular forms* (or property tropes, in our terminology) qualifying particular substances; e.g. the cold of this ice cube, which ceases to be when the ice cube melts.
ii *Universal forms* recurring in many particular substances at once, which explain qualitative resemblance among substances; e.g. being cold.
iii *Powers or dispositions*, whose metaphysical job is to account for the capacity of their possessor to bring about or suffer change;[9] e.g. the power of cold of an ice cube, to make other material things around it cold.
iv *The substantial form*, which defines what type of substance a given particular is;[10] e.g. being a human being.

v *Functional forms*, which are constituted by underlying states; e.g. the nutritive capacity of a living animal body, or the capacity to remember scents or beliefs.
vi *Perceptible forms* of substances, which affect the 'mean' of our perceptual organs 'without matter';[11] e.g. the coldness we perceive of an ice cube.
vii *Intelligible forms*, which enable us to cognise external objects; i.e. the forms of an external object present in the passive mind of the cognisor, such as of the cold ice cube.[12]
viii *Mental forms*, which cause, *qua* mental content, physical changes in the agent's body, by affecting her *pneuma*; e.g. the thought of a lion causes a person's shivering.[13]

The list could be expanded; but what matters for present purposes is that different acts of abstraction may be performed on a given substance, and in particular, more than one type of form may be abstracted away from the substance. Abstracting more than one type of form from a substance does not however 'leave behind'[14] each time the same type entity, only more depleted of qualities. Abstracting more than one form, where the forms are of different types, 'leaves behind' different types of underlying substratum (as different between them as e.g. a person is from their body). Hence, I submit, there are different kinds of entity playing the role of 'matter' in an Aristotelian substance. They roughly correspond to any type of entity that can play the role of an underlying substratum in a substance, underlying in any metaphysical kind of way (as subject possessing properties, or as substratum constituting the substance, or as body grounding functional states, etc.). Some of these 'matters' are correlative to types of form, as per above; but some are introduced to solve specific metaphysical problems (e.g. explaining the non-recurrence of abstract entities, e.g. of a perfect geometrical figure I am thinking of this very moment). I will here give an indicative, but not exhaustive list of different types of matter, to illustrate the point. In a substance there is:

a *Body*, namely the concrete physical substratum of a substance that underlies organisational or functional form in the substance; e.g. the marble that underlies the form of a statue.[15]
b *Proximate matter*, which is matter at the penultimate stage of a transformation in the generation of a substance; e.g. the flesh and bones that underlie the form of a tiger.[16]
c *Quantity of matter*, namely the particular matter that survives in a transformation, e.g. the quantity of gold in the ring that is reshaped into a seal; or even when the type of the matter does not survive, e.g. as in the case of the quantity of bread one eats that survives and becomes flesh. The quantity is quantitatively specific, e.g. 100 grams, and is non-recurrent and particular.[17]
d *Physical potentiality*, namely what underlies the cyclical transformation of the elements (water, fire, earth and air) into one another; e.g. what is common between water and air, when water transforms into air and then into fire, i.e. when from the wet and cold comes the dry and cold, and then the dry and hot.[18] This is a type of ultimate material substratum, where what remains is not body, but rather, *physical potentiality* underlying the elemental powers such as cold or dry; and yet, it is not quite propertiless, *pure potentiality*.
e *Pure potentiality*, namely the material substratum that underlies *any* type of matter, to which all the properties of matter belong. This is the potentiality that the so-called 'stripping-off' thought experiment of *Metaphysics* Z.3 reaches.
f *Intelligible matter*, which underpins the particularity of abstract entities, e.g. when a mathematician is proving a theorem, thinking of a particular perfect triangle,[19] or a philosopher is contemplating, e.g. Plato's Form of the Good.

There are also further types of matter that one can individuate in Aristotle's system, by abstracting further types of form from a substance; for instance, by abstracting all of a substance's perceptible forms, or all its intelligible forms. The list I have given above is intended to be illustrative rather than exhaustive. The point I want to make is that, on account of the multiplicity of forms and of matters there are in a given substance, a substance is *not* a compound of *a* matter and *a* form, contrary to the received view; rather, I submit, a substance is *a cluster of hylomorphic compounds*. (I will return to this issue in the final section of the chapter.)

The potentiality of matter

Aristotle attempts to account for the 'relation' between matter and form by analogy to the 'relation' between potentiality and actuality. There are two kinds of potentiality in his ontology, as he notes in *Metaphysics* IX.6 but does not explain, save for saying that the one is analogous to the other: "as someone seeing is to a sighted person with his eyes closed, as that which has been shaped out of some matter is to the matter[20] from which it has been shaped [e.g. as statue is to the marble it has been made from]" (1048b1–3). The two kinds of potentiality Aristotle distinguishes are the potentiality of causal powers, such as the potentiality of a fire to heat; and the potentiality of the material substratum to receive form and be formed by it, such as the potentiality of marble to be a statue. I call the first *causal potentiality* and the second *metaphysical potentiality*; or C-Pot and M-Pot respectively for brevity. Aristotle's thought in *Metaphysics* IX.6, as I reconstruct it, is that we come to understand the metaphysical potentiality of matter in a substance by studying the physical potentiality of causal powers. Aristotle, however, does not articulate what is, and what is not, analogous between e.g. the heat of a fire and the marble of a statue. I will show that the analogy is somewhat misleading.

What is the conceptual relation between C-Pot and M-Pot? They differ in the following respects (of relevance to the present argument): in their ontological status; in their incompleteness; and in their directedness. With respect to their ontological status, M-Pots (such as the marble of this statue) are *abstract* entities, because they are conceived of (by abstraction) as 'missing' the forms of the respectively generated substances. By contrast, C-Pots (e.g. the causal powers of the fire to heat) exist *concretely* as such – as entities in the world (even if they are inactivated potentialities).

Further, it might seem *prima facie* that C-Pots and M-Pots share a certain type of 'incompleteness', each missing a form: a C-Pot is missing the form of its activation, namely the form its potentiality is aiming at (e.g. when the power of heat in a hot metal is not heating any other object); whilst an M-Pot is missing the form that qualifies it in the substance (e.g. the wood is missing the form of table). The analogy would be that activation, in the first case, and qualification, in the second, 'complete' each of the respective potentialities. However, a metaphysical analysis of the two cases reveals that C-Pot and M-Pot do *not* share any incompleteness. M-Pots are indeed incomplete in the sense that they are what is left behind when forms are taken away by abstraction from substances. This is the sense of incompleteness that division by abstraction generates.[21] However it would be incorrect to think that C-Pots, too, are thus incomplete. Causal powers are defined, essentially, by the form of their manifestation or exercise (e.g. heating).[22] However, their definition does *not* express any incompleteness of their potential states; what the definition does is only to describe the nature of the potential. This is worth emphasising, because it is easily misunderstood: causal powers in potentiality are not incomplete; defining them in terms of their unactualised forms does not express what is missing from them, but only helps us epistemologically to describe that the nature of the potentiality is of the kind that manifests by attaining such and such actualised form. There is nothing incomplete about e.g. a magnet that is not interacting with any magnetic charge, any more than there is anything incomplete about a car that is not running.

Third, C-Pots are thought (by most ancient and contemporary metaphysicians alike) to have some kind of directedness towards their manifestation or exercise (e.g. the hot is directed towards heating).[23] By contrast, M-Pots do not have any tendency or propensity to fulfil themselves as potentialities; they are not directed towards a form in the way that C-Pots are directed towards their exercise or manifestation. For instance, the bare substratum of *Metaphysics* VII.3, as an example of an M-Pot, does not tend to acquire form; this is in contrast to the power of e.g. a magnet, which is directed towards its activation, namely, towards attracting metal objects (when in appropriate conditions).

Finally, M-Pots do not interact in any way, and do not get activated. This is in contrast to C-Pots which interact with their partner potentialities (e.g. two electromagnetic powers),[24] and activate each other. Aristotle says "matter, which is potentially all the particulars" (*DA* III.5, 430a11), but this cannot be understood as matter somehow tending towards becoming these particulars. With the benefit of this analysis, pointing to the disanalogies between the two types of potentiality, can we understand the use of 'potentiality' as the 'metaphysical glue' of hylomorphic composite substances, as Aristotle suggests?

Matter:form = potentiality:actuality

We saw in the previous section that Aristotle talks about the hylomorphic 'relation' between matter and form in a substance as analogous to the 'relation' between the potential and the actual. How can we understand his claim that matter in a substance is the potentiality whose actuality is the form? In what sense is the matter (e.g. marble) potentially the form (a statue)? In *Metaphysics* VIII.6 Aristotle makes some important remarks, which I report here in full:

> ... people look for a unifying formula, and a difference, between potentiality and actuality. But, as has been said, the proximate matter and the form are one and the same thing, one potentially, the other actually. Therefore to ask the cause of their being one is like asking the cause of unity in general [including the unity of an underlying material substratum and form]; for each [hylomorphic] thing is a unity [where each thing consists of material substratum and form], and the potential and the actual [including material substratum and form] are somehow one. Therefore there is no other cause here unless there is something which caused the movement from potentiality into actuality.
>
> *(1045b7–22)*

Clearly, when Aristotle writes in the above passage that we are here discussing unity in general, he means that we are discussing the unity of all the types of example he mentions above, including that of proximate matter and form, and also extending this to the underlying material substrata of individual things and their forms (e.g. marble and the form of a statue, or bodily elements and the form of animal). He explains the cause of their unity, in general, by explaining that the potential and the actual are in some sense one. What then is the potential here? It is the underlying material substratum of a substance; which, in the case of artefacts like the bronze triangle, is identical to the proximate matter, i.e. to the quantity of matter from which the artisans made the artefacts. The potentialities of the underlying matter and of the proximate matter are M-Pots, as per our discussion in the previous section of this chapter. We saw that these potentialities differ from C-Pots in that they do not tend towards their fulfilment. The definition of the potentiality of a material substratum, e.g. marble, does not mention the form of the substance, e.g. statue, by contrast to the definition of a causal potentiality, e.g. of magnetic power, which mentions 'attraction (of metal objects)' in the definition of the potentiality. At most, the potentiality of marble will mention that the marble can receive form – be carved; yet, even then, the marble does *not* have a tendency to be carved.

Aristotle's account of hylomorphic unity in terms of the potentiality of matter for the form is therefore developed in two stages. First, he assumes that the actualisation-form is already in the C-Pot in potentiality, and so there is no need to *unify* the form to the potential; and second, he assumes that matter and form in a composite substance are like the potential and actual in C-Pot, and so the same account of unity applies. In other words, Aristotle's account of the hylomorphic unity of the potential and the actual is that there is only one form, the form of the actual, which is embodied in different ways in the potential and in the actual. It is embodied in the nature of the potentiality, at first, and then, upon manifestation, it is embodied in the form of the actuality. So there is nothing to unite, since there is a form in embodiments. I have already argued against the latter, namely, against the parallel between C-Pot and M-Pot. (M-Pot is not directed towards the form; it is incomplete without the form; and does not interact with other potentialities.) I am here showing that the former point Aristotle appeals to does not apply either in this case; the actuality-form is not present in C-Pot in any way that could illuminate the unity of matter and form in a composite substance. The form of e.g. magnetic-attraction is not present in a magnetic-power-in-potentiality in a way that can help us understand how the form of a statue is present in the marble. I conclude that M-Pot does not explain hylomorphic unity. I find the differences between M-Pot and C-Pot to be too fundamental to justify seeing anything comparable to C-Pot in a material substratum. And yet, it is only the nature of C-Pot that makes a strong claim to the 'presence' of the form of its actuality (its manifestation) in the very nature of the potentiality, on account of its definition in terms of the actuality it is aiming at. Thus the question of what makes matter and form an ontological unity is, from my standpoint, still open. In the next section I will examine whether it is the unity of the definition that accounts for Aristotle for the unity of substance.

The unity of a definition: *double hylomorphism*

When Aristotle talks of the definition of a substance being one, this could be read as meaning that a definition is a single string of words; but read this way it would not be relevant to his (and ours) metaphysical concerns. Rather, I take him as meaning, consistently throughout his work, that the definition refers to a single individual entity, rather than to the many entities that the terms in the definition refer to.[25] How then does Aristotle show that the definition refers to a single individual entity? In *Metaphysics* VII.12, he argues that a definition can be reduced to two terms, the genus and the ultimate difference (e.g. a human is a rational animal). He further argues that the unity of a *definition* is hylomorphic; the genus (e.g. animal) is as matter of the species-difference (e.g. the rational) as form:

> If then the genus absolutely does not exist apart from the species which it, as genus, includes, or, if it exists but exists as *matter* (for the voice is genus and *matter*, but its differentiae make the *species*, i.e. the letters, out of it), clearly the definition is the formula which comprises the differentiae [i.e. the differentiae are the species-determinates of the genus-determinable that is like matter].
>
> (1038a5–9, my emphasis)

The relation between genus-to-species (e.g. animal to tiger) is that of determinable-to-determinate. Yet, genus-to-species is *not* like matter-to-form (quantity of flesh and bones to the form of tiger), contrary to what Aristotle claims. Abstracting e.g. crimson (species form) from red (genus form) is a different type of division from abstracting crimson (species form) from the matter it belongs to (e.g. a red apple, or from material pure potentiality). A determinate (e.g. crimson) does not relate to the determinable (e.g. red) in the same way as form qualifies matter (e.g. crimson qualifies this surface); this surface is not to crimson as determinable is to determinate. What Aristotle is trying to convey here, even if his simile does not tell us this, is that e.g. red does not add any qualification to crimson, just as matter does not add any qualification to the form that it embodies. The latter is more obvious than the former; one would have thought that the genus adds qualification – being – to the species; but Aristotle is trying to tell us that there is no relation that could possibly relate, ontologically, red to crimson, just as there is no relation that could relate the potential to its actuality, or matter to the form it embodies.[26]

This is important, because it blocks the following Aristotelian line of explanation of the oneness of a substance: a hylomorphic composite substance is one because its form is one; the form is one because its definition is one; the definition is one because it is hylomorphically unified under a single differentia (as per *Metaphysics* VII.12). This is what I call *double hylomorphism*. This line of explanation is circular.[27] (I had previously thought (in Marmodoro 2013) that there was a mechanism through which the definition of a substance accounted for its oneness. The mechanism would have been that of *substantial holism* (as per Scaltsas 1994 and 2001), according to which the unity of a substance is holistic, and grounded on the holistic definition of the substance.[28] In the next two sections I will however argue that holism unifies, but does not deliver oneness of substance.)

Aristotle's two accounts of hylomorphic unity

Looking at Aristotle's effort to account of the unity of an individual substance, which is the overarching question driving his search in *Metaphysics* books VII, VIII and IX, we can identify two accounts that can be extracted from these chapters. One account is his *substantial holism* (Scaltsas 2001) where, as Aristotle says, each part of a substance is well defined when defined in terms of the form of the substance, as per *Metaphysics* VII.12. The substance is unified by the re-identification of each physical component of a substance in terms of its contribution to the constitution of the substance, as determined by the substantial form of the substance. Thus, matter and form are unified under the conceptual/functional 'umbrella' of the essence, and hence, of the definition of the substance.

Scaltsas (1994 and 2001) assumed that this is what Aristotle was referring to when in *Metaphysics* VIII.6 he claims that the potential is one and the same with the actual; namely, that Aristotle is there referring to the re-identification of the potential in terms of the actuality given in the form. However, Aristotle writes that,

> What is it, then, that makes man one; why is he one and nt many, e.g. animal – biped, ... if, as we say, one element is matter and another is form, and one is potentially and the other actually, the question will no longer be thought a difficulty. ... *the proximate matter and the form are one and the same* thing [ἡ ἐσχάτη ὕλη καὶ ἡ μορφὴ ταὐτὸ καὶ ἕν], *the one potentially, and the other actually.* Therefore to ask the cause of their being one is like asking the cause of unity in general; for each thing is a unity, and *the potential and the actual are somehow one.* Therefore there is no cause other here unless there is something which caused the movement from potentiality to actuality.
>
> *(1045a14–b22, my emphasis)*

I understand Aristotle's text differently from Scaltsas (1994 and 2001). What Aristotle is presenting here is a different solution for the unity of matter and form in an individual substance than the re-identification of the components of a substance in terms of the form of the substance. What he tells us, explicitly, here, is that the cause of their unity is that 'the potential and the actual are in a sense one'. What does he mean by this? What is it for the potential to be one with the actual? I submit that what Aristotle is claiming here is that the matter and the form in a substance are unified into one, not because of any type of connection between them but because they are somehow one. I submit that the potential is one with the actual because potentiality is defined in terms of actuality; that is, the potential and the actual are in a sense *the same form*. The definition of a potentiality is given by the form of its manifestation (its actuality); e.g. the potentiality of a magnet is defined as a capacity to *attract or repel* other electromagnetic charges when manifesting. The potentiality of a magnet that is not interacting with any other charge is neither attracting nor repelling, and yet, we understand the type of potentiality it is as what can produce a manifestation of attraction and repelling.

I argued elsewhere (Marmodoro 2017a, 2017b) in connection with my neo-Aristotelian account of causal powers that the role of the presence of the form of the manifestation (the actuality) of a potentiality in the definition of that potentiality is *epistemological*, not ontological. When we say that e.g. a magnet can potentially attract and repel other charges we mean that attraction and repelling *informs* us as to what type of potentiality this is. I want to further explicate my stance here. Although the mention of the form of the actuality of a potentiality in the definition of the potentiality (e.g. potentiality to attract/repel) does not introduce the ontological presence of the actuality (attracting/repelling), as it serves only epistemologically to direct us to the type of potentiality this is, nevertheless, the type of potentiality it is *embodies* the form of the actuality or manifestation (e.g. the form of repelling/attracting). This *embodiment is ontological*. It is as if the form of the manifestation, the actuality, of the potentiality, is present in the potentiality itself, because it 'shapes' the potentiality's kind into the type of potentiality it is.

This is what Aristotle is explaining in the quotation above: the potentiality is one with the actuality because the potentiality is already 'shaped' (en-formed) by the actuality's form, even though the potentiality has not manifested. I want to emphasise that Aristotle is not claiming that a relation holds between two entities, the potential and the actual; e.g. that the potential is (mysteriously) inextricably connected with the actual.[29] Rather, Aristotle claims that the potential and the actual are somehow *one and the same*, which we can understand in terms of the potentiality's ontology. What Aristotle is saying, concerning the unity of the potential and the actual, is that the potentiality is already 'shaped' by the form of the actuality, by the kind of potentiality it is. In this sense, if the potentiality happens to manifest, the form of the actuality comes to be present, not only as embodied in the type of the potentiality, but also present in actuality, by instantiation.

This is a different kind of solution for the hylomorphic unity of the potential and the actual in an individual substance than the solution of substantial holism. Substantial holism accounts for the unity of an individual substance by the re-identification of the components of the substance in terms of the form of the individual substance – it delivers what we can call the *horizonal unity* of the substance. The present account of the unity of a substance, deriving from *Metaphysics* VIII.6, which we may call the *inter-determination* account, delivers *vertical unity*. I call it *inter-determination* because this is how Aristotle thinks that the potential and the actual are somehow one – they determine each other: "There is no other reason why the potential sphere becomes actually a sphere, but this was the essence of either" (1045a33–34). The form of the *actual* (sphere) determines the type of the potentiality the actual sphere is; and the *potential* (sphere) determines the type of actuality that results from the potentiality's manifestation. I call it *vertical unity* because, as Aristotle says, it unifies layers of organisation: "the proximate matter and the form are one and the same thing; the one potentially, and the other actually" (1045b19). Thus, if we analyse a substance, by abstraction, in levels of proximate-matter for the actuality they each ground, then we are 'descending' the scale of unities in an individual substance. The most elemental level of matter will be the level of the four elemental powers, of the hot, cold, wet and dry (that is, physical potentiality).[30]

I find that there is a problem with this alternative account of the unity of individual substance Aristotle offers us, because, in brief: the form (or essence) of a potentiality is a spectrum of possibilities of which the form (or essence) of the actuality that is manifesting at a given moment is only one. Hence, if the form (or essence) of the actuality cannot serve as the form of the kind of potentiality that generated it, but only be a constituent of the potentiality's form (or essence). For example, the marble of a statue can be en-formed by many more forms than the form of this statue; similarly for organisms. I will not give here an argument for my stance (which would be a version of the Swapping Argument used e.g. by Alexander Bird (2007: 74–75) in relation to quiddities and potencies.)[31] Showing what is problematic in Aristotle's position on the dependence of the definition of the potentiality's kind on the form of the potentiality's manifestation (actuality) would be long and complex, and cannot be done here.

The account of substantial holism, for the unity of substance, differs from the interdetermination account of such unity, and does explain the unity of substance. However, the unity substantial holism establishes is only the unity of the definition of a substance, and therefore, the number of this unity is the number of the *definiendum*. I have shown above (in the section "The unity of a definition: *double hylomorphism*") that Aristotle's account of the number of the *definiendum* is circular; Aristotle ultimately justifies the number of the definiendum with an explanation that uses the number of a hylomorphic model, which is explained in terms of the number of the definiendum. For this reason I will conclude that substantial holism accounts for the hylomorphic *unity* of a substance, but does not determine the *number* of the substance. I will argue that the determination of the number of a substance depends on a further conceptual operation that reconceptualises the parts of a substance into properties of the substantial subject. To this I turn in the next section.

My account of the unity of substance

I have argued earlier in this chapter that there are many different kinds of forms at a time (synchronously) in a substance, and many kinds of matter; they make up multiple hylomorphic composites in the substance. Each hylomorphic composite explains a different type of problem in the constitution of a substance (see the lists of types of form and of matter above). The question that I address is here is: How do all these hylomorphic composites (with different kinds of form and different kinds of matter) make up *one* single substance? Does Aristotle succeed in explaining their unity into a single entity?

Plato was the first to set out the main dichotomy of composition that challenged ancient mereologists (and still challenges us). In what I call the *Dilemma of Composition*, he claims that what is complex falls under either one or the other of the following two accounts: either the whole is identical to its constituent parts, and therefore it is many; or the whole is not identical to its constituent parts, but is unified into one single entity by a form, in which case the unified constituents are no longer parts of the whole.[32] In the former case, the complex whole is a cluster of many parts; in the latter, it is a single entity. Plato however has not been recognised and given due acknowledgement of the fact that he is the first metaphysician in history to describe the phenomenon of the *emergence* of form. However, Plato is explicit that the emergent form *comes into being* from the parts that make it up; his is a clear statement of the relation of constitution, which we traditionally associate with Aristotle's substantial forms. Plato gives an example of a substance-like entity, a syllable (which is an example Aristotle, too, uses), which emerges from its constituents, the letters. Of the syllable, Plato says that "Maybe we should have supposed the syllable not to be the letters, but rather some one kind of thing which has come into being out of them: something which has a form of its own, and is different from the letters" (*Theaetetus* 203e3–5). The thought is that, when the letters come together into a structure, something new emerges, the syllable, which is *different in number* from the letters, and a *different kind* of thing from the letters that make it up. Plato has thus identified two key features that distinguish substances from aggregates: a qualitatively new form unites the constituents; and the whole is numerically different from the totality of its constituents. The question that remains open is this: how does the new form unite the constituents? What kind of metaphysics can account for the change of number as well as for the being of a new kind/quality, in relation to the components to the composite? We do not find such a metaphysical account in Plato, but we do find it in Aristotle. He holds that a substance is unified *holistically* by its form, and if holistically unified, its constituent physical parts lose their discreteness in the whole. So far, so good. However, Aristotle assumes that, thereby, the substance becomes one, a single individual.

I distinguish here between two ways to account for substantial unity. On one account the unity of substances is grounded on *functions/capacities* – I call this *functional holism*; on the other account it is grounded on *individuating concepts*.[33] The type of unity that functional holism yields may be illustrated with the example of a set of devices making up a security system. The overall function – that of being a security system – places requirements on the interrelations, and thus in some way the types of unity of the devices in the system as a whole: if arranged and connected thus-and-so, they can serve a certain function (which the same devices would not serve if scattered in space). The overall function emerges from the system as something novel that the devices taken individually or discretely could not perform. The unity of the system thus derives from the oneness of its overall function. Functions are instantiated when objects, or systems of them, are in contexts where they implement these functions. Objects can be found and used in functional contexts; or they can be designed for such contexts (e.g. a stone used as a weapon; a piece of metal as a coin; a bicycle for riding, etc.). There may be variation as to how deeply into the nature of an object a function 'descends'.[34] However, importantly for the present argument, functional holism does *not* determine the properties of any object 'all the way down'; for different objects, it determines their properties at different levels. Contrast, for instance, the case of a stone, which acquires the function of being a weapon, to the case of the neurons that store a memory. In the first case, only the shape of the stone (e.g. being made sharp-edged) was determined by the new function; in the second, the constitution as well as organisation of the neurons was determined by the function (through evolution).

A different way to account for unity is that according to which *individuating concepts* can make things one, e.g. the concept of being human. Given a criterion for the application of the concept (e.g. if rational, then a human being), the concept applies to the object if and only if the object satisfies the criterion (e.g. is rational). However, how does the use of an individuating concept account for the metaphysical unity of what it is applied to? While the question remains at this stage open, it is important for my account to recognise that the mechanism by which a whole is unified by a function is different from the mechanism by which oneness is achieved through an individuating concept.

How then can a holistically unified whole be many, and can it become one? Does holism not unify the whole into one? I submit that the mechanism that Aristotle proposed, as per substantial holism (namely, the mechanism of the re-identification of the parts in terms of their functional role in the whole as determined by its definition) unifies the *physical parts* 'seamlessly' between them, by fitting them, definitionally, into an integrated functional whole. For instance a computer system comprising a printer, a scanner, a copier, etc. has an overall function that serves to re-identify its physical components according to the role they play towards the overall function of being a computer system. Thus, the distinct physical parts are unified seamlessly, by becoming *functional parts* of the kind 'computer system'. A confusion might arise here because of an ambiguity in language: if a printer is connected to a computer system, and can thereby print, it is seamlessly unified with the computer through its function. However, if we sever its connection with the computer, and it can no longer print, we still call it a 'printer'. Aristotle would explain to us that it is a 'printer' in name only,[35] and potentially one, but the definition of 'printer' is not true of it since it is severed from a computer. Thus, if we are to 'speak Aristotelian', the severed printer is a physical part that derives from a printer but is not a part of the printer, but only potentially so. To pay justice to the intuition we have that a computer system is complex and has parts, we need to recognise that what we mean is that somehow, we can divide a computer system *by abstraction* into parts which on the one hand are severed from the computer by abstraction, while on the other they are still unified functionally with the computer.

The *metaphysical parts* of an object are the parts of the definition that determines *what* this object is. If a function defines an object, e.g. being a computer system, the functional parts thereby become the metaphysical parts comprising the essence of the object. However, the metaphysical parts are not, as such, one; they are many. Thus, in our example, the computer system is many physical parts, which are many functional parts, and so, in this case, many metaphysical parts, too, all serving the single function of being a computer system. Serving a single function unites the parts holistically, in the way that the players of an orchestra are holistically unified under the function of the orchestra. However, holistic functional unity is compatible with plurality – the orchestra, just like a class in school, are many, tired, rowdy, etc. What, then, could unify the metaphysically many parts into a single entity?

As we saw, Aristotle addressed this problem in *Metaphysics* VII.12:

> Now let us treat first of definition …: wherein can consist the unity of that, the formula of which we call a definition, as for instance, in the case of man, 'two-footed animal'; … Why … is this one, and not many, *viz.* 'animal' and 'two-footed'?
>
> *(1037b9–14)*

He tried to solve this problem, too, hylomorphically (see 1038a5–8). I propose a different solution to the question of how a holistically unified whole becomes a single entity: a whole that is holistically unified by its function, which is metaphysically many, becomes *one* when the metaphysical (functional) *parts* composing the whole become *properties* of the whole. For instance, a human being can be thought of as being many, consisting of many functional parts; alternatively, she can be thought of as one, who is e.g. sighted, handed, biped, rational, etc. Crucially, the transition to oneness does not depend on any additional, 'tighter' unification of the metaphysical parts into a whole. Aristotle thought it did, and introduced this idea into his explanation of a substance's oneness, allotting the task of delivering a 'tighter' unification to the form.[36] However, I submit a substance is rendered one only by the *reconception of the metaphysical parts of the whole as properties of the whole*. This is achieved conceptually, by reconceiving a holistically unified (by its definition) whole under an individuating concept. To return to our example once again, the computer system is indeed one, under the individuating concept 'computer system', because we think of calculating, printing, copying, scanning as properties qualifying the whole – the computer system. So, to generalise, the unity of a substance is a 'double-act' of *holistic unification* of the many physical parts into seamlessly unified functional parts; and the *conceptual individuation* of the functional parts as an entity that is characterised by these functions.

Functional unity is compatible with the plurality of the functional whole, whether the whole is e.g. a class, an orchestra, a set or a rabbit, each of which can be *many*. Thinking of each of these as a single individual is a further step, a conceptual one, where the functional parts become qualifications of the whole. This tells us nothing about what evolution has favoured in human history. It may be that we cannot help but think of a physical unity as one individual, especially if its parts move together. It may be that agency of change, which functionally unifies the agent(s), is irresistibly (as it were) perceived as one by us. However, the issue that I am addressing here is whether we can think of an individual as many, when we discover its internal complexity of any kind (physical, functional, metaphysical), and if not, why can we not think of it as many? And if we can think of it as many, does anything change in the way we think of it? I have argued that although physical unity and functional unity may suggest the oneness of the unified, we may still think of the unified as many. We move from the conception of the many to the conception of the individual by thinking of the many distinct parts of an individual as qualifying the individual, rather than as articulating it into parts. Rendering a part of an individual into a property of the individual is a conceptual operation, which alters the metaphysical constitution of the unified whole.

Conclusion

I should like to conclude with a claim, which summarises the essence of this chapter's argument: there is no primary relation of matter to form in a hylomorphic composite substance. Aristotle suggested there is such a relation, which is responsible for the unity and oneness of a composite substance. His explanation of this relation is that the matter of a substance is unified by the form acting as a principle, e.g. through the metaphysical mechanism of substantial holism. However, holding that the form acts as a principle, and granting that substantial holism does the metaphysical work of re-identifying the parts in accordance with their functional role in the substance, does no more, I argue, than ground the unity and oneness of a composite *substance in the unity and oneness of the form* of the substance. This commits Aristotle to what I call *double hylomorphism*. He explains the unity and oneness of the complex essential form of a substance (namely, the unity and oneness of the determinable/determinate composition, of e.g. animal and tiger, expressed by the definition of tiger), in terms of the hylomorphic model of unity and oneness, as matter to form. So, Aristotle's account of the unity of matter and form in a composite substance is circular, explaining hylomorphic unity by hylomorphic unity; this is *double hylomorphism*, combining material hylomorphism and definitional hylomorphism.

This leaves the question of the unity and oneness of a hylomorphic composite substance open. I argue that there are many and varied types of hylomorphic relations in a composite substance, each solving a different metaphysical problem; and further, that none of them is more primary than the rest. On the account I propose, which complements that of Aristotle's, substances comprise clusters of different types of hylomorphic composites, which are united between them by different types of ontological interdependence. Hylomorphic composites are *unified* by holism; however, their *oneness* is not achieved by holism, but only bestowed upon them conceptually, by us.

Notes

1 The first version of this chapter was presented at the American Philosophical Association meeting held in Chicago in February 2018. I am very grateful to Rob Koons who offered insightful criticisms as my commentator on that occasion, and to the audience for the helpful discussion. Subsequent versions benefitted from the feedback received when presenting the paper at departmental seminars at Kings' College London, the University of Padua, the University of Rostock and the summerschool *Aristotle's Metaphysics Yesterday and Today* held at Falconara (Italy).
2 Abstracta are items in the ontology, like numbers, or the colour 'blue' in a magpie. They are not the concrete primary individuals in the ontology, but they can even exist as instantiated and concrete, e.g. salt in seawater, where seawater is a primary concrete individual in the ontology.
3 A full understanding of this last statement would take us deep into Aristotle's metaphysics, and his *Homonymy Principle*; see footnote 28.
4 See *Metaphysics* 1030a5–6.
5 The only wholes that wear their number 'on their sleeves' are the partless wholes – they are one, *qua* simple.
6 On the topic of Aristotle's theory of abstraction see e.g. Allan Bäck (2014: 8).
7 Each type of form or of matter generates its own hylomorphic division of the object.
8 Abstraction is a conceptual division; I will argue in the penultimate section, that the unification of a substance into one entity is a conceptual operation, too.
9 I consider (ii) and (iii) to be co-extensive, because I interpret Aristotle as holding that all properties are powers, as per Marmodoro (2014: chapter 1).
10 On some interpretations, iv) is a functional form; I will discuss this issue further in the penultimate section of the chapter. I don't engage in this chapter with the question of whether the substantial form is a universal or a particular.
11 See *De Anima* 424a19–22.
12 See e.g. *De Anima* III.8, 431b29–432a1; cf. III.4, 429a27; and III.5.
13 See e.g. *De Motu Animalium* 7, 701a12–16.
14 I am using the metaphor of 'leaving behind' to include under it all variations of division. For example, even if we only individuate a form in an object, e.g. the colour blue, we have thereby divided the object (without needing to literally pry them apart). It is like cutting a piece of a cake and leaving it in the cake; the division has left behind two parts. According to Aristotle's *Homonymy Principle*, abstraction divides by transformation. The reason it transforms is that it severs, by abstraction, the functional dependence of the component on the whole; for example, cutting a finger (or individuating as severed) from its hand transforms the finger into mere matter, because it deprives the finger of its functional form, since the finger is not serving any role in the hand any more. See also footnote 28.
15 See *Physics* I 7, and elsewhere.
16 The wood, rather than the earth, water, fire, etc. out of which wood is made up. See e.g. *Metaphysics* VIII.7, 1049a18–27.
17 'Quantity of matter' is a concept introduced by Helen Cartwright in Cartwright (1979).
18 See e.g. *De Generatione et Corruptione* II.5, 332a14–18. In this context, Aristotle treats instances of dryness, wetness, heat, etc. as hylomorphic composites of matter (physical potentiality) and form (where the form of e.g. wetness can be replaced by that e.g. of dryness).
19 See *Metaphysics* VII.10, 1036a9–12.
20 This is ambiguous between the block of marble from which the statue is shaped, and the quantity of marble from which it is shaped and which constitutes the statue once it is formed – I don't disambiguate it here, as the ambiguity does not interfere with the argumentation which follows: both types of matter, proximate and underlying, have the potentiality to receive the form of the generated substance.
21 Of course, abstract entities do not exist on their own, e.g. the wood of a chair does not exist on its own, being a-shaped, without the shape of the chair or any other shape. But not existing on their own does not make them incomplete, any more than half a cup of water is incomplete just because it is not a full cup.
22 See *Metaphysics* IX.1, 1046a9–10: "All potentialities that conform to the same type are starting points of some kind, and are called potentialities in reference to one primary kind, which is a starting point of change in another thing or in the thing itself *qua* other."
23 George Molnar calls it 'physical intentionality' and articulates the idea in his *Powers* (2006), chapter 3.

24 Aristotle's examples of partner potentialities include the potentiality of a teacher to teach and that of the pupil to learn, as in *Physics* III.3, or that of something to sound and that of the perceiver to hear it, as in *DA* III.2.
25 Aristotle writes in *Metaphysics* VII.12:

> Now let us treat first of definition ...; for the problem stated ... is useful for our inquiries concerning substance. I mean this problem:-wherein can consist the unity of that, the formula of which we call a definition, as for instance, in the case of man, 'two-footed animal' ... Why, then, is this one, and not many, viz. 'animal' and 'two-footed'?
>
> *(1037b9–14)*

26 Aristotle says

> the proximate matter and the form are one and the same; the one existing potentially, and the other actually. Therefore to ask the cause of their unity is like asking the cause of unity in general; for each individual thing is one, and the potential and the actual are in a sense one.
>
> *(Metaphysics 1045b19–22)*

27 One might want to argue in defence of Aristotle that both hylomorphic unities – material and definitional – are grounded on the unity of the potential to the actual. But we saw above, in the preceding section, that M-Pot lacks the relevant characteristics of C-Pot.
28 Key to this interpretation is Aristotle's *Homonymy Principle*: "it is not a finger *in any* state that is the finger of a living thing, but the dead finger is a finger only homonymously [i.e. same in name, but different in definition]" (*Metaphysics* VII.10, 1035b23–25). Substantial holism, developed to account for the unity of Aristotelian substances, proposes a mechanism to explain how a hylomorphic substance is *holistically unified* into a whole. The physical parts of a substance are holistically unified by being re-identified in terms of their functional (or organisational) role in the whole. Re-identification in the case of organisms is radically different from the case of the artefacts. When a cat is generated by its cat-embryo, every bit of matter that goes into its creation is transformed into flesh, bones, blood, etc. So, is this generation *ex nihilo*? It is not, because what survives is the quantity of matter that was the initial pool of katamenial blood that becomes flesh and bone. How does a quantity of matter become re-identified? In the case of a silver spoon that is recast as a silver fork, the quantity of silver that survives the generation of the fork is re-identified as the handle and the tines of the fork. However, in the case of the quantity of blood that becomes a quantity of bone, the blood does not survive. My claim is that, although the *matter* does not survive, because it is radically identified (i.e. totally transformed), the *quantity of matter* it is does survive, securing continuity, and *it* constitutes the new substance. My understanding of Aristotle is that he thought that e.g. the steam from boiling water is the same matter as was water before losing that form and embodying the new form, of steam.
29 See David Charles' causal-explanatory account (but the issue cannot be discussed further here for reasons of space), in Charles (2009).
30 Or of their proximate matter, namely the potentiality that underlies the elemental transformations, which I discussed elsewhere, e.g. in Marmodoro (2014: chapter 1).
31 Bird's Swapping Argument is an adaptation of Chisolm's (1967) argument.
32 In Plato's words: "if something has parts, the whole, and the sum, will be all the parts ... If it's not the case that a complex is its elements, then isn't it necessarily the case that it doesn't have the elements as parts?", to which question Plato's answer is in the affirmative (*Theaetetus* 205a8–b2).
33 I do not use the term 'functionalism' in my discussion of substantial unity, as the view it references usually runs together functional unity and individuating concepts, which I want to tease apart.
34 I investigated the issue of how forms 'descend' in the matter of a substance in relation to Aquinas' (Aristotelian) account of the difference between natural substances and artefacts in Marmodoro and Page (2016).
35 See Aristotle's *Homonymy Principle*, explained in footnotes 14 and 28.
36 Aristotle's proposal is complex and I will offer it here only in outline. Aristotle thinks it is the 'tightness' of essential properties that makes the substance one. He in fact contrasts this with the 'looseness' of accidental properties, which he does not think unify the entity into one:

> ... since the differentiae present in man are many, e.g. endowed with feet, two-footed, featherless. Why are these one and not many? Not because they are present in one thing [i.e. subject of properties]; for on this principle a unity can be made out of all the attributes of a

thing [i.e. accidents, too – by *reductio ad absurdum*]. But surely all the attributes *in the definition* [i.e. the essential attributes] must be one; for the definition is a single formula [hylomorphically] and a formula of substance, so that it must be a formula of some one thing; for substance means a 'one' and a '*this*' [individuating principle], as we maintain.

(Metaphysics VII.12, 1037b 22–28, my emphasis)

I suggest that Aristotle believes that the *individuating concept* is generated by *essential* predication. I believe he thinks this because, in the *Categories*, he combines two metaphysical functions into one, right from the start: the differentiae, at any level, *constitute* the form at that level, but are also *predicated* of the form; e.g. the species form 'human' consists of the differentiae 'rational' and 'biped', which are also predicated of it, being 'said of' it, e.g. a human is a rational biped. This automatically treats *metaphysical parts* of the definitionally unified whole as *properties* of the whole. However, as we saw, this is not neutral. Parts are not predicated of the whole; they only make it up, and therefore, the *whole of many parts is many*. To convert the many into one, *the whole has to be converted into a subject*; what *constitutes* it (aka the whole) will now *qualify* it (aka the subject). Aristotle assumed this from the start in the *Categories*; but he didn't have to. He wanted to improve on Plato's 'compresence' of immanent Forms in a thing, e.g. where a human participates in the Rational and in Bipedness. Thus, Aristotle generated definitions which did two things: definitions bound forms essentially together; and they treated these forms as qualifications of the definiendum. However, these two need not go together; and only the second introduces, conceptually, oneness, not because of the essential interconnection, but by treating parts as properties of a subject.

References

Aristotle's texts are quoted from:

Barnes, J. (ed.) (1995) *The Complete Works of Aristotle*, Princeton University Press.

Plato's texts are quoted from:

Cooper J. and D. S. Hutchinson (eds.) (1997) *Plato: Complete Works*, Hackett Publishing.

Other works cited:

Bäck, A. (2014) *Aristotle's Theory of Abstraction*, Springer Verlag.
Bird, A. (2007) *Nature's Metaphysics*, Oxford University Press.
Cartwright, H. M. (1979) 'Amounts and Measures of Amount', in F. J. Pelletier, *Mass Terms: Some Philosophical Problems*, Springer, pp. 179–198.
Charles, David (2009) 'Aristotle's Psychological Theory', *Proceedings of the Boston Area Colloquium in Ancient Philosophy*, Volume 24, Issue 1, pp. 1–49.
Chisholm, R. (1967) 'Identity through Possible Worlds: Some Questions', *Noûs*, Volume 1, No 1, pp. 1–8.
Marmodoro, A. (2013) 'Aristotle's Hylomorphism without Reconditioning', *Philosophical Inquiry*, Volume 37, Issue 1/2, pp. 5–22.
Marmodoro, A. (2014) *Aristotle on Perceiving Objects*, Oxford University Press.
Marmodoro, A. (2017a) 'Power Mereology: Structural versus Substantial Powers', in M. P. Paoletti and F. Orilia (eds.), *Philosophical and Scientific Perspectives on Downward Causation*, Routledge, pp. 110–127.
Marmodoro, A. (2017b) 'Aristotelian Powers at Work: Reciprocity without Symmetry in Causation', in J. Jacobs (ed.), *Causal Powers*, Oxford University Press, pp. 57–76.
Marmodoro, A. and Page, B. (2016) 'Aquinas on Forms, Substances and Artifacts', *Vivarium*, Volume 54, Issue 1, pp. 1–21.
Molnar, G. (2003) *Powers: A Study in Metaphysics*, Oxford University Press.
Scaltsas, T. (1994) *Substances and Universals in Aristotle's Metaphysics*, Cornell University Press.
Scaltsas, T. (2001) 'Substantial Holism', in T. Scaltsas, D. Charles and M. L. Gill (eds.), *Unity, Identity, and Explanation in Aristotle's Metaphysics*, Oxford University Press, pp. 107–128.

23
FEMINIST METAMETAPHYSICS

Elizabeth Barnes

A discussion of the metametaphysical commitments of feminist metaphysics must inevitably begin with a series of caveats. The first is that there are many different things we might mean by 'feminist metaphysics.' For the purposes here, I will simply mean metaphysics that is of particular interest to philosophers working on feminist philosophy. This is, of course, an imprecise definition – since it hangs on what is meant by 'feminist philosophy,' 'particular interest,' etc. And there are further questions of whether work must be done from a specifically feminist framework (whatever that means) in order to be 'feminist metaphysics,' rather than simply being metaphysics that is especially interesting and helpful to people working on feminist philosophy. I'm not interested, for the purposes here, in giving an account of what it takes for metaphysics to count as 'feminist metaphysics' – especially since I suspect that classificatory questions like this are rarely helpful in philosophy. Instead, I'll proceed with the assumption that the gist of 'feminist metaphysics,' at least as I'm using the term, can be understood by paradigm cases. What I take to be characteristic examples of feminist metaphysics include discussions of the nature of social kinds and properties (including gender and sexuality),[1] accounts of social structure and structural explanation,[2] and particular metaphysical questions about gender (such as whether gender is a natural kind, whether it depends on or is grounded by other social or natural properties, whether social generalizations about gender easily could have been otherwise, and so on).[3]

But, of course, a very wide range of philosophers – and very different methodologies and background assumptions – are involved in the discussion of such question. This brings me to my second major caveat: there are probably no specific metametaphysical commitments that unify everyone doing feminist metaphysics, or which are characteristic of feminist metaphysics as such. Feminist metaphysics is a colorful party happening in a big tent. In what follows, I'll discuss what I take to be some of the major themes that emerge from analytic feminist metaphysics. But I don't take these themes to be universal. For the purposes here, I'll focus on analytic feminism both because analytic feminism has been the area of feminism most interested in metaphysics, and because readers of a volume on *metametaphysics* are – I hazard a guess – most likely those trained in an analytic tradition.

When looking at contemporary analytic feminism, there are three main themes that can be well-described as metametaphysics: the metaphysical importance of the social, the political significance of metaphysical claims, and the link between metaphysics and social progress.

The bulk of this chapter will focus on the third of these themes, since I think it is the place where feminist philosophy becomes the most metaphilosophically distinctive. But I'll briefly discuss the first two before moving on.

The importance of the social

One main theme of feminist metaphysics is that there are interesting, substantive metaphysical questions about social-level matters. To think that questions like 'what is gender?' are worth approaching with the tools of metaphysics, you need to think that macro-level phenomena like gender might fall usefully within the purview of metaphysics. Many debates in feminist metaphysics thus require a certain kind of *non-dismissiveness* about these kinds of macro-level entities.[4]

It's common, within contemporary metaphysics, to see metametaphysical claims that go something like this: metaphysics is not about what exists, it's about what is fundamental; the interesting or substantive questions of metaphysics are about the fundamental, and questions about the non-fundamental are primarily linguistic or conventional, rather than matters of substantial metaphysical debate; the interesting explanations for why things are the way they are should be given in terms of what's fundamental; and so on. 'Fundamental' here is glossed in various different ways – what's most natural, most joint-carving, what you find at the end of a grounding or in-virtue-of relation, what's most materially or explanatorily basic, what's independent, etc. I won't wade into those debates here. What these different approaches have in common, and what matters for this discussion, is a type of dismissiveness about the non-fundamental, at least qua interesting questions of metaphysics. The non-fundamental might be interesting, it might be politically, socially, and morally important, and there might be lots of substantial linguistic or conventional questions to settle about it. But for the specific context of doing metaphysics, fundamentality-centric approaches tend – sometimes explicitly, sometimes implicitly – to assume that questions about 'higher level' things like gender are unimportant. The big, central questions are about the fundamental.[5]

A hallmark of feminist metaphysics is interest in metaphysical questions about the social. And so a hallmark of feminist metaphysics is, quite understandably, a rejection of this sort of dismissiveness about the metaphysics of the non-fundamental. If you think that the metaphysics of things like gender is something that's interesting, difficult, and important, and things like gender are not fundamental, then you will tend to think that the metaphysics of the non-fundamental can be interesting, difficult, and important.

To be clear, this is not at all the same thing as saying that you think the best account of social kinds like gender is a metaphysical one (rather than a conventional or linguistic one), or that you think that some things fail to be grounded in or made true in virtue of the fundamental, or etc. Plenty of accounts in feminist metaphysics end up with relatively deflationary views about the metaphysics of things like gender, and plenty of positions in feminist metaphysics are perfectly compatible with the idea that everything – including gender – is ultimately grounded in/ explained by/made true by/true in virtue of/[your favorite gizmo here] the fundamental. The unifying feature of feminist metaphysics is not a particular stance about how to answer or approach questions concerning the metaphysics of things like gender. Rather, the unifying feature is that these questions are interesting and complicated and worth approaching with the tools of metaphysics. We might end up with a view that's deflationist about gender, or a view that says that linguistic convention tells us what we need to know about gender, or a view that's error-theoretic about gender. But the unifying assumption is that getting there is complicated, and requires careful argument.

Metaphysical concepts and political power

A second main area of metaphilosophical interest within feminist metaphysics has been the way in which metaphysical claims are, perhaps despite initial appearances, not politically neutral or politically idle. Often, in doing metaphysics, we describe things as 'real,' 'fundamental,' 'natural,' 'essential,' 'objective,' and so on. And quite a lot of feminist work has been done discussing the ways in which such claims — although taken to be purely descriptive — often have significant normative upshots.[6] This is not to deny that there is such a thing as objectivity, or that some properties can be had essentially, or etc. Rather, the goal has been to pay careful attention to the way in which claims about the natural world often have subtle normative ramifications and can be used for political ends. For example, when feminists argue that certain gender norms are socially constructed, they often engage in what Sally Haslanger describes as a 'debunking project' — they don't merely argue *for* the social nature of such gender norms, they argue *against* the assumption that such norms are natural, essential, biological, etc. And pushing back against the assumption that gender norms as we experience them are natural or essential is especially difficult, partly because ideas of naturalness are powerful. Similarly, the argument for gay rights was not merely an argument *for* the rights of a certain set of people, it was also, inevitably, an argument *against* the idea that being gay was somehow 'unnatural.'

Again, in calling attention to the political role of these terms and concepts, feminists don't thereby need to deny that anything is natural or essential or etc. Rather, the goal is to illustrate how these concepts — even when they are intended as part of a purely descriptive project — can often have political and normative consequences.

Perhaps the most distinctive — and most complex — aspect of feminist metaphysics is the way in which metaphysical and political dimensions intersect. Typically, when feminists are interested in analyzing gender, they are not merely interested in giving a descriptive account of gender — they are also interested in the normative and political questions of how gender as it presently exists is unjust or harmful. But how, exactly, these descriptive and normative projects connect is tricky. From the outside, feminist metaphysics is often viewed as overly political insofar as it uses normative claims about what should or ought to be the case to argue for how things are in a way that's illegitimate. In what follows, I argue that this general skepticism is unwarranted, but that complicated issues arise in interpreting how exactly we should view the connection between the metaphysical and the political.

Typically, people doing traditional analytic metaphysics see themselves as engaged in a descriptive project. They are trying to describe the structure of reality — whatever it may be. Figuring out the correct theory of metaphysics may, of course, have normative consequences. But assessing those normative consequences is often thought to be a separate matter. So suppose, for example, that you think that reality is ultimately nothing but atoms in the void. Perhaps this means that God is dead and everything is permissible. Or perhaps it means that all normative facts are grounded in the non-normative. Or perhaps it means that morality is ultimately a matter of social contract, and is unconnected to metaphysics. Typically, we think that the project of metaphysics is just to tell us whether (basic? ultimate? fundamental?) reality is nothing but atoms in the void. After that, the meta-ethicists can tell us what that means for normativity.

There are, of course, no clear boundaries here. In many ways, some parts of meta-ethics are just the metaphysics of normativity. And it's not uncommon to see normative claims invoked in order to support a descriptive metaphysical conclusion. Philosophers have objected to Lewisian concrete modal realism based on the normative consequences.[7] They have argued for a constitution theory of material composition based on the moral properties of persons.[8] And in a particularly vivid example, Kris McDaniel argues against a particularly deflationary version of ways of being by stating:

> When I look at my children and I look at the heaps my children make, it seems impossible for me to view the ontological status of my children as being more like that of the heaps rather than, for example, a fundamental particle. Perhaps I am nothing in the eyes of God, but my kids damn well better not be!
>
> (2017, p. 190)

So the idea that the role of normativity in feminist metaphysics somehow sets it at stark disparity from other areas of metaphysics is a caricature. And yet it's hard to deny that discussions of feminist metaphysics are more deeply imbued with normative content than many other areas of metaphysics. Part of this is purely presentational. A philosopher might defend stage theory in part because it gives her an account of how her aging father is, in some sense, 'the same person' he used to be, even though his mental capacities have changed dramatically. Giving an account of this type of persistence through change might be incredibly important to her – it might well be the reason she became interested in persistence conditions to begin with – but she's unlikely to discuss this in print. In contrast, a philosopher might defend a view according to which there are many genders in part because she wants to be able to give an account of the lived experiences of non-binary people that she knows and loves. This latter philosopher is far more likely to mention this in print, since the norms of feminist philosophy allow that personal and political considerations can be philosophically motivating.

Yet the difference here is not entirely presentational. Perhaps the most striking place to see this illustrated is in the discussion of the *ameliorative project*. Sally Haslanger introduced the term 'ameliorative project' to describe a specific, politically oriented approach to answering the question 'what is x?,' where 'x' is some socially salient category, property, feature, or concept. Haslanger describes an ameliorative approach to answering the question 'what is x?' as an approach which seeks to answer the question by asking (at least in part) what way of understanding x would best serve 'our legitimate political and social goals.'

Haslanger's discussion of the ameliorative project has been both deeply influential to, and accurately descriptive of, much discussion in feminist metaphysics. For example, it's common in discussions of the metaphysics of gender to reference specific political goals – greater recognition of non-binary genders,[9] greater inclusion of trans people,[10] greater awareness of the social dimensions of gender inequality,[11] etc. – as a way of motivating a particular approach to gender, or of evaluating the success of different theories of gender. Claims like this, though, admit of multiple interpretations, as does Haslanger's basic definition of the ameliorative project. And, as we'll see, different interpretations of the ameliorative project give radically different pictures of the basic project of feminist metaphysics. In what follows, I outline options for how we might interpret the broader claims of the ameliorative project. This isn't intended as a piece of Haslanger exegesis – that is, I am making no claim about what Haslanger herself intends by the usage of her term 'the ameliorative project.' Rather, insofar as the ameliorative project characterizes a common feature within feminist metaphysics – and I think it does – these options represent various ways we might interpret the metaphilosophical upshot of that common feature.

i *The ameliorative project as theoretical holism*

On perhaps its simplest interpretation, the ameliorative project is just a claim about theoretical holism. We have some normative claims we're antecedently committed to (gender injustice is real and ought to be eliminated, trans rights are important, e.g.). When developing and evaluating theories of gender, we hold these normative claims as relatively fixed points, since we're more committed to them and more confident in their truth than we are in any particular metaphysical esoterica about the social world. The use of claims like this as a rubric for evaluating theories of gender is thus no different in kind than an antecedent commitment to the existence of ordinary objects motivating a rejection of compositional nihilism, or an antecedent commitment to the idea that time *really does pass* [bangs table] motivating A-theory. Theorizing has to start somewhere, and if you're much more confident that patriarchy exists and is bad than you are that social metaphysics is some particular way or other, that commitment about patriarchy can be a good place to start, and a good way to evaluate theories.

On this reading of the ameliorative project, there is nothing particularly distinctive about the role of normative or political commitments in the practice of feminist metaphysics.[12] At most, what's distinctive is that we're more likely to have antecedent normative commitments about gender than we are about, say, composition.

But this reading, although straightforward, struggles to make sense of Haslanger's claim that, in doing feminist metaphysics, we're asking not merely what kinds like gender are, we're asking what we want them to be – in the sense that we are asking what theory of social kinds will help us achieve our political and social goals. 'Patriarchy exists and is bad,' as an antecedent commitment, might constrain our viable options for theorizing about gender. But there's no obvious connection between that type of holistic constraint and thinking that a plausible theory of gender – one which explains the existence of patriarchy, for example – will somehow help us to make social progress. After all, it's perfectly consistent with thinking that patriarchy exists and is bad that it's also inevitable, or that the best way to combat it is just to ignore metaphysics entirely, or etc.

ii *The ameliorative project as a claim about the connection between social metaphysics and social justice*

We can, however, supplement the basic theoretical holism reading of the ameliorative project with a further claim about the relationship between social metaphysics and social progress: that understanding what social kinds like gender really are is one part of how we make social progress. On the face of it, this may sound overly flattering to metaphysics. But the connection here is not as tenuous as it might first appear. Haslanger, for example, has repeatedly emphasized the 'debunking' nature of many projects in feminist metaphysics.[13] Often what we are doing when we are giving an account of what gender is, for example, is explaining why folk ideas about gender are incomplete or mistaken. Plausibly, one of the major political goals of feminism is to change the hierarchical nature of gender (i.e., change the way in which people gendered as men have systematic advantages and people gendered as women have systematic disadvantages). If we can develop an explanation for why such hierarchy is not inevitable, or not ultimately reducible to biological differences between the sexes, or not due to essential differences in how men and women think, or etc., such an explanation might be a powerful tool in convincing others that our current gendered hierarchy both can and should change.[14]

Consider, for example, the somewhat functionalist spin that Haslanger (2005) puts on the ameliorative project. When engaged in an ameliorative analysis of gender, she argues, we are asking – in part – what the point of having a concept of gender is. That is, we are asking what explanatory work it does, and what advantage a theory that says that genders are real will have over an error theory or eliminativism about gender. Pair this claim with Haslanger's own analysis of gender, according to which gender is a social structure that systematically advantages those who are perceived to be male and systematically disadvantages those who are perceived to be female.[15] If Haslanger is right about gender, then the reason that genders exist – their function in our social world – is to give power to one set of people and constrain another set of people, based on assumptions about roles in biological reproduction. And if that is right, then gender injustice is built into the very nature of genders themselves (at least as they currently exist). Understanding that would be a huge piece of working toward gender justice, since it sets a very specific target (the elimination of gender kinds as they currently exist). And it would, arguably, be difficult to work effectively toward gender justice without understanding what genders really are.

Note, though, that on this reading of the ameliorative project the political dimension of metaphysical analysis is both limited and contingent. It is limited because, while an understanding of what genders really are (or aren't) might plausibly be helpful in working toward gender justice, large swathes of such work carry on just fine without the need for metaphysics. It is contingent, because while giving a metaphysics of gender might help in working toward our legitimate social goals, it also might not. Maybe the best metaphysics of gender is so complicated that it can only be understood and evaluated by a select group already deeply enmeshed in debates about metaphysics. (Hello, dear readers!) Or maybe the best metaphysics of gender is that, ultimately, there are just atoms in the void and everything else is a matter of concepts and language – an interesting claim, but not one where metaphysics per se is doing much to help us combat sexism. Or maybe the truth about gender turns out to be depressing – there really are innate, biologically ingrained differences between men and women that make men better suited to being in charge of stuff. To be clear, I do not think this is a plausible hypothesis! But I also don't think it's something we can rule out a priori. In keeping with the holism discussed previously, we might want to start from a claim like 'sexism exists and is bad.' But we still have to negotiate which things count as sexism, and we still have to figure out what the relationship is between the normative and the natural. Perhaps it really is true that men are superior to women, but also true that utility is maximized if we treat everyone as though they are equals, for example. Again, I don't think this is the actual scenario we find ourselves in! But it's at least epistemically possible.

More strongly, it is – on this reading of the ameliorative project – important not to rule such views out as an epistemic possibility. And that's because part of the social utility of a theory in metaphysics, on this view, is that it helps us explain the (perhaps surprising) structure of the social world. A theory of gender which explains gender in terms of social hierarchies might, in virtue of the social explanations it gives us, help us to combat assumptions about the biological inevitability of gender inequality. But it can only do that if the contingency of our current gender system is not, in a sense, 'baked in' to the analysis. If it's an a priori condition on the truth of any analyses of gender that gender is not a biological kind and that the social inequality of men and women is due to injustice, then analysis of gender lose their explanatory force. If we want them to be tools for social understanding, our theories of gender need to make the 'debunking' of naturalistic assumptions something like a discovery rather than an axiom.[16]

And so, on this reading of the ameliorative project, the claim that giving a theory of gender will help us to achieve gender justice – rather than, say, depress us or make us de-convert from our feminist convictions – is perhaps best understood as an optimistic bet, rather than a metaphilosophical guarantee. It is our *hope* for what we're doing in feminist metaphysics, but it cannot be, on such a reading, constitutive of what we're doing.

iii *The ameliorative project as a type of metaphysical anti-realism*

We can make the connection between the political and the metaphysical tighter, but only by making a much stronger – and more restrictive – metaphilosophical claim. Often, the ameliorative project is invoked in a way that – to varying degrees of explicitness – endorses a form of metaphysical antirealism and interprets the project of metaphysics (at least for social kinds like gender) as a species of conceptual analysis. There are lots of ways we can 'carve up' the world using our schemas and concepts. None is intrinsically better than the others, at least insofar as none does a better job of describing what the world is really like. (Describing what the world is 'really like' doesn't make sense, on such view – there are various conceptual carvings of what there is, and that's it.) Typically, when deciding how to describe the world – whether to use 'blue' or 'grue,' whether to use 'table' of 'simples-arranged-tablewise,' etc. – we make choices based on a range of factors. What fits best with folk practice? What lets us communicate most effectively? What's the simplest? And so on. But when it comes to social kinds like gender, a major factor – perhaps even the only factor – we ought to consider is: what best helps us make the political progress we want to make?[17]

On this reading of the ameliorative project, there are many different ways that we could think and speak about gender. That is, there are quite literally many different ways we could 'construct' gender. None does a better job than the others of somehow describing the structure of the social world. The ameliorative project then asks, for kinds like gender, 'what do we want them to be?' in a way that's quite straightforward. There are lots of different concepts that could serve as 'our' gender concept. Pick the one that does the political work you want it to do. That is what gender is.

For example, suppose you think it's a major political goal that we affirm the gender self-identification of people who are non-binary. Some gender concepts are binary; others are non-binary; the ameliorative project tells you to pick one that is non-binary. Then when you are asked whether gender is non-binary, the answer is an unequivocal 'yes.'

This type of approach to the ameliorative project engages in what is sometimes called 'conceptual engineering.'[18] The project of feminist metaphysics, on such a reading, is not to describe the basic structure of the social world. It is, rather, to make politically informed choices about how we should think and speak about the world.

iv *The ameliorative project as a claim about social salience*

A separate, intriguing way making a strong link between the political and metaphysical has recently been advocated by Robin Dembroff (forthcoming). Rather than endorsing a type of ontological antirealism in order to justify an ameliorative approach, Dembroff instead endorses a type of ontological *permissivism*. There are – really are, out there in the world, not dependent on what conceptual schema we're invoking in a particular context – many, many different, and closely related, social kinds. So, for example, there is no single social kind that is, independent of context, the uniquely best candidate for being the reference of our gender term 'woman.' Rather, there are many social kinds which differ in their inclusion and exclusion criteria – some based on visual appearance, some based on biological characteristics, some based on self-identification, and so on. And our usage of gender terms doesn't map neatly onto this (incredibly complex) social reality. In some contexts, it might be the case that none of these kinds is uniquely socially salient, leaving the reference of 'woman' multiply ambiguous in that context. In others, it might be the case that one such kind is made particularly salient, and thus is the unique referent of 'woman,' but this is due entirely to contingent (and often unjust) facts about what we've decided to pay attention to in that particular context.

The ameliorative project then becomes a question of social salience: for all of these overlapping social kinds, which should we choose to give social salience and social significance?[19] For example, trans-inclusive and non-binary gender kinds are often not salient in many social contexts – we ignore them, we're unaware of how they function in queer communities, we don't use them as the basis for our terminology or our laws, etc. But the very same gender kinds are far more salient in the context of, e.g., queer communities. The question 'what do we want gender to be?' can thus be interpreted as a question about which social kinds we want to *make* salient. The metaphysical question, on Dembroff's view, is about which social kinds exist (lots) and in virtue of what they exist (the collective norms, beliefs, and social practices of communities, including queer and alternative communities). The political question is about which such kinds we should pay most attention to, use as the basis for terminology and legislation, and so on. There's a close link here between the political and the metaphysical in both directions. The underlying metaphysics (commitment to a plurality of overlapping social kinds) is part of what explains the legitimacy of arguing for a change in our norms about gender. At the same time, our collective norms about gender are part of what create and sustain individual social kinds, such that politically motivated changes in those norms can alter what gender is (over and above altering how we think or speak about it).

v *Some concerns for the ameliorative project*

Let's turn now to more general questions about how to interpret the sort of connection between political goals and metaphysical inquiry that the ameliorative project endorses. The first and most obvious question is how we should think about *politically effective falsehoods*. It seems quite clear, when we examine the history of social movements, that truth and political effectiveness can easily come apart. Consider, for example, the history of the gay rights movement. The medicalization of homosexuality and the claim that being gay was a type of sexual deviance was originally a claim championed by many gay rights activists – and it's quite plausibly a claim that, in its original social context, did a lot to help people.[20] In a context in which gay people were being imprisoned, chemically castrated, beaten, and killed, the idea that they were simply suffering from a medical condition – and thus deserved pity rather than blame and scorn – was considered a step in the right direction by many gay rights activists. But, of course, this was a limited strategy, and while it might've made things a little better, it didn't make them *good*. It's, of course, false that being gay is a mental disorder, and even if it might have had some limited good effects, it's ultimately wrong and unhelpful to medicalize it. And so the mantra of 'born this way' became central to the cause of de-medicalizing sexuality. The idea that being gay is 'natural,' and that people are simply born gay or straight, was a powerful way of combatting the pathologization that gay people faced, and was uncontroversially politically effective. But it's also – at least as it was widely understood in popular culture – probably false. That is, people probably don't come into the world stamped 'gay' or 'straight' and human sexuality is probably a complex mixture of genetic predisposition and socialization, which for some people (especially women) can be very fluid over their lifetimes.

I doubt very much that examples like this, in which we say something that has politically good effects, but which is false, are rare. I've argued elsewhere, for example, against strong versions of the Social Model of disability – views which says that disability is entirely constituted by social prejudice against people with impairments, and that in the absence of such prejudice impairments would be little more than a minor nuisance. But in arguing that such views of disability are false, I'm of course not denying that they've been remarkably politically effective. The widespread adoption of the Social Model within disability rights communities was a huge part of the political push toward the Americans with Disabilities Act, for example. That doesn't – at least in my view – make it *true* though. And it's important to note that, while it's had tremendous political benefit, it's also had some political downsides. People with medically complex, degenerative, or painful disabilities have often found the 'mere nuisance' claim both implausible and alienating, for example, and some have argued that, in (rightly) attempting to emphasize the social dimensions of disability, it's actually swung the pendulum too far in the other direction, making disability seem overly disembodied and the physical aspects of disability harder to talk about.

There are two key points to emphasize here. One is that things which are false can nevertheless be politically effective. The other is that 'political effectiveness' is rarely an all-or-nothing matter. Often, social progress is incremental and limited political strategies – strategies which have some political downsides, don't manage to fully include everyone who ought to be included, etc. – can still be important and worthwhile strategies.

With these points in mind, it becomes tricky to assess what, exactly, ought to be involved in considering 'our legitimate political and social goals' as part of a process of theorizing what there is and how it is. This is less of concern for the weaker readings of the ameliorative project. But for stronger readings, things quickly become complicated. If political effectiveness is part of what makes a view in metaphysics *true* or part of what determines what social kinds *exist*, then the possibility of a politically effective falsehood becomes obscured, and the vagueness in what counts as 'political effective' becomes a serious – and metaphysically very complicated – problem.

A second, related, concern for ameliorative approaches to metaphysics involves the difference between pragmatic and theoretical virtues. Vagueness, for example, is often very useful when making laws, policies, or mandates about social kinds, and such policies are often written in ways that leave them intentionally vague. This is unsurprising, given that social kinds are complicated, language is imprecise, and it's unlikely that any particular set of policy makers will be able to anticipate all scenarios. Highly specific policies often inadvertently and unexpectedly place constraints on their implementation in ways that make them less effective, and it's often better to have vaguely worded policies that allow for discretion, interpretation, and contextual variation in their implementation. This, by itself, however, is not a good argument that social kinds *really are* vague. (Though they might be!) It's politically effective for us to think and speak about them in a way that's vague – especially when making laws and policies. But that might be a simple function of our own cognitive and moral limitations, rather than a deep insight into what the social world is like.

In other cases, laws and policies are forced – for simple pragmatic reasons – to create precision. Suppose for the sake of argument that human sexuality is in fact a continuous spectrum and that discrete 'orientations' or 'orientations categories' don't actually correspond to anything in the world. Were this the case, it might nevertheless be highly politically effective to write laws and policies in a way that refers to orientations like 'gay' and 'bi.' In this scenario, there's no metaphysical distinction between straight and queer – there's just a continuous spectrum of human sexual attraction – but there might nevertheless be *political* use for a distinction between straight and queer. The simple fact is that, given the way the world is, some aspects of human sexual attraction face more social stigma than others, and as a result need social and legal protections. The simplest and most effective way to do this might be to reify distinct categories – to draw a sharp line between the straight and the queer where, in reality, there's only a spectrum – because it might be the most straightforward way of writing policies that serve to protect vulnerable people.

Similarly, I find it plausible that the distinction between the disabled and the non-disabled is riddled with vagueness. And yet, given the way the world is, there are plenty of cases in which it is politically effective – perhaps even politically necessary – to talk about disability as though this were not the case. Often, when it comes to disability, we need to draw a sharp line where in reality things are probably very fuzzy, simply because we need to make practical decisions about who receives the limited resources we have for accommodations, who gets to participate in the Paralympics, and so on. Creating a line between the disabled and the non-disabled and sorting people onto either side of it is, in many contexts, very politically important. But it doesn't follow, of course, that in doing this we're limning the structure of social reality. Precision might, in such contexts, be a pragmatic virtue in thinking and speaking about disability, but it doesn't follow that precision in a theory of disability as a social kind would likewise be a virtue.

What cases like these highlight is that often, in talking about social kinds, things which help our understanding or further our social goals do so because of both our own limitations and the limitations of the social context we find ourselves in. It can be pragmatically – perhaps even epistemically – useful to speak out about social kinds in ways that, in principle, come apart from what those kinds are really like. And one worry for the ameliorative project is that, depending on how strongly we construe the relationship between our political goals and our metaphysical commitments, we might lose track of this contrast.

Notice, however, that the extent to which these worries are pressing depends which interpretation of the ameliorative project we endorse. On weaker readings, in which the ameliorative project is more or less a species of theoretical holism, we have straightforward responses. Because it's no part of what makes our metaphysical theories *true* that they have a specific political role, there's no problem allowing for politically effective falsehoods. Likewise, such views can easily distinguish between pragmatic and theoretical virtues, since the claim is simply that a better theoretical understanding of social kinds might – by enhancing our understanding of those kinds – be one small part of improving our work to justice. So there's no claim that, for example, we can 'read off' what social kinds exist or what they are like from what is in fact the most pragmatically effective way of speaking about such kinds.

But for stronger versions like (iii), where the connection between the metaphysical and the social is much tighter, these questions loom large. If we make political effectiveness (which will inevitably include pragmatic constraints) part of the truth conditions for our theories, then it becomes harder to explain politically effective falsehoods, more pressing to say exactly what counts as 'politically effectiveness,' and harder to distinguish between pragmatic and theoretical virtues.

Finally, though, there is a worry for all versions of the ameliorative project that is worth thinking about explicitly: elitism. Because most of us who do feminist philosophy care very much about gender and gender justice, we all would like to think that our work matters to these efforts for justice. But there is something uncomfortable about saying that a group of elite and privileged scholars who are attempting to wade through the esoterica of what gender is and how it works are thereby making the world a better place. It is not obvious, for example, that grassroots efforts for gender justice need the work that we're doing, and in many cases our efforts seem (quite appropriately) reactive to the views and norms of those efforts, rather than the other way around.

Feminist philosophy is valuable and important – as, I firmly believe, is all philosophy. But there is a concern that, at least in some ways of framing the ameliorative project, we risk overstating its *political* importance, and thereby straying into somewhat aggrandized claims about our own role in social progress. The value of feminist metaphysics might primarily be for helping the esoterically inclined to understand and reason about the rich, complex, and fascinating world of gender. And we can, of course, hope that this will be politically influential for people like us. (That is, people inclined to read philosophy papers, which to be fair is not most people.) And we can also hope for the gradual osmotic effect that philosophical ideas can often have in communities. But the political upshot of all of this might be relatively mild. And even if it is, the value of philosophical exploration, for topics like gender, is value in its own right.

Notes

1 See, inter alia, Haslanger (2005), Alcoff (2006), Dembroff (2016), Ásta (2018).
2 As in, for example, Lugones (1990), Narayan (1998), Haslanger (2016), Thomasson (2018), Ritchie (2018).
3 Some representative examples (from among very, very many) might include Haslanger (2000a), Calhoun (2001), Narayan (2004), Alcoff (2006), Witt (2011), Bach (2012), Bettcher (2013), Jenkins (2016).
4 For extended discussion of this topic as it relates to broader debates in metaphysics, see Barnes (2014, 2017), Mikkola (2016), Sider (2017).
5 On the interpretations of 'fundamental' typically employed in fundamentality-centric metaphysics, it seems fairly clear that genders and the like are not fundamental. There are, of course, other interpretations of 'fundamental,' some with rich philosophical pedigrees, according to which gender could plausibly be said to be fundamental. I'm not attempting to adjudicate which reading of fundamental is better – rather, I'm just using 'fundamental' as it's often used in contemporary analytic metaphysics, and assuming that on that reading genders (and other social kinds and properties) are not fundamental. See Barnes (2014, 2017).
6 See, inter alia, Anderson (1995), Haslanger (2000b, 2003, 2005), Alcoff (2006) and see Epstein (2010) for a helpful overview.
7 For discussion of this objection to concrete modal realism see especially Adams (1974), Lewis (1983, ch. 3), Heller (2003).
8 Baker (1999).
9 Dembroff (2016), Ásta (2018).
10 Bettcher (2014), Jenkins (2016).
11 Haslanger (2000a), Alcoff (2006), Witt (2011).
12 See Mikkola (2015) for a defense of something like this view. It's also relatively similar to the version of the ameliorative project I endorse in Barnes (2016).
13 Haslanger (2003).
14 See Barnes (2017).
15 Haslanger (2000a).
16 Thanks to Liam Kofi Bright for this framing.
17 See Diaz Leon (2018) and Haslanger (2018) for discussion of this particularly as it relates to Haslanger's own use of the ameliorative project. See also, especially, Dutilh Novaes (2018) and Thomasson (forthcoming).
18 See Burgess and Plunkett (2013) for a helpful overview.
19 Although she is less concerned with framing it as a debate about metaphysics, Bettcher (2013) defends a somewhat similar view.
20 See Minton (1996).

References

Adams, R. M. (1974). Theories of Actuality. *Noûs*, 8(3), 211–231.
Alcoff, L. M. (2006). *Visible Identities: Race, Gender, and the Self*. New York: Oxford University Press.
Anderson, E. (1995). Knowledge, Human Interests, and Objectivity in Feminist Epistemology. *Philosophical Topics*, 23(2), 27–58.
Ásta (2018). *Categories We Live By*. Oxford: Oxford University Press.
Bach, Theodore (2012). Gender Is a Natural Kind with a Historical Essence. *Ethics*, 122(2), 231–272.
Baker, L. R. (1999). What Am I? *Philosophical and Phenomenological Research*, 59(1), 151–159.

Barnes, E. (2014). Going Beyond the Fundamental: Feminism in Contemporary Metaphysics. *Proceedings of the Aristotelian Society*, 114(3pt3), 335–351.
Barnes, E. (2016). *The Minority Body: A Theory of Disability*. Oxford: Oxford University Press.
Barnes, E. (2017). Realism and Social Structure. *Philosophical Studies*, 174(10), 2417–2433.
Bettcher, T. M. (2013). Trans Women and the Meaning of 'Woman.' In *Philosophy of Sex: Contemporary Readings*, A. Soble, N. Power, and R. Halwani (eds.), 6th ed., Lanham, MD: Rowan & Littlefield, pp. 233–250.
Bettcher, T. M. (2014). Trapped in the Wrong Theory: Rethinking Trans Oppression and Resistance. *Signs*, 39(2), 383–406.
Burgess, A., and Plunkett, D. (2013). Conceptual Ethics I&II. *Philosophy Compass*, 8(12), 1091–1101.
Calhoun, C. (2001). Thinking About the Plurality of Genders. *Hypatia*, 16(2), 67–74.
Dembroff, R. A. (2016). What Is Sexual Orientation? *Philosophers' Imprint*, 16(3), 1–27.
Dembroff, R. (forthcoming). Real Talk on the Metaphysics of Gender. *Philosophical Topics*.
Diaz Leon, Esa (2018). Kinds of Social Construction. In *Bloomsbury Companion to Analytic Feminism*, Pieranna Garavaso (ed.), London: Bloomsbury Academic, pp. 103–122.
Dutilh Novaes, C. (2018). Carnapian Explication and Ameliorative Analysis: A Systematic Comparison. *Synthese*, 1–24.
Epstein, B. (2010). History and the Critique of Social Concepts. *Philosophy of the Social Sciences*, 40(1), 3–29.
Haslanger, S. (2000a). Gender and Race: (What) Are They? (What) Do We Want Them to Be? *Noûs*, 34(1), 31–55.
Haslanger, S. (2000b). Feminism and Metaphysics: Negotiating the Natural. In *The Cambridge Companion to Feminism in Philosophy*, M. Fricker and J. Hornsby (eds.), Cambridge: Cambridge University Press, pp. 107–126.
Haslanger, S. (2003). Social Construction: The "Debunking" Project. In *Socializing Metaphysics*, Frederick Schmitt (ed.), Lanham, MD: Rowman & Littlefield.
Haslanger, S. (2005). What Are We Talking About? The Semantics and Politics of Social Kinds. *Hypatia*, 20(4), 10–26.
Haslanger, S. (2016). What Is a (Social) Structural Explanation? *Philosophical Studies*, 173(1), 113–130.
Haslanger, S. (2018). Social Explanation: Structures, Stories, and Ontology. A Reply to Díaz León, Saul, and Sterken. *Disputatio* 10(50), 245–273.
Heller, M. (2003). The Immorality of Modal Realism, Or: How I Learned to Stop Worrying and Let the Children Drown. *Philosophical Studies*, 114(1–2), 1–22.
Jenkins, K. (2016). Amelioration and Inclusion: Gender Identity and the Concept of Woman. *Ethics*, 126(2), 394–421.
Lewis, D. K. (1983). *On the Plurality of Worlds*. Oxford: Wiley-Blackwell.
Lugones, Maria C. (1990). Structure/Antistructure and Agency under Oppression. *Journal of Philosophy*, 87(10), 500–507.
McDaniel, Kris. (2017). *The Fragmentation of Being*. Oxford: Oxford University Press.
Mikkola, M. (2015). Doing Ontology and Doing Justice: What Feminist Philosophy Can Teach Us About Meta Metaphysics. *Inquiry: An Interdisciplinary Journal of Philosophy*, 58(7–8), 780–805.
Mikkola, M. (2016). On the Apparent Antagonism Between Feminist and Mainstream Metaphysics. *Philosophical Studies*, 174(10), 1–14.
Minton, H. (1996). Community Empowerment and the Medicalization of Homosexuality: Constructing Sexual Identities in the 1930s. *Journal of the History of Sexuality*, 6(3), 435–458.
Narayan, Uma (1998). Essence of Culture and a Sense of History: A Feminist Critique of Cultural Essentialism. *Hypatia*, 13(2), 86–106.
Narayan, Uma (2004). Undoing the 'Package Picture' of Cultures. *Signs: Journal of Women in Culture and Society*, 25(4), 1083–1086.
Ritchie, K. (2018). Social Structures and the Ontology of Social Groups. *Philosophy and Phenomenological Research*, 100(2).
Sider, T. (2017). Substantivity in Feminist Metaphysics. *Philosophical Studies*, 174(10), 2467–2478.
Thomasson, Amie L. (2018). Changing Metaphysics: What Difference Does it Make? *Royal Institute of Philosophy* Supplement 82, 139–163.
Thomasson, Amie L. (forthcoming). The Ontology of Social Groups. *Synthese*, 1–17.
Witt, C. (2011). *The Metaphysics of Gender*. Oxford: Oxford University Press.

24
SOCIAL ONTOLOGY

Rebecca Mason and Katherine Ritchie

Traditionally, social entities have not fallen within the purview of mainstream metaphysics. For example, very few original research articles on social metaphysics have been published in top philosophy journals.[1] Moreover, only one metaphysics textbook includes social metaphysics as a topic.[2] This is particularly striking in view of the fact that there has been work on social ontology for decades.[3] Here, in addition to surveying the field of social ontology, we consider whether the exclusion of social entities from mainstream metaphysics is philosophically warranted or if it instead rests on historical accident or bias.

We examine three ways one might attempt to justify excluding social metaphysics from the domain of metaphysical inquiry. Metaphysical inquiry, as we construe it, includes both first-order metaphysics and metametaphysical questions about the nature of metaphysics. Given this construal, we argue that each of the arguments fails and conclude that the exclusion of social entities, properties, kinds, and facts from mainstream metaphysics is unjustified. Further, we show that broadening the scope of metaphysics to include a focus on the social requires us to rethink some commonplace metaphysical assumptions.

The chapter is structured as follows. In Sections I–III, we outline arguments from the literature that might be used to provide justification for the view that metaphysics need not focus on social entities. With respect to each argument, we show that it fails to justify the exclusion of social entities from metaphysical inquiry. In Section I we consider whether eliminativism or reductionism about social entities is true. If either is true, one might think that metaphysics ought not focus on the social, for there are no (irreducible) social entities on which to focus. In Section II we consider whether metaphysics ought to focus exclusively on fundamental entities. Since social entities are plausibly not fundamental, this could be used to justify the exclusion of social entities. In Section III we consider the view that metaphysics should focus on natural kinds or properties. Finally, in Section IV, we consider what metaphysical inquiry that includes social entities as central examples would look like. We gesture towards the view that starting from examples of social entities leads us to rethink the assumption that describing reality in terms of intrinsic, independent, and individualistic features is preferable to describing it in terms of relational, dependent, and non-individualistic features.

Before turning to our main arguments, we outline two general ways social entities might depend on social factors and set out a range of examples of social entities. Social entities might depend on social factors causally or constitutively. The following theses (adapted from definitions of forms of social construction from Haslanger 2003 and Mallon 2014) give a feel for the difference:

> *Causal Dependence*: X (being F) is *causally* dependent on social factors if and only if social factors (partially) cause X to exist (as F).
>
> *Constitutive Dependence*: X (being F) *constitutively* depends on social factors just in case (i) in defining what it is to be X (or for X to be F) reference must be made to some social factors or (ii) social factors are metaphysically necessary for X to exist (as an F) or (iii) social factors ground the existence of X (or the fact that X is F).[4]

An entity might causally depend on social factors like human behavior, practices, beliefs, and so on without constitutively depending on social factors. For instance, compounds that are synthesized in a chemistry lab are causally dependent on social practices. Yet, such compounds are plausibly not constitutively dependent on social factors. That is, they are definable without reference to social factors, social factors are not metaphysically necessary for the compound to exist, and the existence of the compound is not grounded in social factors. In contrast, universities, gender, and inflation might be both causally and constitutively dependent on social factors.

When we use "social entity" we intend for it to be broadly construed so as to include entities of the following sorts:

> *Social properties and relations* (e.g., being married, being a U.S. citizen, being a manager, having more buying power than);
>
> *Social facts* (e.g., that the Supreme Court exists, that Kai has four hundred dollars in her bank account, that the United States has a larger military than Gambia);
>
> *Social kinds* (e.g., money and marriage, war and women, capitalists and cartels, races, recessions, and refugees);
>
> *Social groups* (e.g., racial and gender groups, the Minnesota Twins, the Supreme Court, Migos);
>
> *Social institutions* (e.g., universities, corporations);
>
> *Social structures* (e.g., capitalist power structures, oppressive gender structures, the structure of the U.S. government).

In what follows, we note when arguments apply to some and not other social entities.

I Eliminativism and reduction

Our world appears to include primates, pears, and paper, trees, telephones, and tapirs. It also seems to include ethnic groups, parliaments, nations, bands, and sports teams. Common sense and our everyday experiences seem to confirm that there are social entities. Are appearances misleading? That is, are there any social entities?

Eliminativists about Fs hold that there are no Fs. Eliminativism about social entities can be argued for in three main ways. First, one might be motivated to reject social entities given social analogs to puzzles about composition (e.g., Sorites paradoxes, Ship of Theseus puzzles, the puzzle of the statue and the clay). If one is inclined to hold that ships, cups, and dogs do not exist due to these puzzles, one might be similarly inclined to argue that social groups and at least some other social entities do not exist (van Inwagen 1990; Unger 1980).

However, many philosophers reject drawing nihilist or eliminativist conclusions from these puzzles. For instance, some argue that puzzles about composition motivate accepting mereological universalism (Lewis 1991; Sider 2001). Others argue that the puzzles can be solved while maintaining a restricted, "common sense" ontological view (Korman 2015; Markosian 1998, 2008). Whichever view one is inclined to adopt in the realm of non-social material objects one will plausibly be inclined to adopt for material social entities as well.

Second, one might hold eliminativist views about specific social entities. Some argue that there are no racial groups because there are no biological racial essences (Appiah 1996; Zack 1993, 2002). For instance, Appiah argues that there are not racial groups whose members share "fundamental, heritable, physical, moral, intellectual, and cultural characteristics with one another" (1996, 54). He argues that the existence of races relies on there being biological racial essences. Thus, the lack of shared, biological racial essences entails that there are no races. Similar arguments might be posed for gender, ethnic, sexuality, and other sorts of social groups.

The second eliminativist argument is also widely disputed. Many argue that racial, gender, and other groups exist, but maintain that the existence of such groups does not rely on there being shared, biological essences. On some views concepts for social kinds are taken to be cluster concepts, so that there need not be *an* essence to a kind (Corvino 2000; Hale 1996; Heyes 2000; Outlaw 1996; Stoljar 1995, 2011). Much of the work on social construction in philosophy involves arguing that there are social kinds, but they are dependent on social practices not on shared biological essences (Alcoff 2006; Blum 2010; Diaz-Leon 2015a, 2015b; Haslanger 2000, 2003, 2012; Jeffers 2013; Mallon 2006, 2016; Mills 1997, 1998; Sundstrom 2002; Taylor 2004). For instance, Blum argues that "racialized groups are characterized by forms of experience they have undergone and a sociohistorical identity that they possess *because of* the false attributions to them ... of innate biobehavioral tendencies" (2010, 300). Similarly, Haslanger argues that a view of gender groups should acknowledge "the causal impact of classification" (2003, 315). Hacking's (1996, 1999) notion of looping effects on human kinds is one way to understand the effect of classification. He takes classifying to potentially elicit changes in behavior and self-conception thereby modifying the properties instantiated by kind members, which leads to another "loop" that can elicit further changes.

Moreover, it is possible that although racial, gender, and other social groups do not have shared, biological essences, they have shared, non-biological essences (Witt 1995, 2011a, 2011b). Charlotte Witt argues that we should not conflate essentialism with biologism because biological descriptions are only one way of specifying the essence of these groups. Further, she argues that social constructionism is compatible with essentialism (1995). It follows that by demonstrating that social groups are not unified by shared, biological essences one does not thereby demonstrate that those groups do not exist.

Third, individualists in the social sciences (Hayek 1955; Popper 1966; Weber 1922/2013) have argued that social entities are not required for explanation and that there are no irreducible social entities. Explanatory or methodological individualists hold that explanations can be given solely in terms of individuals and their actions rather than "spooky" social forces. Ontological individualists argue that there are no irreducible social entities.

Arguments relying on ontological and methodological individualism are distinct from the other eliminativist arguments we considered because they potentially eliminate *all* social entities.

In contrast, arguments based on puzzles of composition apply only to material social objects (e.g., groups). They do not apply, for instance, to social facts or properties.[5] Arguments based on essentialism apply only to racial, gender, and other social kinds that are loci of oppression and possible sites for social justice projects. However, an explanatory individualist aims to show that all social entities – including social facts, properties, and so on – are explanatorily superfluous. An ontological individualist might argue that there are no social entities whatsoever.[6]

In their strongest forms, both individualist theses rely on reduction.[7] On a reductive view about Fs, there are Fs, but they are nothing "over and above" some other things G. For instance, according to identity theory in philosophy of mind there are mental states, but they just are brain states (i.e., they are identical to brain states). According to individualists, social facts or other social entities are nothing over and above non-social facts or entities.

The success of explanatory individualism relies on the reduction of social facts to non-social facts. The success of ontological individualism (which need not be paired with explanatory individualism) requires that all purported social entities can be successfully reduced to individuals. Ruben (1985) considers various non-social reduction bases for nations and finds each wanting. Epstein (2009, 2015b) argues that even a weaker version of individualism relying on supervenience rather than reduction fails. He cites cases of supervenience failing for certain social groups.

Note further that a non-circular definition of "social" appears to be required if individualist theses are to get off the ground. We will not canvass an array of definitions of "social" that could be offered, but a successful definition will need to be broad enough to capture the range of social entities like those listed above, as well as social beliefs, actions, habits, desires, and so on.

Several philosophers have expressed skepticism about drawing a sharp distinction between social and non-social individualistic facts. For instance, Haslanger holds that the possibility of giving a non-circular definition of "social" is unlikely (2016, fn 8). Similarly, Epstein (2015b, 102) states that he is "not confident" that a clear distinction can be drawn. He notes that individualism is viable only if there is a clear distinction "[o]therwise, it is pointless for [an individualist] to assert that the social facts are exhaustively 'built out of' the individualistic ones" (ibid.). If no sharp distinction can be drawn, the potential for the success of individualism is undermined.

While we have offered criticisms against eliminativist and reductionist strategies, notice that even if a reductionist argument is successful, the exclusion of social metaphysics from metaphysical inquiry is not warranted. For instance, suppose social entities are identical to some non-social entities (i.e., a social version of identity theory is true). On this view, social entities are still part of metaphysics. The claim is that they are identical to non-social entities, not that they fail to exist. Further, even if one posits that certain social kinds do not exist, that does not mean that questions about *whether* they exist fall outside of metaphysical inquiry. For instance, the view that there are no racial groups is a metaphysical view. When engaged in inquiry about what exists, one is engaged in ontological and metaphysical inquiry. So, if one wants to argue for eliminativism about social entities one is doing both metaphysics and social ontology. Therefore, eliminativism does not provide a reason to exclude questions of social ontology from broader metaphysical inquiry, even if ultimately one wants to argue that there are no social entities.

While ontological questions about the existence of certain sorts of entities have been common in metaphysics, they have fallen somewhat out of favor in contemporary metaphysics, which favors asking questions about fundamentality, grounding, and dependence, rather than existence. Thus, we next consider views according to which social entities exist but are not fundamental or independent.

II Fundamentality and mind-dependence

Another reason for excluding social entities from metaphysics is the view that metaphysics is, or should be, solely concerned with entities that are metaphysically fundamental. For example, one might characterize metaphysics as the study of the *fundamental structure* of reality (Sider 2011).[8] If metaphysics is the study of the fundamental structure of reality, and social entities are not fundamental, then they have been excluded from metaphysical inquiry for good reason. There are, however, several problems with this line of argument.

First, many traditional metaphysical disputes are about *whether* some phenomenon is fundamental (Bennett 2017, 232). If metaphysics is only concerned with fundamentalia, then the question of whether these debates have a metaphysical subject matter depends on who is correct about the fundamentality of the entities in question. This is extremely counterintuitive. For example, consider the question of whether grounding is a fundamental relation. It is not the case that grounding falls within the domain of metaphysical inquiry only if the answer to this question is yes! The dispute itself is metaphysical either way. In other words, when engaged in metametaphysics about the features of the entities in one's ontology or the nature of a particular entity one is engaged in metaphysics.

However, let's set that consideration aside for the moment. Instead, suppose that the aforementioned characterization of metaphysics is correct. That is, suppose that metaphysics is the study of the fundamental structure of reality. It does not immediately follow that social entities can be excluded from the domain of metaphysics on these grounds because it is possible that social entities *are* metaphysically fundamental. That is, if metaphysics is the study of the fundamental structure of reality, but social entities turn out to be fundamental, then social entities are a proper topic of metaphysical investigation.

To illustrate this possibility we consider two views according to which social entities can be fundamental. First, Barnes (2012) develops a view of ontological emergence according to which emergent entities are those that are dependent, but fundamental. While this view of emergence is controversial, it demonstrates the possibility that social entities are fundamental even though they are dependent. Barnes applies her characterization of ontological emergence to debates about minds, living beings and persons, composite objects in gunky ontologies, tropes, and certain quantum phenomena (e.g., quantum entanglement). However, she indicates that entities and phenomena other than these could be characterized as ontologically emergent in ways that are both unmysterious and theoretically useful. Indeed, elsewhere Barnes suggests that one way of understanding Haslanger's constructionist account of social structures is that some of their properties are emergent (Barnes 2017, 2424). On an emergentist interpretation of Haslanger's view, certain properties of social structures emerge from, and depend on, our thoughts and practices. That is, they are constructed from complex patterns of social interaction – but the structures themselves are ontologically fundamental.

Second, Sara Bernstein (forthcoming) argues that "middleism" (the thesis that some middle level is fundamental), is at least as plausible as "topism" (the thesis that the top-most level is fundamental, e.g., the cosmos) and "bottomism" (the thesis that the bottom-most level is fundamental, e.g., mereological atoms). She develops middleism with respect to middle-sized dry goods, but notes that her arguments apply *mutatis mutandis* to any entities which do not inhabit the top-most or bottom-most level. Since social entities occupy neither the top-most or the bottom-most level, then, according to middleism, they are candidates for being fundamental. If either Barnes's or Bernstein's views are viable, then social entities fall within the purview of metaphysics even when it is defined as the study of the fundamental structure of reality.

Nevertheless, there are other reasons why we should reject the idea that metaphysics is the study of the fundamental structure of reality. First, suppose that priority monism is true. In that case, the only fundamental entity is the entire cosmos (Schaffer 2010). It would not follow that the cosmos in its entirety is the only proper subject of metaphysical investigation. Presumably, there would still be a wide variety of phenomena for metaphysicians to investigate other than this maximally inclusive whole – namely, its proper parts, and their relationships to each other and the whole.

Second, it is at least epistemically possible that there are no entities that are metaphysically fundamental. That is, it's possible that there are no fundamentalia (Lewis 1991; Schaffer 2003; Sider 1993). If metaphysics is the study of the fundamental structure of reality, and there are no fundamentalia, then metaphysics lacks a subject matter. But presumably the discovery that reality lacks a fundamental structure would not thereby eliminate the subject matter of metaphysics. Karen Bennett puts the point vividly: if it turns out that there are no fundamentalia "metaphysicians will certainly take notice – but they will not *give* notice, and resign their jobs" (2017, 231). Thus, metaphysics should not be characterized as the study of the fundamental structure of reality.

Furthermore, it is plain that metaphysics is *not* solely concerned with entities that are metaphysically fundamental. Many traditional metaphysical questions obviously concern non-fundamentalia. For example: Do persons have free will? What is the relationship between a statue and the lump of clay that constitutes it? Are biological species individuals or kinds? Are mental states identical to physical states of the brain? These questions concern entities that are on many accounts non-fundamental, i.e., persons, statues, biological species, and mental states. If metaphysics actually answers questions about non-fundamental entities then metaphysics does not exclusively concern the fundamental structure of reality.

Moreover, there are various alternative conceptions of metaphysics on offer that include social entities. For example, building on a tradition going back to Aristotle, Jonathan Schaffer argues "metaphysics is about what grounds what" (2009, 347). On his view, grounded – that is, non-fundamental – entities are not outside of the metaphysical domain. Similarly, Bennett argues that "the proper topic of metaphysics is the fundamental structure of reality, whether there are any less fundamental entities, how they are built from the fundamental, *and* at least some of those nonfundamental entities themselves" (2017, 214). On both of these views of metaphysics, social entities do not fall outside of the metaphysical domain.

A related reason for excluding social entities from the domain of metaphysics rests on the view that metaphysics should be concerned with describing the mind-independent nature of reality. According to this proposal, the subject matter of metaphysics does not exclude all dependent entities; rather, it excludes only those that depend on our mental states in particular. Plausibly, all social entities are mind-dependent in some sense or other. Therefore, on this proposal, metaphysics does not include them.

However, we argue that there are problems with this characterization of metaphysics as well. Mental states are mind-dependent. According to intuitionists (e.g., Brouwer 1981), mathematical entities are too. Yet, inquiry into these entities clearly falls within the domain of metaphysics. Furthermore, secondary qualities and response-dependent properties are mind-dependent, but are likewise proper subjects of metaphysical inquiry.

We conclude that social entities cannot be justifiably excluded from metaphysics because they fail to be fundamental. First, it is possible that social entities are fundamental. Second, and more importantly, it is not the case that metaphysics is merely the study of the fundamental structure of reality. Traditionally, metaphysicians have investigated both fundamental and non-fundamental entities, and their relations. Moreover, the question of whether some entities are fundamental is a properly metaphysical one. Finally, we argued that social entities cannot be justifiably excluded on the basis of being mind-dependent. Metaphysicians have long been concerned to investigate mind-dependent entities. In the next section, we consider whether social entities can be justifiably excluded because they fail to be natural.

III Naturalness

So far we have argued that social entities cannot be excluded from the domain of metaphysics on the grounds that they are not fundamental or mind-independent. Eliminativism and reductionism about social entities also fail to justify the exclusion of social ontology from metaphysical inquiry. In this section, we consider whether social entities can be justifiably excluded because they fail to be *natural* in some sense (Khalidi 2015; Thomasson 2003). We argue that this strategy also fails to justify the exclusion of social entities from the domain of metaphysics.

A naturalistically inclined philosopher might argue that social entities are not legitimate subjects of metaphysical inquiry because metaphysicians should be concerned to investigate all and only those entities to which our best-confirmed scientific theories are ontologically committed (Colyvan 2001; Putnam 1972; Quine 1969). On this view, the question of whether social entities fall within the purview of metaphysics turns on the question of which scientific theories are best confirmed. Now, if physics is the only scientific theory which passes muster, then this would give us a reason to exclude social entities from metaphysical inquiry. However, if social scientific theories are sufficiently well-confirmed, then metaphysicians should be concerned with social entities after all. This is because social scientists clearly theorize about them. For example, economists, sociologists, and anthropologists theorize about social entities such as money, marriage, and refugees.

However, even if social scientific theories are sufficiently well-confirmed, and even if they are about social entities, there is reason to believe that we should not delimit metaphysics in this way. There is no well-confirmed scientific theory that is ontologically committed to the existence of God, and yet whether God exists is a metaphysical question if there ever was one. More generally, metaphysicians ought to be engaged in the project of determining which entities exist, whether or not the best-confirmed scientific theories are committed to them.

Yet another way to exclude social entities from metaphysics is by arguing that metaphysicians should restrict their attention to natural properties or kinds.[9] If social kinds fail to be natural, then perhaps metaphysicians should not be concerned with them. We think that this proposal likewise fails to justifiably exclude social entities for two reasons.

First, this strategy only purports to exclude social *kinds* from metaphysical theorizing, and not other social entities (e.g., social groups, social facts, or social events). Moreover, a clear contrast between social and natural kinds is difficult to draw. It is obvious that social kinds are not found "in nature" so to speak. However, many paradigmatically natural kinds are not found "in nature" either, e.g., synthetically produced chemical compounds like polyethylene and PTFE (Teflon).

Furthermore, social kinds do not contrast with natural kinds in the sense that the former are *super*natural. Social kinds, like the human beings who create them, are occupants of the natural world, and are subject to the same physical laws that govern the behavior of everything from planets to protons.[10] Moreover, social kinds are susceptible to empirical investigation. Indeed, they form the subject matter of a wide variety of scientific disciplines including sociology, anthropology, history, economics, and psychology.

Finally, if natural kinds are just those kinds which enable us to successfully predict and explain empirical phenomena (that is, if natural kinds are those kinds that license inductive inferences, warrant empirical generalizations, and feature in fruitful explanations), there is reason to believe that social kinds are natural in the relevant sense (Bach 2012; Boyd 1999; Griffiths 1999; Khalidi 2013, 2015, 2018; Mallon 2003, 2016; Mason 2016). For example, Ron Mallon argues that social constructionist explanations are a species of causal explanation. On his view, social roles like being a stay-at-home father or being a CEO are loci of predictive and explanatory potential. For instance, the fact that individuals occupy these roles enables us to predict and explain many of the properties they instantiate.

We conclude that social entities should not be excluded from metaphysics on the grounds that they fail to be natural in any of the aforementioned ways.

IV Conclusion

We have argued that there is no good reason to think that social entities do not fall within the purview of metaphysics. In particular, they cannot be excluded on the basis of being non-fundamental, mind-dependent, non-natural, or reducible. And even if there are not social entities, it is still an metaphysical question whether certain social entities exist.

Certainly, some topics in social metaphysics lie at the intersection of metaphysics and other areas of philosophy, for example political philosophy, feminist philosophy, or ethics (e.g., the metaphysics of race and gender). But, as Bennett puts it, "lying at the intersection of A and B does not mean lying in neither A nor B, but in both" (2017, 233). Moreover, philosophers have more permissive attitudes with respect to other metaphysical intersections. For example, many questions concerning the nature of mental phenomena fall at the intersection of metaphysics and philosophy of mind. Indeed, many questions lie at the intersection of metaphysics and empirical disciplines like chemistry and physics. But no one seriously thinks that these questions *really* belong solely to these other disciplines, and not to metaphysics. For instance, many metaphysicians hold naturalistic views informed by and intersecting with various scientific disciplines (Ladyman and Ross 2007; Ney and Albert 2013).

Our view is that the exclusion of social entities from metaphysics has more to do with the interests and preoccupations of metaphysicians than the unsuitability of social entities themselves. We hold that there is no compelling philosophical reason, for example, why metaphysics should include an investigation of the existence and nature of holes and time but not corporations and races. It is just that most metaphysicians, for whatever reason, have been more interested in investigating the former than latter.

By way of concluding, we would like to consider what attending to social metaphysics means for the nature of metaphysics. In particular, we consider to what extent our metaphysical intuitions have been conditioned by a historically contingent focus on certain types of entities (e.g., fundamentalia), to the exclusion of others (e.g., social groups). What would our metaphysical commitments be like if metaphysicians focused on social entities rather than the more traditional targets of metaphysical inquiry?

First, focusing on social entities indicates that the metaphysician's preference for describing reality in terms of features that are independent, individualistic, intrinsic, universal, ahistorical, and non-normative is not well founded. The preference for describing reality in terms of these features is bound up with the sorts of metaphysical commitments addressed in the previous sections of this chapter. If one has a radically eliminativist ontology, or if one endorses a fundamentalist conception of metaphysics according to which metaphysics is *really* about describing the fundamental structure of reality, which consists exclusively of simples or spacetime points, then perhaps there is good reason to prefer a characterization of reality exclusively in terms of these features. But we have argued that metaphysics should not be characterized in these ways and that plenty of plainly metaphysical subjects cannot be so characterized. Today metaphysics has a greater focus on dependence or "building" relations or on "what grounds what." Social metaphysics highlights the importance of dependent and relational features of reality, positioning it squarely within contemporary metaphysics.

Social metaphysics leads us to describe reality in ways that are dependent, anti-individualistic, relational, particular, historical, and normatively laden. For instance, many social entities are spatiotemporally restricted, and so are neither ahistorical nor universal. It is plausible that races and genders came into existence at a particular period in time, and may be found only in the small corner of the universe inhabited by human beings. On some views they are historical or sociohistorical kinds (Bach 2012; Du Bois 1897/1996; Diaz-Leon 2015b; Jeffers 2013; Mallon 2003, 2016; Taylor 2000, see Appiah 1985 for criticism of race as a historical kind). Focusing on social entities also reveals that some entities could have normative natures. For instance, being part of a social group might depend on being bound by particular social conventions or norms, perhaps as part of an overarching social structure (e.g., Ásta 2018; Ritchie 2020; Thomasson 2019).

Social metaphysics also helps to show how normative considerations are relevant to theory choice. For example, feminist metaphysicians argue that metaphysical theories should not simply explain how big things are built from little ones, or how one event causes another. They also ought to incorporate moral and political values (Barnes 2014, 2016, 2017; Haslanger 2012; Haslanger and Ásta 2017; Mikkola 2015, 2017). That is, we should prefer theories that not only describe the world correctly, but can perform the relevant normative work. For example, an adequate theory of race and gender ought to enable us to accomplish our goal of eliminating racial and gender injustice and oppression. One motivation for an intersectional theoretical framework is that a single-axis analysis of oppression (e.g., an analysis focused just on sexism) can mask and reinforce oppression (Collins 2000; Crenshaw 1989, 1991; Spelman 1988). A focus on social entities allows one to see ways that descriptive and normative projects function in tandem.

The inclusion of social metaphysics also has important diversifying effects. By excluding the investigation of social entities from metaphysics we thereby exclude the work of many philosophers who are women, transgender, non-binary, and people of color in an area of philosophy that is particularly white and male-dominated.

Social entities not only have a place in metaphysical inquiry, they deserve a place of prominence. Just as feminist and critical race theory have encouraged philosophers of science and epistemologists to reconsider their starting assumptions, social metaphysics forces metaphysicians to reconsider the vantage points and inherent biases in certain metaphysical and theoretical predilections.

Notes

1 This claim is easily verified by searching keywords such as "social," "social kind," "social group," etc., in top philosophy journals (e.g., *Philosophical Review, the Journal of Philosophy, Mind, Philosophy and Phenomenological Research, Noûs, Australasian Journal of Philosophy, Philosophical Studies*, etc.).
2 For example, see Carroll and Markosian 2010; Conee and Sider 2015; Crane and Farkas 2004; Effingham 2013; van Inwagen 2014; van Inwagen and Zimmerman 2008; Kim et al. 2011; Koons and Pickavance 2015, 2017; Loux 2008; Loux and Crisp 2017; Lowe 2002; Mumford 2012; Sider et al. 2007; Tahko 2016. Most of these titles were published well after the inception of social metaphysics in analytic philosophy. NB: Ney's (2014) introductory metaphysics textbook includes a chapter on the metaphysics of race.
3 For instance, Burman (Andersson) 2007; Gilbert 1989; Hacking 1996, 1999; Haslanger 1995; Ruben 1985; Searle 1990, 1995, 2010; Tuomela 1989; in addition to more than 30 years of work on the metaphysics of race and gender (e.g., Appiah 1985). Other work on collective intentionality also focuses in part on social ontology. See Schweikard and Schmid (2013) for an overview.

4 There are various ways one can spell out constitutive dependence. We do not wish to take a stand on the issue here, hence the disjunctive definition. For discussion of various ways social entities might be "held together" see Epstein (2015a).
5 This is because arrangements of simples could instantiate social properties and thereby ground social facts. For instance, atoms arranged shell-wise could have the property of being money and this could ground the truth of the social fact that atoms arranged shell-wise are money.
6 See Greenwood (2003) and Epstein (2015b) for discussion of forms of individualism and social explanation.
7 A weaker version of either thesis might rely on supervenience.
8 Trenton Merricks (2013) expresses skepticism about Sider's claim that this is what metaphysics is about: "metaphysics is not – not even 'at bottom' – about only one thing, and so not – not even 'at bottom' – about only the fundamental structure of reality" (722). Elizabeth Barnes (2014, 2017) argues against the fundamentalist conception of metaphysics on the grounds that it illegitimately rules out feminist metaphysics. Sider (2017) responds to the worry.
9 Another way of developing this objection is by appeal to perfectly natural properties (Dorr and Hawthorne 2013; Lewis 1983; Schaffer 2004).
10 It is a separate question whether social kinds figure in laws of nature *qua* social kinds, or whether all social scientific laws are reducible to physical laws.

References

Alcoff, L., 2006. *Visible Identities: Race, Gender, and the Self.* Oxford: Oxford.
Appiah, A., 1985. The Uncompleted Argument: Du Bois and the Illusion of Race. *Critical Inquiry*, 12 (1), 21–37.
Appiah, K.A., 1996. Race, Culture, Identity: Misunderstood Connections. In: K.A. Appiah and A. Gutmann, eds. *Color Conscious: The Political Morality of Race*. Princeton: Princeton University Press, 30–105.
Ásta (Sveinsdóttir), 2018. *Categories We Live By: The Construction of Sex, Gender, Race, and Other Social Categories*. Oxford: Oxford University Press.
Bach, T., 2012. Gender Is a Natural Kind with a Historical Essence. *Ethics*, 122 (2), 231–272.
Barnes, E., 2012. Emergence and Fundamentality. *Mind*, 121 (484), 873–900.
Barnes, E., 2014. Going Beyond the Fundamental: Feminism in Contemporary Metaphysics. *Proceedings of the Aristotelian Society*, 114 (3), 335–351.
Barnes, E., 2016. *The Minority Body: A Theory of Disability*. Oxford: Oxford University Press.
Barnes, E., 2017. Realism and Social Structure. *Philosophical Studies*, 174 (10), 2417–2433.
Bennett, K., 2017. *Making Things Up*. Oxford: Oxford University Press.
Bernstein, S., forthcoming. Could a Middle Level Be the Most Fundamental? *Philosophical Studies*.
Blum, L. 2010. Racialized Groups: The Sociohistorical Consensus. *The Monist*, 93 (2), 298–320.
Boyd, R., 1999. Homeostasis, Species, and Higher Taxa. In: R.A. Wilson, ed. *Species: New Interdisciplinary Essay*. Cambridge: MIT Press, 141–186.
Brouwer, L.E.J., 1981. *Brouwer's Cambridge Lectures on Intuitionism*, D. van Dalen, ed. Cambridge: Cambridge University Press.
Burman (Andersson), Å., 2007. *Power and Social Ontology*. Lund: Bokbox Publications.
Carroll, J. and Markosian, N., 2010. *An Introduction to Metaphysics*. Cambridge: Cambridge University Press.
Collins, P.H., 2000. *Black Feminist Thought: Knowledge, Consciousness and the Politics of Empowerment*. 2nd ed. New York: Routledge.
Colyvan, M., 2001. *The Indispensability of Mathematics*. Oxford: Oxford University Press.
Conee, E. and Sider, T., 2015. *Riddles of Existence: A Guided Tour of Metaphysics*. 2nd ed. Oxford: Oxford University Press.
Corvino, J., 2000. Analyzing Gender. *Southwest Philosophy Review*, 17 (1), 173–180.
Crane, T. and Farkas, K., 2004. *Metaphysics: A Guide and Anthology*. Oxford: Oxford University Press.
Crenshaw, K., 1989. Demarginalizing the Intersection of Race and Sex: A Black Feminist Critique of Antidiscrimination Doctrine, Feminist Theory and Antiracist Politics. *The University of Chicago Legal Forum*, 1989 (1), 139–167.
Crenshaw, K., 1991. Mapping the Margins: Intersectionality, Identity Politics, and Violence against Women of Color. *Stanford Law Review*, 43 (6), 1241–1299.

Diaz-Leon, E., 2015a. In Defence of Historical Constructivism About Races. *Ergo: An Open Access Journal of Philosophy*, 2 (21), 547–562.
Diaz-Leon, E., 2015b. What Is Social Construction? *European Journal of Philosophy*, 23 (4), 1137–1152.
Dorr, C. and Hawthorne, J., 2013. Naturalness. *In*: K. Bennett and D. Zimmerman, eds. *Oxford Studies in Metaphysics*. Oxford: Oxford University Press, 3–77.
Du Bois, W.E.B., 1897/1996. The Conservation of Races. *In*: E.J. Sundquist, ed. *The Oxford W.E.B. Du Bois Reader*. Oxford: Oxford University Press.
Effingham, N., 2013. *An Introduction to Ontology*. Cambridge: Polity Press.
Epstein, B., 2009. Ontological Individualism Reconsidered. *Synthese*, 166 (1), 187–213.
Epstein, B., 2015a. How Many Kinds of Glue Hold the Social World Together? *In*: M. Gallotti and J. Michael, eds. *Perspectives on Social Ontology and Social Cognition*. Dordrecht: Springer, 41–55.
Epstein, B., 2015b. *The Ant Trap: Rebuilding the Foundations of the Social Sciences*. Oxford: Oxford University Press.
Gilbert, M., 1989. *On Social Facts*. Princeton: Princeton University Press.
Greenwood, J.D., 2003. Social Facts, Social Groups and Social Explanation. *Noûs*, 37 (1), 93–112.
Griffiths, P.E., 1999. Squaring the Circle: Natural Kinds with Historical Essences. *In*: R.A. Wilson, ed. *Species: New Interdisciplinary Essay*. Cambridge: MIT Press, 209–228.
Hacking, I., 1996. The Looping Effects of Human Kinds. *In*: D. Sperber, D. Premack, and A.J. Premack, eds. *Causal Cognition: A Multidisciplinary Debate*. New York: Clarendon Press, 351–394.
Hacking, I., 1999. *The Social Construction of What?* Cambridge: Harvard University Press.
Hale, J., 1996. Are Lesbians Women? *Hypatia*, 11 (2), 94–121.
Haslanger, S., 1995. Ontology and Social Construction. *Philosophical Topics*, 23 (2), 95–125.
Haslanger, S., 2000. Gender and Race: (What) Are They? (What) Do We Want Them to Be? *Noûs*, 34 (1), 31–55.
Haslanger, S., 2003. Social Construction: The "Debunking" Project. *In*: F. Schmitt, ed. *Socializing Metaphysics*. Oxford: Rowman & Littlefield, 301–325.
Haslanger, S., 2012. *Resisting Reality: Social Construction and Social Critique*. Oxford: Oxford University Press.
Haslanger, S., 2016. What Is a (Social) Structural Explanation? *Philosophical Studies*, 173 (1), 113–130.
Haslanger, S. and Ásta (Sveinsdóttir), 2017. Feminist Metaphysics. *Stanford Encyclopedia of Philosophy*.
Hayek, F.A., 1955. *The Counter-Revolution of Science*. New York: The Free Press.
Heyes, C.J., 2000. *Line Drawings: Defining Women Through Feminist Practice*. Ithaca: Cornell University Press.
van Inwagen, P., 1990. *Material Beings*. Ithaca: Cornell University Press.
van Inwagen, P., 2014. *Metaphysics*. 4th ed. New York: Routledge.
van Inwagen, P. and Zimmerman, D., eds., 2008. *Metaphysics: The Big Questions*. Hoboken: Wiley-Blackwell.
Jeffers, C., 2013. The Cultural Theory of Race: Yet Another Look at Du Bois's "The Conservation of Races." *Ethics*, 123 (3), 403–426.
Khalidi, M.A., 2013. *Natural Categories and Human Kinds: Classification in the Natural and Social Sciences*. Cambridge: Cambridge University Press.
Khalidi, M.A., 2015. Three Kinds of Social Kinds. *Philosophy and Phenomenological Research*, 90 (1), 96–112.
Khalidi, M.A., 2018. Natural Kinds as Nodes in Causal Networks. *Synthese*, 195 (4), 1379–1396.
Kim, J., Korman, D.Z., and Sosa, E., eds., 2011. *Metaphysics: An Anthology*. 2nd ed. Oxford: Blackwell Publishing.
Koons, R. and Pickavance, T., 2015. *Metaphysics: The Fundamentals*. 1st ed. Oxford: Blackwell Publishing.
Koons, R. and Pickavance, T., 2017. *The Atlas of Reality: A Comprehensive Guide to Metaphysics*. 1st ed. Oxford: Blackwell Publishing.
Korman, D.Z., 2015. *Objects: Nothing Out of the Ordinary*. Oxford: Oxford University Press.
Ladyman, J. and Ross, D. 2007. *Every Thing Must Go: Metaphysics Naturalized*. Oxford: Oxford University Press.
Lewis, D., 1983. New Work for a Theory of Universals. *Australasian Journal of Philosophy*, 61, 343–377.
Lewis, D., 1991. *Parts of Classes*. Oxford: Wiley-Blackwell.
Loux, M., 2008. *Metaphysics: Contemporary Readings*. New York: Routledge.
Loux, M. and Crisp, T., 2017. *Metaphysics: A Contemporary Introduction*. New York: Routledge.

Lowe, E.J., 2002. *A Survey of Metaphysics*. Oxford: Oxford University Press.
Mallon, R., 2003. Social Construction, Social Roles, and Stability. In: F. Schmitt, ed. *Socializing Metaphysics: The Nature of Social Reality*. Lanham: Rowman & Littlefield, 65–91.
Mallon, R., 2006. "Race": Normative, Not Metaphysical or Semantic. *Ethics*, 116 (3), 525–551.
Mallon, R., 2014. Naturalistic Approaches to Social Construction. *Stanford Encyclopedia of Philosophy*.
Mallon, R., 2016. *The Construction of Human Kinds*. Oxford: Oxford University Press.
Markosian, N., 1998. Brutal Composition. *Philosophical Studies*, 92 (3), 211–249.
Markosian, N., 2008. Restricted Composition. In: T. Sider, J. Hawthorne, and D. Zimmerman, eds. *Contemporary Debates in Metaphysics*. Malden: Blackwell Publishing, 341–364.
Mason, R., 2016. The Metaphysics of Social Kinds. *Philosophy Compass*, 11 (12), 841–850.
Merricks, T., 2013. Three Comments on Theodore Sider's *Writing the Book of the World*. *Analysis*, 73 (4), 722–736.
Mikkola, M., 2015. Doing Ontology and Doing Justice: What Feminist Philosophy Can Teach Us About Meta-Metaphysics. *Inquiry: An Interdisciplinary Journal of Philosophy*, 58 (7–8), 780–805.
Mikkola, M., 2017. On the Apparent Antagonism Between Feminist and Mainstream Metaphysics. *Philosophical Studies*, 174 (10), 2435–2448.
Mills, C., 1997. *The Racial Contract*. Ithaca: Cornell University Press.
Mills, C., 1998. *Blackness Visible: Essays on Philosophy and Race*. Ithaca: Cornell University Press.
Mumford, S., 2012. *Metaphysics: A Very Short Introduction*. Oxford: Oxford University Press.
Ney, A., 2014. *Metaphysics: An Introduction*. London and New York: Routledge.
Ney, A. and Albert, D.Z., eds., 2013. *The Wave Function: Essays in the Metaphysics of Quantum Mechanics*. Oxford: Oxford University Press.
Outlaw, L., 1996. Conserve' Races? In Defense of W.E.B. Du Bois. In: B.W. Bell, E.R. Grosholz, and J.B. Stewart, eds. *W.E.B. Du Bois on Race and Culture*. New York: Routledge, 15–38.
Popper, K., 1966. *The Open Society and Its Enemies*. London: Routledge & Kegan Paul.
Putnam, H., 1972. *Philosophy of Logic*. London: George Allen & Unwin.
Quine, W.V.O., 1969. *Ontological Relativity and Other Essays*. New York: Columbia University Press.
Ritchie, K., 2020. Social Structures and the Ontology of Social Groups. *Philosophy and Phenomenological Research*, 100, 402–424. doi:10.1111/phpr.12555.
Ruben, D.H., 1985. *The Metaphysics of the Social World*. London: Routledge & Kegan Paul.
Schaffer, J., 2003. Is There a Fundamental Level? *Noûs*, 37 (3), 498–517.
Schaffer, J., 2004. Two Conceptions of Sparse Properties. *Pacific Philosophical Quarterly*, 85 (1), 92–102.
Schaffer, J. 2009. On What Grounds What. In: D. Manley, D.J. Chalmers, & R. Wasserman, eds. *Metametaphysics: New Essays on the Foundations of Ontology*. Oxford University Press, 347–383.
Schaffer, J. 2010. Monism: The Priority of the Whole. *Philosophical Review*, 119 (1), 31–76.
Schweikard, D.P. and Schmid, H.B., 2013. Collective Intentionality. *Stanford Encyclopedia of Philosophy*.
Searle, J., 1990. Collective Intentions and Actions. In: P.R. Cohen, J. Morgan, and M.E. Pollack, eds. *Intentions in Communication*. Cambridge: MIT Press, 401–416.
Searle, J., 1995. *The Construction of Social Reality*. New York: The Free Press.
Searle, J., 2010. *Making the Social World: The Structure of Human Civilization*. Oxford: Oxford University Press.
Sider, T., 1993. Van Inwagen and the Possibility of Gunk. *Analysis*, 53 (4), 285–289.
Sider, T., 2001. *Four-Dimensionalism*. Oxford: Oxford University Press.
Sider, T., 2011. *Writing the Book of the World*. Oxford: Oxford University Press.
Sider, T., 2017. Substantivity in Feminist Metaphysics. *Philosophical Studies*, 174 (10), 2467–2478.
Sider, T., Hawthorne, J., and Zimmerman, D., eds., 2007. *Contemporary Debates in Metaphysics*. Oxford: Blackwell Publishing.
Spelman, E., 1988. *Inessential Woman: Problems of Exclusion in Feminist Thought*. Boston: Beacon Press.
Stoljar, N., 1995. Essence, Identity, and the Concept of Woman. *Philosophical Topics*, 23 (2), 261–293.
Stoljar, N., 2011. Different Women. Gender and the Realism-Nominalism Debate. In: C. Witt, ed. *Feminist Metaphysics: Explorations in the Ontology of Sex, Gender and the Self*. Dordrecht: Springer, 27–46.
Sundstrom, R., 2002. Race as a Human Kind. *Philosophy and Social Criticism*, 28 (1), 91–115.
Tahko, T., 2016. *An Introduction to Metametaphysics*. Cambridge: Cambridge University Press.
Taylor, P.C., 2000. Appiah's Uncompleted Argument: W.E.B. Du Bois and the Reality of Race. *Social Theory and Practice*, 26 (1), 103–128.
Taylor, P.C., 2004. *Race: A Philosophical Introduction*. Cambridge: Polity Press.
Thomasson, A.L., 2003. Realism and Human Kinds. *Philosophy and Phenomenological Research*, 67 (3), 580–609.

Thomasson, A.L., 2019. The Ontology of Social Groups. *Synthese*, 196 (12), 4829–4845.
Tuomela, R., 1989. Collective Action, Supervenience, and Constitution. *Synthese*, 80 (2), 243–266.
Unger, P., 1980. The Problem of the Many. *Midwest Studies in Philosophy*, 5 (1), 411–468.
Weber, M., 1922/2013. *Economy and Society*. Berkeley: University of California Press.
Witt, C., 1995. Anti-Essentialism in Feminist Theory. *Philosophical Topics*, 23 (2), 321–344.
Witt, C., ed., 2011a. *Feminist Metaphysics: Explorations in the Ontology of Sex, Gender and the Self*. Dordrecht: Springer.
Witt, C., 2011b. *The Metaphysics of Gender*. Oxford: Oxford University Press.
Zack, N., 1993. *Race and Mixed Race*. Philadelphia: Temple University Press.
Zack, N., 2002. *Philosophy of Science and Race*. New York: Routledge.

Further reading

Ásta (Sveinsdóttir), 2015. Social Construction. *Philosophy Compass*, 10 (12), 884–892.
Epstein, B., 2019. Social Ontology. *Stanford Encyclopedia of Philosophy*.
Mallon, R., 2007. A Field Guide to Social Construction. *Philosophy Compass*, 2 (1), 93–108.
Mason, R., 2016. The Metaphysics of Social Kinds. *Philosophy Compass*, 11 (12), 841–850.
Mikkola, M., 2017. Feminist Perspectives on Sex and Gender. *Stanford Encyclopedia of Philosophy*.
Ritchie, K., 2015. The Metaphysics of Social Groups. *Philosophy Compass*, 10 (5), 310–321.

25
NATURAL LANGUAGE ONTOLOGY

Friederike Moltmann

Natural language ontology is the study of the ontology (ontological categories, structures, and notions) reflected in natural language. It is a subdiscipline of both philosophy and linguistics. More specifically, natural language ontology is part of both natural language semantics and metaphysics.

Natural language ontology is a new discipline that has emerged with the development of natural language semantics over the last decades. It has been suggested as a discipline first by semanticists (Bach 1986).[1] Research in natural language semantics falls under natural language ontology when it deals with semantic issues that involve metaphysical notions, such as reference to entities of particular ontological categories (e.g., events, tropes, facts, kinds), plurals, the mass-count distinction, tense, aspect, and some modals. Research in philosophy (or the philosophy-linguistics interface) falls under natural language ontology when it deals with metaphysics as reflected in linguistically manifest intuitions.

Natural language ontology, however, is not just an emerging discipline. It has also been a practice throughout the history of philosophy. Philosophers throughout history, at times more often than others, have appealed to natural language to motivate an ontological view or notion, and when they did so, it is fair to say, they practiced natural language ontology. Such an appeal to natural language can be found already in Aristotle and very explicitly in medieval metaphysics (Ockham, Aquinas, Buridan), in the phenomenological tradition (Brentano, Husserl, Meinong, Bolzano, Twardowski), as well as in early analytic philosophy (Frege, Strawson, Austin, Vendler, Ryle).

The appeal to natural language in the history of philosophy had often been based on the assumption that natural language just reflects reality. More recently, though, the view has established itself among philosophers that natural language does not in fact reflect the ontology of what there really is, but rather comes with its own ontology, an ontology that may be quite different from the ontology of the real.

Natural language ontology as a subdiscipline of both linguistics and philosophy raises the following general questions that this chapter will address:

1 How does the semantics of natural language involve ontology and thus in what sense is natural language ontology part of linguistics?
2 How does natural language ontology situate itself within metaphysics and thus is a branch of metaphysics?

3   What sorts of linguistic data reflect the ontology implicit in language, and how is that ontology itself to be understood?
4   What is distinctive about the ontology of natural language and what sorts of conditions does this impose on an ontological theory?

1 The role of ontology in the semantics of natural language

1.1 Ways of the reflection of ontology in natural language

How does natural language reflect ontology? The semantics of natural languages involves entities of various ontological categories, ontological structures, and ontological notions on the basis of syntactic roles of expressions, syntactic categories and features, and lexical words.

First of all, entities may play various roles in the semantic structure of natural language sentences, though, of course, in what way exactly may depend somewhat on particular semantic theories about relevant constructions or expressions. Most importantly, entities play a role as the semantic values of referential noun phrases (NPs) as well as the things that quantificational NPs range over. Moreover, entities play a role as arguments of predicates. Natural language contains a wealth of expressions referring to or quantifying over entities, and it comes with a wealth of expressions that express properties of entities (or relations among them). Thus, in *John owns the building*, the referential NPs *John* and *the building* stand for entities, and *own* is a predicate expressing a relation among entities that is attributed to them in that sentence.

The notion of a referential NP is equally important in linguistics and in philosophy. Referential NPs generally are considered occurrences of NPs in sentences that have the function of standing for objects. Proper names and definite NPs can serve as referential NPs, as can specific indefinites and certain determinerless (bare) plurals and mass nouns. There are various syntactic and semantic criteria for referential NPs. For philosophers, since Frege, they include (very roughly) the ability of an NP to support anaphora, to be replaceable by quantificational NPs, and to serve as arguments of ordinary (i.e., extensional) predicates (Frege 1892; Hale 1987). For syntacticians, referential NPs also must satisfy certain syntactic conditions (having the more complex structure of a DP, that is, a determiner phrase, rather than just an NP, the category of predicative NPs) (Abney 1987; Borer 2005).

The notion of a referential NP (or 'name' as it was called at the time) already played a central role in Frege's (1892) philosophy of language and provided *a syntactic criterion for objecthood*: for Frege, an object is what can be the semantic value of a referential NP (using the contemporary term). Standing for an object is the contribution of a referential NP in the context of a sentence (Frege's Context Principle) (Wright 1983; Hale 1987).

Entities may play also the role of implicit arguments (that is, as arguments of predicates without at the same time being the semantic values of referrential NPs). Thus, on Davidson's (1967) influential analysis, the sentence *John walked slowly* states that there is an event which is an argument of *walk* (together with John) and of which *slowly* is true (*slowly* now being treated as a predicate of events). The very same arguments that lead Davidson to posit events as implicit arguments apply to adjectives and motivate tropes (particularized properties) as arguments of adjectives.[2] *John is profoundly happy* will then state that there is manifestation of happiness (a trope) that (together with John) is an argument of *happy* and of which *profoundly* is true (Moltmann 2009, 2013a).

Another important semantic role of entities in the semantic structure of natural language sentences is that of a parameter of evaluation. The standard semantic view takes a sentence to be true or false not absolutely, but relative to a time and a (possible) world. Possible worlds are generally treated as parameters of evaluation for the semantics of modals and conditionals, and times often for the semantics of tenses and temporal adverbials. In the more recent development of truthmaker semantics (Fine 2017b), situations play somewhat similar roles for the semantics of conditionals and modals, but now as exact truthmakers of sentences.[3]

Natural language also reflects ontological categories, in particular some of its syntactic categories or features. Thus, verbs are generally taken to reflect the category of events (Szabo 2015). Adjectives generally reflect the category of qualities or tropes (that is, particularized properties or concrete property manifestations) (Williams 1953; Woltersdorff 1970; Moltmann 2009). The category singular count noun conveys unity or singularity, the category plural noun plurality. Natural language moreover reflects metaphysical notions of various sorts, such as part-whole relations (Moltmann 1997), constitution (Fine 2003), causation (Swanson 2012; Ramchand 2018), (time- and space-relative) existence (Fine 2006; Moltmann 2013b), and existence of the past (the presentism debate) (Szabo 2007). Besides syntactic categories or features, natural language displays particular types of expressions conveying metaphysical notions, such as metaphysical (*may, must*), existence or ways of being (*exist, occur, obtain*), ontological dependence (*have*), part-whole relations (*part of, whole, partially, completely*), causation (*make*), and truth (*true, correct*).

1.2 The connection between ontology and compositionality

The ontology that natural language reflects is intimately linked to compositionality, the chief tenet of natural language semantics. Whether and how entities play a role in the semantic structure of natural language depends very much on the way the contribution of occurrences of expressions to the composition of the meaning of the sentence is conceived. Generally, the contribution of referential NPs is taken to be that of standing for an object and the role of expressions acting as predicates that of taking objects as arguments and yielding truth values. Without positing entities as semantic values of referential NPs and without positing properties of entities or relations among them as semantic values of predicates, compositionality is hardly possible, or so it seems.[4] The same predicates should (generally) express the same property with different referential NPs, and the same referential NP should (generally) stand for the same entity with different predicates.

1.3 Derivative and language-driven entities as semantic values of referential NPs

The ontology reflected in referential NPs has been subject to the most controversy, raising questions whether natural language could possibly be viewed as a guide to ontology and whether referential NPs even refer to entities. The general observation is that referential NPs display a great range of highly derivative entities, many of which philosophers may not be willing to accept. Yet, NPs that appear to stand for such entities satisfy the very same criteria of referentiality as NPs standing for less controversial entities or 'ordinary referential NPs,' as one may call them. In particular, they go along with the same sorts of predicates as ordinary referential NPs, support anaphoric pronouns, and can be replaced by quantificational and pronominal NPs.

First of all, referential NPs display various sorts of *derivative entities* that are ontologically dependent on others such as artifacts of various sorts, collections that come with a structure or function (classes, groups, teams, orchestras), kinds (with definite NPs of the sort *the Siberian tiger*), as well as 'disturbances' such as shadows, holes, folds, and tropes (particularized properties). Natural language displays a particularly rich ontology of tropes or trope-related entities, which include complex manifestations of non-natural properties (John's happiness, Socrates' wisdom), tropes such as strength and weakness as distinct (order-constituted) tropes, and quasi-relational tropes such as John's tallness as a trope distinct from the ordinary quantitative trope John's height (Moltmann 2009, 2013a).

Derivative entities of this sort also appear to be part of the naïve ontology of ordinary speakers (non-philosophers) (see Section 3.1). This may not be so, however, for another part of the ontology that natural language displays, namely what one may call a *language-driven ontology*, which consists of entities that go along with the constructional semantics of particular natural-language constructions. Referential NPs for entities in that ontology may include, for example:

(1) definite plurals, which stand for (unrestricted) pluralities of entities:
 the students, Quine, and the Eiffel Tower
(2) definite mass NPs, which stand for (unrestricted) quantities:
 the water and the wine, the water in this area
(3) bare (determinerless) plurals and mass nouns that stand for unrestricted kinds:
 empty seats, clean water
(4) definite NPs that stand for variable objects:
 the increasing amount of water in the container, the book John needs to write
(5) definite intentional NP, which may stand for merely conceived (nonexistent) entities:
 the building mentioned in the guide, the trip John is planning

The dominant view about the definite plural *the students* is that it stands for the sum of the contextually restricted extension of the noun *students* (Link 1983; Moltmann 1997; Champollion/Krifka 2016). Similarly, *Quine and the Eiffel Tower* stands for the sum of the two individuals. Definite plurals exhibit criteria of referentiality in that they share predicates with singular NPs and come with plural-specific collective predicates, support anaphora, and can be replaced with ordinary quantificational NPs.[5] The definite mass NP *the water in this area* likewise satisfies criteria of referentiality and thus, on the standard view, stands for the sum of the extension of *water in this area*. Sum formation as involved in the semantics of plural and mass NPs is unrestricted since, given the standard view, any definite plural or mass NP and any conjunction of definite NPs will stand for a sum of entities.

It is a widely accepted view in semantics that bare plural and mass NPs stand for kinds (for example in *empty seats are rare* or *clean water is important*) (Carlson 1977). Again formation of kinds in this sense is rather unrestricted since any bare nominal, whatever its conceptual content, can act as kind-referring term in that sense.

Definite NP of the sort *the increasing amount of water in the container*, it has been argued, stand for variable quantities that have different manifestations as particular quantities at different times (Fine 1999). Similarly, *the book John needs to write* will stand for a variable object that has different manifestations as particular books John has written in possibly only counterfactual situations Moltmann (2013a, forthcoming a). Finally, the construction *the building mentioned in the guide* may have a semantic value that is a merely intentional (nonexistent) entity (*the building in the guide does not exist*) (Moltmann 2015).

The NPs in (1)–(5) all act as referential NPs. But they stand for entities in a language-driven sort of constructional ontology, an ontology ordinary speakers are not likely to accept when thinking about what there is. Such cases make particularly clear that natural language involves its own ontology, an ontology that is in part language-driven and clearly distinct from the ontology of what there ultimately is.

2 How can natural language ontology be situated within metaphysics?

The observation that natural language ontology involves a rich ontology of highly derivative entities has led many philosophers to reject natural language as a guide to ontology. The subject matter of metaphysics, on their view, is fundamental reality, not the ontology reflected in language. Given such a view, natural language ontology no longer has a place within metaphysics.

This is not the only way, however, of understanding metaphysics. There are alternative conceptions of metaphysics which are not just focused on fundamental reality and within which natural language ontology can find its place.

First of all, there are older traditions of metaphysics that are not focused on the ontology of the real. One of them is the Kantian tradition, which dealt with ontological categories, for example, but as preconditions for accessing the world, rather than as categories of how things really are. Another is the phenomenological tradition (Brentano, Husserl), where ontology was pursued based on how things appear, rather than assumptions about a subject-independent reality (the way things appear being taken to be constitutive of the things themselves).

In the mid-twentieth century Strawson (1959) introduced the distinction between descriptive and revisionary metaphysics, a distinction best understood as one based on whether ontology is reflected in particular 'data' or not. Descriptive metaphysics thus concerns itself with the ontology that is reflected in our shared intuitions or ordinary judgments, or in fact natural language. By contrast, revisionary metaphysics pursues a 'better' ontology, not reflected in such data (but more suited, say, for the development of the natural sciences). Given that distinction, natural language ontology clearly is a branch of descriptive metaphysics.

A somewhat related, yet different distinction has recently been made by Fine (2017a), who distinguishes between naïve metaphysics and foundational metaphysics. Fine's notion of naïve metaphysics is basically the same as that of descriptive metaphysics: it is metaphysics whose subject matter is the ontology reflected in our ordinary judgments as well as, more specifically, natural language. Naïve metaphysics is contrasted, however, with foundational metaphysics. The subject matter of naïve metaphysics, the *metaphysics of appearances*, is reflected in our ordinary judgments; the subject matter of foundational metaphysics is the ontology of what there really is. What is important according to Fine is that naïve metaphysics cannot be skipped in favor of foundational metaphysics. Rather foundational metaphysics must take naïve metaphysics as its starting point: foundational metaphysics must start out with the notions that naïve metaphysics deals with, in order to possibly explain them in more fundamental terms. Naïve metaphysics itself, Fine argues, should be pursued without foundational considerations.

Natural language ontology then has a place within metaphysics, as part of naïve metaphysics, or as I will call it, staying with the better established and less misleading Strawsonian term, 'descriptive metaphysics.'[6]

Fine's notion of 'metaphysics of appearances' might suggest that entities in the ontology of natural language are to be viewed as mere appearances. But this is not what the notion is meant to convey. Rather a distinction between actual and merely conceived entities is to be made. Generally only the former contribute to the truth of sentences. Ordinary predicates of natural language (for example, sortals) are existence-entailing and presuppose that the objects they take as arguments be existent. Only certain non-ordinary predicates can be true or false also of entities that fail to exist. (They include intentional verbs such as *think about, refer to, mention*, as well as the verb *exist*.)

3 Recognizing natural language ontology as a discipline of its own

Natural language ontology has faced serious challenges being recognized as a discipline of its own. There are three reasons for that.

The first is the foundationalist orientation of contemporary metaphysics. While metaphysics has enjoyed a significant revival in the twentieth century, it has to a great extent been focused on foundational metaphysics, pursuing questions of ultimate reality in line with physics and some of the other natural sciences. The pursuit of descriptive metaphysics, as metaphysics focused on the ontology reflected in our ordinary judgments or linguistic intuitions, has not been given the same importance.

The second reason is the formal orientation of Montague Grammar ('English as a formal language'), which had dominated linguistic semantics for the last decades (Thomason 1974). In the Montagovian tradition, the purpose of formal semanticists has been considered that of developing logical analyses of parts of natural language that can explain intuitively valid inferences. With that as its main aim, there was little concern in the semantics of language semantics as to the ontological-cognitive status of the formal notions used in model-theoretic semantics (say, the entities and notions posited in the models).

The third reason for the difficulty for natural language ontology being recognized as a discipline is Chomsky's rejection of referentialist semantics. Chomsky (1986, 1998, 2013) took an entirely skeptical stance as to whether language involves reference to entities, and thus did not encourage setting out a research agenda for natural language ontology within generative linguistics. Chomsky's skepticism was based on the view that ontology concerns only mind-independent reality (with the relation being able to relate only to objects in such a reality).

Chomsky's rejection of the involvement of ontology in natural language also had to do with his exclusive focus on referential NPs. For Chomsky, natural language terms, including artifact terms, terms for cities, terms like *the typical student*, cannot stand for objects in a mind-independent reality (or even objects on any standard understanding of the term, including conceived objects, Chomsky p.c.). Chomsky's examples generally involve property attributions associated with a referential NP that violate standard conditions on objecthood (e.g., one can paint a door, but also walk through it).

Chomsky's conclusion that natural language does not involve ontology is not warranted, given the distinction between foundational and descriptive metaphysics ('the metaphysics of appearances') and given the fact that ontology is reflected in many other parts of language than just referential NPs. Moreover, there are promising recent ontological and lexical approaches that permit a reanalysis of Chomsky-style examples, which Chomsky did not take into consideration (variable objects, multi-faceted objects, the generative lexicon, approaches in applied ontology).

Natural language ontology has a well-defined subject matter, a cognitive ontology reflected in natural language, and as such is part of linguistics. This alone makes it an important project to pursue. However, there are also specific reasons for a philosopher to pursue it.

First, as Fine (2017a) argues, descriptive metaphysics is presupposed by foundational metaphysics, which has as one of its aims the clarification of the notions that descriptive metaphysics deals with. At least for some of those notions linguistic data may be particularly relevant, which thus requires the pursuit of natural language ontology.

Second, natural language ontology may shed a new light on longstanding philosophical puzzles. A great range of philosophical views, for example about ontological categories, about propositions, about truth and truthbearers, about numbers, and about the constitution of material objects, have been motivated, at least in part, by appeal to natural language. Often, however, such an appeal turns out to be based on a naïve, incomplete, or mistaken analysis of linguistic data. It is hence important to analyze the full range of relevant linguistic data properly in order to uncover the ontology they in fact involve. A deeper linguistic analysis may then provide new philosophical solutions or perspectives regarding the philosophical issues. An example is difficulties for the notion of an abstract proposition and for abstract objects generally. Philosophers generally take natural language to involve a rich ontology of abstract objects, including propositions. However, a closer examination of the linguistic facts indicates that natural language does not in fact involve abstract propositions in its ontology and that reference to abstract objects generally is highly restricted (Moltmann 2013a) (Section 5).

4 What sorts of linguistic data reflect the ontology of natural language and how is the ontology of natural language to be characterized?

Natural language ontology has as its subject matter the ontology implicit in natural language, or the *ontology of natural language*. This raises two central questions:

1   How is that ontology to be understood, as it cannot be the ontology of fundamental reality?
2   What sorts of linguistic data do reflect that ontology?

Let us address these two questions in turn.

4.1 Natural language ontology and folkmetaphysics

As a first suggestion one might say that the ontology of natural language is just the ontology that ordinary people, i.e., non-philosophers, accept. This cannot be right, however: the ontology implicit in natural language cannot be the ontology ordinary people (non-philosophers) *naïvely accept* when thinking about what there is. That is, natural language ontology is to be distinguished from *folkmetaphysics*. Folkmetaphysics takes different sorts of data into account than natural language ontology. One difference consists in that, just like folkphysics and folk biology, folkmetaphysics takes into account assertions such as:

(6) a   There are artifacts.
     b   Objects are not events.

Metaphysics assertions play no role for natural language ontology. No philosopher or linguist would appeal to assertions such as (6a) when arguing that natural language reflects an ontology of artifacts or assertions such as (6b) when arguing that natural language reflects an ontology of objects being distinct from events. What matters for natural language ontology are not metaphysical assertions, but presuppositions, for example presuppositions of ontological categories carried by referential NPs, quantifiers or predicates.

There are also linguistic data that natural language ontology takes into account, but not folkmetaphysics, for example sentences that involve ontological commitments not accessible to ordinary speakers, such as sentences containing silent syntactic elements with ontological content, as would be the case according to the sorts of syntactic structures posited in generative syntax.[7]

The ontology of natural language thus should be understood as an ontology that speakers *implicitly accept*, not as an ontology speakers naïvely accept when thinking about what there is:

(7) Characterization of the ontology implicit in natural language (1st version)
     The ontology of natural language is the ontology speakers implicitly accept.

This notion of implicit acceptance is special in that it is a particularly robust one. It is a form of acceptance that resists rejection upon reflection. Ordinary speakers may reject entities in the language-driven ontology that natural language displays, unrestricted pluralities, unrestricted kinds, variable objects, or conceived objects, say. Yet anyone that uses the relevant NPs will use them with such entities as semantic values and thus accept those entities. Implicit acceptance of the ontology implicit in natural language is mandatory for users of the language. In that sense, the notion of implicit acceptance is rather different from the notion of implicit acceptance in ethics. In the context of ethics, what is implicitly accepted (bias) permits rejection upon reflection.

The resistance to revision for the ontology of natural language indicates that the ontology reflected in language has the very same status as universal grammar, being implicit knowledge that cannot be subject to revision. What distinguishes ontology from syntax, of course, is that ontology is also the subject matter of a particular branch of philosophy and as such should in principle be subject to reflection and revision. However, as the ontology of natural language, it is not.

4.2 Natural language ontology and cognitive ontology

The ontology implicit in natural language cannot just be understood as our implicitly accepted cognitive ontology. This would be too broad and in part not correct. The mass-count distinction may illustrate that. Natural language appears to distinguish the semantic values of *the rice, the rice grains,* and *the heap of rice* in terms of their properties, and thus ontologically: the rice grains can be (internally) distinguished, compared, listed, or counted, but not so for the rice or the heap of rice (Moltmann 1997). Carrying different properties, 'the rice,' 'the rice grains,' and 'the heap of rice' thus appear to be distinct entities. In perception, by contrast, the distinction between 'the rice,' 'the rice grains,' and 'the heap of rice' hardly matters. Mass nouns such as *water, rice, police force, footwear* appear to stand for distinct entities from those denoted by its plural counterparts such as *water quantities, rice grains, policemen,* and *shoes* (the latter can be counted, distinguished, and enumerated, for example, but not the former). Moreover, unrestricted sum formation does not seem plausible as part of our perception-related cognitive ontology, where formation of sums appears restricted by 'gestalt conditions,' perceivable conditions of integrity. The ontology of natural language is at least in part an ontology closely related to language (its language-driven part) and should thus be characterized in this way:

(8) Characterization of the ontology implicit in natural language ontology (2nd version)
The ontology of natural language is the ontology a speaker implicitly accepts *by way of* using natural language.

Of course not all of the ontology of natural language is language-driven. There are certainly parts of it that belong to our cognitive ontology in general, as is plausible for artifacts and various ontologically dependent entities mentioned earlier.

4.3 Natural language ontology and the core-periphery distinction

The characterization of the ontology of natural language in (8) is still not correct. The ontology of natural language is not reflected in *all* of natural language. Throughout history, when appealing to natural language for motivating a particular ontological view, philosophers made use of certain types of expressions or uses of expressions and not others. Thus, philosophers' technical terms or other terms whose use requires a degree of philosophical reflection are not considered indicative of the ontology of natural language, for example *the property of being happy* or *the truth value true*. Philosophical terms and non-ordinary, philosophical uses of natural language expressions, even though they are part of the legitimate use of natural language, do not reflect the ontology implicit in natural language. Otherwise natural language could reflect any ontology whatsoever that someone may come up with. For example, a particular philosopher may just introduce a technical term for some ontological category, say, that of a platonic universal or that of 'the nothing,' but this does not make that category (platonic universals or the nothing) part of the ontology implicit in language.

The distinction that needs to be made is that between the *core* of language and its *periphery* (Moltmann 2013a, 2017, 2019, forthcoming b). Only expressions in the core reflect the ontology of natural language, not expressions in the periphery. The core-periphery distinction is essential for natural language ontology. It has been relied on by philosophers throughout history when making appeal to linguistic examples, and it likewise guides the practice of contemporary semanticists and philosophers pursuing natural language ontology.

The periphery includes *reifying NPs*, that is, NPs of the sort *the number eight, the property of being happy, the proposition that it is raining,* or *the truth value true*. Reifying NPs introduce objects on the basis of a sortal and possibly non-referential material (*eight, being happy, true, that* S (on a view on which *that*-clauses are non-referential)) (Moltmann 2013a, Chap. 6). Clearly, reifying NPs may introduce entities that need not be considered part of the ontology of natural language (truth values, abstract properties, propositions, and numbers). Philosophers in fact have generally stayed away from reifying NP when appealing to natural language for motivating an ontological category. For example, Frege (1884) did not motivate numbers as objects appealing to the presence of the construction *the number eight* in natural language, and he did not motivate truth values as objects by appealing to *the truth value true*. Rather he used expressions like *the number of planets* and *eight* from the core of language when arguing for numbers being objects, and his motivations for truth values to have the status of objects did not come from particular natural language sentences at all. Hale (1987) did not argue for properties being objects on the basis of terms like *the property of mercy*, but simple terms like *mercy* (from the core of language). Link (1983) did not motivate sums being part of the ontology of natural language on the basis of terms like *the sum of the students*, but simple definite plurals like *the students* (which clearly belong to the core of language).

Clearly then reference to the core-periphery distinction is indispensable for the right characterization of the ontology implicit in natural language (Moltmann 2017, 2019):

(9) Characterization of the ontology of natural language (final version)
The ontology of natural language is the ontology a speaker *implicitly accepts* by way of making use of the *core* of language.

5 Universals of natural language ontology

The core-periphery distinction is also essential for the quest for universals of natural language ontology. Clearly, only the core in the present sense, not the periphery (in that sense) can represent a form of universal cognitive language-related ontology. The existing work in natural language ontology certainly incorporates an implicit restriction to the core of language for generalizations meant to be universal. The core-periphery distinction is used explicitly in the general hypothesis about reference to abstract objects in natural language in Moltmann (2013a):

(10) The Abstract-Objects Hypothesis
Natural language does not involve reference to abstract objects in its core, but only in its periphery.

Given that hypothesis, what appeared to be expressions in the core of language referring to abstract objects (numbers, properties, properties, propositions, degrees, expression types) are in fact expressions referring to particulars, pluralities of (actual or possible) particulars, or variable objects, or expressions that fail to have a referential function in the first place (numerals, clausal complements, predicative complements, complements of intensional transitive verbs). The particulars include tropes, to which natural language displays pervasive reference. (Tropes include quantitative tropes such as John's height or the number of planets (a number trope).) Only in the periphery is reference to abstract objects possible, for example through the use of reifying terms such as *the number eight*, *the property of being happy*, or *the proposition that it is raining*.

The periphery raises questions of its own. The periphery is a legitimate part of natural language, or a legitimate extension of it. As such, it has a semantics and hence comes with an ontology. But that ontology may diverge from the ontology of the core. The question then is, how should the ontology of the periphery be understood, in particular, when it is not part of the ontology of the core of natural language? For answering this question, it is important to keep in mind that ontology in the context of descriptive metaphysics includes ontologies of appearances. Ontology for the purpose of compositional semantics may include merely conceived entities, philosophical entities on some philosopher's conception, not necessarily the ontology of actual entities, let alone real entities.

6 The syntactic core-periphery distinction

The core-periphery distinction raises an important question, namely whether there is a linguistic basis for the distinction. That is, are there syntactic or lexical conditions that determine which expressions (or uses of expressions) will be part of the periphery rather than the core?

The core-periphery distinction recalls the core-periphery distinction that Chomsky's (1981, 1986) introduced for syntax (and which Chomsky (p.c.) still thinks is essential for syntax). For Chomsky, very roughly, the core of the syntactic system of a language represents regularities and in fact universal grammar, whereas the periphery involves exceptions and parts of language added on from outside influences. The question then is whether two separate core-periphery distinctions should be made for syntax and for the ontology of natural language.

At first, the core in the syntactic sense and the core in the ontological sense do not seem to coincide. For example, *the number eight* belongs to the periphery in the present sense, but at the same time seems to belong to the core in Chomsky's sense. However, a different view becomes plausible when the focus is not on entire constructions, but on more elementary parts of language. For Yang (2016), who more recently revived and defended the core-periphery distinction in syntax, functional categories (syntactic categories and features) belong to the syntactic core, but the lexicon to the periphery. Given that, the peripheral status of the reifying NP *the number eight* can be attributed to the occurrence of the sortal *number* in that construction, rather than the construction as such.

Further support for a single core-periphery distinction comes from the ability or inability for an expression or syntactic element to allow for a non-ordinary use or 'conceptual engineering' (Eklund 2015). Lexical categories such as sortal nouns allow for non-ordinary uses, but not so, it seems, functional categories (e.g., (overt or empty) determiners, morpho-syntactic categories (plural, tense), or syntactic constructions).

While it is plausible that functional and structural meaning side with the core (in the present sense), it is not obvious that all of the lexicon belongs to the periphery (in the present sense). For example, the verb *exist* appears to belong to the core (not permitting a non-ordinary use), whereas the non-relational bare noun *existence* belongs to the periphery (Moltmann 2019, forthcoming b).

Thus, while philosophers and non-philosophers are likely to consider 'existence' a univocal notion applying to every actual thing, the predicate *exist* in fact applies only to material and abstract objects, not to events (*the rain still exists* or *the accident existed yesterday* are unacceptable, cf. Hacker 1982; Cresswell 1986; Moltmann 2013b). That holds regardless of a language user's (naïve or not so naïve) philosophical views about existence. By contrast, the bare nominalization *existence* (on its non-relational use) can easily be used to convey any notion of existence a language use may subscribe to (as in the sentence *existence is a univocal notion*).[8]

7 The ontology of natural language and other ontologies

Given the perspective of descriptive metaphysics, there is not a single ontology of the real, but rather different ontologies reflected in different ranges of data can coexist, including different ontologies reflected in different peripheries of natural language and different ontologies for different cognitive domains.[9] Such a pluralist view of ontology recalls Goodman's (1978) 'ways of worldmaking,' and it requires new formats for ontologies theories. Ontology can now no longer be based on a fixed set of categories and their characteristic properties and relations. Rather, it goes along better with a constructional ontology (Fine 1991), where various ontological operations may lead to different ranges of entities for different ontologies.

A constructional ontology is also particularly suited for complex expressions in natural language that serve the introduction of entities by some form of abstraction, namely reifying NPs of the sort *the number eight* or *the truth value true* (Hale 1987; Wright 1983) or the introduction of pleonastic entities (Schiffer 1996). Reifying terms require a distinction between the acceptance of an ontological operation interpreting complex expressions and the acceptance of the outcome of applying the operation. Syntactic knowledge of the construction of reifying terms will go along with acceptance of the ontological operation of reification interpreting that construction, but not with the acceptance of the outcome of that operation, that is, the application of the operation in particular cases.

By contrast, syntactic knowledge of the constructions in (1)–(5) will go along with the acceptance of the ontological operations interpreting them as well as the acceptance of the outcome. Thus, knowledge of English, displaying constructions as in (1)–(5), goes along with a mandatory implicit acceptance of the entities the constructions in (1)–(5) stand for. Their acceptance cannot be subject to revision upon reflection, just as syntactic knowledge cannot be revised. For the language-driven part of the ontology, then, ontology, based on ontological operations introducing derivative entities, can be considered on a par with syntax.[10]

8 Conclusion

Natural language ontology is a branch of descriptive metaphysics whose subject matter is the ontology implicit in natural language. As such it is a new discipline that is both part of philosophy and linguistics, as well as a practice that had been pursued throughout the history of philosophy. Natural language ontology as a discipline on its own not only has a well-defined subject matter. It may also set its own ambitions on a par with that of generative syntax, aiming for a universal ontology associated with the core of language, not subject to revision.

Acknowledgments

I would like to thank Ricki Bliss, Bob Matthews, Ian Dunbar, Gaetano Licata, Benjamin Nelson, Tristan Tondino, and an anonymous referee for comments on an earlier version of this chapter, Kit Fine for numerous conversations on the topic, and Noam Chomsky and Matti Eklund for relevant exchanges. The chapter has benefitted from audiences at talks at the conference *The Language of Ontology*, Trinity College Dublin, 2017, the conference, *Quo Vadis Metaphysics*, Center for Formal Ontology (ICFO), Warsaw, 2017, the conference *Metaphysics and Semantics*, University, New Haven, 2017, the *Biolinguistic Conference on Interface Asymmetries*, NYU, New York, 2017, and the *Oasis 1 Conference*, Paris 8, Paris, 2018 as well as the compact seminars at the University of Duesseldorf, the University of Munich, and ESSLLI in Sofia in 2018.

Notes

1. Bach (1986) uses the term 'natural language metaphysics' for natural language ontology. (See also Kratzer 1989; Asher 1993; Pelletier 2011; Chao/Bach 2012). This is in a sense more adequate in that ontology is generally taken to be narrower than metaphysics, dealing just with what there is and not with the nature of things (and both are dealt with by metaphysics). However, 'ontology' is increasingly used in the broader sense of metaphysics as well, in particular when it has an empirical connection ('applied ontology'). Also 'ontology' is more usable when talking about the subject matter of a discipline, in particular since it comes with a plural: *the ontology of natural language* is the subject matter of natural language ontology, and there are different *ontologies* that are the subject matter of different branches of metaphysics.
2. For the notion of a trope, see Williams (1953) and Woltersdorff (1970).
3. See Moltmann (2018) and Ramchand (2018) for further linguistic applications of truthmaker semantics.
4. There is a recent alternative approach to compositional semantics, though, which aims to do without objects and truth, namely the one of Pietroski (2018). Here semantic composition is based solely on conceptual 'instructions.'
5. There is no universal agreement, however, that definite plurals stand for pluralities conceived as single entities that are sums. Thus, it has been pointed out that that account fails to distinguish the one (singular count) and the many (plural) (Yi 2005, 2006; Moltmann 2016). For example, the semantic value of *the children* below could not be counted as a single entity:

(i) John counted the children and the adults.

Sentence (i) fails to have a reading on which John counted two: the sum of the children and the sum of the adults. Another argument against the mereological account is the reading of the predicate *exist*. *Exist* below cannot apply to 'the children,' stating the existence of the plurality independently of the existence of the individual children:

(ii) The children exist.

Such considerations have led to the exploration of alternative views, on which definite mass NPs plurally refer to each child at once (Yi 2005, 2006; Oliver/Smiley 2013; Moltmann 2016).

6. Naïve metaphysics may be taken to be the metaphysics ordinary people (non-philosophers) pursue when naïvely reflecting upon what there is, namely folk metaphysics. However, folkmetaphysics need to be sharply distinguished from natural language ontology and thus part of descriptive metaphysics (Section 4.1).
7. An example is the silent noun theory of Kayne (2005, 2015), which posits various silent nouns such as age, number, height, etc. as part of apparently simpler syntactic structure suggesting that those structures involve reference to ages, numbers, and heights.
8. Another question that arises is whether a separate core-periphery distinction should be made in the conceptual domain, with an invariant conceptual core not permitting conceptual engineering (the 'conceptual fixed points' of Eklund 2015).
9. There is of course also the issue of the universality of the ontology of natural language, touching upon the Sapir-Whorf hypothesis and the controversy surrounding it (Pinker 1994; Hespos/Spelke 2004).
10. Note that on that view the ontology is not representational, since semantic values of referential NPs are not considered representations. Rather it is a constructional ontology of actual, though derivative and possibly mind-dependent entities.

References

Abney, S. P. (1987): *The English Noun Phrase in its Sentential Aspect*. Ph.D. thesis, MIT, Cambridge, MA.
Asher, N. (1993): *Reference to Abstract Objects*. Kluwer Academic Publishers, Dordrecht.
Bach, E. (1986): 'Natural Language Metaphysics.' In R. Barcan Marcus et al. (eds.): *Logic, Methodology, and Philosophy of Science* VI, North Holland, Amsterdam et al., 573–595.
Borer, H. (2005): *Structuring Sense. Volume I. In Name Only*. Oxford University Press, Oxford.
Carlson, G. (1977): 'A Unified Analysis of the English Bare Plural.' *Linguistics and Philosophy* 1, 413–457.
Champollion, L./M. Krifka (2016): 'Mereology.' In P. Dekker/M. Aloni (eds.): *Cambridge Handbook of Formal Semantics*. Cambridge University Press, Cambridge.
Chao, W./Bach, E. (2012): 'The Metaphysics of Natural Language(s).' In R. Kempson et al. (eds.): *Philosophy of Linguistics, 14*. North Holland, Elsevier, Amsterdam, 175–196.
Chomsky, N. (1981): *Lectures on Government and Binding. The Pisa Lectures. Studies in Generative Grammar*, no. 9. Foris Publications, Dordrecht.
Chomsky, N. (1986): *Knowledge of Language: Its Nature, Origin, and Use*. Praeger, Westport, CT and London.
Chomsky, N. (1998): *New Horizons in the Study of Language and Mind*. Cambridge University Press, Cambridge.
Chomsky, N. (2013): 'Notes on Denotation and Denoting.' In I. Caponigro/C. Cecchetto (eds.): *From Grammar to Meaning: The Spontaneous Logicality of Language*. Cambridge University Press, Cambridge.
Cresswell, M. (1986): 'Why Object Exists, but Events Occur.' *Studia Logica* 45, 371–375.
Davidson, D. (1967): 'The Logical Form of Action Sentences.' In N. Rescher (ed.): *The Logic of Decision and Action*. Pittsburgh University Press, Pittsburgh, 81–95.
Eklund, M. (2015): 'Intuitions, Conceptual Engineering, and Conceptual Fixed Points.' In Christopher Daly (ed.): *Palgrave Handbook of Philosophical Methods*. Palgrave Macmillan, London.
Fine, K. (1991): 'The Study of Ontology.' *Noûs* 25.3, 262–294.
Fine, K. (1999): 'Things and Their Parts'. *Midwest Studies of Philosophy* 23, 61–74.
Fine, K. (2003): 'The Non-Identity of a Material Thing and Its Matter.' *Mind* 112, 195–234.
Fine, K. (2006): 'In Defense of Three-Dimensionalism.' *Journal of Philosophy* 103.12, 699–714.
Fine, K. (2017a): 'Naïve Metaphysics.' *Philosophical Issues* 27.1, 98–113, edited by J. Schaffer.
Fine, K. (2017b): 'Truthmaker Semantics.' In B. Hale/C. Wright (eds.): *Blackwell Philosophy of Language Handbook*. Blackwell, New York.
Frege, G. (1884): *Die Grundlagen der Arithmetik* ('Foundations of Arithmetic').
Frege, G. (1892): 'Funktion und Begriff.' Reprinted in G. Patzig (ed.): *Funktion, Begriff, Bedeutung*, Vandenhoeck and Ruprecht, Goettingen.
Goodman, N. (1978): *Ways of Worldmaking*. Hackett, Indiana.
Hacker, P. M. S. (1982): 'Events, Ontology, and Grammar.' *Philosophy* 57, 477–486.
Hale, B. (1987): *Abstract Objects*. Blackwell, Oxford.
Hespos, S./E. Spelke (2004): 'Conceptual Precursors to Language.' *Nature* 430, 453–456.
Kayne, R. (2005): *Movement and Silence*. Oxford University Press, New York.
Kayne, R. (2015): 'The Silence of Heads.' Ms NYU, available on Academia.
Kratzer, A. (1989): 'An Investigation into the Lumps of Thought.' *Linguistics and Philosophy* 12, 608–653.
Link, G. (1983): 'The Logical Analysis of Plurals and Mass Nouns.' In R. Baeuerle et al. (eds.): *Semantics from Different Points of View*. Springer, Berlin, 302–323.
Moltmann, F. (1997): *Parts and Wholes in Semantics*. Oxford University Press, New York.
Moltmann, F. (2009): 'Degree Structure as Trope Structure.' *Linguistics and Philosophy* 32.1, 51–94.
Moltmann, F. (2013a): *Abstract Objects and the Semantics of Natural Language*. Oxford University Press, Oxford.
Moltmann, F. (2013b): 'The Semantics of Existence.' *Linguistics and Philosophy* 36.1, 31–63.
Moltmann, F. (2015): 'Quantification with Intentional and with Intensional Verbs.' In A. Torza (ed.): *Quantifiers, Quantifiers, Quantifiers*. Springer: Synthese Library, Dordrecht, 141–168.
Moltmann, F. (2016): 'Plural Reference and Reference to a Plurality: Linguistic Facts and Semantic Analyses.' In M. Carrara et al. (eds.): *Unity and Plurality: Philosophy, Logic, and Semantics*. Oxford University Press, Oxford, 93–120.
Moltmann, F. (2017): 'Natural Language Ontology.' *Oxford Encyclopedia of Linguistics*.
Moltmann, F. (2018): 'An Object-Based Truthmaker Theory for Modals.' *Philosophical Issues* 28.1, 255–288.

Moltmann, F. (2019): 'Natural Language and its Ontology.' In A. Goldman/B. McLaughlin (eds.): *Metaphysics and Cognitive Science*, Oxford University Press, Oxford.

Moltmann, F. (forthcoming a): 'Variable Objects and Truthmaking.' To appear in M. Dumitru (ed.): *The Philosophy of Kit Fine*, Oxford University Press, New York.

Moltmann, F. (forthcoming b): 'Abstract Objects and the Core-Periphery Distinction in the Ontological and Conceptual Domain of Natural Language.' In J. L. Falguera and C. Martínez (eds.): *Abstract Objects: For and Against*. Springer: Synthese Library, Dordrecht.

Oliver, A./T. Smiley (2013): *Plural Logic*. Oxford University Press, Oxford.

Pelletier, F. J. (2011): 'Descriptive Metaphysics, Natural Language Metaphysics, Sapir–Whorf and All That Stuff: Evidence from the Mass-Count Distinction.' *The Baltic International Yearbook of Cognition, Logic and Communication* 6, 1–46.

Pietroski, P. (2018): *Conjoining Meanings: Semantics without Truth Values*. Oxford University Press, Oxford.

Pinker, S. (1994): *The Language Instinct*. Harper Perennial Modern Classics, New York.

Ramchand, G. (2018): *Situations and Syntactic Structures: Rethinking Auxiliaries and Order in English*. MIT Press, Cambridge, MA.

Schiffer, S. (1996): 'Language-Created and Language-Independent Entities.' *Philosophical Topics* 24.1, 149–167.

Strawson, P. (1959): *Individuals: An Essay in Descriptive Metaphysics*. Methuen, London.

Swanson, E. (2012): 'The Language of Causation.' In D. Graff Fara/G. Russell (eds.): *The Routledge Companion to the Philosophy of Language*. Routledge, London, 716–728.

Szabo, Z. (2007): 'Counting Across Times.' *Philosophical Perspectives* 20: *Metaphysics*, 399–426.

Szabo, Z. (2015): 'Major Parts of Speech.' *Erkenntnis* 80, 3–29.

Thomason, R. H. (ed.) (1974): *Formal Philosophy: Selected Papers of Richard Montague*, Yale University Press, New Haven, CT.

Williams, D. C. (1953): 'On the Elements of Being.' *Review of Metaphysics* 7, 3–18.

Woltersdorff, N. (1970): *On Universals*. Chicago University Press, Chicago.

Wright, C. (1983): *Frege's Conception of Numbers as Objects*. Aberdeen University Press, Aberdeen.

Yang, C. (2016): *The Price of Linguistic Productivity. How Children Learn to Break the Rules of Language*. MIT Press, Cambridge, MA.

Yi, B.-Y. (2005): 'The Logic and Meaning of Plurals. Part I.' *Journal of Philosophical Logic* 34, 459–506.

Yi, B.-Y. (2006): 'The Logic and Meaning of Plurals. Part II.' *Journal of Philosophical Logic* 35, 239–288.

26
PHENOMENOLOGY AS METAPHYSICS

Dan Zahavi

The relationship between phenomenology and metaphysics is controversial.[1] One initial source of contention concerns the meaning of both terms, which is far from univocal. Consider, for instance, how the term 'phenomenology' is currently used in parts of cognitive science and analytic philosophy of mind to denote the qualitative character of experience. On this reading, phenomenology is a dimension of experience and one way to raise the question concerning the link between phenomenology and metaphysics is by asking whether the fact that it seems to us that we have subjective experiences warrants the conclusion that there really are subjective experiences. This is an interesting (if perhaps also slightly absurd) question, but it is not one I will pursue in what follows. Rather, my concern will be with a different kind of phenomenology, namely phenomenology understood as a distinct philosophical tradition. It was founded by Edmund Husserl and continued by thinkers such as Max Scheler, Martin Heidegger, Jean-Paul Sartre, Maurice Merleau-Ponty and Emmanuel Levinas. This list of names already suggests the next complication, however. Even though they all qualify as phenomenologists, they were independent thinkers with divergent and conflicting views on various issues, including the relation between phenomenology and metaphysis. The later Heidegger, for instance, was critical towards metaphysics. He described it as a thinking of identity, i.e. as a thinking that attempted to annul the ontological difference between being and beings, and in various of his writings, Heidegger sought to substitute the conceptual apparatus of metaphysics for what he took to be a more authentic type of thinking (Heidegger 1998: 185, 188, 224). Levinas, by contrast, defined metaphysics as an openness to otherness, as an acknowledgement of the infinite. In fact, for him, metaphysics was nothing but a movement of transcendence, namely the very relation to the absolute other (Levinas 1969: 43). In *Totalité et infini*, Levinas consequently criticized Heidegger for his disregard of metaphysics and for promoting a totalizing ontology that absorbed and reduced the foreign and different into the familiar and identical (Levinas 1969: 42–43). In the conclusion of Sartre's *L'être et le néant*, we find yet another set of reflections on the relation between phenomenology, ontology and metaphysics. Sartre writes that whereas ontology describes the structure of a being, metaphysics seeks to explain an event, namely the upsurge of the for-itself, and he consequently argues that metaphysics is to ontology as history is to sociology (Sartre 2003: 639, 641). One reason for these divergent answers is obviously that the term 'metaphysics' is being used so differently. And indeed, one of the main reasons why the relation between phenomenology and metaphysics remains controversial to this day is precisely due to the ambiguity of the latter term.

Given that it will be impossible to discuss the positions of all the phenomenologists in a short text like the present and given that some of the most divisive views on the relation between phenomenology and metaphysics can be found in discussions of Husserl's phenomenology, that is what I will focus on in the following.[2]

The metaphysical neutrality of phenomenology

In *Logische Untersuchungen* (1900–1901), a work Husserl considered his 'breakthrough' to phenomenology, Husserl described his overall project as an attempt to establish a new foundation for pure logic and epistemology (Husserl 2001: I/2). He argued that the cardinal task of a theory of knowledge is to determine the conditions of possibility for objective knowledge and not to determine whether (and how) consciousness can attain knowledge of a mind-independent reality. In fact, Husserl explicitly rejected this very type of question, as well as the question of whether there is an external reality, as *metaphysical questions* that have no place in epistemology (Husserl 2001: I/178). What did Husserl have in mind, when talking of metaphysics? In this early work, Husserl considered metaphysics a discipline whose main task was to answer questions concerning the nature and existence of external reality.

Logische Untersuchungen contains numerous passages affirming the difference between phenomenology and metaphysics. In the introduction to the second part of the work, for instance, Husserl described phenomenology as a *neutral* investigation (Husserl 2001: I/166) and claimed that epistemological concerns precede every metaphysics (Husserl 2001: I/178). He then emphasized that all of the six ensuing investigations were characterized by their metaphysical presuppositionlessness (Husserl 2001: I/178–179). Later, in the 5th Investigation, he again explicitly stressed the difference between the metaphysical and the phenomenological endeavour, and went on to say that the descriptive difference between the act of experience and the object of experience is valid regardless of one's take on the question concerning the nature of the being-in-itself. In fact, it is a difference that precedes every metaphysics (Husserl 2001: II/106). Husserl also insisted that phenomenology should disregard the question of whether or not the intentional object has any mind-independent reality. As Husserl phrased it, the *existence* of the intentional object is phenomenologically irrelevant, since the intrinsic nature of the intentional act is the same regardless of whether or not its object exists (Husserl 2001: II/99).

Logische Untersuchungen is typically taken to exemplify a pre-transcendental, descriptive type of phenomenology, and as this brief overview suggests, Husserl's pre-transcendental phenomenology was not only avowedly metaphysically neutral, but also seemingly committed to a form of *methodological solipsism*. How the external world is, makes no difference to one's mental states. What about his later work? One significant difference between *Logische Untersuchungen* and Husserl's next major work *Ideen I* (1913) is that Husserl in the intervening years came to realize that certain methodological steps – the notorious epoché and transcendental reduction – were required if phenomenology were to accomplish its designated task. Whereas both notions were absent in *Logische Untersuchungen*, they came to play a decisive role after Husserl's transcendental turn, i.e. after he explicitly conceived of and presented phenomenology as a form of transcendental philosophy. Did the introduction of these notions (which will be explained shortly) change anything vis-à-vis the relation between phenomenology and metaphysics? According to one deflationary interpretation of Husserl's transcendental idealism, when it comes to the issue of metaphysics, Husserl's view remained basically unchanged. Husserl remained committed to the idea that phenomenology is metaphysically neutral, since his interest and focus was on *meaning* rather than *being*. This is a view shared by people whose Husserl interpretations otherwise diverge wildly. It is a view defended not only by Dreyfus and Carman, but also by Carr, Crowell and Thomasson.

Dreyfus and Carman both consider Husserl an arch-internalist, i.e. as somebody who defended the view that the mind contains a multitude of internal representations that all have the function they have regardless of how the world is, and they take Husserl's phenomenology to be a futile attempt to investigate consciousness from a strictly internal perspective, i.e. a perspective that removes all external components from consideration. As Dreyfus puts it, Husserlian phenomenology is an enterprise that is exclusively interested in the mental representations that remain in consciousness after the performance of the reduction has bracketed the world and any concern with existence (Dreyfus 1982: 108; 1991: 50). It is the same interpretation we find in Carman, who writes:

> The transcendental reduction ... consists in methodically turning away from everything external to consciousness and focusing instead on what is internal to it. The reduction thus amounts to a special kind of reflection in which the ordinary objects of our intentional attitudes drop out of sight, while the immanent contents of those attitudes become the new objects of our attention.
> *(Carman 2003: 80)*

Although his general assessment and appraisal of Husserl is very different from Dreyfus' and Carman's, Carr has also defended the view that a crucial part of Husserl's transcendental methodology is to exclude the actual existence of the world from consideration (Carr 1999: 74). As Carr puts it, all reference to the being of transcendent reality is excluded in order to focus instead on its sense or meaning (Carr 1999: 80). This is also why, according to Carr, Husserl is not engaged in a metaphysical project at all. Transcendental phenomenology is not a metaphysical doctrine but must on the contrary simply be understood as a critical reflection on the conditions of possibility for experience (Carr 1999: 134). Related interpretations have been defended by both Thomasson and Crowell. Whereas Thomasson writes that Husserl's method is not concerned with reality but only with an analysis of meaning that has no immediate bearing on metaphysical questions about "what really exists" (Thomasson 2007: 91), Crowell has emphasized the non-metaphysical direction of transcendental thought. According to Crowell, phenomenology is first and foremost a philosophy of meaning (Crowell 2001: 5) and must ultimately be viewed as a metaphilosophical or methodological endeavour rather than as a straightforwardly metaphysical doctrine about the nature and ontological status of worldly objects. Husserl's transcendental idealism can precisely be distinguished from metaphysical idealism in that the latter, but not the former, makes first-order claims about the nature of objects (Crowell 2001: 237). From this point of view, a metaphysical interpretation of Husserl's transcendental phenomenology entails a dramatic misunderstanding of what phenomenology is all about.[3] It misunderstands the notion of reduction, and it overlooks the decisive difference between the natural attitude and the phenomenological attitude.

The metaphysical implications of phenomenology

Many interpreters have taken Husserl's methodology, his employment of the epoché and the reduction, to involve a bracketing of questions related to existence and being and have for that very reason also denied that phenomenology has any metaphysical implications. If one construes Husserlian phenomenology in such a way, it is quite natural to conclude that important topics are simply missing from its repertoire; being and reality are topics left for other disciplines. But this interpretation neither respects nor reflects Husserl's own assertions on the matter. As Husserl declares in § 23 of *Cartesianische Meditationen*, the topics of existence and non-existence, of being and non-being, are all-embracing themes for phenomenology, themes addressed under the broadly understood titles of reason and unreason (Husserl 1960: 56). Indeed, Husserl is quite unequivocal in his rejection of any non-metaphysical interpretation of transcendental phenomenology:

> [T]ranscendental phenomenology in the sense I conceive it does in fact encompass the universal horizon of the problems of philosophy ... including as well all so-called metaphysical questions, insofar as they have possible sense in the first place.
> (Husserl 1989a: 408)

> Phenomenology is anti-metaphysical insofar as it rejects every metaphysics concerned with the construction of purely formal hypotheses. But like all genuine philosophical problems, all metaphysical problems return to a phenomenological base, where they find their genuine transcendental form and method, fashioned from intuition.
> (Husserl 1997: 101)

> Our monadological results are *metaphysical*, if it be true that ultimate cognitions of being should be called metaphysical. On the other hand, what we have here is *anything but metaphysics in the customary sense*: a historically degenerate metaphysics, which by no means conforms to the sense with which metaphysics, as 'first philosophy', was instituted originally. Phenomenology's purely intuitive, concrete and also apodictic mode of demonstration excludes all 'metaphysical adventure', all speculative excesses.
> (Husserl 1960: 139)

Statements like these strongly suggest that it might be wrong to opt for a deflationary non-metaphysical interpretation of Husserl's transcendental project. This, however, is not to say that it is by any stretch straightforward to interpret Husserl's pro-metaphysical statements, since the term 'metaphysics', as already pointed out, is notoriously ambiguous, and can be understood and defined in a variety of diverse ways. Here are three quite different definitions:

- Metaphysics is a philosophical engagement with question of facticity, birth, death, fate, immortality, the existence of God, etc.
- Metaphysics is a fundamental reflection on and concern with the status and being of reality. Is reality mind-dependent or not, and if yes, in what manner?
- Metaphysics is a theoretical investigation of the fundamental building blocks of reality.

Although Husserl in his later works came to the conclusion that phenomenology had to engage with the problem of facticity, and although one can also find Husserl working on a kind of 'philosophical theology' (cf. Hart 1986), I think Husserl's primary interest concerned the second type of metaphysis, i.e. metaphysics understood as pertaining to the realism-idealism issue, that is, to the question of whether reality is mind-independent or not. Indeed, I would claim that Husserl's interest in this particular form of metaphysics was essential to his conception of transcendental philosophy. As I have argued repeatedly over the years, this commitment is not in conflict with Husserl's phenomenological methodology, his employment of the epoché and the reduction, but on the contrary follows directly from it.

On one interpretation, the real purpose of the phenomenological bracketing is to limit the scope of the investigation. There are certain issues that are excluded from consideration, certain questions that phenomenologists are not supposed to engage with. Phenomenologists might believe that they are directed at something extra-mental, something transcendent, something that is not contained in consciousness, and as phenomenologists, they should investigate this belief and our experiences of natural objects, artefacts, other people, works of arts, social institutions, etc., but they are not entitled to say anything about the being of these entities themselves. As a phenomenologist, I can claim that I experience a lemon, that a lemon is appearing, that it seems as if there is a lemon in front of me, but I cannot as a phenomenologist affirm that there really is a lemon. To do the latter would be to make an illegitimate transition from phenomenology, from a concern with how things appear and what meaning they have for me, to metaphysics, to a concern with reality and real existence. This line of reasoning, however, is based on a misinterpretation (cf. Zahavi 2003b, 2017). The purpose of the epoché was never to bracket either the world or true being from consideration. The epoché does not involve an exclusion of reality, but rather a suspension of a particular dogmatic *attitude* towards reality, an attitude that is operative not only in the positive sciences, but which also permeates our daily pre-theoretical life. Indeed, the attitude is so fundamental and pervasive that Husserl calls it the *natural attitude*. What is the attitude about? It is about simply taking the world we encounter in experience for granted, and to assume that it also exists independently of us. Regardless of how natural and obvious it might be to think of reality as a self-subsisting entity, if philosophy is supposed to amount to a radical form of critical elucidation, it cannot simply take this kind of realism for granted. If we are to adopt the phenomenological attitude and engage in phenomenological philosophizing, we must take a step back from our naive and unexamined immersion in the world and suspend our automatic belief in the mind-independent existence of that world. By suspending this attitude, and by thematizing the fact that reality is always revealed and examined from some perspective or another, reality is not lost from sight, but is for the first time made accessible for philosophical inquiry. Rather than making reality disappear from view, the epoché and reduction is precisely what allows reality to be investigated philosophically. This is why Husserl in the lecture *Phänomenologie und Anthropologie* from 1931 repeatedly writes that the only thing that is excluded as a result of the epoché is a certain naivety, the naivety of simply taking the world for granted, thereby ignoring the contribution of consciousness (Husserl 1989b: 173). And as Husserl repeatedly insists in this 1931 lecture, the turn from a naive exploration of the world to a reflective exploration of the field of consciousness does not entail a turning away from the world, rather, it is a turn that for the first time allows for a truly radical investigation and comprehension of the world (Husserl 1989b: 178).

The position eventually reached by Husserl is clearly stated in *Cartesianische Meditationen*:

> The attempt to conceive the universe of true being as something lying outside the universe of possible consciousness, possible knowledge, possible evidence, the two being related to one another merely externally by a rigid law, is nonsensical. They belong together essentially; and, as belonging together essentially, they are also concretely one, one in the only absolute concretion: transcendental subjectivity. If transcendental subjectivity is the universe of possible sense, then an outside is precisely nonsense.
>
> (Husserl 1960: 84)

> Carried out with this systematic concreteness, phenomenology is *eo ipso* 'transcendental idealism', though in a fundamentally and essentially new sense. It is not a psychological idealism, and most certainly not such an idealism as sensualistic psychologism proposes, an idealism that would derive a senseful world from senseless sensuous data. Nor is it a Kantian idealism, which believes it can keep open, at least as a limiting concept, the possibility of a world of things in themselves.
>
> (Husserl 1960: 86)

Phenomenology is the science of the phenomena. In much of the philosophical tradition, the phenomenon has been defined as the way the object appears to us, as seen with our eyes (and thought with our categories) and has been contrasted with the object as it is in itself. The assumption has then been, that if one wishes to discover and determine what the object really is like, then one has to go beyond the merely phenomenal. Had it been this concept of phenomenon that phenomenology was employing, phenomenology would have been the study of the merely subjective, apparent or superficial. But this is not the case. As Heidegger points out in § 7 of *Sein und Zeit*, phenomenology is drawing on and employing a very different and more classical conception of phenomenon, according to which the phenomenon is that which shows itself, that which reveals itself (Heidegger 1996: 25). Phenomenology is consequently not a theory about the *merely* apparent. As Heidegger also pointed out in a lecture course given a few years before *Sein und Zeit*:

> It is phenomenologically absurd to speak of the phenomenon as if it were something behind which there would be something else of which it would be a phenomenon in the sense of the appearance which represents and expresses [this something else]. A phenomenon is nothing behind which there would be something else. More accurately stated, one cannot ask for something behind the phenomenon at all, since what the phenomenon gives is precisely that something in itself.
>
> *(Heidegger 1985: 86)*

In its radical exploration of the structure and status of the phenomenon transcendental phenomenology cannot permit itself to remain neutral or indifferent to the question concerning the relationship between the phenomenon and reality. But by having to take a stand on this relationship, phenomenology also by necessity has metaphysical implications.

Whereas some might claim that the phenomenon is something merely subjective, a veil or smoke screen, that conceals the objectively existing reality, phenomenologists reject what can be called a two-world doctrine, i.e. the proposal that we have to make a principled distinction between the world that presents itself to and can be understood by us and the world as it is in itself. This is certainly not to deny the distinction between mere appearance and reality – after all some appearances are misleading – but for phenomenologists, this distinction is not a distinction between two separate realms (falling in the province of phenomenology and science, respectively), but a distinction between two modes of manifestation. It is a distinction between how the objects might appear at a superficial glance, and how they might appear in the best of circumstances, for instance as a result of a thorough scientific investigation.

What then about the third type of metaphysics? Well, in one sense this was certainly also something Husserl was engaged in, though he usually spoke of it as a matter of ontology rather than metaphysics. In works such as *Cartesianische Meditationen*, *Ideen III*, *Erste Philosophie II* and *Formale und transzendentale Logik*, Husserl distinguishes formal ontology from material (or regional) ontology. Formal ontology is the name for the discipline that investigates what it means to be an object. It is considered a formal enterprise, for it abstracts from all considerations concerning content. It is not concerned with the differences between stones, trees and violins, it is not concerned with the differences between various types of objects, but with that which holds true for any object whatsoever. Formal ontology is consequently engaged in an analysis of such categories as quality, property, relation, identity, whole, part and so on. In contrast, material (or regional) ontology examines the essential structures belonging to a given region of objects and seeks to determine that which holds true with necessity for any member of the region in question. For instance, what is it that characterizes mathematical entities as such in contrast to social acts or mental episodes? Given this definition of ontology, ontological analyses (of both material and formal nature) are ubiquitous in Husserl's phenomenological writings, and there is no question that Husserl would have agreed with Heidegger's statement that "There is no ontology *alongside* a phenomenology. Rather, *scientific ontology is nothing but phenomenology*" (Heidegger 1985: 72).

Phenomenology and idealism

But where does all of this leave us vis-à-vis the metaphysical implications of Husserl's transcendental phenomenology? Is it simply a form of metaphysical idealism? This was the conclusion reached by quite a few of Husserl's early followers. After the publication of *Logische Untersuchungen*, they had enthusiastically joined Husserl in Göttingen in order to study with him. But many became quite unhappy with Husserl's subsequent development. They took his explicit endorsement of transcendental philosophy in *Ideen I* to constitute a betrayal of the core ideas of phenomenology and to ultimately amount to an unpalatable form of idealism. Ingarden, for instance, claims that for Husserl "every being (real or ideal or purely intentional) is to be deduced from the essence of the operations (acts) of pure consciousness" (Ingarden 1975: 12). On this reading, Husserl is basically committed to the view that the object is created by the subject (Ingarden 1975: 58). Similar interpretations have more recently been defended by Philipse and Smith. Whereas Philipse has claimed that Husserl was a reductive idealist, who considered the world to be nothing but a mentalistic projection (Philipse 1995: 266), A. D. Smith has defended the view that Husserl is an absolute idealist. One who would claim that nothing would exist in the absence of consciousness (Smith 2003: 179). More specifically, Smith suggests that Husserl's idealism amounts to the claim that physical facts and entities supervene on consciousness, they are nothing over and above experiential facts (Smith 2003: 183–185). Even if this doesn't entail that physical objects are simply mental states, or constructions out of these, the way they depend upon mental states is still such that Smith might be said to ascribe a form of sophisticated phenomenalism to Husserl.

Interpretations like these are, however, difficult to reconcile with Husserl's persistent rejection of phenomenalism. Already in *Logische Untersuchungen*, Husserl was very explicit in his criticism:

> However we may decide the question of the existence or non-existence of phenomenal external things, we cannot doubt that the reality of each such perceived thing cannot be understood as the reality of a perceived complex of sensations in a perceiving consciousness.
>
> *(Husserl 2001: 2/342)*

As Husserl points out, if we carefully analyse a physical object, it will not eventually dissolve in consciousness, but in atoms and molecules (Husserl 2003: 28). As he also writes in texts from 1908 and 1923:

> The objects of nature are obviously true objects; their being is true being, and nature is real in the true and full sense of the word. ... And to say that natural science has nothing to do with nature, that the true objects it engages with are mere sensations, and that what we call things, atoms, etc., are merely symbols, abbreviations in our economy of thought, for sensations and complexes of sensations, is the pinnacle of wrongness.
>
> *(Husserl 2003: 70–71)*

> An idealism that so to speak beats matter to death, which explains away experienced nature as a mere illusion and which only admits truth to the being of the psyche, is nonsensical.
>
> (Husserl 2002: 276)

Husserl would never propose that statements about botanical or geological states of affairs are ultimately to be reinterpreted as statements about mental processes. Indeed, whatever Husserl meant by idealism, is certainly didn't entail a denial of the difference between mind and world, subject and object. As he pointed out in *Ideen I*, a material thing "is by essential necessity not a mental process but a being of a wholly different mode of being" (Husserl 1982: 70). Or as he would put it later in the same work, the distinction between the being of consciousness and transcendent being is the "most radical of all ontological distinctions" (Husserl 1982: 171).

Despite his commitment to transcendental idealism, Husserl was not opposed to empirical realism. Indeed, not unlike Kant, Husserl did not merely think that transcendental idealism and empirical realism were compatible; he thought that the latter required the former. This is, of course, not to deny that there are significant differences between Husserl and Kant. By rejecting the Kantian notion of *das Ding an sich*, Husserl also removed any reason to demote the status of the reality we experience to being 'merely' for us. For Husserl, the world that can appear to us – be it in perception, in our daily concerns or in our scientific analyses – is the only real world. To claim that in addition to this world there exists a world behind-the-scenes, which transcends every appearance and every experiential and theoretical evidence, and to identify this world with true reality, is for Husserl an empty and countersensical proposition. Husserl consequently embraces a this-worldly conception of objectivity and reality, and thereby dismisses the kind of scepticism that would argue that the way the world appears to us (even under optimal conditions) is compatible with the world really being completely different.

I think the interpretation of Husserl as a metaphysical idealist is mistaken, but as should also be clear by now, I think the wrong way to counter that misinterpretation is by endorsing a deflationary non-metaphysical interpretation. Husserl's transcendental phenomenology is not neutral in the sense of being in principle compatible with a variety of different metaphysical views, such as objectivism, eliminativism or subjective idealism. Husserl's idealism is not transcendental rather than metaphysical, because it lacks metaphysical impact. It is not transcendental rather than metaphysical, because Husserl retained the notion of *das Ding an sich*. No, Husserl's idealism is transcendental rather than metaphysical, because of the way in which he interprets the dependency relation. The mind-dependency of reality is not due to the fact that worldly objects can be reduced to or supervene on the 'stuff' that mental states are made of. This is also why I earlier argued that Husserl in only one sense was engaged in the third type of metaphysics. There is another way of conceiving of that kind of metaphysics, such that it by no means was something Husserl was concerned with. Husserl's transcendental idealism is not participating in or contributing to the debate between monists and dualists. Husserl's opponent is not the dualist, but the objectivist, who claims that reality is absolute in the sense of being radically mind-independent.

On Husserl's view, the fit and link between mind and world isn't merely external or coincidental: "consciousness (mental process) and real being are anything but coordinate kinds of being, which dwell peaceably side by side and occasionally become 'related to' or 'connected with' one another" (Husserl 1982: 111). By insisting on the interdependence and inseparability of mind and world, Husserl is not reducing the world to intramental modifications or constructions. Reality does not literally exist in the mind but is always a reality for us. He is combining the claim that the world is different from the mind with the claim that world is also related to the mind, and vice versa. Indeed, on one account, the very focus of phenomenology is precisely on this intersection between mind and world, neither of which can be understood in separation from each other. I consequently think Sartre was quite right, when he in a text from 1939 offered the following interpretation of Husserl's theory of intentionality:

> Against the digestive philosophy of empirico-criticism, of neo-Kantianism, against all 'psychologism', Husserl persistently affirmed that one cannot dissolve things in consciousness. You see this tree, to be sure. But you see it just where it is: at the side of the road, in the midst of the dust, alone and writhing in the heat, eight miles from the Mediterranean coast. It could not enter into your consciousness, for it is not of the same nature as consciousness. ... But Husserl is not a realist: this tree on its bit of parched earth is not an absolute that would subsequently enter into communication with me. Consciousness and the world are given at one stroke: essentially external to consciousness, the world is nevertheless essentially relative to consciousness.
> *(Sartre 1970: 4)*[4]

Husserl's transcendental idealism entails the view that worldly objects are constitutively dependent upon transcendental subjectivity. Transcendental subjectivity, however, does not create the objects it constitutes. Nor is it their source, in the sense that they can somehow be deduced from or explained by its operations. It is not as if the fact that water is composed of hydrogen and oxygen, rather than helium and xenon, is somehow to be explained with reference to consciousness. To speak of constituting transcendental subjectivity is not to speak of a mind that shapes the world in its own image; rather, the constitutive process must be understood as a process that permits that which is constituted to appear, unfold, manifest and present itself as what it is (Husserl 1973a: 47, 1973b: 434). It is therefore no coincidence, that Husserl frequently spoke of how the world 'constitutes itself' ('konstituiert sich') in consciousness (Husserl 1973b: 19), and as Levinas once observed in a commemorative piece on Husserl fittingly entitled 'The ruin of representation':

> Intentionality means that all consciousness is consciousness of something, but above all that every object calls forth and as it were gives rise to the consciousness through which its being shines and, in doing so, appears.
> *(Levinas 1998: 119)*

Rather than amounting to an unprecedentedly strong version of internalism – a version that emphasizes the self-sufficiency of the mind to such an extent that it basically eliminates the world from the picture – Husserl's transcendental idealism was (1) motivated by an attempt to save the objectivity and transcendence of the world of experience, (2) characterized by its rejection of phenomenalism and global scepticism and (3) committed to the essential interdependence of mind and world. In addition, and this is a central aspect of his theory that I cannot develop further in this context, Husserl also (4) holds the view that the world is correlated with an intersubjective community of embodied subjects. Jointly these features happen to make his transcendental idealism far less marginal than one might initially have expected. By endorsing the view that the only justification obtainable and the only justification required is one that is internal to the world of experience and to its intersubjective practices, Husserl offers a view on the transcendental that points forward in time rather than backwards to Kant. In that sense, and to that extent, Husserl's conception of the transcendental is distinctly modern, and might even be said to have quite a presence in twentieth-century continental and analytic philosophy.

Notes

1 I have been writing on the relation between phenomenology and metaphysics for more than 20 years. The following chapter draws on and summarizes claims made in more detail in previous publications including Zahavi 2002, 2003a, 2003b, 2008, 2010 and 2017.
2 For discussions of the metaphysical positions of Heidegger, Sartre, Merleau-Ponty and Derrida, see, for instance, Wilson 2000, Braver 2007 and James 2018.
3 It should come as no surprise that Carr and Crowell have sympathies for Allison's Kant-interpretation (Carr 1999: 108–111; Crowell 2001: 238). For Allison, transcendental idealism must be appreciated as a metaphilosophical outlook rather than as a straightforward metaphysical doctrine (Allison 1983: 25). It investigates the epistemic conditions under which we have an experience of the world and does not make any first-order claims about what there is in the world. By contrast, I see a number of intriguing parallels between my own Husserl-interpretation and the Kant-interpretation recently proposed by Allais, who suggests that transcendental idealism should be interpreted as a position situated between strong metaphysical idealism (read phenomenalism) and a deflationary non-metaphysical reading (Allais 2015: 3).
4 This is not only Husserl's view. Heidegger and Merleau-Ponty also deny the self-contained nature of the mind and argue that it is intrinsically world-involved. In addition, however, they also defend the reverse claim, and argue that the world is tied to the mind. As Heidegger writes in the lecture course *The Basic Problems of Phenomenology* from 1927:

> World exists – that is, it is – only if Dasein exists, only if there is Dasein. Only if world is there, if Dasein exists as being-in-the-world, is there understanding of being, and only if this understanding exists are intraworldly beings unveiled as extant and handy. World-understanding as Dasein-understanding is self-understanding. Self and world belong together in the single entity, the Dasein. Self and world are not two beings, like subject and object, or like I and thou, but self and world are the basic determination of the Dasein itself in the unity of the structure of being-in-the-world.
>
> *(Heidegger 1982: 297)*

A similar commitment to the interdependence of mind and world is found in Merleau-Ponty, who towards the end of *Phenomenology of Perception* declares that

> The world is inseparable from the subject, but from a subject who is nothing but a project of the world; and the subject is inseparable from the world, but from a world that it itself projects. The subject is being-in-the-world and the world remains 'subjective'. since its texture and its articulations are sketched out by the subject's movement of transcendence.
>
> *(Merleau-Ponty 2012: 454)*

References

Allais, L. 2015. *Manifest Reality: Kant's Idealism and His Realism* (Oxford: Oxford University Press).
Allison, H. E. 1983. *Kant's Transcendental Idealism: An Interpretation and Defense* (New Haven, Conn.: Yale University Press).
Braver, L. 2007. *A Thing of This World: A History of Continental Anti-Realism* (Evanston, Ill.: Northwestern University Press).
Carman, T. 2003. *Heidegger's Analytic: Interpretation, Discourse and Authenticity in Being and Time* (Cambridge: Cambridge University Press).
Carr, D. 1999. *The Paradox of Subjectivity: The Self in the Transcendental Tradition* (Oxford: Oxford University Press).
Crowell, S. 2001. *Husserl, Heidegger and the Space of Meaning* (Evanston, Ill.: Northwestern University Press).
Dreyfus, H. L. 1982. Husserl's Perceptual Noema. In H. L. Dreyfus and H. Hall (eds.): *Husserl, Intentionality and Cognitive Science* (Cambridge, Mass.: MIT Press), 97–123.
Dreyfus, H. L. 1991. *Being-in-the-World* (Cambridge, Mass.: MIT Press).
Hart, J. G. 1986. A Precis of a Husserlian Phenomenological Theology. In S. C. Laycock and J. G. Hart (eds.): *Essays in Phenomenological Theology* (Albany, NY: SUNY), 89–168.
Heidegger, M. 1982. *The Basic Problems of Phenomenology*, trans. A. Hofstadter (Bloomington: Indiana University Press).

Heidegger, M. 1985. *History of the Concept of Time*, trans. T. Kisiel (Bloomington: Indiana University Press).
Heidegger, M. 1996. *Being and Time*, trans. J. Stambaugh (Albany, NY: SUNY).
Heidegger, M. 1998. *Pathmarks*, ed. W. McNeill (Cambridge: Cambridge University Press).
Husserl, E. 1960. *Cartesian Meditations: An Introduction to Phenomenology*, trans. D. Cairns, Husserliana 1 (The Hague: Martinus Nijhoff).
Husserl, E. 1973a. *Zur Phänomenologie der Intersubjektivität: Texte aus dem Nachlass. Zweiter Teil: 1921–1928*, ed. I. Kern, Husserliana 14 (The Hague: Martinus Nijhoff).
Husserl, E. 1973b. *Zur Phänomenologie der Intersubjektivität: Texte aus dem Nachlass. Dritter Teil: 1929–1935*, ed. I. Kern, Husserliana 15 (The Hague: Martinus Nijhoff).
Husserl, E. 1982. *Ideas Pertaining to a Pure Phenomenology and to a Phenomenological Philosophy. First Book: General Introduction to a Pure Phenomenology*, trans. F. Kersten (The Hague: Martinus Nijhoff).
Husserl, E. 1989a. *Ideas Pertaining to a Pure Phenomenology and to a Phenomenological Philosophy. Second Book: Studies in the Phenomenology of Constitution*, trans. R. Rojcewicz and A. Schuwer, Husserliana 4 (Dordrecht: Kluwer Academic).
Husserl, E. 1989b. *Aufsätze und Vorträge (1922–1937)*, ed. T. Nenon and H. R. Sepp. Husserliana 27 (Dordrecht: Kluwer Academic Publishers).
Husserl, E. 1997. *Psychological and Transcendental Phenomenology and the Confrontation with Heidegger (1927–1931)*, ed. and trans. T. Sheehan and R. E. Palmer (Dordrecht: Kluwer Academic).
Husserl, E. 2001. *Logical Investigations I–II*, trans. J. N. Findlay (London: Routledge).
Husserl, E. 2002. *Einleitung in die Philosophie: Vorlesungen 1922/23*, ed. B. Goossens, Husserliana 35 (Dordrecht: Kluwer Academic).
Husserl, E. 2003. *Transzendentaler Idealismus: Texte aus dem Nachlass (1908–1921)*, ed. R. Rollinger, Husserliana 36 (Dordrecht: Kluwer Academic, 2003).
Ingarden, R. 1975. *On the Motives Which Led Husserl to Transcendental Idealism*, trans. A. Hannibalson (Den Haag: Martinus Nijhoff).
James, S. P. 2018. Merleau-Ponty and Metaphysical Realism. *European Journal of Philosophy* 26/4: 1312–1323.
Levinas, E. 1969. *Totality and Infinity: An Essay on Exteriority*, trans. A. Lingis (Pittsburgh, Penn.: Duquesne University Press).
Levinas, E. 1998. *Discovering Existence with Husserl*, trans. R. A. Cohen and M. B. Smith (Evanston, Ill.: Northwestern University Press).
Merleau-Ponty, M. 2012. *Phenomenology of Perception*, trans. D. A. Landes (London: Routledge).
Philipse, H. 1995. Transcendental Idealism. In B. Smith and D. W. Smith (eds.): *The Cambridge Companion to Husserl* (Cambridge: Cambridge University Press), 239–322.
Sartre, J.-P. 1970. Intentionality: A Fundamental Idea of Husserl's Phenomenology. *Journal of the British Society for Phenomenology* 1(2): 4–5.
Sartre, J.-P. 2003. *Being and Nothingness*, trans. H. E. Barnes (London: Routledge).
Smith, A. D. 2003. *Routledge Philosophy Guidebook to Husserl and the Cartesian Meditations* (London: Routledge).
Thomasson, A. L. 2007. In What Sense Is Phenomenology Transcendental? *Southern Journal of Philosophy* 45(S1): 85–92.
Wilson, J. 2000. Metaphysical Questions in Sartre's Phenomenological Ontology. *Sartre Studies International* 6(2): 46–61.
Zahavi, D. 2002. Metaphysical Neutrality in Logical Investigations. In D. Zahavi and F. Stjernfelt (eds.): *One Hundred Years of Phenomenology: Husserl's Logical Investigations Revisited* (Dordrecht: Kluwer Academic Publishers), 93–108.
Zahavi, D. 2003a. Phenomenology and Metaphysics. In D. Zahavi, S. Heinämaa, H. Ruin (eds.): *Metaphysics, Facticity, Interpretation: Contributions to Phenomenology* (Dordrecht: Kluwer Academic Publishers), 3–22.
Zahavi, D. 2003b. *Husserl's Phenomenology* (Stanford, Calif.: Stanford University Press).
Zahavi, D. 2008. Internalism, Externalism, and Transcendental Idealism. *Synthese* 160/3: 355–374.
Zahavi, D. 2010. Husserl and the 'Absolute'. In C. Ierna, H. Jacobs, F. Mattens (eds.): *Philosophy, Phenomenology, Sciences: Essays in Commemoration of Husserl* (Dordrecht: Springer), 71–92.
Zahavi, D. 2017. *Husserl's Legacy: Phenomenology, Metaphysics, and Transcendental Philosophy* (Oxford: Oxford University Press).

PART IV

The epistemology of metaphysics

27
A PRIORI OR A POSTERIORI?

Tuomas E. Tahko

1 The role of the distinction in metametaphysics[1]

The distinction between the a priori and the a posteriori is important both in epistemology and in metaphysics. What makes the distinction important in *meta*metaphysics is the fact that many questions in metametaphysics are closely related to the source of metaphysical knowledge and our epistemic access to that source. For instance, since many topics in metaphysics concern the realm of *abstract* objects such as sets and numbers, the question about our epistemic access to entities of this type is relevant. This issue is closely tied to *Benacerraf's Problem* (Benacerraf 1973), which concerns the causal isolation of the realm of abstract entities, especially mathematical entities (for discussion, see Horsten 2018). Another area of knowledge that seems causally isolated from us is metaphysical modality – obviously a very important area of inquiry in metaphysics.[2] If we do have knowledge about metaphysical possibility and necessity, then what is the source of that knowledge? One important reason why this question is pressing in metametaphysics is that many contemporary metaphysicians strive for a naturalistic understanding of metaphysics, but the realm of metaphysical modality may be considered especially problematic in this regard, because there does not seem to be a simple way to acquire knowledge about it by a posteriori means.[3]

However, it should be noted right at the outset that there is no particular reason to think that a priori reasoning or the question about the relationship of the a priori and the a posteriori is a special problem for metaphysics. As Karen Bennett (2016) has convincingly argued, all areas of philosophy will face similar issues, since a priori reasoning is employed all over philosophy. Nevertheless, a priori knowledge is often thought to be a challenge for, if not incompatible with, naturalism, so the fact that metaphysics sometimes seems to deal with knowledge accessible only by a priori means poses a problem for naturalistic metaphysicians (see Jenkins 2013). We will discuss all these issues in more detail in what follows but let us first consider some preliminaries.

There are at least four different areas to which the a priori-a posteriori distinction can be applied. These are knowledge, justification, reasoning, and methodology. There is obviously a very large literature on each of these areas, so we will not be discussing them all in detail.[4] One way to understand the difference between knowledge and justification is in terms of truth – one may be justified in believing a certain proposition p without it being true, but knowledge that

p arguably also requires that p is true. We need not dwell on the details of the analysis of knowledge, but it is arguably a priori knowledge in particular that is of interest in metaphysics, whereas a priori justification could be considered to fall under the remit of epistemology. This is, very roughly, what Albert Casullo's analysis of the distinction might suggest: we may distinguish between a non-reductive and a reductive approach to a priori knowledge, where the first is concerned with the analysis of the concept of a priori knowledge, while the latter is concerned with a priori justification (cf. Casullo 2003: 10). Reasoning and methodology are of course closely tied to justification, but methodology is a somewhat broader notion. For instance, if we say that metaphysics generally involves a priori methods, this could mean that metaphysicians justify their claims about the subject matter of metaphysics in terms of conceivability arguments, intuitions, logic, or other similar tools that are typically (but not without exception) considered non-empirical.

2 How can we distinguish between a priori and a posteriori knowledge?

It may seem relatively easy to give a simple definition of a priori and a posteriori knowledge and distinguish them on that basis. A priori knowledge is simply knowledge acquired by non-experiential or non-empirical means, whereas a posteriori knowledge is acquired via experience or the senses, empirically. This does not yet specify whether it's necessary that either type of knowledge was acquired by a priori or a posteriori means though. If a computer provides us with information about a complicated mathematical issue, then we acquired that information by a posteriori means, assuming that the computer is a reliable source of this type of mathematical information. But a skilled mathematician might be able to acquire the same information by a priori means. Yet, it seems that there may be some forms of metaphysical knowledge that we could not have arrived at by empirical means at all. If we have knowledge about any metaphysical necessities or metaphysical possibilities that are nomologically impossible, then this type of knowledge might be a case in point.[5] For our purposes, it is this type of knowledge – knowledge that seems necessarily a priori – that makes for the most interesting case study.

So, we have started with the assumption that a priori and a posteriori knowledge can be distinguished via the notion of *experience*. But there are many well-known problems concerning this. For instance, Laurence BonJour (1998: 7–11) mentions two of the most apparent problems concerning the a priori and experience: the problem of how we define 'experience' itself and how the a priori is supposed to be 'independent' of it. In the first case, the problem is to determine the correct scope of experience. Do mental processes count as experience? How about mathematical or philosophical reasoning that relies on certain learned patterns? Should only perceptual information count as experiential? How does memory fit in with all of this? The second traditional problem involves issues concerning concept acquisition as a precondition for a priori knowledge. We need concepts to formulate our beliefs, and it seems that those concepts must be learned experientially before we can even formulate any beliefs.

Problems with the distinction remain, even assuming that we can overcome these issues and define 'experience' in such a way that a clear-cut distinction can be made and agree that concept acquisition does not contaminate a priori knowledge with a posteriori elements. One central issue is that it is exceedingly difficult to find *pure* examples of a priori or a posteriori knowledge. Consider an example that one might take to be particularly easy:

> [T]he scattering of birds causes you, via the belief that birds have scattered, to infer, with the help of a number of other beliefs, that there is a cat in the vicinity.
>
> *(McGinn 1975–76: 199)*

This example of perceptual information concerning birds, which is used to infer a further proposition, does seem to be relatively easy to classify as a posteriori knowledge. However, there is inference involved, presumably based on inductive information concerning previous cases of bird scattering. But inference is a form of reasoning, regardless of what the initial premises are (i.e. whether the premises are themselves a priori or a posteriori). Based on an appropriate set of premises, we can deduce that the vicinity of a cat is one likely explanation for the scattering of the birds. What is the nature of this form of reasoning? Can it be accurately described as 'a posteriori reasoning', or is that even a sensible notion? Would we not be inclined to say that all forms of *reasoning* are a priori? One could of course suggest that only pure perceptual information, whatever that may be, is truly a posteriori and anything that we might deduce from that perceptual information is in fact a priori. This way we would end up with the curious result that most of our knowledge must be a priori, since we have arrived at it via a priori means, using our capacity to reason deductively. However, this is not a common understanding of a priori knowledge. It is more common to think that whenever the acquisition of a piece of knowledge involves a posteriori elements, that piece of knowledge should be classified as a posteriori. After all, if there was an a posteriori element that was required at some point, then the piece of knowledge in question is not fully independent of experience. So, it would only be pure a priori knowledge in the sense of complete independence of experience that counts as a priori knowledge.

Setting aside the issue of concept acquisition, we may then end up, instead, with the result that only logic counts as a priori, in virtue of it being a purely deductive science (McGinn 1975–76: 199–200). Now, if deduction is a mark of the a priori, and we need deduction to be able to form a proposition concerning a certain state of affairs, such as there being a cat in the vicinity at a location where bird scattering has been observed, then it would appear that any kind of inferred propositional knowledge like this will include a priori elements, even if the present proposal would classify it as a posteriori. Much, if not all, of the underlying information may have originated in our senses, but at some point, reasoning will enter the picture.

So, we have a bit of a dilemma. We use a priori methods all the time and we regard them as reliable. But since there is almost always some preliminary information involved, except perhaps in formal logic, it appears that there isn't much pure a priori knowledge. The issue is further complicated by the fact that interesting types of a priori knowledge, such as that concerning metaphysical modality, are not usually thought to be available to us via logic. Instead, many think that acquiring this type of knowledge requires resorting to methods of inquiry such as conceivability or intuitions. If these methods of inquiry are considered a posteriori, then we end up with the somewhat surprising result that even knowledge of metaphysical modality is a posteriori.

There is a further issue. Since it seems that often we can come to know the same thing both by a priori and a posteriori means, we would need to refine the distinction if we don't want the same piece of knowledge to be both a priori and a posteriori. After observing this issue, Timothy Williamson concludes that:

> Perhaps the best fit to current practice with the term is to stipulate that a truth is a posteriori if and only if it can be known a posteriori but cannot be known a priori.
> *(Williamson 2013: 293)*

Now, we have been applying the distinction to *ways* of knowing and justification rather than truths, and Williamson's suggestion here would result in a similar picture, because he effectively suggests that it is the way that we *can* come to know a truth that determines its status, so that's what we should focus on. However, as he goes on to note, even this approach faces a problem if we accept Kripke's (1980) famous case in favour of contingent a priori truths and necessary a posteriori truths. If there are contingent a priori truths, such as Kripke's example concerning the standard metre, then it seems that they must also be knowable a posteriori – this is in fact how most of us come to learn about the length of the standard metre, even if it was a priori for those who initially proposed the definition. If Kripke is right, the link between apriority and necessity is severed and we lose a seemingly easy way to identify a priori truths with necessary truths and contingent truths with a posteriori truths.

What is the upshot regarding the a priori-a posteriori distinction? Given that it is very difficult to properly distinguish a priori and a posteriori knowledge, one might think that the distinction is not very significant at all. As we have seen, even if we do find a way to draw the distinction without too much vagueness, we may end up with the result that there is very little knowledge of one or the other type. Williamson (2013) has suggested something similar, arguing that even though the distinction can be drawn, the differences between a priori and a posteriori knowledge are superficial. The reason for this is that in both cases, experience does play a role and according to Williamson this role is more than 'purely enabling' – like it would perhaps be in the case of concept acquisition – but also less than 'strictly evidential', like it might be in the case of pure perceptual information.

The problem with distinguishing the enabling and evidential roles of experience is related to our understanding of 'experience' itself. We have noted some of the problems surrounding this already, many of which arise from having to account for 'inner' as well as 'outer' experience: not all experience is perceptual, we often appeal to experience that is purely internal to us as well, such as when we perform a calculation. But it's not clear whether the process of calculation is playing an evidential role here, since we might think that the inner experience of calculating just enables us to access certain mathematical truths. Yet, a similar story can be told about outer experiences as well: there is a certain experience that we associate with perception, but the knowledge that we gain via this process concerns external matters, so once again it could be argued that experience is only mediating our access to certain external facts, hence playing an enabling rather than an evidential role. So, it won't do to draw the a priori-a posteriori distinction in terms of inner and outer experiences, because either type of experience can be interpreted as evidential or enabling.

This rather negative result does not mean that we couldn't find a reasonable way to draw the a priori-a posteriori distinction. The problem, rather, is that all reasonable ways to draw the distinction will be somewhat stipulative and even supposedly clear cases of one or the other type of knowledge may end up being classified differently. The consensus in contemporary epistemology, insofar as there is one, seems to be that the coherence and significance of the distinction are under serious threat.[6] But perhaps this does not mean that the distinction is useless. In the next section we will outline an approach that embraces the vagueness of the distinction and also explains why it may not be fruitful to attempt to classify a certain truth outright as either a priori or a posteriori.

3 Bootstrapping and cyclical processing

Motivated by the difficulty of demarcating a priori and a posteriori knowledge, some philosophers have abandoned the project of trying to find a clear definition of either type of knowledge. Instead, it is suggested that there is a very subtle interplay between different types of inquiry and both are needed in order to acquire any knowledge at all (Chakravartty 2013; Lowe 2011, 2014; Morganti and Tahko 2017; Tahko 2008, 2011).

Another source of motivation for a revised view is more closely related to the role that a priori knowledge is sometimes thought to play in metaphysics. For instance, a priori methods in metaphysics are often associated with intuitions or conceivability, but many metaphysicians are quite sceptical of these methods. E.J. Lowe is one such philosopher. He argues that a view taking intuitions as evidential in metaphysics is 'fundamentally misguided and leads inexorably to an anti-realist conception of metaphysical claims' (Lowe 2014: 256). However, Lowe himself is not against a priori knowledge or a priori methods in metaphysics; he just thinks that they have little to do with intuitions or conceivability. Instead, Lowe argues that it is knowledge of essence that has a central role in metaphysics and that the process of acquiring this knowledge should not be considered completely independent of experience, but rather as proceeding in a 'cyclical manner, by alternating stages of a priori and a posteriori inquiry' (2014: 257).

Those sympathetic to the idea that a priori and a posteriori inquiry go hand in hand in metaphysics often also consider the same to be true of scientific inquiry. We will return to science and naturalistic metaphysics in section 5, but it may be helpful to consider an example from scientific inquiry here to demonstrate the degree to which a priori and a posteriori methodologies can be intertwined. I borrow this example from Tahko (2011: 157), where the 'cyclical' relationship between the a priori and the a posteriori is referred to as 'bootstrapping relationship'.

Consider the phenomenon of gravitational redshift, which refers to the change in the wavelength of light and other electromagnetic radiation when it travels from a stronger gravitational field to a weaker one. This effect of gravity on light was predicted already by Newton, but Newton's results relating to the phenomenon were partly inaccurate, as he relied on the corpuscular theory of light. When light is conceived as an electromagnetic wave instead, the phenomenon of gravitational redshift needs to be reconsidered, as it appears that the wavelength of light could only change from one place to another if the flow of time also changes. This mystery was of course solved by Einstein's special theory of relativity, which models how the flow of time can indeed change, relative to a given frame of reference – the famous example of twins ageing at different rates because one of them is travelling close to the speed of light is a case in point. So, it was Einstein's theory of general relativity that correctly predicted the gravitational redshift phenomenon. Einstein's work, however, was not empirical. It was only the Pound-Rebka experiment in 1959 which correctly measured gravitational redshift, and this experiment is often also considered to have verified Einstein's theory.

There is of course a lot more detail in this example and it should be noted that we are here dealing at the level of complete theories about light and electromagnetic radiation rather than a single proposition. However, we could focus just on the simple proposition <Gravity bends light>, which is effectively what is responsible for the phenomenon of gravitational redshift. Now, this phenomenon is not something that we would have been likely to look for if we didn't have some theoretical reasons to do so. These reasons were already apparent in Newton's theory, which predicted that gravity would influence light. But Newton's theory was based on false assumptions about the nature of light. Later, experiments with light showed that it behaves much like an electromagnetic wave, and the original theory concerning gravitational redshift could not accommodate this. Accordingly, since Einstein's theory predicted different values for the phenomenon, it became an important test case for his theory.

We can now simplify a little and reconstruct the long history of gravitational redshift in terms of the proposition <Gravity bends light>. Newton's theory represents the first a priori step in the bootstrapping relationship. Newton of course had empirical information about gravity that he used to build his theory on, but there are good reasons to think that extending that theory to light was quite independent from experience, since we don't observe the influence of gravity on light without very sensitive equipment. Moreover, Newton's corpuscular theory of light was mistaken, as experiments later confirmed. Thus, we have an a posteriori falsification of Newton's original theory concerning the interplay between gravity and light even though the proposition <Gravity bends light> is true. So, Newton was actually partly correct when he asserted (let us imagine), the proposition. Einstein's work represents a new a priori step from the established a posteriori framework, which had falsified Newton's theory about gravitational redshift. After his theory was empirically verified by the Pound-Rebka experiment (and others), it also became a part of the a posteriori framework. On this basis, it seems wrong to say that the proposition <Gravity bends light> would be either purely a posteriori or purely a priori. Depending on which aspects of this story we emphasize and how strictly we define the a priori and the a posteriori, either result could be derived. The upshot is that it is often more accurate to say that a priori and a posteriori methods are intertwined in a quite intimate relationship, which is better described as cyclical or as bootstrapping.

Let us now move on to some applications of the a priori-a posteriori distinction in (meta)metaphysics.

4 Connection to modal epistemology

We have already noted that one area where one might think that a priori reasoning is needed is knowledge of metaphysical modality and indeed modal epistemology more generally. This is a topic where conceivability, for instance, has been traditionally employed as a source of evidence. But there have been recent developments in the area of modal epistemology as well that point toward a need to re-evaluate the role of a priori and a posteriori methods. Specifically, while it may once have been common to regard all knowledge of metaphysical modality as a priori, there are now many philosophers sympathetic to versions of *modal empiricism* (e.g. Fischer and Leon 2017), where attempts have been made to account for at least some of this knowledge in terms of a posteriori methods. So, we can divide the accounts of the source of modal knowledge into two rough categories: modal rationalism and modal empiricism.

Modal rationalism encompasses intuition- and conceivability-based approaches of the type defended, for instance, by George Bealer (e.g. 2004) and David Chalmers (e.g. 2002), and also the essence-based account of E.J. Lowe (e.g. 2012). Bealer defends an intuition-based account, whereas Chalmers defends a conceivability-based account as a part of a broader rationalist picture. Lowe's approach is slightly different, since he thinks, following Kit Fine, that metaphysical modality is grounded in essence, so modal epistemology becomes a special case of the epistemology of essence.

We do not need to dwell on the details of the various approaches to modal epistemology.[7] But it will be interesting to briefly discuss one key question regarding the relationship of a priori and a posteriori methods in modal epistemology (we continue to work with the – admittedly problematic – assumption that if a proposition is knowable only with the help of a priori methods, then we should classify that proposition as a priori). This is the question of unification, that is, are any of the various approaches to modal epistemology able to account for all modal knowledge or do we need to resort to various different kinds of methods to account for all modal knowledge? The starting point here should presumably be that, all other things being equal, a unified account, a single explanation for all modal knowledge, should be preferred. Two ways to unify the account would be to argue that all modal knowledge is acquired by a priori means or that it is all acquired by a posteriori means (and perhaps also in terms of the very same method, such as conceivability).

However, the unified approach has been forcefully challenged in recent work, especially by modal empiricists. A partial motivation here may be naturalistic, i.e. to avoid a commitment to a priori methods insofar as possible. Indeed, it seems right that there is some modal knowledge that can be reached by a posteriori means, such as via perception. Moreover, as Carrie Jenkins (2010) has suggested, modal empiricism can be understood as a view according to which experience ensures the reliability of our modal knowledge. On Jenkins's version of the idea, experience provides an epistemic grounding for our concepts. Interestingly, since concepts are also the basis of our conceptual abilities, conceivability itself would appear to need this experiential basis. This issue is obviously related to the distinction between experience as enabling and as evidential, so we end up with the same problems as above.

Yet, there are also reasons to think that not all modal knowledge can be acquired by a posteriori means, such as modal knowledge concerning abstract objects or nomologically impossible yet metaphysically possible matters. So, either way, proponents of a uniform account of all modal knowledge will face great difficulties, at least insofar as this uniformity is expected in terms of the a priori-a posteriori distinction (there are more fine-grained ways to understand uniformity, as specified in Wirling 2019, but our focus here is on the a priori-a posteriori distinction, for obvious reasons). Perhaps this constitutes another reason to avoid using this distinction as the basis for distinguishing different positions such as modal rationalism and modal empiricism (cf. Tahko 2017).

5 Connection to science and naturalistic metaphysics

One reason to be interested in the a priori-a posteriori distinction in the context of metametaphysics is the desire to strive for the most naturalistic and scientifically respectable epistemology for metaphysics as possible. Assuming that the a priori-a posteriori distinction can be made in the first place, one might think that it provides us with at least a rough tool to distinguish between naturalistic and non-naturalistic approaches to the epistemology of metaphysics. Unfortunately, this is a naïve attitude. Not only is the distinction itself unlikely to be sharp enough to provide any useful input on this question, but a closer look quickly reveals that even the most 'naturalistic' area of science will need input from methods of inquiry that can be reasonably classified as a priori. We have already seen a rudimentary example of this in section 3, with reference to Newton and Einstein. Here's Anjan Chakravartty's take on the matter:

> The degree to which and the ways in which the many domains of investigation that come under the heading of 'the sciences' are empirical is highly variable. As a consequence, the distinction here between a priori and a posteriori methodology cannot simply be superimposed unproblematically on metaphysics and the sciences, respectively.
>
> *(Chakravartty 2013: 33–34)*

Chakravartty brings this issue up precisely in connection to the prospects of naturalized metaphysics, arguing that the distinction between non-naturalistic and naturalistic metaphysics cannot be simply made on the basis of the a priori-a posteriori distinction. This is of course the expected result given what we have already learned about the distinction. But maybe there are more subtle ways to draw the difference between non-naturalistic and naturalistic metaphysics? Chakravartty (2013: 32) speculates that it may be the idea of a priori theorizing 'with no significant empirical tethering' that is responsible for the hostility towards some, apparently non-naturalistic, approaches to metaphysics. Of course, if what we have observed in earlier sections is correct, there simply is no such area of metaphysics, indeed, there seems to be no area of human inquiry whatsoever which would not take advantage of both a priori and a posteriori resources. It is important to recognize that this by no means entails that all areas of metaphysics would be unproblematic, epistemically speaking. It only means that no area of metaphysics should be considered problematic just because it employs a priori methods.[8] Or if it is, then the same problems propagate to other areas of philosophy as well (cf. Bennett 2016), and to the sciences.

This is not the place to pursue arguments to the effect that the sciences, to variable extents, employ a priori methodology. Instead, it may be worthwhile to take an entirely different angle on the issue, namely, why do we think that the methods used in metaphysics and philosophy are a priori at all? This has been recently discussed by Daniel Nolan (2015), who argues that the 'armchair' methods used in philosophy could just as well be classified as a posteriori. Nolan suggests that these methods generally involve the senses playing a role that is not merely enabling. So, once again we encounter the distinction between enabling and evidential roles. Here, the role of the senses is taken to be something more than a necessary part of the acquisition of concepts. Nolan identifies four possible tasks of this type for the senses. We will discuss each of them briefly.

The first is *Assembling and Evaluating Commonplaces*. This task amounts to the analysis of such stories as the one about (the statue of) Goliath and Lumpl, the piece of clay of which Goliath is made.[9] Nolan suggests that we know only a posteriori that there are statues made of clay and that such statues can be smashed without destroying the clay. But the real interest of the suggestion is that philosophers can make surprising 'discoveries' on the basis of such commonplaces, which are seemingly available to everyone on the basis of a posteriori knowledge. For instance, the realization that Goliath and Lumpl have different persistence conditions has important upshots for metaphysical debates about composition. However, there may still be room to argue that the relevant philosophical work is nevertheless a priori. For one might insist that we need to have some grasp of the *kind* of thing that statues and lumps are before we are in any position to draw the philosophically important conclusions (cf. Lowe 2012).

Nolan's second a posteriori armchair task concerns *New Theoretical Alternatives*. The fairly uncontroversial idea behind this task is that once a set of theoretical alternatives is already known, philosophers can come up with new versions of these theories or combine different theoretical frameworks to the effect that a unified theory can be constructed – all from the armchair but apparently without resorting to a priori inquiry. Nolan mentions David Lewis's neo-Humean framework as an example; indeed, Lewis made a tremendous effort to combine laws, causation, counterfactuals, chance, and dispositions into a unified theoretical framework, the value of which is undeniable regardless of whether Lewis is correct about all the details. This certainly seems like a reasonable and valuable task to be conducted from the armchair and since all the data is, in a sense, already available in previous theories, the task itself would seem to require no further inquiry, a priori or otherwise.

Perhaps it might still be objected that even once all the data is available it is a complicated task to determine all the implications of bringing together different theoretical alternatives. So, one might think that such a task will not be possible without further *interpretation* of the alternatives; thus, if the interpretative work requires metaphysical a priori inquiry, then the task cannot be completed simply in terms of a posteriori armchair inquiry. Be that as it may, since this task is in some sense secondary (given that metaphysical inquiry surely doesn't *start* from it) it is perhaps of less importance whether it is truly a priori or a posteriori.

A related, third task discussed by Nolan is *Integrating Past A Posteriori Investigation*. This important task focuses on determining which discoveries of various disciplines should inform our overall world view as well as the relations between these disciplines. But Nolan readily admits that this task apparently involves much more than just armchair work: we may need to conduct actual scientific experiments to determine what the exact link is, say, between psychology and linguistics, or physics and chemistry. Nolan does point out that at least in certain cases, the information needed to engage in the integration task may be considered 'commonplace' in the sense discussed regarding the first task. One particularly striking example might be quantum mechanics and its apparent violation of determinism: while the specifics of the situation are probably far from commonplace, it is now so widely known that quantum mechanics causes problems for determinism and classical physics more generally that the integration task can effectively proceed from the armchair. Indeed, much of the seminal work in this area was done already in the early 1900s and the philosophical debate about the interpretation of the experimental data has been ongoing ever since. Still, one might again argue that the interpretation of the relevant theories, which is surely required before the integration task can begin, could just as well be considered a priori. The core idea, in any case, is that work on theories or ideas that are widely spread and established could count as armchair work *even if* the original work that resulted in those theories was not armchair work.

The fourth and final task that Nolan proposes is *Applying Theoretical Virtues*. By 'theoretical virtue', Nolan means things like internal consistency, external coherence, simplicity, explanatoriness, fertility, unificatory power, and other such comparative as well as internal virtues. It should be noted that it is not untypical in the literature discussing theoretical virtues to assume that they are part of a priori methodology (see Paul 2012). But Nolan may very well be correct in questioning this assumption, because the process of 'applying theoretical virtues' is rarely elaborated on. He suggests that it's epistemically better to accept a theory that better satisfies these theoretical virtues. An obvious explanation for this is that such theories are more likely to be true. Yet, other attempts to justify theoretical virtues can be made; Nolan mentions that they can sometimes be justified with reference to *other* theoretical virtues. For instance, unificatory power may often promote further simplicity and unifying two theories may also increase explanatory value. Hence, the appeal to unificatory power could be justified in terms of these other virtues, if *they* are considered valuable. Perhaps a more direct justification could be drawn from predictive success, as simpler theories may provide accurate predictions. Nolan's suggestion is that this assessment can be conducted in the armchair since it relies on established evidence rather than direct empirical work.

Interestingly, if Nolan is correct about the first three tasks, then even many of the 'inputs' of this fourth task can be considered to derive from a posteriori armchair work. If the commonplaces, new theoretical alternatives, and integration are all armchair activities, then the results they produce, the more sophisticated theories, are already largely a product of armchair work. We can then further compare these sophisticated theories in the armchair, by applying theoretical virtues.

The picture that emerges from Nolan's proposal may seem attractive, since it corroborates much of philosophical methodology without resorting to the controversial notion of apriority. Of course, some philosophers would likely consider this to be a disadvantage, since if the a priori is to have any place in philosophy, then its best defence is exactly that a priori inquiry is needed in cases such as those described by Nolan. A more deflationary reaction to this discussion would be to say that we have a mere terminological debate in our hands. Indeed, does it even matter whether the armchair methods discussed are a posteriori or a priori, as long as they are reliable?

6 Concluding remarks

The conclusion of our discussion regarding the a priori–a posteriori distinction and its role in metaphysics is somewhat negative. The distinction can be drawn and there are various ways to avoid its arbitrariness, but none of the reasonable accounts seem to help to settle issues such as the status of modal knowledge, the demarcation of science and (different areas of) philosophy, or the prospects for naturalized metaphysics. This does not mean that the distinction could not have its uses, but we should be modest in our attempts to use it when it comes to addressing the above issues. It may be more promising to examine the interplay of a priori and a posteriori elements in certain areas of reasoning, perhaps with the goal of improving our overall methodology.

Notes

1 The discussion in this entry follows and develops on material in Tahko 2011, 2015.
2 For discussion about modal epistemology and the role of modal knowledge in metaphysics, see Sonia Roca-Royes's and James Miller's entries in this volume.
3 For discussion on versions of naturalistic metaphysics, see Matteo Morganti's entry in this volume.
4 See for instance Williamson 2000 for an extensive discussion.
5 However, the issue is not quite as simple as that. There is a relatively new trend in modal epistemology to develop modal empiricist accounts, some of which attempt to provide an empirical basis also for knowledge about metaphysical necessity and possibility. See the articles in Fischer and Leon 2017.
6 For a good overview of some of these challenges, see Casullo 2015. See also the essays in Boghossian and Peacocke 2000.
7 Modal epistemology is covered in more detail in Sonia Roca-Royes's entry in this volume.
8 Again, for further discussion on various accounts of naturalistic metaphysics, see Matteo Morganti's entry in this volume.
9 See Gibbard 1975 for the original example.

References

Bealer, G. (2004) 'The Origins of Modal Error', *Dialectica* 58 (1): 11–42.
Benacerraf, P. (1973) 'Mathematical Truth', *Journal of Philosophy* 70 (19): 661–679.
Bennett, K. (2016) 'There Is No Special Problem with Metaphysics', *Philosophical Studies* 173: 21–37.
Boghossian, P. and Peacocke, C. (eds.) (2000) *New Essays on the A Priori*, Oxford: Oxford University Press.
BonJour, L. (1998) *In Defense of Pure Reason: A Rationalist Account of A Priori Justification*, Cambridge: Cambridge University Press.
Casullo, A. (2003) *A Priori Justification*, Oxford: Oxford University Press.
Casullo, A. (2015) 'Four Challenges to the A Priori–A Posteriori Distinction', *Synthese* 192 (9): 2701–2724.
Chakravartty, A. (2013) 'On the Prospects of Naturalized Metaphysics', in D. Ross, J. Ladyman, and H. Kincaid (eds.) *Scientific Metaphysics*, Oxford: Oxford University Press, pp. 27–50.
Chalmers, D. (2002) 'Does Conceivability Entail Possibility?' in T.S. Gendler and J. Hawthorne (eds.) *Conceivability and Possibility*, Oxford: Oxford University Press, pp. 145–200.
Fischer, B. and Leon, F. (eds.) (2017) *Modal Epistemology After Rationalism*, Cham: Springer.
Gibbard, A. (1975) 'Contingent Identity', *Journal of Philosophical Logic* 4: 187–221.
Horsten, L. (2018) 'Philosophy of Mathematics', *The Stanford Encyclopedia of Philosophy* (Spring 2018 Edition), E.N. Zalta (ed.) URL = https://plato.stanford.edu/archives/spr2018/entries/philosophy-mathematics/.
Jenkins, C.S.I. (2010) 'Concepts, Experience and Modal Knowledge', *Philosophical Perspectives* 24: 255–279.
Jenkins, C.S.I. (2013) 'Naturalistic Challenges to the A Priori', in A. Casullo and J.C. Thurow (eds.) *The A Priori in Philosophy*, Oxford: Oxford University Press, pp. 274–290.
Kripke, S. (1980) *Naming and Necessity*, Cambridge, MA: Harvard University Press.
Lowe, E.J. (2011) 'The Rationality of Metaphysics', *Synthese* 178: 99–109.
Lowe, E.J. (2012) 'What Is the Source of our Knowledge of Modal Truths?' *Mind* 121: 919–950.
Lowe, E.J. (2014) 'Grasp of Essences vs. Intuitions: An Unequal Contest', in A.R. Booth and D.P. Rowbottom (eds.) *Intuitions*, Oxford: Oxford University Press, pp. 256–268.

McGinn, C. (1975–76) '"A Priori" and "A Posteriori" Knowledge', *Proceedings of the Aristotelian Society*, New Series, 76: 195–208.
Morganti, M. and Tahko, T.E. (2017) 'Moderately Naturalistic Metaphysics', *Synthese* 194: 2557–2580.
Nolan, D. (2015) 'The A Posteriori Armchair', *Australasian Journal of Philosophy* 93 (2): 211–231.
Paul, L.A. (2012) 'Metaphysics as Modeling: The Handmaiden's Tale', *Philosophical Studies* 160 (1): 1–29.
Tahko, T.E. (2008) 'A New Definition of A Priori Knowledge: In Search of a Modal Basis', *Metaphysica* 9 (2): 57–68.
Tahko, T.E. (2011) 'A Priori and A Posteriori: A Bootstrapping Relationship', *Metaphysica* 12 (2): 151–164.
Tahko, T.E. (2015) *An Introduction to Metametaphysics*, Cambridge: Cambridge University Press.
Tahko, T.E. (2017) 'Empirically-Informed Modal Rationalism', in R.W. Fischer and F. Leon (eds.) *Modal Epistemology After Rationalism*, Synthese Library, Dordrecht: Springer, pp. 29–45.
Williamson, T. (2000) *Knowledge and Its Limits*, Oxford: Oxford University Press.
Williamson, T. (2013) 'How Deep Is the Distinction Between A Priori and A Posteriori Knowledge?' in A. Casullo and J.C. Thurow (eds.) *The A Priori in Philosophy*, Oxford: Oxford University Press, pp. 291–312.
Wirling, Y. (2019) *Modal Empiricism Made Difficult: An Essay in the Meta-Epistemology of Modality*, Doctoral thesis, University of Gothenburg, *Acta Philosophica Gothoburgensis* 33.

28
THE EPISTEMOLOGY OF MODALITY

Sonia Roca-Royes

1 Background

This chapter is on the epistemology of modality. To provide the right context and focus, however, some preliminaries are in order. To start, the term 'modality' refers to the phenomenon of necessities and possibilities (as well as related ones such as impossibilities and contingencies) and it can be understood in different senses. In its most general sense, it refers to the variety of different modalities.[1] In this chapter, however, our focus will be *metaphysical* modality and, although I will not make it explicit each time, this is how my use of modal vocabulary should be understood.

Metaphysical modality is a type of *alethic* modality and, as such, it tracks modes in which propositions (or sentences) can be true or false.[2] Consider for instance the following propositions:

1. There is gold on Earth
2. There are seven pebbles in my pocket
3. Two plus two equals five
4. Socrates is human

Of these, only (1) and (4) are true – I do *not* have pebbles in my pocket. Do they enjoy the same mode of truth? Many would say no; that while (4) is *necessarily* true and thus Socrates could only exist as a human being, the presence of gold on Earth is *contingent* and thus there could be none. Something similar happens with the falsity of (2) and (3). As most would agree, the falsity of (2) is contingent while that of (3) is necessary: I could easily have seven pebbles in my pocket but there's no way two plus two could be five; (3) is an impossible proposition.

One way – a particularly relevant one for current purposes – in which metaphysical modality can be distinguished from other alethic modalities is by means of its strong link to *essential truths*. Essentialism studies the nature of things and is concerned with questions such as 'what makes a thing the thing it is?' and 'what properties or relations are essential to which entities?'. In the words of Bob Hale, the close tie between essentialism and metaphysical modality gets expressed thus: "Metaphysical necessities have their source in the natures of things, and metaphysical possibilities are those left open by the natures of things" (Hale 2013: 253). This quotation is not neutral among the different views available on what exactly the link between metaphysical modality and essential truths is. Rather, it is a salient representative of a (widely accepted) family of Finean views, unified by the thesis that essential truths imply necessities but not the other way around. Thus, for instance, Socrates being essentially human entails that, necessarily, he is human if he exists. But also, there would be propositions that, while being necessary, their necessity doesn't imply the corresponding essential truth. Kit Fine suggests the following as one among a battery of examples:

(5) Socrates belongs to {Socrates}

According to the most salient opponents to the Fineans in this context – the modalists – the tie between metaphysical modality and essentialism is even more intimate than Hale suggests, in that necessary truths also imply essential ones: something being essential to a given entity *just is* it being necessary of it.[3]

It will also be relevant for the purposes of this chapter that we distinguish between mind-dependent accounts of modality and mind-independent ones. The realm of the imaginable is mind-dependent in that it depends on what the mind can imagine. Some accounts of modality have claimed the truth-conditions of modal claims such as 'donkeys could talk' to be dependent on what we can imagine, thereby making the modal realm a mind-dependent one too. Crudely, on these accounts, to say that donkeys could talk comes very close to just saying that we can imagine that they talk. It is thus not an intrinsic statement about donkeys (generalizing: about the extra mental world) but rather one about our mental lives (what we can imagine). At most, it is then a relational statement about donkeys: about what our mental capacities can do with them; namely, imagining them talking. In opposition to mind-dependent accounts, we find the mind-independent ones. According to them, the truth-makers for modal claims are to be found in the extra-mental world.[4] The claim 'donkeys could talk' gets its truth-value irrespective of what goes (or could go) on in the minds of those who think modally; namely us. As we will see later, some modal mind-independentist theorists take it that imaginability/conceivability provides an *epistemic route* to modal knowledge. This position, however, is not to be confused with the mind-dependent one. The difference is sharp: between taking imaginability/conceivability facts as *constitutive of* modal truth, and taking them as an epistemic route to (mind-independent) modal truth.

The dominant stand in the literature is modal mind-independence. This has also been a default position: modal mind-dependence is usually seen as something we would need to be forced into. As we shall see later on, the challenge that modal knowledge poses is motivated under the assumption of mind-independence, to the point that going mind-dependent can be seen as a way of meeting *that* challenge.

Now, the label 'the epistemology of modality' is usually received as making reference to the epistemology of *metaphysical* modality, and there is a literature-based explanation for this: for the most part, the literature on the epistemology of modality has been focused on the metaphysical type of modality. When doing epistemology, the *rationale* for focusing on one specific type of modality – rather than, for instance, on the more general phenomenon – lies, at least partly, in the fact that not all types of modalities will pose the same pressing epistemological questions and, even when there might be overlap on this level, it is not to be taken for granted that analogous questions will call for analogous answers. A case-by-case approach appears thus to be methodologically adequate. The present chapter, in being focused on metaphysical modality, contributes to the existing tendency in the literature. Also, by way of witnessing the methodological recommendation: it will emerge in later sections that the close link between metaphysical modality and essential truths has been taken by many modal epistemologists as suggesting a specific kind of solution to the epistemic challenge that metaphysical modal knowledge raises; a solution that, by the nature of the case, is not transferable to other types of modalities.

In §2, I present the bare bones of the challenge that modal knowledge raises. After that, I will focus on two meta-epistemological distinctions that are becoming more and more central in the literature. The first one (the focus of §3) classifies the epistemologies of modality on the basis of the amount and nature of the suggested routes to modal knowledge. The second one (the focus of §4) classifies epistemologies on the basis of their suggested epistemic priority relations. The final section (§5) contains final remarks.

2 The epistemic challenge in modality

The existence of modal knowledge is puzzling and its explication challenging. In addition, for reasons that van Inwagen (1998) makes clear, and that are accepted well beyond the cohort of modal epistemologists, having available an epistemology of modality with the required explanatory power is of interest to the whole philosophical community. It is on the grounds of what van Inwagen (1998) calls 'a possibility argument', for instance, that Chalmers (1996) rests his case for his type of dualism in the Philosophy of Mind; an argument that contains, as one of its premises, the possibility of zombies. It is also on the grounds of a possibility argument that the existence of God has been argued for, appealing to the possibility of a perfect being. To the extent that a great deal of philosophical arguments involve modal premises, the epistemology of modality has an important role to play in providing epistemic foundations for philosophy (and likely beyond).

Over the past recent decades, the developments in the epistemology of modality have centred on a puzzlement the dissolution of which has generated – and continues to generate – controversy. The puzzle is formulated under the assumption of modal mind-independence and it owes a lot to Edward Craig and Simon Blackburn. Under that assumption – plus a further epistemological assumption to be spelled out shortly – the problem is that there seems to be no room for the satisfaction of the possibility conditions for modal knowledge.

Here is E. Craig's transcendental question for the case of *necessity* where he already suggests a *causal affection constraint* on any potentially satisfactory answer:

> How are we to know of necessity in reality unless it affects us, and in some way differently from mere truth?
>
> *(Craig 1985: 104)*

This constraint requires commitment to a necessity-sensitive faculty and, in the terms of S. Blackburn, the problem is that 'we do not understand our own must-detecting faculty'.

> *(Blackburn 1986/1993: 52)*

To elaborate on why not:

> Think of the analogy: what affects my senses is the fact of the tree's being there; it wouldn't affect them any differently if its being there were necessary.
>
> *(Craig 1985: 104)*

Taking stock, the problem can be formulated as the unwelcome inconsistency among the following three claims:

a   We have knowledge of extra-mental modal facts.
b   Any knowledge of the extra-mental world is grounded on causal affection.
c   Any knowledge grounded on causal affection is bound to be knowledge of *mere* (as opposed to modal) truths.

Claim (a) denies modal scepticism while incorporating the assumption of mind-independence. And yet, claims (b) and (c) imply modal scepticism; they capture the Blackburn/Craig stand that modal knowledge requires causal affection and that causal affection (by actuality) won't inform us about the modal realm.

Formulating the challenge as the unwelcome (a)(b)(c) tension helps explain the different paths that recent literature on the epistemology of modality has taken. Especially when noting that, while jointly inconsistent, the (a)(b)(c) claims are two-to-two consistent: that is, any two of them are jointly consistent. On these bases we could identify three straightforward paths to solving the challenge: deny (a) and become a modal sceptic; deny (b) and become a modal rationalist; or deny (c) and become a modal empiricist. Less simple paths involve combinations of those strategies.

As a reaction to the increasingly recognized importance of modal epistemology, a lot of effort has been put in the last three decades towards finding a satisfactory theory of modal knowledge that meets the Blackburn/Craig challenge. These years, after the impact of the Kripkean *a posteriori* necessities (1972/1980), have witnessed what Bealer diagnosed, already in 2002, as a *rationalist renaissance*, and the literature is now witnessing a turn partially away from rationalism. Among the witnesses of the rationalist renaissance, one finds prominent accounts such as Bealer's (2002, 2004), Chalmers' (2002), Lowe's (2008, 2012), Menzies' (1998), Peacocke's (1999), Yablo's (1993), and Sidelle's conventionalism (1989). Salient instances of the latter include Biggs's (2011), Elder's (2004), Fischer's (2017), Jenkins' (2010), Leon's (2017), Miščević's (2003), Nolan's (2017), Roca-Royes's (2017), Strohminger's (2015), Tahko's (2017), Vetter's (2016) and Williamson's (2007).[5] Against this dialectics, a salient view that fights the current turn away from rationalism is McLeod's (2009), who defends that modal mind-independence requires modal rationalism.

The three paths to solving the challenge identified above – scepticism, rationalism or empiricism – are neat theoretical options. Still, the literature isn't always as neat as this, and there are good theoretical reasons why not. A bold denial of (a), for instance, is exceedingly implausible: we do have at least *some* modal knowledge – e.g. that you *could* break a bone, that Croatia *could have won* World Cup 2018 – and this knowledge calls for an explanation that requires the denial of either (b) or (c). But at the same time, one might want to put limits on the amount of modal knowledge we have, thereby embracing *partial* modal scepticism and *partial* modal rationalism or empiricism, depending on whether it is (b) or (c) that one denies in their explanations of the modal knowledge we have.[6] And to add complexity, one might choose to explain some modal knowledge in a rationalistic manner and some other modal knowledge in an empiricist way, the overall picture resulting in a highly heterogeneous epistemology of modality.

In the following two sections I will focus on two meta-epistemological distinctions – one centred on the variety of sources of modal knowledge and the other one on an epistemic priority question – that are becoming more and more central to the epistemology of modality and that, at the same time, contribute to explaining the partial turn away from rationalism we are witnessing these days.

3 Uniformity vs non-uniformity

Uniform epistemologies will be characterized here as answering the 'how do we know?' question in a one-fold manner. Until recently, uniformism has been an implicit default among rationalist epistemologies. For instance, 'we know by determinately understanding' is Bealer's reliabilist answer (2002, 2004); 'we know by conceiving' is Chalmers's (2002) conceivabilist one, as well as Yablo's (1993) and Sidelle's (1989), the latter from within conventionalism; 'we know by reasoning and comprehensive thought' is Lowe's (2008, 2012); and 'we know by reasoning from the Principles of Possibility' is Peacocke's (1999).

Such answers have *pro tanto* the virtue of uniformity – and the elegance that comes with it – in their favour. Unfortunately, these accounts have been found wanting on several grounds, some of which have to do, precisely, with their uniformity. A salient complaint is that their proposed (epistemological) explanations are not generalizable far enough beyond their running examples. And, arguably, an explanation of this deficit is that they seem to neglect – by not making any explicit mention of it, and by not incorporating it into their strategies – the potential relevance, for its epistemology, of the heterogeneity of the subject matter of modality. Nominalist worries aside, the subject matter of modality includes both *concreta* and *abstracta*. It is widely agreed that an epistemology of *categorical* truths about concrete entities should be notably different from an epistemology of such truths about abstract entities: the role of causal affection being less, if existent at all, in the latter. The hypothesis that emerges from these considerations is that this necessitates a non-uniform epistemology of modality, according to which there are different sources of (first-hand) modal knowledge depending on their aboutness. For an articulation of the reasons supporting the hypothesis, see Roca-Royes (2007) and Mallozzi's work (2018a, 2018b), who calls for non-uniformism with the slogan of 'putting modal metaphysics first'.

To frame the discussion on the epistemology of modality in terms of the distinction between uniform and non-uniform is a fairly recent phenomenon, but it is increasing fast. Still, and despite the intuitiveness behind it, the literature hasn't quite settled on any one precise way of understanding the distinction between uniform and non-uniform modal epistemologies. One can conveniently distinguish between a coarse-grained and a fine-grained understanding of the distinction. Both understandings can be glossed in the same way: the uniformism/non-uniformism distinction has to do with how many sources of modal knowledge one thinks there are, "the *uniformity view* hold[ing] that there is only one single route to modal knowledge at the most fundamental level of explanation" (Vaidya 2015). Still, different ways of individuating *routes* will yield to (extensionally) non-equivalent understandings of the distinction.[7]

The coarse-grained understanding is tied to the *a priori/a posteriori* distinction. On this understanding, an epistemology of modality for which all of our modal knowledge was acquired by *a priori* means would be uniform irrespective of whether different epistemic tools were involved, at a basic level, in explaining different pieces of modal knowledge. Wirling (2019: footnote 68) is right in attributing to Tahko (2015) the coarse-grained understanding of the distinction.

The fine-grained understanding (the one implicit in Roca-Royes 2007), by contrast, takes 'routes' more narrowly. More needs to be said about what exactly is taken to constitute a route but, roughly, and on the basis of what the literature has offered so far: conceiving, rational insight, determinate understanding, reflection from Peacocke's Principles of Possibility, induction (whether *a posteriori* or *a priori*), abduction (again, *a posteriori* or *a priori*), entitlements and counterfactual reasoning would be among the epistemic tools that, either by themselves or in combination, can constitute routes to modal belief. On this understanding, an epistemology of modality will be uniform provided that one single route (thus individuated) is appealed to in order to explain all instances of modal knowledge.[8,9]

I turn now to identify representatives of these categories that one finds in the literature. As indicated above, the accounts by Bealer, Chalmers, Lowe, Peacocke, Sidelle and Yablo are all arguably uniform in the fine-grained sense. They are also all rationalist accounts – with a qualification that allows them to accommodate, rather than neglecting, knowledge of *a posteriori* necessities within their theories – and, thus, each comes out as uniform in the coarse-grained sense too.[10] Jenkins' conceivability account (2010) also comes out as uniform in both the fine- and the coarse-grained understandings, with the difference that hers is an empiricist one.

Hale's account, in being also – with the same qualification as above – a rationalist one, also counts as uniform in the coarse-grained sense. Yet, it cannot be classified as uniform in the fine-grained one. Hale defends that reflection on the meanings of words is a route to *some* modal knowledge. It is this route that is at play in, for instance, our acquisition of essentialist knowledge about (the property of) *being a square*.

> Clear and straightforward cases in which knowledge of meaning suffices for knowledge of essence are those in which we are able to give an explicit definition of a word. … For example, a plane figure is correctly described as 'square' iff it is made up of four straight sides of equal length, meeting at right-angles. Precisely because such a definition gives [analytically] necessary and sufficient conditions for the word 'square' to apply, there is no mystery how we can know that there is no more (and no less) to being square than satisfying those conditions.
>
> (Hale 2013: 255)

Yet, this route doesn't generalize. It wouldn't generate knowledge of what Hale labels 'general principles of essence' – "principles asserting, schematically, that such-and-such a property is essential to its instances" (Hale 2013: 269) – paradigm examples of which are *the essentiality of origins* and *the essentiality of kind membership*. For such principles, Hale thinks and I agree, an abductive case, with somewhat less probative force than meaning-based arguments, is the strongest non-question-begging type of argument we're in a position to offer (Hale 2013: 271). The resulting epistemology of modality, one that is as wide in scope as the explanandum requires, will thus be non-uniform in the fine-grained sense. As mentioned above though, and because of the rationalist nature of Hale's variety of explanations, it still counts as uniform in the coarse-grained sense.

Things are otherwise with the epistemology of modality I favour (as presented in Roca-Royes 2017 and 2018), which comes out as non-uniform in both the coarse- and the fine-grained senses. In a nutshell, three partial theses set the grounds of the account. First, that while (rationalist) meaning-based explanations are best to explain certain instances of modal knowledge, they are limited in scope: again, and in agreement with Hale, they don't generalize. Second, that our knowledge of ordinary possibilities about concrete entities requires induction-based explanations of an empirical sort that, yet again, don't generalize. Third, that *that* inductive knowledge will not figure among the epistemic grounds for necessity/essentialist knowledge and, as such, an altogether independent route to such knowledge will be needed. The upshot epistemology is, consequently, non-uniform in the two senses.

One might have expected to see Williamson's counterfactual-based account (2007) classified as a uniform one, at least in the fine-grained sense. For Williamson's answer to the 'how do we know?' question can be neatly glossed with 'by reasoning counterfactually'. Yet, such classifying would be inadequate. To begin with, already in 2007 Williamson adverts that "[t]here is no uniform epistemology of counterfactual conditionals. In particular, imaginative simulation is neither always necessary not always sufficient for their evaluation, even when they can be evaluated" (Williamson 2007: 152). There is therefore no reason to expect that our epistemic access to those counterfactuals that are relevantly involved in the acquisition of modal knowledge is uniformly explainable. If anything, there is reason to expect the opposite. In addition, in his *Modal Logic as Metaphysics* (2013) Williamson engages with different routes to knowing objective modal facts, abduction being a prominent one. Saliently, such facts include, as Strohminger and Yli-Vakkuri stress (2017), the most general modal facts.

The issue of generalizability from running examples (illustrated all along this section) has made it into the collective awareness of modal epistemologists. Thus, for instance, Biggs suggests an abduction-based explanation to cover modal knowledge involving scientific entities but, at the same time, he's very cautious (2011: §4.4) not to rush *general* conclusions about the explanatory power of his abduction-based explanations. Similarly, Strohminger offers a perception-based epistemology of our knowledge of *some* non-actual possibilities and, like Biggs, she's also cautious not to generalize hastily:

> I have argued that some claims about nonactual possibilities can be, and often are, known by perception. I am not suggesting that perception has the ability to explain *all* of our knowledge of nonactual possibilities, even when supplemented with inference and testimony.
>
> *(Strohminger 2015: 369)*

The same is true of Hanrahan's imagination-based epistemology, aimed at explaining the justification we have for "a certain set of our modal beliefs" (Hanrahan 2007: 142) while being declared limited beyond this; and of Rasmussen's *continuity*-based guide to possibility (2014), which he claims to be "restricted to modal claims of a very specific sort" (526). To provide one last example, Nimtz (2012) offers an epistemology of modality explicitly confined to explaining knowledge of those metaphysical necessities that are (epistemically) grounded in knowledge of conceptual necessities.

It is not surprising that this collective awareness about the generalizability issue goes hand in hand with, and in fact explains, our increasing interest towards non-uniform epistemologies of modality. Indeed, to put forward an epistemology of modality that one already assesses as not being able to explain all cases of modal knowledge calls for *additional* explanations to cover those instances of knowledge thus far left unexplained. Non-uniformism will be the upshot.

4 The structure of modal knowledge

The second distinction I want to focus on turns on the structure of modal knowledge, and it owes a lot to Bob Hale. Already in 2003 he distinguished between symmetric and asymmetric accounts of modal knowledge and, within the asymmetric ones, he distinguished between the necessity-first and possibility-first ones. Crudely, the key trait of an asymmetric *necessity*-first epistemology of modality is the existence of a base class of known *necessities* such that all other modal knowledge (of remaining necessities and of possibilities) ultimately depends on our knowledge of the elements in that base class. There is thus an asymmetry in the epistemic priority order. Similarly, an asymmetric possibility-first account will claim the existence of a base class containing only known possibilities. By contrast, on a symmetric account one will find both possibility- and necessity-knowledge at the most fundamental level.

Bob Hale (2003) offered powerful reasons against an asymmetric possibility-first account. At that time, he already leaned towards an asymmetric necessity-first one but left the door open to a symmetric epistemology; indeed, his main aim there was to compare the explanatory strengths of the two asymmetric accounts. Things are different in his most recent development of the view (2013) where, building up from his earlier arguments, he endorses a necessity-first approach. His main motivation for doing so stems from his own metaphysics of modality: *the essentialist theory of necessity*; a member of the Finean family of views, as referred to in §1. As mentioned there, Hale thinks that the intimate link between essential truths and the metaphysical realm is to be understood as an ontological priority of the essential over the modal: modality is grounded in essence. At an ontological level, therefore, the essential facts about a given entity imply the *de re* necessities involving it and those, in turn, determine the space of possibilities for it (no reverse implications). Hale finds it that a *mirroring* epistemology of modality is strongly suggested by this ontological picture. As he says, given the ontological priority order, "one might expect an explanation of how we can have knowledge of the nature or essence of things to play a fundamental and central part in explaining knowledge of necessity" (Hale 2013: 254). The full mirroring epistemology thus includes, as a proper part, an asymmetric necessity-first approach. On the resulting picture, knowledge of essence would come first, as most fundamental, followed by knowledge of directly derived *de re* necessities – the elements of the base class – followed in turn by knowledge of *de re* possibilities (as those determined to be compatible with the *de re* necessities) and possibly of further necessities too.[11]

Although the literature has widely endorsed Hale's distinction(s) when it comes to classifying the different epistemologies of modality in the market, there hasn't been much published discussion about Hale's specific arguments for the convenience of an asymmetric necessity-first approach. An excellent exception is Fischer (2016). There, apart from casting doubts on the felicity of Hale's arguments, Fischer stresses the importance of scrutinizing them:

> Hale's goal is to motivate a necessity-based approach. If his arguments work, then we shouldn't accept views like Stephen Yablo's (1993). ... This is a possibility-based approach: I acquire a well-founded belief that there are no conflicting necessities by objectually imagining the relevant world; I don't need to begin with a well-founded belief that there aren't any conflicting necessities. The same point probably applies to most conceivability- or imaginability-based views – including, e.g., Hart (1988), van Woudenberg (2006), Hanrahan (2007), and Kung (2010).
>
> *(Fischer 2016: 77)*

I have my own doubts that conceivability accounts are possibility-first accounts (despite appearances) and have expressed them in Roca-Royes (2011). Regardless of those, Fischer's point is a structural one. If there are general, probative enough reasons to prefer one type of account to the other two, those reasons should be informing our search for an adequate epistemology of modality. For, as Fischer adds later on the same page, "it's hard to know where to begin in the epistemology of modality, so it's worth exploring any way to whittle down our options".

Both Hale's arguments for the inadequacy of a possibility-first account and Fischer's reasons to think that they fall short of establishing their target claim thus deserve proper scrutiny. Rather than engaging in this task, however, the way I want to contribute (with this chapter) to this debate is by doing two other things. First, by arguing, independently, against the adequacy of a necessity-first account. But second, by agreeing with Hale that some knowledge of necessities is basic. In other words, I will contribute to this debate by suggesting the need of a *symmetric* account.

As for the first task: I have already expressed the key rubric of the argument in Roca-Royes (2017, 2018). It is also quite a simple argument, which I will present here by means of a running example:

i   To establish *that my office desk can break* in a necessity-first manner, we first need access to a suitably strong body of relevant known necessities/essential truths about my office desk and determine that they don't rule out its breaking.

(i) is just a statement of the view applied to our running example. 'Suitably strong' means here whatever it needs to rule out the epistemic adequacy (in this context at least) of *arguments from ignorance*. Suitable strength is compatible with the body of necessities being empty, so long as we *know* it to be empty. The literature takes it that plausible candidates to belong to the suitably strong body will include, at some level of generality, the principles of *essentiality of origin* and of *essentiality of kind*. But regardless of which and how many elements are in there, the requirement about suitable strength dictates that we should know them all and know them to be all. As Fischer puts it, absence of evidence (about the existence of a necessity) is not enough; we need evidence of absence. (See Fischer 2016: 78.)

ii   Given (i), on necessity-first accounts, our rational degree of confidence *that my office desk can break* cannot be higher than our degree of confidence about the basic (*de re*) necessities relevantly involved, and about the non-existence of further necessities.

Thus, in the running example, our (alleged) *basic* knowledge that unbreakability is not essential to (and thus not necessary of) the desk, that the desk is essentially made of wood and that the desk is essentially a desk, epistemically grounds our piece of knowledge that the table can break (on the basis of this possibility being determined not to be ruled out by *any of* the table's essential properties). But not only this: given that *that* is the priority order, our rational degree of confidence in the latter is upper bound by our degree of confidence in the former pieces of knowledge. And yet, I contend:

iii   Our confidence that my office desk can break is higher than our confidence about my desk's essential properties. And this difference in degrees of confidence is (at least as per our current epistemic situation) the rational one.

Hale (2013) distinguishes different levels of probative force of arguments for modal claims. As we saw in §3, he believes that the meaning-based arguments that can be given to establish essential truths about, for instance, (the property of) *being square* are more conclusive than the abductive cases that can be offered in favour of, precisely, "general principles of essence". (See in particular Hale 2013: §11.3).

I cannot agree more with Hale on this. And I want to add to it: the inductive arguments that can be given in favour of the desk's breakability are stronger – in terms of probative force – than the abductive arguments supporting essentialist claims. This, I contend, supports the rationality of the different degrees of confidence exhibited. Looking at things with the lens of a social epistemologist, this difference in probative force is, likely, what explains the asymmetric distribution of peer-agreements and peer-disagreements: claims of ordinary possibility are significantly more widely peer-agreed than essentialist claims. It is incompatible with what is pictured by this radiography that our ordinary possibility knowledge is grounded in (basic) necessity/essentially knowledge of the sort required by a necessity-first account.

5 Final remarks and further reading

It is important to note that the two distinctions in sections 3 and 4 cut across each other. As such, results centred on one of them will not per se inform research centred on the other. There's nothing inherent to, for instance, a necessity-first approach that requires uniformity in any of the two senses distinguished in §3. Hale's very own account partially witnesses this claim. For, despite being a necessity-first approach, it is also, for the reasons provided in §3, a non-uniform account in the fine-grained sense. And there's nothing that would impede an account from being *also* non-uniform in the coarse-grained sense. For instance, an account in the spirit of Hale's but with the difference that, for the case of some essentialist principles, it endorsed Elder's empiricist *confirmationism* (rather than Hale's rationalist abduction-based explanations) would arguably result in a necessity-first account that would be non-uniform in both senses.

Despite, therefore, there being general reasons for thinking that a comprehensive modal epistemology must be symmetric (thus prioritizing neither knowledge of necessity nor knowledge of possibility), these reasons do not in turn favour, per se, a non-uniform account. Reasons for non-uniformism and reasons for symmetry will likely be found independently of one another.

There are other pieces in the literature that provide an overview of the epistemology of modality, each with its distinct focus, and that, jointly, render a very rich insight into the topic. These include Evnine (2008), Fischer (2017), McLeod (2005), Strohminger and Yli-Vakkuri (2017) and Vaidya (2015).

Notes

1 These include epistemic (relating to knowledge and belief), deontic (relating to norms) and alethic (relating to truth). See Kment (2017) for an excellent survey on the varieties of modality.
2 Other paradigmatic types of *alethic* modalities include *nomological* (constrained by natural laws), *logical* (constrained by logical laws) and *conceptual* (constrained by conceptual relations). But also modalities such as *historical modalities* (what is possible/necessary at a given point in time given the worlds' past) count as alethic. See Kment (2017: §3) for a thorough discussion on the structure of the metaphysical modal realm.
3 We do not, for current purposes, need to take a stand on the matter. For, even according to the Fineans, who hold the weaker relation between essence and modality, the link is strong enough. For representative literature on the debate between modal and non-modal accounts of the notion of *essence* see Correia (2007), Fine (1994) and Romero (2019).
4 Exceptions to this bold claim include modal claims like ['It is possible that X'] where X involves a mental activity, like 'It is possible that I imagine a talking donkey.' A refined claim would need to take this class of statements into account but, for current purposes, it won't harm to stick to the bold claim.
5 Williamson would probably not be happy having his account classified as either rationalist or empiricist. He thinks that such categories are of not much epistemological significance. Still, without doing *that*, I think it is not unfair to the view's spirit to receive it as witnessing the turn away from rationalist epistemologies of modality.
6 Van Inwagen (1998) is a foundational representative of a partial sceptic. While he is not clear what explanation is to be provided for the cases of modal knowledge that we have, he is absolutely clear both that we have some modal knowledge and that there are some modal facts that are beyond epistemic access.
7 These two ways of understanding the distinction (and the different ways of individuating routes) were first suggested to me by Ylwa Wirling, to whom I also thank for letting me use her convenient labelling.
8 This doesn't mean that the advocates of uniform accounts would only allow *one* and always the same route to modal knowledge. (There isn't always enough explicit discussion on the issue.) It *does* mean, however, that they would believe in the existence of one route capable of providing access to all knowable modal truths; and it does mean that they make this route – whichever that is on each occasion – the centre of their epistemological explanations. Quite likely, for instance, the advocates of the uniform (one-fold) answers mentioned above will also allow for testimony to be a route to modal knowledge and, perhaps more pressingly, they would also agree to actuality being a proof of possibility. This doesn't undermine the claim that their accounts are uniform. What makes them so is the fact that their privileged route *also*, according to them, provides epistemic access to those alternatively accessible modal truths.
9 The remark in the previous endnote makes it convenient to distinguish, for the purpose of leaving it aside, a further way in which the uniform/non-uniform distinction could be understood. On this further understanding, an epistemology of modality would be non-uniform if it endorses more than

one possible route to knowing the same modal truth(s) (for at least some modal truths). This is the understanding Vaidya is making room for when he writes that "[t]he *non-uniformity* view maintains either that different people can come to know the same modal truth through different routes or that at the fundamental level of investigation there must be more than one route to modal knowledge" (Vaidya 2015).

10 The qualification is that they claim modal knowledge to be *fundamentally a priori*. The term 'fundamentally' is here intended to allow for the inclusion of knowledge of *a posteriori* necessities like, perhaps, that *Socrates is necessarily human*. And the way these can be subsumed under a rationalist programme is by factoring them out into an essentialist principle — supposedly (*purely*) *a priori* — like *all humans are necessarily human*, and an *a posteriori* non-modal truth, like *Socrates is human*. For more on how *a posteriori* necessities can be subsumed under a rationalist programme by claiming them to be *fundamentally a priori*, see Peacocke (1999: 168–169).

11 There is room for more and less extreme variants of asymmetric epistemologies depending, respectively, on whether they claim *all* necessity (or else possibility) knowledge to be epistemically prior to *any* possibility (or else necessity) knowledge or not. Hale himself endorses a moderate version. (For more on this, see Hale 2013: 253.)

References

Bealer, G. (2002) 'Modal epistemology and the rationalist renaissance' in Gendler, Tamara, and Hawthorne, John (eds.), *Conceivability and Possibility*, Oxford: Oxford University Press, 71–125.
Bealer, G. (2004) 'The origins of modal error', *Dialectica*, 58(1): 11–42.
Biggs, S. (2011) 'Abduction and modality', *Philosophy and Phenomenological Research*, 83(2): 283–326.
Blackburn, S. (1986/1993) 'Morals and modals', in *Essays in Quasi-Realism*, Oxford: Oxford University Press, 52–74.
Chalmers, D. (1996) *The Conscious Mind*, Oxford: Oxford University Press.
Chalmers, D. (2002) 'Does conceivability entail possibility', in Gendler, Tamara, and Hawthorne, John (eds.), *Conceivability and Possibility*, Oxford: Oxford University Press, 145–200.
Correia, F. (2007) '(Finean) essence and (priorean) modality', *Dialectica*, 61/1: 63–84.
Craig, E. (1985) 'Arithmetic and fact' in I. Hacking (ed.), *Essays in Analysis*, Cambridge: Cambridge University Press, 89–112.
Elder, C. (2004) *Real Natures and Familiar Objects*, Cambridge, MA: MIT Press.
Evnine, S. (2008) 'Modal epistemology: our knowledge of necessity and possibility', *Philosophy Compass*, 3: 664–684.
Fine, K. (1994) 'Essence and modality', *Philosophical Perspectives*, 8: 1–16.
Fischer, B. (2016) 'Hale on the architecture of modal knowledge', *Analytic Philosophy*, 57/1: 76–89.
Fischer, Bob (2017) 'Modal empiricism: objection, reply, proposal', in Fischer, B. and Leon, F. (eds.), *Modal Epistemology After Rationalism*, Berlin: Springer (Synthese Library), 263–279.
Jenkins, C. (2010) 'Concepts, experience, and modal knowledge', *Philosophical Perspectives*, 24: 255–279.
Hale, B. (2003) 'Knowledge of possibility and necessity', *Proceedings of the Aristotelian Society*, 103: 1–20.
Hale, B. (2013) *Necessary Beings: An Essay on Ontology, Modality, & the Relations Between*. Oxford: Oxford University Press.
Hanrahan, R. (2007) 'Imagination and possibility', *The Philosophical Forum*, 38/2: 125–146.
Kment, B. (2017) 'Varieties of modality', *The Stanford Encyclopedia of Philosophy* (Spring 2017 Edition), Edward N. Zalta (ed.), URL = https://plato.stanford.edu/archives/spr2017/entries/modality-varieties/.
Kripke, S. (1972/1980) *Naming and Necessity*, Cambridge, MA: Harvard University Press.
Leon, F. (2017) 'From modal skepticism to modal empiricism' in Fischer, B. and Leon, F. (eds.), *Modal Epistemology After Rationalism*, Berlin: Springer (Synthese Library), 247–261.
Lowe, J. (2008) 'Two notions of being: entity and essence', *Royal Institute of Philosophy Supplements*, 83(62): 23–48.
Lowe, J. (2012) 'What is the source of our knowledge of modal truths', *Mind*, 121/484: 919–950.
Mallozzi, A. (2018a) 'Putting modal metaphysics first', *Synthese*, forthcoming.
Mallozzi, A. (2018b) 'Two notions of metaphysical modality', *Synthese*, forthcoming.
McLeod, S. (2005) 'Modal epistemology', *Analytic Philosophy*, 46/3: 235–245.
McLeod, S. (2009) 'Rationalism and modal knowledge', *Crítica*, 41/122: 29–42.
Menzies, P. (1998) 'Possibility and conceivability: a response-dependent account of their connections' in R. Casati (ed.), *European Review of Philosophy*, 3: 255–277.

Miščević, N. (2003) 'Explaining modal intuition', *Acta Analytica*, 18(30–31): 5–41.
Nimtz, C. (2012) 'Conceptual truths, strong possibilities, and our knowledge of metaphysical necessities' in Drapeau-Contim, Filipe, and Motta, Sébastien (eds.), *Modal Matters*, special issue of *Philosophia Scientiae*, 16(2): 39–59.
Nolan, D. (2017) 'Naturalised modal epistemology' in Fischer, B., and Leon, F. (eds.), *Modal Epistemology After Rationalism*, Berlin: Springer (Synthese Library), 7–27.
Peacocke, C. (1999) *Being Known*, Oxford: Oxford University Press.
Rasmussen, J. (2014) 'Continuity as a guide to possibility', *Australasian Journal of Philosophy*, 92/3: 525–538.
Roca-Royes, S. (2007) 'Mind-independence and modal empiricism' in Penco, C., Vignolo, M., Ottonelli, V. and Amoretti, C. (eds.), *Proceedings of the 4th Latin Meeting in Analytic Philosophy* (LMAP-2007), Genoa, Italy, 117–135.
Roca-Royes, S. (2011) 'Conceivability and de re modal knowledge', *Noûs*, 45/1: 25–49.
Roca-Royes, S. (2017) 'Similarity and possibility: an epistemology of de re possibility for concrete entities' in Fischer, B., and Leon, F. (eds.), *Modal Epistemology After Rationalism*, Berlin: Springer (Synthese Library), 221–245.
Roca-Royes, S. (2018) 'Rethinking the epistemology of modality for abstracta' in Fred-Rivera, Ivette, and Leech, Jessica (eds.), *Being Necessary: Themes of Ontology and Modality from the Work of Bob Hale*, Oxford: Oxford University Press, 245–265.
Romero, C. (2019) 'Modality is not explainable by essence', *Philosophical Quarterly*, 69/274: 121–141.
Sidelle, A. (1989) *Necessity, Essence, and Individuation*, Ithaca, NY: Cornell University Press.
Strohminger, M. (2015) 'Perceptual knowledge of nonactual possibilities', *Philosophical Perspectives*, 29, Epistemology.
Strohminger, Margot, and Yli-Vakkuri, Juhani (2017) 'The epistemology of modality', 77/4: 825–838.
Tahko, T. (2015) *An Introduction to Metametaphysics*, Cambridge: Cambridge University Press.
Tahko, T. (2017) 'Empirically informed modal rationalism' in Fischer, B., and Leon, F. (eds.), *Modal Epistemology After Rationalism*, Berlin: Springer (Synthese Library), 29–45.
Vaidya, A. (2015) 'The Epistemology of Modality', *The Stanford Encyclopedia of Philosophy* (Winter 2017 Edition), Edward N. Zalta (ed.), URL = https://plato.stanford.edu/archives/win2017/entries/modality-epistemology/.
Van Inwagen, P. (1998) 'Modal epistemology', *Philosophical Studies*, 92: 67–84.
Vetter, B. (2016) 'Williamsonian modal epistemology, possibility-based', *Canadian Journal of Philosophy*, 46/4–5: 766–795.
Williamson, T. (2007) *The Philosophy of Philosophy*, Oxford: Oxford University Press.
Williamson, T. (2013) *Modal Logic as Metaphysics*, Oxford: Oxford University Press.
Wirling, Y. (2019) *Modal Empiricism Made Difficult: An Essay in the Meta-Epistemology of Modality*, Acta Universitatis Gothoburgensis, Gothenburg. (http://hdl.handle.net/2077/57967).
Yablo, S. (1993) 'Is conceivability a guide to possibility?', *Philosophy and Phenomenological Research*, 53: 1–42.

29
IDEOLOGY AND ONTOLOGY

Sam Cowling

Introduction

Theories place demands upon what the world is like. If a theory is true, these demands are met. Consider a theory like arithmetic. If true, it seems to require at least the existence of numbers. In philosophical parlance, arithmetic therefore involves an *ontological commitment* to numbers. Roughly put, the ontology of arithmetic includes numbers in virtue of the fact that numbers must exist in order for arithmetic to be true. So, upon pain of inconsistency, it seems that we cannot accept the truth of arithmetic while, at the same time, denying that there are numbers.

Some philosophers have sought to show that the truth of arithmetic is, rather surprisingly, compatible with the view that there are, in fact, no numbers.[1] The prospects for such views are highly controversial, but, regardless of their merits, these efforts are often part of a more general inquiry that aims to determine how, if at all, we can extract specific ontological commitments from various theories. (See Quine 1948; Bricker 2016.) The pursuit of a comprehensive account of the ontological commitments of theories is, in turn, a central ambition of contemporary metaphysics. Among other things, it seeks to provide a way to move from questions about which theories we endorse to an inventory of what we ought to believe exists. This focus on ontological commitments and how theories acquire them is understandable enough, but it has arguably led some philosophers to ignore a related and perhaps equally important question about the demands theories place on the world: are the commitments of theories exclusively ontological?

An increasingly prevalent view in metaphysics is that the commitments of theories are not limited to ontology but also include what has come to be called "theoretical ideology." The most ready examples of the ideological commitments of theories include primitive predicates and operators. For example, on certain views, the predicate 'instantiates' is an ineliminable commitment of our best metaphysical theories, given the apparent truth of sentences like 'Plato instantiates humanity.' But, while we might abide an ontological commitment to both Plato and the property *humanity*, some views deny that there is any ontological correlate of 'instantiates.'[2] According to these views, while the predicate 'instantiates' does not express a relation, we are nevertheless required to accept into our theory the predicate, notion, or concept of "instantiation" as a theoretical primitive – a basic, undefined piece of ideology and perhaps an irreducible aspect of metaphysical structure.

According to what we can call *ideological realism*, theories place ideological demands on the world and seek to capture the metaphysical structure of reality in a different but no less objective or "worldly" way than ontological commitments. For this reason, ideological realists hold that there is an objective sense in which the ideological commitments of theories might "match" (or fail to "match") the non-ontological structure of the world to a greater or lesser extent. As Kment (2014: 150) puts it,

> It shouldn't be assumed that all ingredients of reality must be individuals, properties, or relations – or entities of any kind, for that matter. For example, it's possible that in order to describe reality completely, we need to use some primitive piece of ideology that relates to some aspect of reality that doesn't belong to one of these three ontological categories, and which may not be an entity at all.

If ideological realism of this kind is correct, then maximizing fit between the primitive ideology of a theory and the metaphysical structure of reality ought to be a central aim in devising our metaphysical theories. (See Sider 2009, 2011.)

While the division between ontology and ideology as well as the nature of ideological commitment is controversial, it is difficult to deny that theories are often compared or evaluated with respect to the variety or intelligibility of their primitive predicates or operators. In at least this limited respect, theories seem evaluable for their ideological economy in a manner roughly comparable to their ontological economy. But, if we are to evaluate ideological commitments in parallel to ontological ones, we require an adequate account of precisely how theoretical ideology places demands upon the world. A clearer conception of the metaphysical nature of theoretical ideology is therefore needed. In this regard, the work of W.V. Quine, David Lewis, and, more recently, Ted Sider has done much to bring the ideology-ontology distinction to the forefront of metaphysical inquiry and helped to map out the ways in which theory evaluation hinges on assessing ideological commitments alongside ontological ones. This entry sketches the recent history of the metaphysics of ideology while noting some of the central questions that arise in understanding ideological commitment. Throughout, remarks and running examples draw from what we can call *metaphysical* ideology – roughly, the distinctive ideological commitments of metaphysical theories – though much of what has been said regarding ideology has been intended to generalize to theoretical ideology regardless of subject matter.

Ideological commitments

Let us return to the case of arithmetic. If we grant, as most do, that arithmetic requires an ontological commitment to infinitely many numbers, what other demands does it impose on reality? Consider, for example, the arithmetical claim that zero has a unique successor. Does this claim require that we accept, along with zero and its unique successor, an additional entity: *the successor relation*? Put differently, does the truth of the claim that zero has a unique successor mandate ontological commitments that go beyond numbers alone to a domain of other kinds of entities like relations or operators?

If we answer in the negative, it is natural to hold that, while this claim does not require the existence of a successor relation, it nevertheless commits us to reality being a certain way – after all, we are deploying the notion, *successor*, which other theories might reject as unnecessary, incoherent, ambiguous, or whatever. To this end, we must distinguish our ontological commitment to numbers from our *ideological commitment* to the predicate 'is a successor of,' which is employed in expressing the theory of arithmetic.

When broadly construed, ideological commitment concerns whatever predicates or concepts occur within a theory, regardless of whether they are defined or primitive notions. More narrowly construed, ideological commitment concerns only the undefined or primitive predicates or concepts that figure into our theories. Given that the predicate 'is a successor of' occurs in arithmetical sentences like 'Zero has a unique successor,' it is among the ideological commitments of arithmetic. And, if we can provide no reductive analysis or definition of this predicate, we must treat it as a primitive notion and therefore count it as an ideological commitment in the narrow sense. In the construction of our theories, the judicious choice of primitives enhances expressive and explanatory power and permits the analysis and reduction of other theoretical terms – e.g., we might analyze the notion of *is greater than* in terms of *successor*. Conversely, if we introduce primitive predicates willy-nilly, we obscure systematic analytic connections within our theories and induce gratuitous complexity. Since theories are typically thought to improve when the stock of basic notions is minimized (without reducing their perspicuity or explanatory credentials), the natural stance is that, when it comes to ideology in the narrow sense, less is more.

Although there is no clear sense in which one might reject ideological commitments as just defined, one could, in principle, oppose the ideological realist thesis that ideological commitments can do a better or worse job of matching the structure of reality. One way of opposing ideological realism is to hold that putative ideological commitments are little more than ontological commitments in disguise. Such a view would hold that there is no way in which we might admit 'is a successor of' into our theory without positing a relation correlated to the predicate. Similarly, the introduction of primitive modal and temporal operators would be held to require distinctive sorts of presumably abstract ontological commitments expressed by these operators. Below, we'll consider why assimilating all metaphysical disagreement to ontological disagreement implausibly distorts debates in the metaphysical of modality. Before doing so, it is worth explaining why ideological commitments seem both fruitful and arguably unavoidable in metaphysics.

Consider the familiar case of properties and instantiation. For almost all who accept an ontology of properties, talk of "instantiation" plays a key role in understanding the nature of properties. But, when pressed to elaborate upon such a view, problems quickly arise and are exacerbated if "instantiation" is held to express an abstract relation.[3] Very roughly: if we attempt to explain the fact that *a* instantiates *F* as holding in virtue of the fact that *a* and *F* stand in the *instantiation* relation, then we must explain the fact that *a*, *F*, and *instantiation* are so related. To this end, we might posit some additional relation to do this work, but this merely generates another structurally similar explanatory burden. Upon pain of an infinite regress of posited relations, we seem required to treat our talk of instantiation as quite different from our talk of "is five feet from" or "loves." An attractive option is therefore to introduce 'instantiates' as a primitive predicate in our theory of properties – one required for describing reality but that has no ontological correlate.

The case of instantiation illustrates the appeal (perhaps even the necessity) of introducing primitive ideological commitments. It is worth noting, however, that the relation between primitives, ideology, and ontology is a complex affair. Notice that primitive notions are distinguished by their theoretical status. Where other theoretical terms admit of reductive definition, primitives do not. And, upon pain of circularity or a kind of infinite conceptual "descent" (i.e., a non-terminating chain of notions defined in terms of yet other notions), all theories must abide at least some primitives. There is, however, nothing about being a primitive notion that, in principle, precludes it from having an ontological correlate. One might, for example, take "goodness" to be a primitive and insist that it expresses a fundamental property, while others might take the predicate as a primitive and deny it has any ontological correlate. So, although philosophers often talk of taking something as a primitive as a shorthand way of communicating that it incurs no ontological commitment, it would distort the connection between definability and the ideology-ontology distinction to conflate these equally important but notably different questions.

Lewis' contribution

While inquiry into ideology and its interface with ontology is implicit in a range of metaphysical debates, the prevalence of inquiry into metaphysical ideology owes greatly to its role in discussions regarding modality and its analysis. Moreover, the nature and significance of ideology is perhaps made clearest by attending to the case of modal operators.

Possible worlds theorists hold that our modal thought and talk is properly analyzed in terms of quantification over a specific kind of entities: *possible worlds*. Necessary truths obtain in all possible worlds. Impossibilities obtain at none. Contingent truths obtain at some but not all possible worlds. The backbone of possible worlds theory is a pair of biconditionals, which analytically connect the operators of modal logic with quantification over possible worlds: (i) $\Box P$ iff, at all possible worlds, P is true; and (ii) $\Diamond P$ if and only if, at some possible worlds, P is true.

Setting aside certain exceptions, possible worlds theorists purport to analyze modality by taking on an ontological commitment to worlds. Possible worlds theory is, however, far from mandatory as a metaphysics of modality. (See Divers 2002.) One competing view, modalism, holds that our modal thought and talk is in perfectly good standing, but denies that it is rightly analyzed in terms of possible worlds or any other additional ontological commitments. (See Forbes 1989 and Melia 1992.) One kind of modalism rejects the above biconditionals and contends that facts like $\Box P$ and $\Diamond P$ admit of no further analysis. On such a view, reality has an irreducibly modal aspect that cannot be assimilated into a special kind of ontological commitment to possible worlds. A second competing and considerably more radical view, modal eliminativism, denies that our modal thought and talk is in working order. (See Sider 2003: 185.) Instead, modal claims are either meaningless or systematically mistaken. It is neither true or false that $\Box P$, since reality has no modal aspect whatsoever.

Possible worlds theory, modalism, and modal eliminativism are markedly different views and aptly characterizing their differences is a matter of much importance for modal metaphysics. Notice, though, that it seems quite possible for the modalist and the modal eliminativist to agree on all ontological matters since each denies there are possible worlds. Upon close inspection, we would badly distort the character of their disagreement if we assimilate it to a disagreement about what things exist. The modalist's view is not distinctive for believing that there are special entities – the box and diamond of modal logic – that the modal eliminativist does not believe in. Rather, the modalist believes reality exhibits modal structure that is aptly expressed using the '$\Box$' and '$\Diamond$' of modal logic. In stark contrast, the modal eliminativist denies reality has any modal structure. The difference between these views is therefore not an ontological one. Instead, it concerns metaphysical ideology – in this case, modal ideology pertaining to necessity and possibility.

According to Lewis, the pitfalls of primitive modal ideology are best avoided by endorsing modal realism, a view on which quantification over concrete possible worlds furnishes us with a reduction of our modal notions. It is in the context of this project that Lewis offers arguably the most influential remarks on ideology. In setting the stage for his defense of modal realism, Lewis speaks of the ideological commitments of set theory, their importance, and their relation to ontology with a casualness that belies the paucity of preceding philosophical inquiry into ideology. Summarizing the case in favor of set theory, Lewis (1986: 4) says:

> Set theory offers the mathematician great economy of primitives and premises, in return for accepting rather a lot of entities unknown to *Homo javanensis*. It offers an improvement in what Quine calls ideology, paid for in the coin of ontology. It's an offer you can't refuse. The price is right; the benefits in theoretical unity and economy are well worth the entities.

A parallel is then swiftly drawn with the hypothesis of modal realism and its vast ontology of concrete possibilia. And, in that regard, Lewis notes that the ontological commitment to a plurality of concrete worlds provides "the wherewithal to reduce the diversity of notions we must accept as primitive." (*Ibid.*) The philosophical result is an improved, more unified, and more parsimonious theory, not just regarding modality, but in "the theory that is our professional concern – total theory, the whole of what we take to be true." (*Ibid.*)

The picture of theory choice Lewis points to in these early pages goes largely unelaborated in *On the Plurality of Worlds*. There is no sustained investigation into the metaphysics of ideology. There is no careful defense of the assumptions regarding ontology and ideology. There is no examination of other methodological orientations that might discount the value of minimizing ideology. But, despite the sparse character of Lewis' remarks on ideology, it is hard to understate the influence these passages have exerted on subsequent metaphysics. They have prompted others to follow methodological suit and make explicit the practice of weighing ontological commitments against ideological ones. (See, e.g., Rodriguez-Pereyra 2002: 199–210; Melia 2008: 112–113; and Van Cleve 2016.) And, although the pursuit of fewer basic notions in our theories is by no means novel, the rise of this Lewisian picture of theory choice has changed how metaphysicians describe and evaluate competing theoretical alternatives. Talk of "bloated ontology" is now regularly set against the appeal of a "lean" or "minimal ideology."

For all its influence, the ideological "turn" inaugurated by Lewis' remarks has unfolded with conspicuously little in the way of explicit comment. For better or worse, this is because Lewis' remarks presuppose several intuitively plausible theses about ideology that have met with wide, albeit largely implicit, approval.

Ideological Virtues: Metaphysicians frequently appeal to theoretical virtues like conservatism, simplicity, and fertility that extend beyond the constraint of empirical adequacy. The most familiar of these appeals are Ockhamist ones that caution us to dispense with unnecessary ontological commitments. Lewis' remarks, in *On the Plurality of Worlds* and elsewhere, presuppose that attending to these virtues (and correlative vices) while assessing ideological commitments is a route to delivering better theories. Indeed, the reduction of primitives is a recurring aim in Lewis' work on metaphysics. By applying these theoretical standards to both ontology and ideology, Lewis' methodology marks them, not only as comparable concerns, but as among our better grounds for preferring certain theories over rivals that differ only in ideological terms. (See Sider 2013 for an application of ideological parsimony to the metaphysics of mereology.) As a consequence, differences in ideology cannot be discounted or ignored as merely notational variants. These differences bear upon what theories we ought to accept since they have direct implications for what reality is ultimately like.

Comparative Ontology and Ideology: Throughout Lewis' work on metaphysics, we are often invited to consider competing theories that differ in how they allocate their theoretical spending. Where some theories take on vast ontological costs and spend little on ideology, others minimize ontological commitments in favor of a wealth of primitive notions. For instance, in the metaphysics of time, we can contrast eternalist views that posit non-present entities with presentist views that reject such entities while introducing primitive tense operators. (See

Cowling 2013.) In these and other cases, explanatory work needs to be done and we ought to seek out those theories that maximize their theoretical efficiency. Quite often, the space of competing theories includes pairs of theories that can be seen to differ in whether to invoke ontological or ideological resources to achieve a given explanatory end. In certain cases, this induces epistemic indecision. Memorably, Bennett (2009: 65) describes theory choice under these circumstances as follows:

> At this point, it starts to feel as though we are just riding a see-saw – fewer objects, more properties; more objects, fewer properties. Or perhaps – small ontology, larger ideology; larger ontology, smaller ideology. Either way, it starts to feel as though we are just pushing a bump around under the carpet.

A standard Lewisian strategy for addressing theoretical impasse (apart, of course, from simply abiding it) is to note that the construction of our metaphysical theories is a global affair with different ontological and ideological commitments generating interlocking theoretical virtues and vices. For example, we can usefully ask how our modal metaphysics ought to align with our temporal metaphysics. It might, for instance, seem unprincipled to eschew primitive modal operators and accept merely possible entities while, at the same time, rejecting non-present entities and introducing primitive temporal operators. In hopes of selecting from among theoretical rivals, Lewis' work on ideology often points towards an expansive and daunting task: discerning what cumulative ideological commitments yield a principled and internally consistent view of reality and its structure.

Quine's contribution

Although Lewis' remarks on ideology have exerted greater influence on recent metaphysics, the relation between ontology and ideology as well as the term 'ideology' owe to Lewis' predecessor, Quine. At the same time, the distinctively Quinean conception of ideology is tethered to Quine's controversial account of theories and meaning (an account too broad to summarize here). Quine offers an early statement on this front while clarifying his treatment of ontological commitment:

> Given a theory, *one* philosophically interesting aspect of it into which we can inquire is its ontology: what entities are the variables of quantification to range over if the theory is to hold true? Another no less important aspect into which we can inquire is its *ideology* (this seems the inevitable word, despite unwanted connotations): what ideas can be expressed in it?
>
> *(Quine 1951: 14)*

On this Quinean conception of ideology, theories are to be viewed as comprising (at least) ontology and ideology.[4] The divide between these notions owes the irreducibility of ideological commitments and ideological distinctions to exclusively ontological matters. For, as Quine (*Ibid.*) notes, "the ideology [of the theory of real numbers] embraces many such ideas as sum, root, rationality, algebraicity, and the like, which need not have ontological correlates in the range of the variables of quantification of the theory." Given that these are separable domains of commitment and inquiry, Quine (*Ibid.*) concedes the possibility of ideological variation even in the face of ontological agreement:

> Two theories can have the same ontology and different ideologies. Two theories of the real numbers, for example, may agree ontologically in that each calls for all and only the real numbers as values of its variables, but they may still differ ideologically …

The prospects for ideological variation in the face of ontological agreement owe to the amorphous connection between these two aspects of theories. Quine (*Ibid.*) notes that the "ontology of a theory stands in no simple correspondence to its ideology," but adds that ideological matters arguably constrain ontological commitments in virtue of the kinds of admissible ideas or predicates it permits.[5] Quine (*Ibid.*) says:

> The ideology of a theory is a question of what the symbols mean; the ontology of a theory is a question of what the assertions say or imply that there is. The ontology of a theory may indeed be considered to be implicit in its ideology; for the question of the range of the variables of quantification may be viewed as a question of the full meaning of the quantifiers.

Quine's distinction between ontology and ideology is of particular significance for understanding his account of ontological commitment, but his subsequent remarks make clear either an inability or unwillingness to hammer the notion of ideology into clearer or more tractable terms. Quine (1951: 15) says:

> I have described the ideology of a theory vaguely as asking what *ideas* are expressible in the language of the theory. Urgent questions of detail then arise over how to construe 'idea.' Perhaps, for what is important in ideological investigations, the notion of ideas as some sort of mental entities can be circumvented … Now the question of the ontology of a theory is a question purely of a theory of reference. The question of the ideology of a theory, on the other hand, obviously tends to fall within the theory of meaning; and, insofar, it is heir to the miserable conditions, the virtual lack of scientific conceptualization, which characterize the theory of meaning.

Some years later, Quine abandoned the invocation of "ideas" in characterizing ideology and suggests that, alongside ontology, "what may be called ideology [is] the question of admissible predicates" (Quine 1957: 17.) In doing so, Quine moves the subject matter of ideological comparison away from what seem to be shadowy mental entities to the more familiar realm of theoretical vocabulary. The resulting strategy for evaluating ideology continues even during Quine's later flirtations with Pythagoreanism in Quine (1976). In mapping out the prospects for a pythagorean ontology comprising only sets but replete with a rich ideology, Quine simply identifies ideological commitment with a theory's proprietary predicates and functors:

> We must note further this triumph of hyper-Pythagoreanism has to do with the values of the variables of quantification, and not with what we say about them. It has to do with ontology and not with ideology. The things that a theory deems there to be are the values of the theory's variables, and it is these that have been resolving themselves into numbers and kindred objects – ultimately into pure sets. The ontology of our systems of the world reduces thus to the ontology of set theory, but our system of the world does not reduce to set theory; for our lexicon of predicates and functors still stands stubbornly apart.
>
> *(Quine 1976: 503)*

Quine's insights into theories extend to recognizing the need for something beyond his conception of ontology and to tethering this additional category to what is expressible within a theory. Understandably, for Quine, what is expressible proves to be a matter of roughly which predicates are admitted within a theory. And, while this notion is useful to facilitate the rough evaluation of theories, it falls short of a viable metaphysics of ideology in a number of ways. Not only is it unclear whether ideology is inherently tied to the syntactic types Quine points to (i.e., predicates and functors), there is a clear temptation in Quine to tie ideological commitments to linguistic items. There is, however, nothing metaphysically special about uninterpreted word types or linguistic items, so, when we find Lewis picking up Quine's ontology-ideology distinction, he often speaks loosely (or perhaps equivocally) of linguistic items like operators and predicates, "notions," "concepts," or simply of "primitives." And, while there seems to be a variety of ways to regiment ideology within a broader metaphysical picture, Lewis, like Quine, offers us rather little to go on.

Sider's contribution

Sider (2011) stands out as the leading effort in articulating a full-fledged metaphysics of ideology. Where Lewis' methodological commitments require an implicit commitment to ideological realism, Sider (2011: 13) is explicit on this point:

> The term 'ideology,' in its present sense, comes from Quine. It is a bad word for a great concept. It misleadingly suggests that ideology is about ideas – about *us*. This in turn obscures the fact that the confirmation of a theory confirms its ideological choices and hence supports beliefs about structure. A theory's ideology is as much a part of its worldly content as its ontology.

For Sider, ideological commitments "are as much commitments to metaphysics as are ontological commitments" (Sider 2011: 230.) Moreover, Sider's variety of ideological realism is one that extends the realm of ideological commitment beyond predicates to operators, quantifiers, and an unspecified variety of other syntactic categories. Sider defends this expanded conception of ideology by drawing upon Lewis' influential view that certain properties are metaphysically privileged or perfectly natural in virtue of inducing objective resemblance, figuring into natural laws, and occurring in analyses of intrinsicality and other notions.[6] On Lewis' view, there is an objectively privileged structure to reality and the predicates deployed in our theories can do a better or worse job of capturing it. For example, a theory that invokes a predicate 'is a shmelectron' that applies to electrons and shuttlecocks is worse than one that invokes a predicate 'is an electron' that applies to all and only electrons. For, while either theory might (with some encumbrances) be put to work, the former is inferior for its lack of perspicuity or, as it is often put, for failing to carve reality at its (objectively distinguished) joints. A chief aim in introducing our theoretical vocabulary – in this case, predicates – is to accord with reality's underlying structure. As Sider (2011: 12) puts it:

> A good theory isn't merely likely to be *true*. Its ideology is also likely to carve at the joints. For the conceptual decisions made in adopting that theory – and not just the theory's ontology – were vindicated; those conceptual decisions also took part in a theoretical success, and also inherit a borrowed luster. So we can add to the Quinean advice: regard the ideology of your best theory as carving at the joints. We have defeasible reason to believe that the conceptual decisions of successful theories correspond to something real: reality's structure.

Sider argues that objective metaphysical structure is not exhausted by the kind of structure which predicates do a better or worse job of carving at reality's joints. Rather, Sider (2009: 404) says "[w]e should extend the idea of structure beyond predicates, to expressions of other grammatical categories, including logical expressions like quantifiers. (Interpreted) logical expressions can be evaluated for how well they mirror the logical structure of the world."

If Sider is correct, our pursuit of concordance between theory and reality requires, not merely an apt choice of predicates, but also the judicious choice of quantifiers, operators, and perhaps other sorts of theoretical vocabulary. And, in each of these domains, we are faced with a choice that runs parallel to our above decision between introducing predicates 'electron' or predicates like 'shmelectron.' We can, according to Sider, obscure or illuminate metaphysical structure by making errant or apt conceptual choices. This raises the exceptionally difficult question of what, if any, quantifiers might "carve at reality's joints." But it also provides a backdrop to frame the earlier debate between the modalist and eliminativist about modality: at bottom, this disagreement concerns whether the world has any objective modal structure and, in turn, whether the box and diamond aptly render this structure.

Sider's conception of metaphysical inquiry increases the breadth of ideological inquiry and compounds its theoretical importance. It also forecloses any hope of extracting ideological commitments simply by scrutinizing a theory's stock of primitive predicates. For, if Sider is correct, our ideological choices are reflected across the diversity of grammatical categories we deploy in our theorizing. Notice, also, that if the kinds of operators and quantifiers we introduce constitute substantial ideological decisions, we seem required to provide reasons in favor of their adoption and, in turn, against rival theoretical structures. Mapping out these rival views has required increased attention to heterodox metaphysical options, often with noteworthy ontological consequences. To take an extreme example, we might opt for metaphysical theory that dispenses with quantification altogether in favor of predicate functors. Such a view requires an expanded ideology, but, on a roughly Quinean conception of ontological commitment, dispenses with any ontological commitments whatsoever. (See Quine 1960; Dasgupta 2009; and Turner 2011.) While there is little to recommend such a radical view, utilizing ideological resources like predicate functors in more modest ways generates some promising alternatives.[7] At the same time, considerable methodological issues arise when trying to determine how to evaluate competing options that differ in increasingly radical respects. But, if Sider is right, the differences among these options are not merely notational or representational; they concern "worldly" structure and so ontology and ideology require equal attention. As Sider (2011: 14) puts it,

> We often face a choice between reducing our ontology at the cost of ideological complexity, or minimizing ideology at the cost of positing new entities. If ideology is psychologized, the trade-off is one of apples for oranges; whether to posit a more complex world or a more complex mode of expression. But on the present approach, both sides of the trade-off concern worldly complexity. A theory with a more complex ideology posits a fuller, more complex, world, a world with more structure. Thus ideological posits are no free lunch.

Open questions about ideology

Inquiry into the nature of the non-ontological commitments of theories is a metametaphysical project in its nascent stages. In fact, it remains an open question whether we are best served by using "ideology" as a catch-all label for whatever commitments are non-ontological or for some more narrow domain of theoretical aspects. There is also little consensus how, if at all, we might most usefully taxonomize ideological commitments or even explicate the notion of ideology except by appeal to a carefully tailored range of examples.[8] If, for example, a puzzled metaphysician asks for a general characterization of an "ideological disagreement," it is unclear whether an informative answer – one that illuminates the notion of ideology without simply invoking it – can be offered. Perhaps, then, the notion of ideological structure must itself be taken as primitive. (Whether this leaves it in better or worse shape than the notion of ontological structure is yet another open question.) Fortunately, our grasp on the notion of ideology seems to be improved through our examination of competing metaphysical options that differ in their deployment of ideological and ontological resources.

If a comprehensive account of ideology and its connection to metaphysical structure can be provided, the central challenge for ideological inquiry is likely to be a species of a more general challenge for metaphysics: accounting for the epistemic significance of theoretical virtues. The absence of a consensus rationale for this pervasive methodology is a bit of a scandal in its own right, but, for ideological realists like Lewis who rely upon theoretical virtues, the story is bound to be worryingly complex. This is because the diversity of ontology and ideology seems to rule out any straightforward account of how we ought to compare theoretical costs across the ontology-ideology divide. For example, should we quantify ideological commitments by grammatical category, conceptual kind, or via some other means? And, once we can count up ideological commitments, how do they compare with ontological ones? Is every primitive worth exactly one fundamental entity or twelve derivative entities or what? Absent a recipe for comparing ontology and ideology or calculating overall theoretical cost, it remains mysterious how we might make principled choices from among competing "package deals" of ontology and ideology. Articulating some principles for theory choice that attend to the ideology-ontology interface is therefore a central task for the broader challenge of understanding the role of theoretical virtues in metaphysics.

Notes

1 Nominalists deny the existence of abstract entities and therefore typically reject the existence of numbers. On nominalist options, see Hellman (1989), Burgess and Rosen (1997), and Cowling (2017).
2 On instantiation and various metaphysics options, see Strawson (1959: 167–173) and Loux (1998: 30–36).
3 On Bradley-style regresses, see MacBride (2011) and Maurin (2012). On ideological resolutions of the regress, see Armstrong (1978: 109–111), Nolan (2008), and Cowling (2017: 120–129).
4 On Quine's on ontology and ideology, see Burgess (2008) and van Inwagen (2008).
5 The connection between ideology and ontology proves yet more complicated in light of the interaction of ideology with identity and indiscernibility. For discussion, see Geach (1967) and Kraut (1980).
6 On Lewis' conception of naturalness and related theoretical options, see Lewis (1983). Sider (2013) demarcates objective structure via a "structure" operator that attaches to expressions of various grammatical types to single them out as structural.
7 A largely unexplored option is to more selectively apply the tools required by ontological nihilism – namely, a suitable powerful language of predicate functors – in order to replace certain bodies of quantified discourse – e.g., theories concerning merely possible, non-present, or abstract entities. For discussion of nominalist options of this kind, see Burgess and Rosen (1997: 185–188) and Cowling (2017: 238–242).
8 For discussion of ideological kinds and the individuation of ideological commitment, see Finocchiaro (forthcoming).

References

Armstrong, D.M. 1978. *Nominalism and Realism: Universals and Scientific Realism*. Cambridge: Cambridge University Press.
Bennett, Karen. 2009. "Composition, Colocation, and Metaontology" in David Chalmers et al. (eds.) *Metametaphysics*, Oxford: Oxford University Press: 38–76.
Bricker, Phillip. 2016. "Ontological Commitment" in Edward N. Zalta (ed.) *Stanford Encyclopedia of Philosophy*.
Burgess, John. 2008. "Cats, Dogs, and so on." *Oxford Studies in Metaphysics* 4: 56–78.
Burgess, John, and Rosen, Gideon. 1997. *A Subject with No Objects*. Oxford: Oxford University Press.
Cowling, Sam. 2013. "Ideological Parsimony." *Synthese* 190: 889–908.
Cowling, Sam. 2017. *Abstract Entities*. London: Routledge.
Dasgupta, Shamik. 2009. "Individuals: An Essay in Revisionary Metaphysics." *Philosophical Studies* 145: 35–67.
Divers, John. 2002. *Possible Worlds*. London: Routledge.
Finocchiaro, Peter. (forthcoming) "The Explosion of Being: Ideological Kinds in Theory Choice." *Philosophical Quarterly*.
Forbes, Graeme. 1989. *Languages of Possibility*. Oxford: Blackwell.
Geach, Peter. 1967. "Identity." *Review of Metaphysics* 21: 3–12.
Hellman, Geoffrey. 1989. *Mathematics without Numbers*. Oxford: Oxford University Press.
Kment, Boris. 2014. *Modality and Explanatory Reasoning*. Oxford: Oxford University Press.
Kraut, Richard. 1980. "Indiscernibility and Ontology." *Synthese* 44: 113–135.
Lewis, David. 1983. "New Work for a Theory of Universals." *Australasian Journal of Philosophy* 61: 343–377.
Lewis, David. 1986. *On the Plurality of Worlds*. Oxford: Blackwell.
Loux, M. 1998. *Metaphysics: A Contemporary Introduction*. London: Routledge.
MacBride, Fraser. 2011. "Relations and Truthmaking." *Proceedings of the Aristotelian Society* 111: 161–179.
Maurin, Anna-Sofia. 2012. "Bradley's Regress." *Philosophy Compass* 7.11: 794–807.
Melia, Joseph. 1992. "Against Modalism." *Philosophical Studies* 68: 35–56.
Melia, Joseph. 2008. "A World of Concrete Particulars." *Oxford Studies in Metaphysics* 4: 99–124.
Nolan, Daniel. 2008. "Truthmakers and Predication." *Oxford Studies in Metaphysics* 4: 171–192.
Quine, W.V. 1948. "On What There Is." *Review of Metaphysics* 2: 21–38.
Quine, W.V. 1951. "Ontology and Ideology." *Philosophical Studies* 2: 11–15.
Quine, W.V. 1957. "The Scope and Language of Science." *British Journal for the Philosophy of Science* 8.29: 1–17.
Quine, W.V. 1960. "Variables Explained Away." *Proceedings of the American Philosophical Society* 104.3: 343–347.
Quine, W.V. 1976. "Whither Physical Object?" in R.S. Cohen et al. (eds.) *Essays in Memory of Imre Lakatos*, Dordrecht: Reidel: 497–504.
Rodriguez-Pereyra, Gonzalo. 2002. *Resemblance Nominalism*. Oxford: Oxford University Press.
Sider, Theodore. 2003. "Reductive Theories of Modality" in M.J. Loux and D.W. Zimmerman (eds.) *The Oxford Handbook of Metaphysics*, Oxford: Oxford University Press: 180–208.
Sider, Theodore. 2009. "Ontological Realism" in David Chalmers et al. (eds.) *Metametaphysics*, Oxford: Oxford University Press: 384–423.
Sider, Theodore. 2011. *Writing the Book of the World*. Oxford: Oxford University Press.
Sider, Theodore. 2013. "Against Parthood." *Oxford Studies in Metaphysics* 8: 237–293.
Strawson, P.F. 1959. *Individuals: An Essay in Descriptive Metaphysics*. London: Methuen.
Turner, Jason. 2011. "Ontological Nihilism." *Oxford Studies in Metaphysics* 8: 3–54.
Van Cleve, James. 2016. "Objectivity without Objects: a Priorian Program." *Synthese* 193: 3535–3549.
van Inwagen, Peter. 2008. "Quine's 1946 Lecture on Nominalism." *Oxford Studies in Metaphysics* 8: 125–143.

30
PRIMITIVES

Jiri Benovsky

Theories and their primitives – some examples

Generally speaking, primitives are often understood as axioms of a given theory, and we encounter them in any field in any science. In short, axioms are postulates that a theory takes to be true without argument – they are starting points for arguments instead of being results of argumentative work. We encounter axioms as explicitly stated in mathematics and logic, but any theory about anything always involves – explicitly or implicitly – basic postulates which serve as the foundation for the work that the theory is doing. In this chapter, the main point of interest will concern metaphysics and the role primitives play in metaphysical theories. Indeed, metaphysical theories are no exception and they all contain primitives. Furthermore, their primitives play a central role in the way the theories work – as we shall see, primitives actually do the most important part of the job. In this chapter, I am going to try to better understand the nature of primitives and the role they play in metaphysical theories. The best way to start is to have some clear examples in mind, so let us begin by comparing some traditional well-known metaphysical theories about the nature of material objects and about the way objects have their properties, and see what kind of primitives they contain and how these primitives behave.

A first such theory is the Armstrongian view (see Armstrong 1978) according to which material objects are bare particulars (substrata) that instantiate spatio-temporal multiply locatable (repeatable) universals. In this view, two objects *a* and *b* are both F ('they share the same property') iff *a* and *b* both instantiate the numerically same universal F-ness. An alternative solution to this traditional 'problem of attribute agreement' is the bundle theory with tropes (see Williams 1953), according to which objects are bundles of compresent non-repeatable, non-multiply locatable properties: two objects *a* and *b* are both F ('they share the same property') iff *a* and *b* both have among their constituents numerically different but exactly similar F-tropes. Yet another alternative view, namely resemblance nominalism (see Rodriguez-Pereyra 2002), claims that objects are basic – here, the objects themselves are not further analysed, rather they are taken as being primitive. In this view, *a* and *b* are both F ('they share the same property') iff they are both members of the same resemblance class.

These three views have this in common: when it comes to the question about how two objects can 'share the same property', they all answer the question in a primitive way. The instantiation of the same universal, the relation of exact resemblance between tropes, or the fact that *a* and *b* resemble each other are primitives postulated by the theories that use them, and these primitives are the tools that provide an answer to the problem of attribute agreement. If one asks in virtue of what *a* and *b* are both F (in virtue of what 'they share the same property'), one theory answers by saying that it is in virtue of instantiating the same universal – and if one asks further in virtue of what two instances of F-ness are instances of the same universal, then the answer is that this is a primitive. Similarly for trope theory: *a* and *b* are both F in virtue of containing exactly similar tropes, and tropes are exactly similar in a primitive way. Resemblance nominalism also answers the question in a primitive way since 'being in the same resemblance class', or for *a* and *b* to 'resemble each other' is primitive.

We thus realize that the main job – the very reason why these theories are there in the first place – is done by the theories' primitives. This is really not very surprising, and even less is it a problem. Indeed it's entirely adequate to introduce primitives that do heavy-duty jobs, for otherwise there would be little justification to introduce them in the first place! Why would we need to postulate the existence of an entity such as a non-relational instantiation tie (to take only one example) if it weren't for some important theoretical job to be done? Primitives are acceptable – and welcome – in metaphysical theories precisely because they do an important job.

This situation is not specific to the theories I just quickly mentioned, indeed, it is commonplace in metaphysics. The bundle theory, for instance, can come in a version with tropes, as we have just seen, but it can also be combined with the view that properties are immanent universals. This version of the bundle theory is often rejected because of worries concerning the principle of Identity of Indiscernibles. In short, here is the objection: in Max Black's (1952) possible world containing nothing but two homogeneous spheres that are perfect qualitative duplicates, how can this version of the bundle theory explain the numerical diversity of the two spheres? Since the spheres are perfect qualitative duplicates, they are bundles of the very same universals, and so they turn out to be *one and the same bundle*, instead of *two*. Thus, this version of the bundle theory is committed to endorse the principle of Identity of Indiscernibles which says that if two objects have all exactly the same properties they are identical – 'they' are one object – since if the 'two' objects are bundles of properties and if the properties are numerically identical (i.e. they are universals), the two bundles are numerically identical as well. But, the objection goes, given Black's world, this principle is false, and so the bundle theory (with universals) is false. Of course, there are several ways to defend this version of the bundle theory here, for instance one can reject Black's world as being a genuine possibility, or one can try to account for the numerical diversity between the two spheres by appealing to haecceistic properties or to location properties. But this is not the point of interest for us now. We are interested in the way the primitives do their job here. To see this, consider how the competition works: the main rival view in this debate is the 'bare particulars' substratum theory – let us ask: how does *this* view account for numerical diversity of the two spheres in Black's world? Easily: the substratum theory uses a primitive to do that, namely, the numerical diversity of the substrata. In this view, a substratum, or a 'bare particular', is the entity that instantiates the properties of an object. In the case of the two indiscernible spheres, if the substratum theory is combined with the view that properties are universals, the spheres can neither be qualitatively distinguished nor numerically distinguished by their properties as it would be the case if these were tropes. But the substratum theory can solve the problem with its primitive substratum: substrata are primitively responsible for the particularity of particulars, and they are also responsible for primitive numerical difference between them – indeed, it just is a primitive postulate of the substratum theory that two substrata are numerically different even if they are not qualitatively different (and the two spheres are then numerically different in virtue of the numerical difference between their substrata).

"Problem-solvers"

Such primitives are what I call *problem-solvers*. A problem-solver is a primitive that is there to solve a problem. The theories we have seen above all answer the question of attribute agreement (i.e. the question about how two objects can 'share the same property') by appealing to their primitives: the relation of exact resemblance between tropes, the instantiation of the same universal, or the fact that *a* and *b* resemble each other. In the same crucial places, all three views introduce a primitive with the same function: *primitively answer the question* ("In virtue of what are *a* and *b* both F?"). Problem solved. With a problem-solver. Problem-solvers are commonplace in metaphysics, and in philosophy in general; without them we would not get very far. Primitive problem-solvers are the pillars that sustain the structures of our theories.

If, then, this is how the substratum theory explains numerical diversity and the particularity of particulars, the bundle theory (with universals) can do the same. According to the bundle theory, there are no substrata, rather there are only properties tied together by a special variably polyadic relation often called 'compresence'. Here, compresence, instead of a substratum, plays then the functional role of particularizing particulars: it is a 'unifying device' (like the substratum, functionally speaking) whose role in the theory is to take properties to make up objects. Concerning the problem with the two indiscernible spheres, the relation of compresence does not contribute to the *qualitative* nature of objects and so it can very well be a numerically different relation in different objects without spoiling the two objects' qualitative identity. It can thus account for *numerical* diversity of the two *qualitatively* identical spheres – in this way, the two spheres are numerically different in virtue of the numerical difference between the (instances of) relations of compresence that tie together their properties. Similarly to the substratum, the relation of compresence is then not only a primitive problem-solver when it comes to the problem of particularity of particulars, but it can also very well primitively solve the problem with qualitatively indiscernible but numerically distinct objects, simply by postulating, exactly as the substratum theorist does it with *her* problem-solver, that this primitive unifying device is such that different objects are bundled together by qualitatively indistinguishable but primitively numerically distinct (instances of) relations of compresence. This does not make the bundle theorist's answer to the objection unpalatably ad hoc. Indeed, the bundle theorist is allowed to make such a move, since this is exactly what her opponents do: she has at her disposal a primitive problem-solver, that she introduced herself in her theory because she needed to perform a theoretical job, and she can postulate it to be as she needs it to be, and to be able to do whatever she needs to be done. Both the bundle theory and the substratum theory contain a *primitive unifying device* to make up objects, and if one side claims her device to have the additional primitive capacity to be numerically distinct albeit qualitatively identical in different objects, there is no reason why the other side could not claim the same thing as well.

Let us now turn our attention to the primitives themselves. They seem to be tools, problem-solvers, that do a theoretical job, and that's what defines them – indeed, they are individuated by their functional role. In the situation described above (the Black's world scenario and the particularization of particulars), their job is the same and they do it in the same way. It appears then that they just turn out to be one and the same thing under different names. One can ask: what difference does it make to call a primitive problem-solver a "substratum" rather than "compresence" if the job it does is the same? The difference only seems to be terminological. If we were to stick to a neutral vocabulary, like "unifying device", and if we reformulate the two views and how they behave with respect to the problem of Identity of Indiscernibles, we get a situation where both theories will be able to claim that the two spheres can be distinguished in virtue of there being a primitively distinguished unifying device for each sphere. What we realize here is that not only are primitives like pillars that support the weight of our theories, by doing most of the work, but that they also are 'points of contact' between the theories. Keep in mind the three answers above to the problem of attribute agreement ('sharing the same property'): the three theories all contain a primitive problem-solver at a crucial place in the theory that allows them to solve the problem. Their primitives are, of course, clearly different, but they have the same overall function within a given theory, and the role to explain how *a* and *b* can share the same property. In a very general sense, all three views do answer the question *in the same way* – a primitive way. One can then raise the meta-metaphysical and methodological question: when it comes to the problem of attribute agreement, what difference does it really make to pick one theory rather than another? The current debate in metametaphysics provides a lot of discussion in the neighbourhood of this question, even if the discussion typically does not focus on primitives. For instance, Hirsch (2005, 2007, 2008) defends a claim of equivalence between endurantism and perdurantism; in his view, these debates are merely verbal disputes. They seem to say different things but they are in fact making the same claims, merely formulated in different ways. Another example is Bennett (2008) who focuses on theories of composition and argues that there is little reason to embrace one side of the debate rather than the other, even if they are *not* just terminological variants, for, in her view, it is epistemically under-determined which one we should choose. To come back to our examples above, in Benovsky (2008), I argue for an equivalence claim between (various versions of) the bundle theory and the substratum theory. In my view, the question that arises is: at the end of the day, if competing theories answer their theoretical challenges by appealing to their primitives, what difference does it make to embrace one theory rather than another? To push the point even further, one can ask: is there any non-verbal difference between the theories at all? If not, are these theories then somehow equivalent? If one suspects that this might be so, in what sense can one speak here of equivalence? Metaphysical equivalence? Theoretical equivalence? Epistemic equivalence? To begin to answer these questions, let me try to say more about the nature of primitives.

The functional view

In my way of talking about primitives above, I have taken on board what I will refer to as the 'functional view' of primitives. This is the idea of a problem-solver, as a tool that is introduced in a theory to do a job – its main reason to be there is its function. Indeed, primitives are typically introduced in a theory by a philosopher who needs them to perform a theoretical job. When introducing them, the author of the theory describes what they *are* by saying what they *do* – for instance, when introducing in his view the notion of primitive instantiation, Armstrong described it as a tie that has the capacity to relate universals with substrata in a non-relational way. This says almost nothing about the *nature* of instantiation, but it clearly stipulates what its *function* is. With a good reason: it's what it *does* that counts – it's the very reason for it being in the theory in the first place. One can then think that the primitive *is what it does* – this is the view that primitives are individuated by their functional role, the 'functional view'. And this is where claims of metaphysical equivalence can be made – indeed, if two primitives do the same theoretical job, and if they are individuated by their functional role, they are then equivalent not just for all theoretical purposes but metaphysically equivalent as well. At the end of the day, they thus turn out to *be* just one and the same thing, only called different names.

The alternative view to the functional view of primitives can be called 'the content view'. This view embraces the idea that primitives not only have a function in a theory but that they (also) have a *nature* – they not only have a function, but they also have a content. The idea is that primitives may very well *do* the same thing, but that they *are not* the same thing. In this view, then, even if two sides of a debate use problem-solvers that are functionally equivalent, they still have a different nature/content and so they are not *metaphysically* equivalent even if they are functionally/theoretically equivalent. In our example above, the idea here is that compresence is a relation and a substratum is not a relation, so they are not the same thing, even if their functions are so similar (or even entirely the same).

The functional view seems to behave better than the content view. Consider the idea that the problem-solver used by one theory is a substratum, while the problem-solver of another theory is a relation, and so they have a different nature. What is it that this idea captures? Does it say something about the substratum or the relation of compresence, like for instance the often-cited difference that properties cannot 'float free' while a substratum can 'stand alone' and support properties? But whatever this alleged difference between the two primitives means, it would be a functional difference – a difference in what a substratum can do, while compresence doesn't. So, here again, the functional view seems to be the adequate story. Here, we would have a difference between the two primitives, and so they would not be equivalent, but they would not be equivalent for functional reasons. To repeat: this is no surprise at all – primitives being there to do a job, it's only natural that they are individuated in terms of their functions. For this reason, any difference between primitives will always be a functional one – their functions are the very reason for introducing them in the first place. In the content view, we would be forced to say that in addition to their functions primitives have a non-functional content, but this would objectionably mean that there is *a difference that makes no difference* – a somewhat prejudiced attitude towards one of the primitives and against the other, where one would stick too heavily to the terminology one uses, namely to words like 'substratum' or 'relation'.

To sum up the suggestions I have been making above: primitives are problem-solvers, that are central to any theory, and that do most, if not all, of the theoretical job. They are also 'points of contact' between theories, and they are individuated by their functional role. In this sense, they can lie at the heart of a possible equivalence claim between theories. (My equivalence claim between the bundle theory and the substratum theory, that I quickly hinted at above and that I defend in detail in Benovsky 2008, is construed in this way).

On explanatory power

We have seen above that most of the work of metaphysical theories is often done by their primitives. Does this mean, then, that the explanatory power of our theories is just primitively postulated? If so, one can raise a sceptical question about how a primitive can really explain anything, and consequently about how our theories can explain anything, since they answer the questions we ask by primitively postulated problem-solvers. In this last section, I am going to try to say more about what an explanation is and about where explanatory power comes from.

Many explanations involve causality. If, for instance, we ask "Why did an avalanche occur this morning on the north face of Mont Blanc?", we get an answer that exhibits a complex causal chain of events including snowfall, wind, and perhaps unwise skiers. Of course, this is not the type of explanation we are looking for when we ask how and what a theory and/or a primitive explains anything – the explanation here is not causal. Perhaps this other kind of explanation is closer to what we need here:

> Why does lightning occur just when there is an electric discharge between clouds or between clouds and the ground? Because lightning simply *is* an electric discharge involving clouds and the ground. There is here only one phenomenon, not two that

are correlated with each other; and what we thought were distinct correlated phenomena run out to be one and the same. Here the apparent correlation is understood as *identity*.

(Kim 2006, p. 85)

In this example, the relation between the explanadum and the explanans is simply identity. We have a similar situation in the case of our example concerning attribute-agreement: for instance, we can say that the explanandum is the sharing of the same property, while the explanans is the instantiating of the same immanent universal. In this case, we can say that sharing the same property *just is* instantiating the same universal, exactly as lightning *just is* atmospheric electric discharge. To have another example, according to the substratum theory, numerical diversity of the two spheres in Black's world *just is* or *consists in* their having a numerically different substratum. So, can we say that the relation between a primitive problem-solver and the phenomenon it explains is identity? There are two problems with this view.

First, as Ruben (1990, p. 219) claims, not all identities are explanatory. The identity claim "lightning is lightning" is not explanatory, while the identity claim "lightning is atmospheric electric discharge" is, because in the second case, even if there is only one phenomenon involved, it is conceptualized in two different ways. We see here that explanation is (and identity isn't) an irreflexive relation.

Second, perhaps even the claim "lightning is atmospheric electric discharge" is *not* explanatory. This is the sceptical challenge: if the relation between 'sharing the same property' and 'instantiating the same universal' is identity, how does this *explain* anything? We encounter this worry in many places; for instance when discussing the psychoneural identity theory, Kim (2006, pp. 97–98) says:

> Our conclusion, therefore, has to be that both forms of the explanatory argument are open to serious difficulties. Their fundamental weakness lies in a problematic understanding of the role of identities in explanation, an important topic that has not received much attention in the literature. ... We do not have to say that identities have no role to play *in* explanations. For they can help *justify* explanatory claims – the claim that we have explained something. ... It is only that identities do not generate explanations on their own.

Kim's worry seems to be close to the one I raised above: since a primitive problem-solver (the explanans) is actually the very same thing as the phenomenon we wanted to understand (the explanandum), it is not clear that we have really gained anything by such an explanation – i.e. by providing an explanation of what we wanted to understand in terms of a primitive that's actually the same thing as what we wanted to have a better understanding of.

What we learn here is that in the cases we are interested in the relation of explanation is close to identity but it is not identity. Not only is it irreflexive, it is also asymmetrical. Granted, lightning *is* atmospheric electric discharge, and the phenomenon of sharing the same property *is* (say) the phenomenon of instantiating the same universal, but the explanation does not consist just in pointing to this fact, it also points to the fact that the explanans is *more fundamental* than the explanandum, and this is what we get, this is what we learn from a good explanation. In an explanation of the kind we are interested in here, one of the two sides of the explanation relation is more fundamental than the other. One could perhaps understand the situation here as involving the growingly familiar notion of grounding: as Bricker (2006), DeRosset (2010), and Schaffer (2009) claim, if a is grounded in b, a is nothing over and above b. a, in other words, is an "ontological free lunch" in Armstrong's (1997) sense; the "ontological price" you pay for *a* and *b* is just whatever you would pay for *b* alone. Only in this particular sense can one talk about identity between *a* and *b*. This kind of relation is found in many cases of metaphysical theories. For instance, according to a version of perdurantism, *a*'s persisting through time *just is a*'s having temporal parts at different times, where the latter is taken to be a more fundamental phenomenon than the former. Quite often, the terminology that is used in such cases appeals to the locution "in virtue of": *a* persists through time *in virtue of* having temporal parts at different times.

If we took this idea involving the notion of grounding on board, we would then be in a situation where we step on a primitive in our effort of understanding the nature of explanation – indeed, the notion of grounding is typically taken as being primitive itself (as well as other related notions such as the notion of 'being nothing over and above', the notion of an 'ontological free lunch' and similar). In the case of lightning and the explanation in terms of atmospheric electric discharge, the chain of explanation goes on until the most fundamental level is reached – the most fundamental level being largely dependent here on empirical matters. The situation is different in the case of explanations found in metaphysical theories involving problem-solvers – we are not limited here by empirical matters. Rather we reach the bottom of our metaphysical inquiry when we arrive at a notion that is unanalysable any further without circularity. The examples we have seen include a substratum, a non-relational instantiation, resemblance, and others. These notions are taken by our theories to be too fundamental to be further explained.

In this sense, what is at stake in metaphysical inquiry is to find out which is the best primitive, and what is more fundamental than what; what explains what. The metaphysician's work can thus be best understood as work on primitives – the pillars that sustain the structure of our metaphysical theories and that do the central part of the job. Explanatory power is often the main criterion that we use to evaluate our theories. This is only natural since the very reason to build a metaphysical theory in the first place is to provide an explanation of phenomena that we want to better understand, such as the particularity of particulars, the sharing the same property, persistence through time, and so on. The picture we get here of what metaphysics does is, close to what Schaffer (2009) argues for, not to tell us *what there is* but to tell us "what grounds what" – as I suggest to formulate it, to discover what are the most fundamental notions, which are primitive and which are not. In this view, the task of metaphysics is not to provide us with a *list*, a kind of inventory of what there is; rather, the point is to provide a top-bottom structure of relations of 'grounding' or 'explanation' and to say which concepts and entities are primitive, which are fundamental and which are not.[1]

Note

1 I would like to thank Baptiste Le Bihan for his useful comments that helped me to improve this chapter.

References

Armstrong, D. M. 1978. *Nominalism and Realism*. Cambridge University Press.
Armstrong, D. M. 1997. *A World of States of Affairs*. Cambridge University Press.
Bennett, K. 2008. Composition, Colocation, and Metaontology. In D. J. Chalmers, D. Manley, and R. Wasserman, eds., *Metametaphysics: New Essays on the Foundations of Ontology*, Oxford University Press, pp. 38–76.
Benovsky, J. 2008. The Bundle Theory and the Substratum Theory: Deadly Enemies or Twin Brothers? *Philosophical Studies* 141:175–190.

Black, M. 1952. The Identity of Indiscernibles. *Mind* 61:153–164.
Bricker, P. 2006. The Relation Between the General and the Particular: Entailment vs. Supervenience. In Dean Zimmerman, ed., *Oxford Studies in Metaphysics*, Vol. 2, Oxford University Press, pp. 251–287.
De Rosset, L. 2010. Getting Priority Straight. *Philosophical Studies* 149 (1):73–97.
Hirsch, E. 2005. Physical-Object Ontology, Verbal Disputes, and Common Sense. *Philosophy and Phenomenological Research* 70:67–97.
Hirsch, E. 2007. Ontological Arguments: Interpretive Charity and Quantifier Variance. In John Hawthorne, Theodore Sider, and Dean Zimmerman, eds., *Contemporary Debates in Metaphysics*, Blackwell, pp. 367–381.
Hirsch, E. 2008. Ontology and Alternative Languages. In D. J. Chalmers, D. Manley, and R. Wasserman, eds., *Metametaphysics: New Essays on the Foundations of Ontology*. Oxford University Press, pp. 231–258.
Kim, J. 2006. *Philosophy of Mind*. Second Edition. Westview Press.
Rodriguez-Pereyra, G. 2002. *Resemblance Nominalism, A Solution to the Problem of Universals*. Oxford University Press.
Ruben, D. 1990. *Explaining Explanation*. Routledge.
Schaffer, J. 2009. On What Grounds What. In D. J. Chalmers, D. Manley, and R. Wasserman, eds., *Metametaphysics: New Essays on the Foundations of Ontology*. Oxford University Press, pp. 347–383.
Williams, D. C. 1953. On the Elements of Being. *Rev. Metaphysics* 7:171–192.

31
CONCEPTUAL ANALYSIS IN METAPHYSICS

Frank Jackson

1 Introduction

What role does conceptual analysis play in metaphysics? No easy question to answer in the absence of accounts of metaphysics and conceptual analysis. And there's the rub. Accounts of metaphysics and, especially, of conceptual analysis are inevitably controversial. Skirting controversy is not an option. I start by explaining how I will understand metaphysics and conceptual analysis in this chapter.

2 What is metaphysics?

By metaphysics, I will mean the study of what there is and what it is like. What this comes to is best grasped via examples. For one example, objects change over time: a tree gets taller; a snooker ball moves, i.e. changes its position over time. But what, exactly, is change over time? We see it and measure it, but what are we seeing and measuring? One answer many like is: change over time is a matter of a single thing enduring through time and having different properties at different times. When a tree gets taller, the *very same thing* has one height early on and a greater height later on. When the ball moves, the very same thing, the ball, is at different places at different times. But there is a different answer that also has many supporters. It holds that objects have temporal parts, that they are extended in time as well as in space. On this view, change over time is a matter of an object's temporal parts having different properties. When a tree gets taller, later temporal parts of the tree are taller than earlier temporal parts of the tree. When a snooker ball moves, it is because different temporal parts of the ball are at different locations.[1]

Here's a second example. We have beliefs. I here and now believe that I am in front of a laptop; many people believe that Federer is the GOAT. What kind of state is belief, and how is it related to the brain states which science has told us are necessary conditions for believing? Some insist that the state of belief is literally identical with some brain state or other, playing a certain kind of functional role.[2] This is a (debated) view about the metaphysics of belief.

My final example is existence. Some things exist, our earth for example; and some things do not exist, Vulcan, for example. Does this mean that existence is a property that some things have and some things do not have?[3] This is a dispute in the branch of metaphysics known as ontology.

3 What is conceptual analysis?

How does conceptual analysis impact on the kinds of questions we have just been talking about? On the face of it, the question about the nature of change discussed above would seem to be one settled by empirical investigation. Indeed, many who favour the view that we should understand change in terms of different temporal parts of an object having different properties do so because they think that relativity theory in physics tells us that this is the correct account.[4] The answer is that, on one understanding of conceptual analysis – the understanding that informs this chapter – conceptual analysis is in the business of clarifying and explicating what it takes, or if you like, how something has to be, to fall under some concept or other. If we have a word for the concept in question, a conceptual analysis will tell us what it takes to be correctly described by that word.

What do I mean by 'what it takes' and 'how something has to be' to fall under some concept or other? The example of change tells us. The dispute over change is in part a dispute over the concept of change. One party holds that what it takes for object, O, to change – how O has to be to change – is for O to be the very same thing with different properties at different times. The other party holds that what it takes is for O to have different temporal parts with different properties.

What should those who hold that what it takes to change is to be the very same thing with different properties at different times say about the claim that physics favours the account of change in terms of temporal parts? They have four ways to go. The first is to urge that the physics has been misunderstood. Although one understanding of relativity theory thinks of our universe as a huge thing extended through Space-Time and the objects in it as having temporal parts (Smart 1963, ch. 7 and Nerlich 1979), other understandings are possible. The second is to grant that physics has indeed shown that it is never the case that the very same thing has different properties at different times, that what happens is always that different temporal parts of things have different properties. It follows that, *strictly speaking*, nothing ever changes. Surprising no doubt, but why shouldn't physics surprise us? According to this second way to go, I went wrong when I introduced the example of change to illustrate what metaphysics is concerned with. I presupposed that things change and proceeded to contrast two views about its nature. My presupposition was mistaken. The third way to go is to insist that it is incontrovertible that things do sometimes change and that the concept of change is incompatible with relativity theory in physics. The upshot is that we should reject relativity theory. In the early days of relativity theory responses of this kind were not unknown – some did seek to refute relativity theory by reflecting on our concepts of time and change – but I think it is fair to say that almost no-one nowadays goes down this path. The fourth way to go is to urge that we should modify our concept of change. Yes, physics has shown that nothing changes according to our extant concept of change, but the sensible response is to fashion a new concept. Those who go down this path may well describe their opponents as being wrong about the nature of change, but right about the nature of what we should put in its place and about the nature of what happens in our world which we have been calling 'change' – wrongly according to the old concept but rightly according to the replacement concept.

A version of the fourth kind of response arguably happened when the atomic theory of matter came along. Many responded to the atomic theory by saying that we have discovered that nothing is solid. For we have, they said, discovered that what we thought were solid objects are in reality very 'gappy' arrays of atoms. However, they went on to note, the objects we thought were solid resist intrusion on the space they occupy – that's why we can safely sit on some of them – so we should change our concept of solidity to one that counts something as solid when it resists intrusion.

4 How much does conceptual analysis matter to metaphysics?

What counts as change is an interesting question. So is, Does what counts as change mean that physics teaches us that nothing changes? All the same, one might well think that they are interesting but not central to metaphysics. Surely one could simply distinguish two concepts of change, tag them change$_1$ and change$_2$, and leave it to science to rule on which concept is exemplified given the way things actually are?

That would be too quick. Suppose it turns out that the only sense in which change occurs (be it right to call it 'change' or not, be it change according to our concept or not) is the sense on which it is to have different temporal parts with different properties. This has implications for punishing people for past wrongs. It means that the thing you punish is, strictly speaking, a different thing from the thing that acted wrongly: the temporal part that gets punished is a different, later temporal part from the earlier temporal part that acted wrongly. The implications of this point for the justification of punishment are far from trivial.

However, this much is true. There is a sense in which the debate over how to analyse change (in the sense of what it takes to count as change) does not affect the metaphysical question of most interest. The debate bears on which kinds of happenings are rightly called 'change', which kinds do and which do not fall under the concept of change, or maybe on which do and which do not fall under an old concept of change and whether it would be sensible to adopt a new concept. But, surely, the issue of most interest *for metaphysics* is the nature of what kinds of events, objects, processes, etc. are to be found in our world, not whether they do or do not fall under this or that concept, or under some sensible revision of this or that concept.

This may well be the right attitude to take to the debate over the concept of change – and the debate over the concept of solidity we mentioned briefly in passing – but there are cases where conceptual analysis bears directly on important issues in metaphysics as such. One kind of case is where a conceptual analysis opens our eyes to a property that we would not otherwise have known about and whose coherence we might well have doubted. I give examples in the next section, section 5. A second kind of case is where a conceptual analysis, or sometimes simply a plausible claim about the nature of a concept, tells us about interconnections between theses in metaphysics in a way that helps us make important choices about which of those theses we should accept, and sometimes bears on what is required to resist scepticism about one or another property. I give examples in section 6.

5 Conceptual analysis and the discovery of properties

Perhaps the simplest example of the first kind of case is the discovery that there are higher orders of infinity.

We distinguish finite sets from infinite sets. There are finitely many people over 1.8 metres in height and finitely many trees. There are infinitely many positive integers (the natural numbers: 1, 2, 3 ...), and there are infinitely many prime numbers. Some finite sets are different in size than others. For example, the set of trees is different in size (bigger in fact) than the set of people over 1.8 metres in height. An obvious question is, Are some *infinite* sets different in size from other infinite sets? Although this is an obvious question, it is far from obvious how we might answer it, and indeed what it could mean for one infinite set to be different in size than another. Isn't to be infinite to be as big as big can be, and does that come in more than one size?

The key to answering these questions lies in the famous analysis of what it takes for two infinite sets to be the *same* size. It goes as follows: infinite set S is the same size as infinite set T if and only if it is possible to put the members of the two sets into one-one correspondence – every member of T is paired with a distinct member of S, and every member of S is paired with a distinct member of T. That tells us what it takes for one infinite set to be the same size as another infinite set (and means that, e.g., the set of natural numbers is the same size as the set of prime numbers). But now we know how to make sense of one infinite set *not* being the same size as another – their members cannot be put in one-one correspondence – and can approach the question of proving, for instance, that the set of real numbers is not the same size as the set of natural numbers. The best-known proof goes back to Georg Cantor, as does our understanding of the concept of one infinity being or not being the same size as another.[5]

Cantor's analysis of the concept of one infinity being the same size as another informed us about that property and, thereby, the property of one infinity *not* being the same size as another – a property we might well have thought incoherent otherwise, and allowed us (Cantor, in the first instance) to prove that it is sometimes instantiated.

Here's a second example. Consider the harmonic series: $1 + 1/2 + 1/3 + 1/4 + \ldots$ When a mathematician tells us that it *diverges*, what are they telling us? We understand what they are telling us when they explain that for any positive number, N, there is a positive integer, n, such that $1 + 1/2 + \ldots + 1/n$ is greater than N. And we then have at least the glimmer of an idea of how a proof that the harmonic series diverges might go. This example is probably best described as one where a conceptual analysis clarifies and makes precise a property, rather than as one where it tells us about a property we might not otherwise have believed in. Presumably, we all have a rough grasp of the idea of a series diverging, independently of the analysis.

6 The bearing of conceptual claims and analyses on issues in metaphysics: three case studies

Our first case study is presentism and its relationship to the special theory of relativity (STR).

Presentism is a thesis in metaphysics that holds that what exists and what exists now are one and the same. According to it, what will exist does not exist – what is true is that it will exist. Similarly, what did exist does not exist – what is true is that it did once exist. The contrast is with the picture of reality we discussed earlier that thinks of our universe as a huge thing extended through Space-Time. On that picture, things in the past, present and future exist equally, while being differently located in time. Harold Wilson existed. On the extended universe view that's to exist – in the past; whereas according to presentism, Harold Wilson does not exist in any sense at all, although he did once exist.[6]

STR is a thesis in metaphysics. This is because it is a thesis about the nature of the world we occupy. Among other things, STR holds that there is no absolute simultaneity at a distance. A victory salute in a soccer stadium in Brazil cannot be simultaneous with one in Italy in any absolute sense. The salutes can be simultaneous relative to some given inertial frame, but when they are, they will fail to be simultaneous relative to some other inertial frame. The reason this is incompatible with the possibility of the events being simultaneous in any absolute sense is that there is and can be no sense in which the first frame is privileged.[7]

A plausible thesis about the concept of existence implies that presentism and STR cannot be true together. The plausible thesis about the concept of existence is that it is an absolute one: something either does or does not exist. There is no sense to something existing relative to such and such, but not existing relative to so and so. But then presentism implies that there is absolute simultaneity at a distance: distant events E_1 and E_2 will be simultaneous if and only if they both exist. For according to presentism, everything that exists, exists at exactly the same time, namely, the present. We have, therefore, a plausible example of a conceptual claim (about existence)

Conceptual analysis in metaphysics

telling us that one thesis in metaphysics (STR) implies the falsity of another thesis in metaphysics (presentism).

Our first case study concerned the connection between, on the one hand, the nature of a concept and, on the other, two theses in metaphysics. Our second case study concerns the connection between conceptual analyses of concepts and how to resist scepticism when better theories of some subject matter are developed.

We explain the behaviour of gases in terms of temperature, pressure and volume. For example, we explain why the pressure in tyres goes up on hot days in terms of the increase in temperature. (Thus, the advice in car handbooks to check tyre pressures when the tyres are cold.) We explain why balloons expand when we blow into them in terms of the interconnection between pressure and volume. And so on. This is a bit of folk theory in the sense of being pretty much common knowledge. The thermodynamic theory of gases makes things precise and specifies the relationships between temperature, pressure, volume and the behaviour of gases in terms of equations.

The kinetic theory of gases is an established thesis in metaphysics about the nature of gases. It holds that gases are aggregations of very loosely bound atoms or molecules, and, in virtue of this fact, have various atomic and molecular kinetic energy properties. The kinetic theory explains the behaviour of gases in terms of these atomic and molecular kinetic properties; the behaviour which is explained in the folk theory and its precisification, the thermodynamic theory, in terms of temperature, pressure and volume is explained in terms of these atomic and molecular properties.

This suggests that we should conclude that gases lack the properties of temperature, pressure and volume. The explanations of the thermodynamic theory that appeal to these properties have been superseded by explanations in terms of the properties of the kinetic theory, explanations which famously do a much better job of explaining a number of important details – what is special about absolute zero and why gas behaviour changes markedly when gases are highly compressed, for example. The suggestion would be that we should think of the thermodynamic theory of gases as like the phlogiston theory of combustion, the theory which was superseded by the oxidation theory. Much as we concluded that there is no property of being phlogisticated, so we should conclude that there is, for example, no such property as having such and such a temperature. The explanatory job that once warranted our believing in temperature has been taken over by mean atomic or molecular kinetic energy.

I think that there is only one way to resist this suggestion – and we should resist, who wants to have to say that winds are never hot? It is to defend an analysis of the properties of the thermodynamic theory of gases in terms of functional roles.[8] The terms 'temperature', 'pressure' and 'volume' pick out functional roles. We know how to articulate those roles – that is in fact what the laws of the thermodynamic theory do – and when we do this, we can resist the sceptical suggestion. Far from refuting the thermodynamic theory, the kinetic theory tells us which properties play the roles definitive of temperature, pressure and volume, and thereby what those properties are. Here is how it looks, spelt out for the case of temperature:

> Being at so and so a temperature is having the property playing such and such a functional role – the role filled by 'temperature' in the thermodynamic theory of gases. (Analysis of the concept of temperature in a gas)
>
> Having thus and so a level of mean atomic or molecular kinetic is the property playing such and such a functional role. (Discovery of the kinetic theory)
>
> Therefore, being at so and so a temperature is having thus and so a level of mean atomic or molecular kinetic energy.

When I say above that, for example, 'temperature' picks out a certain functional role, and treat that as equivalent to holding that the concept is the concept of something that plays a certain functional role, I mean that what it takes to be temperature is to be that which plays that functional role. And my reason for holding this is that this is what is needed in order to make the equations of the thermodynamic theory of gases true. The equations are in fact implicit definitions of temperature, volume and pressure.

Could we say something similar about being phlogisticated? Should we hold that it is a property defined by its role in phlogiston theory and that what happened when the oxidation theory came along was that being phlogisticated was shown to be a kind of receptivity to being oxidized? If that's right, the conventional wisdom that the oxidation theory refuted the phlogiston theory is a mistake; instead, it told us what being phlogisticated is. However, a crucial role for being phlogisticated in phlogiston theory is that of containing stuff (phlogiston in fact) that gets given off in combustion. And there is no such stuff; the functional role assigned to being phlogisticated in the phlogiston theory isn't occupied.

The important methodological message that comes from the above example is independent of whether or not the functional analyses of temperature, pressure and volume sketched above are correct. The message is that we need analyses – answers to what it takes to be temperature, pressure and volume – to resist scepticism in the face of the explanatory success of the kinetic theory. Some may argue that the functional story I sketched above should be replaced by an account that treats 'temperature', 'pressure' and 'volume' as natural kind terms, and that what the kinetic theory did was to reveal the natural kinds in question. That would be another way to resist scepticism, but it would not alter the fact that we need analyses. Having said that, I should record that it has sometimes been argued that ontological austerity alone is enough to resist scepticism, for example by Block and Stalnaker (1999). The thought is that identifying, for example, temperature with mean atomic or molecular kinetic energy reduces the number of properties we need to countenance and that's reason enough to affirm the identity. And if temperature is identical to mean atomic or molecular kinetic energy, temperature exists. However, austerity is not to the point. The choice is not between believing in two properties and believing in one. It is between believing in one, which is mean atomic or molecular kinetic energy and *not* temperature, and believing in one, which is mean atomic or molecular kinetic energy and *also* temperature, as 'they' are one and the same property.

Our third case study concerns the inter-connections between the famous problem of finding an analysis of knowledge, on the one hand, and the metaphysics of belief and knowledge, on the other. Traditionally, knowledge has been thought of as a species of true belief. The metaphysics of knowledge has accordingly been thought of as the metaphysics of a special kind of true belief. The major driver of this view has been the conviction that knowledge entails true belief. You cannot know something that is false, and you cannot know something without believing it. Our discussion will take this for granted (though it has been contested).

If knowledge is a special kind of true belief, what it takes to be knowledge is to be that special kind of true belief. So if we think of a conceptual analysis as answering a 'what it takes' question, we should be able to give an analysis of knowledge in terms of true belief combined with whatever it is that makes true belief special in the way needed to count as knowledge. It is a matter of record that this has proved to be extremely difficult. For what is to come, we need to give a bit of the unhappy history.

Many thought, plausibly enough, that what is special about true beliefs that count as knowledge is that they are justified. Accordingly, they suggested the following analysis of knowledge: S knows that p if and only if (i) p, (ii) S believes that p, and (iii) S's belief that p is justified.[9] Nowadays, thanks to Gettier (1963), the great majority of analytical philosophers accept that cases like the following refute this analysis. A sample of my blood is sent to a famously reliable pathology laboratory. Two days later I receive an email from the laboratory saying that the blood sugar level in the sample is 100mg/dl. As a result, I believe that the blood sugar level in the sample of my blood sent to the laboratory is 100mg/dl. This belief is justified. I have every reason to trust the laboratory. However, by a one in a million chance, my blood sample was confused with a sample from someone with the same name as me, and the result that was emailed to me was an accurate report of the blood sugar level in his sample. However, as it happens, the blood sugar level in his sample is the same as in mine. This means that my belief that the blood sugar level in my sample is 100mg/dl is true as well as being justified. It seems clear that my true justified belief that the blood sugar level in my sample is 100mg/dl is not a case of knowledge; I do *not* know that the blood sugar level in my sample is 100mg/dl.[10]

When philosophers first became aware of this kind of counter-example to the true justified belief analysis of knowledge, many thought it would be easy to make the needed repair – adding a clause to rule out 'flukiness' or some such would do the trick. Others thought that instead of adding a clause, we should replace the requirement that the belief be justified with X, with different authors giving different accounts of X. It all turned out to be much more difficult than expected and there is still no generally accepted analysis of knowledge as a special kind of true belief. Our question concerns the significance of this fact for metaphysics.

The significance is that it makes it obscure how to meet a certain sceptical challenge concerning knowledge. I do not mean the traditional sceptical challenge, What if anything is known? I mean the challenge that asks *who* knows things, not *what* they know. We all agree that tables do not know things. We all agree – traditional sceptics aside – that often people know things, their mobile phone number, for example. What justifies the belief that often people know things, including in particular their mobile phone numbers? Part of the answer lies in what they do; indeed what they do is often crucial. Perhaps they cite a certain number when we ask them what their number is. We then note what happens when we dial the number they cite – their phone rings. Information of this kind tells us what the subject believes and gives us a reason to hold that their belief is true. We may also have information bearing on the evidence they have for their belief. Perhaps they themselves dialled the number from another phone and noted that their own phone rang. Perhaps they opened *Settings* on their phone and noted the number recorded there. We may also have information that tells us about the causal origins of their belief, how reliable beliefs of this kind are, that chance played no role in their belief being true, and so on and so forth. But how might we be justified in moving from this body of evidence to the conclusion that they know their mobile phone number? Our evidence supports, strongly supports we may suppose, that they have a true belief of so and so a kind about what the number is. But how does this bear on whether or not they know what the number is? What could warrant taking the extra step, the step from all the information about what they believe to their knowing the number?

One response to this challenge is that there is no extra step. Somewhere in all the information about the nature of their belief about the phone number – the information about causal origin, reliability, justification, truth, etc. – lie the clause or clauses we need to add to true belief to obtain the sought for analysis of knowledge. It is, after all, possible to analyse knowledge as a special kind of true belief. The unhappy history does not tell us that no analysis is possible. It tells us that articulating one is hard, and maybe that those seeking an analysis have made their task unnecessarily hard. They have sought to analyse a single concept of knowledge, whereas in fact there are a number of concepts – knowledge is not a single species of true belief. Or maybe the problem is that they have sought precision about a matter that is inherently vague to one extent or another. Or maybe the problem is that they have overlooked the fact that what counts as knowledge is context dependent. Etc.

What should we say if this response is a big mistake? We might argue that knowledge is a fundamental category that resists reductive analysis, perhaps expressing the thought by saying that it is *sui generis*.[11] The interest of a view of this kind is clear, but there is a problem with it, one that is really a corollary of the sceptical worry we have been discussing. Why should we believe that knowledge so understood is ever possessed? What explanatory work does it do that is not done by one or another species of belief? By explanatory work, I mean here explanatory work in explaining and predicting our interactions with the world around us, and our inferential processes in arriving at opinions about how things are. If knowledge is *sui generis*, it is hard to resist eliminativism about knowledge.

The two final sections are concerned, in turn, with a sense in which conceptual analysis is unavoidable, and the relevance to metaphysics of analyses of sentences as opposed to concepts.

7 The inevitability of conceptual analysis

Any view about the role of conceptual analysis in metaphysics is inevitably controversial. There is, however, one role for it which is not controversial, or, I urge, should not be.

The world is a complex place. That makes predicting and explaining what happens hard. Explaining and predicting share movements and house prices are no easy tasks. And even when we make things easier by restricting ourselves to cases where relativistic effects can be ignored, explaining and predicting the movements of collections of physical bodies often requires help from powerful computers.

We make things more tractable by finding patterns in complex bodies of information, the aim being to find the right patterns for predicting and explaining. This is why economists look at changes in average house prices, rent to house price ratios, graphs of general inflation versus housing price inflation, etc. when seeking to explain and predict the next housing bubble. A much discussed issue of current concern is finding the right patterns in past weather data to allow well-founded judgements of global temperatures in the future. Conservation principles in physics often concern the conservation of a pattern in facts about the masses, positions and motions of collections of physical bodies. For example, in Newtonian physics the conservation of kinetic energy says that $\Sigma_i 1/2 m_i . v_i^2$ is conserved.

These patterns are a priori determined by the data they supervene on and can be given analyses in terms of that data. In that sense, we should all be hospitable to conceptual analysis.[12] It is good to ask after the right concepts to use in explanation and prediction. Economists are right to explore the merits of one or another measure of economic activity, and when they do this, they are considering which a priori determined properties are best for one or another job, or, as we can say it, they are wondering which *concepts* are best for the task at hand.

8 Conceptual analysis and rewriting sentences

We have mainly focused on the bearing of conceptual analysis thought of as making explicit what it takes to be a so and so – or, if you like, what is needed to fall under the concept of so and so – to questions in metaphysics. Before we finish, we should note that sometimes the focus is on *sentences*, the aim being to show how to avoid a metaphysical commitment held to be, for one reason or another, undesirable. For example, a metaphysician may want to avoid a commitment to the existence of limps, despite the fact that people have limps. Our metaphysician insists that 'S has a limp' should be rewritten as (*analysed as*, they may say) 'S limps'. Or a philosopher of mind may insist that although I may say that I have a pain in my hand, there is no pain in my hand, literally speaking. After all, if my hand is in my pocket, I will not say that I have a pain in my pocket. This philosopher may insist that 'I have a pain in my hand' should be rewritten as 'My hand hurts.' Or, finally, consider our earlier example of Vulcan. Some metaphysicians are happy to allow that Vulcan is, quite literally, something that does not exist, and so that there are things that do not exist. But the majority side with, e.g. Quine (1961), insisting that what there is and what exists are one and the same question. They urge that 'Vulcan does not exist' should be rewritten as 'There is no such thing as Vulcan.'

9 A quick summary

Conceptual analysis, understood as the clarification and explication of what it takes for a concept to apply to something, bears on issues in metaphysics in the following ways: it tells us when explanations and predictions in terms of one set of concepts are or are not in potential conflict with explanations and predictions in terms of a different set of concepts; it can alert us to properties that might otherwise have escaped our attention, or ones we might have thought incoherent; it can alert us to patterns in data that are important for prediction and explanation, and, relatedly, suggest when it would be wise to make one or another modification to the concepts we employ in prediction and explanation; and, in its 'rewriting sentences guise', can show us how to avoid unwanted ontological commitments.

Notes

1 Quine (1960) is a prominent supporter of the second view, Geach (1967) of the first.
2 Usually this view is held along with the view that mental states in general are states of the brain playing certain roles, see, e.g., Armstrong (1968).
3 Quine (1961) is perhaps the best-known advocate of a negative answer to this question; he holds (as do most but not all metaphysicians) that what exists and what there is are one and the same question.
4 Smart (1963, ch. 7) for example, but this is not his only reason for favouring the view.
5 See pretty much any history of mathematics, e.g. Courant and Robbins (1941). The set of real numbers is bigger than the set of natural numbers.
6 Thus defenders of the extended through Space-Time picture, e.g. Smart (1963) and Nerlich (1979) cited earlier, are opposed to presentism. A supporter of presentism is Bigelow (1996).
7 This comes from the standard reading of STR and I do not dispute it, but it has been disputed, e.g. by Tooley (1997, ch. 11).
8 This would be a special case of the approach to theoretical terms in, e.g., Lewis (1970).
9 For a view of this kind, see Ayer (1956). Those who defend analyses of this kind are not thinking of them as merely descriptions of how we use 'knowledge'. The analyses are in part accounts of how we use 'knowledge' and in part accounts of how it would be good to use the word.
10 It 'seems clear' to me and very many that this kind of case is not a case of knowledge. What should we say to those who disagree? And what should we say to those who agree that this kind of case is not a case of knowledge but insist (against what I say in the text) that it isn't a case of justified belief either? These are interesting questions I set aside here, but see Jackson (2011) and references therein.
11 This is one way of reading one theme in Williamson (2000).
12 For more on the many issues raised by the way we use patterns in seeking to make sense of our world, see Jackson (2017).

References

Armstrong, D. M. (1968) *A Materialist Theory of the Mind*, London: Routledge & Kegan Paul.
Ayer, A. J. (1956) *The Problem of Knowledge*, London: Macmillan.
Bigelow, John (1996) 'Presentism and Properties', *Philosophical Perspectives*, 10: 35–52.
Block, Ned and Robert Stalnaker (1999) 'Conceptual Analysis, Dualism and the Explanatory Gap', *The Philosophical Review*, 108: 1–46.
Courant, Richard and Herbert Robbins (1941) *What Is Mathematics?* London: Oxford University Press.

Geach, P. T. (1967) 'Identity', *Review of Metaphysics*, 21: 3–12.
Gettier, Edmund (1963) 'Is Justified True Belief Knowledge?' *Analysis* 23: 121–123.
Jackson, Frank (2011) 'On Gettier Holdouts', *Mind and Language*, 26, 4: 468–481.
Jackson, Frank (2017) 'Armchair Metaphysics Revisited: The Three Grades of Involvement in Conceptual Analysis', in Giuseppine D'Oro and Søren Overgaard (eds.) *The Cambridge Companion to Philosophical Methodology*, Cambridge: Cambridge University Press, pp. 122–140.
Lewis, David (1970) 'How to Define Theoretical Terms', *Journal of Philosophy*, 67: 427–446.
Nerlich, Graham (1979) 'What Can Geometry Explain?' *British Journal for the Philosophy of Science*, 30: 69–83.
Quine, W. V. (1960) *Word and Object*, Cambridge, Mass.: MIT Press.
Quine, W. V. (1961) 'On What There Is', reprinted in *From a Logical Point of View*, Cambridge, Mass.: Harvard University Press, pp. 1–19.
Smart, J. J. C. (1963) *Philosophy and Scientific Realism*, London: Routledge & Kegan Paul.
Tooley, Michael (1997) *Time, Tense, and Causation*, Oxford: Clarendon Press.
Williamson, Timothy (2000) *Knowledge and Its Limits*, Oxford: Oxford University Press.

32
CONTINGENTISM IN METAPHYSICS[1]

Kristie Miller

1 Introduction

Let us distinguish two kinds of contingentism: *entity contingentism* and *metaphysical contingentism*. Here, I use 'entity' very broadly to include anything over which we can quantify – objects (abstract and concrete), properties, and relations. Then *entity contingentism* about some entity, E, is the view that E exists contingently: that is, that E exists in some possible worlds and not in others.[2] By contrast, *entity necessitarianism* about E is the view that E exists of necessity: that is, that E exists in all possible worlds. We can distinguish two views: *global entity contingentism* and *global entity necessitarianism*. Global entity contingentism is the view that for any possible E, E exists contingently. Global entity necessitarianism is the view that for any possible E, E exists necessarily.

While entity contingentism and entity necessitarianism are views about the modal status of entities, *metaphysical* contingentism and necessitarianism are views about the modal status of metaphysical principles. *Metaphysical contingentism* about some metaphysical principle, P, is the view that P is contingent: it is true in some worlds, and false in others. *Metaphysical necessitarianism* about P is the view that P is necessary: either P is true in every world, or false in every world. We can then distinguish two views: *global metaphysical contingentism* and *global metaphysical necessitarianism*. Global metaphysical contingentism is the view that for any internally coherent[3] metaphysical principle P, P is contingent: P is true in some worlds and false in others. Global metaphysical necessitarianism is the view that for any internally coherent metaphysical principle, P, P is necessary: it is either true in every world, or false in every world.

This chapter principally focuses on metaphysical rather than entity contingentism, though §2 briefly discusses the latter. As we will see (§2), both global entity contingentism and global entity necessitarianism are controversial views, and most philosophers fall somewhere between these two extremes, holding that some, but not all, entities exist necessarily – a view we can call *entity moderatism*. By contrast, global metaphysical necessitarianism has, until recently, largely been the default view. It is only recently that some philosophers have argued that we should be metaphysical contingentists about at least some metaphysical principles – a view we can call *metaphysical moderatism*. §3 considers why global metaphysical necessitarianism has hitherto been so persuasive, then evaluates (§3.1) some arguments for the view before considering more recent arguments in favour of the contingency of (at least some) metaphysical principles (§3.2).

2 Entity contingentism

Neither global entity contingentism nor global entity necessitarianism have proved particularly attractive, with most philosophers accepting entity moderatism. In part this is because there are two apparently compelling counterexamples – one to global entity contingentism and one to global entity necessitarianism – that jointly suggest that some view between the two extremes must be right. The first counterexample, to global entity contingentism, arises in the philosophy of maths. Mathematical Platonists hold that there exist (abstract) mathematical objects, while mathematical nominalists deny that said objects exist. Whether Platonist or nominalist, though, it is almost universally agreed that mathematical objects either exist of necessity (Hale and Wright 1992; Schiffer 1996; Resnik 1997; Shapiro 1997) or necessarily fail to exist (Balaguer 1998 and 2018; Rosen 2001; Yablo 2005). Hence almost all parties to the dispute are entity necessitarians about mathematical objects.

Entity necessitarianism about mathematical objects is partly motivated by the widely shared belief that mathematical claims are either true, or false, of necessity. If mathematical objects exist in *any* world, then, surely they are what make (true) mathematical claims true in that world. But on the plausible assumption that mathematical claims are made true *by the same kind of thing* in every world, it follows they must be made true by mathematical objects in every world. Moreover, since mathematical claims are true of necessity, it must be that the very same mathematical objects exist in every world (and make those claims true). Likewise, if there is a world in which mathematical objects do not exist, then it must be that *something else* makes (true) mathematical claims then true in that world. Hence, by similar reasoning, since whatever makes mathematical claims true in a world that lacks mathematical objects will make them true in all worlds, we should conclude that mathematical objects necessarily fail to exist.

The second counterexample, this time to global entity necessitarianism, is the existence of ordinary medium sized dry goods such as you and I (and goats, toasters, and cars). In all these cases, it seems, things would not need to have gone very differently for the relevant entities to have failed to exist. But if that is right then entity contingentism is true of a whole range of entities, and hence global entity necessitarianism is false.

There are, however, those who resist these counterexamples. There are entity contingentists about mathematical objects (Field 1993; Colyvan 1998, 2001) who think we ought to posit mathematical objects in our world if and only if they are indispensable to the best scientific theories of our world. Moreover, they argue, there are possible worlds in which mathematical objects *are* dispensable to the best scientific theory – worlds in which we can do science without numbers (Field 1980) – and worlds in which we cannot do science without numbers (Colyvan 2001). Hence we ought to conclude that there exist numbers in some, but not all, possible worlds. If the most compelling counterexample to global entity necessitarianism is the existence of mathematical objects, then resistance to this counterexample could deliver global entity contingentism.

On the other side of the modal fence are those who reject the second purported counterexample: views that deny the contingency of ordinary objects. There are two routes to this conclusion. On the first there is only one possible way things could be – the way they are – and so whatever actually exists, trivially exists of necessity. This view entails both global entity necessitarianism and global metaphysical necessitarianism. On the second, the same objects, properties, and relations *exist* in every world, but what differs between worlds is which objects are concrete, which properties are instantiated, and which relations obtain (Linsky and Zalta 1994; Williamson 2010, 2013). Here is a helpful analogy. According to the view that there is only one way things could be, there exists a single theatre, with a single (small) cast of characters, performing the only possible play. By contrast, on the second view there exist many theatres, and backstage in each theatre there exists a single, very large, cast of abstract characters. But each theatre has a different play running, and which characters come out on stage – and hence which are concretely realised – varies depending on which play is running. (Strictly speaking, for Williamson there are three classes of objects: concrete ones (those on stage), non-concrete ones (those that are back-stage, but could come out on stage (but don't), and abstract ones (those that never come out on stage, and always remain behind the scenes).

If the only real threat to global entity necessitarianism is the apparent contingency of ordinary objects, then this view represents a way of defending global entity necessitarianism.

Nevertheless, despite resistance to each counterexample, by and large philosophers have endorsed entity moderatism. As we will see, however, the same reasoning has not prevailed in the case of metaphysical principles.

3 Metaphysical contingentism

When we ask whether we ought be contingentists or necessitarians about some, or all, metaphysical principles, which principles do we have in mind? In some sense the principles at issue are a motley crew covering a wide variety of areas of metaphysics. They include, but are not limited to, the following:

- *composition principles* – principles that tell us under what conditions some xs compose a y (van Inwagen 1990).
- *property principles* – principles that tell us what it is to be a property (whether properties are immanent universals (Armstrong 1978), tropes (Campbell 1997; Maurin 2002), or whether nominalism is true (Lewis 1983; Rodriguez-Pereyra 2002)).
- *persistence principles* – principles that tell us what it is for an object to persist through time (whether objects perdure (Lewis 1986; Sider 2001) or endure (Thomson 1983; Johnston 1987)).
- *time principles* – principles that tell us about the nature of time (whether time is characterised by the instantiation of irreducibly tensed A-properties (Zimmerman 2005; Bourne 2006), or merely B (or C)-relations (Mellor 1981; Price 1996)).
- *law of nature principles* – principles that tell us what it is to be a law of nature (whether laws are generalisations that feature in the most virtuous true axiomatisation of all the particular matters of fact (Lewis 1986) or are relations of necessity that hold between universals (Tooley 1977; Armstrong 1983).[4]
- *object principles* – principles that tell us what it is to be an object (whether objects are bundles of properties (Ehring 2011), or substrata with properties attached; whether objects are identical with regions of space-time (Schaffer 2009a) or distinct from them (Gilmore 2014); whether composite objects are identical with pluralities of simples (Bohn 2009; Cotnoir 2014) or distinct from said pluralities (Sider 2007a; Cameron 2014).
- *priority principles* – principles that tell us what ontologically depends on what (see Schaffer 2009b) (for instance whether wholes depend on their parts (Sider 2007b) or parts depend on the whole (Schaffer 2007)).

We can think of many (or perhaps all) such principles as answering the question, for particular metaphysical kinds of thing, *what is it to be that kind of thing?* Alternatively, we might think of them as *metaphysical laws* (so long as (for now) we leave open the modal status of said laws). In what follows I will take a 'we known 'em when we see 'em' approach to these principles:

for while there is some disagreement, around the edges, as to which principles ought to count as metaphysical principles, there is enough agreement for us to proceed without a rigorous characterisation. Moreover, doing so heads off an argument for a very uninteresting kind of global metaphysical necessitarianism. That argument proceeds as follows. It is analytic that metaphysical principles are principles of complete generality – i.e. principles that are true of necessity – and so any purported metaphysical principle that is not true of necessity is not really a metaphysical principle at all. Accepting this characterisation of metaphysical principles guarantees the truth of global metaphysical necessitarianism, but only at the cost of entailing that should it turn out that all of the principles just introduced are contingent, then none of them is, in fact, a metaphysical principle. Indeed, it is consistent with this characterisation that it turns out that there are no metaphysical principles at all! This seems like an objectionably cheap way to purchase global necessitarianism, and so I set it aside.

Until relatively recently something close to global metaphysical necessitarianism has been the overwhelming orthodoxy.[5] It might seem puzzling that global metaphysical necessitarianism has fared so much better than global entity necessitarianism. In part, it has fared better because global metaphysical necessitarianism is not as ontically committing as global entity necessitarianism. Metaphysical principles tell us what it is for there to be a certain kind of thing, and so can be true in worlds in which there are none of those things. It can be true (and necessarily so) that what it is to be a dog is to be a *Canis familiaris* (say) and it also be true that dogs exist contingently. Likewise the metaphysical principle that says that laws of nature are relations of nomic necessitation between universals can be true, of necessity, even though there are worlds with no such relations – for there can be worlds in which there are no laws.

But that can't be the whole story. There has to be a positive *reason* to endorse global metaphysical necessitarianism. The following section considers a number of arguments in favour of the view. To the extent that these arguments fail we have little reason to accept global metaphysical necessitarianism.

3.1 Arguments for global metaphysical necessitarianism

Few explicit arguments in favour of global metaphysical necessitarianism have been made. Recent arguments about the modal status of metaphysical principles have typically focused on trying to show that metaphysical contingentism is true of at least some of the principles (and hence global metaphysical necessitarianism is false). Nevertheless, it is worth trying to articulate arguments for metaphysical necessitarianism that are implicit in the literature to see whether any is plausible.

3.1.1 The argument from ground

Let's begin with the argument from ground: an argument that appeals to the nature of ground to show that metaphysical principles hold of necessity. Roughly, the idea here is that metaphysical principles are best understood as general claims about what grounds what.[6]

1. Metaphysical principles are best understood as claims about what it is to be a G.
2. Claims about what it is to be a G are best understood as claims of the form: being F is what it is to be G.
3. If being F is what it is to be G, then necessarily, x is F iff x is G.
4. Therefore, necessarily, x is F iff x is G.[7]

Then the question arises: how should we interpret (2)? What does it mean to say that being F is *what it is* to be G? The most natural interpretation is either that F-ness is identical with G-ness, or that F-ness is the essence of G-ness. In either case we don't really need to appeal to ground per se. So I consider each of these options in the next section.

3.1.2 The arguments from identity and essence

Let's consider two arguments in which we interpret the claim that what it is to be F is to be G, in terms of identity (first) and essence (second).

The argument from identity

1. The metaphysical principles are best understood as identity statements.
2. Identities hold (or fail to hold) of necessity.
3. Therefore, the metaphysical principles are necessary.

The argument from essence

1. The metaphysical principles are best understood as claims about essences.
2. Claims about essences hold (or fail to hold) of necessity.
3. Therefore, the metaphysical principles are necessary.

Here, the supposition is that metaphysical principles are best understood as expressing claims of the form 'X is Y', but according to the identity argument we interpret 'X is Y' as a claim about identity (i.e. that X = Y) and in the essence argument as the claim that X is essentially Y, or that Y is part of the essence of X. Since both identity statements and essential truths hold of necessity, (if at all),[8] in either case we discover that necessary, X is Y. As we will see, it doesn't much matter, for present purposes, which picture we go with. To see this, let's start with the former, identity, picture. Before we do so, however, it is important to notice something that both approaches share and which will become important in what follows. Consider the case of water and H_2O.

Notice that whether it is necessary that water is H_2O because water = H_2O, or because water is essentially H_2O, that tells us nothing about whether there are worlds in which there is something very like water except for its chemical composition. Indeed, it is typically assumed that there is such a possible substance – twater – which is superficially and functionally just like water except it is made of, say, XYZ. In discovering that water is H_2O we discover that XYZ is not water. By parity, then, suppose we discover that there are actual tropes, and so 'properties are tropes' is actually true. It does not follow that there are no possible entities that play the property role – which, is, say, grounding objective similarities and differences, grounding causal powers, determining natural laws or regularities, and connecting (in some way) to predicates in natural language – and are not tropes. If it should turn out that 'properties are tropes' ought to be interpreted either as 'properties = tropes' or as 'properties are essentially tropes', this tells us that if there are possible immanent universals, *then they are not properties*, just as if there is XYZ, it is not water. It doesn't tell us that there are no immanent universals.

Let's return to the identity interpretation. According to a broadly Humean tradition, the discovery of identities involves two components: empirical discoveries about the actual world, combined with *a priori* claims about the semantics of the relevant term.[9] In the case of 'water' the *a priori* component is something like the following: 'water' refers, *in every world*, to whatever is *actually* the watery stuff. It is part of the semantics of 'water' that 'water' rigidly refers to whatever it actually refers to. Prior to knowing the microstructure of the actual watery stuff, we could have known, *a priori*, that *whatever* that structure is, water is, necessarily, that stuff.

The virtue of this Humean view is that we don't need to know which metaphysical principles are true to know their modal status: we only need to know the reference fixing descriptions of the terms that feature in the principles, and these can be known *a priori*. Consider a principle of the form 'X is Y'. There are a number of different reference fixing descriptions that would entail that X = Y. Here are two that are (at least in the abstract) plausible. Option 1: it is *a priori* that 'X' refers, of necessity, to whatever 'X' actually refers to. Option 2: it is *a priori* that if 'Y' possibly refers, the 'X' refers to what 'Y' refers to, otherwise 'X' refers to Z. In the former case if 'X' actually refers to Y, then necessarily, X is Y, and in the latter if 'Y' possibly refers, then necessarily, X is Y.[10]

Is it plausible that any of the metaphysical principles have some such *a priori* semantic claim as part of their referencing fixing description?

Consider the principle that says that laws of nature are relations of nomic necessitation between immanent universals. Might it be *a priori* true that 'laws of nature' refer to whatever metaphysical kind of things are actually the laws of nature, and necessarily so? Suppose that rather than discover that there actually exist relations of nomic necessitation between immanent universals, instead we had discovered that the actual laws are the best systematisation of particular matters of fact. On the assumption that there are possible worlds with relations of nomic necessitation between universals, it seems unlikely that we would have concluded that, necessarily, laws are best systematisations. For then even in worlds with relations of nomic necessitation between universals, the laws are best systematisations, not those relations of necessitation. But it's not overly plausible that if there are worlds with relations of nomic necessitation, that those relations are not laws of nature. That's because, in general, the reason philosophers think that laws are not relations of nomic necessitation is because they think there are no relations of nomic necessitation (and, perhaps, no universals to be thus related) not because they think there are, or possibly are, such things, but they aren't laws of nature! So if there are, possibly, such things, we'd be tempted to say that even if *actually*, laws are best systemisations (because actually there are no relations of nomic necessitation) nevertheless, laws are relations of nomic necessitation in worlds where there are such relations.

So consider the second referencing fixing option. Perhaps it is *a priori* that 'laws of nature' necessarily refer to relations of nomic necessitation if relations of nomic necessitation are possible, and otherwise refer to some other metaphysical kind. One might think this if one thinks that relations of nomic necessitation are the best deserver to be what laws of nature are, and so if there are possibly such things, then those are the laws; nevertheless, if there are necessarily no such things then the laws must be something else (perhaps, say, the best systematisation of matters of fact) assuming we think there is some second-best deserver. If not, we have to conclude that error theory about laws is true. Since a best systems theory of laws seems like a good enough deserver, likely we will say that if, of necessity, there are no relations of nomic necessitation, then laws are just best systematisations. But if there are, possibly, relations of nomic necessitation, then those relations are, of necessity, the laws.

In fact, something like that sounds plausible. If there are worlds with relations of nomic necessitation we might be inclined to say that worlds that lack such relations don't *really* have laws at all. Perhaps there *appear* to be laws because there are regularities, but in such a world these are *mere* regularities. Nothing is really governing the dynamical unfolding of said worlds. If that is right, then the principle 'the laws of nature are relations of nomic necessitation' is necessary (if said relations are possible).

The problem for the global metaphysical necessitarian is that if she thinks the necessity of every metaphysical principle issues from its being an identity, then she needs to show that *all* metaphysical principles contain expressions with referencing fixing description that function like these, and it is far from clear that this is plausible.

Consider the principle that says that properties are immanent universals. In order for that principle to be necessary, something like either of the following must be true *a priori*: 'properties' refer to whatever metaphysical kind of thing actually plays the property role and necessarily so, or 'properties' refer, of necessity, to immanent universals if they are possible, and if not, refer to things of metaphysical kind K. Neither of these looks very plausible because properties just are the things that do certain jobs – the things that ground objective similarities and causal powers, and determine natural laws or regularities, and connect to predicates in natural language. It's hard to see how something could do all those things and not be a property. In fact, it seems as though something like the following might be true *a priori*: whatever plays the property role is a property.[11] If so, then if different kinds of thing play the property role in different worlds, it follows that any property principle is contingent.[12]

The same sort of reasoning holds if we think that the metaphysical principles are best interpreted as claimed about essences. For if we think, for instance, that whatever plays the property role in some world is a property, then we must reject the claim that if actually immanent universals play that role then, essentially, properties are such universals. In general the essentialist interpretation of the metaphysical principles will seem implausible whenever more than one kind of metaphysical thing *can* equally well do some metaphysical job, and we cannot rule out that all those metaphysical kinds are possible. Since there are such cases, it is plausible that neither the argument from essence nor from identity will secure the conclusion that global metaphysical necessitarianism is true. So let us consider some further arguments for that conclusion.

3.1.3 *The argument from* a priority

Let's now consider an argument from *a priority*. Roughly, the idea is that since metaphysics is an *a priori* endeavour, and since *a priori* reason can only furnish us with necessary truths, metaphysical principles must be necessary.

1   We use *a priori* reason to determine which metaphysical principles are true.
2   *A priori* reason can only determine the truth (or falsity) of necessary claims.[13]
3   Therefore, the metaphysical principles are necessary.

Premise (2) says, in effect that if it's *a priori* then it's necessary. There are good reasons to deny (2) in its full generality (see Kripke 1980) since there are *a priori* truths involving indexicals (such as 'I am here') that are clearly contingent.[14] Still, if indexical claims are the only exception to the necessity of *a priori* truths, then an amended version of (2) might hold, to wit (2★): *a priori* reason that does not involve indexicals can only determine the truth (or falsity) of necessary claims. Then if we add (3★): the metaphysical principles do not involve indexicals, we can reach the conclusion (3), therefore, the metaphysical principles are necessary. Not everyone accepts that the only instances of *a priori* contingencies arise from indexicals (see Williamson 1986 and Hawthorne 2002), but even amongst those who think there is deeply contingent *a priori* knowledge, it seems plausible that knowledge of metaphysical principles is not this kind of knowledge. So if our knowledge of metaphysical principles is entirely *a priori*, we might still make a case for a version of the argument from *a priority*.

The problem is that while a lot of reasoning about metaphysical principles is *a priori*, a good deal of it also appeals to a range of empirical facts, including facts about the physics of our world, folk intuition and semantics, the content of our phenomenal states, and so on. That is because often we evaluate competing metaphysical principles by determining which best meets a complex set of desiderata that include the extent to which a principle preserves or explains our folk intuitions, perceptions and phenomenologies, and folk semantics, is simple, parsimonious, explanatory, and consistent with our best science. Such reasoning might afford us *a priori* knowledge of some range of conditional claims: if things are thus are so, then P (where P is some metaphysical principle) for a range of ways that things can be. Perhaps these conditionals are necessary. But even if they are, it does not follow that the metaphysical principles themselves are necessary. For it can be necessary that if things are thus and so, then P is true, and yet, if it is contingent that things are thus and so, it can be contingent that P is true. Quite generally then, since *a priori* reasoning *alone* does not, in general, yield the metaphysical principles, then we have little reason to suppose that (2) is true. Perhaps sometimes *a priori* reasoning alone yields knowledge of metaphysical principles, (and so perhaps in those cases we have reason to think that the principles hold of necessity) but not all knowledge of these principles is achieved by *a priori* reasoning alone, as (2) requires.

Still, if this is indeed how metaphysical theorising goes, then there is another argument to be considered, which I call the argument from virtue:

3.1.4 *The argument from virtue*

The next argument seeks to show that the metaphysical principles are necessary by appealing to a certain methodology used in metaphysis: namely that we appeal to certain virtues when determining which theories are true. Here is that argument:

1 We determine which metaphysical principles are true by determining which are most virtuous.
2 Whichever metaphysical principles are actually most virtuous are, necessarily, most virtuous.
3 Therefore, whichever metaphysical principles are actually true are, necessarily, true.

The problem, however, is that there seems little reason to think that (2) is true. It may often be that since what needs explaining varies between worlds, which principles are the most virtuous also varies. Consider competing principles about time. Some A-theorists argue for the A-theory on the grounds that because it better explains our temporal phenomenology (see Baron *et al.* 2015) it is overall most virtuous. Even if that is so, it is plausible that there are worlds in which temporal phenomenology is very different, and hence better explained by some other metaphysical principle. Or suppose that the B-theory is actually most virtuous because it best accommodates empirical facts about our world, namely the truth of general relativity.[15] Assuming there are worlds in which general relativity is false, however, there may be worlds in which the A-theory best accommodates empirical facts and is overall more virtuous. Given the variability of what needs explaining in different worlds it seems unlikely that (2) is true.

But suppose it were true that metaphysical principle P is most virtuous in every world. Should we conclude that P is necessarily true?[16] Not obviously. We hope that in our world virtue is a guide to truth. Perhaps we even have inductive reason to think it is. But unless we posit a necessary connection between virtue and truth, we should expect that in some worlds the virtuous will come apart from the true. It's hard to see, though, what could ground the presence of such a necessary connection. If being true is distinct from being the most virtuous theory, as surely it is, then the presence of such a necessary connection is mysterious. Given this, we should conclude that P's being most virtuous in *w* is merely *evidence* of P's being true in *w*. But then even if P is most virtuous in every world, this is not reason to suppose P to be true of necessity; instead, it is reason to conclude that, likely, in some worlds the evidence is misleading. Of course, the same reasoning holds true for contingency: the fact that *different* competing principles are most virtuous in different worlds does not, if the previous reasoning is right, give us reason to conclude that those principles are contingent (and so the analogous reasoning wouldn't give us reason to be entity contingentists about mathematical objects). So consideration of virtue doesn't seem to tell us anything about the modal status of the metaphysical principles.

3.1.5 *The no contingent difference-maker argument*

The next argument appeals to the idea that metaphysical principles do not have contingent difference-makers. Difference-makers, are, very roughly, things in the world that make a difference. They are often appealed to in the context of causation, where the idea is that a cause is something that makes a difference to what happens (the effect) so that what did happen wouldn't have happened without it. So what makes something, say P, a difference-maker to Q, is that without P, Q wouldn't have happened. We expect contingent things to have contingent difference-makers. After all, if what makes the difference between P obtaining, and P not obtaining is something necessary (say Q), then Q obtains in every world, and so we'd expect P to obtain necessarily. So if P is contingent, we'd expect to find that that it has a contingent difference-maker. Here is the argument:

1   A metaphysical principle is contingently true only if it has a contingent difference-maker.
2   No metaphysical principle has a contingent difference-maker.
3   Therefore, no metaphysical principle is contingent.[17]

Since (1) is plausible, the weight of the argument rests on (2). Why accept (2)? Hale and Wright (1994) offer something like this argument (appropriately amended) for entity necessitarianism about mathematical objects. Whether (2) is plausible depends on how one is thinking about difference-makers. Hale and Wright clearly suppose that a contingent difference-maker (for the existence of mathematical objects) needs to be a contingent *non-mathematical* thing whose existence in some worlds *grounds* or *explains* the existence of mathematical objects in just those worlds. But one might hold that the existence of mathematical objects is not grounded in some further (non-mathematical) contingency, and that the contingent existence of mathematical objects is its own difference-maker.

In the case of metaphysical principles, then, one might look for a contingent difference between worlds that *grounds*, or *explains*, a particular principle's being (contingently) true. So, for instance, if objects persist by perduring in some worlds, and enduring in others, then one is looking for a difference between worlds that explains this difference in the way objects persist. By contrast, one might simply look for a contingent difference between worlds in which objects endure and worlds in which objects perdure, and *that* difference can simply be the fact that objects endure in one world, and perdure in another.

While it is plausible that whether or not mathematical objects exist might not be grounded in some further contingent features of a world, it is less plausible that the truth of metaphysical principles is not grounded in further facts. To think otherwise is to think that it is a brute matter *which* principles are true in a world. While one could think this, let's suppose we don't. Then we can understand the call for contingent difference-makers as a call for an explanation of the contingent truth of a metaphysical principle – an account of what grounds the contingent truth of said principle.

Let's consider an example. Suppose principles of composition – which specify the conditions under which composition occurs – are contingent. One might initially think that there *is* something that can vary from world to world and ground the truth of different composition principles: namely the distribution of simples. Perhaps in some world simples are arranged in such a way that they ground its being true that every plurality of disjoint objects composes some object (i.e. unrestricted composition is true),[18] while in some other world simples are arranged in such a way that they ground its being true that no plurality of disjoint objects composes any object (i.e. mereological nihilism is true).[19]

Quite generally, one might think, if metaphysical principles are principles about what grounds what, and if grounding relations are super-internal[20] – if facts about what grounds what are grounded in the grounds themselves – then worlds with different grounds will generate different grounding relations, and different derivative facts, and hence in at least some such cases different metaphysical principles will be true at different worlds. But if so, (2) is false.

On further reflection, however, this reasoning is tenuous. We can grant that in one world the arrangement of simples is such that every plurality of disjoint objects composes some object, and that in some other world the arrangement of simples is such that no plurality of disjoint objects composes some object. Call the way the simples are arranged in the first world the *R-way*. The reason the simples compose in that world is because they are arranged in the R-way. Suppose, further, that in the closest world in which the simples *aren't* arranged in the R-way, those simples fail to compose anything. In fact, that's why they don't compose anything in the second world. Then even though in first world unrestricted composition is true and in the second nihilism is true, the nihilistic metaphysical principle (to wit that disjoint objects never compose) is not true in the second world, and the unrestricted composition principle (to wit, that every set of disjoint objects composes) is not true in the first world. Rather, it seems as though the principle that says that things must be arranged in the R-way to compose, is true in both worlds. It is just that in one world the simples are arranged in the R-way, and in the other they are not. In each case the *principle* is the same, but what is *generated* via that principle – which composites there are – varies.

This suggests that if composition principles really are contingent, then it must be because there exists a pair of worlds, w_1 and w_2, such that (i) in w_1 there exists a plurality of simples arranged X-wise, and those simples compose something and (ii) in w_2 there exists a plurality of simples arranged X-wise, such that that plurality is an intrinsic duplicate of the plurality in w_1,[21] and such that in w_2 those simples fail to compose anything. But if *that* is what contingentism about compositional principles requires, then it is difficult to see what could make for the difference between w_1 and w_2. But it is plausible that whether or not some plurality of simples composes something should be a function of the internal relations between, and intrinsic properties of, those simples. While one might well think that *what* is composed is in part a function of extrinsic properties and relations, it seems more difficult to suppose that whether or not anything is composed is a function of these properties and relations.[22] If so, it is difficult to see what could ground the difference between composition occurring in w_1 and failing to occur in w_2.

If all the metaphysical principles are relevantly like composition principles in this respect, then (2) is plausible. Are they? Not obviously. If there being, say, tropes in one world and not in another can be a brute matter not explained by some other contingent difference-maker (which seems plausible) then there being tropes in some worlds, and not others, is enough of a difference-maker to explain why properties are tropes in some, but not all, worlds. At the very least, it is far from obvious that (2) is true for all metaphysical principles.

3.2 Arguments for global metaphysical contingentism

So far, then, we have not found any compelling arguments for global metaphysical necessitarianism. In what follows I consider arguments for global metaphysical contingentism.

3.2.1 The argument against necessity

The following argument aims to show that there is no plausible source of necessity for metaphysical principles, and so they must be contingent. The argument proceeds as follows:

1   There are only two sources of necessity: analyticity and *a posteriori* necessities of identity or essence.
2   The metaphysical principles are neither analytic, nor are they *a posteriori* necessities of identity or essence.
3   Therefore the metaphysical principles are contingent.[23]

Let's grant (1) (though one might deny it). Roughly, the idea is that a proposition is necessarily true, this must either be because it is true as a matter of meaning (analyticity) or because although it is true *a posteriori*, it is a matter of identity or essence (and hence is true of necessity). Instead, I will focus on (2). It certainly looks implausible that *all* metaphysical principles are analytic: it surely cannot be a matter of meaning that laws are relations of nomic necessitation, or that the things that ground similarities and causal powers are, say, tropes. But perhaps certain mereological and locational principles[24] are analytically true and, more controversially, perhaps it is analytic that only some ways of arranging simples compose something.[25] If some metaphysical principles are analytic then (2) is false.

In the previous section we considered whether the metaphysical principles are *a posteriori* necessary claims of identity or essence. While we concluded that *some* principles are not plausible candidates to be claims of identity or essence, we did not rule out that no principles can plausibly be interpreted as claims of identity or essence. And if some can, then (2) is false.

So while it is plausible that at least some (and perhaps most) principles are neither analytic nor *a posteriori* necessary, it is not obvious that none are. If so, the argument fails. At best, an argument like this shows that *some*, or perhaps *most*, principles are contingent, and hence supports moderate metaphysical contingentism.

3.2.2 The argument from conceivability

The following argument appeals to conceivability as a method for determining what is possible, and aims to show that since we can conceive both of any metaphysical principle being true in some world and of it being false in some other world, it follows that the principles must be contingent.

1   Appropriate conceivability is a good guide to possibility.
2   For each metaphysical principle, we can appropriately conceive of that principle being true in some world, and appropriately conceive of it being false in some other world.
3   Hence for each principle, it is possible that the principle is true and possible that it is false.
4   Therefore, each principle is contingent.[26]

Though it is controversial exactly what role conceivability ought to play in modal epistemology, almost everyone agrees that it will feature in some form or other. Chalmers (2002) and Rosen

(2006) both defend the view that x is possible just in case x is in some way *appropriately* conceivable. So if both x and not x are appropriately conceivable then both x and not x are possible and x is contingent. Chalmers spells out appropriate conceivability in terms of *ideal positive* conceivability, and Rosen in terms of *correct* conceivability. Roughly speaking, both proposals hold that appropriate conceiving is conceiving that is consistent with a full specification of the way the actual world is (Chalmers) or with the natures of the kinds concerned (Rosen). So suppose we are trying to appropriately conceive of water being XYZ. We will find ourselves unable to do this once we have specified the microstructure of actual water. For having so specified, it will no longer be consistent with this specification that water is something other than H_2O.

This brings us back to the arguments from identity and essence. Insofar as a metaphysical principle appears to be a good candidate to be an *a posteriori* necessity of identity or essence, this will be because one cannot appropriately conceive of both its truth and its falsity. So the argument from conceivability is really the flip side of the arguments from identity and essence: the former claims that for each principle we *can* appropriately conceive of it, and its negation; the arguments from identity and essence claim that for each principle we *cannot* appropriately conceive of both it, and its negation.

As we have seen, though, it not clear that the principles stand or fall together. Plausibly, some are claims of identity or essence, and it is impossible to appropriately conceive both their truth and their negation, and others are not, and it is possible to appropriately conceive both their truth and their negation. So the argument from conceivability is not sound, though a similar, amended, argument may show that *some* metaphysical principles are contingent.

4 Conclusion

Consideration of these arguments suggests that if global metaphysical necessitarianism is true, then it is not because there is a single source of necessity for all the metaphysical principles. Instead, it must be because some are necessarily true because they are analytic, and some because they are identity claims or claims about essences, and some for other reasons. If, however, the source of this necessity varies, it is unlikely that it will turn out that all metaphysical principles are necessary. So, at least at this stage of the investigation, the arguments considered here, jointly, suggest that metaphysical moderatism is the most plausible view. Then there is work to be done in establishing the modal status of each metaphysical principle.

Notes

1 With thanks to Michael Duncan, Dana Goswick, Naoyui Kajimoto, Dan Korman, Shang Lu, Kris McDaniel, James Norton, Jon Simon, Rory Torrens, and Jennifer Wang for helpful comments on an earlier draft.
2 By talk of possible worlds I intend to pick out the broadest sphere of 'real' worlds: worlds that represent ways things *really could be*. Sometimes this sphere is known as the metaphysically possible worlds, and sometimes as the logically possible worlds.
3 That is, the principle itself is not internally contradictory.
4 This is to be distinguished from the claim that the laws of nature are contingent in the sense that there are different laws of nature in different worlds (see Schaffer 2005 for a defence of that widely held view).
5 See Sider (1993) and Schaffer (2007) on composition; Sider (2001) on persistence; McTaggart (1908); Bigelow (1996); Markosian (2004); Cusbert and Miller (2018) on the modal status of theories of time; Miller (2013) on the modal status of the nature of properties; and Schaffer (2007) on the modal status of priority principles.). Recent notable dissenters from this orthodoxy have been Parsons (2013), Cameron (2007), and Balaguer (2018) who defend the contingency of accounts of composition; Rosen

(2006) who defends the contingency of the nature of properties; Lewis (1999 p227) who defends the contingency of accounts of persistence; and Bourne (2006), Balaguer (2014), and Le Bihan (2014) who defend the contingency of time's being tensed and Miller (2009) who offers a general defence of contingentism. Finally, Wildman (2018) develops a contingentist account of fundamentality itself.

6 I'm taking grounding to an asymmetric, transitive, irreflexive relation of ontological dependence that obtains between world facts (though for these purposes it doesn't really matter if the relata of grounding relations are, instead, objects and properties rather than facts). See Trogdon (2013) for an overview of grounding.
7 With thanks to Dan Korman for this suggestion.
8 One might deny this in some cases. See for instance McDaniel (2017, chapter 9). But I will suppose, for present purposes, that it is so.
9 Jackson (1998, 2004); Chalmers (2004).
10 This is an example of what is sometimes known as a conditional analysis. See Hawthorne (2001) and Braddon-Mitchell (2003) for views of this kind applied to the analysis of phenomenal concepts. There the idea is, roughly, that it might be that our concept of qualia is such that if actually there are dualistic mental states to which we have certain direct introspective access, then those states are qualitative states and necessarily so, but if there are no such states (because physicalism is true) then some kind of functional states turn out to be the qualitative states.
11 See Miller (2013) for an argument of this kind.
12 This leaves open that the necessitarian might try to show that although, say, tropes exist, they cannot play the property role because (for instance) there are aspects of that role that they are ill-equipped to play.
13 Notice this is consistent with there being necessary claims that are not *a priori*.
14 What are sometimes known as superficially contingent *a priorities*.
15 See Putnam (1967) and Savitt (2000).
16 For consideration of this worry see Miller (2009).
17 Hale and Wright (1994) offer something like this argument in the context of the modal status of mathematical objects.
18 See Rea (1998), Sider (2003), and van Cleve (2008) for a defence of unrestricted composition.
19 See Sider (2013) and Contessa (2014) for a defence of mereological nihilism.
20 Bennett (2011) and deRosset (2013).
21 Where plurality, x, is an intrinsic duplicate of plurality, y, if and only if every one of the xs has an intrinsic duplicate in y, and no one of the ys fails to have an intrinsic duplicate in the xs, and the arrangement of the ys is the same as the arrangement of the xs (i.e. the internal relations that obtain between the xs are duplicated in the internal relations that obtain between the ys).
22 Of course, the contingentist could deny this, and offer an account of why composition is not (or is not necessarily) a function of intrinsicality in this way.
23 Arguments like this can be found in Cameron (2007); Miller (2009); Parsons (2013); and Balaguer (2014).
24 See Saucedo (2011) and Kleinschmidt (2016) for consideration of this idea.
25 See Thomasson (2007).
26 See for instance Balaguer (2018 and 2020).

References

Armstrong, D. (1978). *A Theory of Universals: Universals and Scientific Realism, Volume II*. Cambridge: Cambridge University Press.
Armstrong, D. (1983). *What Is a Law of Nature?* Cambridge: Cambridge University Press.
Balaguer, M. (1998). *Platonism and Anti-Platonism in Mathematics*. Oxford: Oxford University Press.
Balaguer, M. (2014). 'Anti-Metaphysicalism, Necessity and Temporal Ontology'. *Philosophy and Phenomenological Research* 89(1): 145–167.
Balaguer, M. (2018). 'Why the Debate about Composition Is Factually Empty'. *Synthese* 195(9): 3975–4008.
Balaguer, M. (2020). 'Why Metaphysical Debates Are Not Merely Verbal'. *Synthese* 197(3): 1181–1201.
Baron, S., Cusbert, J., Farr, M., Kon, M., and Miller, K. (2015). 'Temporal Experience, Temporal Passage and the Cognitive Sciences'. *Philosophy Compass* 10(8): 560–571.
Bennett, K. (2011). 'By our Bootstraps'. *Philosophical Perspectives* 25(1): 27–41.

Bigelow, J. (1996). 'Presentism and Properties'. In J. Tomberlin (ed.), *Philosophical Perspectives*, Volume 10. Malden, MA: Blackwell, pp. 35–52.
Bohn, Einar (2009). 'Composition as Identity: A Study in Ontology and Philosophical Logic'. *Dissertations* 92. URL=http://scholarworks.umass.edu/open_access_dissertations/92.
Bourne, C. (2006). *A Future for Presentism*. Oxford: Oxford University Press.
Braddon-Mitchell, D. (2003). 'Qualia and Analytical Conditionals'. *Journal of Philosophy* 100(3): 111–135.
Cameron, R. (2007). 'The Contingency of Composition'. *Philosophical Studies* 136: 99–121.
Cameron, Ross P. (2014). 'Parts Generate the Whole but They Are Not Identical to it'. In Aaron J. Cotnoir and Donald L. M. Baxter (eds.), *Composition as Identity*. Oxford: Oxford University Press, pp. 90–107.
Campbell, K. (1997). 'The Metaphysics of Abstract Particulars'. In D. H. Mellor and A. Oliver (eds.), *Properties*. Oxford: Oxford University Press, pp. 125–139. First published 1981, *Midwest Studies in Philosophy* 6: 477–488.
Chalmers, D. (2002). 'Does Conceivability Entail Possibility?' In T. Gendler and J. Hawthorne (eds.), *Conceivability and Possibility*. Oxford: Oxford University Press, pp. 145–200.
Chalmers, D. (2004). 'Epistemic Two Dimensional Semantics'. *Philosophical Studies* 118: 153–226.
Colyvan, M. (1998). 'In Defence of Indispensability'. *Philosophia Mathematica* 6(1): 39–62.
Colyvan, M. (2001). T*he Indispensability of Mathematics*. New York: Oxford University Press.
Contessa, G. (2014). 'One's a Crowd: Mereological Nihilism Without Ordinary-Object Eliminativism'. *Analytic Philosophy* 54(4): 199–221.
Cotnoir, Aaron J. (2014). 'Composition as Identity: Framing the Debate'. In Aaron J. Cotnoir and Donald L. M. Baxter (eds.), *Composition as Identity*. Oxford: Oxford University Press, pp. 3–23.
Cusbert, J. and Miller, K. (2018). 'The Unique Groundability of Temporal Facts'. *Philosophy and Phenomenological Research* 97(20): 410–432.
deRosset, L. (2013). 'Grounding Explanations'. *Philosophers' Imprint* 13(7): 1–26.
Ehring, Douglas. (2011). *Tropes: Properties, Objects and Mental Causation*. Oxford: Oxford University Press.
Field, H. (1980). *Science Without Numbers*. Princeton, NJ: Princeton University Press.
Field, H. (1993). 'The Conceptual Contingency of Mathematical Objects'. *Mind* 102(406): 285–299.
Gilmore, Cody (2014). 'Building Enduring Objects Out of Spacetime'. In Claudio Calosi and Pierluigi Graziani (eds.), *Mereology and the Sciences*. Springer: New York, pp. 5–34.
Hale, B. and Wright, C. (1992). 'Nominalism and the Contingency of Abstract Objects'. *The Journal of Philosophy* 89(3): 111–135.
Hale, B. and Wright, C. (1994). 'A Reduction Ad Surdum? Field and the Contingency of Mathematical Objects'. *Mind* 103(410): 169–184.
Hawthorne, J. (2001). 'Intrinsic Properties and Natural Relations'. *Philosophy and Phenomenological Research* 63(2): 399–403.
Hawthorne, J. (2002). 'Deeply Contingent A Priori Knowledge'. *Philosophy and Phenomenological Research* 65(2): 247–269.
Jackson, F. (1998). *From Metaphysics to Ethics: A Defence of Conceptual Analysis*. Oxford: Oxford University Press.
Jackson, F. (2004). 'Why We Need A-Intensions'. *Philosophical Studies* 118: 257–277.
Johnston, M. (1987). 'Is There a Problem about Persistence?' *The Aristotelian Society* Supp 61: 107–135.
Kleinschmidt, S. (2016). 'Placement Permissivism and the Logics of Location'. *Journal of Philosophy* 113(3): 117–136.
Kripke, S. (1980). *Naming and Necessity*. Cambridge, MA: Harvard University Press.
Le Bihan, Baptiste (2014). 'No-Futurism and Metaphysical Contingentism'. *Axiomanthes* 24(4): 483–497.
Lewis, D. (1983). 'New Work for a Theory of Universals'. *Australasian Journal of Philosophy* 61: 343–377.
Lewis, D. (1986). *On The Plurality of Worlds*. Oxford: Basil Blackwell.
Lewis, D. (1999). *Papers in Metaphysics and Epistemology*. Cambridge: Cambridge University Press.
Linsky, B and Zalta, E. N. (1994). 'In Defense of the Simplest Quantified Modal Logic'. *Philosophical Perspectives* 8: 431–458.
Markosian, N. (2004). 'A Defense of Presentism'. In D. Zimmerman (ed.), *Oxford Studies in Metaphysics: Volume 1*. Oxford: Oxford University Press, pp. 47–82.
Maurin, A. S. (2002). *If Tropes*. Dordrecht: Kluwer Academic Publishers.
McDaniel, K. (2017). *The Fragmentation of Being*. Oxford: Oxford University Press.
McTaggart, J. E. (1908). 'The Unreality of Time'. *Mind* 17(68): 457–474.

Mellor, D. H. (1981). *Real Time*. Cambridge: Cambridge University Press.
Miller, K. (2009). 'Defending Contingentism in Metaphysics'. *Dialectic* 62(1): 23–29.
Miller, K. (2013). 'Properties in a Contingentist's Domain'. *Pacific Philosophical Quarterly* 94(2): 225–245.
Parsons, J. (2013). 'Conceptual Conservatism and Contingent Composition'. *Inquiry* 56(4): 327–339.
Price, H. (1996). *Time's Arrow and Archimedes' Point: New Directions for the Physics of Time*. New York: Oxford University Press.
Putnam, H. (1967). 'Time and Physical Geometry'. *Journal of Philosophy* 64: 240–247.
Rea, M. (1998). 'In Defense of Mereological Universalism'. *Philosophy and Phenomenological Research* 58(2): 347–360.
Resnik, M. (1997). *Mathematics as a Science of Patterns*. Oxford: Oxford University Press.
Rodriguez-Pereyra, G. (2002). *Resemblance Nominalism: A Solution to the Problem of Universals*. Oxford: Clarendon Press.
Rosen, G. (2001). 'Nominalism, Naturalism, Epistemic Relativism'. In J. Tomberlin (ed.), *Philosophical Topics* XV (Metaphysics), pp. 60–91.
Rosen, G. (2006). 'The Limits of Contingency'. In F. McBride (ed.), *Identity and Modality*. Oxford: Oxford University Press, pp. 13–38.
Saucedo, R. (2011). 'Parthood and Location'. In Dean Zimmerman and Karen Bennett (eds.), *Oxford Studies in Metaphysics: Volume 6*. Oxford: Oxford University Press, pp. 225–287.
Savitt, S. (2000). 'There's No Time Like the Present (in Minkowski Space-Time)'. *Philosophy of Science* 67: S563–S574.
Schaffer, J. (2005). 'Quiddistic Knowledge'. *Philosophical Studies* 123: 1–32.
Schaffer, J. (2007). 'From Nihilism to Monism'. *Australasian Journal of Philosophy* 85: 175–191.
Schaffer, Jonathan (2009a). 'Spacetime the One Substance'. *Philosophical Studies* 145(1): 131–148.
Schaffer, J. (2009b). 'On What Grounds What'. In D. Chalmers, D. Manley, and R. Wasserman (eds.), *Metametaphysics: New Essays on the Foundations of Ontology*. Oxford: Oxford University Press, pp. 347–383.
Schiffer, S. (1996). 'Language Created, Language Independent Entities'. *Philosophical Topics* 24: 149–167.
Shapiro, S. (1997). *Philosophy of Mathematics: Structure and Ontology*. New York: Oxford University Press.
Sider, T. (1993). 'Van Inwagen and the Possibility of Gunk'. *Analysis* 53(4): 285–289.
Sider, Theodore (2001). *Four-dimensionalism*. Oxford: Oxford University Press.
Sider, T. (2003). 'Against Vague Existence'. *Philosophical Studies* 114: 135–146.
Sider, Theodore (2007a). 'Parthood'. *Philosophical Review* 116(1): 51–91.
Sider, T. (2007b). 'Against Monism'. *Analysis* 67(1): 1–7.
Sider, T. (2013). 'Against Parthood'. *Oxford Studies in Metaphysics* 8: 236–293.
Thomasson, A. (2007). *Ordinary Objects*. Oxford: Oxford University Press.
Thomson, J. J. (1983). 'Parthood and Identity Across Time'. *Journal of Philosophy* 80: 201–220.
Tooley, M. (1977). 'The Nature of Law'. *Canadian Journal of Philosophy* 7: 667–698.
Trogdon, K. (2013). 'An Introduction to Grounding'. In Miguel Hoeltje, Benjamin Schnieder, and Alex Steinberg (eds.), *Varieties of Dependence: Ontological Dependence, Grounding, Supervenience, Response-Dependence (Basic Philosophical Concepts)*. Munich: Philosophia Verlag, pp. 97–122.
van Cleve, J. (2008). 'The Moon and Sixpence: A Defense of Mereological Universalism'. In Theodore Sider, John Hawthorne, and Dean W. Zimmerman (eds.), *Contemporary Debates in Metaphysics*. Oxford: Blackwell, pp. 321–340.
van Inwagen, P. (1990). *Material Beings*. Ithaca, NY: Cornell University Press.
Wildman, N. (2018). 'On Shaky Ground?'. In Ricki Bliss and Graham Priest (eds.), *Reality and its Structure*. Oxford: Oxford University Press, pp. 275–291.
Williamson, T. (1986). 'The Contingent A Priori: Has it Anything to do with Indexicals?' *Analysis* 46(3): 113–117.
Williamson, T. (2010). 'Necessitism, Contingentism, and Plural Quantification'. *Mind* 119(475): 657–748.
Williamson, T. (2013). *Modal Logic as Metaphysics*. Oxford: Oxford University Press.
Yablo, S. (2005). 'The Myth of Seven'. In Mark Eli Kalderon (ed.), *Fictionalism in Metaphysics*. Oxford: Clarendon Press, pp. 88–115.
Zimmerman, D. (2005). 'The A-theory of Time, the B-theory of Time and "Taking Tense Seriously"'. *Dialectica* 59(4): 401–457.

Further reading

Kristie Miller in 'Defending Contingentism in Metaphysics', *Dialectica* (2009) 62(1): 23–29, and in 'Three Routes to Contingentism in Metaphysics', *Philosophy Compass* (2010) 5(11): 965–977 are accessible general introductions to the idea of metaphysical contingentism, and include presentation of a range of arguments for the view. Gideon Rosen, in 'The Limits of Contingency', in F. McBride (ed.), *Identity and Modality*, Oxford: Oxford University Press (2006), pp. 13–38 is a more detailed examination of contingentism. Mark Balaguer, in 'Anti-Metaphysicalism, Necessity and Temporal Ontology', *Philosophy and Phenomenological Research* (2014) 89(1): 145–167, and Mark Colyvan in *The Indispensability of Mathematics*, New York: Oxford University Press (2010) are both discussions of contingentism in the philosophy of mathematics. Ross Cameron, 'The Contingency of Composition', *Philosophical Studies* (2008) 136: 99–121 and Josh Parsons, 'Conceptual Conservatism and Contingent Composition', *Inquiry* (2013) 56(4): 327–339, are both discussions of contingency in the domain of composition.

33
IS METAPHYSICS SPECIAL?

Thomas Hofweber

What is the question?

The question whether metaphysics is special can be understood in a number of different ways. Being special involves an implicit contrast class: special compared to what? And being special involves an implicit dimension: special in what way? Thus there are many questions one might ask when one asks whether metaphysics is special. And metaphysics might be special in several of these ways: it might have the highest dropout rate among philosophy graduate students over 30. That would make it special with respect to dropout rate compared to other sub-disciplines in philosophy. But this is clearly not what is intended by the question as it would occur in a handbook on metametaphysics. I will focus on it being special in a way that gets at what has been a longstanding debate about the status of metaphysics. In this sense philosophers have long thought that metaphysics was special in one of two ways: either especially grand and glorious, or especially confused and problematic. Metaphysics could be special in a good way, or in a bad way. It could be special in some way compared to other parts of philosophy, or compared to other parts of inquiry in general. Philosophers have defended one or the other option throughout the history of philosophy, including to this day. Some have thought that metaphysics has a distinguished place in inquiry, one that gives it a special standing among all other parts of inquiry. Metaphysics is the queen of the sciences, the discipline that truly reveals what reality is like, our best guide to ultimate reality, and so on. Others have held that metaphysics is special in a purely negative way: especially confused, or misguided, a mere pretender among the sciences. On this line, metaphysics is not really a legitimate part of inquiry at all, but rather something like an outgrowth of an illegitimate attempt to try to find out what the world is like with pure speculation, while having the guts to claim that this is possible. Both of those attitudes have been around for a long time, and it is hopeless and impossible to survey the history of the debate about whether metaphysics is special in a good or a bad way in this short chapter, and so I won't even attempt to do so. Instead of discussing the literature, I will highlight and discuss a few ways in which one might naturally think that metaphysics is special in the relevant senses. I will refer to a few publications that push a particular line discussed in the main text in more detail, but I can't hope to give anything like a survey of the literature. To get started, I will first introduce a little bit of terminology about how we should think of metaphysics and what it is supposed to do, and then look at whether metaphysics might be special in a glorious way, and finally whether it might be especially problematic or dubious.

Metaphysics

What metaphysics is supposed to do, if anything, is controversial, and I won't try to answer this question here. But even without having settled what metaphysics is supposed to do more precisely, we can nonetheless distinguish various aspects of what it might do. Most philosophers, even those critical of metaphysics and sympathetic to the idea that metaphysics is special in a bad way, do not think that there is nothing for metaphysics to do and that it should just go away. Many think that there is something useful for metaphysics to do, although they disagree on what that is. Some might think that although metaphysics itself does not rise to the status of a science or a real partner of the sciences in inquiry, it nonetheless has a useful auxiliary role to play: it might clarify concepts, or point to confusions, or raise problems that are not yet addressed in the sciences, and so on. Let us call *the task* of metaphysics whatever metaphysics is supposed to do. This leaves open whether metaphysics has a place in inquiry alongside the sciences, or possibly even above the sciences as their queen. Metaphysics would have a task if it has some job to do. And even anti-metaphysical philosophers often think that it has something to do, even if it isn't settling questions of fact, and thus has a task. The question remains whether that task is anything like the task of other parts of inquiry. Inquiry in general tries to find out what the world is like, what is true, and what the facts are. Does metaphysics do this as well?[1] If so, and thus if metaphysics properly should aim at determining what the world is like and which facts obtain, then we can say that metaphysics has *a domain*. The domain of metaphysics consists of those facts, which it should figure out whether they obtain. Or equivalently, the domain of metaphysics consists of those questions of fact that metaphysics should aim to answer. We can also call those *a question for metaphysics*. Biology has such a domain: the facts concerning living things, or something near enough. Physics has a domain, and so on. Does metaphysics have one too? If so, which facts or which kinds of facts are in it? One straightforward answer is that metaphysics has a domain and that its domain is captured by the question: what is reality like? This is fair enough, but it also could be seen as the answer that any other part of inquiry could also give. Biology could take its domain to be articulated by the question what reality is like, with special emphasis on living things, and so on. Everyone who engages in inquiry ultimately has reality in mind and ultimately wants to find out what it is like. So, even if metaphysics has a domain, the next question is whether it has *its own domain*: some facts that metaphysics in particular is supposed to investigate, or some questions of fact that are properly addressed by metaphysics and especially metaphysics. If metaphysics has its own domain the issue will be what distinguishes the facts in its domain from those in other domains and how do they all relate to each other to form one totality of facts.

Besides the issue of the task of metaphysics and the domain of metaphysics, there is the question of the method of metaphysics. Are the questions in the domain of metaphysics to be addressed with a distinct method, one that applies to metaphysics, but not to other parts of inquiry? Or are the methods of metaphysics the same as those properly employed in other parts of inquiry?

Metaphysics could be special with regard to all three of these things: the task, the domain, or the method. It could be that the task of metaphysics is rather different than the task of other parts of inquiry, be it other parts of philosophy more narrowly or other parts of inquiry more broadly. Maybe what metaphysics is supposed to do is very different, or rather similar, to what philosophy in general, or science in general, is supposed to do. It could be that the domain of metaphysics, if there is one, is rather similar or rather different than the domains of other parts of inquiry. Maybe metaphysics aims at a distinct class of facts, or maybe it is just concerned with the same kinds of facts as the sciences, but possibly in a more general or abstract way? And it could be that the methods of metaphysics are rather similar or rather different than the methods employed in the sciences. Maybe metaphysics tries to find out what reality is like a priori, and the sciences try to do the same a posteriori? Any one of these ways might make metaphysics special, and we should now look more closely at how one might think that metaphysics is special, be it in a good way or a bad way.

Is metaphysics especially glorious?

There is a long tradition within philosophy to give metaphysics a special status among all the sciences. This tradition holds that metaphysics is glorious: it is in the business of fact-finding, and it has a special role in that business overall. Not only does metaphysics have a domain, and thus some facts to investigate, just like biology or other sciences, but metaphysics has a special place among all fact-finding parts of inquiry. I want to consider two ways in which this might be: first, metaphysics is special, since the facts in its domain are more central than the facts in any other domain, for example because all other domains depend on metaphysics somehow. Second, metaphysics is special, since the facts in its domain are the most revealing of reality: they show what reality is really or ultimately like. Either way, metaphysics would be especially glorious. It could be the queen of the sciences, or the true revealer of reality. We will look at them in turn.

Metaphysics as the queen of the sciences

To think of metaphysics as the queen of the sciences can be motivated in at least two main ways: metaphysics as the great unifier, and metaphysics as the discharger of presuppositions. On the first, metaphysics, or a related larger part of philosophy, is seen to unify the results of the other sciences in a way that none of the other sciences do, but that is required to bring all of science together. Only in metaphysics do we see the larger picture of what reality is like, although smaller parts of this picture are painted in the individual sciences. Metaphysics is a central hub where all the pieces are being put together. The sciences each deliver their results to this hub, and philosophy and metaphysics put them all together into one overall picture of reality, possibly with some additions and augmentations. So understood, metaphysics has a domain and it has its own domain: it concerns certain large-scale facts about how the different domains investigated in the particular sciences come together and relate to each other. Metaphysics thus has a special place in inquiry, it is the central place where the puzzle pieces are being put together.

This picture, however, is rather problematic, both regarding science as well as metaphysics. It isn't the case the individual sciences merely locally look at their areas with no regard to what is going on elsewhere. Science itself is not that local, and although no single science has as its domain the totality of facts, all sciences have the totality of facts in mind, at least implicitly, and how their domains relate to other domains, in particular nearby ones. In addition, metaphysics as it is commonly practiced is not at all like gathering facts from various sciences and putting them together into one picture. Some aspects of metaphysics are certainly related to this issue and to how the different sciences relate to each other (when considering questions of emergence, for example), but metaphysics in general is not like that. Just think of any one of the standard metaphysical problems: freedom of the will, the ontology of numbers, essence and modality, and so on. It would be hard to see what to do with these areas of metaphysics if one held that the domain of metaphysics concerns facts about unifying the sciences.

A second attempt to give metaphysics a special status among all parts of inquiry, and to support the "queen of the sciences" metaphor, is to think of metaphysics as the discharger of presuppositions, and the discipline that puts the sciences on a firm footing. To illustrate this approach, consider questions about change in science and in metaphysics. The sciences try to find out, amongst other things, how things change over time: the dynamical laws of physics, the evolution of species, the change of materials under pressure, and so on. But, the argument goes, the sciences simply presuppose that change is possible at all, and then aim to determine what changes happen, given this presupposition. This assumption that change is possible at all is an assumption made at the outset, an assumption that is never cashed in within the sciences. Without cashing it in the results of the sciences are only conditional: if change is possible at all then things change this way. And this is where metaphysics comes in: it shows how change is possible at all, and with showing this, metaphysics puts the sciences on solid ground. It takes the queen of the sciences to establish unconditional versions of the results of the sciences. The sciences depend on metaphysics for having unconditional results, and only metaphysics can help the sciences to achieve them. The special domain of metaphysics so understood can thus be seen as facts about what is possible: the possibility of change, the possibility of time, and so on.[2]

But is it true that the sciences simply assume that change is possible at all, and then do their work conditional on this assumption? An alternative way of looking at it is that the sciences establish various results, including things like that the polar ice caps are melting, and that these results imply that things change and that change is thus possible. Why think that the possibility of change is assumed or presupposed by the sciences, rather than implied by their results? To compare the situation to a simpler case: the sciences have found out that pure plutonium is silvery-gray in color. Did they assume or presuppose that it is colored at all, and then, under that assumption, establish that it has that particular color? Or did they establish empirically what the color is, and then conclude from that that it is colored at all? In that case I would think that it is the latter. That pure plutonium, something never observed before 1940 or so, has a color at all might be an open question, although, of course, it is reasonable to predict that it is not translucent. But was it assumed or presupposed that it has a color, or established empirically, by seeing that it is silvery-gray? Again, the latter. And so it seems with change. That things change in the world was established empirically, although maybe not with certainty, but nonetheless similarly as most other things are established empirically. If so, then the results of the sciences hold unconditionally, although not with certainty, and they do not need final vindication from philosophy and metaphysics. Thus it can seem rather dubious to try to establish that metaphysics is the queen of the sciences along those lines.

Overall then, to hold that metaphysics is special among other parts of inquiry in a glorious way, since it is the queen of the sciences, is not an easy line to defend.

Metaphysics as the true revealer of reality

Another way in which metaphysics might be special in a glorious way is that it might go deeper than the sciences. The sciences uncover what reality is like, what is true, and what is the case. But metaphysics goes further. There are different ways of trying to articulate how it goes further, with some prominent options including the following: metaphysics tries to find out what reality is ultimately like, what fundamental reality is like, what is ultimately true, or what is true in reality. And this is claimed to be especially revealing of reality. To pick just one example how this might go, the sciences might give a description of the empirical world: what is in it, how it changes, etc. Metaphysics, on top of that, might determine that ultimately empirical things are ideas in the mind of God, or bundles or tropes, or what have you. This would not conflict with the sciences, but add another level of description to what reality is like. And that level is in a sense the deepest one, one that shows what really is the case, behind the appearances, or what reality is fundamentally like, or something similar. Metaphysics is special in this sense, since it uncovers a description of reality that is most revealing, or at least more revealing than those of the sciences. It does not merely describe the appearances, but uncovers what reality is ultimately like.

This way of taking metaphysics to be special relies on a distinction between what is the case and what is ultimately the case, or a similar version of it. Can this distinction be made sense of? There are several distinctions in the neighborhood that do make sense, but that won't help here. For example, there clearly is a legitimate distinction between what seems to be the case and what is the case. This is the distinction between appearances and reality, understood in a particular way: how reality appears to us, and how it in the end is. But so understood the distinction does not help carve out a special place for metaphysics, since the sciences clearly aim to find out how things are, not just how they appear to be. This alone won't help.

But maybe there is a different way of understanding something like the appearance-reality distinction, or the distinction between a more superficial and a deeper description of reality. And maybe metaphysics can claim for itself the deeper description as its domain. But what is this distinction? And why should we think that the level that belongs to metaphysics is deeper or more revealing than the one that belongs to the sciences, however the distinction is drawn? There are two main ways to try to approach this: one is to hold that there is a primitive notion of being "ultimate" or "really" or "reality," a notion that can't be defined or spelled out, but that we grasp and can make sense of. Using this notion one can then distinguish what is true from what is really true or ultimately true or true in reality. On such an approach it might then be a primitive, unexplained insight that what is ultimately true is deserving of greater attention and is more revealing than what is merely true. Second, one could hold that such a notion can be spelled out, propose a way to spell it out, and then explain why so spelled out the ultimately true is special. Both approaches face serious obstacles, although versions of both approaches are popular in metaphysics. The first approach was championed in recent work by Kit Fine, in particular (Fine, 2001), who argues that we need to assume a notion of *reality* in metaphysics, and that this notion should be accepted even if it cannot be spelled out any further. In particular, this notion of reality is to be contrasted with an ordinary, naive notion of reality, where reality contrasts with fiction or things that are not the case. Instead, what is true in reality, in Fine's sense, contrasts in particular with what is merely true, not only with what is false. What is true in reality is of special concern for metaphysics, maybe even the distinct domain of metaphysics, whereas the sciences primarily aim at what is true. But do we have a grasp of such a metaphysical notion of reality, and do we understand when we ask about what is true in reality, as opposed to what is merely true? This way of understanding the domain of metaphysics, as concerning what is ultimately the case, opens itself up to a common criticism of metaphysics, one that is connected to the position that metaphysics is special in a problematic way. This criticism will be discussed in more detail below, and it concerns that metaphysical questions rely on meaningless expressions, and the project of metaphysics is based on an illusion of meaning. Whether this criticism applies here is, of course, not clear so far. Nonetheless, it might seem suspicious that we would have a primitive notion of reality or being ultimately the case, where this notion is distinct to metaphysics. If metaphysics concerns what is true in metaphysical reality, and that notion can't be spelled out or explained to others, then this can give rise to the suspicion that metaphysics is an esoteric discipline: you need to be a metaphysician to know what the discipline is even supposed to be about.

The second option, to spell out the relevant notions of metaphysical reality or being ultimately the case, has a different set of problems. Although many attempts have been made to get clearer on what such a distinction might be, they have generally not been very succesful.

One could try to say what being ultimately the case consists in. And there are several options that one could pursue: maybe it consists in being fundamental in some sense, or being a supervenience base for the rest, or being a final truth-maker of other truths, or something along those lines. But in general one faces a dilemma in such attempts at making these notions explicit: either they are too weak and not distinctly metaphysical enough to do the work they are intended for, or they rely on other primitive metaphysical notions that are just as problematic as the notion of metaphysical reality or being ultimately the case. The former horn of the dilemma is widely seen to apply to attempts that heavily rely on the notion of supervenience. Although this notion can be defined precisely in several different ways, relying on modal notions, like what is necessarily the case, the result is often considered weak and unsuitable to carve out a domain for metaphysics with proper significance. To spell out a notion like metaphysical reality that is suitable to give metaphysics a special place can easily push one into the first horn of our main dilemma, since such attempts could rely on a primitive metaphysical notion that is just as problematic as a primitive notion of metaphysical reality. For example, the proposal that metaphysical reality is the totality of all facts that give the *ultimate* explanation of the other facts is in danger of being in that position, unless it is made clear what ultimate explanation is and how it relates to just plain old explanation. If it is taken to be a primitive kind of metaphysical explanation then little progress would have been made. And similar worries apply to the other options one has, relying on notions like fundamentality, reality, grounding, and so on.

Whether the relevant notion of, for example, being ultimately the case is primitive or can be spelled out or not, it remains unclear why what is ultimately the case is more revealing than what is the case. Simply because we use the grand sounding word "ultimately" for this notion does not thereby give it special significance. Anyone who holds that metaphysics is especially glorious in this way will have some work to do to motivate this claim for specialness. Why should we especially care and attach special value to what is ultimately the case? There are some clear strategies that might answer this: what is ultimately the case is tied to what is fundamental, which is tied to what explains the rest, and there is value in explanation. And similarly for other attempts that spell out being ultimately the case. But it is not clear how such an account would work for a primitive notion of being ultimately the case, and it is not clear if it would work even if it is spelled out in terms of fundamentality and explanation, since the relevant notion of fundamentality and explanation might be of a distinctly metaphysical kind, one that isn't obviously tied to fundamentality and explanation as it is used in the sciences. All these issues deserve further and much more detailed discussion and they have been widely discussed in the literature.[3] But overall we can say that it is not clear whether metaphysics is especially glorious, since it is the true revealer of reality.

Is metaphysics especially problematic?

There is a long tradition in philosophy to think that there is something profoundly wrong with metaphysics. There is no single reason why metaphysics seems to be especially problematic, but there are two that stand out: worries about meaning and worries about epistemology. On the former the charge is that metaphysics relies on terminology that is not fully meaningful or that is falsely taken to have a different meaning when used in metaphysics than it normally has. On the latter, epistemological worry, metaphysics is charged to be pure speculation, with no basis for settling the questions it wants to answer one way or another. We will look at them in turn.

Is metaphysics special?

Metaphysics as the babbler in the meaningless

Metaphysics can be seen as often making meaningless or semantically confused assertions, and as an attempt to answer not well-defined questions. The suspicion that there is a deep confusion about meaning at the heart of many metaphysical projects can be supported by a more careful look at some of the assertions that metaphysicians are hoping to make. This can range from criticizing particular sentences, like the famous "The nothing nothings" example, to criticizing the central reliance on expressions like "ultimately," "really," or "fundamentally," as used above, where metaphysics was understood as being concerned with what is ultimately the case. Furthermore, one might argue that a particular feature of metaphysical assertions, namely that it is not clear how they could be tested or verified empirically, shows that these assertions are devoid of meaning, since all meaning must be tied to experience and verification. Such a criticism of metaphysics, which is based on some general criterion of being meaningful, might seem rather dated these days, although it was prominent during the days of logical positivism. But the more general worry remains without such a criterion. For example, one might point to a pattern in how metaphysical projects are motivated that can be accused of relying on a confusion about meaning. Here is one example how this can go.

In the debate about the ontology of numbers there is a standard motivation for the significance of this project that comes from thinking about mathematics as a whole and from the outside. Does mathematics aim to describe a part of reality which is there independently of our mathematical practice, or is it a game with certain rules that we figure out as we go along? Is mathematics discovery or invention, etc.? One way to support the discovery side is to claim that there is an ontology of mathematical objects, things which exist and which are talked about in mathematics. This side accepts a certain picture of mathematics as describing parts of reality, and in particular that there are these parts which are described this way. So, this side accepts the existence of mathematical objects, objects like numbers, functions, and so on. And there being such objects is a central part of this picture. So, one might think that the question whether there are such mathematical objects is a great candidate for being in the domain of metaphysics. It seems to be just the kind of question that pushes you metaphysically one way if answered one way, and another way if answered another.

But this is problematic since it sure seems that the question whether there are numbers and functions is settled within mathematics itself. Many theorems of mathematics imply that there are numbers and functions. And thus they imply that there are mathematical objects. Mathematics itself has shown that there are infinitely many prime numbers, so there are infinitely many numbers, so there are numbers. But this is not how it was supposed to go. The question whether there are numbers was supposed to be a question that is central to understanding mathematics from the outside, a question that divides two main sides of how mathematics as a whole is supposed to be understood philosophically. It was supposed to be a metaphysical question, one about the ontology of numbers and one in the domain of metaphysics not in the domain of mathematics.

This can be taken to show that the metaphysical question is based on a confusion about meaning. Metaphysics hopes to ask a question with certain grand metaphysical features, and it articulates the question in a particular way ("Are there numbers?"), but that question with that meaning doesn't seem to have any of the grand features. Instead of a grand metaphysical question the metaphysician asked a trivial mathematical question. They were simply confused about this, maybe hoping that the question they were aiming to ask is a different question than the trivial mathematical one, but this hope is futile. The metaphysical project is thus confused, and the confusion is one about meaning. The question the metaphysician is asking is trivial, not deep.[4]

Similar criticisms can be brought up against many other metaphysical projects. Are we free? Are there tables? And so on. In each case there are some ordinary meanings associated with the relevant words "free" and "table," and with those meanings it is trivial that we are free and that there are tables. What the metaphysician is trying to do is to ask other, nontrivial questions with those same words, but there are no other questions in the neighborhood here. The project of metaphysics is, or is largely, based on a confusion about meaning, or so goes the worry.

If this is to be accepted then metaphysics would not have a domain in the above sense, but it might still have a task. Its task might well be understood as unearthing these misuses of language and making sure that they do not happen any more. Metaphysics thus turns away from reality, and focuses on language, and the confusions that arise from language. The task so understood can be largely critical, making sure no further errors happen, or more constructive, proposing a better language that is less likely to lead to confusion. Either way, metaphysics is far from the queen of the sciences, and in fact not a real part of inquiry at all.

How should one assess this situation? One reaction can simply be that metaphysics does not have its own domain. It asked a question – are there numbers? – which is answered in mathematics: yes. Similarly, metaphysics might ask some general questions about the material world, and those questions are then answered in the sciences. Metaphysics would thus be more in the business of asking questions than answering them, and it would ask very general questions, maybe too general, or too trivial, for the sciences themselves to ask. But if this is all there is to metaphysics, this would certainly warrant seeing it in a negative light and not an equal partner in inquiry. But there are also other options.

First there is an issue about whether we are asking the same question when we ask whether there are numbers in mathematics and in metaphysics. Although it is natural to express the question in each case with the words "Are there numbers?," it is not clear that these utterances of those words express the same question. This general possibility should be clear, since in many other, ordinary, cases the same words can be used to ask different questions. The issue is whether this indeed applies here as well. One could get a ring of a difference when one considers that normally when asking this question one is after examples. So, when I ask "Are there prime numbers between 10 and 15?" I ideally want to have an example of such a prime number as an answer. But when I ask whether there are numbers in metaphysics then I am not satisfied with an example of a number, but I am asking about what kinds of things exist, what the world is made from, or something along those lines. The question remains why uttering these words in these different situations leads to different questions. How does this difference arise from those words? Are some of them ambiguous or polysemous? Is this difference tied to philosophy, or does it arise in ordinary communication and then lead to a philosophically significant consequence? These are not easy questions to answer, but it should be clear that the general possibility of two questions expressed with the same words should be considered an option.[5]

Second, there is an issue about whether the metaphysical question was properly expressed with the words "Are there numbers?" Maybe, contrary to the proposal just discussed, those words can only be used properly to ask one and the same question, but the question we wanted to ask in metaphysics is not the one expressed in those words. Instead, the question properly articulated should be loaded up with more words, including more metaphysical terminology. So, the mathematical question might well be "are there prime numbers between 10 and 15?" and with it "are there numbers?," both of which have a trivial affirmative answer. But the metaphysical question instead is "are numbers among the fundamental things?" or "are numbers real?" or "are there really numbers?" or "are numbers part of reality?" So understood asking the metaphysical question is not making a confusion about language. The only confusion was to articulate the metaphysical question incorrectly as "are there numbers?" That confusion, the proposal goes, can be overcome quite easily: just ask the better articulated question instead. All one would have to explain is why we articulated the question we wanted to ask in this incorrect way.

Just as the proposal discussed above about metaphysics being the true revealer of reality, this proposal faces a similar worry. To repeat a question discussed in some more detail above, how is the distinction between there being numbers and there being numbers in reality to be understood? After all, anyone who thinks that there are numbers would think that they are part of reality. What else would they be part of? This is not to say that such a distinction cannot be drawn, but certainly more needs to be said.[6]

Overall then it is not so clear whether metaphysics is especially problematic, since it is based on a confusion about meaning.

Metaphysics as unjustified speculation

Even if metaphysical theses and questions are fully meaningful, a worry remains that attempting to answer them is pure speculation, carried out on the basis of vastly insufficient evidence. As such, metaphysics is highly problematic, as it attempts to answer questions that we are in no position to answer, not just now, but in principle. Let me illustrate this issue with the debate about the existence of ordinary objects. Consider the question about composition: do the particles that are arranged like a table compose a table? That is, is there simply an array of particles, and nothing else, or is there a table in addition to the particles? Suppose, as many, but not all in this debate think, that there is a real, meaningful, and nontrivial question about whether there is a table in addition to the particles arranged table-wise. How could we tell the difference? One attempt to answer it is to insist that one can see the table. But would things not look exactly the same to us if there was no table, but only particles arranged table-wise? In both cases, one might insist, would I form the belief that there is a table, but only in one of them would this belief be true. We seem to see a table, but we can't tell by just opening our eyes whether this seeming is correct. But then, how else could we decide? Could science decide it? Do we have evidence from science that there is a table? It might seem that it would be question begging to describe the evidence in terms of ordinary objects, as evidence that the table is this heavy, or the like. Why not describe it as evidence that the particles arranged table-wise are collectively this heavy? Maybe the issue thus can't be settled empirically. That doesn't mean it can't be settled. There are numerous other considerations that could come into play. One set of arguments is that one side or the other is not coherent: tables are incoherent, somehow, or just particles arranged table-wise with no composition of tables is incoherent. But this seems pretty tough to defend, and I won't try to pursue it. Instead it is more popular to focus on considerations about theoretical virtues: by considerations concerning simplicity or fruitfulness or parsimony or the like can we see that one side is better than the other. No tables leads to a simpler or more parsimonious theory, and so we should favor that composition does not occur. Or tables leads to a more fruitful theory, and so we should favor composition occurring. But this puts all the weight and all the sources of evidence on theoretical virtues. How much weight can we really give theoretical virtues alone, and how much of their appeal comes from our desire for simplicity and parsimony for practical reasons, rather than there being any grounds that the world is simple and parsimonious. In light of this one might conclude that the whole debate is pointless, not because the question is meaningless, but because the answer is too elusive.

This line of reasoning can be challenged in numerous ways. First, it is not at all clear that we do not have empirical evidence for the existence of tables, and thus for the occurring of composition. That perception presents the world to me in terms of objects might well be evidence for them, even if I would have this belief falsely in cases where there are no objects like tables, but only particles. There is a further issue about whether the evidence we originally have gets defeated in light of metaphysical considerations, and thus the evidence goes away in light of further thought. Either way, the issue is there whether we do have such evidence, and the answer is not completely clear.[7] Second, it is unclear whether considerations about theoretical virtues alone carry much weight without the addition of other evidence. Maybe metaphysical considerations are closely tied to overall theoretical virtues in purely non-empirical theorizing. That should not be taken as a bad thing, or something that shows that metaphysics is pointless, but instead as a somewhat distinctive methodology of metaphysics: relying on theoretical virtues alone.[8]

These issues naturally do not only arise for the debate about composition. Many other debates can be understood along similar lines: the evidence that is being presented for one view or the other is one largely about which overall picture makes the most sense and is the easiest to digest. But the considerations brought in up in favor of one side over the other can be seen as not presenting evidence or giving reasons, but rather as expressing a desire or preference: this would be simpler, and thus preferable, that would be more appealing to us in some other way, and thus preferable, etc. But pointing to the truth is another matter, and here, the worry goes, we have little to show for. If considerations in metaphysics indeed were of this kind, and if the theoretical virtues are not themselves to be taken as reasons for truth then maybe there is little left of metaphysics. But whether this is indeed so is far from clear. Thinking of metaphysics as purely being based on theoretical virtues is not exactly plausible as a reconstruction of actual metaphysical debates, and thinking of theoretical virtues as merely expressions of wishful thinking is not exactly how theoretical virtues are relied on in other parts of inquiry.

Thus overall we can say that here, as in all of our other cases before, it is not clear which side in this debate is correct.

Conclusion

Whether metaphysics is special, either when compared to the natural sciences or to other parts of philosophy, is unclear and subject to a longstanding and ongoing debate and discussion. Much of this debate concerns the nature of metaphysics and how it compares to other parts of inquiry with respect to its task, domain, and method. As was the case throughout the history of philosophy, to the best of my knowledge, so is the case these days that some philosophers take metaphysics to be special in a negative way, and others take it to be special in a positive way. Although much progress has been made in this debate, and many of the positions have been developed and defended in much greater detail and with much force, the debate is ongoing and it seems fair to say that nothing resembling consensus on the special status of metaphysics has been reached.

Notes

1 Here we should more properly distinguish facts about what is the case, and other facts, like facts about what languages we should use, or what confusions we should avoid. Metaphysics, on the anti-metaphysical line, would only be concerned with facts of the latter kind, while the sciences are concerned with facts of the former. I will gloss over this in the following.
2 A position of this kind is defended by E.J. Lowe in Lowe (1998).
3 The literature on these issues is vast, and hopefully other chapters in this handbook discuss some of it in more detail. Still, here is a small selection of relevant work: For a defense of a place of a primitive notion of reality in metaphysics, see Fine (2001, 2005). For another primitivist version, see Sider

(2011). For criticism of such approaches, see chapter 13 of Hofweber (2016) and Dasgupta (2018). For a discussion of supervenience, see Kim (1993). For truth-making, see Cameron (2008).
4 A famous objection to metaphysics along those lines is due to Rudolf Carnap in Carnap (1956). A contemporary version is defended by Amie Thomasson, in particular in Thomasson (2015).
5 I have defended this answer in Hofweber (2005) and in particular Hofweber (2016). There I argue that the two readings of the question arise from two functions that polysemous quantifiers have in ordinary communication and that the affirmative answer to the mathematical question does not answer the metaphysical question.
6 And, of course, more has been said, for example in Fine (2009), Schaffer (2009), and many others.
7 See Merricks (2001) for an argument that the evidence gets defeated, and Hofweber (2016) for an argument that it does not.
8 See Sider (2013) for more on theoretical virtues in the debate about composition, Korman (2015) for more about objects in general, and Bennett (2009) for a defense that the issue is elusive and Bennett (2016) for an argument that metaphysics is not distinctly different than other parts of philosophy.

References

Bennett, K. (2009). Composition, coincidence, and metaontology. In Chalmers, D., Manley, D., and Wasserman, R., editors, *Metametaphysics: New Essays on the Foundations of Ontology*, pages 38–76. Oxford University Press.

Bennett, K. (2016). There is no special problem with metaphysics. *Philosophical Studies*, 173:21–37.

Cameron, R. (2008). Truthmakers and ontological commitment: or how to deal with complex objects and mathematical ontology without getting into trouble. *Philosophical Studies*, 140:1–18.

Carnap, R. (1956). Empiricism, semantics, and ontology. In *Meaning and Necessity*, pages 205–221. University of Chicago Press, 2nd edition.

Dasgupta, S. (2018). Realism and the absence of value. *Philosophical Review*, 127(3):279–322.

Fine, K. (2001). The question of realism. *Philosophers' Imprint*, 1(1):1–30.

Fine, K. (2005). Tense and reality. In *Modality and Tense*, pages 261–320. Oxford University Press.

Fine, K. (2009). The question of ontology. In Chalmers, D., Manley, D., and Wasserman, R., editors, *Metametaphysics: New Essays on the Foundations of Ontology*, pages 157–177. Oxford University Press.

Hofweber, T. (2005). A puzzle about ontology. *Noûs*, 39:256–283.

Hofweber, T. (2016). *Ontology and the Ambitions of Metaphysics*. Oxford University Press.

Kim, J. (1993). Concepts of supervenience. In *Supervenience and Mind*, pages 53–78. Cambridge University Press.

Korman, D. (2015). *Objects: Nothing Out of the Ordinary*. Oxford University Press.

Lowe, E. J. (1998). *The Possibility of Metaphysics*. Oxford University Press.

Merricks, T. (2001). *Objects and Persons*. Oxford University Press.

Schaffer, J. (2009). On what grounds what. In Chalmers, D., Manley, D., and Wasserman, R., editors, *Metametaphysics: New Essays on the Foundations of Ontology*, pages 347–383. Oxford University Press.

Sider, T. (2011). *Writing the Book of the World*. Oxford University Press.

Sider, T. (2013). Against parthood. In Bennett, K. and Zimmerman, D., editors, *Oxford Studies in Metaphysics*, volume 8, pages 237–293. Oxford University Press.

Thomasson, A. L. (2015). *Ontology Made Easy*. Oxford University Press.

PART V

Science and metaphysics

34
SCIENCE-GUIDED METAPHYSICS

Kerry McKenzie

Introduction

Though its very right to exist is regularly called into question by those proudly identifying as 'scientistic,' it seems that metaphysics is very much alive and kicking in today's philosophy of science. For the contemporary field is teeming with attempts to mine scientific theories for metaphysical conclusions – thus creating what is sometimes called 'naturalistic' metaphysics, 'scientific metaphysics,' or (most explicitly) 'science-guided metaphysics' (SGM). Philosophers have dug deep into our best contemporary physics, for example, in search of the world's fundamental category, whether it be that of objects, structures, or events (Čapek 1984; French 2014); and as to which, if any, of the traditional conceptions of the nature of properties give the best interpretation of the fundamental quantities (Maudlin 2007, Chapter 4; Kuhlmann 2010). Those same theories have been invoked to argue that the world is fundamentally holistic (Maudlin 2007, Chapter 3), that it exhibits fundamental indeterminacy or vagueness (Bokulich 2014), and even that it is devoid of any fundamental entities at all (McKenzie 2011). The debate over whether modality is a fundamental or derivative aspect of reality – surely one of the most central questions in contemporary metaphysics – has in recent years largely morphed into questions of the interpretation of quantum mechanics, with some arguing on this basis that spacetime should no longer be regarded as fundamental (see Ney and Albert 2013 for discussion). And those are just some of the metaphysical conclusions that have been reaped from physics: modern chemistry and biology have likewise been brought to bear on deep questions of identity, composition, and persistence.

This resurgence of metaphysics in the philosophy of science is in many ways only to be expected, given the developments the twentieth century witnessed in both science and philosophy. For what made these changes in science truly revolutionary is the fact they entailed grand changes of world-view – changes in the sort of basic categories and concepts that metaphysics is typically tasked with articulating. And surely one enduring legacy of Quine's philosophy is his thesis of the 'continuity' between metaphysics and science – the thesis that has probably done most to restore metaphysics to intellectual legitimacy from its positivistic nadir.[1] No wonder, then, that metaphysics is back in vogue and often being developed in close contact with the sciences. However, since this fact about what kind of work is being done and with what frequency is presumably just an empirical fact about philosophers, the distinctively *metametaphysical* interest lies in the conceptual and normative aspects of the practice. What is 'science-guided metaphysics,' precisely? And what value, epistemic or otherwise, pertains to it – both intrinsically and in comparison with other approaches? The latter question seems particularly pressing in light of the rash of papers that have appeared in the last decade or so arguing that contemporary metaphysics has gone badly awry in its *failure* to utilize the guidance of science. By choosing instead to 'go it alone' and produce 'the kind of metaphysics that floats entirely free of science' (Ladyman and Ross 2007, 9), the bulk of metaphysics has, in the eyes of the most vocal supporters of SGM, become 'irrelevant' (Ladyman and Ross op. cit., vii), 'frivolous' (French and McKenzie 2015, 28), 'pseudoscientific' (Ladyman and Ross op. cit., 17), 'sterile or even empty' (Callender 2001: 34) – in sum, an activity lacking in seriousness, self-awareness, or accountability to anything outside itself. Some have even been so impolitic as to state that it 'fails to qualify as part of the enlightened pursuit of objective truth, and should be discontinued' (Ladyman and Ross op. cit., vii), and – in the circles I move in at least – has largely been applauded for doing so.

This resurgence of science-guided metaphysics, then, has been accompanied by a deeply antagonistic meta-metaphysics – namely, the explicit contention that an alternative metaphysics conducted in a scientific vacuum isn't worth wasting our time with. Let us, rather crudely, call this latter metaphysics – that operating without the guidance of science – 'armchair' metaphysics.[2] But since the most vocal critics of armchair metaphysics themselves engage in science-guided metaphysics, we can only assume that *that* activity, by contrast, *is* taken to have value. Let us express the conjunction of these views as the 'normative claim of SGM' (NCSGM):

NCSGM: Metaphysics has value, and as such ought to be conducted, iff it is guided by science.

Our focus in what follows will be on this normative claim. The central moral will be that, for all of this fighting talk, it is much harder to sustain a normative distinction between science-guided and armchair metaphysics than many advocates of the former would seem to believe. Three arguments will be advanced to this end. The first, concerning we might call the *conceptual problem*, addresses the content of the normative claim. Here the objection will be that the notion of 'science-guided metaphysics' is at present so nebulous that we are not able to state what it is that the normative claim prescribes, let alone evaluate it. The second, concerning what we will call the *practice problem*, is that science-guided metaphysics routinely employs the deliverances of armchair metaphysics in the course of fleshing out its claims. As such, pending some major change in how they go about it, proponents of science-guided metaphysics cannot assert the normative claim without running into performative contradictions. The third concerns we will call the *progress problem*, which focuses on the fact that the science upon which contemporary SGM relies is overwhelmingly likely to be false, meaning that a metaphysics based on it is likely to be false also. Given that – unlike in science itself – there is also no clear sense in which metaphysical claims can at least be said to be 'making progress,' the epistemic value of a present-day metaphysics that is based in current science becomes very difficult to discern.

That in any case is what I will argue here. It might help to state at the outset however that I myself think it would be bizarre to choose not to employ relevant, well-supported science in the course of engaging in metaphysical theorizing – indeed in the course of any sort of theorizing whatsoever. Thus while the thrust of the chapter will be negative, my intention is more to direct science-guided metaphysicians as to where we might usefully place our efforts rather than to score points against the rival view (assuming that such dichotomies even make sense – something surely undermined by the arguments to follow). Furthermore, it strikes me as only appropriate that science-guided metaphysicians engage in self-criticism of this sort, given that the lack of self-reflection has formed the core of so much of their own critiques of armchair metaphysicians.

Since this debate is normative through and through, we begin by outlining in more detail the normative presumptions of science-guided metaphysics. From then on we assess the normative claim.

The norms of 'science-guided metaphysics'

The normative claim of science-guided metaphysics embeds a pair of propositions: that SGM is an activity worth engaging in, and that any other metaphysics is not. Clearly each of these claims needs supporting. Let us call the argument for the former claim the 'positive' argument for SGM, and that for the latter the 'negative.' While the negative argument has dominated the recent literature on naturalistic metaphysics, far less has been said regarding the positive. To my mind this imbalance is somewhat curious, given that the modern discipline has its roots in a movement whose whole political point was the elimination of metaphysics. However, the one positive argument that *is* offered, if only occasionally and in passing, is the most obvious one – namely, that which appeals to the sheer success of science. Insofar as we think that science is a reliable guide to the world, the thought goes, we can expect a metaphysics guided by it to be a reliable guide in turn. While we can take it as implicit in the very fact that SGM exists, that this is the motivating thought is occasionally made explicit in the literature.[3]

As noted, however, much more ink – certainly much more memorable ink – has been spilled attacking the 'armchair' alternatives to SGM, and hence in outlining the negative argument. Now it must be said that the existing literature here is somewhat ambiguous as to the status of metaphysical inquiry on matters upon which science is silent (assuming that there are such things). For present purposes, we will assume that NCSGM addresses, at least in the first instance, only metaphysical theses whose subject matter 'overlaps' with that of a well-established science. So if metaphysicians are writing about the nature of fundamental properties, typically assumed to be physical in character, according to NCSGM the scientific theory describing those properties must be consulted in the course of our theorizing if it is to generate work of any value. Similarly for work theorizing relations of material composition, the persistence conditions of objects, the nature of colors and moral properties, and so on (and on). So the claim is that, where there is 'overlap' in this sense, metaphysics that fails to incorporate the relevant science deserves only to be 'cast to the flames.'

Why is this? The basic complaint is methodological in character – namely, that there is no reason to think that the methods employed in armchair metaphysics are well-equipped to reveal the relevant facts about the world, and plenty of reason to think they do not. To understand these objections, it is helpful to have some examples of *a priori* approaches to some of the questions outlined in the opening paragraph (thus establishing overlap).[4]

The existence of fundamental objects.[5] Ted Sider and Ross Cameron have argued that the world contains a fundamental level composed of mereologically fundamental objects. Despite the widespread assumption that the fundamental is physical, at no point is any physical theory consulted in the course of arguing for this claim. Rather, the views are defended on the grounds that greater 'ideological parsimony' or 'unity' would be exhibited by the resulting metaphysical theories than would be exhibited by their anti-fundamentalist rivals. In each case it is stated without argument that such 'virtues' of theories enhance the likelihood of their truth.[6]

The nature of fundamental objects.[7] Ned Markosian defends the idea that what makes 'simples' – the 'basic building blocks that are meant to be combined in order to form composite objects' – truly 'simple' and hence fundamental is the fact that they are *maximally continuous* bits of matter

(Markosian 1998). Roughly speaking, this means that they 'occupy the largest matter-filled, continuous regions of space around.' Despite clarifying that it is, in particular, *physical* objects that are the target of the thesis, there is no consultation with physics as to whether space is continuous, whether fundamental matter is continuous, or indeed with anything that physics might have to say about what fundamental physical objects are like. Rather, the principal reason that this account of the 'basic building blocks' is to be preferred is that the main rival theory – according to which the fundamental objects are fundamental in virtue of being 'point-sized' – is inconsistent with our ability to 'imagine a possible world in which there is only one physical object, a perfectly solid sphere made of some homogeneous substance.' Thus substantive questions about the nature of fundamental physical objects are addressed not by consultation with physics, but rather with the content of our imaginations, together with the modal intuitions we experience concerning what our imaginations present to us.[8]

The denial of holism. In his statement of Humean supervenience – needed to ground the success of his reductive modal programme – Lewis postulates that the world resolves into a mosaic of fundamental physical properties all of which are pointwise instantiated and intrinsic. Little is offered in support of locality beyond that it is 'inspired by classical physics' (Lewis 1994, 474); and all that is said regarding their intrinsicality is that it is 'plausible' or 'seems right' (Lewis 1986, 61; Lewis 1983, 375). But whatever fudges might be buried in this statement of what 'inspired' the view – think here of a cable TV drama 'inspired' by a true story – classical mechanics is certainly *not* a true story and subsequent theories strongly suggest that both assumptions are highly questionable. And while one could certainly object that the arguments concerning quantum non-locality, etc. only came to prominence in the philosophical literature after Lewis had laid down his system, it remains that substantive claims about the nature of fundamental physics properties are simply asserted and this just seems problematic – not least because these assumptions lie at the fulcrum of the success of his system.

While this list is obviously not exhaustive, these are all clear cases in which metaphysicians confidently argue, or even simply assert, that fundamental entities *assumed to be physical in character* are a certain way without ever bothering to consult any physics. Rather, an alternative methodology is employed, two features of which stand out. The first is the explicit appeal to intuition, or to what intuitively 'seems right' to the person making the claim. The second is a crucial appeal to 'theoretical virtues.' There are furthermore reasons to think these examples are relevantly representative of methodology in metaphysics, for the literature is replete with further examples and quotations from metaphysicians stating that these strategies *just are* the core strategies of the discipline (Bryant 2017; Kriegel 2013; Ladyman 2017). The claim of advocates of SGM is that there is little reason to believe either strategy is reliable, and plenty to suggest that it is not.

Let's take the reliability of intuitions first. It seems, first of all, that there is *no reason to think* that our impressions regarding what is possible – intuitions that, it is held, were acquired in our environment of adaptation – would be reliable in the unobservable domain.[9] Nobody today thinks that the claims of fundamental physics – such as regarding color confinement or mass oscillation – could be reliably revealed by intuition (thus explaining why we shell out huge sums for experiments), and there seems no good reason to think things are any different when it comes to the fundamental metaphysics of this physics. Indeed, if physics has any metaphysical implications whatsoever (and I'm not sure anyone would go on the record denying that), then any agreement between our metaphysical intuitions and the implications of a deeply unintuitive physics would have to be a mere matter of chance; following Gettier, then, they would presumably fall short of knowledge. More strongly still, it seems we have ample historical evidence to claim definitively that intuition is *not* a good guide to aspects of the world transcending our immediate experience. Whether it concerns the necessity of determinism, the logical properties of parallel lines or simultaneity relations, or the necessity of immutable objects, it seems that 'appeal to intuition has a really, really bad track record' (Howard ms, 10). But while physics can employ experimental methodologies to circumvent our lack of reliable instincts when it comes to the deep structure of the world, it is not obvious that a metaphysics that is impervious to the deliverances of the lab has any analogous alternative.

While the objections to the evidential role of intuition (at least in the relevant domains) seem straightforwardly compelling, the strength of the argument against the appeal to virtues is less clean-cut. One certainly gets the impression from the critical literature that the appeal to virtues such as 'simplicity' and 'unity' by armchair metaphysicians is at the very least highly under-theorized and not subject to any measure of quality control, and that it has in effect become an in-house methodology lacking any independent rationale (Ladyman and Ross op. cit., 16–17). Now, admittedly one does find frequent appeals by armchair metaphysicians to the fact that such virtues are routinely invoked in the process of scientific theory selection; as such, the claim goes, they may be legitimately appealed to in metaphysical theorizing too.[10] However, many critics of metaphysics have taken pains to deny that this 'continuity gambit' works, largely on account of relevant disanalogies between scientific and metaphysical theories.[11] For example, Michael Huemer has argued that while we can show that 'simpler' theories with fewer free parameters may be better supported by actual data than those with more – on the grounds that those extra parameters could be tuned to agree with different configurations of properties than those given in the data, resulting in that data giving less of a probability 'boost' – it is hard to see how this argument could generalize to theories in which the 'data' to be explained is the far more general fact that 'things have properties' (Huemer 2009, Section II.4). Others have claimed that metaphysical – unlike scientific – theories are unlikely to differ significantly in the degree of virtue they exhibit, so that the question of their role in conferring differential likelihood does not even arise. Thus Kriegel (op. cit., 21), for example, claims that metaphysical theories are unlikely to exhibit significant parsimony differences, owing to the fact that they routinely make as many denials as they do assertions (Humeanism not being mere 'quietism' about modality, and so on). Others still have argued that the role of virtues in scientific theory selection is grossly over-stated by metaphysicians, for it is in fact only empirical features that the scientific community regards as rational criteria for theory selection (Ladyman 2012, Section 4). If this is the case, then there isn't even a 'continuity' case to argue for.

This completes our whistlestop tour of the positive and negative arguments for SGM. In sum, the positive argument is that SGM may be expected to inherit the epistemic success of science. The negative argument is that the methods employed by the alternative are not fit for purpose. If these are sound, it would indeed seem that producing SGM is a valuable way to spend one's time while engaging in armchair enquiry is not. However, both what SGM is, and why exactly it is more valuable than the alternative, turn out to be extremely difficult to articulate, and this for several reasons. What I will call the 'conceptual' problem and the 'practice' problem both exploit the fact that the methodology of SGM is not cleanly quarantined from that of armchair metaphysics, meaning that the negative arguments applying to the latter would seem to infect the former. The 'progress' problem, by contrast, questions the positive argument for SGM, by denying that the success science enjoys in describing the deep portions of the world can be expected to 'percolate up' to a metaphysics guided by it.

The conceptual problem

In order for the normative claim to be well-defined, we clearly need to be able to say something meaningful about the relata involved – namely 'science' and 'metaphysics' – and the 'guiding' relation that SGM advocates claim ought to obtain between them. But arguably nothing here is as clear as it needs be to bring about wholesale change in the discipline, and this has not gone unnoticed by apologists for *a priori* metaphysics. Thus in one attack on the naturalistic critique Williamson (2013) focuses on the relata concerned and the difficulties involved in defining 'science' in particular. Chakravartty (2017), by contrast, presses that the 'guiding' relation is hopelessly ill-defined, and that any attempt to sharpen it up will likely sanction either both forms of metaphysics or neither.[12]

Consider first of all Williamson's objection – an objection which in effect resurrects the demarcation problem against the advocate of NCSGM. For in order for their injunction to be well-defined they must specify what it is that they mean by 'science.' But here a narrow course must be steered. Construe 'science' too selectively, and they risk ruling out logic and mathematics – clearly bodies of theory indispensible to metaphysics. Construe it too widely, however, and naturalism 'loses its bite' (30). All things considered it is better, he says, not to 'be implicated in an equivocal dogma.' As such, rather than committing to adhering to 'scientific theories' in the course of producing theories of metaphysics, we should resolve to conduct philosophy with 'a scientific spirit' – meaning that it is *virtues*, such as honesty and rigor, that we elect to adhere to instead.

The advocate of NCSGM will no doubt here protest that there is in fact something deeply *dishonest* about the way that many armchair metaphysicians go about their work (in the way that they pay 'mere lip-service' to science and so on).[13] But regardless of whether they are right about that, this challenge must be addressed by those committed to the normative claim if they are to have a claim at all. However – as philosophers of science will be well aware – the long and not entirely productive debate over the 'problem of demarcation' suggests there is no obvious way to do so. The naturalist might hope to avoid tackling the problem head-on by (a) finding some principled way to make an exception for fields, such as logic and mathematics, which are essential to reasoned enquiry in general and yet not obviously classifiable as 'science,' or (b) insisting that we do not in fact need a general definition of 'science' to make the point that they want to make. For since metaphysics is very often concerned with the *fundamental* in particular, in most cases it is simply fundamental physics that is to be consulted – and that *physics* qualifies as science if anything does, has been a fixed-point in the demarcation debate.[14] But this, alas, will not do: Callender's argument against the 'flowing present' view of time, for example, draws almost as heavily on psychology as it does on physics, and psychology is a field whose scientific status has perennially been regarded with suspicion (Callender 2017). At this point the naturalist might reach for Laudan's response to the demarcation problem, holding that we should abandon the attempt to identify 'science' in favor of determining *that for which we have good evidence* (Laudan 1983). But it's not clear that this helps at all, for what counts as evidence in metaphysics is surely a core part of what is here in dispute.

To see this, consider now the 'guidance' relation advocates of NCSGM insist must hold between science and metaphysics, and that forms the focus of Chakravartty's concerns. How are we to understand it? If all 'guidance' means is that a bit of science must be thumbed through prior to arguing for a metaphysical claim, SGM runs the risk of reducing to a sort of 'corporate listening exercise' – an exercise in which some science is ostensibly 'consulted' prior to the drawing of a conclusion that would have been drawn either way. More than this is presumably required by the normative claim, with the 'guidance' of science in some way *constraining* what can legitimately be inferred. However, it is widely believed – including by some of the leading protagonists in this debate – that science will generically *underdetermine* the relevant metaphysics.[15] If this is right, then some kind of 'Goldilocks principle' must operate here, such that science must have some non-trivial role in constraining the relevant metaphysics without any expectation that it will be uniquely constraining. But then what must the science be admixed with before we are entitled to draw metaphysical conclusions from it? What, in other words, counts as evidence in metaphysics, in addition to that which supports the scientific theories that have guided it?

A standard lore here is that 'theoretical virtues' – 'nice-making' features transcending empirical adequacy and logical consistency, such as 'simplicity,' 'elegance,' and so on – must at this point be appealed to.[16] Now clearly, if this is the case then we immediately face the question of what makes science-guided metaphysics in better epistemic shape than its armchair counterpart. For the criticisms of the role of virtues in metaphysical theory selection outlined above are, by and large, insensitive to whether the metaphysics involved is armchair or naturalistic; indeed, since at this point the science is presumed to no longer be playing any role, it seems there is no distinction that could be made here in principle. And if such appeals are viewed as deeply problematic in the case of armchair metaphysics, then the significance of the fact that in the science-guided case the theorizing *started out* as based in science is far from obvious. For is not a chain of reasoning only as strong as its weakest link?

If science routinely underdetermines metaphysics, then, whatever does the work in arguing for metaphysical claims is by definition outside of science, and the defender of NCSGM faces the question of what makes SGM kosher and yet not its armchair counterpart. As I have argued elsewhere, however, it is unclear to me that the methodology of SGM does of necessity involve the weighing up of theoretical virtues at crucial inferential junctures (McKenzie 2018). To take just a couple of examples, Maudlin (Maudlin 2007, Chapter 4) has argued that the idea that properties are universals is *inconsistent* with the local gauge structure of fundamental physics; I myself have argued that the thesis of ontic structuralism is *implied* by the same sort of gauge-theoretic considerations (McKenzie 2016). While these arguments have (predictably) been criticized, it seems that the methodology employed by naturalistic metaphysicians does not, as a matter of fact, inevitably make appeal to 'virtues' over and above the compulsory empirico-logical constraints. Furthermore, it is often admitted that science can at least *delimit* the number of live metaphysical options (see e.g. Bigaj and Wüthrich op. cit). But if it can reduce this number, what in principle prevents it from decreasing to one?

These, then, are the questions we face. What does it mean to say 'science guides metaphysics'? Can the relationship be determinative, or not? If not, what else must be added? And what would make the resulting metaphysics in better shape than that resulting from the armchair? Without a clear policy on these issues, the normative claim doesn't have a clear meaning, let alone a justification.

The practice problem

The conceptual problem is posed to the very notion of 'science-guided metaphysics.' But the meaning attached to metaphysical claims themselves is notoriously difficult to pin down, and there is *prima facie* little reason to think that the claims of SGM are in a different boat here. Take for example the core claim of structuralist metaphysics – the view that '*relational structure is ontologically fundamental*' (Ladyman and Ross op cit., 145). The doctrine proposes that if we take modern physics – principally, quantum theory and relativity – sufficiently seriously, then we will have to regard the *category of physical objects* as a *derivative category*, in contrast to that of *structure*; or at the very least, that it can no longer be regarded as a category *ontologically prior* to relations and structure. Its adherents typically also hold that *identity facts* are *derivative* and that the world is profoundly *holistic* in that objects *ontologically depend* either on each other or on the relations between them. Needless to say, few of the terms just highlighted have an especially straightforward meaning. Nor is there anything remotely straightforward about how one would go about arguing for them. So what sort of strategies have structuralist metaphysicians invoked to sharpen up and support their claims?

The short answer to this question is that structuralists have largely adopted a strategy of *mining the existing metaphysics literature* for appropriate resources, followed by a process of *applying* or *adapting* those resources so as to suit their particular needs. Thus in order to articulate the core claim that structure is ontologically fundamental I have drawn heavily on Kit Fine's writings on ontological dependence, and to argue that we can make sense of physicists' ambitions to explain the fundamental I have plundered *a priori* work on 'grounding.'[17] Conceptual work by Jessica Wilson on determinates and determinables has been employed by Steven French to articulate the fundamentality of structure over kind properties, and he has also invoked Ross Cameron's theory of truthmaking to communicate how radical structuralists interpret physicists' talk about objects while denying that they really exist. Similarly, Simon Saunders has appropriated Leibniz's principle of the identity of indiscernibles, revamping it à la Quine and extending to allow discernibility with respect to relations, to demonstrate the identity dependence of objects on relations in the context of quantum mechanics. And David Lewis' notion of 'elite' or 'perfectly natural' properties has been invoked in the service of showing that the core claim of epistemic structuralism is non-trivial.

There are many other examples that could be cited in this connection. But the key point for present purposes is that all these metaphysical packages that have proved useful to appropriate in structuralism were not only created independently of structuralism, but were moreover (by and large) developed independently of *any scientific considerations whatsoever*. Cameron's version of truthmaker theory, for example, was developed to make sense of talk of tables and chairs, and Leibniz' principle of the identity of indiscernibles was (obviously) articulated too early on to be informed by the quantum mechanics that it subsequently helped to illuminate. Furthermore, presumably naturalistic metaphysicians proceed in this way only because this division of labor is in some way advantageous to them. But whatever those advantages are, as a methodology it clearly poses problems for the advocate of NCSGM. For how can it be that armchair metaphysics has no value if it repeatedly proves highly useful for constructing the science-guided metaphysics that ostensibly *does*? So long as the metaphysics of science takes its concepts 'off the peg' instead of making everything it needs 'to order,' it seems the normative claim of science-guided metaphysics cannot be asserted without performative contradiction.[18]

The progress problem

What I have called the 'practice problem' took aim at the negative argument for SGM – the argument to the effect that 'there is no alternative' to science-guided metaphysics. The progress problem, on the other hand, sets its sight on the positive support that can be given for it. This, recall, was the simple but *prima facie* compelling thought that since science succeeds in guiding us to facts about the deep structure of the world – or so we will assume – a metaphysics that is based on it may be expected to successfully describe the world also. In making such an assumption we of course take up the standpoint of scientific realism.[19] But as everyone versed in the realism debate knows, our realist statements must be appropriately hedged and qualified if they are to be defensible. For nobody thinks our current scientific theories are anything other than *approximately* true at best. Indeed, giving the vicissitudes of scientific history it isn't far-fetched to think that they may still be *very far from the truth*. But constitutive of the realist stance is the idea that as we produce theories that are more and more empirically adequate, we are at least making *better approximations* to the truth, and hence *making progress* toward it. It is belief in this sequence of *ever better* approximations, and hence in *progress* transcending the empirical, that defines the thesis of 'convergent realism.' And since 'progress' is itself a normatively loaded term, it is hard to deny that the truth of this belief would constitute a good, something of epistemic value.

How does this thesis about progress in science connect with the issue of the value of metaphysics? First and most obviously, if our belief in the truth-content of our current scientific theories is hedged in some way (as all contemporary realists will hold it must be), then presumably our belief in the truth-content of any metaphysics that is based in these theories must be hedged in some way also. But on the assumption that science is at least *making progress* towards a true description, it seems natural to assume that a metaphysics that is firmly based in that science may be expected to be making progress too. As Ladyman and Ross (op. cit., 35) put it, while it must be conceded that 'it could turn out that the best current metaphysics is substantially wrong' since we still expect future scientific change, 'to the extent that metaphysics is closely motivated by science, we should expect to make progress in metaphysics *iff* we can expect to make progress in science.'[20]

The increasing explicitness in the metaphysics of science about the likely falsity of our theories is surely a laudable development. But it also raises urgent questions about what the value of that metaphysics is. While there is much to say here, I will press only that the notion of progress in metaphysics remains gravely under-developed, as do the reasons for thinking that progress in science would render it likely.[21] For it seems, once again, that there are relevant disanalogies between each description of nature that would problematize such inferences. Recall in this connection that the thesis of 'convergent realism' is that science is making progress via its production of successively *better approximations* to the truth. And presumably progress may be inferred from science to metaphysics only if there are analogous senses of progress involved in each case. However, it seems that metaphysical theses are just not the sort of thing to which the language of approximation happily applies.

To see this, let us return – albeit briefly – to some of the metaphysical debates guided by current physics mentioned at the outset. As is typical, these are debates over what is *fundamental*. Now, given that we do not think that the theories involved are truly fundamental, we expect our metaphysics based in those theories to undergo change as well. (After all, if our metaphysics were not sensitive to such changes, it would be hard to even motivate SGM.) Thus the hope is that our current metaphysics can at least be viewed as an 'approximation' to that to which we will be guided in future. Suppose, then, that we are Humeans on the basis of current scientific theories. On standard renderings, this requires the belief that *all* the fundamental properties are *categorical*. Thus their opponents, the Anti-Humeans, believe that at least *some* fundamental properties are *not* categorical (see e.g. Bird 2007, 45). But what does any of this mean? Well standardly, to say that a property is categorical is to say it 'has no essential or other non-trivial modal character' (Bird op. cit., 67), and hence is 'free from any modal commitment' (Schrenk 2016, 71). And 'modal commitment' will be modeled in terms of the presence of a modal operator (box or diamond) in the analysis of the property. Now this presence clearly cannot be 'approximated,' for such operators either appear in a definition or they do not. And should *any properties whatsoever* turn out not to be categorical then the anti-Humean – i.e., directly opposing – position is automatically sanctioned. What, then, could it mean for the position taken by a Humean to be an 'approximation' to some other view?

It seems to me that the root of the problem here is that metaphysical theses tend to have a totalizing, all-or-nothing character, both in terms of their extension and in terms of how their core concepts are defined. Given this feature, the notion of approximation simply does not make any obvious sense as applied to claims of metaphysics – unlike the claims formed in the characteristically much less general and much more discriminating language of physics. For this reason, there is simply no obvious way to infer optimistic doctrines of progress in metaphysics from the presumed fact of progress in the sciences, *even if* that metaphysics is created in close consultation with the sciences. As such, what the value is in engaging in metaphysics at any point prior to the emergence of a final theory – that is, at times like now – emerges as distinctly unclear. At the very least, to my mind, it is an issue that cries out for further theorizing.

Conclusion

Contemporary metaphysics of science is replete with claims that metaphysics constitutes a legitimate activity if and only if it is guided by science – claims that seem initially compelling. But a closer look reveals that science-guided metaphysics is implicated in armchair metaphysics and its problems as well. Furthermore, in spite of their close proximity it is hard to see how science-guided metaphysics participates in the epistemic value we associate with the sciences themselves. To be clear, my claim is *not* that the issues discussed here have gone unnoticed by metaphysicians of science; nor is it the case that I regard it as settled that any of them are in principle intractable. The point is rather that there is no consensus on these issues and no obvious way out of the ensuing problems. In a sense this is heartening, for it shows that there is still much valuable work to be done in a field whose value has been put into question. But the lack of obviousness about how to respond in the face of these problems is hardly consonant with the supposed obviousness of NCSGM. Perhaps we naturalists ought to reflect more on this before writing manifestos, for the *n*th time, on the illegitimacy of armchair philosophy.

Notes

1 See Callender 2001 for a nice discussion of this rehabilitation and many other themes discussed below.
2 See however Guay and Pradeu 2017 for how such loose ways of characterizing the intended counterpoint to 'science-guided metaphysics' can mask important distinctions.
3 See e.g. Ladyman and Ross 2007: 7, 35, and Chapter 1 *passim*; Maudlin 2007, Chapter 3 *passim*; Baker 2013: 265.
4 Note that in talking about 'a priori metaphys*ics*' I am not thereby implying that this is produced by 'a priori metaphysi*cian*.' Sider and Lewis, for example, have all engaged deeply with science at many junctures in their theorizing (cf. Nolan 2015).
5 This discussion has primarily assumed a problematic mereological conception of fundamentality, although the structure of the argument would seem to generalize.
6 I should point out that Sider, in a footnote, admits that appeals to simplicity are questionable, and especially so in philosophy. He states he 'suspect[s] that principles of parsimony cannot be derived from more fundamental epistemic principles' (Sider 2013, 239). While the admission is in a sense laudable, it is not clear what is actually gained from it.
7 Again, the question has been primarily discussed in mereological terms, but here I cannot see how the argument would generalize.
8 Note that Markosian explicitly acknowledges that such intuitions are 'notoriously difficult to defend.' While (again) this concession is in a sense laudable, it is this difficulty that is precisely the issue.
9 This has long been a theme in philosophy of science: see e.g. Shapere 1990.
10 See Sider et al. 2008, 7 and Paul 2012 for particularly explicit examples.
11 This terminology is from Saatsi 2017 – a paper which criticizes the idea that inference to the best explanation may be relied on in metaphysics given the assumption it is employed in science.
12 Note that this is rather a simplification of Chakravartty's richer, voluntarist position.
13 See Bryant op. cit. for a powerful reflection on metaphysicians' 'false consciousness.'
14 Of course, the status of string theory may reasonably be thought to undermine this claim.
15 E.g. Bigaj and Wüthrich (2015, 13) state that such underdetermination is exemplified in 'virtually all cases where attempts are made to draw metaphysical lessons from physical theories.'

16 See Chakravartty 2017 for a recent statement of this sort of view.
17 See French and McKenzie 2012, 2015 for references to many of these examples of appropriation.
18 This argument is defended in more detail in French and McKenzie 2012, 2015.
19 For some discussion of how metaphysics of science and the presumption of scientific realism intersect, see again Guay and Pradeu op. cit.
20 Ladyman and Ross op. cit., 35.
21 The ensuring argument is outlined in more detail in McKenzie 2020, forthcoming.

References

Baker, David. 2013. Identity, Superselection Theory, and the Statistical Properties of Quantum Fields. *Philosophy of Science*, 80(2), 262–285.
Bigaj, Tomasz and Wüthrich, Christian. 2015. Introduction. In *The Metaphysics of Contemporary Physics* (Poznan Studies in Philosophy of Science and the Humanities, Vol. 104), ed. Tomasz Bigaj and Christian Wüthrich. Leiden: Brill, pp. 7–24.
Bird, A. 2007. *Nature's Metaphysics*. Cambridge: Cambridge University Press.
Bokulich, Alisa. 2014. Metaphysical Indeterminacy, Properties, and Quantum Theory. *Res Philosophica*, 91(3), 449–475.
Bryant, Amanda. 2017. Keep the Chickens Cooped: The Epistemic Inadequacy of Free Range Metaphysics. *Synthese*, 1–21.
Callender, Craig. 2001. Philosophy of Science and Metaphysics. In *The Continuum Companion to the Philosophy of Science*, ed. S. French and J. Saatsi. London: Continuum, pp. 33–54.
Callender, Craig. 2017. *What Makes Time Special?* Oxford: Oxford University Press.
Čapek, Milič. 1984. Particles or Events? *Boston Studies 82*. D. Reidel Publishing Company (1984), 1–28.
Chakravartty, Anjan. 2017. *Scientific Ontology: Integrating Naturalized Metaphysics and Voluntarist Epistemology*. New York: Oxford University Press.
French, Steven. 2014. *The Structure of the World: Metaphysics and Representation*. Oxford: Oxford University Press.
French, Steven and McKenzie, Kerry. 2012. Thinking Outside the Toolbox: Towards a More Productive Engagement between Metaphysics and Philosophy of Physics. *European Journal of Analytic Philosophy*, 8(1), 42–59.
French, Steven and McKenzie, Kerry. 2015. Rethinking Outside the Toolbox: Reflecting Again on the Relationship between Metaphysics and Philosophy of Physics. In Bigaj and Wüthrich 2015, pp. 25–54.
Guay, Alexandre and Pradeu, Thomas. 2017. Right Out of the Box: How to Situate Metaphysics of Science in Relation to other Metaphysical Approaches. *Synthese*, 1–20.
Howard, Don. ms. The Trouble with Metaphysics. Unpublished manuscript.
Huemer, Michael. 2009. When Is Parsimony a Virtue? *Philosophical Quarterly*, 59(235), 216–236.
Kriegel, Uriah. 2013. The Epistemological Challenge of Revisionary Metaphysics. *Philosophers' Imprint*, 13.
Kuhlmann, M. 2010. *The Ultimate Constituents of the Material World – In Search of an Ontology for Fundamental Physics*. Frankfurt: ontos Verlag.
Ladyman, James. 2012. Science, Metaphysics and Method. *Philosophical Studies*, 160(1), 31–51.
Ladyman, James. 2017. An Apology for Naturalized Metaphysics. In *The Metaphysics of the Philosophy of Science: New Essays*, ed. Matthew H. Slater and Zanja Yudell. Oxford: Oxford University Press, pp. 141–161.
Ladyman J. and Ross, D. 2007. *Every Thing Must Go*. Oxford: Oxford University Press.
Laudan, Larry. 1983. The Demise of the Demarcation Problem. In *Physics, Philosophy and Psychoanalysis: Essays in Honor of Adolf Grünbaum*, ed. Robert S. Cohen and Larry Laudan. Dordrecht: D. Reidel, pp. 111–127.
Lewis, D. 1983. New Work for a Theory of Universals. *Australasian Journal of Philosophy*, 61, 343–377.
Lewis, D. 1986. *On the Plurality of Worlds*. Oxford: Blackwell.
Lewis, D. 1994. Humean Supervenience Debugged. *Mind*, 103(412), 473–490.
Markosian, Ned. 1998. Simples. *Australasian Journal of Philosophy*, 76, pp. 213–226.
Maudlin, Tim. 2007. *The Metaphysics within Physics*. Oxford: Oxford University Press.
McKenzie, K. 2011. Arguing Against Fundamentality. *Studies in the History and Philosophy of Physics*, 42(4), 244–255.

McKenzie, Kerry. 2016. Looking Forward, Not Back: Supporting Structuralism in the Present. *Studies in History and Philosophy of Science (Part A)*, July 2016 doi:10.1016/j.shpsa.2016.06.005.
McKenzie, Kerry. 2018. Review of 'Scientific Ontology: Integrating Naturalized Metaphysics with Voluntarist Epistemology.' *BJPS Review of Books*: https://bjpsbooks.wordpress.com/2018/05/01/mckenzie-anjan-chakravartty/.
McKenzie, Kerry. 2020. A Curse on Both Houses: Naturalistic versus A Priori Metaphysics and the Problem of Progress. *Res Philosophica*, 97(1), 1–29.
McKenzie, Kerry. forthcoming. The 'Philosopher's Stone': Physics, Metaphysics, and the Value of a Fundamental Theory. In *Beyond Spacetime: The Philosophical Foundations of Quantum Gravity* (Volume II), ed. Nick Huggett, Baptiste Le Bihan, and Christian Wüthrich. Oxford: Oxford University Press.
Ney, Alyssa and Albert, David Z. 2013. *The Wave Function: Essays on the Metaphysics of Quantum Mechanics*. Oxford: Oxford University Press.
Nolan, Daniel. 2015. Lewis's Philosophical Method. In *A Companion to Lewis*, ed. B. Loewer and J. Schaffer. Oxford: Wiley-Blackwell, pp. 25–39.
Paul, L. A. 2012. Metaphysics as Modeling: The Handmaiden's Tale. *Philosophical Studies*, 160(1), 1–29.
Saatsi, Juha. 2017. Explanation and Explanationism in Science and Metaphysics. In *Metaphysics and the Philosophy of Science: New Essays*, ed. Matthew Slater and Zanja Yudell. Oxford: Oxford University Press, pp. 163–192.
Schrenk, Markus. 2016. *Metaphysics of Science*. London and New York: Routledge.
Shapere, Dudley. 1990. The Origin and Nature of Metaphysics. *Philosophical Topics*, 18(2), *Philosophy of Science* (Fall), pp. 163–174.
Sider, Theodore. 2013. Against Parthood. In *Oxford Studies in Metaphysics* 8, ed. Karen Bennett and Dean Zimmerman. Oxford: Oxford University Press, pp. 237–293.
Sider, Theodore, Hawthorne, John, and Zimmerman, Dean W. (eds.) 2008. *Contemporary Debates in Metaphysics*. Oxford: Blackwell.
Williamson, Timothy. 2013. What is Naturalism?. In *Philosophical Methodology: The Armchair or the Laboratory?*, ed. Matthew C. Haug. London: Taylor & Francis Group, pp. 29–31.

35
METHODS IN SCIENCE AND METAPHYSICS

Milena Ivanova and Matt Farr[1]

1 What are science and metaphysics?

Can we distinguish science from metaphysics? Traditionally, metaphysics is defined as the most general study of reality, concerned with the actual and possible, essences and potentialities, identities and priority relations. Science, on the other hand, is qualified as the study of the natural or physical or actual world. Already we face problems here: what is 'natural' and 'physical' if not just that which can be scientifically studied? Take any candidate for something non-physical or non-natural (e.g. ghosts, souls, spirits, and the like): if there were some way of reliably measuring it, then it would presumably be of relevance to science. Thoughts such of these raise scepticism as to whether metaphysics and science can be distinguished solely in terms of their subject matter. Although the subject matters of science and metaphysics may indeed overlap, it is reasonable to suppose that the disciplines may be distinguished in terms of their methodologies, particularly with respect to the empirical nature of the scientific method that has no obvious analogue in metaphysics. In the present chapter, we go further and make the case against a clear-cut methodological distinction between science and metaphysics. One might think that metaphysics is a purely conceptual, a priori, or 'armchair' discipline, whereas scientists are in the lab or the field, employing a distinctive empirical method that all and only sciences share, but the reality is much less clearly structured, with a variety of non-empirical methods shared by both metaphysicians and scientists in attempts to answer similar foundational questions about the world. In what follows we investigate different methodologies employed within science and metaphysics, both analytic and naturalistic, and argue that when it comes to methodology, there is a substantial overlap between science and metaphysics that undermines a sharp set of demarcation criteria between the two disciplines.

1.1 What is science?

Science is standardly understood to differ from non-scientific activities in terms of its method. Much of early twentieth-century philosophy focused on identifying the scientific method and offering necessary and sufficient conditions for a discipline to constitute science as opposed to non-science or pseudo-science. The logical positivists employed a verificationist criterion of meaning, arguing that scientific statements are meaningful insofar as they are empirically verifiable, as opposed to metaphysical statements. Karl Popper (1963) argued that the scientific method is falsificationist as opposed to verificationist; although some scientific statements, such as universal generalisations, are not clearly verifiable through empirical means, it should be possible to falsify them through some empirical test. Thomas Kuhn (1970) offered a broader perspective on the demarcation problem and scientific methodology, taking genuine science to offer its own puzzle-solving tools that advance the scientific paradigm. Imre Lakatos (1977) appealed to the ability of scientific theories to make novel predictions to be what demarcates science from pseudo-science and to be a genuinely distinctive feature of progressive science. In attempting to resolve the demarcation problem and overcome objections raised against existing accounts, Paul Thagard (2013[1978]) offered a historical and social perspective on the question by focusing on the very practitioners of science. For Thagard, it is not only that pseudoscientific theories are less progressive than their competitors, but the very community of practitioners make little attempt at resolving problems endemic to their positions. On this account, demarcation is no longer an absolute matter but becomes contextual: evaluating the methods and approaches a community follows when addressing certain questions can make a theory scientific at one time and pseudoscientific at another.

In the contemporary debate in philosophy of science, it is widely acknowledged that all such attempts to offer a set of necessary and sufficient conditions to demarcate science from non-science face a series of foundational problems. Many philosophers of science instead take science to adopt a plurality of methods, and that the relationship between science and non-science is resolved only when a more contextual approach is taken. What about metaphysics? Can we provide demarcation criteria for metaphysics, and is its methodology a helpful place to start looking?

1.2 What is metaphysics?

While once taken to constitute a single activity, science and metaphysics are now taken to be two very different disciplines. While science aims at making precise predictions about the physical world, metaphysics is taken to study questions of broader significance and generality. For instance, we turn to physics to predict where our planet will be with respect to the sun in a month's time; we turn to biology to understand the evolutionary difference between a sugar glider and a flying squirrel; we turn to metaphysics to address questions like "What is the essence of X?", "Are there universal properties?", "How do we understand actuality and possibility?", "What are the fundamental ontological categories?", etc. While many questions of metaphysics are clearly outside the scope of science, some central questions, such as "Does time pass?" and "How does the mental relate to the physical?" are taken to fall under the study of both science and metaphysics, paving the way for what is commonly called *naturalistic metaphysics*. When it comes to questions of composition, finding a fundamental level to reality, or understanding time, modern science – whether physics, empirical psychology, or biology – can contribute towards highly non-trivial answers. Naturalistic metaphysicians take science to provide the basis for their investigations, prescribing that we should read our metaphysics from our contemporary science, and answer questions about the fundamental nature of the world by appealing to fundamental physics (Ladyman and Ross 2007; Maudlin 2007; French and McKenzie 2012, 2015; Ney 2012; Morganti and Tahko 2017). In this way, naturalistic metaphysics takes scientific theories to play a primary role in addressing many paradigm metaphysical questions.

On the other hand, *analytic metaphysicians* often see the relationship between science and metaphysics to be the other way around, with metaphysics being an autonomous area of study that determines the conceptual background that makes science possible. For instance, the metaphysician E.J. Lowe (2005: vi) holds that "metaphysics goes deeper than any merely empirical science, even physics, because it provides the very framework within which such sciences are conceived and related to one another". On the positive side, one can see an analogy here between, on the one hand, metaphysics and science and, on the other hand, pure and applied mathematics, with the metaphysician's job being to provide a kind of conceptual apparatus required for undertaking scientific study. On the negative side, one cannot neglect to take seriously the fact that metaphysics invariably has epistemic aims, such as establishing facts about the nature of how the universe operates (whether it be the function of laws of nature, or the nature of time, space, causation, or even existence itself), and it is highly controversial to hold that we can achieve such epistemic aims from the armchair, divorced from empirical findings.

We examine the metaphor of one discipline being 'prior' to the other in greater detail in section 3. Beforehand, in the next section, we shall see how the methodologies of science and metaphysics overlap though the shared use of non-empirical evaluative factors such as theory virtues, appeals to intuition, and the related use of modelling and inference to the best explanation.

2 What are the methodologies of science and metaphysics?

Traditionally, science was taken to proceed by observation, intervention through experimentation, and logical forms of inference in order to form and test hypotheses. This positivist picture of science can hardly be considered adequate in describing science today, since these considerations cannot be seen as either necessary or sufficient for science. Scientists often need to choose between competing explanations of the same observations and the only way they can do so is by employing non-empirical factors in their decision making, such as aesthetic considerations like simplicity and elegance. As Pierre Duhem (1954[1906]) pointed out, when it comes to choosing between competing (empirically equivalent) hypotheses on non-empirical grounds, one runs into a meta-problem: having to choose which aesthetic virtue is to be prioritised and how it is to be defined. What ultimately determines the choice is the scientist's intuition or 'good sense' (Ivanova 2010).

Furthermore, hypotheses are often accepted within the scientific community despite the unavailability of empirical confirmation. This point is best illustrated within contemporary high energy physics in which theories are entertained by the community despite either making predictions that cannot be tested due to technological constraints, or moreover failing to make specific predictions due to an overabundance of free variables within the theories, such as with multiverse cosmology and string theory (see Ellis and Silk 2014). Such problems have led some to propose non-empirical accounts of theory assessment and confirmation within physics (most notably Dawid 2013, though see the collection of papers in Dawid et al. 2019). Less controversially, physicists have historically placed a high degree of trust in scientific theories and hypotheses prior to empirical confirmation, such as with the Higgs mechanism prior to its famously complex empirical confirmation, and the atomic hypothesis prior to J.J. Thomson's cathode ray experiments. In each case, the restricted availability of empirical tests has prompted the use of alternative non-empirical grounds for supporting a theory. Beauty, elegance, simplicity, unity, and coherence with other frameworks are among considerations widely employed by scientists in this regard, with not only pragmatic weight placed upon these factors, but also epistemic significance, since such considerations are often taken to justify belief in the truth of theories.[2]

2.1 Use of theory virtues

Simplicity and parsimony have long been considered important elements in scientific reasoning. For instance, when discussing scientific methodology, Isaac Newton offered several Rules of Reasoning (or methods), with the first rule committing to parsimony:

> Rule I. No more causes of natural things should be admitted than are both true and sufficient to explain their phenomena. As the philosophers say: Nature does nothing in vain, and more causes are in vain when fewer suffice. For nature is simple and does not include in the luxury of superfluous causes.
>
> *(Newton 1999[1687]: 794)*

Simplicity was also used by Poincaré (2001[1902]) in his defence of Euclidean geometry in light of the underdetermination of physical geometry by experience. In the context of the measurement problem of quantum mechanics, simplicity is often appealed to by defenders of the Everett interpretation insofar as unlike rival interpretations, such as collapse theories and the De Broglie–Bohm theory, it does not add extra mathematical structure or postulates to the orthodox quantum mechanics formalism. On the contrary, defenders of rival interpretations standardly dismiss the Everett interpretation on the grounds of failing to be *ontologically* parsimonious, since despite its lack of extra postulates, it is standardly taken to entail a branching multiverse with every possible measurement outcome actually occurring.

Theory virtues play a central methodological role within metaphysics, with competing pictures of the world standardly evaluated with reference to their simplicity, parsimony, and fit with other metaphysical theories. For example, in ontology, trope theorists (such as Campbell 1990) take their theory to be preferable to rival theories on grounds of simplicity, since it holds that there exist only tropes. Objects are understood as ontologically derivative bundles of tropes, being nothing over and above their constituting properties. For a trope theorist, there is nothing more to a tree than its particular colour, shape, length, weight, and mass. Contrary to the minimal ontology postulated by trope theorists, Armstrong (1993) defends a two-category ontology, which postulates both particulars and universals, and Lowe (2006) defends a four-category ontology, which holds there to be two fundamental categories of particulars (objects and tropes) and two categories of universals (kinds and properties). Both Armstrong and Lowe argue that the various theoretical virtues of their theories with respect to trope theories, such as their explanatory power, outweigh the perceived vice of their respective lack of simplicity.

The fact that both science and metaphysics employ theory virtues has been used as a reason to defend the legitimacy of metaphysics on methodological grounds as a means of establishing truths about the world. As L.A. Paul claims:

> The theoretical desiderata we use to choose a theory include simplicity, explanatory power, fertility, elegance, etc., and are guides to overall explanatory power and support inference to the truth of the theory. ... [I]f the method can lead us to closer to the truth in science, it can lead us closer to the truth in metaphysics.
>
> *(Paul 2012: 21)*

On the other hand, Ladyman and Ross (2007) argue that the similarity of methods here instead has the consequence of making metaphysics pseudoscientific:

> Some metaphysicians have realized that they can *imitate science* by treating their kind of inquiry as the search for explanations. ... Taking the familiar explanatory virtues of

unity, simplicity, non-ad hocness, and so on, they ... argue with each other about whose particular metaphysical package scores highest on some loosely weighted vector of these virtues and requires the fewest unexplained explainers. On the basis of such reasoning, metaphysics is now often regarded *as if it were a kind of autonomous special science*, with its explananda furnished by the other sciences.

(Ladyman and Ross 2007: 17; emphasis added)

Ladyman (2012) is dismissive of such an approach at least in part because he sees theory virtues, in the spirit of Van Fraassen (1980), as being merely pragmatic devices for theory choice and not being concerned with truth per se, whereas metaphysics explicitly does aim at truth. An obvious rejoinder here is simply to hold that theory virtues *are* guides to truth, this being an attitude found not only within metaphysics but also in science. However, such a claim requires serious justification. Scientific realists have offered a number of arguments in defence of the idea that theory virtues can be indicators of truth, both a priori (Swinburne 1997) and empirical arguments, often based on inferences from the history of science (McMullin 2009; Schindler 2018), but these are not without problems. In particular, the a priori arguments often assume that nature itself is simple, making them circular, while empirical arguments suffer from being inconclusive, since inductive arguments from the history of science can be offered both in support of and against realism (Ivanova 2020).

2.2 Use of intuitions

A further salient point of overlap between the methodologies of science and metaphysics is the use of intuition. In addition to weighing competing theories with respect to theory virtues, metaphysicians also appeal heavily to their intuitions, with 'intuitiveness' commonly taken as a key desideratum in itself. For instance, adherents of the A-theory of time take it to be intuitive that time passes and that the distinction between past, present, and future are mind-independent, with the B-theory's rejection of these claims being highly counter-intuitive. Scientists are not immune to talking about the role of their intuitions, their aesthetic sensibility, and their intuitive sense. As mentioned above, Duhem claims that what resolves theory choice in science often is the 'good sense' of the scientists. The mathematician and scientist Henri Poincaré similarly argues that scientists use their aesthetic sensibility as a "delicate sieve" to select "the most elegant and beautiful combinations" that the mind produces (2001[1902]: 397). But can we reasonably claim that intuitions play the same role in science as in metaphysics?

At first glance, intuitions appear to play a more central role in metaphysics than they do in science; after all, the scientist may appeal to experiment to test theories. Taking this viewpoint, French and McKenzie (2015) claim there is a key asymmetry between the use of intuition in science and metaphysics:

[I]n the scientific case, and arguably in [naturalistic metaphysics], the intuitions are functioning only as a starting point, a guide to what to try and justify by other means; by contrast in the [case of metaphysics] intuition itself has an essential justificatory role.

(ibid: 29)

Though there is certainly a lack of symmetry between the two cases, we take the asymmetry to be far less clear-cut; in particular, intuition is demonstrably used as a tool of justification within science. Being distrustful towards intuitions as a philosopher of science is of course well placed, since the history of science shows intuitions to routinely run counter to scientific discovery.

But can we really appeal to the traditional distinction between context of discovery and context of justification to claim that intuitions, when used in science, are only relevant in the former context, while in metaphysics they are also crucial in the latter? We think that such an attitude does not do justice to much of what happens in science.

Scientists, for better or worse, appeal to intuition not only to come up with hypotheses but also to justify their projects. They do so primarily in cases where there is insufficient empirical data to confirm or disconfirm the relevant theory. Poincaré claimed that the aesthetic intuition of scientists leads them to select the hypotheses that are most likely to be successful (Ivanova 2017). Pierre Duhem argued that theory virtues cannot resolve theory choice, but rather it is the scientist's good sense that selects the most fruitful theories. The Nobel laureate Subrahmanyan Chandrasekhar held that aesthetic intuition can play an epistemic role in science, noting that "we have evidence ... that a theory developed by a scientist, with an exceptionally well-developed aesthetic sensibility, can turn out to be true even if, at the time of its formulation, it appeared not to be so" (1987: 64).[3] Aesthetic intuitions are commonly invoked in contemporary physics where theories are compared on grounds of naturalness, elegance, and beauty (Chandrasekhar 1987; McAllister 1996; Green 1999; Ivanova 2017). While it is doubtful whether our intuitions are any good at picking the right theory in advance of decisive empirical data, we cannot overlook the fact that scientists routinely employ intuitions to justify belief in a theory in the absence of empirical confirmation, and commonly take such intuitions to be a guide to the likelihood of a theory's truth. In order to do justice to scientific practice, it is important to acknowledge that intuitions within science demonstrably go beyond the context of discovery.

It is, of course, important to distinguish the descriptive aspect here from the normative. It is a matter of fact that many scientists do use intuition as a means of justification, but *should* they do so? A traditional approach to addressing this question is to look at the history of science and establish the track record of intuitive thinking: have intuitions led us down the right track or mostly taken us astray? It is well known that much of contemporary science is highly counterintuitive, and that many revolutionary theories were initially poorly received due to their lack of fit with received wisdom, whether this being because of the entrenchment of certain concepts within science (such as quantum mechanics overturning the determinism of classical physics) or the social restriction of religious dogma (such as with the Copernican revolution and Darwin's theory of evolution by means of natural selection), but such examples are also exceptional cases that stand out for their conceptually revolutionary status; it does not follow that intuitions are of no use in the standard inferential practices central to the construction of theories and hypotheses.[4] We take such track-record arguments to be inconclusive in resolving the normative question but take it that an independent justification is needed if any epistemic import is to be placed on one's intuition.

2.3 Modelling, thought experiments, and modes of inferences

While the use of intuition and theory virtues are two very clear examples of the overlapping methodologies of science and metaphysics, they are not the only cases. Metaphysicians and scientists also use similar forms of inference and modelling, such as the use of the imagination and thought experiments, the use of inference to the best explanation, and modelling by abstraction and idealisation. Both scientists and metaphysicians employ the imagination and the construction of thought experiments. Descartes' evil demon argument is a thought experiment that asks us to imagine what our experiences would be like were there no external world. Einstein's elevator thought experiment asks us to imagine what our experience would be like were we subjected to (a) uniform upward acceleration in an elevator or (b) the gravitational force in a stationary elevator, in order to argue for the equivalence principle. Poincaré's heat disc thought experiment (1902) asks us to imagine what our experience would be like were we living in a non-Euclidean disc exposed to non-uniform heat forces in order to argue for the underdetermination of geometry by experience and to establish the legitimacy of non-Euclidean geometries. The use of the imagination to generate hypotheses and justify them is common practice in both science and metaphysics.[5]

Furthermore, scientists as well as metaphysicians use inference to the best explanation (IBE) as a reliable form of reasoning. Scientists often explain the success of science by invoking IBE: an example is the predictive success of the atomic theory being explained by the fact that atoms are real, rather than fictitious, entities. Similarly, in metaphysics, Platonists invoke IBE to argue for the existence of abstract objects. The raspberries and strawberries in my breakfast bowl share something in common – they are all red. This fact cannot be explained, one may argue, unless we posit the property – redness – shared by these objects. Platonists argue that the best explanation for this resemblance is the existence of abstract objects.[6]

A further related point of methodological overlap is the use of abstraction and idealisation in the creation of models. Abstraction and idealisation are common in science; we omit and simplify the systems we study in order to model them and make predictions, and we also introduce elements to the system that are factually incorrect. For instance, when we try to predict economic behaviour we use models that make certain assumptions about agents, by simplifying the parameters involved in making choices and assuming certain patterns of behaviour that do not correctly describe individual human behaviour. Paul (2012) sees the job of the metaphysician to be continuous with that of the scientist insofar as each uses abstraction and idealisation to model different ways the world could be. The advantage that science holds over metaphysics is that experience will eliminate a good number of competitors, while the metaphysician needs to rely on theoretical desiderata to choose between these competing models.

In summary, insofar as the practice of science routinely goes far beyond a strictly empirical method, there are many key points of overlap between the methodology of science and that of metaphysics.

3 The relationship between science and metaphysics

Since there is a substantial overlap of the methodologies of science and metaphysics, appealing to methodology itself fails to provide a sharp demarcation between science and metaphysics. As with Venn diagrams, partial overlap does not entail indistinguishability; the empirical aspects of scientific methodology are sufficient to make key aspects of science distinct from metaphysics. However there is an asymmetry here insofar as the central methodological tools of metaphysics are also central to science. In light of this methodological overlap, this section considers the relationship between science and metaphysics in more detail. Metaphysics and science are often spoken of using the metaphor of a disciplinary hierarchy. First, it is often held by analytic metaphysicians (such as in the above quoted passage of E.J. Lowe) that metaphysics is conceptually prior to science through its wider, more general scope, and its concern with the basic conceptual tools used by science. Second, many naturalistic metaphysicians take the opposing point of view, such that metaphysics is secondary to science through its need to 'keep up' or 'fit' with the latest received wisdom within science, with science-free metaphysics being deemed as either limited to analytic questions that are not about the world per se, or else being an illegitimate means of addressing questions that ought to be informed by science.

Can either of these offer a clear means of demarcating science from metaphysics? In this section, we see that the methodological overlap of science and metaphysics undercuts both avenues. First, the use of non-empirical methods within science includes the kind of self-reflective conceptual analysis common to metaphysics, undermining the idea that metaphysics is where concepts are formed and science where they are applied or tested. Second, science's usage of non-empirical methods poses a problem for those inclined towards positivist, falsificationist, or pragmatist criticisms of metaphysics as meaningless, pointless, or simply less legitimate than science due to its non-empirical nature.

3.1 *Is metaphysics conceptually prior to science?*

Consider the characterisation offered by Paul (2012) that "metaphysics tries to tell us what laws, naturalness, properties, objects, persistence, and causal relations fundamentally *are* ... and science tries to discover *which* entities there are or how these natures are exemplified" (ibid: 6), and Morganti and Tahko's (2017) related claim that metaphysics differs from science "in its greater generality and perhaps conceptual priority" and that "science represents at least an indirect 'testing ground' for metaphysical hypotheses, which thus get fleshed out, as it were, in the same process that employs them to provide an interpretation of our best scientific theories" (ibid: 2560–2561). This idea of a clear hierarchy of metaphysics as prior to science implies that science is not properly reflective on its own conceptual tools. To pick on just one example from many, we see that this is patently not the case with respect to causal relations in the case of physics. Much of the focus in quantum foundations since the publication of Bell's theorems (1964, 1976) has been to reflect on what causal structure *is*, in order to understand *how* the non-local quantum correlations are to be explained, and indeed what is ultimately sufficient to provide a causal explanation in general, with the ultimate goal of utilising quantum causal relations for new technologies. Similarly, the task of unifying general relativity and quantum mechanics has forced theoreticians to query what space and time are, and whether spatiotemporal relations are properly fundamental or derivative of some deeper concepts. We could suppose that the researcher employed by the physics department becomes a de facto metaphysician when carrying out such inquiries, but this would appear to spin the facts to fit an inappropriate demarcation criterion.

Rather, the general idea of the scientist requiring a set of well-defined metaphysical concepts and tools before getting to work overlooks the central task in foundational work within science, namely how to reflect upon and adjust key central theoretical terms, such as 'time', 'space', and 'simultaneity' (in the case of classical to relativistic physics), 'motion' (in the case of Aristotelian to Galilean physics), 'cause', and indeed even 'reality' and 'identity' (in the case of classical to quantum mechanics). These are all examples from physics, but that is not to pick out physics as a special case; one can find analogous examples throughout the sciences, such as 'living', and 'individual' in the case of biology.[7] What we see routinely in the foundations of the sciences is a reflection of its practitioners upon the basic terms and concepts of inquiry in order to deal with well-confirmed anomalies in the data that fail to fit existing theory, or else deal with the problem of unifying accepted but prima facie incompatible theories (as with general relativity and quantum theory). As such, the conceptual priority of metaphysics is in danger of being overstated: it is central to scientific methodology to perform the kind of key conceptual analysis that metaphysicians aim to do, and this is invariably done in response to empirical findings.

This is not simply a point about the tendency of scientists to effectively 'do metaphysics', but more importantly concerns the epistemology of the kind of concepts used in science. Very often, physics (to pick again on a preferred example) requires us to consider possibilities not seriously entertained or explored by metaphysics, not least due to their unintuitiveness, such as the inertial-frame-dependence of simultaneity employed to understand the nature of light in special relativity, the lack of definiteness of properties of subatomic particles between measurements in order to explain their motion in quantum mechanics, and the use of non-Euclidean geometry to understand the gravitational force in general relativity. To pick on the latter point in a bit more detail: there was a long debate within philosophy concerning the status of Euclidean geometry given that some of its axioms, particularly the 'parallel postulate', fail to have the status of logical truths. On this point, Schopenhauer (1966[1819]: 130) held that "this truth is supposed to be too complicated to pass as self-evident, and therefore needs a proof; but no such proof can be produced, just because there is nothing more immediate", comparing the parallel postulate to the principle of contradiction. The fact that not only were consistent non-Euclidean geometries formulated but that they later proved useful in accounting for physical forces demonstrates that armchair reasoning about metaphysical possibilities can be less general and more restrictive than it may initially appear. It is often the empirical method of science, namely the existence of data that fails to fit existing theories, that forces us to take seriously possibilities that may not have been entertained as metaphysically possible, with this empirical grounding of conceptual analysis playing a key epistemic and motivational role in our understanding of central concepts within scientific theories. As such, when conceptual apparatus is most needed within science, namely in the foundations of new theories, the data often plays a crucial role in the development and application of concepts, and as such this undermines the idea of metaphysics alone putting in the 'background' needed for science to proceed.

3.2 Is metaphysics less legitimate than science?

Finally, metaphysics has historically been demarcated from science through being a less legitimate discipline, whether it being due to having the hopeless epistemic aim of achieving synthetic *a priori* knowledge (as the empiricists would have it), of being concerned with meaningless pseudo-questions (as the positivists would have it), of being merely untestable (as the falsificationists would have it), or merely of having no practical relevance (as the pragmatists would have it). More recently, we see Ladyman and Ross (2007: vii) hold that much of analytic metaphysics "fails to qualify as part of the enlightened pursuit of objective truth, and should be discontinued" through its failure to meet the science-first ideal of naturalistic metaphysics. But there are clearly questions studied by metaphysicians on which science doesn't reasonably bear, and which we have no particularly strong reason to regard as somehow constituting illegitimate intellectual inquiry. So long as the metaphysician does not have immodest epistemic aims and is not under the illusion that their inquiry is sufficient to uncover some deep, synthetic facts about the world, then it seems reasonable to consider scientifically uninformed metaphysical inquiry into issues like the special composition problem (van Inwagen 1990), or the conceivability of backwards causation (Dummett 1954, 1964; Black 1956), as analogous to pure mathematics.

Such a picture is endorsed by French and McKenzie (2012), naturalistic metaphysicians who defend the legitimacy of analytic metaphysics, in part because the latter can and often does prove useful to the former. This way of seeing the role of metaphysics is analogous to other theoretical endeavours: just as mathematicians come up with different theoretical frameworks, some of which can find their application in science (non-Euclidean geometries, knot theory), such might be the fate of many projects of analytic metaphysics, making them legitimate as potential scientific tools. Of course, much of the subject matter of this kind of metaphysics may be of little to no intrinsic interest to the scientist, the philosopher of science, nor indeed the naturalistic metaphysician, but this lack of intrinsic interest does not amount to a lack of legitimacy of analytic metaphysics.[8]

Potential usefulness, however, is a low bar. Though legitimate, many of the questions tackled in the field of analytic metaphysics may be reasonably deemed 'pointless', as Ladyman and Ross

(2007: 29–30) note, in the spirit of the pragmatist C.S. Peirce, on the grounds that they venture far outside what can in principle be tested. In contrast, scientific hypotheses, in the spirit of Popper, are standardly characterised as in-principle testable, or at least are sufficiently connected to the concept of experimental testing, and as such are not pointless. Can we then suppose that this is a key mark of discernibility between metaphysics and science; that the kinds of questions asked by the latter are *not* pointless? Ultimately, the methodological overlaps between science and metaphysics we have highlighted, specifically the widespread usage of non-empirical methodologies in science, once more undercut this putative means of demarcation. This turns out to be a key issue in the case of so-called 'post-empirical' science, namely those research programmes of theoretical physics that fail to make empirically testable predictions, with string theory and multiverse cosmology being prime examples (Ellis and Silk 2014). A central question in this regard is whether post-empirical science deserves its name: has theoretical physics reached the point at which large parts of it no longer constitute science at all? What exactly is the alternative: to dismiss such research as non-scientific, or as merely pointless, speculative science, or simply as metaphysics? There is the significant worry here that one is merely left playing with semantics. What is relevant is that empirically driven scientific discourse invariably has and does venture outside what can ultimately be tested, that non-empirical methods are for that reason a key tool within science, and hence that scientific methodology can overlap significantly with that of metaphysics, making it hard to delegitimise metaphysics on grounds of methodology without also condemning much of science.[9]

Notes

1 Authors are listed in reverse alphabetical order; this work is fully coauthored.
2 For a historical exploration of such attitudes, see Chandrasekhar (1987), McAllister (1996), Ivanova (2017), and Hossenfelder (2018).
3 Chandrasekhar goes on: "[a]s Keats wrote a long time ago, 'what the imagination seizes as beauty must be truth – whether it existed before or not'" (1987: 64).
4 The idea that intuition plays a largely reliable role in making scientific inferences from the data and forming new hypotheses, is of course distinct from the evidently false claim that contemporary scientific theories give a picture of the world that is 'intuitive', taken to mean 'aligned with common sense'.
5 For further reading on the role of thought experiments in science, see Stuart (2016) and Salis and Frigg (2020).
6 For a recent examination on the relationship between IBE in science and metaphysics, see Saatsi (2017).
7 For 'living', see Machery (2012); for 'individual', see Clarke (2010).
8 Indeed, Ladyman (2012) acknowledges this epistemically modest picture of metaphysics as not intrinsically worthless, but rather as less sure-footed than other a priori disciplines such as pure mathematics or logic due to the metaphysics' lack of an analogous notion of proof.
9 The authors thank Steven French for helpful comments. This chapter was written during the quiet hours, amidst the haze of new parenthood; we dedicate it to our daughter, Cailyn.

References

Armstrong, D.M. (1993) A World of State of Affairs. *Philosophical Perspectives*, 7, Language and Logic, 429–440.
Bell, J.S. (1964) On the Einstein-Podolsky-Rosen Paradox. *Physics*, *1*(RX-1376), 195–200.
Bell, J.S. (1976) The Theory of Local Beables. *Epistemological Letters*: March 1976, 11–24, reprinted as chapter 7 of Bell's 1987 *Speakable and Unspeakable in Quantum Mechanics*, Cambridge: Cambridge University Press.
Black, M. (1956) Why Cannot an Effect Precede Its Cause? *Analysis*, 16(3), 49–58.
Campbell, K. (1990) *Abstract Particulars*, Oxford: Blackwell.
Chandrasekhar, S. (1987) *Truth and Beauty: Aesthetics and Motivation in Science*, Chicago: The University of Chicago Press.

Clarke, E. (2010) The Problem of Biological Individuality. *Biological Theory*, 5(4), 312–325.
Dawid, R. (2013) *String Theory and the Scientific Method*, Cambridge: Cambridge University Press.
Dawid, R., Thébault, K., and Dardashti, R. (Eds.). (2019) *Why Trust a Theory?: Epistemology of Fundamental Physics*, Cambridge: Cambridge University Press.
Duhem, P. (1954[1906]) *The Aim and Structure of Physical Theory*, Princeton, NJ: Princeton University Press.
Dummett, M. (1954) Can an Effect Precede Its Cause? *Proceedings of the Aristotelian Society*, 28, 27–44.
Dummett, M. (1964) Bringing about the Past. *The Philosophical Review*, 73(3), 338–359.
Ellis, G., and Silk, J. (2014) Scientific Method: Defend the Integrity of Physics. *Nature News*, 516(7531), 321.
French, S., and McKenzie, K. (2012) Thinking outside the (Tool)Box: Towards a More Productive Engagement between Metaphysics and Philosophy of Physics. *The European Journal of Analytic Philosophy*, 8, 42–59.
French, S., and McKenzie, K. (2015) Rethinking outside the Toolbox: Reflecting again on the Relationship between Philosophy of Science and Metaphysics. In T. Bigaj and C. Wüthrich (eds.), *Metaphysics in Contemporary Physics*. Poznan Studies in the Philosophy of the Sciences and the Humanities. Amsterdam: Rodopi, pp. 25–54.
Greene, B. (1999) *The Elegant Universe*, New York: W.W. Norton and Company.
Hossenfelder, S. (2018) *Lost in Math: How Beauty Leads Physics Astray*, New York: Basic Books.
Ivanova, M. (2010) Pierre Duhem's Good Sense as a Guide to Theory Choice. *Studies in the History and Philosophy of Science*, 41, 58–64.
Ivanova, M. (2017) Aesthetic Values in Science. *Philosophy Compass*, 12, DOI: 10.1111/phc3.12433.
Ivanova, M. (2020) Beauty, Truth and Understanding. In M. Ivanova and S. French (eds.), *The Aesthetics of Science: Beauty, Imagination and Understanding*, Routledge.
Kuhn, T. (1970) Logic of Discovery or Psychology or Research? In Imre Lakatos and Alan Musgrave (eds.), *Criticism and the Growth of Knowledge*, Cambridge: Cambridge University Press, pp. 4–10.
Ladyman, J. (2007) Does Physics Answer Metaphysical Questions. *Royal Institute of Philosophy Supplements*, 61, 179–201.
Ladyman, J. (2012) Science, Metaphysics and Method. *Philosophical Studies*, 160(1), 31–51.
Ladyman, J., and D. Ross (with D. Spurrett and J. Collier) (2007) *Every Thing Must Go*, Oxford: Oxford University Press.
Lakatos, I. (1977) *Philosophical Papers, vol. 1*, Cambridge: Cambridge University Press.
Lowe, E.J. (2006) *The Four-Category Ontology*, Oxford: Oxford University Press.
Machery, E. (2012) Why I Stopped Worrying about the Definition of Life … and Why You Should as Well. *Synthese*, 185(1), 145–164.
Maudlin, T. (2007) *The Metaphysics within Physics*, Oxford: Oxford University Press.
McAllister, J.W. (1996) *Beauty and Revolution in Science*, Ithaca, NY: Cornell University Press.
McMullin, E. (2009) The Virtue of a Perfect Theory. In Martin Curd and Stathis Psillos (eds.), *The Routledge Companion to Philosophy of Science*, New York: Routledge, pp. 561–571.
Morganti, M., and Tahko, T.E. (2017) Moderately Naturalistic Metaphysics. *Synthese*, 194(7), 2557–2580.
Newton, I. (1999[1687]) *The Principia: Mathematical Principles of Natural Philosophy*. I.B. Cohen and A. Whitman (trans.), Berkeley, CA: University of California Press.
Ney, A. (2012) Neo-positivist Metaphysics. *Philosophical Studies*, 160(1), 53–78.
Paul, L.A. (2012) Metaphysics as Modeling: The Handmaiden's Tale. *Philosophical Studies*, 160(1), 1–29.
Poincaré, H. (2001[1902]) *Science and Hypothesis: Essential Writings of Henri Poincaré*, ed. Stephen Gould, New York: Modern Library.
Popper, K. (1963) *Conjectures and Refutations*, London: Routledge and Kegan Paul.
Saatsi, J. (2017) Explanation and Explanationism in Science and Metaphysics. In M. Slater and Z. Yudell (eds.), *Metaphysics and the Philosophy of Science: New Essays*, Oxford: Oxford University Press, pp. 162–191.
Salis, F., and Frigg, R. (2020) Capturing the Scientific Imagination. In P. Godfrey-Smith and A. Levy (eds.), *The Scientific Imagination*, Oxford: Oxford University Press.
Schindler, S. (2018) *Theoretical Virtues in Science: Uncovering Reality Through Theory*, Cambridge: Cambridge University Press.
Schopenhauer, A. (1966[1819]) *The World as Will and Representation*, Vol. II, E.F.J. Payne (trans.), New York: Dover Publishing.
Stuart, M.T. (2016) Taming Theory with Thought Experiments: Understanding and Scientific Progress. *Studies in History and Philosophy of Science Part A*, 58, 24–33.

Swinburne, R. (1997) *Simplicity as Evidence of Truth*, Milwaukee, WI: Marquette University Press.
Thagard, P. (2013[1978]) Why Astrology is a Pseudoscience. In Martin Curd, J.A. Cover, and Christopher Pincock (eds.), *Philosophy of Science* (2nd ed.), New York: W.W. Norton and Co., pp. 27–36.
Van Fraassen, B. (1980) *The Scientific Image*, Oxford: Oxford University Press.
Van Inwagen, P. (1990) *Material Beings*, Ithaca, NY: Cornell.

36
THING AND NON-THING ONTOLOGIES

Michael Esfeld

Discrete objects: from Aristotelian metaphysics to ontic structural realism

Thing ontologies are ontologies that consider the universe to be made up of a plurality of discrete objects. Non-thing ontologies can take the form of ontologies of discrete objects, too, but not necessarily so: they can also be conceived as ontologies of one continuous object. Let us consider the central versions of ontologies of discrete objects first, starting with thing ontologies, moving from there to non-thing ontologies of discrete objects and finally ontologies of one continuous object.

The paradigm example of a thing ontology is atomism, which can be traced back to the Presocratics. Thus, Democritos is reported as maintaining

> That substances infinite in number and indestructible, and moreover without action or affection, travel scattered about in the void. When they encounter each other, collide, or become entangled, collections of them appear as water or fire, plant or man.
> *(Fragment Diels-Kranz 68 A57, translation taken from Graham 2010, p. 537)*

At our time, Feynman says in the introduction to the *Feynman lectures on physics*:

> If, in some cataclysm, all of scientific knowledge were to be destroyed, and only one sentence passed on to the next generations of creatures, what statement would contain the most information in the fewest words? I believe it is the *atomic hypothesis* (or the atomic *fact*, or whatever you wish to call it) that *all things are made of atoms – little particles that move around in perpetual motion, attracting each other when they are a little distance apart, but repelling upon being squeezed into one another*. In that one sentence, you will see, there is an *enormous* amount of information about the world, if just a little imagination and thinking are applied.
> *(Feynman et al. 1963, vol. 1, ch. 1–2)*

The atoms are things, notably because they are discrete and thus countable (even if there were an infinite number of them) and because they are permanent – they do not come into existence and they do not go out of existence.

In virtue of the atoms being permanent, they can furthermore be considered as things in the sense of substances. Over and above that, for them to be substances, one can follow Aristotle in demanding that they exist independently of each other (cf. *Categories*, ch. 5). That is to say, no atom is ontologically dependent on other atoms – the atoms may be ontologically dependent on an absolute space, as in Democritean and Newtonian physics, or on God, as in Christian metaphysics; but if they claim an independent existence, that means that no atom is dependent for its existence on there being other atoms. If they exist independently of each other, each atom has some intrinsic features that characterize it as the atom that it is – that is, features that belong to each atom independently of whether or not there are other atoms (cf. the definition of intrinsic properties in Langton and Lewis 1998; see Hoffmann-Kolss 2010, part 1, for an extensive discussion of intrinsic properties). According to Aristotelian metaphysics, these intrinsic features are a form (eidos) that makes each thing the thing that it is (see *Metaphysics*, book 7) – at least as far as its being a thing of a certain kind is concerned, if not its being a certain individual (see Frede and Patzig 1988 for the latter view).

However, Aristotle does not have atoms in mind. His first and foremost candidate for substances are living organisms. Aristotelian forms then are intrinsic kind properties in the first place, namely those intrinsic properties that make it that a thing is a thing of a certain kind – such as tigers, lions or elephants in the case of living beings, or electrons, protons, neutrons viz. quarks in the case of candidate objects for atoms in the literal sense of indivisible particles. In the latter case, properties such as a certain value of mass and charge (and spin in quantum physics) are the candidates for kind-making properties. In any case, one does not find in physics candidates for intrinsic properties that go beyond kind-making properties. For instance, all electrons have the same value of mass and charge. In other words, there are no intrinsic properties that distinguish and hence individuate the things that belong to the same natural kind.

If one searches for a property that individuates, the first and foremost candidate is position. That is to say, its position relative to other things individuates each thing in a configuration of things at any given time, since no two things can be at the same place at the same time. Moreover, the history of position – that is, the trajectory of each thing – provides for its identity in time, since no two things can have the same trajectory. However, position is not an intrinsic property. It is a relation, namely either the occupation relation between things and points or regions of space on absolutism about space, or the distance relation between things on relationalism about space. If one enquires into an intrinsic feature that individuates things, there is no qualitative property available. Nonetheless, one can go for a primitive thisness (haecceity) of each thing, namely the feature of being that thing conceived as primitive. But one has to pay a high metaphysical price for haecceitism: on this view, possible worlds are distinct simply by permutation of haecceities, without there being any qualitative difference between them (see Adams 1979).

Position is not only the first and foremost candidate for the feature that individuates things. It also is the first and foremost candidate for the world-making relation. If there is a plurality of things, there has to be something that relates these things so that they make up a world. In other words, if one considers a plurality of worlds, there has to be something that relates the things in each world, by contrast to a relation that establishes a comparison between things across different possible worlds (such as e.g. the counterpart relation). At least as far as the actual world is concerned, position in the sense of spatial relation (distance) is what unites the world (cf. Lewis 1986, ch. 1.6). Ontologies that admit a plurality of things, but no relations (see Heil 2012, ch. 7; Lowe 2016) face the problem that there then is no relation that binds the many things together so that they make up a world. On this basis, one can make a case for putting relations before intrinsic properties: a relation of a certain type is indispensable as world-making relation if one admits a plurality of things, and only a relation can individuate these things (if one dismisses haecceities). It turns out that position in the sense of the distance relation can fulfil both these tasks.

Moreover, taking physics into account, it is by no means obvious that the kind-making properties are indeed intrinsic properties. Although parameters like mass and charge (as well as spin in quantum physics) are attributed to microphysical objects taking individually, this fact does not warrant the conclusion that they are intrinsic properties of the physical objects. The reason is that in physics, these parameters are exclusively considered in terms of their dynamical role in the laws of a given theory, namely in terms of what they do for the motion of the physical objects. Thus, commenting on Newton's *Principia*, Mach says that "the true definition of mass can be deduced only from the dynamical relations of bodies" (1919, p. 241). The same goes for charge, spin, etc. On this basis, one can make a case for considering these parameters to be relations rather than intrinsic properties, namely as expressing dynamical relations as encoded in the laws of a given physical theory (see Esfeld and Deckert 2017, ch. 2).

This case is strengthened when one turns to quantum physics: in brief, in quantum physics, also the classical parameters such as mass and charge are situated on the level of the wave-function, which is the central dynamical parameter in quantum physics. The wave-function cannot be considered as being or encoding intrinsic properties of physical objects due to its entanglement. That is to say, in brief, there is a wave-function only for a configuration of objects – in the last resort, the configuration of the objects of the entire universe. Entanglement cannot be reduced to (or be replaced with) additional local variables of the physical objects that could be candidates for intrinsic properties, as was proven by Bell's theorem (1964, reprinted in Bell 1987, ch. 2). And if one admits additional variables in order to solve the famous measurement problem of quantum physics, the first and foremost candidate for such variables is position, which is, again, a relation. Thus, basing oneself on physics, one can make a case for relations instead of intrinsic properties.

The metaphysical stance that makes this case is known as *ontic structural realism*, with structures being conceived as concrete physical relations among physical objects that do not have any intrinsic properties. This does not imply that one either has to resort to haecceities as that what individuates objects or to come to the conclusion that there is nothing that individuates physical objects (so that there are no objects at all). Dynamical relations such as entanglement in quantum physics or mass and charge in classical physics cannot individuate objects – in the formalisms of physical theories, they presuppose (individual) objects as that to which they are applied and whose evolution they constrain. But position can individuate objects – as mentioned above, the objects in a configuration can be conceived as being individuated by the distance relations in which they stand. Thus, on the stance known as *moderate* ontic structural realism, there still are objects, but all there is to these objects are the relations in which they stand (see Esfeld 2004; Esfeld and Lam 2011).

By contrast, on the stance known as radical ontic structural realism, the objects are dissolved in the structures in the sense that any putative object is supposed to turn out to be constituted by relations (see French and Ladyman 2003). The main objection to this stance is that structures are commonly defined over a set of objects. That is to say, there cannot be relations without relata that these relations relate. Even if this objection is sound, note that it is no argument for intrinsic properties, for it does not say anything against the relata being individuated by the relations in which they stand. In any case, radical ontic structural realism is not a thing ontology: it poses structures in the guise of relations instead of things. Moderate ontic structural realism, on the contrary, is a thing ontology, with its specific feature being the view that the things are individuated by the relations in which they stand and do not have intrinsic properties. Hence, in any case, ontic structural realism is opposed to Aristotelian metaphysics with its stress on intrinsic properties; in particular, it is diametrically opposed to the above-mentioned view that there are no relations. Quite to the contrary, relations are all there is or at least all there is when it comes to the fundamental features of things and to what individuates things.

If relations individuate things, the things are no longer substances in the sense of independent things, since they are no longer ontologically independent of each other. If position in the sense of distance is the relation that individuates, any object is ontologically dependent on there being at least two further objects for it to be individuated by the distance relations that it bears to the other objects with which it makes up a world. The failure of ontological independence of the things in the world is the reason why Schaffer (2010) maintains ontological monism in the sense of the ontological priority of the whole: only the whole universe is a substance according to this criterion. But note that this is only a verbal issue (i.e. a matter of definition): by adopting the criterion of ontological independence for something to be a substance, we automatically get to the stance of ontological monism in the sense of the view that there is only one substance as soon as we take relations to be essential to the objects in the world (e.g. because they are individuated by relations). However, by thus getting to monism, we have not moved away from an ontology of a plurality of things. We only refuse to grant the things the status of substances. This monism light so to speak is to be contrasted with a substantial monism so to speak according to which there is only one object. In other words, what is at issue here, in the context of (moderate) ontic structural realism, is still atomism with a plurality of individual things, albeit considered as ontological monism of only one substance, in contrast to the non-thing ontology of there being only one continuous stuff.

Discrete objects: events instead of substances

Another change to the traditional thing ontology inspired by contemporary physics consists in representing the configuration of matter of the universe as being inserted not in three-dimensional space, but in four-dimensional space-time. In this case, the configuration does not change in time; it is wholly present in space-time. In other words, everything that there is in space-time simply exists. Existence is not tied to a mode of time, such as only the present, or the present and the past existing and the future not existing, coming into existence (becoming) as time passes. The shift from space to space-time hence rules out in the first place ontologies that tie existence to a specific mode of time, such as presentism (see e.g. Saunders 2002); but it thereby also has implications for thing ontologies.

If things are inserted in space-time instead of in space, each material object, including its entire evolution, is conceived as occupying a certain region in space-time, namely what is known as a worldline (instead of a point or a region in space, with the occupied points or regions changing as time passes). This view implies the following shift in metaphysics: if things are inserted in three-dimensional space and change in time, they persist by enduring. They are wholly present at each point of time. Consequently, they do not have temporal parts; they only have spatial parts (and thus occupy regions of space), unless they are atoms in the sense of things that are located at points of space. By contrast, if things are inserted in four-dimensional space-time, they persist by perduring: they occupy a region in space-time, namely a worldline. Consequently, they do not only have spatial parts (unless they are atoms), but also temporal parts. The parts of the worldline of a thing are its temporal parts, and the existence of the thing as a whole is its entire worldline.

More precisely, the shift in ontology then is the following one: one can consider discrete, enduring things as substances, now taken in the sense of objects that are wholly present at each moment of their existence. By contrast, if one identifies things with their worldline in four-dimensional space-time, one considers them to be processes in the sense of continuous sequences of events, because they have temporal parts. Events and processes are four-dimensional entities, whereas substances in the sense of endurants are three-dimensional entities. An ontology of such four-dimensional entities still is a thing ontology; it still is an ontology of discrete objects, but these are sequences of events (processes) instead of enduring substances (as regards four-dimensionalism, see e.g. Heller 1990; Sider 2001; and Benovsky 2006, first part).

The physics that suggests this shift in ontology is the special as well as the general theory of relativity. The reason for this shift is that in the special theory of relativity, in contrast to Newtonian physics and consequent upon the velocity of light being independent of a reference frame and thus absolute, there no longer is a unique temporal order of all the events in the universe. Spatial and temporal distances between events are relative to a reference frame, and there is no universally privileged reference frame. Only the four-dimensional, spatio-temporal interval between any two events is independent of the choice of a reference frame. This is the reason for considering all the events in the universe as being inserted in a four-dimensional space-time, resulting in what is known as the block universe metaphysics. According to this metaphysics, the universe is a single four-dimensional block with everything that there is in space and time simply existing in that block, that is, being located in that block by occupying a certain region in that block, namely a worldline. The absence of a privileged reference frame then means that there is no unique, privileged way of slicing the four-dimensional space-time block into three-dimensional, spatial hypersurfaces that are ordered in one-dimensional time (see e.g. Balashov 2010 for the case for the block universe metaphysics based on relativity physics).

The block universe metaphysics is not committed to the existence of a four-dimensional, absolute space-time into which the material things are inserted. It can also be conceived as a relationalism about space-time, namely in terms of four-dimensional, spatio-temporal relations between events and their continuous sequences, that is, the worldlines. The main challenge that the block universe metaphysics faces is to establish a distinction between variation within a given configuration of things and change of that configuration. In brief, the differences in the four-dimensional, spatio-temporal relations between events and continuous sequences of events (worldlines) provide for variation within the block universe, but they do not change, since they all exist at once; *a fortiori*, there is no becoming in the block universe (Geach 1965, in particular p. 323, is the main source for this objection).

The block universe metaphysics is inspired by the physical theories of relativity, but it is not imposed upon us by the physics. In the first place, one can be a scientific realist about that physics and still adopt the attitude that is also available for the Leibnizian relationalist when it comes to Newtonian mechanics, which is formulated in terms of an absolute space and time: the relationalist can regard the absolute space and time and all the absolute quantities that appear in Newtonian mechanics as the means to represent the (Leibnizian) distance relations among point particles throughout the history of the universe, instead of subscribing to an ontological commitment to them (see Huggett 2006). By the same token, the relationalist can regard the Minkowskian geometry of special relativity and the Riemannian geometry of general relativity as the means to represent the basic ontology of (Leibnizian) distance relations among elementary discrete objects and the change in these relations throughout the history of the universe (see Esfeld and Deckert 2017, chs. 5.2–4). Furthermore, both in the classical, Newtonian case as well as in the relativistic case, there is an alternative formulation of the physics available that in the end yields the same empirical results based on an ontology of Leibnizian distance relations, namely Barbour's shape dynamics (see e.g. Barbour and Bertotti 1982; Barbour 2012; Gomes and Koslowski 2013; and Gryb and Thébault 2016).

Be that as it may, as regards an ontology of discrete objects in terms of a thing ontology, conceiving the things as being inserted into a three-dimensional space with change in one-dimensional time or conceiving them as being inserted into a four-dimensional block of space-time is no big issue. The things still are substances in the sense of being permanent and the considerations about their ontological dependence or independence apply in the same manner as mentioned in the preceding section. In the case of four-dimensionalism, the things are not substances only in the sense that they are perdurants instead of endurants.

The big issue, if there is a big issue here, is the possibility to maintain an ontology of discrete objects, but to abandon the assumption that the objects persist. If they do not persist, they no longer are things – and *a fortiori* not substances – but there still is a plurality of discrete objects that are individuated by their spatio-temporal location. These then are single events. On this ontology, there are no continuous sequences of events and hence no worldlines and no processes. There just are discrete, single events. If there are enough of such events and if they are close enough to each other, it may be possible to explain our experience of macroscopic objects that persist.

These events are ephemeral. If one formulates this ontology in terms of three-dimensionalism, these events come into being out of nothing and they are annihilated instantaneously – there is no change in anything, since there is nothing that persists. If one formulates this ontology in terms of four-dimensionalism (the block universe metaphysics), then some points of space-time are occupied by material events, and these events are distributed in a discrete manner in continuous space-time – there are unoccupied space-time points between any two material events, hence no continuous sequences of events.

Such an ontology is not only a possible position in the logical space of philosophical ontologies. It can also be based on physics, in this case a formulation of quantum physics with a dynamics of wave-function collapse. In brief, on this view, whenever the quantum mechanical wave-function collapses in the mathematical space on which it is defined, there occurs a single event at a point in physical space; these events are known as flashes (see Tumulka 2006, p. 826, and see Bell 1987, ch. 22, for the original proposal, as well as Allori *et al.* 2008, section 3.2, and 2014, section 2.3). The claim then is that these discrete flashes are all there is in physical space. As any ontology based on quantum physics, this ontology is in dispute (see Arntzenius 2012, ch. 3.15, for an endorsement and Maudlin 2011, ch. 10, and Esfeld and Deckert 2017, ch. 3.3, for a critical discussion). The interest for metaphysics is that this ontology illustrates the possibility of subscribing to a genuine event ontology – that is, an event ontology that cuts philosophical ice by admitting only single, discrete events that hence are discrete objects, but not substances because they do not persist (by contrast to a reformulation of a substance ontology in terms of continuous sequences of events, as in the case of the block universe metaphysics based on relativity physics).

Continuous stuff

The alternative to an ontology of a plurality of discrete objects is an ontology of one continuous object. There hence is only one substance that is permanent and that, since it is the only object in the universe, is not ontologically dependent on any other object. This ontology can also be traced back to the Presocratics, more precisely to the first Presocratic natural philosophers such as Thales, Anaximander and Anaximenes, who can all be interpreted as searching for the one stuff out of which the universe is made. On this ontology, there is, in contrast to a plurality of discrete, finally indivisible objects (atoms), just one continuous stuff that fills all of space. This stuff is known as gunk. However, like the ontology of discrete objects, this ontology is not committed to subscribing to an absolute background space: there can just be one extended, continuous stuff, but there does not have to be a space or space-time that is distinct from that stuff and into which it is inserted. This stuff can be conceived as being infinitely divisible. By contrast to the ontology of atoms, which are indivisible and therefore point particles, the stuff ontology does not have commit itself to points, neither point-objects (atoms), nor points of space (see Arntzenius and Hawthorne 2005).

If there is just stuff, the only variable is the density of that stuff. That is to say, in order to account for variation, one has to admit different degrees of density of stuff at different points or in different regions of space, with these degrees of density changing in time. The primitive stuff is a bare substratum that, moreover, admits of different degrees of density in different points or regions of space as a primitive matter of fact: there is nothing that accounts for the difference between the degrees of density of the stuff in different regions of space (mass, for instance, would, again, be a dynamical variable, not an intrinsic property of the stuff – see Allori *et al.* 2014, pp. 331–332). Apart from the commitment to a bare substratum of matter, the main challenge for this ontology is to formulate a dynamics for the evolution of the density of stuff such that stuff is concentrated in certain regions of space in such a way that the experience of discrete objects is accounted for. In other words, our experience of discrete objects down to molecules and atoms in the form of the chemical elements does not impose an ontology of a plurality of discrete objects on us. There may be just one object in the guise of a continuous stuff on the fundamental level. But any such ontology then has to accomplish the formulation of a dynamics for that stuff that allows it to be distributed in a way as if there were not only discrete macroscopic, but also discrete microscopic objects down to the chemical elements, if not the quantum particles.

Whereas the ontology of discrete objects is supported by particle physics, the ontology of a continuous stuff can draw on fields in physics. However, fields enter into modern physics as mediators of particle interactions, like the electromagnetic field in the Maxwell-Lorentz theory of the electrodynamic interaction in the nineteenth century. Nonetheless, one can claim that in the subsequent development of physics, fields have become ever more important and gained an autonomous status, such as the metrical field in general relativity theory, which is also the gravitational field. That notwithstanding, physics has not replaced particles with fields: the fields in quantum physics (quantum field theory) are not fields that have values at the points of space (or space-time), and quantum field theory is employed to formulate today's standard model of particle physics (one can even go as far as maintaining an ontology of permanent particles for quantum field theory, see Esfeld and Deckert 2017, ch. 4).

Nevertheless, there is a formulation of quantum physics available that works with a dynamics of the collapse of the wave-function and that regards the quantum mechanical wave-function as referring to a matter density field in physical space and as describing the evolution of this field – which is then such that, upon collapse of the wave-function, the matter density concentrates itself at certain points or regions of space so that precise measurement outcomes and, in general, the experience of discrete objects is accounted for (see Ghirardi *et al.* 1995 and Allori *et al.* 2008, section 3.1, and 2014, section 2.2). However, it is in dispute whether this proposal really succeeds in accommodating precise measurement outcomes. Moreover, in order to account for correlated measurement outcomes across space, this proposal has to postulate that the matter density can be instantaneously delocated across arbitrary distances in physical space (corresponding to the instantaneous collapse of the wave-function all over space) (see Egg and Esfeld 2015). These reservations notwithstanding, this worked-out proposal for an ontology of quantum physics in terms of a matter density field shows that a non-thing ontology in the guise of an ontology of one continuous stuff is a live option in contemporary metaphysics.

Such an ontology does not necessarily have to be conceived of in terms of only one continuous stuff that is extended all over space with different degrees of density. Rovelli (1997) proposes in the context of general relativistic physics an ontology of a plurality of fields that overlap and interact with each other, such as the gravitational field and the electromagnetic field both conceived as fields of primitive stuff that stretch all over space. In this case, the difference between these fields consists in differences in their dynamics, that is, in differences in their characteristic evolutions of concentrating and stretching out.

Conclusion

There is evidence of a plurality of discrete objects, in common sense as well as in science, where the theory of the atomic constitution of matter is the basis of the success of modern science from physics to molecular biology. If one takes this evidence to be a good argument for endorsing an ontology of a plurality of discrete objects, the formulation of that ontology then depends on the view that one adopts with respect to space and time: a plurality of enduring substances, or a plurality of worldlines (processes) that are continuous sequences of events, being perduring objects. In both cases, one defends an ontology of a plurality of things. One gets to a non-thing ontology if one admits instead of continuous sequences of events only a plurality of single, discrete events. These still are discrete objects, but they are ephemeral (flashes). The main version of a non-thing ontology, however, is the ontology of a continuous stuff that stretches all over space. Both this non-thing ontology and the ontology of a plurality of discrete objects can be traced back to the first Presocratic philosophers. They remain live options for an ontology that takes today's physics into account.

References

Adams, Robert M. (1979): "Primitive thisness and primitive identity". *Journal of Philosophy* 76, pp. 5–26.
Allori, Valia, Goldstein, Sheldon, Tumulka, Roderich and Zanghì, Nino (2008): "On the common structure of Bohmian mechanics and the Ghirardi-Rimini-Weber theory". *British Journal for the Philosophy of Science* 59, pp. 353–389.
Allori, Valia, Goldstein, Sheldon, Tumulka, Roderich and Zanghì, Nino (2014): "Predictions and primitive ontology in quantum foundations: a study of examples". *British Journal for the Philosophy of Science* 65, pp. 323–352.
Arntzenius, Frank (2012): *Space, time and stuff*. Oxford: Oxford University Press.
Arntzenius, Frank and Hawthorne, John (2005): "Gunk and continuous variation". *The Monist* 88, pp. 441–465.
Balashov, Yuri (2010): *Persistence and spacetime*. Oxford: Oxford University Press.
Barbour, Julian B. (2012): "Shape dynamics: an introduction". In: F. Finster, O. Müller, M. Nardmann, J. Tolksdorf and E. Zeidler (eds.): *Quantum field theory and gravity*. Basel: Birkhäuser, pp. 257–297.
Barbour, Julian B. and Bertotti, Bruno (1982): "Mach's principle and the structure of dynamical theories". *Proceedings of the Royal Society A* 382, pp. 295–306.
Bell, John S. (1987): *Speakable and unspeakable in quantum mechanics*. Cambridge: Cambridge University Press.
Benovsky, Jiri (2006): *Persistence through time, and across possible worlds*. Frankfurt (Main): Ontos.
Egg, Matthias and Esfeld, Michael (2015): "Primitive ontology and quantum state in the GRW matter density theory". *Synthese* 192, pp. 3229–3245.
Esfeld, Michael (2004): "Quantum entanglement and a metaphysics of relations". *Studies in History and Philosophy of Modern Physics* 35, pp. 601–617.
Esfeld, Michael and Deckert, Dirk-André (2017): *A minimalist ontology of the natural world*. New York: Routledge.
Esfeld, Michael and Lam, Vincent (2011): "Ontic structural realism as a metaphysics of objects". In: A. Bokulich and P. Bokulich (eds.): *Scientific structuralism*. Dordrecht: Springer, pp. 143–159.

Feynman, Richard P., Leighton, Robert B. and Sands, Matthew (1963): *The Feynman lectures on physics. Volume 1*. Reading (Massachusetts): Addison-Wesley.

Frede, Michael and Patzig, Günther (1988): *Aristoteles 'Metaphysik Z'. Band 2: Kommentar*. München: Beck.

French, Steven and Ladyman, James (2003): "Remodelling structural realism: quantum physics and the metaphysics of structure". *Synthese* 136, pp. 31–56.

Geach, Peter (1965): "Some problems about time". *Proceedings of the British Academy* 51, pp. 321–336.

Ghirardi, Gian Carlo, Grassi, Renata and Benatti, Fabio (1995): "Describing the macroscopic world: closing the circle within the dynamical reduction program". *Foundations of Physics* 25, pp. 5–38.

Gomes, Henrique and Koslowski, Tim (2013): "Frequently asked questions about shape dynamics". *Foundations of Physics* 43, pp. 1428–1458.

Graham, Daniel W. (2010): *The texts of early Greek philosophy. The complete fragments and selected testimonies of the major Presocratics. Edited and translated by Daniel W. Graham*. Cambridge: Cambridge University Press.

Gryb, Sean and Thébault, Karim P. Y. (2016): "Time remains". *British Journal for the Philosophy of Science* 67, pp. 663–705.

Heil, John (2012): *The universe as we find it*. Oxford: Oxford University Press.

Heller, Mark (1990): *The ontology of physical objects: four-dimensional hunks of matter*. Cambridge: Cambridge University Press.

Hoffmann-Kolss, Vera (2010): *The metaphysics of extrinsic properties*. Frankfurt (Main): Ontos.

Huggett, Nick (2006): "The regularity account of relational spacetime". *Mind* 115, pp. 41–73.

Langton, Rae and Lewis, David (1998): "Defining 'intrinsic'". *Philosophy and Phenomenological Research* 58, pp. 333–345.

Lewis, David (1986): *On the plurality of worlds*. Oxford: Blackwell.

Lowe, E. Jonathan (2016): "The are (probably) no relations". In: A. Marmodoro and D. Yates (eds.): *The metaphysics of relations*. Oxford: Oxford University Press, pp. 100–112.

Mach, Ernst (1919): *The science of mechanics: a critical and historical account of its development. Fourth edition. Translation by Thomas J. McCormack*. Chicago: Open Court.

Maudlin, Tim (2011): *Quantum non-locality and relativity. Third edition*. Chichester: Wiley-Blackwell.

Rovelli, Carlo (1997): "Halfway through the woods: contemporary research on space and time". In: J. Earman and J. Norton (eds.): *The cosmos of science*. Pittsburgh: University of Pittsburgh Press, pp. 180–223.

Saunders, Simon (2002): "How relativity contradicts presentism". In: C. Callender (ed.): *Time, reality and experience*. Cambridge: Cambridge University Press, pp. 277–292.

Schaffer, Jonathan (2010): "Monism: the priority of the whole". *Philosophical Review* 119, pp. 31–76.

Sider, Theodore R. (2001): *Four-dimensionalism. An ontology of persistence and time*. Oxford: Oxford University Press.

Tumulka, Roderich (2006): "A relativistic version of the Ghirardi-Rimini-Weber model". *Journal of Statistical Physics* 125, pp. 825–844.

37
MODERATELY NATURALISTIC METAPHYSICS

Matteo Morganti

Introduction

In a rough, generic connotation, philosophical naturalism consists in the request that philosophy be in some sense continuous with our best current science. At least since the work of Quine, this is a rather popular doctrine. And naturalism has become even more fashionable as of late, more and more philosophers attempting to systematically ground their claims in the empirical input coming from the special sciences. However, naturalism is in fact a tricky concept, and distinct claims to the effect that philosophy should be naturalistic, or naturalized, may mean many different things, and even lead to conflicting conclusions. How should the notion of 'continuity' be understood? Does philosophy have to become an empirical enterprise? Does naturalism imply the elimination of (some) philosophical questions? The situation is particularly complex when one looks at metaphysics, the part of philosophy that allegedly inquires into the fundamental structure and nature of being – that is, the part of philosophy which studies reality in its most general and basic aspects. Indeed, given its characterization, metaphysics appears paradigmatic of the a priori nature of philosophical research, hence to be clearly something that it is hard, if not altogether impossible, to render continuous with science – whatever this may exactly mean. Indeed, when it comes to metaphysics naturalism might be, and has been, intended in a *radical*, essentially eliminativist sense – the demand for the attribution of priority to science leading to the abandonment of metaphysics.[1] Alternatively, the naturalist may contend that metaphysics can be accepted, but only to the extent that it can be made dependent on science. It is, however, possible to endorse a more *moderate*, yet substantial, form of naturalism about metaphysics.

This chapter will indeed illustrate the meaning and motivation of various forms of naturalized metaphysics, and then provide arguments in favour of moderate naturalism about metaphysics. In particular, section 1 will briefly rehearse various ways of understanding metaphysics, and present reasons for adopting a science-oriented approach to it. Section 2 will first consider arguments that have been put forward in favour of radical naturalism, leading to either the elimination of metaphysics or its subordination to science, and then challenge the view on the basis of a trilemma. A case study concerning the ontological nature of quantum entities will be used to guide the reader by means of an example. Section 3 will argue that (a form of) moderately naturalistic metaphysics sidesteps the trilemma and should in general be preferred by philosophers. Section 4 describes in more detail the sort of moderately naturalistic metaphysics which promises to strike the best balance between science and philosophy. A brief concluding section follows.

1 Kinds of metaphysics, and the naturalistic option

As mentioned, metaphysics can roughly be characterized as the study of the most general and fundamental features of everything that exists, independently of what is the case, or seems to be the case, in any specific subdomain of reality. Defining science as the search, based on empirical methods, for the explanation of particular matters of empirical fact based on putative laws of nature, a contrast immediately emerges – essentially due to the testability of scientific predictions as opposed to the seeming lack of empirical import of metaphysical claims.

Indeed, metaphysics is often conceived as an independent, purely a priori discipline and science as an independent, purely a posteriori discipline, with no overlap in their methods and basically different subject matters. For instance, one might understand metaphysics as conceptual analysis, focused on clarifying certain basic notions and primarily based on linguistic considerations (see Jackson 1998). On a slightly different note, even conceding that metaphysics and empirical science have a significant amount to share in terms of object of study, one may insist that they differ drastically in terms of methodology, since metaphysics is an essentially a priori enterprise (see Lowe 2011).

Given the undisputable successes of science, and the lack of agreement, if not of objective progress, in metaphysics, a completely different approach may however appear advisable. One, that is, that demands metaphysics to be continuous with science also in terms of methodology. This leads to at least three alternative options.

The first is the so-called 'experimental metaphysics' championed, for instance by Goldman (see, e.g., Goldman 2007, 2015). Basically, Goldman suggests that the empirical study of the human mind should be a fundamental part of metaphysical investigation. He shows that many presuppositions that we may regard as obvious, for example concerning the individuation of events, actually depend on the specific way in which our mind works with respect to the environment. Thus, he concludes from this, rather than doing traditional philosophy on the basis of these presuppositions, we should instead make the latter our primary object of inquiry. While it is certainly right that we should study via empirical methods everything that we can, however, the sort of naturalization proposed by Goldman seems to actually imply a significant change of perspective, since experimental metaphysics understood in this way is a sui generis sort of enterprise, not really metaphysical, but rather *about* metaphysics. For, it leads to the formulation of claims about humans as the subjects of metaphysical inquiry, not about the world as the object of such an inquiry. For this reason, we will not discuss it further here.[2]

The other two options will instead constitute the focus of the rest of this chapter: as we did in the introduction, we will call them 'radical' and 'moderate' naturalistic metaphysics, respectively. We will discuss the former in the next section, and the latter in the following two.

2 Radically naturalistic metaphysics

Radical naturalism about metaphysics has a more or less clear, and more or less explicit, debt towards the logical empiricists' rejection of metaphysical statements as meaningless.

Ritchie (2008), for instance, defends a sort of 'deflationary naturalism' based on the general principle that philosophers should only pursue philosophical projects that can be carried out through a detailed investigation of science. Similar ideas are expressed by Maddy (2007), who defines and defends what she calls 'second philosophy', a way of doing philosophy that she regards as a radical and austere form of naturalism. Essentially, second philosophy ignores big philosophical questions and systems, and recommends instead the use of "what we typically describe with our rough and ready term 'scientific method'" (Ib.: 2).

The most illustrative example of radical naturalism, however, is probably constituted by van Fraassen. On the way towards defining and refining his own constructive brand of empiricism, van Fraassen mounts a sustained attack against metaphysics, especially in his (2002). There, van Fraassen argues that:

- science is constantly and harshly tested, and often falsified, but this doesn't affect, but rather grounds, its practical relevance; metaphysics, instead, seeks the truth, but is never in a position to establish whether what it says is actually true or false, and therefore turns out to be a merely formal exercise;
- metaphysical questions are irredeemably context-dependent and such that they lack well-defined answering strategies. He uses the example of the question "Does the world exist?" (see also Putnam's 2004 discussion of the mereology-related question "How many objects are there in a universe with only three particles?");
- metaphysics accounts for "what we initially understand [in terms of ...] something hardly anyone understands" (Van Fraassen 2002: 3), and consequently turns out to be a superfluous addition to the indications coming from empirical science.

These may well be sensible criticisms, especially from a strictly empiricist viewpoint aiming to revive the neopositivist idea of sharply demarcating between what is verifiable and amenable to scientific inquiry and what is not. However, without entering a detailed discussion of each point (for which, see Morganti 2013, esp. sec. 2.2) and empiricism more generally, we will just call attention to two basic problems for this sort of approach.

First of all, as is well-known since the time of the late neopositivist movement and Popper, the sharp divide between science and metaphysics just mentioned is very problematic. And this clearly entails that sweeping claims about metaphysics as opposed to science inevitably rest on shaky ground.[3]

Second, empiricists often employ a distinction between the observable and the unobservable – most notably, in the context of the debate concerning the epistemic import of scientific theories and the status of theoretical entities. However, besides the fact that this distinction too is fuzzy, it seems to lend insufficient support to the empiricists' anti-metaphysical claims. Some philosophers (see Maclaurin and Dyke 2012), for instance, contended that metaphysics should be naturalistic in the sense that it shouldn't be in principle unable to have observable consequences. However, as McLeod and Parsons (2013) argued drawing a parallel with the failure of Ayer's criterion of 'factualness', every theory can be made to have observable consequences by aptly adding auxiliaries to it. If one responds (Dyke and Maclaurin 2013) that auxiliary hypotheses themselves should only be admitted if they are supported by current science and able to lead to novel predictions, the risk of begging the question becomes evident. For, again a clear distinction between science and non-science is merely presupposed. Moreover, the vague notions of 'support' and 'prediction' definitely need to be made more precise.

We will get back to this shortly. Before that, it is important to notice that the foregoing considerations invite one to divide the empiricism-oriented, radical naturalistic approach to metaphysics into two typologies. The first, that we can call 'eliminative naturalism about metaphysics' (ENM), recommends the elimination of metaphysics based on its methodological shortcomings. The second, which we will dub 'reductive naturalism about metaphysics' (RNM), urges instead that metaphysics be made dependent on (current) science. Other than in the abovementioned papers by Maclaurin and Dyke, RNM can be traced in the widely discussed Ladyman and Ross (2007). Ladyman and Ross urge that metaphysics not be eliminated, but practised only to the extent that it can be put to the service of science. In particular, these authors argue against what they call 'neo-Scholastic metaphysics', which they connote as an activity that puts forward abstract claims and hypotheses that only pay lip service to science and are, in fact, grounded almost exclusively in commonsense intuition. As an alternative, Ladyman and Ross defend a conception of metaphysics as the search for, and promotion of, unification among scientific theories on the basis of physics. Naturalized metaphysics should, in their opinion, follow two basic principles:

- the 'Principle of Naturalistic Closure', according to which

    Any new metaphysical claim that is to be taken seriously at time t should be motivated by, and only by, the service it would perform, if true, in showing how two or more specific scientific hypotheses, at least one of which is drawn from fundamental physics, jointly explain more than the sum of what is explained by the two hypotheses taken separately

    (Ib.: 37);

- the 'Primacy of Physics Constraint', stating that

    special science hypotheses that conflict with fundamental physics, or such consensus as there is in fundamental physics, should be rejected for that reason alone. Fundamental physical hypotheses are not symmetrically hostage to the conclusions of the special sciences. This, we claim, is a regulative principle in current science, and it should be respected by naturalistic metaphysicians.

    (Ib.: 44)

One might object to the assumption of the primacy of physics; to the idea that the use of intuition is bad and is not present in physics; to the claim that metaphysics is only based on intuition and basic knowledge of science; or to the statement that non-science is useful only if it leads to unification in science (and not, say, explanation and understanding in a more general sense). Here, however, we will not discuss Ladyman and Ross' views, nor RNM more generally, any further. Rather, we will tackle the issue from a slightly different perspective. To do so, we will first introduce a case study, against the background of which we will critically assess both ENM and RNM.

2.1 Identity in quantum mechanics

The case study that we will briefly look at concerns the ontological status of the entities described by quantum physics. For our present purposes, it is sufficient to stick to standard textbook quantum mechanics.

The things that we interact with most of the time at the level of common sense are no doubt individual objects: they have a unique identity both in the sense that they are numerically unique and distinct from everything else at any instant of time, and in the sense that they persist as the same thing at different times. Focusing on the first, 'synchronic', sense of identity, it seems to be accompanied by three basic features: first, every object occupies a well-defined region of space-time; second, every object is qualitatively unique, in that it differs from everything else not only numerically, but also with respect to some qualitative feature – at the very least, its position in space at a given time – consequently obeying Leibniz's *Principle of the Identity of the Indiscernibles*; third, when we consider their collective behaviour (their *statistics*), objects combine in familiar and intuitive ways – four equally probable combinations being available, for instance, for two non-biased coins.

In the quantum domain, however, these three general facts do not hold anymore – or so it seems. The occupation of a well-defined region cannot be taken for granted (especially so in the relativistic case); the properties of things correspond to objective probabilities, and it is possible for two or more things to have all their probability assignments in common, including those attributing a spatio-temporal location to them (the Identity of the Indiscernibles is violated); and, lastly, quantum statistics is very different from classical statistics – roughly, the quantum counterpart of our system of two classical coins is such that only three combinations, or even just one in some cases, are available to 'particles' of the same type in the same system, and for which two states are possible.[4]

Now, the supporter of RNM could interpret this as lending decisive support to their stance. A bit in the spirit of 'experimental metaphysics' understood in the sense of Shimony (1980),[5] one could say that quantum mechanics provides us with a clear (negative) answer to the question 'Are fundamental physical objects individuals?'. Indeed, it is remarkable that empirical evidence now available to us can be used to evaluate philosophical principles such as Leibniz's Identity of the Indiscernibles, initially introduced on purely a priori grounds, and to say relevant things about synchronic identity and non-identity.[6] Unfortunately, though, things are not as straightforward as one may think. For, it is in fact possible to resist the above conclusion that quantum entities are non-individuals. For instance, one could claim that being an individual object does not require occupying a definite location (or, alternatively, argue in favour of quantum theories/interpretations, such as Bohmian mechanics, that explicitly attribute precise locations to particles); next, s/he could argue that the Identity of Indiscernibles is not a general truth, and quantum entities may qualify as distinct individuals in spite of their being exactly similar in their intrinsic properties (their numerical identity and diversity being a primitive fact);[7] lastly, one could claim that the 'weird' statistics obeyed by quantum entities is not due to their non-individuality, but rather to the peculiar properties that they exemplify, or the specific ways in which arrangements of many things emerge out of collections of individual things at the microscopic level.[8]

We don't need to get into the details. The mere logical possibility of the alternatives just mentioned suffices to show that the naturalist's hope of reading off metaphysical consequences from our best current theories is ill-posed.[9] The case of the ontological status of quantum entities is arguably paradigmatic of a more general fact. Namely, that the empirical input *underdetermines* metaphysics, hence it is simply impossible to reduce metaphysics to science in the sense of becoming able to reach clear-cut philosophical conclusions simply on the basis of empirical data and scientific theories.

One may reply that this is bad news for the radical naturalists endorsing RNM, hence reduction as 'reading off', but not for those who take naturalism about metaphysics to lead to ENM, i.e. to full-blown eliminativism about metaphysics. Van Fraassen, for one, would be happy with the idea that empirical data underdetermine our general claims about alleged fundamental facts: from it, he would conclude that we should simply stop seeking such putative general truths. While it may be a consistent stance for the empiricist to reject all claims going beyond the observable (provided, of course, that the observable-unobservable divide can be drawn on the basis of well-defined criteria, which is, as mentioned earlier, notoriously debatable), the problem is that, on a closer look, ENM also turns out to be untenable. For, the problem is that the empirical input underdetermines our hypotheses, and it is a well-known fact that this is the case for science as well, not only metaphysics! One may respond that metaphysical hypotheses are 'more' underdetermined than scientific ones.[10] However, this is not clearly decisive: as illustrated by our case study above, at least some metaphysical hypotheses are in some way connected to the empirical data, hence can be evaluated on an at least partly a posteriori basis.[11] Thus, even granting that there is a difference between underdetermination in science and metaphysics, it is one of degree, not of kind. Consequently, a differentiated treatment of science and metaphysics, leading to the elimination of the latter, does not appear warranted on the basis of underdetermination alone. To the contrary, it seems fair to say that, if taken seriously, the problem of underdetermination of hypotheses by the empirical data should (or, at least, could) affect our attitude towards science itself, and consequently to the rejection of ENM.

Summing up in the form of an explicit argument, we get the following:

1. The relevance and empirical success of science advises us to be naturalists about philosophy (assumption)
2. Metaphysicians must ground their claims and hypotheses on the best available science (from 1.)
3. Scientific input underdetermines metaphysical hypotheses (fact)
4. There is no way to pick out in a precise and univocal manner metaphysical hypotheses (from 3.)
5. Metaphysics doesn't follow directly from science, hence cannot be reduced to it – RNM should be rejected too (from 4.)
6. Metaphysics cannot be pursued (from 2. and 5.)
7. Naturalist philosophers cannot discard only metaphysics, as scientific hypotheses are also underdetermined by the empirical data – ENM should be rejected too.

What should we do, then? Is naturalistic metaphysics a hopeless project?

3 Moderately naturalistic metaphysics

There seem to be three possible reactions to the above argument: i) to consider underdetermination fatal to scientific hypotheses too (so simply accepting the consequences of 7. above); ii) to provide further arguments to demarcate between (underdetermination in the case of) scientific and metaphysical hypotheses (so rejecting 7.); iii) to give up naturalism itself (so giving up on 1.). It is easy to see that each one of these alternatives is not appealing in the context of naturalistic metaphysics, and we are consequently led into a problematic trilemma.

Indeed, option i) does not simply entail scientific anti-realism but a stronger form of scepticism: for, the key point here seems to be that underdetermination suffices for questioning not simply the truth of certain hypotheses, but their rational acceptability. It is clear, however, that the conclusion that scientific hypotheses are in fact not selected rationally is very implausible at least from a naturalistic perspective. Of course, as Hume has taught us, we should strictly speaking always refrain from making general claims based on experience, for the latter never constitutes a sufficient basis for obtaining the former by means of deduction. But, as Hume also argued, this does not mean that non-deductive hypotheses are not rational. To the contrary, on a sufficiently comprehensive understanding of the way in which our beliefs are formed and ground our actions, empirical underdetermination seems to just be a fact we need to accept. And this is what we in fact do all the time, especially in science.

As for option ii), if one were able to (re-)establish a principled differentiation between science and metaphysics, be it based on the kind of underdetermination occurring in the two domains or some other criterion, underdetermination could stop representing a problem for the scientifically minded philosopher regardless of its consequences for the naturalist about metaphysics. This, however, was one of our key points in the preceding section: if the credibility of a naturalistic stance is essentially connected to the fate of classical empiricism and the demarcation problem, given the unquestionable fact that the latter has so far proven to be insoluble, perhaps one should lose faith in the naturalistic project itself before attempting once again to revive a somewhat outdated philosophical agenda. As a matter of fact, while examples of metaphysical questions that appear not to have any substantial connection with the empirical input can certainly be provided, this falls short of grounding a general argument of the sort we are currently seeking, and it is unclear what else could be added so as to effectively substantiate the case against metaphysics.

Moving on, it goes without saying that option iii) is definitely unattractive as well. First, we are discussing the prospects of naturalistic metaphysics, and abandoning naturalism is, in this sense, a complete non-starter. Second, it is important to point out that even if one were ready to contend that science has nothing to contribute whatsoever to philosophical problems, the issue of underdetermination would not thereby be avoided. To the contrary, it would persist almost untouched: for, even in a fully anti-naturalistic context, in which the empirical input is ignored in the context of a priori philosophical analysis, the metaphysician would in any case have to choose among many different, logically possible alternative explanations of things. The only difference would be that s/he, unlike the naturalist, would be forced to choose exclusively on the basis of non-empirical considerations.

The trilemma just discussed may seem fatal for the naturalistic metaphysician. The notions of underdetermination, explanation and non-empirical factors, however, at the same time suggest a possible way out. In the rest of the chapter, we will look more closely at the details of this way out – which, it will be argued, constitutes the right way of characterizing moderately naturalistic metaphysics.

The basic argument in favour of moderately naturalistic metaphysics is pretty straightforward: if, as argued, underdetermination is ubiquitous, then both in science and in metaphysics the choice among rival hypotheses is of an abductive nature. That is, it presents itself as an inference to the best explanation among the many available explanations compatible with the empirical data. Consequently, both in science and in metaphysics the choice is made on the basis of a) an empirical input and b) non-empirical factors. Therefore, science and metaphysics should be developed in parallel, without any claim of reduction or elimination, exploiting their differences as well as their points of contact.[12] Let us see this in a bit more detail.

As for a), the above discussion of quantum objects and experimental metaphysics has hopefully given an idea of what is at stake. As mentioned, it may be granted that scientific hypotheses are 'closer' to the data and more directly testable than their metaphysical counterparts. However, the latter can be tested nonetheless, at least to the extent that they concern, say, the nature of material entities rather than that of angels, gods or numbers; and to the extent that one employs a sufficiently loose notion of empirical testing. More specifically, the notion of 'indirect testability' can be introduced: metaphysical hypotheses and theories can be put to empirical test in the sense that one can systematically derive consequences from them (or, better, from the conjunction of them and a given scientific theory), compare them with the data and, on the basis of this, establish how far they 'square', as it were, with the world out there. In our earlier example, for instance, Leibniz's principle of the Identity of the Indiscernibles is indirectly tested on the basis of what quantum theory and observation tell us about the properties of microscopic entities.[13] As for b), if it is correct that our hypotheses about reality always have an abductive nature and are underdetermined by the empirical input (there are no 'experimenta crucis', either in science or in philosophy), then it seems plausible to think that both scientists and metaphysicians will evaluate various hypotheses on the basis of well-known pragmatic, or 'theoretical', factors: simplicity, unification, consistency, minimization of established beliefs, etc.[14] In our example, for instance, one may insist that quantum particles are non-individuals based on the belief that the view of the world as ultimately constituted by purely qualitative facts is very explanatory.

Overall, this leads one to realize that the above trilemma can in fact be avoided, by denying that 6. actually follows from 5., that is, that the underdetermination of hypotheses of kind X by the empirical data is a sufficient reason for not asking X-questions at all. Indeed, if underdetermination is inevitable in any kind of abductive (and, more generally, non-deductive) reasoning, then we should simply accept the fallibility of our hypotheses, and consequently be ready to change our beliefs, theories and models as we keep inquiring into the nature of things. Thus, in contrast to both RNM and ENM – but also anti-naturalism – what we should correctly say is that there may well be a difference between science and metaphysics, but not one that warrants an eliminativist or at least reductionist stance towards metaphysics.

This point is crucial, as it grounds the claim that the resulting view of metaphysics is moderately naturalistic. Indeed, since it is explicitly based on the acceptance of underdetermination as pervasive, instead of elimination or subordination it leads to the *parallel development* of science and metaphysics. On the one hand, once the limits and fallibility of our (means to achieve) knowledge of the empirical domain are acknowledged, it becomes fair to consider some elements of metaphysics to be prior to science in that metaphysics explores a basic possibility space in such a way that the grounds for the *interpretation* of scientific theories are laid. (Consider the quantum example again: what does metaphysics do there other than providing conceptual categories and schemes for interpreting quantum entities as individuals or non-individuals of particular sorts? The Leibnizian view, for instance, leads us to certain conclusions; insisting on primitive identities or irreducible relations licenses different inferences; and other options are available). On the other hand, some elements of science remain prior to metaphysics in that science not only contributes to the definition of the basic possibility space itself, but also gathers the indications coming from the actual world that are necessary for fleshing out the various metaphysical hypotheses and selecting the most appropriate (i.e. informative, explanatory, simple, etc., perhaps also likely to be true) among them.

The picture that emerges (for more details, see Morganti 2013 and Morganti and Tahko 2017) is that the methodologies of the two disciplines are in fact significantly different, in that metaphysics is strictly speaking purely a priori while (natural) science is almost entirely a posteriori. This, together with the abovementioned difference in terms of 'distance' from the empirical input, surely marks a distinction between the two disciplines. However, crucially, this is perfectly compatible with the view, defended earlier, that there is no sharp divide here to be found. As a matter of fact, it seems fair to say that, in spite of their mutual dissimilarities, science and metaphysics are nonetheless so intertwined that we cannot properly pursue one without the other if we want to describe and understand the structure of reality at the deepest and most general level. As for their subject matter, it seems fair to grant the fundamental unity and uniqueness of the domain that scientists and metaphysicians explore, at least to the extent that one focuses on the subset of metaphysical questions that are indeed at least potentially amenable to indirect empirical testing as defined a moment ago.[15]

4 Further remarks

Having outlined the main features of a seemingly viable form of moderately naturalistic metaphysics, let us now add a few brief remarks before closing.

A   *Modality.* Ladyman and Ross support their own form of radically reductive naturalism by claiming, among other things, that philosophers have often been wrong in deeming something possible or impossible, and it is thus best to learn directly from scientists. Relatedly, Callender (2011) also laments the lack of a clear definition of the sui generis conceptual space that metaphysics is supposedly concerned with, supporting instead the view that it is ultimately physics that determines what we regard as metaphysically possible, necessary or impossible. However, there is first of all a case to be made that metaphysical modality outstrips physical/nomological modality – the debate on this is still wide open. Moreover, and more importantly, what we said about moderately naturalistic metaphysics in fact requires no presupposition whatsoever about modality. Metaphysics seeks the most fundamental and general truths, and therefore has to employ peculiar concepts and categories, and explore a space of possibilities (probably characterized by dependence and priority relations connected to the essential natures of things) defined in terms of these concepts of categories. But the use of these general notions by no means presupposes the existence of a sui generis domain of purely metaphysical possibilities and necessities. To the contrary, it is perfectly compatible with the idea that there is just one world and one sets of facts governed by one sets of general laws, and metaphysics and science just look at this world from different perspectives, as it were, and using different vocabularies.[16]

B   *Metaphysics and mathematics.* As suggested, even in a moderately naturalistic perspective metaphysics remains, by itself, a purely a priori enterprise. While there are obvious differences (for one, mathematics follows an essentially deductive method, through which it obtains something like cumulative progress, objective results and widespread agreement), this makes metaphysics analogous in important respects with mathematics. For instance, both mathematics and metaphysics proceed a priori and, by and large, independently of natural science; and both mathematics and metaphysics may or may not be then applied to science – the former to translate generic empirical claims into precise quantitative statements, the latter, as suggested, to provide an interpretation of scientific theories. This leads to the question whether moderate naturalism entails that metaphysics should be pursued only to the extent that it is applicable to an actual part of current science with a view to interpreting it; or, instead, metaphysics can be pursued freely, independently of the prospects of application, provided that we keep in mind that application towards interpretation and understanding of empirical science, whenever possible, is essential. Here, naturalism and the idea of preserving some degree of autonomy for metaphysics may pull in different directions, the former leading towards the first approach, the latter pushing towards the second. Be this as it may, as already mentioned (see note 13), in this case too no final stance must be taken in order that moderate naturalism be a credible approach to metaphysics.

C   *Metaphysics and models.* Based on what we said concerning underdetermination, explanation and abductive reasoning, one may agree with, for instance, Paul (2012) and Godfrey-Smith (2006) that – modulo the differences between them – metaphysics is continuous with science in the crucial respect that they both consist in the construction of possible models of reality. If so, the important question of realism must be tackled. In light of our earlier considerations concerning fallibility and underdetermination, one may wish to side with Godfrey-Smith and think that, whatever model we construct about reality, what counts the most is its explanatory efficacy, not its (approximate) truth. On the other hand, here too an alternative stance – according to which the very idea of seeking explanation of a progressively more fundamental nature suggests an essentially realist approach with respect to both science and metaphysics – also seems plausible. Once again, a general characterization of moderately naturalistic metaphysics does not require a final verdict.

D *Moderately naturalistic metaphysics and liberal naturalism.* Lastly, it is worth pointing out that the form of moderately naturalistic metaphysics outlined emerged from considerations that were primarily *methodological*, having to do with the definition of the aims and scope of science and metaphysics and their mutual relationships. These are, at least prima facie, significantly different from the considerations involved in the recent debate concerning whether a 'liberal' naturalism is conceivable, whereby – at the *ontological* level – a naturalistic stance does not automatically entail that a commitment to entities, processes and laws not explicitly posited by science qualifies as 'supernatural', hence unacceptable. It is a very interesting question, and one deserving careful study in the future, whether and how moderately naturalistic metaphysics as defined here meshes with liberal naturalism, what the logical/conceptual links between the two (families of) views exactly are, and what would follow from their joint endorsement (or rejection).[17]

Conclusions

Naturalism requires that science be given centre stage. Applied to philosophy, it has historically led to a more or less explicit subordination of the a priori philosophical enterprise to the a posteriori work of scientists (plus logic and mathematics). When it comes to metaphysics, this has been often accompanied, in recent times as in the golden age of logical empiricism, by a bolder form of eliminativism. Radical naturalism about metaphysics, of either the reductive or the eliminativist sort, is indeed quite popular nowadays. Nonetheless, a rather simple argument against it can be mounted based on the fact that literally every hypothesis about the 'world out there' is underdetermined by the empirical data, and no sharp distinction can be drawn between science and metaphysics. This leads to the identification of a moderately naturalistic alternative, whereby a parallel development of science and metaphysics is recommended, and metaphysics keeps his status of autonomous a priori enterprise while at the same time being connected systematically to science via the notions of indirect empirical testing and philosophical interpretation of scientific theories. Only a brief case study could be discussed here, and many questions (concerning the status of metaphysical models and explanations, the extent to which moderately naturalistic metaphysics should be deemed autonomous, the issue of realism and other things) remain open. Nonetheless, it seems fair to say that moderately naturalistic metaphysics should be regarded as a significant stance in today's meta-philosophical debate.

Notes

1 Note that this does not entail the abandonment of metaphysical realism, the view that reality has objective mind-independent features. Nor does it entail the anti-realist view according to which we have no reason for thinking that we have, or at least can gain, knowledge of the nature of reality.
2 For a different sense of 'experimental metaphysics', see section 2.1 below.
3 In connection to this, it is interesting to notice (following Friedman 2001: 12–13) that supposedly paradigmatic scientific figures such as Helmholtz and even some members of the Vienna Circle didn't (always) see 'scientific philosophy' as necessarily grounded on the sort of deflationist stance outlined a moment ago. Schlick, for instance, argued at one point that we must begin with special problems of the special sciences but then move up "to the ultimate attainable principles ... which, because of their generality, no longer belong to any special science, but rather lie beyond them ... in philosophy" (1978: 335).
4 'Permutation invariance' is said to hold: exchanging two exactly similar entities does not give rise to a new, physically relevant state. In the terminology of the philosophy of modality, this is usually translated by saying that anti-haecceitism holds, and two possible worlds cannot differ merely in the identities of the things that inhabit each one of them.

5 According to whom certain metaphysical conjectures, when conjoined with our best physics, lead to empirical consequences and, therefore, it becomes possible to test metaphysics empirically, and even falsify it via modus tollens.
6 That it is metaphysics that we are talking about nonetheless is, of course, shown by the fact that we are using concepts (identity, non-identity, property, individuality, possible arrangements/modality) that are clearly philosophical, and do not belong to the 'vocabulary' of natural sciences. More on this later.
7 Alternatively, one could i) again have recourse to Bohmian mechanics, where particle positions and trajectories are unique: or ii) claim that quantum entities (may) differ purely in the relations holding among them – being, say, in the relation of having opposite spin in spite of their not having different spin magnitudes taken separately. This is only possible, of course, if the relations in questions are not reducible to separate properties. But it can be argued that this is indeed the case in the quantum domain. The idea of 'weak discernibility', recently applied to quantum entities, is based exactly on this assumption that irreducible, irreflexive, symmetric relations can obtain. For further discussion, and for an overview, see French (2015).
8 For instance, certain properties may only be exemplified, hence attributed, to things collectively. Or, some physical states may just be inaccessible as a matter of physical law. On this, again see French (2015, esp. Sec. 6) for more details.
9 Of course, that a view is dominant and/or has been entertained by authoritative figures (in this case, some of the founding fathers of quantum mechanics – such as, for instance, Schrödinger – who explicitly held that quantum theory entails that quantum particles are not objects in the traditional sense) may instead have little weight, or none at all.
10 See, e.g., Ladyman (2012) and Saatsi (2017).
11 Thus, anti-metaphysical claims merely based on putatively paradigmatic examples – comparing, say, the question how many angels can dance on the head of a pin and questions concerning the existence of Higgs bosons – are not effective.
12 The similarity between science and metaphysics with respect to the fact that both empirical and non-empirical elements play a role in theory-choice is well-known. What is much less common is the use of it with a view to fostering conciliation rather than conflict between the two disciplines.
13 This is analogous to what Shimony's experimental metaphysics, mentioned earlier, recommends, but involves much more than just modus tollens. As for issues that are not (or appear not to be) indirectly testable in the sense just suggested, it can be left open here whether they should be explored anyway, perhaps with a view to putting the results of one's a priori analysis in connection to science if and when this becomes possible; or they should instead be discarded, albeit perhaps temporarily, as uninteresting. In connection to this, see also the remarks in section 4, point B below.
14 The well-known fact that there is no algorithm leading the evaluation of these factors, nor a shared criterion for quantifying each one of them and putting them together is again a problem, if it is a problem at all, both for scientists and metaphysicians.
15 This makes it possible to agree with both those who are sceptical towards free-floating, entirely a priori metaphysics (see, e.g., Bryant forthcoming) and those who emphasize that a correct, or at any rate, sensible understanding of certain empiricist tenets and principles does not entail the elimination or strong reduction of metaphysics (see, e.g., Ney 2012).
16 The difference at the level of vocabulary suffices for explaining the difference in degree of testability discussed earlier. Indeed, this view may well be preferable from the viewpoint of the naturalist; here, however, we are only interested in resisting the argument moving from the putative instability of metaphysical modality to the need to make metaphysics subordinated to science.
17 For nice discussions of liberal naturalism, see De Caro and Macarthur (2004) – where, incidentally, the important role played in this context by Hilary Putnam, whom we quoted as one of the empiricist detractors of metaphysics earlier in the chapter, is illustrated – and Macarthur (2019).

References

Bryant, A., (forthcoming): Keep the Chickens Cooped: The Epistemic Inadequacy of Free Range Metaphysics, *Synthese*.
Callender, C., (2011): *Philosophy of Science and Metaphysics*, in French, S. and Saatsi, J. (eds.): *The Continuum Companion to the Philosophy of Science*, London, Continuum, 33–54.
De Caro, M., and Macarthur, D., (eds.), (2004): *Naturalism in Question*, Harvard, Harvard University Press.

Dyke, H., and Maclaurin, J., (2013): What Shall We Do with Analytic Metaphysics? A Response to McLeod and Parsons, *Australasian Journal of Philosophy*, 91, 179–182.
French, S., (2015): Identity and Individuality in Quantum Theory, in Zalta, E.N. (ed.), *The Stanford Encyclopedia of Philosophy*, https://plato.stanford.edu/archives/fall2015/entries/qt-idind/.
Friedman, M., (2001): *Dynamics of Reason*, The 1999 Kant Lectures at Stanford University, Stanford, CSLI Publications.
Godfrey-Smith, P., (2006): Theories and Models in Metaphysics, *The Harvard Review of Philosophy*, 14(1), 4–19.
Goldman, A.I., (2007): A Program for 'Naturalizing' Metaphysics, with Application to the Ontology of Events, *The Monist*, 90, 457–479.
Goldman, A.I., (2015): Naturalizing Metaphysics with the Help of Cognitive Science, in Bennett, K. and Zimmerman, D.W. (eds.), *Oxford Studies in Metaphysics*, 9, 171–212.
Jackson, F., (1998): *From Metaphysics to Ethics: A Defence of Conceptual Analysis*, Oxford, Clarendon Press.
Ladyman, J., (2012): Science, Metaphysics and Method, *Philosophical Studies*, 160, 31–51.
Ladyman, J., Ross, D., with Spurrett, D., and Collier, J., (2007): *Every Thing Must Go: Metaphysics Naturalized*, Oxford, Oxford University Press.
Lowe, E.J., (2011): The Rationality of Metaphysics, *Synthese*, 178, 99–109.
Macarthur, D., (2019): Liberal Naturalism and the Scientific Image of the World, *Inquiry*, 62, 565–585.
Maclaurin, J., and Dyke, H. (2012): What Is Analytic Metaphysics For?, *Australasian Journal of Philosophy*, 90, 291–306.
Maddy, P., (2007): *Second Philosophy: A Naturalistic Method*, Oxford, Oxford University Press.
McLeod, M., and Parsons, J., (2013): Maclaurin and Dyke on Analytic Metaphysics, *Australasian Journal of Philosophy*, 91, 173–178.
Morganti, M., (2013): *Combining Science and Metaphysics: Contemporary Physics, Conceptual Revision and Common Sense*, Basingstoke, Palgrave MacMillan.
Morganti, M., and Tahko, T., (2017): Moderately Naturalistic Metaphysics, *Synthese*, 194, 2557–2580.
Ney, A., (2012): Neo-Positivist Metaphysics, *Philosophical Studies*, 160, 53–78.
Paul, L.A., (2012): Metaphysics as Modeling: The Handmaiden's Tale, *Philosophical Studies*, 160, 1–29.
Putnam, H., (2004): *Ethics without Ontology*, Cambridge, Harvard University Press.
Ritchie, J., (2008): *Understanding Naturalism*. Durham, Acumen.
Saatsi, J., (2017): Explanation and Explanationism in Science and Metaphysics, in Slater, M. and Yudell, Z. (eds.), *Metaphysics and the Philosophy of Science: New Essays*, Oxford, Oxford University Press, 162–191.
Schlick, M., (1978): *Philosophical Papers I, 1909–1922*, Vienna Circle Collection 11, Heath, P. (trans.), Mulder, H.L. and Van de Velde-Schlick, B.F.B. (eds.), Dordrecht, Reidel.
Shimony, A., (1980): Critique of the Papers of Fine and Suppes, *Philosophy of Science, Proceedings*, 2, 572–580.
Van Fraassen, B., (2002): *The Empirical Stance*, New Haven, Yale University Press.

38
METAPHYSICS AS THE 'SCIENCE OF THE POSSIBLE'

J.T.M. Miller

This chapter considers the view that a central concern of metaphysics is what is possible. That is, the idea is that, unlike science, metaphysics studies not only what is actual, but the ways that reality could be. This view, if right, provides metaphysics with a distinct subject matter from that of science, and, depending on what modal epistemology we adopt, a distinct methodology too. In this chapter, I first provide an overview of the view, before highlighting some of the most prominent objections, and possible routes of response.

1 Stating the view

Conee and Sider, in their introduction to metaphysics, describe a view wherein:

> metaphysics is about what could be and what must be ... Metaphysics is about some actual things, only because whatever is necessary has got to be actual and whatever is possible might happen to be actual. This allows us to say that physics pursues the question of what the basic constitution of reality actually is, while metaphysics is about what it must be and what it could have been.
>
> *(2005: 203)*

Callender puts it as the view that 'whereas scientists excavate dusty field sites and mix potions in laboratories to tell us which states of affairs are actual, metaphysicians are concerned with what is and isn't metaphysically possible' (Callender 2011: 36).

However, despite the fact that, as French and McKenzie state, these 'sentiments pepper the contemporary [metaphysical] literature' (2012: 46), there has been relatively little explicit discussion of this conception of metaphysics. One older statement of a version of the view can be found in the work of Leibniz. Grosholz and Yakira comment that Leibniz takes the 'science of the possible' to be the investigation of the rational grounds of reality. This takes Leibniz to have been interested in the ways that reality could be such that certain types of knowledge, in particular knowledge of arithmetic, are possible (Grosholz and Yakira 1998: 74). That is, given that we know certain things about the world, how could the world be such that we could know those things.

Leibniz's rationalism thus seeks to understand how reality could be such that what we observe and what knowledge we have can be adequately explained. Wolff held a similar position, arguing that philosophy is 'the science of the possible, with the task of showing how and why things are possible' (Frängsmyr 1990: 33). Note that this gives metaphysics a crucial *explanatory* element. The world is a certain way and what metaphysics is trying to do is provide some account as to why it is that way and which is the theory that best explains it.

Russell also, in the final chapter of his 1912 book *Problems of Philosophy*, remarks that:

> Philosophy is to be studied, not for the sake of any definite answers to its questions since no definite answers can, as a rule, be known to be true, but rather for the sake of the questions themselves; because these questions enlarge our conception of what is possible, enrich our intellectual imagination and diminish the dogmatic assurance which closes the mind against speculation.

If philosophy aims are enlarging our conception of what is possible, then metaphysics, as a branch of philosophy, would seem to aim to do likewise.

More recently, one of the few explicit and detailed statements of the view appears in the work of E.J. Lowe.[1] For Lowe, taking metaphysics to be science of the possible takes metaphysics to be 'charged with charting the domain of "objective or real possibility"' (Lowe 2011: 100). That is, metaphysics seeks to tell us what entities are possible and compossible and hence, of the range of ways that we might theorise reality to be, which are genuine ways reality as a whole could be.

This makes metaphysics an inherently holistic or systematic enterprise. Understanding what are the genuinely possible ways that reality could be requires us to consider how certain commitments cohere, or not, with other commitments within our other metaphysical views. That is, to know whether it is a genuine possibility that Xs exist requires us to consider whether Xs are compossible with Ys, and to know whether this is genuinely possible may require us to consider various other claims about Zs and the relations that hold between Xs, Ys, and Zs too.

This also makes metaphysics 'implicitly modal' (Lowe 2011: 106). It is modal in the sense that it can by itself only tell us what there could be, not what there is, and that it is only with experience, added to the metaphysical theorising, that we can arrive at claims about the actual world. We can theorise a myriad of ways that reality as a whole could be, and the role of metaphysical debate is to reduce that range of possibilities by ruling out some views, or combinations of views, as not real or genuine possibilities.

This provides metaphysics with a distinct subject matter from science. Metaphysics, unlike science, takes part of its subject matter to be possible ways reality could be, and not just ways that it actually is. Metaphysics may be more general than the particular sciences, but even if there were an all-encompassing science that took all existing entities to be within its remit, it would still not have the same subject matter as metaphysics, as metaphysics would also include enquiries into what is possible.

An immediate response could be that there is no kind of necessity (and possibility) that is broader or distinct from that of scientific necessity (see Swoyer 1982; Shoemaker 1998; Ellis 2001). How far this is problematic depends on how such claims are cashed out in more detail. On one reading, such views do not deny that there is a notion of metaphysical possibility, but only that it (entirely) overlaps with that of scientific (or physical) possibility. If that is the case, this overlap may limit or constrain the scope of metaphysical inquiry, but it would not mean that it has no subject matter as there is still some sense of 'what are possible ways reality could be' to be studied. However, as I will discuss in section 3, there remains the question of whether this would allow metaphysics to have a *distinctive* subject matter from that of science.

If metaphysics is concerned with the possible, then what about the actual? As noted above in the quote from Conee and Sider, the answer normally given is to point out that what is possible is deeply connected to what is actual. Metaphysics can thus be 'devoted to exploring the realm of metaphysical possibility, seeking to establish what kinds of things could exist and, more importantly, co-exist to make up a single possible world' (Lowe 2006: 4), whilst science (primarily) aims to establish which of the possibilities explored is the actual way reality is. This results in a complementary relationship between science and metaphysics, where both are needed to provide an account of the structure of reality:

> Metaphysics and empirical science are not 'continuous' with each other in any sense which implies that they have the same goals and methods, or that metaphysics is just the extension of empirical science to questions of greater generality than any that are addressed by the so-called 'special' sciences. Rather, when both are conducted fruitfully, metaphysics and empirical science exist in a symbiotic relationship, in which each *complements* the other.
>
> (Lowe 2011: 101–102)

This symbiotic relationship means that advances in one of these fields will influence the work in the other. Breakthroughs in science will allow us to consider possibilities that were previously ignored, or reject theories that had been taken to be genuine possibilities.[2] Similarly, work on the possible ways that reality might be will help to direct our investigations into how reality actually is. Metaphysics therefore progresses by ruling out possible ways reality could be as being non-genuine possible ways, providing insights into possible ways that reality could be that are still 'live'.

This account of the relationship between science and metaphysics that emerges out of this modal conception of metaphysics is one of the main benefits of the view in that the view seems to provide a distinctive subject matter for metaphysics. Furthermore, the view potentially allows for metaphysics to have a distinctive method depending on our modal epistemology. As will be discussed in more detail below, if metaphysics is concerned with the possible, then its methodology will be adopted from how it is that we can gain knowledge of what is possible. If we do so through a priori methods, and assuming that the methods of science are (at least primarily) non a priori, this would make the method of metaphysics distinct from that of science. Any account of what metaphysics is that could do both, or even one, of these would certainly be significant.

2 Some further details

Before moving onto potential problems for this conception of metaphysics, I will comment and clarify a couple of further points.

The first concerns how we should read the notion of 'possible'. The first option is epistemically. It seems plausible that Lewis had something like this in mind in his conception of metaphysics. Lewis endorsed the idea that philosophy sought to arrive at a total reflective equilibrium. This is arrived at by systematising philosophical and pre-existing opinions into an orderly system (Lewis 1973: 88) such that the task of metaphysics is to 'find out what equilibria there are that can withstand examination' whilst being compatible with certain general principles (Beebee 2018: 16; see also Lewis 1986: x). States of reflective equilibria thus seem best understood as epistemic states. They are theories or systems that provide the wanted balance between philosophical views, pre-existing intuitions, and general principles (such as parsimony): different ways, in line with our general epistemic interests, the world might be like.

In contrast, Lowe reads the notion of 'possible' more metaphysically. Competing theories are not just different ways that, for all we know, the world might be like, but rather as competing ways that reality is or could be. This is not epistemic then, or better, not *merely* epistemic. If metaphysics is the science of the possible then it is investigating reality itself, not just how we happen to think, talk, or perceive it.

I will assume this stronger metaphysical reading in what remains of this chapter, noting here that as currently stated it leaves open the question of how we can tell what are the genuinely possible ways that reality could be. I will return to this question and the topic of modal epistemology later in this chapter.

Second, we might ask how this conception relates to other metametaphysical views. As discussed elsewhere in this volume, we might think of metaphysics as being about understanding what grounds what, or asking what is fundamental. If metaphysics is understood as being the investigation of what are the possible ways that reality could be, then all of these are different (perhaps competing) conceptions as to how to delineate those possibilities. Put another way, the metametaphysical claim is that the aim of metaphysics is to investigate possible ways reality could be. The results of such an investigation, however, will depend on other metametaphysical commitments we may have (e.g. that metaphysics should investigate grounding or fundamental relations). This means that, metaphysicians who conceive of metaphysics as the science of the possible, but differ with regard to other metametaphysical (and first-order metaphysical) commitments, may also differ with regard to what they think are genuine ways reality could be.

I wish to take no position on this here. Minimally stated, taking metaphysics to be the science of the possible requires us to take no position on what limits there are on the genuine ways reality could be. Other metametaphysical positions, when combined with this one, may do so.

One potential exception to this concerns neo-Quinean approaches that argue that metaphysics is about asking 'what exists?' (see Quine 1948; Egerton, this volume, Parent, this volume). This is because if we were to endorse an existence based approach, we might think that to take metaphysics to be the science of the possible is to commit ourselves to a Meinongian ontology that accepts the existence of merely possible entities. Science would then study actual entities, and metaphysics would have a broader domain that includes the merely possible. Whilst some might accept merely possible entities into their ontology, most do not, and this would certainly cause some issues with how far we can combine this account of metaphysics with various forms of naturalism. Ney, for example, writes that:

> metaphysicians frequently remark when describing their subject matter that although the sciences are concerned only with what is actually [the] case, or what can happen that is compatible with the actual laws of nature, metaphysicians are concerned too with what is merely possible, including what may be only logically possible and incompatible with actual scientific laws. But note that this doesn't correspond to a broader domain. For there aren't in addition to the actual entities that exist, also any merely possible entities for metaphysics (though not the sciences) to be about. One would have to adopt the modal realism of David Lewis (1986) to think otherwise.
>
> *(Ney 2019: 15)*

However, to think that this requires us to take merely possible entities as existing is to view metaphysics through a neo-Quinean lens that those who endorse the view that metaphysics is the science of the possible need not accept.[3] Rather, the claim being discussed is not that merely possible entities exist, but that debates in metaphysics are about 'conditionals of the form, "If such and such counterfactual situation were to obtain, then so and so would be the case"' (Ney 2019: 15).[4]

Even if we accept that Xs do not actually exist, claims about whether *were* Xs to exist, could Ys also exist, can be true or false, and can reveal something about the nature of Xs, Ys, and the ways that reality actually is. In this way, it would seem that no ontological commitment to the existence of merely possible entities is needed in order to take metaphysics to be the science of the possible.

3 Floating free from science?

A first criticism of this conception of metaphysics is whether it allows metaphysical theorising to 'float free' from science, and in particular physics. The problem is that this way of viewing metaphysics could allow for rampant unrestrained speculation by metaphysicians under the cover of trying to find out what is metaphysically possible, as opposed to what is 'merely physically possible'.

Morganti and Tahko, for example, take up this line of thought, arguing that the 'claim that science can determine which of the possibilities identified by metaphysics is actual falls short of constituting a satisfactory methodological basis, exactly because it seems to allow for totally unconstrained metaphysical theorising that, nevertheless, somehow latches onto reality' (2017: 2567). The concern is that there is nothing that limits our metaphysical speculation or determines the limits of what is *metaphysically* possible, allowing metaphysical theories to float free of science, contra the claim of a close, symbiotic relationship.

Connected concerns are raised by both Callender (2011), and French and McKenzie (2012). Callender diagnoses the clash between science and metaphysics as arising from the metaphysicians' move towards inquiring into a notion of metaphysical possibility that leaves science as no longer relevant to metaphysics:

> Being about what metaphysically must and could be, metaphysics on this conception is forced by the change of target into studying more general abstract principles, such as whether two objects can ever occupy the same place and same time. If the concern is whether this principle holds in the real world, science will be relevant to assessing its truth. But why should science be relevant to assessing its truth in metaphysically impossible worlds wherein science is very different? Plainly it's not: science, after all, is mostly about the metaphysically contingent.
>
> *(2011: 40–1)*

French and McKenzie argue that the way that metaphysicians have approached investigating the possible ways reality could be has rendered science 'peripheral', meaning that 'physics has only an "incidental" or marginalized role within metaphysics *even* if we buy into this conception of metaphysics' (2012: 46). Indeed their aim is to go further and argue that we should reinstate 'physics as the proper point of departure for modal questions concerning the actual', and emphasise 'just how fruitless modal discussions concerning physical ontology are if taken to be divorced from actual physics' (2012: 46–7). The idea therefore is that a conception of metaphysics that is in principle divorceable from physics should be rejected, and metaphysics as the science of the possible is one such conception.

There are at least two points that are correct in these objections. The first is that science is concerned with possibility and necessity. Williamson (2016, 2017), for example, has argued that possibility and necessity are studied in natural sciences. Natural science thus is not a modality-free zone, though it remains an open question as to whether the modalities explored in science are the same as the metaphysical modality invoked above.

The second point is that metaphysics, as a discipline, cannot allow its speculation to entirely float free of science. This is in part because of the above recognition that science does investigate modality, but more centrally because, unconstrained by science, metaphysical speculation would have no plausible route to justify its claim that it is actually getting at how reality is at all.

Perhaps the most natural response, echoing the outline above, is to reassert that we should not think of metaphysics as an isolated or bordered off domain. That is, to argue that the boundary between metaphysics and science is blurred as investigations into how reality could be, and how it actually is, are intrinsically connected. For example, we could accept metaphysical investigation as being about the genuinely possible ways that reality could be, but when it comes to assessing a theory's relative merits, a lack of coherence with empirical data could be taken as a mark against that view as a plausible account of how reality as a whole is. That is, a lack of coherence with empirical data will help to indicate that that account does not really refer to a genuine way that reality could be.

French elsewhere seems to suggest something along these lines, arguing that metaphysical inquiry into possible ways that reality could be will produce a

> kind of 'conceivability spectrum', ranging from unconstrained conceivability, which should perhaps come with a metaphysical 'health warning' as it may include Meinongian objects and inconsistencies in general, to logically constrained conceivability, with the Principle of Non-Contradiction, of course, acting as a significant constraint, to conceivability constrained by intuitions, with regard to which we can draw on recent discussions of the role of such intuitions in philosophy ..., to metaphysically constrained conceivability, which may come before the previous entry depending on what the relevant metaphysical and 'intuitive' constraints are, to, finally, physically constrained conceivability (or 'naturalistic' conceivability, perhaps), within which we might distinguish constraints based on classical physics, quantum physics and so forth.
>
> (2018: 225)

Where we fall on this conceivability spectrum is a matter for debate. It would be wrong to think that there are no metaphysical theories that are at that extreme point, and hence, presumably metaphysicians that might argue for being unconstrained in that way. It would be equally wrong to think that there are no metaphysical theories that are strongly and deeply linked to empirical findings. I have suggested that metaphysicians should avoid the furthest end of metaphysical speculation that is divorced from all empirical consideration. I have not argued directly for that here; however, my view is similar to the 'conditionalized' support for analytic metaphysics recently provided by French and McKenzie (2016). What we can say, at least, in response to the worry that metaphysics will float free of science is that it need not float free, and that we should be vigilant against allowing it to. Admittedly, this is unlikely to assuage some, as it does allow for there potentially being unconstrained metaphysical theorising.[5]

Ultimately, what constraints guide the limits of possibility will turn on questions about how we conceive of the relationship between the various different kinds of possibility, and the scope of those possibilities. For example, if we take metaphysical possibility to be constrained by facts about the laws of nature or the causal powers of fundamental properties and relations, then the empirical inquiry into those aspects of reality will be crucial to understanding what we think is metaphysically possible. These are questions, though, about what are the genuine possible ways reality could be, and not about whether metaphysics studies what are the genuinely possible ways reality could be. If metaphysical possibility is constrained in these ways – if, for example, we conclude that what is metaphysically possible must still be consistent with the laws of nature – this does not undermine the conception of metaphysics under discussion. Rather, it indicates progress made in determining what are the genuinely possible ways reality could be.

4 How distinctive?

Blurring the boundaries between metaphysics and science, though, only reignites questions about whether the subject matter of metaphysics is suitably distinctive, or at least distinct from that of science. The issue is how sharp the distinction between science and metaphysics is to be drawn. Too loosely and we cannot say that metaphysics has a distinct subject matter from science; too sharp and metaphysical reasoning might (always) be too unconstrained. Metaphysics as the science of the possible needs to thread the needle between these two unwanted conclusions. I will suggest a way in which we might do so.

Let us call 'narrow metaphysics' that part of human enquiry done by metaphysicians into the possible ways that reality (as a whole) might be. Narrow metaphysics is not divorced from empirical data as which (types of) entities we are interested in will be influenced by experience or science. But its subject matter is possible ways that reality might be.

We can distinguish this from 'broad metaphysics'.[6] Broad metaphysics aims to provide an account of how reality actually is, drawing upon findings and views within narrow metaphysics and empirical findings in order to try to discover which of the possible ways reality could be is the way that it actually is.

So understood, it is in virtue of the subject matter of narrow metaphysics that we can say that the subject matter of metaphysics is distinctive. Narrow metaphysics is distinctively concerned with how reality could be, whilst broad metaphysics is non-distinctively concerned with how reality actually is. We therefore can avoid too loose or too sharp a boundary between metaphysics and science by recognising that all work labelled 'metaphysics' may not have the same (primary) aims.[7]

This does mean that, in so far as science is not a 'modality-free zone' (Williamson 2018), (parts of) science may also be engaged in 'narrow metaphysics'. Speaking as a philosopher, I cannot say how accepted this consequence might be by scientists, but restricting metaphysics to solely those that work within philosophy departments would seem unhelpfully limiting. A claim that the subject matter of metaphysics is distinctive cannot be based on a desire to restrict where such work takes place as it is unclear why we should expect our conception of metaphysics to neatly fit into contingent discipline boundaries within universities.[8]

Even if we grant this distinction between narrow and broad metaphysics, we might relatedly worry that this does not make metaphysics distinctive enough from other *philosophical* domains. For example, Bennett (2016) has argued that studying the extent of what is possible in an a priori fashion is not distinctive of metaphysics compared to other philosophical areas of inquiry.[9]

The response to this will depend on how one conceives of philosophy more broadly. For some that have defended the view that metaphysics is the science of the possible, a lack of sharp distinction between metaphysics and the rest of philosophy indicates the central role of metaphysical thinking in philosophy (Lowe 2006: ch. 1). If this is the case then it will naturally be the case that metaphysical research exists in a variety of philosophical domains. Whether this is right, though, will depend on a discussion of meta-philosophical issues concerning the relationship between various domains of philosophical theorising that I cannot cover in depth here.

5 Conceptual analysis?

The above focused on the relationship between science and metaphysics, and tried to describe a view wherein metaphysics is distinct from, but closely related to, science. However, by drawing a distinction between metaphysics and science, a related objection arises: if metaphysics is about how reality could be, then is metaphysics 'merely' conceptual analysis? The worry is that metaphysics, under this view, will ultimately just be involved in reflecting on our concepts, thereby again allowing metaphysics to float free from any connection to reality itself.[10]

Certainly under this view, conceptual analysis will likely be a significant part of the method of metaphysics. For example, it is by considering definitions of central terms within our theorising that we arrive at views about what (types of) entities are possible and compossible. One method to try to work out whether it is genuinely possible that Xs, Ys, and Zs all exist is to consider the concept that each of these entities fall under and see whether positing their existence is compossible with each other (and any general principles that we might also endorse).[11]

However, accepting that metaphysicians do engage in conceptual analysis, does not immediately lead to the conclusion that metaphysics is *mere* conceptual analysis. That is, we can hold that metaphysicians when engaging in conceptual analysis are trying to improve our concepts. This is an interest in concepts which is 'normative' and metaphysical: 'it is concerned to improve our concepts, or ways of thinking of things, by making them more accurate, that is, more truly reflective of the essences or natures of the things that we are thinking of when we deploy those concepts' (Lowe 2011: 108). Concepts are not the primary subject matter of metaphysics, but part of the work required to delineate the possible ways that reality as a whole might be may involve honing and improving our concepts.

This can be linked to the notion of 'metalinguistic negotiation' (Burgess and Plunkett 2013a, 2013b; Plunkett and Sundell 2013, 2014; Plunkett 2015). Plunkett describes a metalinguistic negotiation as 'a dispute in which speakers each use (rather than mention) a term to advocate for a normative view about how that term should be used' (2015: 832).

The extent to which metaphysics is engaged in conceptual analysis is the extent to which metaphysicians are engaging in metalinguistic negotiation. For example, when metaphysicians argue about the concept 'property', they are arguing about how the concept should be used and what it should mean. We might, amongst other views, think that the concept 'property' denotes a universal, or a trope, or something that has a dispositional essence, etc. One reason, potentially amongst others, to engage in metalinguistic negotiation is in order to make those concepts more accurate to the natures of the things that the concepts pick out.

This is in line with Plunkett's claim that to engage in metalinguistic negotiation does not mean that we are 'merely' talking about concepts:

> Suppose one argued (as I think is correct) that an important part of communication among biologists involves metalinguistic negotiation. (The different meanings of 'species' is a good place to start with such a proposal, as is the different meanings of 'intelligence'). Would that mean that there aren't facts about animals and their behavior to investigate, and then all biological argument is just about normative issues about word and concept choices? Clearly not.
>
> *(2015: 860)*

The biological argument is not only about word or concept choice, as what settles the issue with respect to concept choice may be 'non-voluntary' (Plunkett 2015: 860–1). Similarly, the different meanings of 'property' are not just about word and concept choices, but about which of the competing concepts best track reality, and the genuinely possible ways that reality could be. To say which does best track reality, though, may involve a substantial element of conceptual analysis to refine the concepts first.

6 Knowing the possible?

If metaphysics is the science of the possible, then we must also provide some account of how it is that we can come to know what is possible. That is, a fully developed version of this view requires a developed modal epistemology. This chapter is not the place for an in depth discussion of modal epistemology, but I will highlight some consequences which may ensue from the adoption of different forms of modal epistemology.[12]

There are a wide spectrum of views that seek to explain how we can come to have modal knowledge, and each of them will provide a way to explain how we acquire such knowledge. If metaphysics is the science of the possible, these views will influence what we take to be the methodology of metaphysics.

This is because if metaphysics is the science of the possible, then its method will be derived from whatever methodology is required to gain knowledge of what is possible. For example, if we adopt a form of modal rationalism, then the methodology of metaphysics will be a priori. If this is right, then metaphysics may have both a distinct subject matter (that of what is possible), and a distinct methodology.[13] If, however, we adopt some form of modal empiricism,[14] holding that empirical methods can lead to knowledge of what is possible, then metaphysics may not have a distinctive method, though the distinctiveness of the subject matter will remain.

There are, though, views in modal epistemology that will cause more serious problems for this conception of metaphysics, such as an extreme form of modal scepticism that argues that we can have no modal knowledge (see van Inwagen 1998 for a discussion of such views). If extreme modal scepticism is right, then this would force us to reject all claims that purport to be about metaphysical possibilities, thereby undermining the claim that metaphysics has a distinctive subject matter from that of science. For these reasons, it seems that combining the strongest form of modal scepticism with the view that metaphysics is the science of the possible cannot be done whilst maintaining a view of metaphysics that has a subject matter that is suitably distinct from science.

Alternatively, we might accept modal conventionalism (Cameron forthcoming), or hold that modal statements are not truth-evaluable. Though very different, these kinds of views argue that claims about necessity (and possibility) are not about the world, holding that modal claims are not true in virtue of the world, but rather due to how we speak about it (in the case of conventionalism), or only reflect rules of use of our terms (Thomasson 2007).

One line of response to these views holds that they fail to adequately distinguish between claims about how we use words, and how the entities that our words refer to are. For example, Yablo (1993), in his response to Sidelle (1989), argues that the fact that we can decide how we use certain terms does not mean that the properties of the entity referred to by those terms are conventional. 'Water' could have been used in various ways, but this does not rule out water – the substance – having various modal features. Along similar lines, Russell (2010) has argued that conventionalist accounts are focused on the meaning of sentences, whilst a metaphysical interest in modality is about the status of the propositions expressed by those sentences, and has argued that the conventionalist has no plausible way to bridge this gap.[15]

If, though, modal claims do turn out to be true only in virtue of our language, then, this would certainly have serious consequences for the view that metaphysics is the science of the possible. At the very least we would be forced to reject the assumption I have made throughout this chapter that modal claims, and hence metaphysical claims, track real features of the world. Perhaps a deflationary version of metaphysics could come from this, whereby metaphysics is useful as a systematic inquiry into the nature of our concepts or language. However, this is certainly not the sort of metaphysics that those in the literature who have defended this conception of metaphysics would wish to accept.

7 Conclusion

The view that metaphysics is the science of the possible has a long, but often implicit, history. Often the view is listed as one of the major positions within the literature, but there are admittedly few that have explicitly defended it. In light of this, this chapter aimed to bring some elements together in a single place, highlighting what I have taken to be the central aspects of the view.

Taking metaphysics to be the science of the possible is, naturally, not without its objections, and there are still a number of ideas to explore and refine in order for such an account to be persuasive. Perhaps, the most central task facing those that would wish to defend the view that metaphysics is the science of the possible is the clear need for an adequate modal epistemology. This task is not easy, but nor should it be surprising that an account of what metaphysics is may require us to combine it with commitments elsewhere. I have also suggested that (minimal) coherence with science is something that should constrain the most extreme forms of metaphysical theorising. However, it has to be noted that as I have outlined it, the view does not in principle rule out unconstrained metaphysical theorising (if, that is, we should want any definition of metaphysics to do so in the first place).

If these objections can be responded to, and the finer details outlined, then taking metaphysics to be the science of the possible potentially allows for a complementary conception of metaphysics and science, without a superiority of metaphysics over the empirical sciences or vice versa, and an inherently holistic conception of metaphysics that recognises the need to understand how putative entities are related to each other in order to assess the plausibility of positing those entities.

Notes

1 For a more detailed reconstruction of Lowe's metametaphysical views, see Miller (2018a).
2 See French and McKenzie (2016: 44) who provide examples that they argue shows that the 'actual can veto crucial assumptions about what [the available] possibilities are'.
3 See, for example, Lowe (2013: ch. 4) in which he argues that '∃' should be analysed as the 'particular quantifier' rather than as an existential quantifier, with the particular quantifier being able to quantify over non-existent objects without implying their existence.
4 For clarity, Ney's focus is on a related but distinct topic to that of this chapter. Ney is discussing the view that metaphysics is more fundamental than science, and arguing against the claim that metaphysical necessity is more fundamental than physical necessity. I am making no claims about what is more fundamental here.
5 This leaves aside the question of whether philosophers and scientists are talking to each other enough. Again, it seems that the answer is that some are; some are not. Some metaphysics should be more in line and engaged with empirical data; some interpretations of the empirical data maybe could benefit from an increased engagement with metaphysical considerations.
6 To distinguish between narrow and broad metaphysics is not to say that they are distinct domains of enquiry. They are linked, given that, as noted above, investigating the ways that reality could be and the ways it is are done alongside each other.
7 I say 'primary' as it is of course possible that some metaphysical work will fall into both categories, as explorations of the genuine ways reality could be *and* how reality actually is. Indeed, it may even be that most work does not fit neatly into one or other of these categories. Even if this is the case, this distinction can still be useful as a way to illustrate the complex aims of metaphysical inquiries.
8 This is not to deny that those trained as metaphysicians within philosophy departments might have developed skills to engage in metaphysics that those trained in empirical scientific methods have not, and vice versa, but this is more a topic for the sociology of metaphysics.
9 I am grateful to an anonymous reviewer for pushing me to consider this point.
10 Though, see Jackson (1998, this volume) and Bealer (1998) for more detailed discussions of metaphysics as conceptual analysis.

11 By 'general principles' here, I mean what elsewhere has been labelled 'ideology' (see Cowling 2013; Miller 2018b).
12 For a more in depth discussion of modal epistemology, see Roca-Royes (this volume) and Vaidya (2015).
13 See Lowe (1998). This also assumed that science does not rely on a priori methods (see Farr and Ivanova, this volume).
14 A lot here will depend on the precise details of the favoured account of modal empiricism. For more detailed discussions about non-rationalistic modal epistemologies, see Fischer and Leon (2017).
15 There is far more that could be said about this, and about other 'neo-conventionalist' views of modality (see Cameron 2009, 2010a, 2010b; Sider 2003).

References

Bealer, G. 1998. 'Intuition and the autonomy of philosophy'. In M. DePaul and W. Ramsey (eds.), *Rethinking Intuition: The Psychology of Intuition and Its Role in Philosophical Inquiry*, Lanham, MD: Rowman & Littlefield, pp. 201–40.
Beebee, H., 2018. 'Philosophical scepticism and the aims of philosophy', *Proceedings of the Aristotelian Society*, 118(1): 1–24.
Bennett, K. 2016. 'There is no special problem with metaphysics', *Philosophical Studies*, 173: 21–37.
Burgess, A., and Plunkett, D., 2013a. 'Conceptual ethics I', *Philosophy Compass*, 8(12): 1091–1101.
Burgess, A., and Plunkett, D., 2013b. 'Conceptual ethics II', *Philosophy Compass*, 8(12): 1102–10.
Callender, C. 2011. 'Philosophy of science and metaphysics'. In S. French and J. Saatsi (eds.), *Continuum Companion to the Philosophy of Science*, London: Continuum, pp. 33–54.
Cameron, R. 2009. 'What's metaphysical about metaphysical necessity?', *Philosophy and Phenomenological Research*, 79(1): 1–16.
Cameron, R. 2010a. 'On the source of necessity'. In B. Hale and A. Hoffman (eds.), *Modality: Metaphysics, Logic and Epistemology*, Oxford: Oxford University Press, pp. 137–52.
Cameron, R. 2010b. 'The grounds of necessity', *Philosophy Compass*, 5(4): 348–58.
Cameron, R. forthcoming. 'Modal conventionalism'. In O. Bueno and S. Shalkowski (eds.), *The Routledge Handbook of Modality*, London: Routledge.
Conee, E., and T. Sider (2005). *Riddles of Existence: A Guided Tour of Metaphysics*, Oxford: Oxford University Press.
Cowling, S. 2013. 'Ideological parsimony', *Synthese* 190(17): 3889–3908.
Ellis, B. 2001. *Scientific Essentialism*, Cambridge: Cambridge University Press.
Fischer, B., and Leon, F. (eds.), 2017. *Modal Epistemology After Rationalism*, Cham: Springer.
Frängsmyr, T., 1990. 'The mathematical philosophy'. In T. Frängsmyr, J. L. Heilbron, and R. E. Rider (eds.), *The Quantifying Spirit in the Eighteenth Century*, Berkeley: University of California Press, pp. 27–44.
French, S. 2018. 'Toying with the toolbox: How metaphysics can still make a contribution', *Journal for General Philosophy of Science*, 49(2): 211–230.
French, S. and K. McKenzie, 2012. 'Thinking outside the toolbox: Towards a more productive engagement between metaphysics and philosophy of physics', *European Journal of Analytic Philosophy*, 8(1): 42–59.
French, S. and K. McKenzie. 2016. 'Rethinking outside the toolbox'. In T. Bigaj and C. Wüthrich (eds.), *The Metaphysics of Contemporary Physics*, Leiden: Brill, pp. 25–54.
Grosholz, E. and E. Yakira. 1998. *Leibniz's Science of the Rational*, Stuttgart: Franz Steiner Verlag.
Jackson, F. 1998. *From Metaphysics to Ethics: A Defence of Conceptual Analysis*, Oxford: Oxford University Press.
Lewis, D. K., 1973. *Counterfactuals*. Oxford: Blackwell.
Lewis, D. K., 1986. *On the Plurality of Worlds*, Oxford: Blackwell.
Lowe, E. J., 1998. *The Possibility of Metaphysics*, Oxford: Clarendon Press.
Lowe, E. J., 2006. *The Four-Category Ontology*, Oxford: Clarendon Press.
Lowe, E. J., 2011. 'The rationality of metaphysics', *Synthese*, 178: 99–109.
Lowe, E. J. 2013. *Forms of Thought: A Study in Philosophical Logic*, Cambridge: Cambridge University Press.
Miller, J. T. M. 2018a. 'E. J. Lowe', *Internet Encyclopedia of Philosophy*.
Miller, J. T. M. 2018b. 'Are all primitives created equal?', *The Southern Journal of Philosophy*, 56(2): 273–92.

Morganti. M., and Tahko, T., 2017. 'Moderately naturalistic metaphysics', *Synthese* 194: 2557–80.

Ney, A. 2019. 'Are the questions of metaphysics more fundamental than those of science?', *Philosophy and Phenomenological Research*, Online First. DOI: 10.1111/phpr.12571.

Plunkett, D., 2015. 'Which concepts should we use?: Metalinguistic negotiations and the methodology of philosophy', *Inquiry*, 58(7–8): 828–74.

Plunkett, D., and Sundell, T., 2013. 'Disagreement and the semantics of normative and evaluative terms', *Philosopher's Imprint*, 13(23): 1–37.

Plunkett, D., and Sundell, T., 2014. 'Antipositivist arguments from legal thought and talk: The metalinguistic response'. In G. Hubb and D. Lind (eds.), *Pragmatism, Law and Language*, London: Routledge, pp. 56–75.

Quine, W. V. O. 1948. 'On what there is', *The Review of Metaphysics*, 2(5): 21–38.

Russell, B., 1912. *Problems of Philosophy*, Home University Library.

Russell, G. 2010. 'A new problem for the linguistic doctrine of necessary truth'. In C. D. Wright and N. J. L. L. Pedersen (eds.), *New Waves in Truth*, New York: Palgrave-Macmillan, pp. 267–81.

Shoemaker, S. 1998. 'Causal and metaphysical necessity', *Pacific Philosophical Quarterly*, 79(1): 59–77.

Sidelle, A. 1989. *Necessity, Essence, and Individuation: A Defense of Conventionalism*, Ithaca, NY: Cornell University Press.

Sider, T. 2003. 'Reductive theories of modality'. In M. J. Loux and D. W. Zimmerman (eds.), *The Oxford Handbook of Metaphysics*, Oxford: Oxford University Press, pp. 180–208.

Swoyer, S. 1982. 'The nature of natural laws', *Australasian Journal of Philosophy*, 60(3): 203–23.

Thomasson, A. L. 2007. 'Modal normativism and the methods of metaphysics', *Philosophical Topics*, 35(1/2): 135–60.

Vaidya, A. 2015. 'The epistemology of modality', *Stanford Encyclopedia of Philosophy*, Edward N. Zalta (ed.).

Van Inwagen, P. 1998. 'Modal epistemology', *Philosophical Studies*, 92(1): 67–84.

Williamson, T. 2016. 'Modal science', *Canadian Journal of Philosophy*, 46(4–5): 453–92.

Williamson, T. 2017. 'Modality as a subject for science', *Res Philosophica*, 94(3): 415–36.

Williamson, T. 2018. 'Spaces of possibility', *Royal Institute of Philosophy Supplements*, 82: 189–204.

Yablo, S. 1993. 'Is conceivability a guide to possibility?', *Philosophy and Phenomenological Research*, 53(1): 1–42.

INDEX

aboutness illusion 176
absolute fundamentality 212–15
absolute generality 130–41
absolute positing 27–8
absolutism: meta-metaphysics 130–1; ontology 130; philosophy of logic 131; philosophy of mathematics 131–2; quantifying over everything 139; role of 130–2; strong 138–9
abstraction: metaontology of 153–5; principles 145, 146
Abstract Objects 144
actual-literature verbalism 125
Adams, R.M. 72
aesthetic intuitions 452
alethic modality 364
Alston, William 241
ameliorative project 303; claim about social salience 306–7; concerns for 307–10; metaphysical anti-realism 306; social metaphysics and social justice 304–5; as theoretical holism 303–4
anachronism 17
analogy, fiction 259–61
analytic judgements 24
analytic metaphysicians 49, 448, 455, 485
Anglo-Saxon monosyllables 50
anti-absolutism 132; beyond one's language 137–8; for empty names 136; language-relative conception, objecthood 135–6; metaphysical conception of language 133–5; question-begging 136–7; recarving 132–8
anti-metaphysicalism 152
anti-realism 67–8, 229–30; *see also* Metaphysical Anti-Realism
anti-sceptical argument 66
Aporia 20
a posteriori knowledge 415; bootstrapping and cyclical processing 356–8; modal epistemology 358–9; a priori *versus* 353–6; science and naturalistic metaphysics 359–61
apparent ontological commitments 236
Appiah, A. 314
a priori knowledge 410–12; bootstrapping and cyclical processing 356–8; modal epistemology 358–9; *versus* a posteriori 353–6; science and naturalistic metaphysics 359–61
Aristotelian metaphysics 459–62
Aristotle 13, 18–22, 213, 245, 250, 284–6, 288–93, 295, 460; enquiry into being 19, 20; semantic-based reading 19, 20; "What is being?" 20
Armstrong, D.M. 72, 387, 390, 393, 450
assertion 274, 275
asymmetric essential predicates 75
atomists 201
Audi, P. 229
Austin, J.L. 174
Azzouni, J. 93

Bach, E. 336n1
backing model 225
Balaguer, M. 121, 124, 125
Barnes, E. 316
Baron, S. 227
Basic Law V 143–5
Bealer, George 358, 367, 368
Bellerophon 86
Bell's theorems 454, 461
Benacerraf, Paul 151
Benacerraf's Problem 353
Bennett, Karen 121, 164, 206, 317, 353, 381, 390
Benovsky, J. 390
Bernstein, Sara 316
big-D Dependence approach 212
Biggs, S. 370

Biglish 105–7; Smallish semantic theory of 106, 107
Bird, Alexander 293
Black, Max 388
Blackburn, Simon 41, 69n12, 270, 366
Bliss R.L. 218
Block, Ned 400
Blum, L. 314
Boghossian, Paul 146
BonJour, Laurence 354
Boolos, George 144, 145
Brandomian inferentialism 274
Bricker, P. 392
Brock, S. 91
Brock-Rosen objection 265–6
Bunsen 119
Button, Tim 62, 64, 67, 68, 68n2

Callender, C. 476
Cameron, Ross P. 21, 148, 167, 186, 234, 437
Canis familiaris 408
Cantor, Georg 398
Caplan, Ben 185
Carman, T. 341
Carnap, Rudolf 32–47, 58, 72, 88–90, 94, 95, 135, 138, 161, 172, 184; no exit, reality and existence 44–5; no return, explained 45–6; ontological relativity 42–4; 'principle of tolerance' 52; problems and complexities 42–6; skepticism 35–7; theory 37–42
Carr, D. 341
'Cartesian Credo' 67
Cartesianische Meditationen 343
Cartesianism 62, 65
Casullo, Albert 354
categorical modalism 247
Categories 18, 284
causal dependence 313
causal-descriptivism 280
causal potentiality (C-Pot) 286, 288–90
ceteris paribus 55
Chakravartty, Anjan 359, 440
Chalmers, David 358, 366, 368, 415
Chandrasekhar, Subrahmanyan 452
charity 176
charity-based argument 103
charity to understanding 125
Chisholm, R. 72
Chomsky, N. 330, 334
Chomsky conjunction test 179
cognition 26, 29
cognitive ontology 332
collateral assistance 154
common-sense ontology 111–12, 314
compositional function 105
compositionalist quantifier 187
compositional semantic theory 105
composition principles 407

conceptual analysis 396, 403; case studies 398–402; claims, bearing 398–402; discovery of properties 397–8; inevitability of 402; metaphysics and 397; and rewriting sentences 402–3
conceptual problem 436
conditional modalism 247, 248
Confessions of a Confirmed Extensionalist and Other Essays 79n1
confirmationism 373
conflicts of charity 115n41
consequential essence 250, 251–3
constitutive dependence 313
constitutive essence 250, 251–3
constructional ontology 335
contingentism 405–17
conversational exchange 120
core-periphery distinction 332–3
Correia, F. 75, 251, 252, 255
correspondence 62
counterfactual conditionals 77
Craig, Edward 366
Crowell, S. 341

Dasgupta, S. 217, 224
Davidson, Donald 55, 72, 326
De Broglie–Bohm theory 450
Deductive-Nomological (DN) model 226
de facto rules 174
deflationary/deflationism 33, 103, 110, 118–19, 125–6, 161, 274, 281, 488
deflationary naturalism 469
De Florio, Ciro 147
Dembroff, Robin 306
Democritos 459
denial of holism 438
de re metaphysical necessity 246
derivative essence 252
derivative truths 241
DeRosset, L. 206, 392
Devitt, Michael 63, 180n1
Die Grundlagen der Arithmetik 144
discourse-domains 173
discrete objects: Aristotelian metaphysics to ontic structural realism 459–62; events and substances 462–4
dispositional essentialists 254
dispositions 286
distinctive ideological commitments 377
Divers, J. 268
domain disagreement 177, 178
Donaldson, Tom 147
Dorr, Cian 101, 107, 167, 255
double hylomorphism 285, 290–1, 296
Douven, I. 64
Dreyfus, H.L. 341
Duhem, Pierre 449, 451, 452
Dummett, Michael 143, 148

Index

Dunn, J.M. 74, 75
Dyke, H. 471

easy arguments: history 160–3; objections to 164–7
easy ontology 159–68; importance of 163–4
egalitarianism 101–3
Einstein, A. 35
Einstein's theory of general relativity 357
Eklund, M. 106, 114n23
Eleatic Visitor 14
eliminative naturalism about metaphysics (ENM) 470, 472, 475
eliminativism 313–15
elite quantifiers 185
emotivism 38, 41, 44
empiricism 36
emptiness 281
entity contingentism 405, 406–7
entity moderatism 405
entity necessitarianism 405
epistemological concerns 164
epistemological holism 54
epistemological priority 53
epistemologies, modality 8, 364–74; uniformity vs non-uniformity 367–70
Epstein, B. 315
equivocality disagreement 177
equivocality ontology 177–9
e-representation 274
essence 245–57; consequential 250, 251–3; constitutive 250, 251–3; definition 250–1; extending 254–5; first wave modalism 247–8; and identity 255–6; modalism and 247–50; modalism bites back 248–50; theories of modality 253–4
essentialism 314, 364
essential truths 366
Everett, Hugh 35
existence 26, 27, 29, 31n15, 37, 44; Carnap and ontological debate 173–5; cogent/serious ontological questioning and debates 172–3; contemporary approach 23; disagreement 178; first-level 23; judgements 27, 28; as modality 24–6; nature of 24–6; ordinary questions 171–2; particular quantifier 24; questions 200; second-level 23–4; of universals 40
experimental metaphysics 469
explanandum 226
explanans 226
explanation 222–3
explanatory power 391–3
expressiveness 109–10
expressivism 6, 39, 45, 270, 277; about meaning 275–7; globalizing 271–3; metasemantics 277–82; *see also* global expressivism
expressivist explanations 46
extensionalism 72

'face value' theory 163
feminist metametaphysics 300–10; importance of social 301; metaphysical concepts and political power 302–10
feminist philosophy 309
fictionalism 259; benefits of 266–8; Brock-Rosen objection 265–6; incompleteness objection 263–5; objections to 263–6
fictionalist paraphrases 261–3
fictionalists 259
fictionalist strategies 259–68; analogy, fiction 259–61; paraphrases 261–3
Fine, Kit 72, 74, 199, 200, 217, 234, 250, 251, 329, 330, 365, 425, 442
first-order existence questions 23
first-order metaphysical claims 246
first-order metaphysics 1
Fischer, B. 371, 372
focus effect 165
folkmetaphysics 331–2
Frege, Gottlob 134, 143, 144, 151, 161, 326, 333
Frege-Geach problems 42
Frege's Conception of Numbers as Objects 144
French, Steven 442, 451, 455
functional-antecedents 271
functional forms 287
functional holism 294
functional unity 296
fundamentality 58, 78, 211–31, 316–18; absolute fundamentality 212–15; needs for 216–19; open issues 219–20; relative fundamentality 215–16; view 211–16; well-foundedness 215
fundamental objects: conceptual problem 439–41; existence of 437; nature of 437

Geach-Kaplan sentence 236, 237
gender 305; inequality 303
generalized identity 255
generic existential dependence 75
generic quantification 188–91
Gettier, Edmund 400, 438
Gibbard, Allan 41
Global-E conception 272–4
global entity contingentism 405
global entity necessitarianism 405
global expressivism 270–82
global metaphysical contingentism 405
global metaphysical contingentism arguments 415; from conceivability 415–16; against necessity 415
global metaphysical necessitarianism 405
global metaphysical necessitarianism arguments 408–14; from essence 409–11; from ground 408–9; from identity 409; no contingent difference-maker argument 413–14; from a priority 411–12; from virtue 412–13
global response-dependence 66, 67

Index

God, concept 28
Godfrey-Smith, P. 476
'God's-Eye' point of view 67
Goldilocks principle 440
Goldman, A.I. 469
Goodman, N. 335
Gorman, M. 251
Gosling, Ryan 120
grounding 199–209, 212; heterodox accounts of 216; location problems 205–7; substantive questions 200–2; theoretical economy 202–4
Grundlagen 150

Hacking, I. 314
Hale, Bob 143, 144, 146, 148, 150–3, 155, 161, 163, 250, 251, 333, 369–73, 413
Hanrahan, R. 370
Haslanger, Sally 302–5, 314, 315
Hawley, Katherine 150
Heidegger, M. 38, 339, 344
Hellman, Geoffrey 144
Hilbert, David 140n4
Hintikka, J. 72
Hirsch, Eli 102, 119, 121–3, 150, 186, 187, 390
Hobbes 77
Hofweber, Thomas 90, 95, 165, 167, 200
holistic unification 296
Homo javanensis 380
Homonymy Principle 297n14, 298n28
Horwich, P. 273, 276
Huemer, Michael 439
Hume 473
Hume's principle (HP) 143, 144, 146, 150, 152, 161, 166
Husserl, E. 339–47
hylomorphic composites 296
hylomorphic divisions 286–8
hylomorphic unity 284–99; Aristotle's accounts of 291–3; double hylomorphism 290–1; hylomorphic divisions 286–8; matter and form 289–90; potentiality and actuality 289–90; potentiality of matter 288–9; unity of definition 290–1; unity of substance account 293–6
hyperintensional notion 71–2, 74, 77–9

idealism 345–7
identity 216, 255–6, 279
Identity of Indiscernibles 388
ideological commitments 377–9
ideological disagreement 385
ideological economy 204
ideological parsimony 437
ideological realism 377
ideological virtues 380
ideology: Lewis' contribution 379–81; open questions 384–5; Quine's contribution 381–3; Sider's contribution 383–4

incompleteness objection 263–5
independence 62
Independence Credo 62
indeterminacy 57; of reference 56, 57; of translation 56, 57
inference to the best explanation (IBE) 453
inheritance 271, 275
instantiated particular forms 286
instantiation 25, 292, 376, 378, 388–90
intellectual inquiry 1
intelligence dispute 119
intelligible forms 287
intelligible matter 287
intended interpretation 64
intentional-dependency 277, 279
intentionality 347
intentionality-dependency 277, 281
inter-linguistic semantics 105–7
internal realism 67
intuitions 30, 31n16, 358, 438–9, 451–2, 471
Inwagen, Peter van 49
i-representation 274

Jaksland, Rasmus 33
Javier-Castellanos, Arturo 186, 187
Jenkins, C.S.I. 127n9, 359, 369
'just more theory' manoeuvre 64, 65

Kanger, S. 72
Kantian meta-ontology 23–31
Kim, J. 205, 392
Kitcher, P. 227
Kment, B. 226
knowledge 104
Korsgaard, Christine 45
Kovacs, D. 231n12
Kriegel, Uriah 164, 439
Kripke, Saul A. 44, 66, 72, 73, 174, 247, 255, 356
Kripkean thesis 136
Kuhn, Thomas 448

Ladyman, J. 443, 450, 451, 455–6, 456n8, 471, 476
Lakatos, Imre 448
La La Land 120
language-driven ontology 328
language-relative conception: of objecthood 135–6
language-transcendent domain 135
law of nature principles 407
Leibniz 472, 474, 480
Leibnizian view 475
Leibniz' principle of the identity 442
Levinas, E. 347
Lewis, C.I. 72
Lewis, David 73, 76, 78, 88, 93, 139, 266, 360, 377, 379–81, 438, 442

495

liberal naturalism 477
Liggins, D. 266
linguistic frameworks 33, 39, 40, 42
linguistic interpretation 121
linguistic science 179–80
Linnebo, Oystein 144, 147, 152, 161
local realisms 61
location problems 5, 205–7
logical positivism 53, 58, 447
logicism 146–8
Logische Untersuchungen 340
The Lord of the Rings 238, 240, 262
Lowe, E.J. 75–6, 245, 357, 358, 368, 449, 450, 481–2, 487

MacBride, Fraser 148, 149, 151
MacFarlane, John 153
Maclaurin, J. 471
Maddy, P. 469
Mandeville, Bernard 45
Marcus, Ruth Barcan 97n19
Markosian, Ned 187, 437
material truth 24
Maudlin, Tim 441
maximalism 150
Maxwell-Lorentz theory 465
McDaniel, Kris 185, 187, 303
McGee, Vann 131
McKenzie, K. 451, 455
McLeod, M. 470
McLeod, S. 367
McX 86, 175
mechanical model 225–6
Meinong, Alexius 90–3, 234
Meinongianism 90
Meinongianism of Priest 91
Meinongian quantifier 92
Meinongians 176
mental forms 287
mereological nihilism 204
mereological nihilist quantifier 187
mereological nihilists 203
mereological universalism 100
Merricks, Trenton 188, 189, 321n8
metaethics 38
meta-level questions 1
metalinguistic pragmatism 39
metametaphysical nihilism 270, 277
metametaphysical verbalism 125
meta-ontology 23–4, 87, 148
'Meta-ontology' (1998) 49
metaphysical anti-realism 67; *see also* realism
metaphysical coherentism 219
metaphysical contingentism 405, 407–16
metaphysical dependence relations 212
metaphysical explanation 78, 222–31; backing model 225; grounding and 224–5; mechanical model 225–6; models of 225–8; pragmatic account 228; realism and anti-realism 229–30; subsumption model 226–7; supporting 223–4; unificationism 227–8
metaphysical grounding *see* grounding
metaphysical ideology *see* ideology
metaphysical infinitism 219
metaphysicalism 147
metaphysical modality 364, 365
metaphysical moderatism 405
metaphysical necessitarianism 405
metaphysical potentiality (M-Pot) 286, 289, 290
metaphysical quietism 67
metaphysical realism 3, 61–2; brains in vats and response-dependence 65–7; model-theoretic argument 63–5; semantics and epistemology 63
Metaphysics (Aristotle) 18, 20, 21, 213
metasemantic naturalism 64, 65
metasemantics 277–82; assumptions 279–82; positive metaphysics 278–9; real-definitions 278
methodological individualism 314
A Midsummer Night's Dream 95
minimalism 154
minimal semantic theory 105
Moby Dick 119
modal dualism 206
modal empiricism 358, 488
modalism 247–50
modality 24–6, 72, 73, 79, 476; epistemic challenge in 366
modal knowledge 366; structure of 370–2
modal logic 72
Modal Logic as Metaphysics 369
modal metaphysics: problems with 74–7
modal myopia 3
modal realism 264, 265
modal representation 29; absolute positing 27–8; problem of 26–8; relative positing 26–7
modal revolution 72–4, 79
model-theoretic argument 63–5
moderately naturalistic metaphysics 473–5, 477
Moltmann, F. 328, 333
monism 15, 16
monists 201
Montague, R. 72
Montague Grammar 330
Morganti, M. 454
Morning Star 223

name principle 136
Naming and Necessity 72, 247, 255
naturalism 54, 55, 468, 477
naturalistic metaphysics 9, 359, 435, 437, 448, 468–77
natural language ontology 325–36; and cognitive ontology 332; compositionality and 327; constructional ontology 335; core-periphery

distinction and 332–3; and folkmetaphysics 331–2; metaphysics 329; recognizing, discipline of own 330–1; referential NPs, semantic values 327–8; semantics 326–8; syntactic core-periphery distinction 334–5; universals of 333–4; ways of reflection 326–7
natural numbers system 39
neo-Aristotelian approach 201, 284
neo-Fregeanism 143–56; content recarving and implicit definition 152–3; easy existence 149–50; neo-Fregean, quietist 148–9; Platonism and logicism 146–8
neo-Quinean approach 160, 164, 165
Newton's corpuscular theory of light 357
Nihilese 100, 101
Nimtz, C. 370
Nolan, Daniel 72, 79n1, 264, 265, 360, 361
non-fact-stating mechanism 41
non-reductive physicalists 206
non-thing ontologies 459–66
non-ultimate grounding 201
'normative claim of SGM' (NCSGM) 436, 437, 440, 441, 444; metaphysics and science 436–7
Norton, J. 227
notational variance 191–3
noun phrases (NPs) 326

object principles 407
objectual essentialist statements 254
obsessive verificationism 35
Occam 202, 278
O'Leary-Hawthorne, J. 265
On the Plurality of Worlds 73, 380
ontic structural realism 459–62
ontological commitments 85–97, 241, 376, 378; Carnap and successors 88–90; Meinong and successors 90–3; Quine and successors 86–8; truthmaking 233–7
ontological conflict 41
ontological debates 51, 52, 101, 104, 167, 171–83, 239
ontological dependence 18, 75, 76, 214, 327, 442, 464
ontological discourse 33, 39, 42, 44
ontological economy 204
ontological languages 177
ontological monism 462
ontological nihilism 242
ontological permissivism 306
ontological pluralist/pluralism 5, 184–94
ontological pluralist 5
ontological relativity 42–4, 49
'Ontological relativity' (OR) (1968) 56–8
ontology 4, 23, 28, 30, 119; Carnap, Rudolf 32–47; continuous stuff 464–6; "deflation" of 39; easy 159–68; easy arguments history 160–3; equivocality 177–9; Lewis' contribution 379–81; pragmatist and expressivist 32–47; quantum mechanics, identity 471–3; Quine's contribution 381–3; Sider's contribution 383–4; of universals 36
'On what there is' (OWTI) (1948) 49–52, 58, 160
operationalism 36
ousia 21

Parent, T. 94, 127n11
Parmenides 14
Parsons, Charles 144
Parsons, J. 470
Paul, L.A. 450, 453, 454, 476
Peacocke, C. 368
Peirce, C.S. 456
perceptible forms 287
permutation invariance 477n4
persistence principles 407
Pettit, Philip 66, 69n14
Phänomenologie und Anthropologie 343
phenomenology 7, 339–48; and idealism 345–7; metaphysical implications of 341–5; metaphysical neutrality of 340–1
philosophical ontologies 464
Philosophy of Logical Syntax 38
phlogiston theory 400
physicalism 15, 206, 207, 208n20
physical potentiality 287
physical reality 36
Plantinga, A. 72
Plato 13, 14–18, 293, 294, 376
Platonism 146–8
Platonism-Parmenideanism 15
pleonastic entities 161, 162
pleonastic entity 163
pluralism 15, 16, 101–3; and generic quantification 188–91; 'shorthand' argument 189–91
pluralism, defined 185–8
plurality 466
Podolsky, B. 35
Poincaré, Henri 450–3
politically effective falsehoods 307
Popper, Karl 448
positive metaphysics 278–9
'post-empirical' science 456
post-modal future 77–8
Pound-Rebka experiment 358
practice problem 436
pragmatic account 228
pragmatic cost-benefit analysis 41
presentism 398
Price, H. 89, 93, 274, 275
Primacy of Physics Constraint 471
primitives 387–93; explanatory power 391–3; functional view 390–1; problem-solvers 389–90; theories and 387–8

Principia 461
principle of charity 55
principle of interpretive charity 102
Principle of Naturalistic Closure 471
Principle of Sufficient Reason (PSR) 219
Principle of the Identity of the Indiscernibles 472
Principles of Possibility 368
Prior, A.N. 72
priority principles 407
The Problems of Philosophy 184
problem-solvers 389–90
progress problem 436
property principles 407
proximate matter 287
pure potentiality 287
pushback strategy 16, 17
Putnam, Hilary 50, 59n7, 61–8, 174, 184
puzzlement 15, 35, 36, 366

quantification 24
quantificational pluralism 185
quantifier egalitarianism: challenges for 109–11; expressiveness 109–10; quantifier naturalness 110–11
quantifier naturalness 110–11
quantifier pluralism 109; challenges for 103–8; collapse arguments 107–8; inter-linguistic semantics demand 105–7
quantifiers 85–97
quantifier variance 100–15, 153, 181n5; arguments for 102–3; claim 101–2; and common-sense ontology 111–12; components of 101; proponents of 100
quantifier variantists 110
quantity of matter 287
quasi-scientific methodology 109
quietism 153
Quine, W.V.O. 43, 44, 49–59, 72, 79n1, 86–8, 95n4, 172, 174, 377, 381–3, 468; Metametaphysics 49–59; 'Ontological relativity' [OR] (1968) 56–8; 'On what there is' (OWTI) (1948) 50–2; resurrection of ontological debate 175–6; truthmaking *versus* 233–7; 'Two dogmas of empiricism' [TD] (1951a) 52–4; view from distance 49–50; *Word & Object* [WO] (1960) 55–6
Quinean quantifier 92

racialized groups 314
radical conception 13
radically naturalistic metaphysics 469–73
radical metametaphysics 16
radical naturalism 470
radical translators 57
rationalist renaissance 367
Raven, M. 224
Rayo, Agustín 147

Razor, Occam's 202, 203, 278
realism 63, 229–30; *see also* Metaphysical Realism
real modalities: cognising 28–9
The Reason's Proper Study 144
reductive naturalism about metaphysics (RNM) 470–2, 475
reference thesis 137
relative fundamentality 215–16
relative positing 26–7
relativity 463
resemblance nominalism 388
rigid existential dependence 75
Ritchie, J. 469
Roca-Royes, S. 371
Rosen, Gideon 147, 246, 251, 263, 267, 415–16
Rosen, N. 35
Rosen-Schwartzkopff principle 147
Ross, D. 443, 450, 455–6, 471, 476
Routley (Sylvan), R. 93
Rovelli, Carlo 466
Ruben, D.H. 315, 392
Russell, Bertrand 50, 86, 143, 184, 190
Russell, Jeffrey Sanford 144

Salmon, W. 225
salva veritate 71, 72
Santa Claus paradigm 175
Sapir-Whorf hypothesis 336n9
Sartre, J.-P. 339
Saunders, Simon 442
Scaltsas, T. 286, 291, 292
Schaffer, Jonathan 199, 200, 203, 217, 317, 392, 393, 462
Scharp, Kevin 168n9
Schiffer, Stephen 161, 162, 164
Schnieder, B. 78
Schopenhauer, A. 455
Schroeder, M. 271
Schwartzkopff, Robert 147
science 447–8; intuitions, use 451–2; metaphysics, relationship 453–6; methodologies of 449–53; modelling and thought experiments 452–3; modes of inferences 452–3; theory virtues, use 450–1
science-guided metaphysics (SGM) 435–45; norms of 437–9; practice problem 441–2; progress problem 442–4
science of being 21
scientific theory 90
second-order logic 40, 41
Second Order Number Theory 41
semantic-conceptual enquiry 13
semantic minimalism 273
semantic under-specification 125–6
sentential truth 36
separatism 224
Shapiro, Stewart 40, 144

Shimony, A. 472
Shmenglish 100
'shorthand' argument 189–91
Sidelle, Alan 119, 124, 126, 368
Sider, Theodore 78, 103, 106, 110, 122–3, 150, 164, 167, 185, 234, 377, 383–4, 437
Skiba, L. 264
Skiles, A. 249, 255
Skyrms, B. 72
small-d dependence ordering 213
Soames, S. 97n20
social entities 312, 313, 315, 317
social factors 313
social facts 313
social groups 313
social institutions 313
socialism 124
social justice 304–5
social kinds 313
social metaphysics 304–5, 320
social ontology 312–21; eliminativism and reduction 313–15; fundamentality and mind-dependence 316–18; naturalness 318–19
social properties and relations 313
social reality 306
social salience 306–7
social structures 313
Socrates 75, 76, 134, 207, 226, 245, 248, 252, 365
Sophist 14, 16–18
special theory of relativity (STR) 398, 463
Stalnaker, Robert 76, 400
Sterelny, K. 180n1
Strawson, P. 329
Strohminger, Margot 370
strong absolutism 138–9
substantial form 286
substantial holism 291, 296
substantial metaphysical questions 165
subsumption model 226–7
syntactic core-periphery distinction 334–5
syntactic knowledge 335
Szabó, Zoltán 184

Tahko, T.E. 357, 368, 454
Tarski-style theory of truth 176
Taylor, E. 229
Thagard, Paul 448
theoretical ideology 376
A Theory in Physics about all Physical Things 18
The Theory of Everything 18
thing language 40, 138
thing ontologies 136, 459–66
thing properties system 40
Thomasson, Amie 33, 89, 125, 162–4, 180n1, 200, 238–40
time principles 407
totality of things 18

transcendental phenomenology 342, 346
transcendental subjectivity 347
translation function 105
triangularity 74
trilaterality 74
triviality 125–6
Trogdon, K. 226
truthmaker theory 233
truthmaking 233–43; easy and hard ontology 237–42; ontological commitment 233–7; Quine *versus* 233–7
"Two Dogmas of Empiricism" (1951/1953) 52–4, 88, 166–7

ultimate-success 278, 279, 281
underdetermines metaphysics 472
unionism 224
Universalese 100, 101
universal forms 286
univocality: approach 177; coupling with neutrality 179; disagreement 177
unproblematic truth 102
unrestricted-quantifier-like expression (UQE) 102, 103, 105, 108, 110, 111; direct specification, demand 103–5
Urelement Set Axiom 131

vagueness 308
value-assertion 271
value of copula 25
Van Fraassen, B. 228, 451, 470, 472
Van Inwagen, P. 87, 88, 119, 168n3, 366, 373n6
verbal deflationism 124–5
verbal disputes 118–28; semantic under-specification 125–6; triviality 125–6
verbalness: accounts of 121–2
Vermeulen, I. 122, 127n2, 127n6

Was ist Metaphysik? 38
well-foundedness 215
Whittle, Bruno 192
Williams, J.R.G. 204
Williamson, Timothy 132, 139, 167, 355, 356, 369, 373n5, 440
Wilson, Harold 398
Wittgenstein, L. 174
Woodward, R. 266, 268
Wright, Crispin 143, 144, 146, 148, 150–3, 155, 161, 163, 273, 413
Wyman 86, 90, 175

Yablo, Stephen 159, 160, 165, 171, 175, 368, 371
Yang, C. 334
Yli-Vakkuri, Juhani 370

Zanetti, Luca 147
Zermelo, Ernst 131, 132